Lecture Notes on Data Engineering and Communications Technologies

Volume 85

Series Editor

Fatos Xhafa, Technical University of Catalonia, Barcelona, Spain

The aim of the book series is to present cutting edge engineering approaches to data technologies and communications. It will publish latest advances on the engineering task of building and deploying distributed, scalable and reliable data infrastructures and communication systems.

The series will have a prominent applied focus on data technologies and communications with aim to promote the bridging from fundamental research on data science and networking to data engineering and communications that lead to industry products, business knowledge and standardisation.

Indexed by SCOPUS, INSPEC, EI Compendex.

All books published in the series are submitted for consideration in Web of Science.

More information about this series at http://www.springer.com/series/15362

Bernard J. Jansen · Haibo Liang ·
Jun Ye
Editors

International Conference on Cognitive based Information Processing and Applications (CIPA 2021)

Volume 2

Springer

Editors
Bernard J. Jansen
Qatar Computing Research Institute
Doha, Qatar

Haibo Liang
School of Mechanical Engineering
Southwest Petroleum University
Chengdu, China

Jun Ye
Hainan University
Haikou, China

ISSN 2367-4512 ISSN 2367-4520 (electronic)
Lecture Notes on Data Engineering and Communications Technologies
ISBN 978-981-16-5853-2 ISBN 978-981-16-5854-9 (eBook)
https://doi.org/10.1007/978-981-16-5854-9

This Springer imprint is published by the registered company Springer Nature Singapore Pte Ltd.
The registered company address is: 152 Beach Road, #21-01/04 Gateway East, Singapore 189721, Singapore

Foreword

Cognition has emerged as a new and promising methodology with the development of cognitive-inspired computing, cognitive-inspired interaction, and systems that enable a large class of applications and has developed a great potential to change our life. However, recent advances in artificial intelligence (AI), fog computing, big data, and cognitive computational theory show that multidisciplinary cognitive-inspired computing still struggles with fundamental, long-standing problems, such as computational models and decision-making mechanisms based on the neurobiological processes of the brain, cognitive sciences, and psychology. How to enhance human cognitive performance with machine learning, common sense, natural language processing, etc., is worth exploring.

2021 International Conference on Cognitive-based Information Processing and Applications includes data mining, intelligent computing, deep learning, and all other theories, models, techniques related to artificial intelligence.

The purpose of CIPA2021 is to provide a forum for the presentation and discussion of innovative ideas, cutting-edge research results, and novel techniques, methods, and applications on all aspects of technology and intelligence in intelligent computing.

At least two independent experts reviewed each paper. The conference would not have been a reality without the contributions of the authors. We sincerely thank all the authors for their valuable contributions. We would like to express our appreciation to all Program Committee members for their valuable efforts in the review process that helped us guarantee the highest quality of the selected papers for the conference.

We would like to express our thanks for the strong support of the Publication Chairs, Organizing Chairs, Program Committee members, and all volunteers.

Our special thanks are also due to the editors of the Springer book series "Advances in Intelligent Systems and Computing", Ramesh Nath Premnath and Karthik Raj Selvaraj, for their assistance throughout the publication process.

Bernard J. Jansen
Haibo Liang
Jun Ye

Welcome Message

Cognition has emerged as a new and promising methodology with the development of cognitive-inspired computing, cognitive-inspired interaction, and systems, which can enable a large class of applications and has great potential to change our lives. However, recent advances in artificial intelligence (AI), fog computing, big data, and cognitive computational theory show that multidisciplinary cognitive-inspired computing still struggles with fundamental, long-standing problems, such as computational models and decision-making mechanisms based on the neurobiological processes of the brain, cognitive sciences, and psychology. How to enhance human cognitive performance with machine learning, common sense, natural language processing, etc., is worth exploring.

2021 International Conference on Cognitive-based Information Processing and Applications includes precision mining, intelligent computing, deep learning, and all other theories, models, techniques related to artificial intelligence. The purpose of CIPA2021 is to provide a forum for the presentation and discussion of innovative ideas, cutting-edge research results, and novel techniques, methods, and applications on all aspects of technology and intelligence in intelligent computing.

At least two independent experts reviewed each paper. The conference would not have been a reality without the contributions of the authors. We sincerely thank all the authors for their valuable contributions. We would like to express our appreciation to all Program Committee members for their valuable efforts in the review process that helped us guarantee the highest quality of the selected papers for the conference.

We want to express our thanks for the strong support of the General Chairs, Publication Chairs, Organizing Chairs, Program Committee members, and all volunteers.

Our special thanks are also due to the editors of the Springer book series "Advances in Intelligent Systems and Computing", Ramesh Nath Premnath and Karthik Raj Selvaraj, for their assistance throughout the publication process.

Jim Jansen
Haibo Liang
Jun Ye

Organization

Conference Committee

Local Organizing Chairs

Tao Liao	Anhui University of Science and Technology, China
Xiaobo Yin	Anhui University of Science and Technology, China

Program Chairs

Jim Jansen	Qatar Computing Research Institute, Qatar
Haibo Liang	Southwest Petroleum University, China
Jun Ye	Hainan University, China

Publication Chairs

Neil Y. Yen	University of Aizu, Japan
Vijayan Sugumaran	Oakland University, USA

Publicity Chairs

Weidong Liu	Inner Mongolia University, China
Sulin Pang	Jinan University, China

Program Committee

Ameer Al-Nemrat	University of East London, UK
Robert Ching-Hsien Hsu	Chung Hua University, China
Neil Yen	University of Aizu, Japan
Meng Yu	The University of Texas at San Antonio, USA

Shunxiang Zhang	Anhui Univ. of Sci. & Tech., China
William Liu	Auckland University of Technology, New Zealand
Mustafa Mat Deris	Universiti Tun Hussein Onn Malaysia, Malaysia
Zaher AL Aghbari	Sharjah University, UAE
Guangli Zhu	Anhui Univ. of Sci. & Tech., China
Raja Al Jaljouli	College of Computer Science and Engineering, Kingdom of Saudi Arabia
Tao Liao	Anhui Univ. of Sci. & Tech., China
Abdul Basit Darem	University of Mysore, India
Xiaobo Yin	Anhui Univ. of Sci. & Tech., China
Vjay Kumar	VIT, India
Xiangfeng Luo	Shanghai Univ., China
Jemal Abawajy	Deakin University, Australia
Ahmed Mohamed Khedr	University of Sharjah, UAE
Xiao Wei	Shanghai Univ., China
Sabu M. Thampi	Indian Institute of Information Technology and Management, India
Huan Du	Shanghai Univ., China
Shamsul Huda	Deakin University, Australia
Zhiguo Yan	Fudan University, China
Rick Church	UC Santa Barbara, USA
Tom Cova	University of Utah, USA
Susan Cutter	University of South Carolina, USA
Yi Liu	Tsinghua University, China
Kuien Liu	Pivotal Inc, USA
Wei Xu	Renmin University of China, China
V. Vijayakumar	Professor & Associate Dean, SCSE, VIT Chennai, India
Abdullah Azfar	KPMG Sydney, Australia
Florin Pop	University Politehnica of Bucharest, Romania
Kim-Kwang Raymond Choo	The University of Texas at San Antonio, USA
Mohammed Atiquzzaman	University of Oklahoma, USA
Rafiqul Islam	Charles Sturt University, Australia
Morshed Chowdhury	Deakin University, Australia

CIPA 2021 Keynotes

Jim Jansen is a Principal Scientist in the social computing group of the Qatar Computing Research Institute. He is a graduate of West Point and has a Ph.D. in computer science from Texas A&M University. Professor Jansen is editor-in-chief of the journal, Information Processing & Management (Elsevier), a member of the editorial boards of seven international journals, and former editor-in-chief of the journal, Internet Research (Emerald). He has received several awards and honors, including an ACM Research Award, six application development awards, and a university-level teaching award, along with other writing, publishing, research, teaching, and leadership honors. Dr. Jansen has authored or co-authored 300 or so research publications, with articles appearing in a multidisciplinary range of journals and conferences. He is author of the book, Understanding Sponsored Search: A Coverage of the Core Elements of Keyword Advertising (Cambridge University Press).

Jemal Abawajy is a faculty member at Deakin University and has published more than 100 articles in refereed journals and conferences as well as a number of technical reports. He is on the editorial board of several international journals and edited several international journals and conference proceedings. He has also been a member of the organizing committee for over 60 international conferences and workshops serving in various capacities including best paper award chair, general co-chair, publication chair, vice-chair, and program committee. He is actively involved in funded research in building secure, efficient, and reliable infrastructures for large-scale distributed systems. Toward this vision, he is working in several areas including pervasive and networked systems (mobile, wireless network, sensor networks, grid, cluster, and P2P), e-science and e-business technologies and applications, and performance analysis and evaluation.

Contents

Cognitive Environment, Sensing and Data

Development of Advertising Art Design Based on Information Technology

Zhinong Liu(✉)

Hangzhou Polytechnic, Hangzhou, Zhejiang, China
lzn@mail.hzrtvu.edu.cn

Abstract. With the rapid development of information technology, the application of new media has become more and more extensive. With the arrival of the information age, great changes have taken place in people's behavior and way of thinking. This paper mainly studies the development of advertising art design based on information technology. Based on the interaction of commercial advertising as the breakthrough point, let people of interaction in the design of commercial advertising have a deeper understanding of thinking, not just focus on information science and technology, more of a combination of technology and visual communication design, and then the reasonable use in commercial advertising, publicity and brand value, at the same time, guide people to better use of interactive advertising.

Keywords: Information technology · Advertising design · Big data · Interactive advertising

1 Introduction

In the digital era with big data as the background, "information" production, "digital" operation and "virtual" technology manufacturing affect the development of the national economy, and penetrate into every aspect of life. The trend of the big data era allows people to analyze more data without relying on random sampling, bringing new ways of communication and communication to the public [1, 2]. The production and life of human beings is not limited to a single operation of availability, but has been expanded to a wider and richer user experience, which will affect the public's consumption consciousness, consumption motivation and consumption ability, and directly lead to the change of their consumption behavior. The development of science and technology has changed people's consumption and living habits, and the informationization degree of modern commercial advertising is constantly expanding. The advertising push mode has changed. Nowadays, as long as there is a mobile network, we can complete the purchase and consumption without leaving home. Therefore, the opportunity to deal with traditional two-dimensional commercial advertisements has been reduced [3]. Internet advertising has become the bridge between producers and consumers, consumers gradually become all kinds of digital terminal controller, can be easily circumvented most commercial advertisement, and the number of consumers choose to deal directly with

B. J. Jansen et al. (Eds.): *International Conference on Cognitive based Information Processing and Applications (CIPA 2021)*, LNDECT 85, pp. 3–10, 2022.
https://doi.org/10.1007/978-981-16-5854-9_1

manufacturers to increase, the phenomenon is a big problem in terms of advertisers, so advertisers to survive in the cracks, Find a new identity.

At the beginning of the 21st century, based on the rapid development of economic globalization and information technology, the nature and forms of commercial advertising have been gradually enriched. Advertisement is no longer limited to the tool used to exchange products and promote sales, and a complex and diverse communication process has been added. Advertisement has also been gradually applied to deal with various social problems [4, 5]. Modern advertising was born in the United States in the 19th century after the First World War. With the rapid development of the overall consumer economy in the United States, modern advertising has become a significant social and cultural phenomenon, attracting much attention from the academic community [6]. Foreign commercial advertising began to increase in the 1930s, and New York, London and Tokyo were the three major advertising development centers in the world. Some scholars have provided more new topics for the study of "digitalization" and analyzed the path and space of advertisement. They believe that under the digital environment, advertisement has a variety of operation space and operation path, which also provides more theoretical and practical reference for advertisement research [7].

The research of this paper is helpful to provide practical guidance for the development of commercial interactive advertising. It provides practical reference for designers to create design works that keep pace with the times, and also provides better visual experience and sensory satisfaction for consumers, so as to further expand the application scope of the commercial advertising market.

2 Problems of Advertising Design under Information Technology

The development of information technology promotes the rapid development of communication, providing more creative space for advertising design, and People's Daily life has also changed. The emergence of interaction design has reduced the distance between people and spawned various forms of advertising. The rise and development of advertising cannot be separated from the progress of society, the development of science and technology, and the improvement of people's material life and spiritual and cultural level [8, 9].

2.1 Content and Form are Too Single, and the Interactivity is Insufficient

The advertising design is too monotonous, the content is uniform, it is difficult to make the public in the visual and content to be satisfied, resulting in the failure to let the audience understand the focus of the information transmitted in a specific space, the communication effect is not good. Advertising should not only consider "what to do", but also grasp the expression form of the message and pay attention to "how to do". Domestic advertising compared with abroad, rise too late, development is relatively lagging, designers study started relatively late, in the advertisement on the creation of a lack of visual aesthetic and creative performance, from the point of view of advertising creative, domestic advertising in the design of simple drab, or the lack of creative, lead to the core design point is actually weakened [10].

The addition of interactive design puts higher requirements on the creative level of designers. Designers should not only have skilled professional ability and keen insight to discover social problems, but also need to have interactive design ability and innovative and creative thinking to keep pace with the times. Clever design does not need too many words to explain, it can walk into the hearts of the audience and resonance. On the contrary, even if the content is clearly expressed in language, the lack of creative design will not attract the audience [11, 12]. Interactive emphasis on "interactive" two words, that is, the audience and advertising interactive communication, not interactive will lose the meaning of interaction. Domestic advertisements still remain at the propaganda level of text output and form output. In addition, advertisements are non-profit advertisements, and the placement of advertisements is economically unprofitable, which indirectly limits creativity and leads to unsatisfactory application effect.

2.2 Overall Use of Scale, Communication Effect is not Good

In the design of advertisement, the effect of communication comes from the link of communication and the medium of expression. At the same time, the effect of communication is also an important factor to determine the effect of advertisement. "By what use" means the concept of the medium, in the absence of a medium, the message of advertising is also unattached. "How to convey" is also very important. The development of interactivity design in China is late, and the development is not too mature. Designers' cognition of interactivity design is insufficient, and interactivity has not been popularized in the current advertising, which also affects the audience's acceptance of new forms of advertising, leading to the poor communication effect.

In the era of "Internet+", cross-media communication has been recognized and applied as a new form of publicity. While commercial marketing is doing its best to use as many channels as possible to interact with a wide range of media audiences, social communication has fallen far short. It is no longer the time to try to enter people's hearts through an advertisement. In this era of information flood, any brand marketing activity can not guarantee the perfect communication with the audience and improve the positive awareness of the audience. In this case, if the needs of the audience cannot be actively met, it is impossible to truly achieve interactive communication.

2.3 Constancy of Thought Solidifies

The mission of advertising is to communicate information to the target audience as a kind of conceptual communication, so that the audience knows "what is being conveyed", so that the audience can accept the information conveyed by advertising, and then persuade the audience and guide the audience, then it will play its role. The most fundamental premise of communication lies in the audience's understanding of the information transmitted in the form of advertisement, and the expression method of advertisement must conform to the language category mastered by the target audience. The most important value of advertising is the result of the interaction between the target audience and the work. The development of information technology, in the transformation of the popularization of network and information technology, in the information technology environment of the advertising is endowed with interactive design, but people often contact in daily life or from traditional media products, such as television, magazines, etc., which makes most of the ads or through traditional single way to mass communication. As the program flashes in front of the public, the advertisement may only have a one-side relationship with the public, and will not go to in-depth understanding and express their own views, and its sense of participation will be greatly reduced. This is what people call customary. The so-called habitual thinking refers to the way that people accumulate relevant information in the process of long-term understanding of nature and natural transformation. Therefore, what people need to pay attention to now is how to improve the form of content, how to attract different target groups to actively accept advertisements, and participate in the communication and interaction.

3 Domestic Internet Advertising Status Quo

3.1 Interdisciplinary Research Method

This paper involves communication, marketing, information technology, art design and other disciplines. Therefore, this research method derives from the overall comprehensive research of this subject through the relevant knowledge, theoretical methods and research results of various disciplines, absorbs the new methods of other disciplines to study the interactive thinking of commercial advertising, and brings some new viewpoints.

3.2 Data Collection

In this paper, the relevant data of China's domestic media advertising market from 2013 to 2017 were analyzed by referring to the periodicals, literature and online materials in recent years.

3.3 Mathematical Statistics

In this paper, SPSS200 data statistics software was used to code and input the collected valid data, and further reliability and validity test, descriptive statistics, t-test, correlation analysis and regression analysis were carried out to provide data support for the quantitative research in this paper. The t-test formula used in this paper is as follows:

$$t = \frac{\overline{X} - \mu}{\frac{\sigma X}{\sqrt{n}}} \tag{1}$$

$$t = \frac{\overline{X_1} - \overline{X_2}}{\sqrt{\frac{(n_1-1)S_1^2+(n_2-1)S_2^2}{n_1+n_2-2}\left(\frac{1}{n_1} + \frac{1}{n_2}\right)}} \tag{2}$$

$$t = \frac{\overline{d} - \mu_0}{s_d/\sqrt{n}} \tag{3}$$

4 Domestic Internet Advertising Status Quo Research Results and Development Trends

4.1 China's Five Major Media Advertising Market Scale

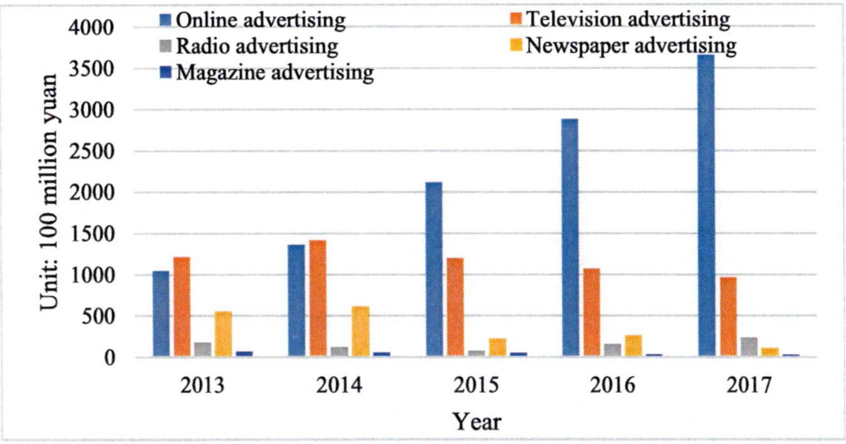

Fig. 1. China's five major media advertising market scale

As shown in Fig. 1, with the acceleration of the digitalization process, mobile Internet advertising has grown rapidly in recent years, and its proportion in traditional Internet advertising is increasing year by year. In 2017, the scale of China's mobile Internet advertising market reached 265 billion yuan, accounting for 66.8% of the Internet advertising market, and the CAGR in the past five years was as high as 66.80%. It is expected that the mobile Internet advertising will maintain a growth rate of more than 50% in the next 5 years and continue to increase its market share in the domestic advertising market as users spend more and more time on mobile devices.

4.2 Scale of Mobile Internet Advertising Revenue under the Background of Information Technology

Table 1. China mobile Internet advertising revenue scale analysis

	2013	2014	2015	2016	2017
Production value	135.6	479.4	971.5	1755.7	2184.6
Year-on-year growth		253%	102%	80.72%	24.42%

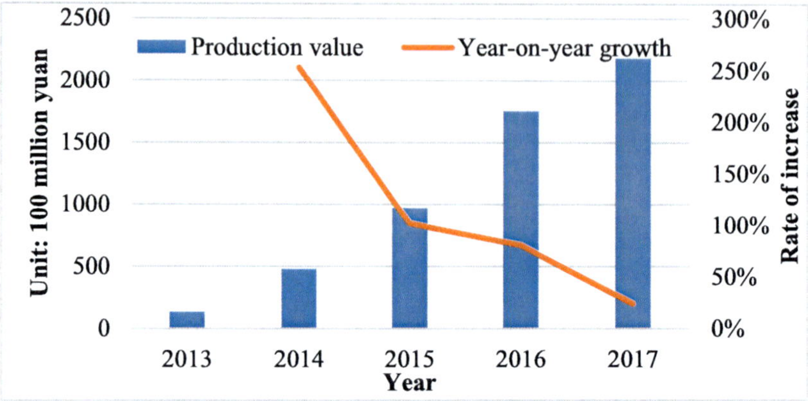

Fig. 2. China mobile Internet advertising revenue scale analysis

As shown in Table 1 and Fig. 2, the number of mobile Internet users in China is growing very rapidly, and the role of mobile Internet advertising in the field of advertising is also obvious to all. From less than 10% of mobile Internet advertising a few years ago to more than 65% in recent years, mobile Internet advertising channels are developing rapidly. Mature mobile Internet technology and e-commerce platform provide a natural and organic soil for commercial advertising design. At the same time, we are also full of various challenges in commercial advertising design.

4.3 Trend of Advertising Design Based on Information Technology

With the continuous progress of the times, people are exposed to more and more things, more and more broaden their horizons, more and more demands, competition is also intensified. The development of information technology promotes the next generation of innovation in the people-oriented, user-oriented, knowledge-based society, and interactive design can better contribute to advertising with the help of "Internet+". Interactive design is of great significance for information dissemination. It makes the information transmitted more intuitively and comprehensively show in front of the audience, which makes the information more powerful, and it is very easy to spread and make the audience take the initiative to participate and accept.

The combination of "Internet+" and traditional industries will collide with new sparks, which will continuously provide new creative inspiration for the interactive design of advertisements. With the development of the times and the innovation of science and technology, the forms of interactive design will be constantly updated, the forms of expression will become more and more rich, and the ways of advertising communication will gradually become rich. The traditional advertising communication mode can only carry out one-way point-to-point communication, and the injection of interactivity can make the advertisement carry out two-way communication, so that the audience can participate in the experience of two-way interaction and communication. Multi-sensory interaction brings more real experience to the audience, while virtual, remote and shared interaction can quickly and accurately convey information. In the future, these technologies will be more perfect and constantly innovate, so as to realize more effective and faster innovation.

The development of interactive design in the future will enrich advertising, so seize the opportunity to invest creative power to innovate content, interact with it and operate flexibly. This can not only create a good communication effect, but also increase the rich experience for people. The changes and applications of interaction design can bring greater changes to society and represent the future development trend of the advertising field.

5 Conclusion

With the development of science and technology, advertising also ushered in the form of a variety of changes, from the advertisement effect, advertising based on information technology mature analysis system for data collection, processing, screening, to avoid the traditional advertising there showing mode of inefficient propaganda, starting from the audience demand, real advertising is practical and entertaining in the first place. Advertising is not only the information output of products, but also the feedback of audience's feelings and aesthetic needs. Only through reciprocal interaction can the publicity effect of advertising be maximized. Based on information technology of the interactive advertising pay more attention to the importance of the audience consumption situation, grasp the current consumption situation and demand, and targeted advertising messages for audience to meet the requirements, in the case of appropriate, appropriate push corresponding demand of advertising information, make advertising push more conform to the audience's cognitive and accept ability in my heart, Thus to bring better interactive advertising experience to the audience.

References

1. Hosseini, Z., Mohammadi, S., Safari, H.: An assessment of the impact of information technology on marketing and advertising. Eng. Technol. Appl. Sci. Res. **8**(1), 2526–2531 (2018)
2. Pitt, J.: Facts, policies, and values? The democratic triad: technology for restoring civil discourse and civic dignity editorial. IEEE Technol. Soc. Mag. **38**(1), 5–8 (2019)
3. Alkan, M., Surmeneli, H.G.: Development of an advertisement tax system based on a geographic information system. In: Proceedings of the Institution of Civil Engineers, vol. 171, no. ME2, pp. 93–104 (2018)
4. Rezaei, S., Emami, M., Ismail, N.: Paid internet advertising (PIA) and frequency of apps-buying decisions: insights from the retail enterprises in an emerging market. J. Enterp. Inf. Manag. **31**(3) (2018)
5. Doland, A.: International agency of the year FF Shanghai. Advertising Age **87**(2), 0046 (2016)
6. Depaula, N., Fietkiewicz, K.J., Froehlich, T.J.: Challenges for social media: misinformation, free speech, civic engagement, and data regulations. Proc. Assoc. Inf. Sci. Technol. **55**(1), 665–668 (2018)
7. Havakhor, T., Sabherwal, R., Steelman, Z.R.: Relationships between information technology and other investments: a contingent interaction model. Inf. Syst. Res. **30**(1), 291–305 (2019)

8. Barabash, O., Shevchenko, G., Dakhno, N.: Information technology of targeting: optimization of decision making process in a competitive environment. Int. J. Intell. Syst. Appl. **9**(12), 1–9 (2017)
9. Rubel, A., Castro, C., Pham, A.: Agency laundering and information technologies. Ethical Theory Moral Pract. **22**(4), 1017–1041 (2019). https://doi.org/10.1007/s10677-019-10030-w
10. Safonov, A.: Discrepancy process control updating on basis of information technology application. Bull. Bryansk State Tech. Univ. **2**(1), 146–152 (2016)
11. Kameneva, N.A.: Analysis of lexical features of the Russian and English languages in the sphere of information technologies. Russ. J. Linguist. **23**(1), 185–199 (2019)
12. The role of social media on enhancing advertising effectiveness. Int. J. Civil Eng. Technol. **8**(9), 1042–1047 (2017)

Micro Animation Design Based on New Media App Interaction

Ping Hu[1] and Ronghui Fu[2](✉)

[1] Department of Games, Software Engineering Institute, Guangzhou 510900, Guangdong, China
[2] College of Computer Science, Neijiang Normal University, Neijiang 641100, Sichuan, China
20200120689@nxmu.edu.cn

Abstract. In this paper, the interaction strategy of micro animation frame prediction is used to learn and complement each other between different modes of micro animation image sequence, so as to improve the performance of the model and its generalization ability. This paper proposes a micro animation recognition method based on dual flow network information interaction. The experiment uses three micro animation databases animebase, move and move II to evaluate the effect of the method. This paper uses 3d-cnn as the basic network, and uses optical flow sequence diagram to assist training. The research shows that although the flow visualization application in the two-dimensional display interactive environment also provides interactive controls, due to the lack of depth information, users need to use the mouse to drag repeatedly to adjust the angle of view in order to put the interactive controls in the correct position. Therefore, the interaction simulation in the new media app can better increase the rendering effect of micro animation. For example, the range of seed location selection and data cutting can be expanded by 35%, the fluency of interaction can be improved by 18%, and the demand for hardware can be reduced by 5%–10%. The system can provide efficient and natural interaction for users in typical flow visualization scenarios.

Keywords: New media app · Interactive micro animation design · Frame prediction algorithm · Virtual reality technology

1 Introduction

With the rapid development of information, communication, data analysis and other technologies, the demand for micro animation interaction in various industries is growing. In the field of new media animation, the typical topology of data interaction between devices is direct access through the network, so it is particularly important to obtain the key data of devices safely, reliably and efficiently. Different from the multi view management method of separating two-dimensional environment visualization view and interactive window, in interactive virtual reality environment, user observation and interaction are completed in the same workspace. At present, the research on micro animation is less, mainly because the app has higher requirements for hardware.

B. J. Jansen et al. (Eds.): *International Conference on Cognitive based Information Processing and Applications (CIPA 2021)*, LNDECT 85, pp. 11–18, 2022.
https://doi.org/10.1007/978-981-16-5854-9_2

Because micro animation has the characteristics of motion coordination and frame predictability (as Fig. 1), many scholars optimize the frame prediction from the perspective of frame.

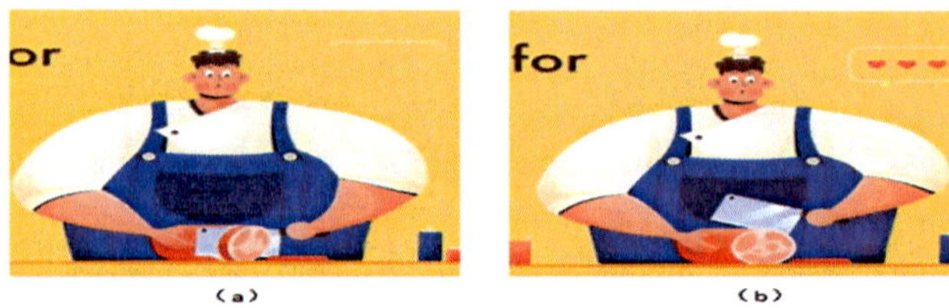

Fig. 1. Micro animation has the characteristics of motion coordination and frame predictability

In the field of micro animation, many scholars optimize the frame prediction from the perspective of frame. Kang n proposed to calculate the top frame of each micro animation sequence, that is, the frame with the largest motion amplitude, and then calculate the vertical and horizontal optical flow diagrams between the top frame and the starting frame respectively, and input them into the convolutional neural networks (CNN) to extract features. The final classification effect is greatly improved than that of using only optical flow method [1]. JD a designed a host computer human-computer interaction software, including receiving and displaying serial data, micro animation real-time decoding, simulation dexterous hand control three modules, realized the accurate and real-time control of micro animation decoding and simulation dexterous hand based on EMG signal [2]. Charalambous C focuses on the realization of human-computer interaction software, and verifies three kinds of micro animation decoding algorithms. In the future research, more kinds of micro animation decoding algorithms can be studied, so that users can use more rich micro animation operation dexterous hand [3]. Liu f proposes various visualization algorithms. Due to different configuration parameters, different interactive panels and controls are needed. For example, users use multiple visualization methods at the same time. The existence of multiple interactive interfaces and interactive controls will increase the number of 3D objects in the user's workspace. A large number of 3D objects will not only cause serious occlusion to the visualization results, it also greatly increases the interaction complexity of users [4]. The above research optimizes the micro animation from the structure of micro animation, and optimizes the demand of its hardware resources. However, in the face of the current graphics card technology improvement, strengthening the interactive rendering effect of micro animation should be worthy of further study.

In this paper, the interaction strategy of micro animation frame prediction is used to learn and complement each other between different modes of micro animation image sequence, so as to improve the performance of the model and its generalization ability. In this paper, a micro animation recognition method based on two stream network information interaction is proposed. Three micro animation databases, animebase, move and move II are used to evaluate the effect of the method. The results show that the method has certain advantages compared with the current methods. In order to better obtain the characteristics of time domain, 3d-cnn is used as the basic network, optical flow sequence diagram is used to assist training.

2 Micro Animation Interaction and Frame Prediction

In addition to space operation, there are a lot of complex interaction requirements such as parameter configuration or color coding configuration in flow visualization applications. In the two-dimensional display interactive environment, users use mouse and keyboard to manipulate the list, slider, text box and other two-dimensional interactive functions to interact [5]. In order to meet such needs, interactive virtual reality environment usually provides panel based on handle ray interaction for class interface operation [6]. However, this method requires users to use the handle for interaction, which requires users to spend more time remembering the operation method of the handle. In the system dominated by micro animation interaction, the switching between the handle and micro animation will also reduce the user's immersion, consume a lot of time, and reduce the user's work efficiency [7]. Therefore, in order to solve the above problems, an interactive panel based on user gaze and micro animation is designed to meet the requirements of parameter configuration, maintain the consistency of system interaction methods, and provide users with an integrated natural interaction solution [8]. Red gaze viewpoint is provided on the interaction panel, which indicates the intersection of user gaze and interaction panel. Users can select interaction functions in the interaction panel by staring. When the red gaze view coincides with the function button, the user can trigger the button function by clicking the micro animation [9]. After the user configures all the parameters on the panel, click OK to complete the visualization algorithm. Users may use a variety of visualization algorithms to analyze the data [10]. Due to different parameters, various visualization algorithms need to be configured by different interactive means. The visualization system should properly manage all kinds of interactive means, so that users can only contact the relevant interactive means when configuring visualization algorithms [11]. In addition, with the deepening of user analysis, the data transfer relationship between visualization algorithms will form a complex topological network, and the system should provide the function of intuitive display and modification of the network, so that users can manage the visualization algorithms and results [12]. According to the above requirements, the framework of interactive flow visualization system is designed, which includes interactive visualization management method based on Visualization scene and visual network management method based on visualization algorithm tree [13]. The visualization scene is used to configure the algorithm nodes in the network, and the visualization algorithm tree is used to manage the visualization nodes in the algorithm network [14, 15].

3 Micro Animation Interactive Experiment Design

3.1 Content

In this paper, the interaction strategy of micro animation frame prediction is used to learn and complement each other between different modes of micro animation image sequence, so as to improve the performance of the model and its generalization ability. In this paper, a micro animation recognition model based on two stream network information interaction is proposed. Three micro animation databases animebase, move and move II are used to evaluate the effect of the method.

3.2 New Media App Micro Animation Interaction Design

This model first initializes the interface, then binds the signal with the slot function, and then initializes the serial port to obtain all the serial port numbers currently open, and adds them to the drop-down menu, then initializes the status display bar, and finally runs the main window display. When you click to open the serial port, the string reading thread will be opened to read the data read by the serial port and display it in the window. When you click the link animation button, the simulation dexterous hand will be started, waiting for the control signal to be received. When you click the decode button, the real-time decoding sub process will be started to decode the EMG signal into micro animation and send out the control signal. At this time, unity 3d simulation dexterous hand receives the control signal, and the finger joint will rotate to make the corresponding micro animation.

Casme, move II and animebase databases are used in this experiment. The samples are from move database, move II database and animebase HS in animebase database. The move database uses 60 frames per second to record 253 spontaneous micro animation sequences with 25 themes. The system organizes and manages the interactive tools of visual algorithm based on the scene. Each algorithm has its corresponding scene, including parameter configuration panel, color coding panel and interactive control. When user interaction occurs, the micro animation and gaze receiver in the head mounted virtual reality device of the system will capture the user's interaction behavior, convert the behavior into data and transfer it to the interactionist module, which will convert it into specific system events, and the system events will operate various controls and interfaces in the current scene of the system. This management method can make the user's workspace only have the interactive panels and controls required by the current algorithm configuration. It can better solve the problem of user's workspace confusion in the interactive flow visualization system, effectively avoid the occlusion of the interactive interface to the visualization results, and improve the user's analysis efficiency and use experience.

4 Results and Analysis

As shown in Fig. 2, because the interactive 3D display environment has no space limitation, the system directly displays the network structure of the visualization algorithm in the 3D interactive panel. The algorithm tree uses nodes to represent the algorithm and its visualization results, and the connection between nodes to represent the data flow relationship between algorithms. Users can interact with the algorithm tree by manipulating the gaze panel to add, delete and configure each algorithm node, and display and hide the visualization results generated based on the algorithm. When the user configures an algorithm node, the system will switch to the corresponding scene, load the parameters that have been set in the algorithm, and make corresponding changes to the interactive controls and interactive panels in the scene. This method can help users quickly locate the target algorithm node in complex algorithm topology network, and effectively improve the analysis efficiency and experience of users.

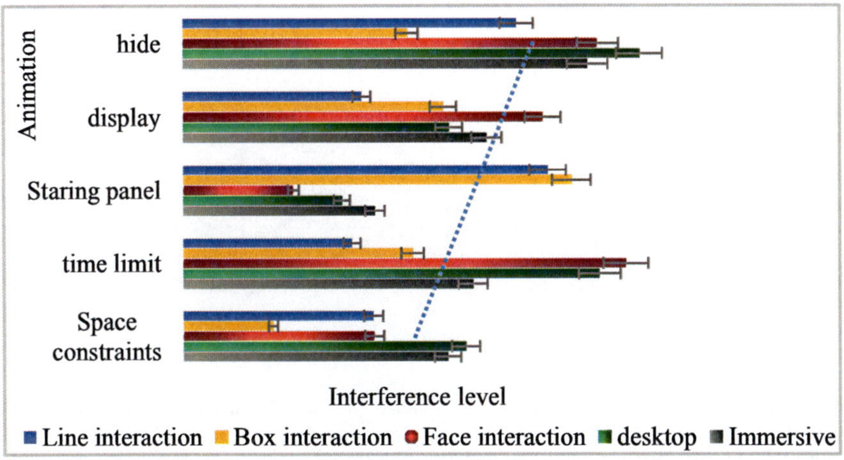

Fig. 2. Limitations of interactive 3D display environment

Table 1. Three-dimensional interactive controls of new media APP

Item	Space constraints	Time limit	Staring panel	Display
Immersive	1.39	2.12	3.22	3.44
Desktop	5.32	3.52	3.53	5.08
Face interaction	1.96	4.64	2.33	1.93
Box interaction	1.32	2.78	3.7	3.24
Line interaction	6.47	3.77	4.94	5.57

As shown in Table 1, the use of interactive controls for spatial operation in 3D space is an important interactive requirement in the application of flow visualization. For example, the selection of seed point location and data cutting range can be expanded by 35%, the interactive fluency can be improved by about 18%, and the demand for hardware can be reduced by 5%–10%. The system can provide efficient and natural interaction for users in typical flow visualization scenarios. Although the flow visualization application in two-dimensional display interactive environment also provides interactive controls, due to the lack of depth information, users need to use the mouse to drag repeatedly to adjust the angle of view in order to put the interactive controls in the correct position. Therefore, the interaction simulation in the new media app can better increase the rendering effect of micro animation. On the other hand, interactive controls in two-dimensional environment use mouse drag to interact, which can not be directly applied to interactive environment. To solve the above problems, a 3D interactive control based on new media app is designed. There are three kinds of interactive controls in the system, which are face interactive control, box interactive control and line interactive control.

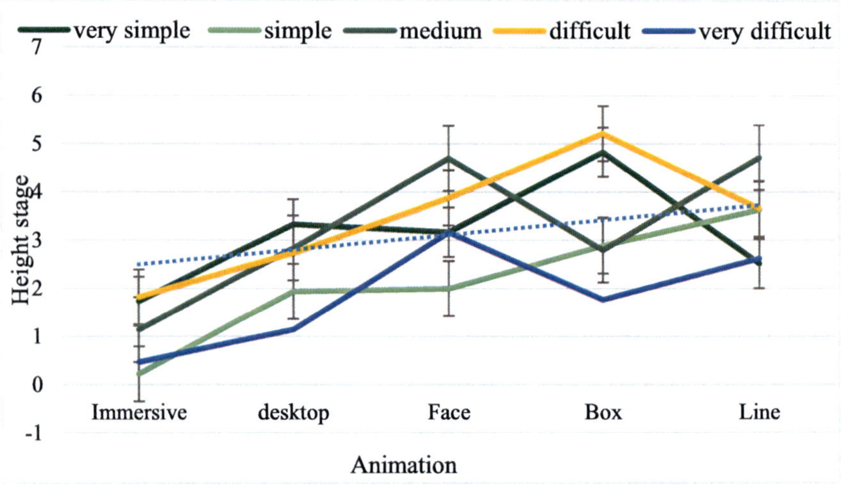

Fig. 3. Line interactive controls are often used in streamline layout

The line interactive control is often used to specify the position of line seed points when arranging streamline, as shown in Fig. 3, including two end handballs and a schematic line segment. Two handballs uniquely determine the spatial position of the schematic line segment, and the schematic line segment refreshes the handball position in real time according to the endpoint. The user can move the handball by grasping the micro animation to specify any line segment in 3D space. Face interaction control is often used to specify the position of line seed points in streamline layout, including three end handballs and a sketch plane. Among them, the three ends of the handball uniquely determine the plane in the space, and the sketch plane refreshes in real time according to the position of the handball. Users can also grasp the micro animation to move the handball and specify any plane in the workspace. Box interactive control is used to specify a box type area when cutting data, which is composed of six stretch handballs, one center handball and six sketch planes.

As shown in Fig. 4, the micro animation information and gaze information are collected and transmitted to the interactive device by the micro animation sensor leapmotion and the head mounted virtual reality device HTC vive Pro respectively; The interactive device transforms the micro animation information into micro animation events, and transmits the micro animation events and gaze information to the interaction interface and micro animation interaction control based on gaze and micro animation; The interaction interface based on gaze and micro animation will transform user gaze into interface UV coordinates, and new media app events into QT mouse click events, thus triggering interface interaction logic; QT interface and interactive controls generate visualization parameters according to user interaction, and transmit them to VTK visualization algorithm to generate visualization results; The visualization results will be transmitted to the HTC vive Pro device via steamvr.

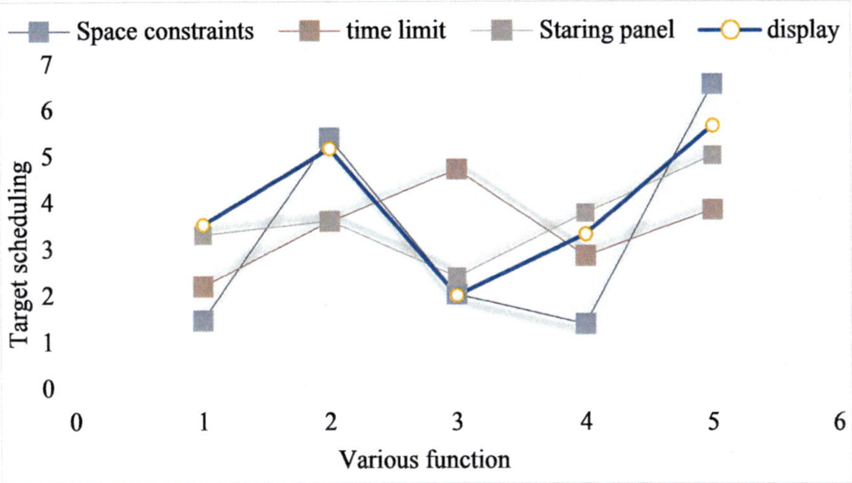

Fig. 4. Micro animation information and gaze information

5 Conclusions

This paper designs and implements micro animation and interactive control based on new media app, and interactive panel based on user gaze and micro animation. The interactive interaction management method based on visual scene and the visual network management method based on visual algorithm tree are designed and implemented. The system interaction and visual algorithm network are systematically managed, which can reduce the complexity of user interaction and provide the whole process support for typical flow visualization operation in interactive environment. Taking the streamline and cutting of typical visualization algorithms in flow visualization application as the scene, the comparative experiment between interactive system and desktop system is designed. The results show that interactive flow visualization virtual reality interactive system can provide more efficient and natural interaction for users in typical application scenarios. In the future, we can design and implement the multi-user collaboration function on the basis of the existing system. In order to optimize the design and implementation of micro animation interaction mode, the natural interaction mode in interactive environment is used to design a more interactive flow visualization interaction method, which can reduce the false trigger frequency and labor intensity of micro animation.

References

1. Kang, N., Bai, J., Pan, J., et al.: Interactive animation generation of virtual characters using single RGB-D camera. Vis. Comput. **35**(6–8), 849–860 (2019)
2. Doyle, J., Glass, K.C., Racz, M., et al.: Student-directed interactive animation for learning cytochrome P450-mediated drug metabolism. Curr. Pharm. Teach. Learn. **10**(12), 1565–1573 (2018)
3. Charalambous, C., Yumak, Z., Frank, V.: Audio-driven emotional speech animation for interactive virtual characters. Comput. Anim. Virtual Worlds **30**(4), 88–92 (2019)
4. Liu, F., Sun, Z., Ma, C., et al.: Context-aware animation data description and interaction method based on sketch. J. Beijing Inst. Technol. **28**(01), 12–20 (2019)
5. Dharmawati, R., Destiana, H.: Interactive animation design of Hijaiyah letters in early age children at Al-Hidayah Kindergarten Bekasi. SinkrOn **3**(2), 89–90 (2019)
6. Zhao, J.: Research on color design in interactive animation design of multiculturalism in the information age. In: Journal of Physics Conference Series, vol. 13, no. 25, pp. 56–61 (2019)
7. Ciccone, L., Oztireli, C., Sumner, R.W.: Tangent-space optimization for interactive animation control. ACM Trans. Graph. **38**(4), 1011–1012 (2019)
8. Ussiph, N., Seidu, H.K.: The impact of using 3D interactive animation tool in teaching computer programming at the senior high school level. Circulation Comput. Sci. **3**(2), 18–28 (2018)
9. Doyle, J., Glass, K.C., Racz, M., et al.: Student-directed interactive animation for learning cytochrome P450-mediated drug metabolism. Curr. Pharm. Teach. Learn. **10**(12), 56–59 (2018)
10. Liu, Z., Zhou, L., Leung, H., et al.: High-quality compatible triangulations and their application in interactive animation. Comput. Graph. **76**(11), 60–72 (2018)
11. Gan, B., Zhang, C., Chen, Y., et al.: Research on role modeling and behavior control of virtual reality animation interactive system in Internet of Things. J. Real-Time Image Process. **5**(1), 1–15 (2020)
12. Şılbır, L., Coşar, A.M., Kartal, Y., et al.: Graphic symbol based interactive animation development process for deaf or hard of hearing students. Int. Electron. J. Elementary Educ. **12**(4), 41–42 (2021)
13. Paier, W., Hilsmann, A., Eisert, P.: Interactive facial animation with deep neural networks. IET Comput. Vis. **14**(3), 995–998 (2020)
14. Li, Y., Li, J.: Interactive system research based on animation visual effects. J. Comput. Methods Sci. Eng. **19**(4), 1–7 (2019)
15. Red cell indexes made easy using an interactive animation: do students and their scores concur. Adv. Physiol. Educ. **42**(1), 50–55 (2018)

The Impact of Digital Economy on Economic Growth Based on Pearson Correlation Test Analysis

Rong Li and C. G. Gospodarik(✉)

Faculty of Economics, Belarusian State University, 220030 Minsk, Belarus
C.G.Gospodarik@bsu.by

Abstract. In the global economic system, the digital economy (DE) has shown a trend of rapid development both internationally and domestically in recent years. In the country, affected by the dividend of my country's population base, the development of the DE is also very rapid. This article takes China, Japan, South Korea, Brunei, and Cambodia in the Asia-Pacific region as the analysis objects to explore the impact of the DE on national economic growth (EG) in the information age. Through the collection and research of related literature on informatization, DE and EG, this article analyzes and studies the relevant theories of the impact of DE on EG in the information age, and lays a theoretical foundation for subsequent research. This article uses factor analysis to measure the DE development index in the Asia-Pacific region when calculating the DE development index in the Asia-Pacific region. The Pearson correlation test, EG two-step cointegration analysis, Granger causality analysis and other methods are used in the empirical research of the DE on the EG of the Asia-Pacific region. The research results show that the DE accounts for an increasing proportion of each country. The absolute value of the Pearson correlation coefficient between the DE development index and their EG in these 5 countries is greater than 0.80, and the P value of the residual unit root test is Both are below 0.025, and the development of the DE has a positive effect on EG.

Keywords: DE · EG · Informatization · National economy

1 Introduction

In the 21st century, the development and wide application of information technology has brought about the birth of the DE, which has caused major changes in social and economic development, and has become the most important force to promote social progress after the industrial revolution [1, 2]. At present, the developed countries in the world are racing to develop the DE, expanding the influence of the DE in the national economy, in order to improve the national economic capacity as soon as possible, thereby enhancing the country's comprehensive factors, so as to be in a leading position in the fierce competition in the 21st century [3, 4].Whether from the perspective of social development or current

B. J. Jansen et al. (Eds.): *International Conference on Cognitive based Information Processing and Applications (CIPA 2021)*, LNDECT 85, pp. 19–27, 2022.
https://doi.org/10.1007/978-981-16-5854-9_3

research, the research on the impact of the DE on EG has important practical significance and theoretical value [5, 6].

The rapid development of the DE has led many scholars to conduct research on it. For example: Patrizia Gazzola, Gianluca Colombo and others believe that the emergence of the DE and competition in the online market have caused new challenges in consumer protection measures; it is expected that online consumer skills will be improved, and consumer awareness and participation. The level will also increase; these are the basic prerequisites for sustainable procurement decisions and should therefore be regarded as the pillars of responsible online consumption [7]. Elena-Madalina Vatamanescu, Bogdan Gabriel Nistoreanu and Andreea Mitan used partial least squares structural equation modeling to investigate subjective and objective measurement standards related to consumer awareness and action levels in the context of fast-growing competition and the enhancement of the DE Technology, found that online shopping decisions are indirectly affected by fierce competition in the DE through consumer protection awareness and consumer protection behavior [8].

This article first uses factor analysis to measure the digital economic development indexes of China, Japan, South Korea, Brunei, and Cambodia from 2011 to 2020, and compares and analyzes the DE development levels of these five countries. Secondly, Pearson correlation test, EG two-step cointegration test, Granger causality test and other methods are used to conduct empirical analysis on the relationship between the DE and EG. Finally, based on the above research content, policy recommendations are made on the impact of the DE on the EG of five countries.

2 Research on the Impact of DE on EG in the Information Age

2.1 Informationization

(1) The connotation of informatization
To a large extent, the concept of "Internet+" has emerged with the need of modern people to understand social development trends [9]. Informatization does not only mean the application of our information science and technology, the comprehensive development and utilization of information resources, the adjustment and optimization of the industrial structure, the expansion of the added value of the national economy and society, and the growth of total employment. A future social life and mode of production should also include the main body in the process of social life, a person's way of thinking, production mode and social lifestyle changes, as well as the determination of other social behavior patterns, social structures, and other social norms and systems [10, 11].

(2) Basic characteristics of informatization
This article summarizes the characteristics of informatization as three characteristics: fast, wide, and deep, and these three complement each other and promote each other to form a joint force to advance the process of informatization. This is the most basic external manifestation of the connotation of informatization. Among them, "fast" refers to the rapid development of informatization; "wide" refers to a wide range of informatization; "deep" refers to the profound influence of informatization.

2.2 DE

(1) The connotation of the DE
This article summarizes the connotation of the DE as: DE is the result of the transformation of information and communication technology, digitization is its key, and information, computers, and communication technology are important aspects of it.

(2) The main characteristics of the DE
From a tax perspective, the DE has the following characteristics: First. Liquidity; second, it is dependent on data, especially "big data"; third, it has a network effect; fourth, adopt a multi-party business model; fifth, in some business models that rely heavily on network effects, the DE has a trend of monopoly or oligopoly; sixth, volatility due to low entry barriers and rapid development of technology: mainly reflected in the rapid update of products and services, and the rapid change of industry-leading companies [12].

2.3 EG

(1) EG theory

1) Classical EG theory
The theory believes that the output in society is mainly the function of human capital. The increase in population and the increase in total social output will promote EG; however, the more land input is not the better, the marginal return of land is indeed diminishing. When the population reaches a certain level, the growth of land cannot meet the needs of its production development and life, population growth slows down, and finally reaches equilibrium after development. At this time, the population growth rate and land growth rate are zero.

2) New EG theory
The new EG theory is also called the endogenous growth theory. The comparison of the new EG theory with classical economics, neo-classical economics and other economic disciplines are mainly manifested in two aspects. One is to treat scientific and technological progress as an endogenous variable; the second is to redefine our labor force emphasizes the importance of the employees in the enterprise for the knowledge reserve and the influence of the knowledge reserve on the rapid EG. The new EG theory advocates that improving its innovation capability will become a very inevitable way for developing countries and underdeveloped regions to implement industrialization and transform economic development models.

(2) EG model
The production function can be expressed by the following mathematical expressions:

$$Y = f(X_1, X_2, \cdots, X_n) \quad (1)$$

Among them, Y is output, X_1, X_2, $\cdots$, X_n are the inputs of various production factors, and f represents the relationship between the input of production factors and the final output.

The meaning of this kind of production function is the input of various production factors. There is only one production factor input, and the specific production factors are not clearly pointed out. Moreover, the correlation formula of this function is not completely given, so this the expression can only be used as a conceptual model between input and output. The manufacturing functions widely used in practical applications mainly include: Cobb Douglas manufacturing function; ces manufacturing function and transcendental logarithmic manufacturing function.

2.4 Recommendations for the Development of the DE to Promote EG in the Asia-Pacific Region

Through the empirical analysis and research conclusions of this article, it can be found that there is a positive correlation between the development of the DE and EG, and the increase of the DE development index can effectively promote EG. In order to promote the economic development of the Asia-Pacific region, this article proposes the following countermeasures: 1) Improve scientific and technological innovation capabilities and promote the development of the DE; 2) training and attracting information technology talents; 3) increasing government funding support and build an environment that promotes the development of the DE.

3 Research on the Impact of DE on EG in the Information Age

3.1 Calculation of the DE Development Index of 5 Countries in the Asia-Pacific Region

(1) DE development indicator system
This paper takes China's Asia-Pacific DE development indicator system as the main reference system, and combines the objectivity of data selection, the operability of data, and the spatiality of data selection of territorial data. The regional DE development indicator system that is in line with the actual situation in the Asia-Pacific region has measured the digital economic development indexes of 5 countries in the Asia-Pacific region. In the DE growth rate indicator system of 5 cities and countries in the Asia-Pacific region that this article needs to build, the primary indicators mainly include government basic indicators, industrial indicators, integration indicators, and environmental protection indicators. The secondary indicators of basic

indicators mainly include Traditional digital infrastructure, new digital infrastructure, environmental protection secondary indicators mainly include government network new media, government network services, and government data governance. The secondary indicators of industrial indicators include industrial scale, industrial body, and secondary indicators of integration. The indicators include the integration of industry and information technology, the digitalization of agriculture, and the digitalization of the service industry.

(2) Measurement method
This article comprehensively considers various factors, based on the consideration of objectivity and operability, selects the factor analysis method to measure the DE development index of 5 countries in the Asia-Pacific region, uses the factor analysis method to reduce the dimensionality of the index system, and then goes through a series of processing, finally, based on the main factors obtained and the contribution rate of each main factor, the required DE development index is obtained by adding them up respectively.

In this part, it mainly involves the calculation of the principal factors in the principal component analysis method model, and finally the calculation of all index indexes. The principal component analysis method model is:

$$\begin{cases} F_1 = a_{11}X_1 + a_{21}X_2 + a_{31}X_3 + \cdots + a_{n1}X_n \\ F_2 = a_{12}X_1 + a_{22}X_2 + a_{32}X_3 + \cdots + a_{n2}X_n \\ \vdots \\ F_n = a_{1n}X_1 + a_{2n}X_2 + a_{3n}X_3 + \cdots + a_{nn}X_n \end{cases} \tag{2}$$

Among them, $a_{1m}^2 + a_{2m}^2 + \cdots + a_{nm}^2 = 1$. $a_{1m}, a_{2m}, \cdots, a_{nm}(m = 1, 2, 3, \cdots, n)$ are the eigenvectors corresponding to the largest eigenvalues of the covariance matrix of X, and $F_i(i = 1, 2, 3, \cdots, n)$ is the main component. The principal components in the analysis, $X_i(i = 1, 2, 3, \cdots, n)$ are the factors in the factor analysis method.

The calculation of the comprehensive index (V) is calculated from the calculated principal component $F_i(i = 1, 2, 3, \cdots, n)$value and its contribution rate:

$$V = \frac{F_1a_1 + F_2a_2 + F_3a_3 + \cdots + F_na_n}{\sum_1^n a_i} \tag{3}$$

Among them, the value calculated by the principal component of $F_i(i = 1,2,3,\cdots,n)$ is the contribution rate of the corresponding principal component.

(1) Data source
Most of the original data in the research process of this article comes from the statistical data published by the governments of various countries on the Internet, and a small part comes from the research data of other scholars.

(2) Data verification
In the process of calculating the DE development index of 5 countries in the Asia-Pacific region, KMO and Bartlett tests will be performed on the original data first to check whether the selected data meet the requirements.

3.2 Empirical Study on the Impact of DE on EG

(1) Model construction
The idea of model construction in this paper is mainly based on the theoretical basis of the contribution of the DE to EG, based on the Cobb-Douglas production function as the model, and the improved Cobb-Douglas production function is selected as the model:

$$GDP = A_0 K^{\alpha} L^{\beta} I^{\gamma} \quad (4)$$

This article is only empirical on the impact of the DE on the EG of 5 countries in the Asia-Pacific region, so it is based on a simple linear model:

$$Y = C + \beta X + \varepsilon \quad (5)$$

(2) Data source
The digital economic development index of 5 countries in the Central Asia-Pacific region calculated in this paper is used as the independent variable, and the GDP of 5 countries in the Asia-Pacific region is the dependent variable. The independent variables are obtained through factor analysis. The DE development index of 5 countries in the Asia-Pacific region can more truly reflect the level of DE development in these 5 countries. Based on the nature of comparability and scientificity, in the empirical process, the original data selected are the relevant data of 5 countries in the Asia-Pacific region from 2011 to 2020, and the data source is the statistical data published by the governments of each country on the Internet.
(3) Empirical test

1) Pearson correlation test
The Pearson correlation coefficient is mainly used to judge whether there is a correlation between the research variable and another variable. It is an indicator of the correlation between the research variables. The Pearson correlation coefficient can be positive or negative. When it is positive, it indicates that there is a positive correlation between the variables, and vice versa, which indicates that there is a negative correlation between the variables. When the absolute value of Pearson's correlation coefficient is 1, it means that the variables are completely positively correlated or completely negatively correlated. It is generally believed that when the absolute value of the Pearson correlation coefficient between one variable and another variable is greater than 0.80, there is a significant correlation between the two variables.
2) EG two-step cointegration test
The EG two-step cointegration test is mainly used to test the cointegration relationship of a single variable, and the EG two-step cointegration test is mainly used to verify that the unit root test is unstable when the variable is level, and the first-order or second-order unit root is stable Under the circumstances, it is used to eliminate the problem of "false regression". The EG two-step cointegration test is based on the same order of the single integer of each variable.

3) Granger non-causality test
   Granger causality test is aimed at the causality of statistical data in the statistical sense. It cannot express the causality in real life, but it still has certain significance for economic forecasting.

4 Research on the Impact of DE on EG in the Information Age

4.1 Analysis of the DE Development Index of 5 Countries in the Asia-Pacific Region

After standardizing the 2011–2020 data of 5 countries in the Asia-Pacific region and using factor analysis methods to measure and calculate the 2011–2020 DE development index of 5 countries in the Asia-Pacific region, as shown in Table 1:

Table 1. Digital economic development index 2011–2020 for five countries in the Asia-Pacific region

	China	Japan	South Korea	Brunei	Cambodia
2020	923.8	902.4	871.6	904.7	857.6
2019	1004.4	974.1	947.6	991.6	934.3
2018	982.7	948.7	901.7	975.6	893.1
2017	427.4	404.4	337.1	437.7	341.8
2016	267.7	247.6	189.3	231.7	127.6
2015	104.6	107.8	61.7	110.4	73.1
2014	60.8	57.9	20.7	59.7	27.3
2013	40.7	37.4	3.4	37.7	6.2
2012	30.3	29.7	−14.7	24.1	−16.4
2011	25.7	20.6	−21.4	14.7	−19.7

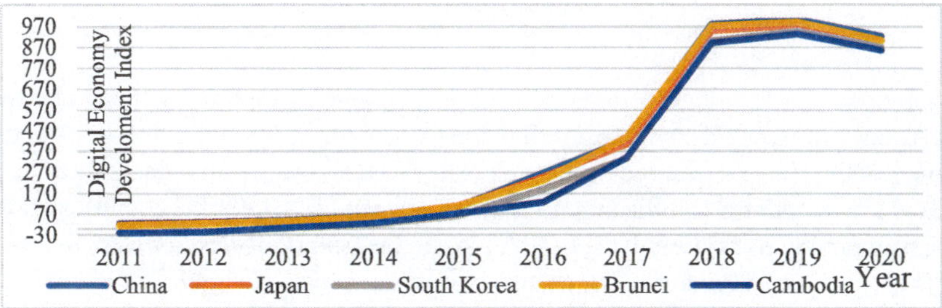

Fig. 1. Digital economic development index 2011–2020 for five countries in the Asia-Pacific region

As shown in Fig. 1, from 2011 to 2020, the DE development index of 5 countries has risen year by year, and the upward trend is getting bigger and bigger. Generally speaking, the DE development of several countries such as China, Japan and Brunei, the index is higher than South Korea and Cambodia. The above results show that the DE accounts for an increasing proportion of each country, and each country also attaches importance to the degree of development of the DE in its own country every year.

4.2 Empirical Study on the Impact of DE on EG

This paper takes the DE development index of 5 countries as the independent variable and the GDP of each country as the dependent variable, and performs Pearson correlation test and EG cointegration test on them. The test results are shown in Table 2:

Table 2. Test result

	China	Japan	South Korea	Brunei	Cambodia
Pearson Correlation	0.978	0.879	0.901	0.897	0.941
Residual Unit Root Test (P)	0.0204	0.0134	0.0245	0.0107	0.0127
GDP Change Ratio	0.0809	0.0762	0.0641	0.0716	0.0743

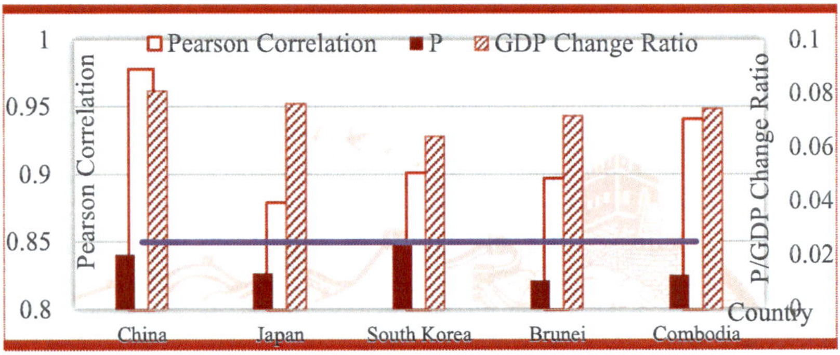

Fig. 2. Test result

As shown in Fig. 2, the absolute value of the Pearson correlation coefficient between the DE development index and their EG in the five countries is greater than 0.80, and the P values of the residual unit root test are all below 0.025. It shows that under the bilateral level of 1%, there is a significant correlation between the GDP of 5 countries and the DE development index during 2011–2020. When the economic development index of 5 countries grows by 1%, their GDP change rate It is of practical significance, and its rate of change is around 0.07. The development of the DE has a driving effect on EG.

5 Conclusions

This article studies the relationship between the DE and EG. The DE is the trend of economic development in the world today and is closely related to EG. It can be said that future EG depends largely on the degree of digital economic development. Based on the collection and sorting of domestic and foreign documents, this article first measures the DE development index of China, Japan, South Korea, Brunei, and Cambodia from 2011 to 2020 by constructing a DE development index system in the Asia-Pacific region. Calculate the comprehensive index index for analysis and research. Secondly, it analyzes the impact of the DE and EG of these five countries, and empirically analyzes the impact of the development of the DE on their EG, using the Pearson correlation test, the EG two-step test, and the granger causality test, an empirical analysis of the impact of the economy on its EG.

References

1. Malyuk, A.A., Gavdan, G.P.: Development and use of national information resources as the basis for DE development. Bezopasnost Informacionnyh Tehnol. **26**(2), 67–85 (2019)
2. Lupak, R.L., Shtets, T.F., Pryshlyak, O.P.: Institutional aspects of the state regulation of the development of DE of Ukraine. Bus. Inform. **1**(504), 271–277 (2020)
3. Ku Lyna, H., Nalukova, N.: Informatization of social providing as a main element in the formation of DE. World Finan. **1**(62), 95–106 (2020)
4. Abrashkin, M.S., Vershinin, A.A.: The impact of the DE on the development of Russian industry. Voprosy Regionalnoj Ekonomiki **34**(1), 3–9 (2018)
5. Zaslavskaya, O.Y.: The impact of global processes of informatization the development of the modern education system in a DE. Rudn J. Informatization Educ. **15**(3), 271–281 (2018)
6. Voronkova, V.H., Nikitenko, V.A., Teslenko, T.V., et al.: Impact of the worldwide trends on the development of the DE. Revista Amazonia Investiga **9**(32), 81–90 (2020)
7. Gazzola, P., Colombo, G., Pezzetti, R., et al.: Consumer empowerment in the DE: availing sustainable purchasing decisions. Sustainability **9**(5), 693 (2017)
8. Vatamanescu, E.M., Nistoreanu, B.G., Mitan, A.: Competition and consumer behavior in the context of the digital economy. Amfiteatru Econ. J. **19**(45), 354–366 (2017)
9. Evseeva, S., Evseeva, O., Kalinina, O.: HR staff leasing: opportunities in the DE environment (case of Russia). In: E3S Web of Conferences, vol. 175, no. 1, p. 13041 (2020)
10. Jiang, H., Cao, Y.: Research on DE boosting the transformation and upgrading of traditional industries in Foshan City. In: E3S Web of Conferences, vol. 235, no. 3, p. 03060 (2021)
11. Zavalko, N.A., Matyunina, O.E., Izmaylova, S.A., et al.: Key components of the financial infrastructure of the de within the fuel and energy complex. Int. J. Eng. Technol. **7**(4.38):672 (2018)
12. Akmarov, P., Gazetdinov, M., Tret'Yakova, E.: Problems of protection of commercial information in the conditions of economy digitalization. Vestnik of Kazan State Agrarin University **15**(2), 133–138 (2020)

Tourism Demand Analysis Algorithm Based on Data Forecast Model Analysis

Meng Chen(✉)

Chengdu Neusoft University, Chengdu, Sichuan, China
chenmeng@nsu.edu.cn

Abstract. Big data provides a good data foundation for accurate analysis and forecasting of tourism and other industries. With the rapid development of tourism industry, tourism demand analysis urgently needs real-time and accurate data as support. Taking Sichuan's inbound tourism demand as the research object, collecting big data for empirical research, then through the establishment of double logarithmic linear model to analyze the main factors affecting the demand of inbound tourism in Sichuan Province. Based on the results of big data analysis, macroeconomic variables (relative CPI, GDP per capita, and exchange rate) and special event factors have varying degrees of impact on the demand for inbound tourism in Sichuan Province. This research is expected to provide a new data-based analysis perspective for the tourism industry in the era of big data.

Keywords: Inbound tourism demand · Double logarithm model · Big data

1 Introduction

Sichuan's inbound tourism has developed rapidly, and the number of inbound tourists and foreign exchange income from inbound tourism have maintained a continuous growth trend. Sichuan Province has rich cultural tourism resources and the largest passenger airport in Central China, and is within the strategic circle of the One Belt One Road. Therefore, the development of inbound tourism in Sichuan Province has the research significance of "regional inbound tourism economy".

In the study of factors affecting inbound tourism demand, scholars generally build a logarithmic linear regression model, which is closely related to the elasticity of the coefficient of the regression model and has a more intuitive economic significance [1]. The establishment of the model can directly reflect the impact of a specific variable on tourism demand.

From the perspective of econometrics, a regression model of Sichuan Province is constructed, and the SPSS statistical analysis software is used to analyze the inbound tourism-related data of Sichuan Province from 2009 to 2018. The value of the independent variable coefficient can be directly obtained from the model result, and the regression equation can be directly written according to the value of the independent variable coefficient. The coefficient before each independent variable in the regression

B. J. Jansen et al. (Eds.): *International Conference on Cognitive based Information Processing and Applications (CIPA 2021)*, LNDECT 85, pp. 28–37, 2022.
https://doi.org/10.1007/978-981-16-5854-9_4

equation refers to the degree of change of the explained variable caused by each unit of the explanatory variable [2–4]. After the parameters of the regression model are estimated, the model should be tested for goodness of fit [5, 6], residual series correlation test [7] and multicollinearity test [3].

2 Data Source and Variable Selection

2.1 Data Source

The main sources of the data used in this article: Sichuan Provincial Statistical Yearbook (2009–2018); Sichuan Provincial Yearbook (2009–2018); National Bureau of Statistics official website; Sichuan Provincial Tourism Official Website; Bank for International Settlements; World Bank Data Website; World Economic and Trade Cooperation Organization, Statistical Database, etc.

2.2 The Main Influencing Factors of Sichuan's Inbound Tourism Demand

Although scholars have different views on the factors affecting the demand of inbound tourism. However, it is generally accepted that the following aspects have an important impact on the demand for inbound tourism, including the per capita income level of the source countries, the consumer price index, the exchange rate and emergencies [8, 9]. This study only analyzes and studies the main influencing factors of inbound tourism demand in Sichuan Province.

(1) Number of Inbound Tourists from Main Source Countries
(2) The number of inbound tourists can reflect the tourist attraction of a country (region) as well as the degree of openness to the outside world [10]. This chapter selects three countries with relatively large number of tourists to Sichuan Province, namely: The United States, The United Kingdom and Malaysia. The above three countries were selected for the research, on the one hand, because these three countries come to Sichuan Province in the majority of tourists; on the other hand, geographical representation is also considered.
(3) Per capita GDP of source countries
(4) The personal income of the residents in the source country is the most important factor that affects the tourists to travel abroad. It is impossible to make statistics on the income of each inbound tourist, and it is even more difficult to analyze them one by one. Therefore, the academic circles usually use per capita GDP to represent the personal income level of residents.
(5) Relative CPI of Tourist Source Countries and Sichuan Province
(6) The price of tourism products has a direct impact on the demand for tourism, which is embodied in the premise that the income level of tourists remains unchanged. Due to the different source countries, these fees vary greatly, and there is no specific statistical data and no unified pricing standard. Therefore, this dissertation takes the relative consumer price index (relative CPI) of tourist source countries and Sichuan Province as an index to measure the tourism price.

(7) The relative CPI formula is expressed as follows:

$$\text{Relative CPI}_t = \frac{CPI_t^{\text{source country}}}{CPI_t^{\text{Sichuan province}}} \tag{1}$$

(8) Exchange Rate
(9) The exchange rates of different source countries have different impacts on inbound tourism demand. With the slow appreciation of Chinese currency in 2009, the national economy of Sichuan Province has been affected to a certain extent, and the corresponding inbound tourism will also be affected. Therefore, Sichuan provincial government and relevant tourism enterprises should pay attention to the change of exchange rate.
(10) The exchange rate is expressed as:

$$ExR = ER_t \frac{CPI_t^{\text{source country}}}{CPI_t^{\text{Sichuan province}}} \tag{2}$$

Refers to how much Chinese currency can be exchanged for a unit of country's currency. In order to eliminate the influence of inflation, the ratio of the CPI of the source country and Sichuan Province is introduced in the formula.
(11) Major or Unexpected Events
Tourism is a very fragile industry, easily affected by special events such as domestic and foreign politics and economy. The 2012 China International Friendship Cities Conference attracted the attention of tourists from all over the world and also increased the number of tourists in Sichuan Province since 2012. As a result, foreign exchange income in 2012 increased by 34.4% compared to 2011. The number of arrivals in 2012 also increased by 38.6% compared to the previous year.

3 Model Analysis of Inbound Tourism in Sichuan Province

3.1 Model Setting

International tourism is essentially equivalent to international trade, and the double logarithmic linear model is derived from the study of changes in the gravity model of international trade [11, 12]. In 1989, the most basic trade gravity model was proposed by Bergstrand, and its formula is as follows:

$$M_j^i = aY_i^b Y_j^c D_{ij}^d A_{ij}^e \tag{3}$$

Among them: M_j^i represents the trade import value of country i from country j in a certain period;

Y_i represents the gross national product of the importing country;
Y_j represents the gross national product of the exporting country;
D_{ij} represents the direct distance between two countries;

A_{ij} represents other factors that promote or hinder the trade between two countries.

The various influencing factors of inbound tourism demand are incorporated into the above formula and converted into a log-linear model. This is to transfer the variables used in the literature to the variables available in this study.

The dependent variables and explanatory variables (excluding dummy variables) are converted to logarithms. The parameters of the log-linear model have elastic interpretations, and assume constant elasticity across all values of the data set. Therefore, the inbound tourism demand model is as follows:

$$\ln(\text{numberofvisitors})_t = \beta + b\ln(\text{GDP per capita})_t + c\ln(\text{relativeCPI})_t + d\ln(\text{exr})_t + eD_t \tag{4}$$

Among them:

$(\text{numberofvisitors})_t$ indicates the number of inbound tourists from a certain country to Sichuan Province during the period t;
$(\text{GDP per capita})_t$ means the per capita gross national product of a source country during t, excluding inflation;
$(\text{relativeCPI})_t$ represents the ratio of the consumer price index of a certain source country and Sichuan Province in period t;
$(\text{exr})_t$ represents the exchange rate of a source country's currency relative to the Chinese currency during the t period, excluding inflation factors;
D_t represents a dummy variable, that is, a major event that has an important impact on inbound tourism in Sichuan Province during the period. In this study, IFConference is dummy variables, which are set to 1 when they occur, and 0 when they do not occur.

The study uses the method of dummy variables to quantitatively analyze the impact of special political and economic events on the inbound tourism industry in Sichuan Province.

Therefore, the above double logarithmic linear formula can be written as the following formula:

$$\begin{aligned}\ln(\text{numberofvisitors})_t = {} & \beta + b\ln(\text{GDP per capita})_t + c\ln(\text{relativeCPI})_t \\ & + d\ln(\text{exr})_t + D_1(\text{IFConference}) \quad t = (2009, 2010, \ldots, 2018)\end{aligned} \tag{5}$$

From the Sichuan Province inbound tourism demand model established by the formula (5), it can be seen that the independent variable coefficients directly represent the elasticity of Sichuan Province's inbound tourism demand to the per capita income of the source country, the elasticity of inbound tourism demand to the relative consumer price index, and The elasticity of inbound tourism demand to exchange rates is in line with the requirements of studying tourism demand rules. The quantitative relationship between each influencing factor and the demand for inbound tourism is established through a double logarithmic linear model.

3.2 Empirical Analysis

Using the Eq. (5) to establish the inbound tourism demand regression model in Sichuan Province, using the statistical software SPSS, analyze the relevant statistical data of British,Malaysian and American's tourism in Sichuan Province. The model parameter estimates, correlation tests and collinearity tests shown in the table below are obtained.

3.2.1 United Kingdom

Model		Unstandardized Coefficients		Standardized Coefficients	t	Sig.	Collinearity Statistics	
		B	Std. Error	Beta			Tolerance	VIF
1	(Constant)	-14.776	6.343		-2.329	0.059		
	ln(GDPpercapita)	4.104	0.572	0.397	7.175	0	0.979	1.021
	ln(ExR)	-7.576	0.49	-0.865	-15.461	0	0.957	1.045
	IFConference	0.245	0.111	0.123	2.21	0.069	0.973	1.027

Fig. 1. Model parameter estimation, correlation test and collinearity test for British tourism in Sichuan Province

It can be seen from Fig. 1 that the regression equation for British traveling to Sichuan Province is:

$$\begin{aligned}\ln(\text{numberofvisitors})_t = & -14.776 + 4.104\ln(\text{GDP per capita})_t - 7.576\ln(\text{exr})_t \\ & + 0.245(\text{IFConference}),\quad t = (2009, 2010, \ldots, 2018) \qquad (6)\end{aligned}$$

British GDP per capita, exchange rates between the two places, and China International Friendship Cities Conference are the main factors affecting British tourism in Sichuan Province. Because the correlation coefficient between ExR and related CPI is relatively large, there is a correlation. In order to reduce the impact of collinearity on model diagnosis, the relative CPI is excluded from the model (Figs. 2, 3 and 4).

Model	R	R Square	Adjusted R Square	Std. Error of the Estimate	Durbin-Watson
1	0.991	0.982	0.973	0.1038655	2.504

Fig. 2. The fitting degree and sequence correlation test of British model for Sichuan inbound tourism

ANOVA						
Model		Sum of Squares	df	Mean Square	F	Sig.
1	Regression	2.496	3	0.832	27.859	0.001
	Residual	0.179	6	0.03		
	Total	2.675	9			

Fig. 3. Model rationality

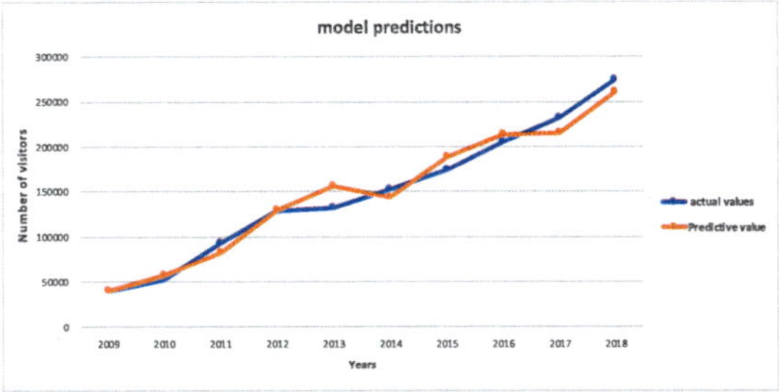

Fig. 4. Fitting graph of model prediction results for British tourists

3.2.2 Malaysia

Model		Unstandardized Coefficients		Standardized Coefficients	t	Sig.	Collinearity Statistics	
		B	Std. Error	Beta			Tolerance	VIF
1	(Constant)	3.307	4.123		0.802	0.453		
	ln(GDPpercapita)	1.115	0.44	0.272	2.536	0.044	0.973	1.028
	ln(ExR)	-3.674	0.654	-0.751	-5.62	0.001	0.625	1.6
	ln(RelativeCPI)	7.892	5.312	0.197	1.486	0.188	0.636	1.572

Fig. 5. Model parameter estimation, correlation test and collinearity test for Malaysian tourism in Sichuan Province

It can be seen from Fig. 5 that the regression equation for Malaysian traveling to Sichuan Province is:

$$\ln(\text{numberofvisitors})_t = 3.307 + 1.115\ln(\text{GDP per capita})_t - 3.674\ln(\text{exr})_t + 7.892(\text{RelativeCPI}) \quad t = (2009, 2010, \ldots, 2018) \quad (7)$$

Malaysia's per capita GDP, the exchange rates of the two places and the relative CPI of the two places are the main factors that affect Malaysians to travel to Sichuan Province. Since the significance of IFConference is much greater than 0.05, indicating that the coefficient is 0, it has almost no effect on inbound tourism, so the variable is eliminated (Figs. 6, 7 and 8).

Model	R	R Square	Adjusted R Square	Std. Error of the Estimate	Durbin-Watson
1	0.966	0.933	0.9	0.1728003	2.232

Fig. 6. The fitting degree and sequence correlation test of Malaysia model for Sichuan inbound tourism

ANOVA						
Model		Sum of Squares	df	Mean Square	F	Sig.
1	Regression	2.496	3	0.832	27.859	0.001
	Residual	0.179	6	0.03		
	Total	2.675	9			

Fig. 7. Model rationality

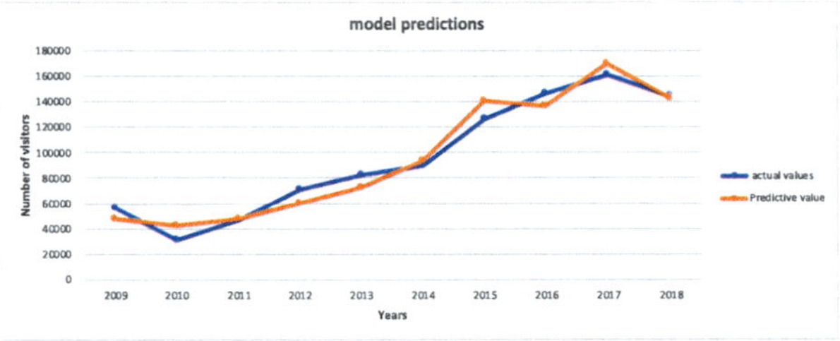

Fig. 8. Fitting graph of model prediction results for Malaysian tourists

3.2.3 United States

Model		Unstandardized Coefficients		Standardized Coefficients	t	Sig.	Collinearity Statistics	
		B	Std. Error	Beta			Tolerance	VIF
1	(Constant)	-45.74	7.121		-6.423	0.001		
	ln(GDPpercapita)	5.865	0.621	0.968	9.447	0	0.967	1.034
	ln(ExR)	-3.203	1.404	-0.236	-2.282	0.063	0.95	1.052
	IFConference	0.233	0.19	0.128	1.228	0.266	0.93	1.076

Fig. 9. Model parameter estimation, correlation test and collinearity test for American's tourism in Sichuan Province

It can be seen from Fig. 9 that the regression equation for American traveling to Sichuan Province is:

$$\begin{aligned} \ln(\text{numberofvisitors})_t = & -45.74 + 5.865\ln(\text{GDP per capita})_t - 3.203\ln(\text{exr})_t \\ & + 0.233(\text{IFConference}) \quad t = (2009, 2010, \ldots, 2018) \end{aligned} \tag{8}$$

U.S. per capita GDP, exchange rates between the two places, and IFConference are the main factors that affect Americans' travel to Sichuan Province. Since the correlation coefficient between per capita GDP and related CPI in the United States is 0.7543, the correlation is strong. In order to reduce collinearity, the relative CPI is excluded from the model (Figs. 10, 11 and 12).

Model	R	R Square	Adjusted R Square	Std. Error of the Estimate	Durbin-Watson
1	0.969	0.939	0.909	0.173831	1.852

Fig. 10. The fitting degree and sequence correlation test of America model for Sichuan inbound tourism

ANOVA						
Model		Sum of Squares	df	Mean Square	F	Sig.
1	Regression	2.791	3	0.93	30.788	0.000485
	Residual	0.181	6	0.03		
	Total	2.972	9			

Fig. 11. Model rationality

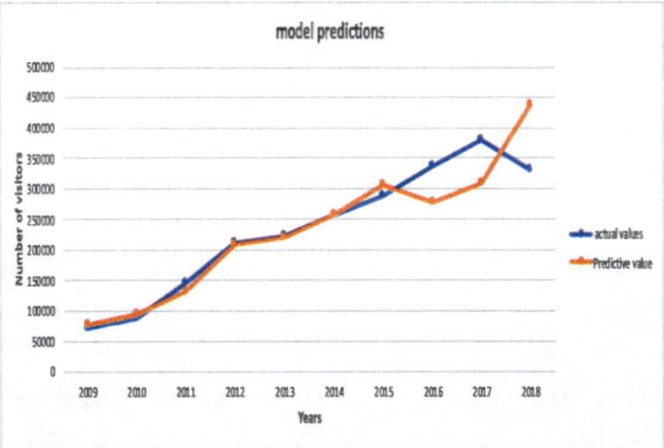

Fig. 12. Fitting graph of model prediction results for American tourists

3.3 Model Diagnosis

1) The coefficient of determination (R^2(UK) = 0. 982, R^2(Malaysia) = 0. 933, R^2(USA) = 0. 939) is very close to 1, indicating that the function of goodness-of-fit test is very effective, and there is a very good linear relationship between the independent variable and the dependent variable.

2) In the ANOVA table, the significance of the three source countries is far less than 0.05, indicating that the coefficient is not all 0, and the equation is reasonable.
3) The Durbin-Watson values of the three source countries are all within the acceptable range (The sample size selected in this article is 10, the number of explanatory variables is 5, and the significance level is set to 0.05. Refer to the DW test critical Value table, it can be seen that the upper critical value dU is 2.822, and the lower critical value dL is 0.243). It shows that there is no serial correlation in the residuals.
4) From the perspective of the variance expansion factor, it can be found that the VIF values are all less than the critical value of 10, indicating that there is no multicollinearity between the independent variables.
5) The prediction results are drawn as a line graph (the two lines are the predicted value and the actual value of the model). It can be seen intuitively that the model fits well and the prediction results are accurate. The predicted value is very close to the actual value, and the trend is consistent.

4 Conclusion

Through the empirical analysis of the three major countries of the United Kingdom, the United States and Malaysia, it can be seen that:

1) The per capita GDP of the three main source countries has a positive impact on the inbound tourism demand of Sichuan Province, and the degree of impact is basically related to the wealth of each country. The United States and the United Kingdom are all developed countries with relatively high-income levels. The per capita GDP of these two countries has a greater impact on the elasticity of tourism demand in Sichuan Province (US 5.865, UK 4.104). The per capita GDP of Malaysia, a relatively backward country, has a relatively small elasticity of demand for tourism to Sichuan Province (1.115). In this way, the results of the empirical analysis are in line with actual facts.
2) The exchange rates of the three main source countries all have a negative impact on the inbound tourism demand in Sichuan Province, and the impact is very significant. The exchange rate coefficients are all negative, indicating that the appreciation of the currency will generally hinder nationals of source countries from traveling to Sichuan Province. The main reason is that the increase in the exchange rate will allow the nationals of the source country to choose more tourist destinations, so the residents of the source country may choose relatively more expensive places to travel.
3) The impact of relative CPI on inbound tourism demand in Sichuan is uncertain. Judging from the empirical results of the main source countries, it is not certain whether the correlation is positive or negative. Among the three source countries, the relative CPI only significantly affected Malaysian nationals' travel to Sichuan Province. The coefficient of Malaysia's relative CPI is positive, indicating that an increase in relative CPI will promote more Malaysians to travel to Sichuan Province.

For the dummy variable introduced in the article: the China International Friendship City Conference. From the empirical results, it can be seen that the China International

Friendship Cities Conference has a greater impact on Americans and British people traveling to Sichuan Province, and the coefficients are all positive, thus promoting the nationals of these two countries to travel to Sichuan Province.

References

1. Ma, J., Liu, S.: Inbound tourism and the marketization of China's institutions: analysis of regional panel data. Nankai Bus. Rev. Int. **7**(4), 542–554 (2016). (in Chinese)
2. Fernandes Costa, B.J., Serrano Fernandes Rodrigues, S.C., Moreno Pacheco, P.: Circular economy and the tourism industry. J. Glob. Bus. Technol. **16**(1), 45–61 (2020)
3. Garín-Muñoz, T., Montero-Martín, L.F.: Tourism in the Balearic Islands: a dynamic model for international demand using panel data. Tour. Manag. **28**(5), 1224–1235 (2007)
4. Garin-Munoz, T., Perez Amaral, T.: An econometric model for international tourism flows to Spain. Appl. Econ. Lett. **7**(8), 525–529 (2000)
5. Bucchianico, A.D.: Coefficient of Determination (R^2). In Wiley StatsRef: Statistics Reference Online (2014)
6. Qu, H., Or, Y.: Determinants of the travel demand model for Canadian tourists to the U.S. Int. J. Hosp. Tourism Adm. **7**(4), 1–19 (2006)
7. Kabaila, P., et al.: The effect of a Durbin-Watson pretest on confidence intervals in regression. Statistica Neerlandica (2020)
8. Crouch, G.I.: The study of international tourism demand: a review of findings. Bull. Econ. Res. 55–72 (1994)
9. Yang, M.: Research on Influencing Factors of Inbound Tourism Demand Based on Grey Relational Analysis. Southwest University (2010). (in Chinese)
10. Ma, A., Cao, R., Guan, B.: Analysis of the main factors influencing the competitiveness of China's inbound tourism. Henan Sci. 874–877 (2008)
11. Pei, Y., Xibo, L., Qian, Z.: Analysis of influencing factors of Beijing inbound tourism demand based on double logarithmic linear model. Math. Pract. Knowl. **41**(11), 15–23 (2011). (in Chinese)
12. Önder, A.Ö., Candemir, A., Kumral, N.: An Empirical analysis of the determinants of international tourism demand: the case of Izmir. Eur. Plan. Stud. **17**(10), 1525–1533 (2009)

The Harm of Social Class Consolidation and Its Solutions from the Perspective of Fairness Based on Big Data Predictive Analysis

Caiyun Ding(✉)

University of Pompeu Fabra, Barcelona, Cataluña, Spain

Abstract. With the development and progress of electronic information professional technology, especially the rapid development of physical information systems, service networks, cloud computing and social service Internet technologies, big data can be seen everywhere, and big data is creating huge use value for people, and using big data can quickly deal with massive data forecasting and analysis problems. "Classification" has become a hot topic in the current public opinion communication in China. It is a concept constructed based on the two meanings of "class mobility blocked" and "risk social amplification". This concept "breaks into" the academic field from the public opinion circle, that is, people have gone beyond the rigorous demonstration of whether the state of "stratum solidification" is "true", and directly enters the stage of analyzing the causes and coping strategies of "stratum solidification". The purpose of this article is to study the harm of social class consolidation and its solutions from the perspective of fairness based on big data predictive analysis. This article first summarizes the basic theory of big data, and then extends the core technology of big data forecasting and analysis, and then combines the problems and shortcomings of the current social class solidification phenomenon in our country, and conducts research and discussion from the perspective of fairness based on big data forecasting and analysis. And conduct in-depth analysis and research on its resolution methods. This study systematically expounds the causes and development of the phenomenon of social stratification, analyzes its harm, and explores the ways to resolve it through experimental analysis. This experiment conducted research on it through research methods such as data collection and interviews. The experimental research shows that nearly one-third of people believe that the conflicts of interests of social groups in our country will intensify.

Keywords: Big data · Predictive analysis · Class consolidation · Discussion on resolution

1 Introduction

At present, our country is undergoing a major social transformation, with profound changes in the pattern of interests, social division of labor, values and lifestyles [1, 2]. Promote the intensification of social differentiation in our country, which has brought

B. J. Jansen et al. (Eds.): *International Conference on Cognitive based Information Processing and Applications (CIPA 2021)*, LNDECT 85, pp. 38–47, 2022.
https://doi.org/10.1007/978-981-16-5854-9_5

serious social consequences and impacts [3, 4]. The original "two classes, one stratum" in our country have now undergone significant changes, gradually turning into a situation where multiple stratums coexist [5, 6]. The current differentiation of social strata, coupled with the increasingly patterned and solidified repulsive system, has made the differentiation and flow between classes weak, which has gradually led to the emergence of social class solidification [7, 8].

Many scholars have achieved good results in the study of the phenomenon of social class consolidation. For example, Hou Liyu believes that the social distribution mechanism of contemporary China makes groups in a dominant position of distribution [9]. Utilize their own superior resources to maintain and pass on the original superior position between generations, leading to the hereditary inheritance of wealth and power, and the solidification of the class [7, 10]. Xu Zhiqiang believes that due to the intervention of rights and capital in China's current education system, it is difficult for the lower class to break through the limits of the "resource poverty" of the parents, and the education system itself cannot bear the responsibility of promoting social mobility [11].

This article adopts the method of combining theory and reality, through the analysis of social class, from the perspective of fairness to analyze the reasons for the solidification of social class in our country and explore ways to resolve it. This article analyzes the phenomenon of social class consolidation on the basis of reading a large number of predecessor documents and the research results of relevant experts.

2 Research on the Harm of Social Class Solidification Phenomenon from the Fair Perspective Based on Big Data Forecasting and Analysis

2.1 Big Data Characteristics

The big data features widely recognized by the industry are the large-scale, diversified, rapid, and value-based proposed by IBM.

(1) Massification

With the development of technology and storage technology, a lot of data in human daily life have been recorded, including data generated by individuals and data generated by business operations. At present, the data of personal computers are all terabytes, and the amount of data of some large enterprises has reached the (Exabyte, EB) level.

(2) Diversification

With the development of the Internet of Things technology, there are more and more types of sensor devices, which can collect various types of data from all walks of life. Big data can not only process structured data of traditional relational types, but also unstructured and semi-structured data such as audio, video, and pictures.

2.2 Hazard Analysis of Stratum Solidification

(1) Harm to economic development
In our country, the decline in the flow of various social strata has led to a reduction in the exchange of resources between the various strata. The upper class with a lot of social wealth can only move between the same strata, causing a series of social injustices. The limited social resources available in the hand, coupled with the inability to obtain wealth from the upper level, will directly affect the social and economic development and cause social conflicts.

(2) Harm to political construction
Class consolidation can cause political crises. Mainly manifested in the various sharp social contradictions brought about by the alienation of power and monopoly. The concentration of rights among the upper class has aggravated social inequality and increased social instability. The first is the disorderly participation in politics. Most of the people at the bottom of society are scattered, the organization of activities is difficult and the disorderly way of appealing for interests, when their own interests are greatly harmed, unscientific behaviors give It will become a factor affecting the political stability of our country. The ruling foundation of the Chinese Communist Party is the masses. Due to the solidification of the strata, the boundaries of the various strata are obvious and the flow is not smooth. The third is the brain drain in political parties. Due to the lack of fair competition, in the selection of party and government officials, elites from the lower levels are increasingly hindering the upward entry of the middle and high levels of society. The government cannot absorb real elites to manage society.

(3) Harm to cultural construction
Class consolidation hinders the progress and development of our country's culture. Culture is the irreplaceable soul of a country, and the development of culture promotes the development of the country and the progress of the nation. The masses of the people are the main body of cultural creation. Different classes and regions of human beings have different cultural inheritance. The upper class and the lower class have their own living habits and customs. These customs and habits are also the class culture and culture in a certain sense. The development of China is inseparable from the cultural exchanges and innovations of various strata. When the various strata of the society stabilize and reach a solid state, the lack of communication between various cultures causes cultural lag and stagnation. From an international perspective, the solidified class is psychologically difficult to accept foreign cultures, and there are feelings of resistance to foreign new ideas. However, the trend of global dissemination of culture is the current trend. Failure to learn from other cultures will inevitably affect the construction and development of my country's cultural undertakings.

(4) Harm to society
The first is to hinder the formation of an olive-shaped society. The olive-shaped society is conducive to the social development of our country. At present, the result of the consolidation of the class in our country is that the entire society presents a pyramidal social structure. At the top are the upper-class groups with a relatively small proportion, and at the bottom are the lower-class groups with a large population but few social resources. The gradual formation of this irrational social structure

is a great threat to social harmony and stability. On the other hand, unfair social competition has led to the psychological imbalance of the bottom social personnel, and abnormal psychology such as "hatred of the rich" and "hatred of officials". Not only is there a sense of despair about their own future, but at the same time they lose confidence in the power holders. When this negative emotion accumulates to a certain extent, it will become a destabilizing factor in society, causing crime and illegal behavior in society, harming the health of others, disrupting public order, and affecting social stability.

2.3 Analysis of Causes of Stratification Consolidation

(1) Economic factors

1) Restrictions on living conditions
People at the bottom of society live at the bottom of society, without a stable income, and lead an insecure life. The living conditions of the following categories of people reflect the current living conditions of the bottom people in contemporary China: First, they can earn a meager income from farming in poor areas. It is impossible to guarantee the basic life, so many people go out to work, which is what we call migrant workers. They do the hardest and most tiring work on the construction site. They eat in the open air and live in simple concierges. They are not protected from wind in winter and rain in summer. Second, vegetable farmers in rural areas keep a small portion of the vegetables they obtain from their homes for their own use, and sell the rest for a meager living allowance. Because they have no money, they can only use their shoulders to carry them to the surrounding county towns to shout and sell. There is a risk of being confiscated or fined when going to the city management. 3. Sanitation workers have to get up early and return late every day, just to make the living environment of the city better. They work hard, but have very low income. Not only that, they have to shuttle on the road every day to face the coming and going vehicles, life is threatened at any time. Fourth, children who go to school in remote areas may be clothed all year round, and may even have no food to eat. Not only do they have to go far in the mountains to get out of school, but they also have to help their parents with housework after school. People living at the bottom of society do not have a stable environment for development and the necessary conditions for development. Even solving the basic living conditions of food and clothing is a problem. In addition, they have to endure discrimination from the upper class. In the continuous development of society, this group of people has gradually been marginalized and stabilized, forming a class that is difficult to move upwards.
2) Disadvantages of market economy
At the beginning of the reform, the state allowed some people to get rich first to achieve the goal of getting rich first and getting rich later. The disadvantage of this policy was the emergence of social polarization between the rich and the poor. With the further development of reform and opening up, it also proposed to insist on "efficiency". With the policy of giving priority to fairness, the gap

between the rich and the poor has widened and brought about a series of social problems. But this situation cannot be changed in a short time.

(2) Political factors

1) Restrictions of the household registration system
In today's society, the social influence of identity cannot be ignored. The household registration of the people at the bottom represents a fixed social identity. Our country implements a dual household registration system that separates urban and rural areas. For the people at the bottom, changing this type of household registration requires funding, contacts and other conditions, but the bottom people are short of these two resources. For the population, many social activities will be restricted. It also created the de facto inequality of the people.
2) Restrictions of the education system
The educational opportunities of different classes of people in our country are not equal. The education system separated from urban and rural areas and the college entrance examination system in my country have different levels of education for all classes. From the perspective of more economically developed areas in remote areas, the educational opportunities available are unequal, and the contrast between urban and rural areas is more obvious. When urban children rush to interest classes during quality education, rural children still worry about not being able to go to school. Another reason is the unfair distribution of educational resources. First, Beijing, Shanghai and other first-tier cities have more educational resources in developed areas than underdeveloped areas, and urban educational resources are also much more saturated than surrounding rural areas. Furthermore, in comparison between schools, key schools have better teaching staff and better hardware and software facilities than ordinary schools.
3) Imperfect power constraints
In recent years, the civil service examinations, public sector recruitment, and monopoly sector recruitment examinations have all aroused doubts about the interference of political power. The recruitment of civil servants in some places is not afraid to use family advantages as a special condition for recruitment, so as to increase the chance of being hired. Party and government officials naturally have to find ways to transfer their power to future generations and the market economy implemented in our country has a relatively concentrated power. A large number of economic activities require government officials to approve. It is inevitable that there will be In order to obtain the collusion of private interests, long-term unchecked political power and imperfect power check-and-balance mechanism have resulted in the dominance of administrative power and has gradually formed a network of heaven and earth to encroach on all areas.

2.4 Analysis of Strategies Solidification and Resolution

(1) Establish a fair and just governing concept
Only by firmly establishing a fair value idea in the ruling concept can a more fair and just social system be established, and only through a fair and just social system can the fairness and justice of the society be effectively maintained, and the emergence of the solidification of social strata can be effectively prevented. In other words, to ensure the justice of the system must first ensure that the system itself fully embodies the principles and spirit of justice. If the content of the various systems formulated by the government lacks justice, so in the process of system implementation, not only can it not guarantee social justice, on the contrary, it will only hinder the formation of a just environment, and ultimately fix the unfairness of society itself.

(2) Break the regional differences in the household registration system
The regional differences in the household registration system hinder the free migration of citizens of the country, and at the same time cause the inequality in the rights of urban and rural residents, and it is not suitable for the current development of our society. Therefore, it is necessary to deepen the reform of the household registration system, break the regional differences in the household registration system, and form a unified household registration system that integrates urban and rural areas.

2.5 Predictive Analysis Methods Based on Big Data

Traditional forecasting models are generally historical periods and time series. Time series are divided into Moving Average, Exponential Smoothing, Kalman Filters, Adaptive Filters and Autoregressive Integration. Autoregressive Integrated Moving Average Model (Autoregressive Integrated Moving Average Model), but hotel revenue management forecasts are different from traditional forecasts. Generally, there are two time variables, which are the scheduled time and the consumption time. Lee et al. used the data generated between the scheduled time and the time of consumption, combined with traditional forecasting models, to divide the revenue management forecast into three categories, historical booking models, advanced booking models, and combined models. (combined models).

1) Historical booking models: In the forecast based on historical data, the final order number on the target date is predicted according to the time series of the final number of orders per day in the past;
2) Improved booking models (advanced booking models): In the forecast based on advance booking data, the final booking volume is predicted based on the observation data. The model infers future booking increments by estimating the growth rate scheduled in each period, and finally predicts the total number of orders on the target date. The difference between the forecast based on advance booking data and the forecast based on historical data is that in the latter forecast, complete historical data is required, while the former has less demanding data requirements and can also be forecasted with incomplete data;

3) Combined models: It is also known as a prediction model that combines historical data with pre-booked data. It combines the predictions of the two through other methods such as regression or weighted average.

2.6 Simple Random Sampling

Suppose that from the sampling frame $A^q = (q = 1, \ldots, Q)$, a simple random sampling method n^q is used to extract samples with a capacity of nq, and the sample size falling into the domain K is n_K^q, denote $f = (f^1, \ldots, f^Q)$ Is a Q-dimensional column vector composed of the sampling ratios in each sampling frame, where $f^q = n^q/N^q$, then

$$n_K^q = \widehat{N_K^q} f^q \tag{1}$$

In the case of multiple boxes, the overall total value can be expressed as

$$Y = \sum\nolimits_{k \in U} Y_k = N^T \overline{Y} \tag{2}$$

3 Experimental Research on the Harm and Resolution of Social Stratification Phenomenon from the Perspective of Fairness Based on Big Data Predictive Analysis

3.1 Subjects

(1) In order to make this experiment more scientific and effective, this experiment investigated the composition of the people's congress of our country. Our Constitution affirmed the status of the ruling class of workers and peasants from a legal point of view. The proportion of city representatives attending the National People's Congress shows the harm of the solidification of the social class in our country.
(2) In order to further investigate the phenomenon of social solidification in our country, this experiment also randomly distributed questionnaires in a certain place. The purpose of the survey is to investigate people's perceptions of the conflict of interests of social groups and whether the conflict of interests of social groups in my country may be intensified.

3.2 Research Methods

(1) Questionnaire survey method
This study set up targeted questionnaires by asking relevant experts, and adopted a semi-closed method to prompt investigators to fill in correctly.
(2) Mathematical Statistics
Use related software to perform statistical analysis on the data obtained from the research.
(3) Interview method
This article conducts face-to-face interviews with relevant experts, and organizes and analyzes the collected data. These data not only provide theoretical support for the topic selection of this article, but also provide data support for the research results of this article.

4 Based on the Analysis of Big Data Prediction and Analysis, the Harm of Social Class Solidification Phenomenon and the Experimental Analysis of Its Resolution from the Perspective of Fairness

4.1 Analysis of Political Participation

In order to investigate the phenomenon of social class consolidation in our country, this experiment investigated the composition of the 10th to 12th National People's Congress. The results are shown in Table 1.

Table 1. Political participation analysis

	Worker ratio	Farmer ratio	Ratio of cadres	Proportion of intellectuals
Tenth	10%	8.4%	32.4%	21.14%
Eleventh	1.77%	4.34%	41.81%	19.22%
Twelfth	7.28%	6.14%	34.88%	20.42%

It can be seen from Fig. 1 that since the reform and opening up, workers and peasant representatives have always been a minority. Even 13.42% of the total number of workers and peasants in the twelfth session, there are many cadre directors, and there are few workers and peasant representatives in the true sense. There are few, which fully shows the solidification of the social class in our country.

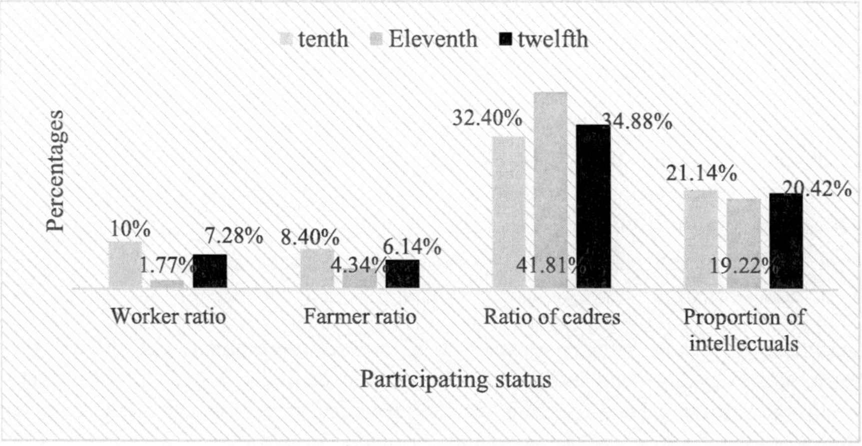

Fig. 1. Political participation analysis

4.2 Analysis of Interest Conflicts of Social Groups

In order to further investigate the conflicts between classes, this experiment launched a research and investigation in the form of questionnaires, and the results obtained are shown in Table 2.

Table 2. Cognitive distribution of interest conflicts of social groups

	Serious conflict	Big conflict	A little conflict	No conflict	Can't tell
Man	2.5%	9.2%	21.7%	7.3%	7.9%
Woman	2.3%	9.0%	23.1%	9.1%	7.9%

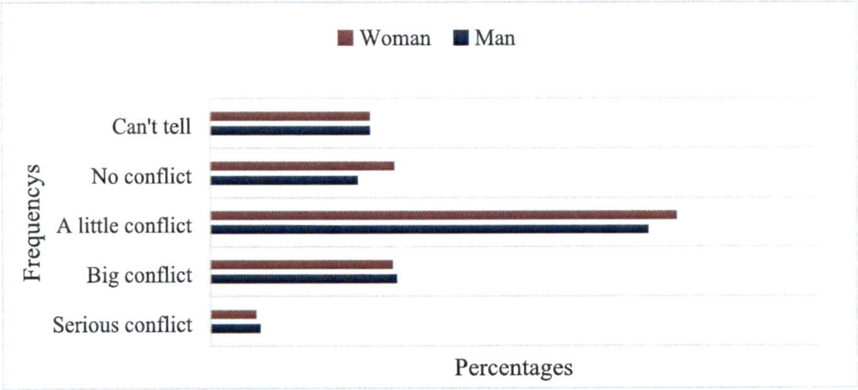

Fig. 2. Cognitive distribution of interest conflicts of social groups

It can be seen from Fig. 2 that 67.9% of the interviewees believe that there is a conflict of interest between various groups in the current society, which is reflected in the serious imbalance of the interest pattern caused by the inclined allocation of social resources to the strong class, and the relative deprivation of the weak class, the stratification consciousness in the process of differentiation is showing more and more conflicts or confrontations.

5 Conclusion

Objectively determine subjective, we should consider the problem and find solutions based on the current social reality in China. At present, China is in a special period of social transformation, and the problems of deepening social class contradictions and imminent solidification have been urgently needed to be resolved. The phenomenon of "second generation of officials" and "second generation of wealth" in recent years shows that the inheritance of wealth and the family solidification of social status have begun. In view of this serious problem, it is necessary to find feasible countermeasures to deal with class solidification.

References

1. Fewell, C.J., Hess, M.E., Lowery, C., et al.: Experiences of stereotyping and tracking in appalachia: a case in consolidation and social class case narrative background and context. J. Cases Educ. Leaders. (7), 12 (2021)
2. Kraus, M.W., Park, J.W., Tan, J.: Signs of social class: the experience of economic inequality in everyday life. Perspect. Psychol. Sci. **12**(3), 422–435 (2017)
3. Harackiewicz, J.M., Canning, E.A., Tibbetts, Y., et al.: Closing achievement gaps with a utility-value intervention: disentangling race and social class. J. Pers. Soc. Psychol. **111**(5), 745–765 (2016)
4. Mouncey, P.: Social class in the 21st century. Int. J. Mark. Res. **58**(2), 335–338 (2016)
5. Francombe-Webb, J., Silk, M.: Young girls' embodied experiences of femininity and social class. Sociology **50**(4), 652–672 (2016)
6. Grabe, M.E.: Tabloid and traditional television news magazine crime stories: crime lessons and reaffirmation of social class distinctions. J. Mass Commun. Q. **73**(4), 926–946 (2016)
7. Bjornsdottir, R.T., Rule, N.O.: The visibility of social class from facial cues. J. Pers. Soc. Psychol. **113**(4), 530–546 (2017)
8. Dubois, D., Rucker, D.D., Galinsky, A.D.: Social class, power, and selfishness: when and why upper and lower class individuals behave unethically. J. Pers. Soc. Psychol. **108**(3), 436 (2016)
9. Hou, L.: The interrelations between social class, personal relative deprivation, and prosociality. Soc. Psychol. Pers. Sci. **8**(6), 660–669 (2017)
10. Callan, M.J., Kim, H., Gheorghiu, A.I., et al.: The interrelations between social class, personal relative deprivation, and prosociality. Soc. Psychol. Pers. Sci. **8**(6), 660–669 (2017)
11. Xu, Z.: The social psychology of the great recession and social class divides. J. Soc. Issues **73**(1), 8–22 (2017)

The Visual Art Design of the Inheritance and Development of the Invisible Embroidery Culture Under the Smart City

Kunkun Yang(✉)

Jiangxi University of Applied Science, Nanchang, Jiangxi, China

Abstract. With the rapid development and continuous innovation of big data technology, a new cultural form constructed with a variety of big data technologies as media and channels has created a brand-new cultural ecology, and big data technology has also been applied to the inheritance of traditional culture. More and more countries and regions are beginning to explore the use of digital methods to protect and inherit their traditional culture, so that traditional culture can make full use of the advantages of big data technology to get a new life. Yuzhang (Nanchang) embroidery culture crafts, as an indispensable and important result of traditional Chinese crafts, is Chinese traditional culture a jewel of China, it is also a cultural heritage that the Chinese people are proud of. Embroidery craftsmanship is also facing the impact and challenge of modern civilization. Many traditional techniques are also at a critical juncture. Therefore, paying attention to embroidery craftsmanship and paying attention to traditional craftsmanship is an urgent task we face. Yuzhang embroidery craftsmanship is gradually being lost, and it is particularly important to protect and inherit it using big data technology. The purpose of this article is to design the visual art of the cultural heritage and development of the intangible Yuzhang under the background of big data. This article firstly summarizes the basic theory of big data, and then extends the core technology of big data, and combines the current situation of Yuzhang embroidery culture inheritance and visual art design research in our country, to give an overview of the inheritance and visual art of Yuzhang embroidery under big data. Carry out design and research. This article systematically expounds the origin, development, technical characteristics and cultural heritage of Yuzhang embroidery, and finally combines big data technology to design and research it. This article uses questionnaire data method, mathematical statistics method, etc. to study the theme of this article. Experimental research shows that this research shows that in modern society, traditional Yuzhang embroidery is difficult to inherit and develop. Therefore, the use of modern technology such as big data is very necessary for the visual art design of the inheritance and development of the intangible cultural heritage of Yuzhang embroidery.

Keywords: Big data technology · Yuzhang embroidery culture · Visual art · Design research

B. J. Jansen et al. (Eds.): *International Conference on Cognitive based Information Processing and Applications (CIPA 2021)*, LNDECT 85, pp. 48–56, 2022.
https://doi.org/10.1007/978-981-16-5854-9_6

1 Introduction

Embroidery culture originated in ancient labor and life, and was inspired by people's production and life. The ancestors of the Chinese nation used embroidery to express their praise of life [1, 2]. The working people use dexterous hands to show the style and artistic achievements of different eras, different regions, different cultural heritages and different folk customs characteristics [3, 4]. Here covers the artistic quintessence of Chinese folk culture for thousands of years. The difference between the embroidery cultures of other regions in China is that the embroidery culture of Yuzhang not only preserves the characteristics of traditional folk paper-cutting, but also has its unique form of expression, whether in modern visual design or in future visual design [5, 6].

In today's social environment with affluent material life, people begin to pay attention to things with individuality and cultural aesthetics [7, 8]. Embroidery, an ancient handicraft with a certain market value and cultural value, is undoubtedly a treasure house for mining. With the development of industrial civilization, there are various forms of embroidery. The emergence of machine embroidery has greatly reduced the cost of embroidery. The decline of traditional handicrafts is inevitable. Both China and the West have protected the inheritance and development of traditional hand embroidery [9, 10]. In the process of the pursuit of excellence in details and the worship of artistic creation and handicraft, the Chinese and Western haute couture provides the best habitat for embroidery craftsmanship. Embroidery craftsmanship makes clothing possess high-level capital and expensive prices. The combination of the two contains profound cultural heritage and humanistic pursuits [11, 12].

Based on the folk customs of Yuzhang, combined with the perspective of cultural inheritance, this article investigates and sorts out the ancient folk art of Yuzhang embroidery culture, analyzes the artistic characteristics and cultural connotations it expresses, and proposes some protective measures.

2 Application Research of Visual Art Design Based on the Inheritance and Development of Intangible Cultural Heritage of Yuzhang Embroidery

2.1 Research on Visual Art Design

(1) Pixel-based visual art
Pixels are the basic elements of digital images, and there are various image processing methods based on pixel color value and gray level. These methods have great application potential in style analysis for statistics and research on image low-level information. Although these simple pixel information is effective, for a work of art, the pixels are related. Obviously, these methods are all effective analytical methods for the study of visual art understanding.

(2) Visual art based on brushes
In addition to the color information of the pixel itself, the information in the image is more contained in the lines and edges. At the same time, images do not have basic elements such as words in literary works, so researchers choose brushes as

the focus of research and analysis. Brush can be considered as the smallest unit with independent structure in the art work, which represents the characteristics of the art work.

(3) Visual art based on spatial statistics
The analysis based on pixels and strokes all start from the details of the image, and there is no discussion on the overall composition method and law of the image. The earlier discussion on the characteristics of spatial statistics was Taylor et al., who used the box dimension method to calculate the fractal dimension of pollock's drip painting. In addition, EMD, orthogonal mirror filter, artificial neural network and other methods have also been applied to researches such as identification and classification, and certain results have been achieved. These studies are based on "category-based judgments", rather than based on image content such as pixels and brushes.

(4) Visual art based on other characteristics
In addition to the above three methods, image processing methods such as frequency domain analysis, sparse coding analysis, controllable golden decomposition, and multi-scale dichotomy have also been widely used in the research of visual art understanding.

2.2 Yuzhang Embroidery Culture

(1) Cultural origin
According to historical records, the Yuzhang embroidery culture first originated in the Qianlong period. Its founder was Qianlong's "god sister" Qiu Huanggu, and his husband was the husband of the old man Ji Xiaolan. Qiu Huanggu has been very talented since she was a child. She has been exposed to embroidery since she was a child, and became obsessed with embroidery when she grew up. She is good at observing details and has a delicate mind. If you embroider the landscape, look at the shape of the mountains and rivers; if you embroider the vegetation, study the fineness of the stems and leaves; if you embroider the animals, observe the texture of the hair. Only in this way can the works of Yuzhang embroidered be lifelike and possess the godlike form. So the special feature of Yuzhang embroidery comes from this. It originated from folk life.

(2) Technical characteristics
The vivid image of Yuzhang embroidery is also inseparable from the unique skills of Yuzhang embroidery. Yuzhang embroidery adopts special needle methods derived from the folk, such as flat needles, hairpin needles, floating needles, etc., which have a particularly strong three-dimensional effect. The unique stitching method of Yuzhang embroidery means that embroidery cannot be made by machine. Every embroidery can only be made by the embroidery mother one stitch by stitch. The embroidered products are all orphans, and it is hard to find the second one in the world. One element uses up to nine colors, and the color is too uniform, which will restore the essence of the material to the greatest extent. It is different from Suzhou embroidery, one of the four famous embroidery, with distinct colors, too exquisite and distorted.

(3) Cultural inheritance
Yu Zhang Embroidery has been passed down through eight generations. After Yu Zhenglian and Zhou Nanying inherited the pioneering and innovation, Yu Zhang Embroidery combined the craftsmanship of embroidery and conveyed painting. His work "Konggu Goshawk" uses silk threads to create the characteristics of light and darkness, highlighting the artistic sense, texture, layering and three-dimensional sense of the work. It vividly expresses the artistic conception and humanistic spirit of traditional Chinese ink and wash. This unique folk embroidery method is unique in China. It perfectly combines the art of embroidery with the art of calligraphy and painting. It pushes embroidery from the traditional field to the field of culture and art, and has become an important cultural business card of Nanchang, a famous historical and cultural city.

(4) Cultural development
Under the leadership of Yu Zhenglian and Zhou Nanying, the quiet folk traditional Yuzhang embroidery began to actively appear in our eyes. They established a productive protection base-Yuzhang Embroidery Textile. All Yuzhang embroidery works are from the embroiderers trained by Yuzhang Embroidery Workshop. The embroidery workshop provides free training to laid-off women workers and the disabled, giving them a chance to re-employ them. Yuzhang embroidery skills have also been inherited and developed. They established the inheritance base of Yuzhang Teachers College and the Provincial Disabled Promotion Association, allowing Yuzhang embroidery to enter the campus and into the lives of more people. At the same time, they also cooperated with the School of Clothing of Jiangxi Normal University to study the derivation of intangible cultural heritage embroidery. In recent years, Yuzhang embroidery has successfully declared the provincial intangible cultural heritage and was selected as a gift from Nanchang. Yuzhang embroidery is not only a representative of Jiangxi, but also a representative of China, which is exported to foreign countries.

2.3 The Application of Visual Art Design Based on the Inheritance and Development of Intangible Cultural Heritage of Yuzhang Embroidery

(1) The perfect expression of imagery
Yuzhang embroidery uses contemporary subjective ideas to explain traditional folk cultural concepts, and through the use of a large number of cultural elements, embroidery is no longer monotonous and boring, and seeks many new modern expressions. This feature is unique to Yuzhang embroidery. form. The concept of "imagery" is a special cultural concept in the field of Chinese classical culture. It is often used extensively in traditional art forms. It is the most true embodiment of natural phenomena and things by the audience. It is usually used in the expression of modern art creation. It is a cultural thinking and design concept subject to fluctuations in subjective consciousness.

(2) The modeling language of free perspective
Another unique form of artistic expression in Yuzhang embroidery is the free perspective effect. This unique modeling language is unique to Yuzhang embroidery.

It is different from paper-cutting in other places and does not follow the conventional perspective principle. It goes beyond In order to express things objectively, the objects that are not on the same plane are created and deformed to a certain extent, so that the food presented is more unique and real.

2.4 Realization of Yuzhang Embroidery Database

(1) Storage of Yuzhang embroidery database resources
Miao embroidery resources should be easy to find, and the rapid dissemination and sharing depends largely on whether the metadata description of the resources is accurate and complete. When constructing the Miao embroidery resource database, a unique Miao embroidery database table must be established according to the specifications to realize video, Import and export resource data such as audio, graphics, images, and text. According to the characteristics of Miao embroidery resources, the structured relational data and unstructured resource files are respectively designed for data storage.
(2) Realization of the foreground function module

1) User browsing module
The user browsing module allows users to understand the production process of traditional handicrafts as a viewer. It mainly includes two sub-modules: work selection and production process demonstration. The work selection module presents thumbnails of handicraft works on the screen in the form of a sorted list for users to select.
2) User experience module
The user experience module allows users to experience the production process of traditional handicrafts as a producer. It mainly includes four sub-modules: material selection, work generation simulation, independent creation, and work printing. The material selection window lists the materials required for making handicrafts, such as pattern modeling, making tools, etc. for users to choose.

2.5 Calibration Estimation Method

On the premise that the sampling design has been determined, S represents all possible sample sets, and s represents a specific sample that has been drawn. The probability that the k-th unit is included in the sample is called the inclusion probability of the k-th unit, denoted as πk, which can be expressed as follows:

$$\pi_k = \Pr(\mathrm{k} \in \mathrm{S}) = \Pr(I_k = 1) = \sum\nolimits_{s \in K} P(s) \tag{1}$$

Where $\sum\limits_{s \in k} P(s)$ represents the sum of the occurrence probabilities of all samples containing the k-th unit. Similarly, the probability of containing units k and l is recorded as π_{kl}, and the given sampling design p(.) is expressed as follows:

$$\pi_{kl} = Pr(k\&l \in S) = Pr(I_k i_s = 1) = \sum\nolimits_{s \in k\&l} p(s) \tag{2}$$

3 Experimental Research on Visual Art Design Based on the Inheritance and Development of Intangible Cultural Heritage of Yuzhang Embroidery

(1) Research object
In order to make the research results of this article more scientific and effective, this article conducted a questionnaire survey in Nanchang. A total of 50 questionnaires were distributed in this survey, 25 of which were male and female, using a ten-point scoring system. "1" means unfamiliar. "10" means familiarity. The degree of familiarity from 1 to 10 is from low to high, and the data obtained is used to obtain a more accurate result by using the analytic hierarchy process. This research will start with the cultural identity of Yuzhang District residents towards Yuzhang embroidery, in order to explore whether the cultural inheritance of intangible Yuzhang embroidery has been affected. In addition, in order to understand the application of Yuzhang embroidery in visual technology, investigation has been carried out so far. The result of the investigation uses the calibration estimation method to obtain a more accurate result.
(2) Research methods

1) Questionnaire survey method
This article investigates the inheritance and protection of Yuzhang embroidery culture by issuing questionnaires to the residents of Yuzhang District. The purpose of this semi-closed questionnaire survey is to make the data more accurate.
2) Field research
This article conducts in-depth visits to Yuzhang Embroidery Workshop, etc. and conducts face-to-face interviews. And to record and organize the collected data, these data provide a reliable reference for the research results of this article.
3) Mathematical Statistics
Use relevant software to perform statistical analysis on relevant data.

4 Experimental Analysis of Visual Art Design Based on the Inheritance and Development of Intangible Cultural Heritage of Yuzhang Embroidery

(1) Analysis on the development and inheritance of Yuzhang embroidery
Folk art has been integrated into our lifeline, and its endangerment is to some extent the danger of the entire national culture. To rescue and protect traditional folk art is to save our cultural lifeline. The results of the questionnaire survey on Yuzhang embroidery in Nanchang area are shown in Table 1:
It can be seen from the picture that the residents of Yuzhang District know little about Yuzhang embroidery culture, especially the positive characteristics of Yuzhang embroidery. This reflects from the side that the inheritance and development of intangible cultural heritage Yuzhang embroidery has been affected by contemporary society to a certain extent. Protecting traditional culture has a long way to go.

Table 1. Analysis on the development and inheritance of Nanchang embroidery

	Cultural origin	Cultural development	Cultural heritage	Technical characteristics
Man	3	4	5	2
Woman	4	6	4	5

(2) Application analysis of Yuzhang embroidery in visual technology
Applying the distinctive artistic features of Yuzhang embroidery to visual art design, traditional culture will show a distinctive new visual experience and bring unprecedented sensory experience. The survey data on the visual technology of Yuzhang embroidery is shown in Table 2 (Fig. 1):

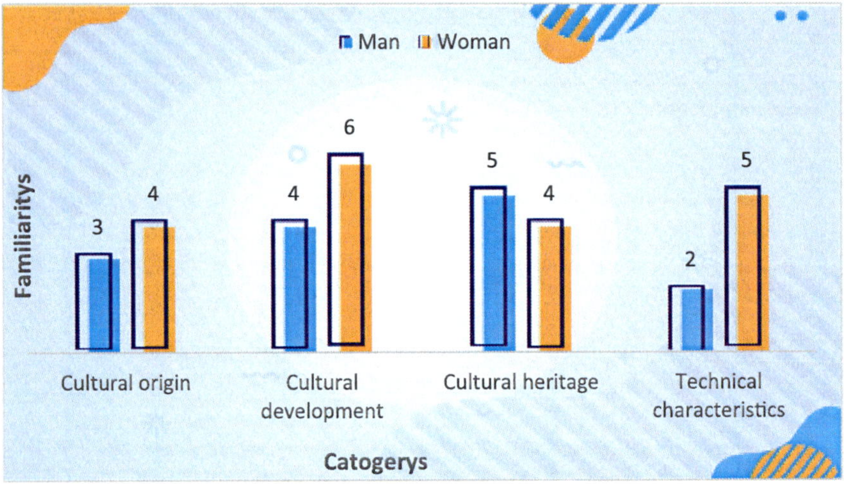

Fig. 1. Analysis on the development and inheritance of Nanchang embroidery

Table 2. Analysis on the application of Nanchang embroidery in visual technology

	Imagery	Free perspective	Allegory	Time-space concept
Man	6 .	8	7	4
Woman	4	6	8	7

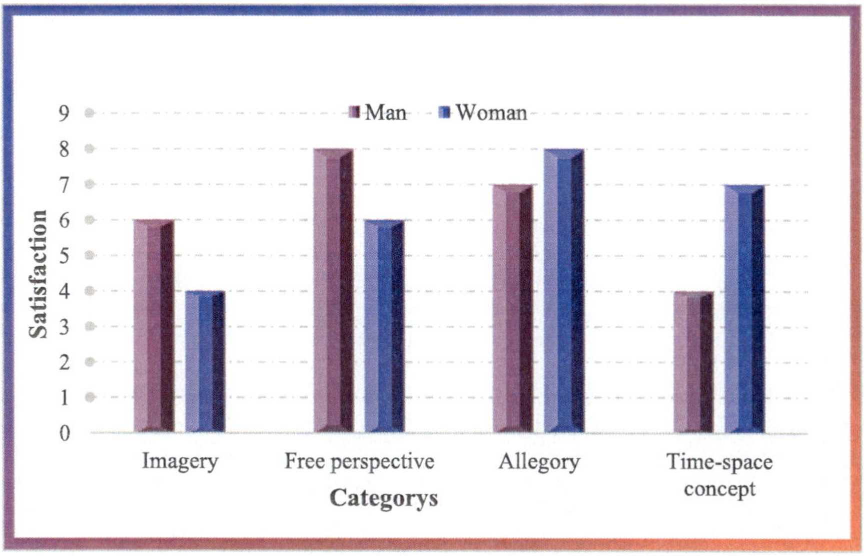

Fig. 2. Analysis on the application of Nanchang embroidery in visual technology

It can be seen from Fig. 2 that in the visual art research based on the inheritance and development of the intangible Yuzhang embroidery culture, compared with the traditional Yuzhang embroidery, the Yuzhang embroidery integrated into the visual art application brings a new visual experience and brings a new visual experience to the traditional culture. The development of China has a very good protective effect.

5 Conclusion

In recent years, the modern spread of Yuzhang embroidery has received more and more attention, and has shined on the world clothing stage. From its cultural significance to its craftsmanship, the scope of research is getting wider and wider. How to better carry on the modern inheritance of traditional embroidery, its research significance and value are very important. Big data technology is an important means of modern scientific research, and its application in the textile and apparel field has become more popular. Digital network technology is a branch of big data technology and a very popular digital platform. The application in the field of intangible cultural heritage inheritance has achieved initial results, but the application in the inheritance of Yuzhang embroidery is very weak. Application in the modern inheritance of traditional embroidery in the Central Plains is a breakthrough for the development of both.

Acknowledgements. Project: 2020 Jiangxi University Humanities and Social Science Project, "Research on visual art design based on Inheritance and development of Yuzhang embroidery culture", compere: Yang kunkun. (Project Number: YS20225).

References

1. Wang, Z., Yao, L., et al.: Innovation in evaluating the cultural value of Bian embroidery heritage. Int. J. Innov. Sci. **8**(3), 217–229 (2016)
2. Kim, K., Lee, M.: Flocking in interpretation with visual art design principles. Wirel. Pers. Commun. **93**(1), 211–222 (2016). https://doi.org/10.1007/s11277-016-3925-1
3. Danielle, M., Iuri, T.D.: Visual art and regulatory fit messages on consumer evaluations. Revista De Administrao De Empresas **56**(2), 152–165 (2016)
4. Page, J.: GOVER, KE art and authority: moral rights and meaning in contemporary visual art. Oxford University Press, 2018, vi+ 188pp. 4 b&w illus. $55.00 cloth. J. Aesthetics Art Criticism **77**(1), 92–96 (2019)
5. Adams, J.: Editorial: visual art-based education research. Int. J. Art Des. Educ. **38**(3), 558–561 (2019)
6. Silaci, I., Ebringerova, P., et al.: New interpretation of public visual art in urban space. In: IOP Conference Series: Materials Science and Engineering, vol. 471, no. 9, pp. 92038–92038 (2019)
7. Mcmahon, J.A.: The ancient quarrel between art and philosophy in contemporary exhibitions of visual art. Curator: Museum J. **62**(1), 7–17 (2019)
8. Harrison, T.: Difficult texts: John 20.16–17a – the Resurrection in visual art. Theology **122**(2), 108–110 (2019)
9. Ibert, O.: Making Value and Career Building in the Creative Economy. Evidence from Contemporary Visual Art. Springer Briefs in Geography. Economic Geography (2018). **94**(3), 327–329
10. Forsler, I.: Image ecologies: infrastructures of visual art education in Sweden and Estonia. Int. J. Educ. Art **14**(2), 239–246 (2018)
11. Slitine, M.: Representing the past for renegotiating the present in a warring context: the case of visual art in Gaza. Neurosci. Res. **65**(3), 69–78 (2018)
12. Smiraglia, C.: The art of listening: listening skill development, classical music appreciation, and personal response through visual art in a middle school program. Res. Issues Music Educ. **14**(1), 2 (2018)

A Diachronic Study of Second Language Mental Lexical Attrition Based on WordNet Computer Data Analysis

Chunxia Zhang(✉)

Foreign Languages School, Jingchu University of Technology, Jingmen 448000, Hubei, China

Abstract. Vocabulary attrition is a very common phenomenon in foreign language study. This paper studies the vocabulary attrition of Chinese learners through an empirical research. It tries to analyze the Chinese non-English majors' vocabulary attrition in different stages of their English study Based on WordNet Computer Data Analysis., which provides a reasonable explanation to maintain the students' ability to learn English as well as the university foreign language reform. The results show that the proportion of semantic reaction words in the first three tests increased with the improvement of the subjects' vocabulary level. Since the fourth test took place one year after the end of English learning, the number of respondents declined significantly, while the number of formal respondents did not show much change. Nonsense words of the four groups of subjects increased significantly in the final test. No matter the level of the subjects' vocabulary, it would be attrited to some extent.

Keywords: Computer data analysis · Mental lexicon · Storage model · Second language · Language attrition · WordNet

1 Introduction

Psycholinguistics is a rapidly developing interdisciplinary subject in recent years, which provides a more scientific perspective and method for the study of the development and origin of human language, and greatly promotes people's understanding of cognition and information processing, making people more clear about the process of language acquisition and guiding teaching.

Researchers have different views on the organizational patterns of L2 mental vocabulary. The first is the phonetic theory. Scholars who hold this view believe that the mental words in a second language are linked together by sounds. Meara believes that bilinguals store words in their second language and their native language in completely different ways [4]. Mental words in the second language are linked mainly by phonetic means, while mental words in the native language are linked mainly by semantic means. The second viewpoint is semantic theory, that is, the storage of second language mental words in the brain is essentially semantic connection. An experimental study by Singleton confirmed this view [7]. The third viewpoint is syntactic theory, which holds that the

B. J. Jansen et al. (Eds.): *International Conference on Cognitive based Information Processing and Applications (CIPA 2021)*, LNDECT 85, pp. 57–63, 2022.
https://doi.org/10.1007/978-981-16-5854-9_7

mental vocabulary of second language speakers is essentially combined by collocation relations, including the collocation of word prefix and suffix, preposition collocation, adjective and adverb collocation, and so on. Wolter found through research that, for the vocabulary stored in the brain, there are obvious differences in the connection ways between the mother tongue and the second language mental vocabulary [9]. The second language mental vocabulary relies on the semantic connection, while the mother language vocabulary relies on the phonetic connection.

Information is stored on the surface of the brain, such as foreign words, which is easy to be attrited, while deep stored language elements, such as pronunciation and syntax, are not easy to be attrited [1]. Therefore, the fundamental attrition of language is the attrition of words. If language users cannot randomly extract words from the brain that constitute the text and syntax, then language learning is undoubtedly a failure. Speech is rarely affected in the process of language attrition and can basically remain stable. To slow down the rate of language attrition is to slow down the process of word attrition.

Weltens found through research that the more similar the mother tongue and foreign language are in pronunciation and syntax, the easier it is to maintain the foreign language proficiency [8]. On the contrary, the greater the degree of interference and abrasion of the foreign language. Monika S. Schmid completed a work in this field, *Language Attrition*. In this work, the author explained in detail the definition of Language Attrition, the theoretical background and the issues that should be paid attention to in the study of Language Attrition. The book also involved a large number of empirical studies. These studies reveal the attrition phenomenon caused by the interaction between the first language and the second language [5]. Schmid & Kopke pointed out that the erosion of mother tongue and second language first starts from the erosion of vocabulary [6]. De Bot argues that language attrition belongs to the category of psycholinguistics. In recent years, some scholars have begun to analyze the attrition of words from the perspective of psycholinguistics [2].

2 The Research Methods

2.1 The Subject of Study

All of the 20 non-English majors from 2 classes in Jingchu University of Technology took part in the study. Before the experiment, the vocabulary level of the subjects was tested by vocabulary level test. According to test results (as shown in Table 1) were divided into four groups, the lowest level 1 subjects vocabulary, grasp the vocabulary by less than 2500 words, level 2 participants grasp the vocabulary between 2500 to 3500 words, which is above the average of level 3, who can reach 3500 to 4500 words. The final group, level 4 subjects, whose vocabulary is the largest, more than 4500 words. Five students from each group of pretests were randomly selected as subjects of this study.

2.2 Research Methods and Means

In this study, 20 non-English majors were randomly selected to conduct vocabulary tests in the first semester, the second semester of the freshman year and their sophomore respectively. 30 high frequency words were selected from the teaching requirements of college English course, and these stimulating words were randomly shuffled. Participants will be given an vocabulary association test, in which they will be given access to a mental lexicon during the test. The test was conducted four times over a period of two years. Before and after the test, the "look-write" free association test was adopted. The test was conducted in the multimedia classroom. The subject looked at the words appearing on the slide at the same time, and wrote down the words that could be thought of after 5 s, which was the response words of the subject. Finally, these response words are statistically analyzed to answer the two questions raised in this study: first, what is the mental lexicon association mode of second language learners? 1. Semantic or formal association? 2. What kind of semantic or formal connection are they? Second, at different stages of learning (i.e., at different levels of English proficiency), what are the changes in the mental lexicon association patterns of English learners? Third, is the rate of word attrition related to the mode of word association, and if so, what is the mode of association?

This study intends to use WordNet to identify the longitudinal aggregation relationship of words. WordNet is a vast vocabulary repository designed by psychologists, linguists and computer engineers at Princeton University. Its characteristic is that it is in accordance with the words in the meaning rather than to organize the information of vocabulary and in WordNet, nouns, verbs, adjectives and adverbs are in accordance with the synonym set organization, each Synset gives a basic vocabulary concept, and has established including synonymous relationships between these concepts (synonymy, antonym relations (antonymy), a relationship (hypernym & hypony), part of the up and down (meronym) and so on many kinds of semantic relations. For the horizontal combination relationship between response words, we can use the corpus tool BNC. BNC can not only use its matching SARA retrieval software, but also support a variety of general retrieval software, and can directly search online, which is convenient and fast.

3 The Results and Analysis

3.1 Results of the First Test

The first test took place at the beginning of enrollment, and the results showed that the total number of response words given by the four groups was 117, 137, 154, 160 respectively. Semantic responses were 54%, 60%, 63%, 70%, and formal responses were 36%, 33%, 34% and 28%, respectively. It can be seen that the semantic response words given by the four groups of subjects are all higher than the formal response words, which indicates that no matter what the level of the second language vocabulary of the subjects is, the semantic association is dominant and the formal association is auxiliary in their psychological lexicon. The results also showed that the proportion of response words provided by the four groups showed an increasing trend from level 1 to level 4, while nonsense words showed a decreasing trend. That is to say, the higher the level of the second language, the more response words the subjects provided. This phenomenon is in line with the vocabulary response characteristics of learners at different English vocabulary levels (Table 2).

Table 1. Test 1

		Level 1 (n = 5)		Level 2 (n = 5)		Level 3 (n = 5)		Level 4 (n = 5)	
		Num.	Percent	Num.	Percent	Num.	Percent	Num.	Percent
Semantics	Synonymy	22	18%	29	21%	32	21%	36	23%
	Antonymy	19	16%	24	18%	31	20%	33	21%
	Hypernym	13	11%	17	12%	19	12%	24	15%
	Meronym	10	9%	13	9%	16	10%	18	11%
Forms	Phonological	25	22%	29	21%	30	19%	26	16%
	Inflect	16	14%	17	12%	23	15%	20	12%
Meaningless response		12	10%	9	7%	5	3%	3	2%
Total		117	100%	137	100%	154	100%	160	100%

Table 2. Test 2

		Level 1 (n = 5)		Level 2 (n = 5)		Level 3 (n = 5)		Level 4 (n = 5)	
		Num.	Percent	Num.	Percent	Num.	Percent	Num.	Percent
Semantics	Synonymy	28	19%	33	20%	34	19%	36	19%
	Antonymy	25	17%	36	22%	31	18%	34	18%
	Hypernym	19	13%	19	12%	27	15%	31	17%
	Meronym	20	14%	15	9%	21	12%	29	16%
Forms	Phonological	25	17%	28	17%	31	18%	25	14%
	Inflect	18	12%	22	14%	27	15%	28	15%
Meaningless response		12	8%	8	5%	4	2%	2	1%
Total		147	100%	161	100%	175	100%	185	100%

3.2 Results of the Second Test

After one year of English study, the subjects were given a second vocabulary response test. The results showed that the total number of response words given by the four groups was 147, 161, 175, 185 respectively. In this study, the proportion of semantic responses and formal responses given by the four groups were 63%, 64%, 65% and 70% respectively. It can be seen from the above results that semantic association is still dominant and formal association is secondary. Compared with semantic and formal reactions, nonsense response still show a decreasing trend. According to the total number of response words provided by the subjects, after one academic year of English learning, the vocabulary level of the four groups of subjects has improved somewhat compared with that of the students who just entered the school (Table 3).

Table 3. Test 3

		Level 1 (n = 5)		Level 2 (n = 5)		Level 3 (n = 5)		Level 4 (n = 5)	
		Num.	Percent	Num.	Percent	Num.	Percent	Num.	Percent
Semantics	Synonymy	24	18%	35	25%	39	22%	37	21%
	Antonymy	20	15%	34	24%	35	20%	35	20%
	Hypernym	21	15%	2	1%	28	16%	26	15%
	Meronym	19	14%	16	11%	20	11%	25	14%
Forms	Phonological	23	17%	25	18%	27	15%	29	16%
	Inflect	15	11%	25	18%	21	12%	25	14%
Meaningless response		15	11%	5	4%	7	4%	2	1%
Total		137	100%	142	100%	177	100%	179	100%

3.3 Results of the Third Test

After two years of learning English, the participants took the third vocabulary response test. The results showed that the total number of response words given by the four groups was 137, 142, 177, 179 respectively. Among them, the proportion of semantic reaction is 61%, 61%, 69%, 69%, and the proportion of formal reaction is 28%, 35%, 27% and 30% respectively. From the above results, it can be seen that the results of the third vocabulary test are not much different from the results of the second vocabulary test. The results also showed that the higher the vocabulary level of the subjects, the less nonsense words and unresponsive words were provided. However, from the total number of words provided by the third and second subjects, there was no significant difference in the vocabulary level between the four groups in the first and second year (Table 4).

Table 4. Test 4

		Level 1 (n = 5)		Level 2 (n = 5)		Level 3 (n = 5)		Level 4 (n = 5)	
		Num.	Percent	Num.	Percent	Num.	Percent	Num.	Percent
Semantics	Synonymy	18	16%	31	21%	31	21%	32	20%
	Antonymy	19	16%	27	18%	29	19%	26	17%
	Hypernym	21	18%	19	13%	23	15%	24	15%
	Meronym	15	13%	15	10%	24	16%	22	14%
Forms	Phonological	13	11%	21	14%	21	14%	26	17%
	Inflect	11	9%	25	17%	16	11%	25	16%
Meaningless response		19	16%	8	5%	6	4%	2	1%
Total		116	100%	146	100%	150	100%	157	100%

3.4 Results of the Fourth Test

The fourth vocabulary test was conducted in the second semester of the third year of college English learning. The results showed that the total number of response words given by the four groups was 116, 146, 150, 157 respectively. The results of this study showed that the proportion of semantic responses given by the four groups was 63%, 63%, 71%, 66%, and the proportion of formal responses was 20%, 31%, 25% and 32%, respectively. The above results show that the total number of responses provided by the subjects is significantly lower than that of the first three times after they stop learning English for one year, but the number of words provided by the subjects with high vocabulary level is significantly higher than that of the subjects with low vocabulary level, and the proportion of semantic association is still significantly higher than that of formal association.

4 Discussions

Mental vocabulary is the mental representation of word knowledge in the brain and the long-term memory of word knowledge in the brain (Gui Shichun 2000) [3]. One of the main contents of mental vocabulary research is how words are stored and extracted in the human brain. As the basic ideographic unit of language, vocabulary plays an important role in language learning. Only paying attention to foreign language learning research and ignoring foreign language attrition can not effectively improve the efficiency of foreign language learning. The study of language acquisition is inseparable from the study of language attrition, which opens up an important new way for language acquisition. From the perspective of psychological lexicon in psycholinguistics, it can provide an important theoretical basis for the study of vocabulary acquisition.

5 Conclusion

How to intervene in foreign language attrition is an effective way to delay the decline of college students' foreign language learning ability. In China, English is only a foreign language, especially for non-English majors, after intensive learning of English in senior high schools, some students have been slack in learning English after entering the university. As a college English teacher, learners will inevitably experience the phenomenon of language skills attrition or even forgetting over time. Therefore, it is of great significance and research value to explore how to maintain foreign language competence and prevent foreign language abrasion in the current context.

Acknowledgments. This work is supported by Humanities and Social Sciences Project of Hubei Education Department (Project No. 19G056).

References

1. Craik, F., Tulving, E.: Depth of processing and the retention of words in episodic memory. J. Exp. Psychol. Gener. **104**, 268–294 (1975)
2. De Bot, K.: Language attrition—where are we and where are we going. In: International Conference on First Language Attrition, Amsterdam (2002)
3. Gui, S.: New Edition of Psycholinguistics. Shanghai Foreign Language Education Press, Shanghai (2000). (in Chinese)
4. Meara, P.: Review article: the rediscovery of vocabulary. Second. Lang. Res. **18**(4), 393–407 (2002)
5. Schmid, M.S.: Language Attrition. Cambridge University, London (1991)
6. Schmid, M.S., Kopke, B.: L1 attrition and the mental lexicon. In: Pavlenko, A. (ed.) The Bilingual Mental Lexicon: Interdisciplinary Approaches, pp. 209–238. Multilingual Matters, Clevedon (2008)
7. Singleton, D.: Exploring the Second Language Mental Lexicon. Cambridge (2001)
8. Weltens, B.: The attrition of foreign-language skills: a literature review. Appl. Linguist. 22–38 (1987)
9. Wolter, B.: Comparing the L1 and L2 mental lexicon. SSLA **23** (2001)

Establishment of Agricultural Economic Management Data Model Based on Block Chain Technology

Wei Liu(✉)

Inner Mongolia Technical College of Construction, Hohhot 010070, Inner Mongolia, China

Abstract. With the investment of large-scale and high-speed informatization under the traditional institutional framework, serious information islands have been caused. Therefore, the establishment of data models for the agricultural economy has become an urgent requirement for the construction of agricultural economic informatization. With the rise of Bitcoin in China, the region Block chain (BC) technology has entered people's sight, providing new possibilities for the establishment of agricultural economic management data models. Based on this, this article carried out a research on the establishment of a data model for agricultural economic management based on BC technology. This article elaborates on the establishment of a data model based on BC technology. Through an overview of BC in technology, it introduces the technical characteristics of the BC itself, and discusses several possible application scenarios of BC technology, and explains the working principle of BC technology used in agricultural economic management shows the characteristics and advantages of decentralized management. This article is aimed at the problems of data sharing in the economic network, combined with the scalability of the information technology on the BC and the privacy protection solution we proposed to use the BC to classify the nodes of each agricultural unit and department connected, data exchange is carried out through the platform, and the authenticity and reliability of the data exchange is ensured and it is not maliciously tampered with. The investigation and research results show that in the experiments of the paillier cryptosystem system in this paper, as the total length of the key generated by it gradually increases, the number of calculations that must be required in the process of generating the key gradually increases, and its increasing time and rate gradually increase by 67%; as the total length of the key decreases, the time spent in the encryption and decryption process will gradually decrease, the amount of calculation in the encryption and decryption process gradually increases by 127%, and the time required for decryption and decryption as the key length changes exponentially.

Keywords: Block chain · Agricultural economy · Data management · Data model

B. J. Jansen et al. (Eds.): *International Conference on Cognitive based Information Processing and Applications (CIPA 2021)*, LNDECT 85, pp. 64–70, 2022.
https://doi.org/10.1007/978-981-16-5854-9_8

1 Introduction

In the new generation of information technology, BC technology will be the key to solving the problem [1, 2]. Currently, BC technology is becoming a highly valuable technological frontier with its unique technical advantages, innovative value concepts, and extensive application scenarios [3, 4]. Many experts even believe that BC technology is expected to become the technology that has the most potential to trigger the next wave of disruptive revolution after steam engine, electricity, information and Internet technology [5, 6].

In the research on the establishment of the agricultural economic management data model based on BC technology, many researchers have conducted research on this and achieved good results [7, 8]. For example, in order to solve the problem of BC storage capacity, Campanile L designed and implemented a BC-based cloud computing environment database system, using BC to ensure data integrity, improve performance, and improve stability [9]. Livraga G proposed a BC naming storage system. The system is designed with a four-layer architecture, which makes full use of the decentralization characteristics of the BC to ensure high data security. The distributed and trusted cloud data of the BC can be used [10].

For better research, this article first briefly introduces the implementation steps of BC-based agricultural economic management data collaboration and sharing, and further analyzes the module composition established by the BC-based agricultural economic management data model, focusing on agricultural economic management data. The model is established, combined with the scalability and privacy protection program on the BC, and a solution is proposed for the collaborative sharing of agricultural economic management data. To this end, in order to verify the effectiveness of the measures taken in this article, we conducted an in-depth analysis of the survey data.

2 Research on the Establishment of Agricultural Economic Management Data Model Based on BC Technology

2.1 Implementation Steps for Collaborative Sharing of Agricultural Economic Management Data Based on BC

(1) Data preparation. Users who participate in the sharing of agricultural economic management data encrypt their prepared data through a homomorphic encryption algorithm and add relevant digital tags. According to the initial data request requirements, they are prepared to transmit relevant data through an encrypted secure channel.
(2) Data transmission. Each data sharing participant transfers data to a secure computing container through a secret channel. After entering the secure channel, the relevant data is automatically encrypted, and when the data reaches the secure computing container, it is automatically decrypted and prepared for related operations.
(3) Collaborative data operation. After each data collaborator transmits the data to the secure computing container, it uses the determined secure multi-party computing algorithm to operate on the received data.

(4) Collaboration results are returned. The collaborative data result calculated by the secure container is transmitted to the requester of the collaborative data through a reverse one-way secure channel under strong supervision. After the collaborative data result is delivered to the requester of the collaborative data, it is immediately eliminated and will not be backed up.
(5) The original data is returned. After the collaborative data calculation is completed and the collaborative data result is returned to the collaborative data requester, the original data used for collaboration should also be returned to the original holder of the data, and no backup will be left in the secure computing container, and the original return of data is also carried out on a one-way secure channel.

2.2 Modules Established Based on the BC-Based Agricultural Economic Management Data Model

Departmental system: multiple departments participating in the sharing of agricultural economic management data, select the data to be shared, and transfer the data in their own database and the data in the office system to the data sharing interface service module through the business adapter in an encrypted manner [11, 12].

Data sharing interface service module: The data sharing interface service module is composed of data collection interface, data sharing interface, data conversion interface, service sharing interface, task scheduling interface and data transmission interface. The received government affairs data to be shared is formatted, different data is scheduled according to the pre-determined mode, and the planned tasks and goals are completed.

BC network service module: The BC network service here mainly includes smart contract modules and distributed ledgers. Distributed ledgers here are broad and include the basic elements of BC, P2P networks, consensus algorithms, signature algorithms, etc. Smart contracts mainly complete data cleaning and data synchronization. The generalized distributed ledger mainly completes the tasks of storing data resources, controlling permissions, and storing and analyzing natural person information. This module is connected with the data sharing interface service module.

Data resource catalog service module: The data resource catalog service module mainly includes data resource catalogs and template catalogs, which are juxtaposed with the BC network service module to provide data indexing, data matching and data query services for the BC network service module.

System module: The system module mainly includes data interface service system, data visualization system and data sharing business system. The system module is connected with the BC network service module and the data resource catalog service module, and the information transfer between them is encrypted and decrypted.

2.3 Encryption Algorithm of BC Technology

Homomorphic encryption allows users to directly perform ciphertext-specific algebraic operations, get the result, and perform the same operation on the same plaintext encryption result. Homomorphic encryption has properties as shown in Eqs. (1) and (2):

$$\mathrm{E}_{pk}(m_1 + m_2) = E_{pk}(m_1) \oplus E_{pk}(m_2) \tag{1}$$

$$E_{pk}(a \cdot m_1) = a \oplus E_{pk}(m_1) \tag{2}$$

PAILLIER cryptosystem has good similar properties and good homomorphism, and can be used to construct many practical and efficient cryptographic algorithms. In the decryption process, each party i needs to use the calculation part of the decryption algorithm as shown in Eq. (3):

$$c_i = c^{2\Delta sk_i} \tag{3}$$

3 Experimental Research on the Establishment of Agricultural Economic Management Data Model Based on BC Technology

3.1 Research Materials and Experimental Design

In this article, we use cloud storage to expand BC storage to realize the privacy protection of data shared by multiple parties using the Paillier cryptosystem. Applying the Paillier cryptosystem to the BC can effectively protect sensitive information and solve the privacy protection problem of the BC.

As mentioned earlier, in the data model, privacy threats mainly come from all parties. Therefore, the goal of the solution proposed in this article is to protect each user's data from being leaked to other parties. Since the scheme of this article is based on the data privacy protection protocol given, this article starts with the privacy analysis of this protocol.

3.2 Experimental Method and Analysis Content

The solution in this article is implemented using Python and Matlab. The user data storage database is SQL Server. Our system runs on 4 servers with 2.4 GHz, Intel i7-5500M CPU and 8 GB RAM.

In the ciphertext operation, the Paillier cryptosystem takes a relatively short time to homomorphize and is more efficient. The safety analysis of the scheme proves that the scheme has high safety.

4 Research and Analysis on the Establishment of Agricultural Economic Management Data Model Based on BC Technology

4.1 Research and Analysis of Key Generation Performance Evaluation

In our Paillier cryptosystem experiment, we generated 64, 128, 256, 512, and 1024-bit keys, and generated the corresponding time and public key length, thereby generating the public and private keys required for the entire encryption and decryption process. The first and second experimental data of time and public key length are shown in Table 1:

Table 1. Paillier key generation time

Public key length	First experiment	Second experiment
64	0.31	0.33
128	0.66	0.64
256	0.88	0.86
512	3.02	3.04
1024	7.01	7.02

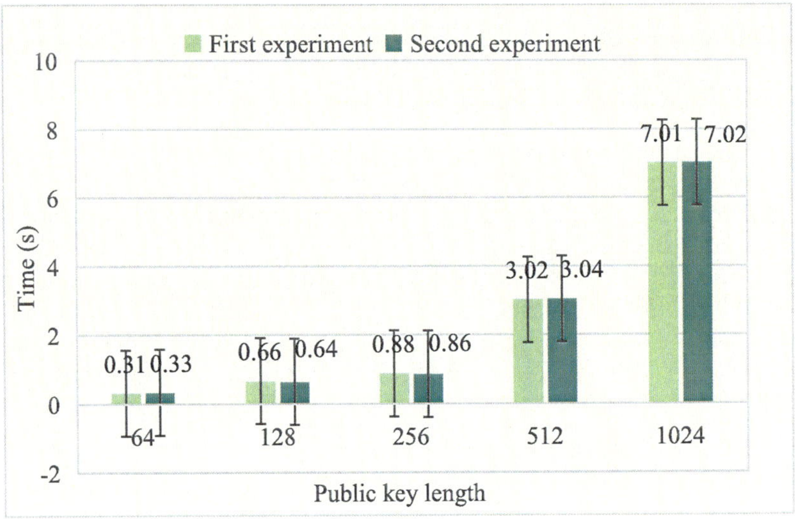

Fig. 1. Paillier key generation time

As shown in Fig. 1, when the key length is 64, the calculation time is 0.32 s, when the key length is 128, the calculation time is 0.65 s, when the key length is 256, the calculation time is 0.87 s, when the key length is 512, the calculation time is 3.03 s, when the key length is 1024, the calculation time is 7.015 s. It can be seen that as the length of the generated key gradually increases, the amount of calculation required in the generation process gradually increases, and the rate of increase gradually increases.

4.2 Encryption and Decryption Performance Evaluation Research Analysis

The experimental plan in this paper gives the time required for the experiment of encrypting 32-bit data when the public key length is 64, 128, 256, 512, and 1024 bits. The experimental data is shown in Table 2:

Table 2. Paillier encryption and decryption into time

Public key length	Paillier encryption time	Paillier decryption time
64	0.11	0.13
128	0.36	0.34
256	0.78	0.76
512	0.92	0.94
1024	4.01	7.02

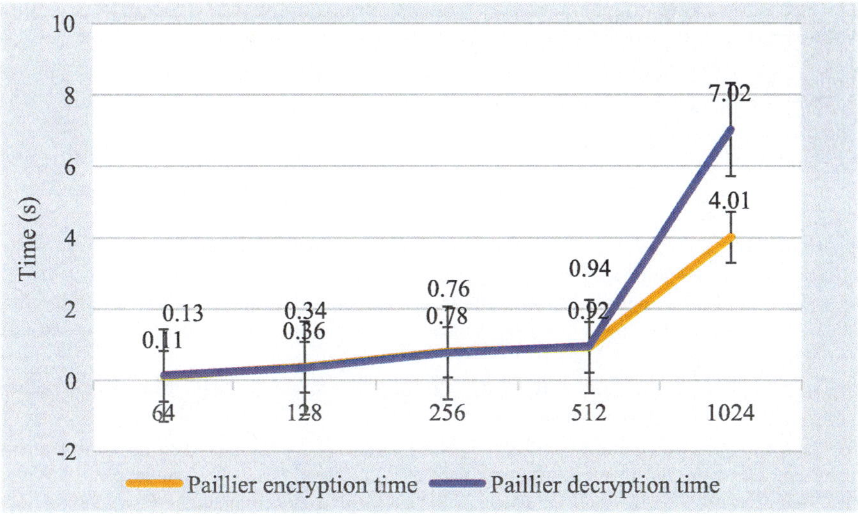

Fig. 2. Paillier encryption and decryption into time

As shown in Fig. 2, as the length of the key increases, the time consumed in the encryption and decryption process gradually increases, and the amount of calculation in the encryption and decryption process gradually increases, and the time required for decryption and decryption increases with the length of the key. The change has increased exponentially. In addition, it can be seen that the time difference between the decryption time and the encryption time of the Paillier algorithm is getting larger and larger.

5 Conclusions

Based on the establishment of the agricultural economic management data model, combined with the scalability and privacy protection scheme on the BC, this paper proposes a solution to the collaborative sharing of agricultural economic management data. According to the characteristics of economic collaboration data, it is possible to achieve collaborative sharing without data collection, while ensuring the security and privacy of data subjects. In the process of realizing data collaboration, it is necessary to pay attention to how to ensure the boundaries of data sources and other related rights, as well as how to ensure the accuracy of economic data.

References

1. Chen, J., Lv, Z., Song, H.: Design of personnel big data management system based on block chain. Futur. Gener. Comput. Syst. **101**(012), 1122–1129 (2019)
2. Cha, H.J., Yang, H.K., Song, Y.J.: A study on access structure management of CP ABTD based block chain for medical information monitoring system. Adv. Sci. Lett. **24**(3), 2026–2030 (2018)
3. Yang, Y., Jia, Z.W.: Application and challenge of block chain technology in the field of agricultural internet of things. Agric. Netw. Inf. (012), 24–26 (2017)
4. Queiroz, M.M., Wamba, S.F.: Block chain adoption challenges in supply chain: an empirical investigation of the main drivers in India and the USA. Int. J. Inf. Manag. **46**(029), 70–82 (2018)
5. Fathi, M.M., Galal, A., Al-Homidan, I., Abou-Emera, O.K., Rayan, G.N.: Residual feed intake: a limiting economic factor for selection in poultry breeding programs. Ann. Agric. Sci. **66**(1), 53–57 (2021)
6. Khiav, L.A., Zahmatkesh, A.: Vaccination against pathogenic clostridia in animals: a review. Trop. Anim. Health Prod. **53**(2), 1–12 (2021)
7. Mo, F., et al.: Environmental and economic benefits of micro field rain harvesting farming system at maize field scale in semiarid east African Plateau. Agric. Water Manag. **56**(206), 102–112 (2018)
8. Garcia Alfaro, J., Navarro Arribas, G.: Searchable encrypted relational databases. Risks Countermeasures **10**(005), 70–85 (2017)
9. Campanile, L.: Designing a GDPR compliant block chain based Io V distributed information tracking system. Inf. Process. Manag. **58**(2), 1–23 (2021)
10. Livraga, G.: Lecture notes in computer science data privacy management. Cryptocurr. Block Chain Technol. **101**(109), 29–30 (2018)
11. Miraz, M.H., Ali, M.: Applications of blockchain technology beyond cryptocurrency. Ann. Emerg. Technol. Comput. **2**(1), 1–6 (2018)
12. Sikorski, J.J., Haughton, J., Kraft, M.: Blockchain technology in the chemical industry: machine-to-machine electricity market. Appl. Energy **195**(JUN.1), 234–246 (2017)

Improvement Design of Information Processing of Urban Municipal Crossing Facilities

Yu Liu[1] and Yu Peng[2,3](✉)

[1] Chongqing Municipal Research Institute of Design Co., Ltd., Chongqing 400020, China
[2] Chongqing Chengtou Road and Bridge Administration Co., Ltd., Chongqing 400060, China
[3] School of Civil Engineering, Chongqing University, Chongqing 400045, China

Abstract. With the continuous development of my country's information technology and the advent of the era of artificial intelligence, in the context of the continuous advancement of knowledge and technology in my country, cities have gradually evolved into an open and innovative model. In order to adapt to the development of the new technological era, the informationization of urban municipal festivals has emerged. Since municipal engineering facilities and public life service facilities, including water supply and power supply, transportation, environmental protection, health education, postal services, etc., cover all aspects of urban construction, can these municipal infrastructures achieve efficient operation, data sharing, interconnection, and intelligent supervision? It has become the focus of the construction of a modern informationized smart city. This article first briefly analyzes the development status of our country's urban municipal crossing infrastructure information industry through an overview, and points out that the construction of standardization is of great significance. By analyzing the construction and application status of municipal infrastructure information standardization, establishing a standard system analysis framework, and putting forward relevant effective opinions and suggestions on the design direction of the next stage of information processing and improvement. Finally, the allowable time of the road section is analyzed, and appropriate evaluation indicators are selected accordingly. Combine the various influencing factors at pedestrian crossings to set simulation conditions, and set a certain range and slope. Establish a simulation plan by crossing various factors and conditions, and run the simulation output results. The experimental analysis results show that this research can give planning guidance in the relevant standard phase of the Internet of Things informatization of municipal crossing infrastructure, and realize the connection and coordination between standards and standards.

Keywords: Municipal administration · Crossing facilities · Information technology · Improvement analysis

1 Introduction

The construction of barrier-free facilities for urban crosswalks is a systematic project, which is a basic condition for the disabled to participate in social and economic activities

B. J. Jansen et al. (Eds.): *International Conference on Cognitive based Information Processing and Applications (CIPA 2021)*, LNDECT 85, pp. 71–78, 2022.
https://doi.org/10.1007/978-981-16-5854-9_9

on an equal basis, and it also provides convenience for the elderly, women and children [1, 2]. Barrier-free facilities are a major event that safeguards the legal rights and interests of the disabled and other special social groups. It is a noble act that reflects social humanistic care, a need to improve the functions of modern cities, and an important symbol of social civilization and social development [3]. At present, China has entered a period of rapid aging. The disabled, the elderly and other disadvantaged groups have an increasing demand for urban barrier-free facilities [4]. The awareness of barrier-free society is growing. Accessibility laws and regulations at home and abroad are constantly updated and improved. The barrier-free facilities of urban crosswalks also show a new development trend [5]. Therefore, it is instructive to analyze the promulgation process of barrier-free laws and regulations at home and abroad, and summarize the development trend of barrier-free crosswalk facilities at home and abroad [6].

At present, the crossing facilities are mainly divided into two types: plane crossing facilities and three-dimensional crossing facilities. Relatively speaking, traditional level crossing facilities have poor safety, low efficiency, and easily cause traffic congestion and other problems. Although three-dimensional crossing facilities can solve this problem, they face huge construction, operation and maintenance costs [7, 8]. The selection of pedestrian crossing facilities should first meet the needs of pedestrians and ensure the safety of pedestrian crossings; secondly, we should consider the convenience of pedestrians. The main problems of household level crossing facilities are: unreasonable location; due to lack of signal control or unreasonable signal timing, pedestrians have a long waiting time, and the interests of pedestrians are not taken into consideration. Due to poor safety, the area of the safety island at the intersection of the passage cannot meet the needs of pedestrians crossing the road, which makes pedestrians occupy the right-turn lane when waiting at a red light, and increases the conflict between right-turning vehicles and the intersection [9, 10]. Pedestrians and traffic are delayed and threaten the safety of pedestrians.

The existing research results mainly focus on the evaluation of the layout of the intersection facilities and the service performance evaluation of the three-dimensional intersection facilities, while the comprehensive evaluation of the three-dimensional intersection facility layout is less, and the evaluation index system is imperfect and needs to be optimized [11]. The determination of index weight coefficients often adopts expert scoring method, which has poor evaluation accuracy, strong subjectivity, and lacks a quantitative comprehensive evaluation model that combines subjectivity and objectivity [12].

2 Method

2.1 The Positioning Algorithm of the Base Path Dynamic Programming

The problem of anchor node movement path planning is a traveling salesman problem. The set of virtual anchor nodes must be equivalent to the set of tourist attractions that the traveller must pass. The mobile anchor node is equivalent to the traveller. The mobile anchor node must pass through each virtual anchor. The node is equivalent to the traveler passing through each tourist attraction. Design the moving route so that the moving anchor node passes through all virtual anchor nodes, and the moving path is the shortest.

At present, scholars have proposed many heuristic optimization search algorithms such as genetic algorithm, ant colony algorithm, tabu search algorithm, and simulated annealing algorithm to solve the TSP problem. The ant colony algorithm simulates the way of ants foraging, and selects the next path through the pheromone released on the path when the ants are foraging.

Obtained must pass virtual anchor node set V_{indis}, store the must pass virtual anchor nodes in arrays A and B, where $a_{ij} = b_{ij}$

$$\mathrm{A} = a_{ij}(i = 1, 2, 3 \ldots\ldots j = 1, 2) \tag{1}$$

$$B = b_{ij}(i = 1, 2, 3 \ldots\ldots j = 1, 2) \tag{2}$$

$$d_{a,b} = \sqrt{(x_a - x_b)^2 + (y_a - y_b)^2} \tag{3}$$

Obtain the weighted matrix D_(i,j) of the virtual anchor node that must pass through formula (4)

$$D_{i,j} = \begin{cases} 0 & i = j \\ \sqrt{\left(a_{i,1} - b_{j,1}\right)^2 + \left(a_{i,2} - b_{j,2}\right)^2} & i \neq j \end{cases} \tag{4}$$

Suppose there are m ants in the entire ant colony, there are n cities, and the distance between each city is D_(i, j) (i, j = 1, 2, 3 … n).

$P_{ij}^{k}(t)$ represents the probability of ant K (K = 1, 2, 3, …, m) moving from city i to city j at time t, and the calculation formula is:

$$P_{ij}^{k}(t) = \begin{cases} o & s \in allow_k \\ \frac{[\tau_{ij}(t)]^{\alpha} \cdot [\varphi_{ij}(t)]^{\beta}}{\sum_{s \in allow_k} [\tau_{ij}(t)]^{\alpha} \cdot [\varphi_{ij}(t)]^{\beta}} & allow_k \not\ni s \end{cases} \tag{5}$$

$$allow_k = \{0, 1, \ldots, n - 1\} - tab_k$$

2.2 Research and Analysis on the Improvement of Urban Crossing Facilities Using Big Data

As the basis of this research, big data is used to analyze the role of urban crossing facilities in the process of improvement and development, and it is more important to the use and analysis of big data.

Analysis of the improvement of urban crossing facilities using big data. The variables of urban traffic rules can be expressed as follows:

$$\mathrm{X} = (w_1, w_2, \ldots, w_n) \tag{6}$$

$$W_k = \{X^{k,1}, X^{k,2}, \ldots, X^{k,N}\} \tag{7}$$

$$X^{k,i} = (X_1^{k,i}, X_2^{k,i}, \ldots, X_d^{k,s})^{Y} \tag{8}$$

2.3 Define Matrix by Sample

$m_k = \frac{1}{N_k} \sum_{x=\partial_k} x$ represents the $m = \frac{1}{N} \sum_{K=1}^{K} \sum_{x=\partial_k} x$ mean of the k-th class of samples, and is the overall mean:

$$P_b = \sum_{k=1}^{k} \frac{N_k}{N} (m_k - m)(m_k - m)^Y = \theta_b \theta_b^Y \tag{9}$$

3 Experiment

3.1 Model Parameter Settings

Occupation parameter: This article sets a value of -1 for a position occupied by other pedestrians. The position of the idle primitive cell at its own station is 0, and the position occupied by other pedestrians is -1.

Safety parameters: In the process of crossing the street, pedestrians often make safety judgments based on the state of the vehicles that will come and go. When pedestrians find that a vehicle is driving toward themselves, they usually decide based on the distance between them and themselves. Whether to take pictographic measures to avoid.

Therefore, this paper sets a probability value based on this phenomenon, and can complete the duplication of safety parameters. The safe parameter is actually a matrix. The initial value of the matrix is 0. When a probability of a horizontal row is activated, the parameter of the horizontal row will be copied to -1, which is used to represent pedestrians.

When the model calculates the safety parameters, we first need to introduce a direction to determine the parameters, so if the vehicle is driving from left to right, the safety parameter is set to 1, otherwise it is set to -1. Therefore, when pedestrians are not included in the scope of safety considerations for vehicles far away from themselves, they can use the following formula to check whether the vehicle is close or far away:

$$\mathrm{D} = D_1 \times D_2 \tag{10}$$

According to the above formula, if D is assigned a value of 1, it means that the passing vehicles are driving towards pedestrians, which may have a possible impact on the safety of pedestrians. On the contrary, if it is -1, it means that the car crossing the street lane is moving away from pedestrians and will not cause a safety impact on people.

According to the social reality, our analysis can conclude that the speed of vehicles on the street is related to the probability of pedestrians crossing the street. When the speed of the vehicles increases, pedestrians will wait for the next time, and the formula for calculating the probability of crossing the street is as follows Show:

$$p_1 = \frac{v_{max} - v_n}{v_{max}} \tag{11}$$

Therefore, the closer the distance, the lower the probability of pedestrian crossing the street. The following formula can analyze a dynamic probability caused by the distance from the nearest vehicle to itself:

$$p_2 = \frac{d}{50} \tag{12}$$

Finally, the probability that the pedestrian will continue to drive in the crossing can use a 3 * 3 matrix. The safety parameters are only calculated and copied for the first three squares of the pedestrian's location. Therefore, there are only two ways for two-way pedestrians to copy. The matrix is shown in Table 1.

Table 1. Matrix of setting values of safety parameters for pedestrians crossing the street

−1	−1	−1
0	0	0
0	0	0

In summary, the safety parameter process of pedestrians can be judged, and finally a profit parameter is established.

Income parameters: The income per step forward for pedestrians is l, the return is − 1, and the immobility is O. If the pedestrian moves sideways, that is, change the direction of travel, the income of moving to the side and forward is 0.85, and the income of moving to the rear is 0.85, as shown in Table 2.

Table 2. Matrix parameter setting values of income parameters

0.85	1	0.85
0	0	0
−0.85	−1	−0.85

4 Results

It can be seen from Fig. 1 that pedestrians wait at the side of the road with the highest frequency in 1s. Pedestrians need to wait for a certain time mainly because pedestrians arrive at the end of the green light or during the red light, and wait for the green phase of the pedestrian in the waiting area. The use of information-based related processing technology to test the two sets of data shows that there is no significant difference between the two at 98% confidence. It can be seen from the figure that the pedestrian crossing time has the highest frequency in 20.25 s. Use information-related processing technology to detect the two sets of data. The test results show that there is no significant difference between the two at 97.3% confidence. The test of pedestrian waiting time and pedestrian crossing time shows that the model can describe the pedestrian crossing behavior of the road section more accurately, which verifies the rationality and accuracy of the model. Motor vehicle delays here refer to motor vehicle delays caused by signal red lights and pedestrians running red lights. When pedestrians cross the street illegally, they may affect the flow of motor vehicles and increase motor vehicle delays. Through

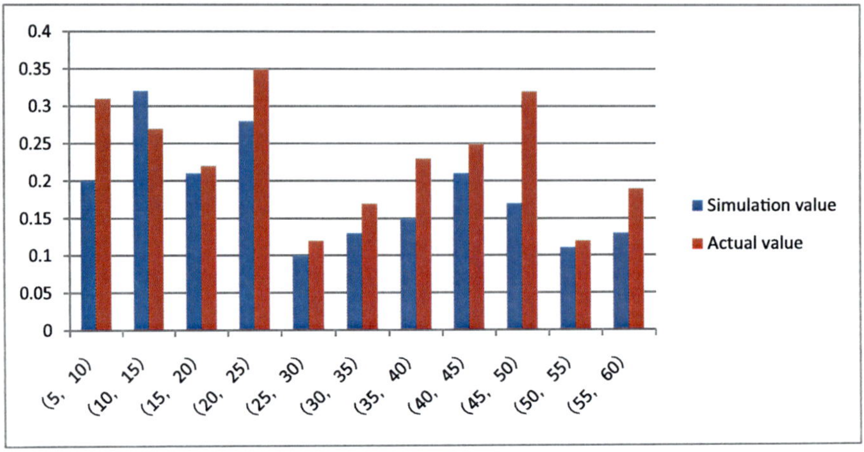

Fig. 1. Frequency analysis chart of pedestrian crossing market in information processing

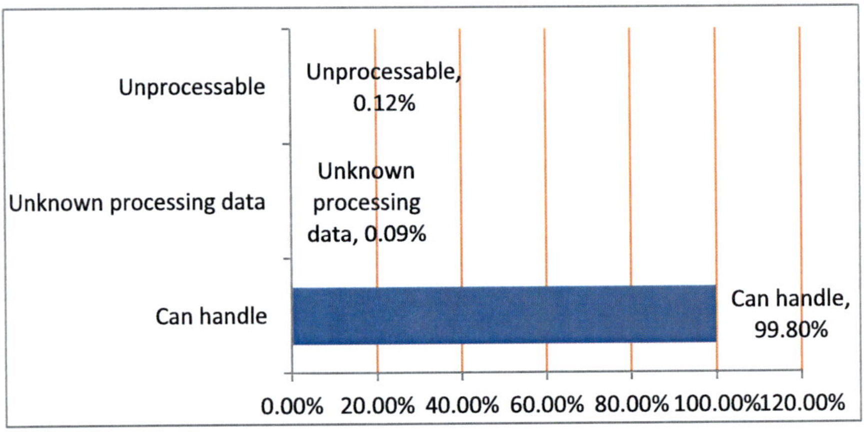

Fig. 2. Data processing degree of informatized municipal crossing facilities

the statistical analysis of motor vehicle delays, the average value of motor vehicle delays is 3.29 s.

Analyzing the data shown in Table 1 and Fig. 2, in the implementation of intelligent analysis of the processing degree of the data in the informatized municipal crossing facilities, and the big data technology for mining and processing, it will lead to the possibility of data Intermittent situations, the consequences of this situation are more serious. Use consistent evaluation indicators to evaluate the performance of the model, and finally select the best performing model to accurately predict the status of the risk rating in computer network communications.

According to the survey, first of all, the establishment of an informatized municipal crossing facility under informatization data has achieved quite effective and satisfactory results for data processing on various crossing facilities, but there are still many problems in the given municipal crossing process. Second, the division of labor is not clear, and the powers and responsibilities of the informatization municipal crossing facility system are unfair. Research shows that the reasons for this phenomenon of inadequate use of informatized municipal crossing facilities are more complicated.

In the previous article, we used information technology to process data in the informatized municipal crossing facilities. After data mining through the improved algorithm, different degrees of support and confidence have been obtained.

5 Conclusion

In summary, in the process of municipal public engineering construction, in order to continuously improve the level and quality of project construction management, it is imperative to strengthen the application of information management mode, which can focus on management and control the defects and loopholes in project construction management. Moreover, with the help of information management, project information resources can also be shared to ensure that construction managers have real-time control of the project construction situation. Municipal facilities information management system has become an inevitable link in urban construction. In the process of system construction, it is necessary to clarify the characteristics of the system according to the actual situation of the city, and to be realistic in all aspects of the construction system to provide support for the effective management of equipment. A successful information management system can monitor people's bad behavior, standardize people's behavior and habits, and improve the overall quality of the masses. This article investigates and evaluates various pedestrian crossing facilities in the city, and finds out the difficulties and bottlenecks that need to be resolved.

References

1. Longoni, R., Wetter, O.: Urban stream works in Central Europe 1200–1700: municipal administration, hydraulic engineering and flood reconstruction. Water Hist. **11**(1–2), 31–57 (2019)
2. Yang, L.: Negative externality of fiscal problems: dissecting the contagion effect of municipal bankruptcy. Public Adm. Rev. **79**(2), 156–167 (2019)
3. Fjeldstad, O.H., Ali, M., Katera, L.: Policy implementation under stress: central-local government relations in property tax administration in Tanzania. J. Financ. Manag. Prop. Constr. **24**(2), 129–147 (2019)
4. Solbakken, R., Bondas, T., Kasén, A.: First-line nurse managers' challenges at the crossroads of Norwegian health care reforms. Nurs. Adm. Q. **44**(3), 205–214 (2020)
5. Griffin, J.: Three parallel crossings in corpus Christi. Undergr. Constr. **74**(6), 32–33 (2019)
6. Patra, M., Perumal, V., Rao, K.: Modelling the effects of risk factor and time savings on pedestrians' choice of crossing facilities at signalised intersections. Case Stud. Transport Policy **8**(2), 460–470 (2020)
7. Cruz, R., Cruz, R.: Purposes and outcomes of information technology education towards sustainable national development. Inf. Technol. Educ. Soc. **17**(1), 5–20 (2020)

8. Tsohou, A., Siponen, M., Newman, M.: How does information technology-based service degradation influence consumers' use of services? An information technology-based service degradation decision theory. J. Inf. Technol. **35**(1), 2–24 (2020)
9. Pan, X., Han, C., Song, M., et al.: The impact of information technology investment on the performance of apparel manufacturing enterprises: based on the moderating effect of equity concentration. IEEE Trans. Eng. Manag. **PP**(99), 1–9 (2020)
10. Konak, A., Kulturel-Konak, S., Cheung, G.W.: Teamwork attitudes, interest and self-efficacy between online and face-to-face information technology students. Team Perform. Manag. Int. J. **25**(5/6), 253–278 (2019)
11. Abdurrahman, L.: Hybrid configuration in information technology value model. IEEE Syst. J. **PP**(99), 1–8 (2020)
12. Davies, N.: Foreign/second language education and technology in the future. IALLT J. Lang. Learn. Technol. **16**(3–4), 5–16 (2019)

Visual Communication of Light in 3D Film and Television Animation

Shuguang Li(✉)

School of Software Engineering, Anhui Vocational College of Electronic Information Technology, Bengbu 233000, Anhui, China

Abstract. In the production of three-dimensional animation, the performance of light is particularly important. Different types of lights can play different lighting effects. Through the artistic treatment of lights, the atmosphere of the scene can be expressed, and the overall tone of the film can be set. This article through the analysis of the visual semantics of light and the different types of light characteristics, proposed the basic theory of the three-point lighting method of main light, auxiliary light and effect light, and realized the artistic expression and communication of the light's "light color" visual semantics.

Keywords: 3D · Animation · Lighting · Rendering

1 Introduction

Light is a vital element of visual expression in the development of 3D animation and virtual reality. Video art can also be called the art of "light and color". For lighting in 3D animation films, animators should freely deploy light sources according to the overall tone of the film, and strengthen the artistic expression of the film through the artistic treatment of lighting.

The primary function of lighting is to improve the brightness of the environment and ensure the necessary lighting conditions for the scene. At the same time, the color saturation of the image is also inseparable from the sufficient lighting conditions during rendering. If there is no sufficient light source, the picture will appear dull and dull. The lack of basic lighting information in the dark part will cause the picture to lose a lot of level of detail.

Through the reasonable adjustment of direct lighting and indirect lighting, the performance of the lighting is greatly enriched, making the main character in the film more vivid and the scene more artistic Expressive.

2 The Visual Semantics of Light

2.1 Units of Measurement of Light

Traditionally, the wattage of commonly used bulbs defines the brightness of light, but with the development of science and technology, the luminous efficiency of bulbs continues to increase, so a more precise and unified definition of brightness is required. In

B. J. Jansen et al. (Eds.): *International Conference on Cognitive based Information Processing and Applications (CIPA 2021)*, LNDECT 85, pp. 79–85, 2022.
https://doi.org/10.1007/978-981-16-5854-9_10

1967, the International Association for Weights and Measures stipulated that the melting point of platinum was 1774 °C. As a light source, its 600,000th of an area (m^2) has a light measurement of 1cd as the basic unit of light brightness [1]. According to this definition, the following light measurement relationship occurred: the brightness of the light source (luminance)-the degree of light received by the object (illuminance)-the reflectance of the object (luminance)-the sensitivity (sensitivity) of the human eye. The luminosity is determined by the energy of the light source, the illuminance is determined by the luminosity and distance, the gray scale is determined by the luminosity and distance and the depth of the object's color, and the hardness is determined by the human physiological condition. In this process, only the luminosity and sensitivity can be determined. The illuminance is related to the distance, and the brightness is related to the color of the object.

2.2 The Color Temperature of Light

Color temperature is a way to describe colored light. Low tones are warmer colors and cooler colors with high color temperature. The color temperature of common light sources is shown in Fig. 1. Lights have different color tendencies, but the human eye has self-adaptive characteristics. Colored light is often regarded as colorless light. For example, the colors of flames of different degrees are red, orange, yellow, and white in order, which shows that the color of light is there is a certain relationship between color and temperature. Usually, the "Kjeldahl method" is used to measure the color temperature of the light, and the symbol is "k". Digital camera menus have color temperature settings, such as 3200 k and 5500 k. In 3D animation movies, a relatively stable colored light source is often needed. There are high color temperature light source 5500 k and low color temperature light source 3200 k. These two standards have become industry standards. During production, you can use the color temperature characteristics to reproduce a specific light source, and add unique light and color effects to the image.

Light Color		Red →			White →			Cyan	
Color temperature (K)	1000	2000	3000	4000	5000	6000	7000	8000	9000
Natural Light		20 minutes of sunlight after sunrise	One hour of sunlight after sunrise		The midday sun		Overcast sky		Blue Light
Artificial light	Candlelight	Incandescent lamp	Halogen lamp	White fluorescent lamp	Fluorescent lamp				

Fig. 1. The color temperature of common light sources

2.3 The Physics and Psychology of Light

There is a two-fold proportional relationship between the physical illuminance of light and the brightness reflected in human psychology [2]. Through the light and dark level, it is actually a measurement method of mental brightness. It is very important in lighting. The brightness level on the screen is twice the illuminance, and the number formula is n square. Different angles, different soft and hard lights will have a certain semantic effect on character description, film narration, and expression of emotions. For example, the character has less side light distribution and hard light, indicating that the character is emotionally agitated; the character has more side light distribution and light perception Softness means that the character's inner mood is calm.

3 Classification of Lights and Shadows

Light travels in a straight line and changes direction when encountering an object. Light is divided into direct light source and indirect light source. In terms of photometry, the light source of direct light is called "primary light". In addition to the vacuum zone, there must be various substances in the real natural scene, such as air, fog, etc. After being irradiated by light, it will produce secondary reflection, which is called "secondary light", and the light reflected by the secondary reflection will be reflected again, which is called "third light", and then form indirect lighting. An excellent lighting system is provided in the 3D software, which can create different lighting and shadow performance effects, simulate the real light changes in the real world, and also simulate the performance effects of indoor or stage lighting, thereby highlighting the film's own performance style.

3.1 Direct Light Source

In 3D software, direct light sources include various types of lights, physical sky lights, and HDR high-dynamic map light sources. The lights mainly include Ambient Light, Directional Light, Point Light, Spot Light, Area Light, Volume Light and other types.

Different lights have some common properties, such as the type of light, the color of the light and the intensity of the light, the range of influence, the way of attenuation, etc. The scope of influence of the light is the unique attribute of the light in the three-dimensional world. It can define the relationship between the light and the objects in the scene, and control the illumination of the light on the diffuse reflection area and the highlight area of the object. The attenuation of light according to different algorithms includes: linear attenuation, square attenuation and cubic attenuation, which simulate the physical propagation properties of light in the real world, and represent the energy attenuation of the light intensity during the propagation process after the light is emitted from the light source, because the light is in the propagation process There is a certain angle between the medium and the light source. The larger the angle, the larger the range of light irradiation, and the light intensity per unit area will also decrease, that is, the lighting effect will become weaker.

Using a professional renderer, you can directly set the physical skylight to simulate real outdoor lighting. The physical skylight is generally linked to the sun node, and the numerical change of the angle of incidence of the light is used to simulate the lighting effects in different time periods. HDR environment lighting uses high dynamic range images directly as a light source, which can simulate real-world lighting for a three-dimensional scene, so as to obtain a more realistic lighting effect.

3.2 Indirect Lighting

Indirect lighting means that light is illuminated by reflection, that is, reflected light. The light quality after the reflected light is diffused is softer than the light diffused by the light through the soft box, which can eliminate shadows well. It is called in the 3D professional renderer. For Global Illumination, the GI lighting method uses light to generate a large number of photons. These photons bounce multiple times in the scene. The GI engine emits sampling points in the scene according to the camera. Each sampling point can emit the color of the rays to the environment, the light intensity, distance and other information are sampled, and then the obtained information will be fed back to each sampling point. These sampling points can determine their own brightness or color based on this information, so that the scene can obtain a more realistic lighting effect of light transmission. The three GI engines in the Vray renderer are shown in Fig. 2.

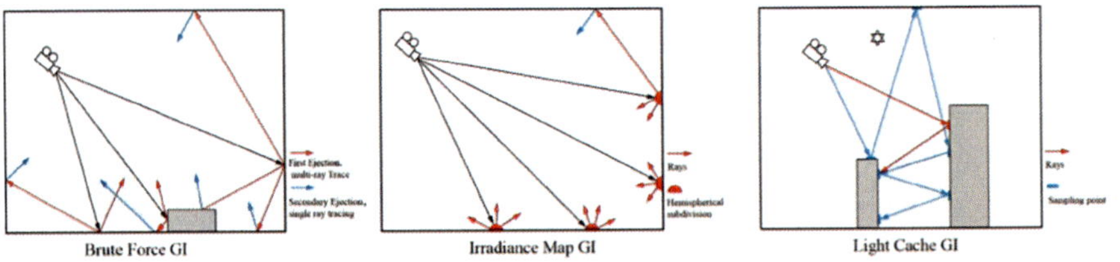

Fig. 2. Three GI engines in the Vray renderer

3.3 Shadow Processing

In a lighted environment, the performance of the shadow effect can increase the realism of the picture. Of course, it is very important to choose a suitable shadow creation method. The light-receiving part and the backlit part of the object usually produce shadows. The purpose of lighting is to make reasonable use of shadows. While eliminating redundant shadows [3]. In animated films, the phenomenon of ghosting is everywhere. The more typical one is the wax shadow light effect, which is the fuzzy area produced by the projection edge. Its artistic value is that the confusing light and shadow will give people endless reverie. Contribute to the performance of specific environmental atmosphere.

In the real world, the virtual reality of the shadow is controlled by the distance between the light source and the object, the distance between the object and the projection surface, and the area of the light source. The factors that control the virtual reality of the shadow are shown in Fig. 3. For example, for objects close to the light source, the larger the shadow, the more blurred the edges, and the corresponding objects away from the light source, the larger the shadow, the sharper the edges. In 3D software, you can directly control the virtual reality of shadows through parameterization. Common methods of casting shadows include depth map shadows and ray tracing shadows. The depth map shadows are the information recorded in the depth map file generated by the software during the rendering process to cast shadows. In this way, the rendering speed of the shadows generated in this way is very fast, and the effect is relatively soft, but it is not as real as the ray tracing shadows. The virtual and real shadows can be controlled by the Filter Size parameterization. The principle of light and shadow tracing shadows is to emit light from the position of the light source to the surroundings of the scene [4]. These light rays will be reflected on the surface of the object. The reflected light can illuminate other objects and then form reflections again. Light and shadow tracing shadows require more rendering. Time is more suitable to show the scattering effect of transparent and semi-transparent objects. Among them, Light Radius is to control the change of the virtual and real shadows. The larger the value, the more blurred the shadow.

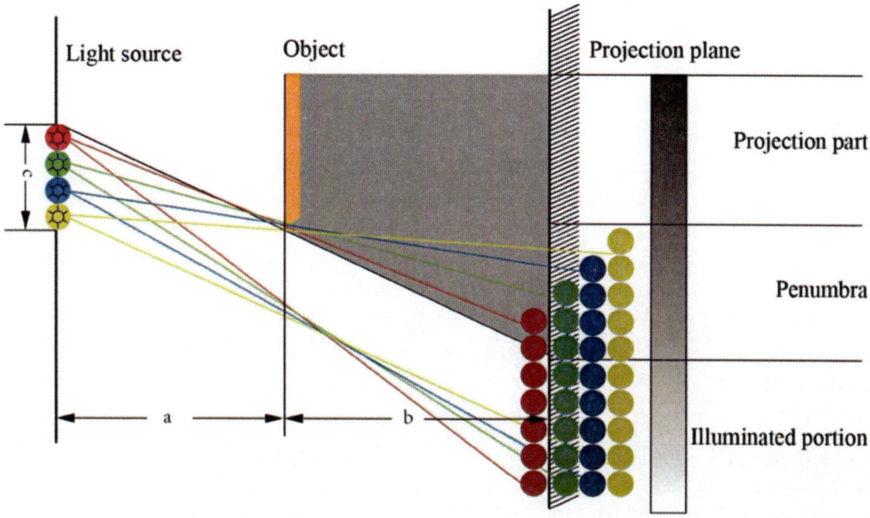

Fig. 3. Factors that control the virtual and real shadows

4 Visual Communication of Light

In the three-dimensional animation film, the light must first deal with the three basic elements of side light, light and dark, and three-dimensional sense. According to the specific requirements of the scene, different light sources can be flexibly set [5]. The visual communication of light mainly includes three-point lighting arrangements such as main light processing, auxiliary light processing, and effect light. Excellent visual transmission is firstly to use natural light to illuminate the pursuit of the reality of the scene, and secondly to express the sense of volume of the object [6]. The reasonable combination of light and shadow ultimately conveys the different emotions of the characters and meets the narrative needs of the film itself.

4.1 Transmission of the Main Light

The task of the main light is to imitate the natural side lighting. In the lighting of the three-dimensional animation film, the main light is the state of the panoramic light receiving direction. The main light defines the lighting rules of the movie scene. Its effect is similar to that of the main lens. Other lighting basis of the film, the main light is generated after setting, which can be processed subjectively. It can be assumed that the main light is the sun, moon, street lamp, indoor artificial light, etc., and the main light is set in the three-dimensional scene. After lighting, the thing to do is to simulate the effect of the main light [7]. As the basic lighting, the position, intensity, and color temperature of the main light should not be changed generally, that is, the same main light should be used in the same scene to ensure the scene the constancy of the lighting effect and the continuity of the lens narrative.

4.2 Transmission of Auxiliary Light

The task of the auxiliary light is to adjust the brightness of the picture. In the three-dimensional scene, the place where the main light is not irradiated is often dark, and the dark part does not have enough details. It is necessary to use auxiliary light to fill the light, so as to get a better balance of light and darkness. The auxiliary light may be the indirect light reflected by the reflector. If the intensity of the secondary ejection is not enough, you can also use a softer light as the auxiliary light. Because the auxiliary light only brightens the dark parts that the camera cannot see, it is not the intensity should not be higher than the main light [8]. In addition, when the auxiliary light is illuminated, try to avoid the generation of shadows, otherwise meaningless shadows will harm the quality of the image.

4.3 Transmission of Effect Light

In order to achieve a certain artistic effect, it is necessary to use some special lighting techniques to modify the picture and enhance the artistic effect, such as increasing the backlight effect of the three-dimensional effect, the gloss and the beam processing method that increase the spatial effect [9]. The basic meaning of the backlighting effect has nothing to do with whether there is a realistic light source. Its function is mainly to emphasize the contour lines of the character, increase the sense of space, and express the crystal clear texture [10]. The black hair and clothing make the character and the dark color of the background blend into one. At this time, it is necessary to outline the outer contour of the character with a backlight effect to obtain a strong three-dimensional effect. In addition, the backlighting effect helps the performance of the smoke and the uneven texture of the object. The environment fog node is often used in the scene, under the action of the environment fog, a clear beam effect can be obtained, but at the same time, the entire scene will become blurred.

5 Conclusion

In the lighting of 3D animation films, light not only stays in reproduction, but its value lies in performance. Light produces light and dark shadows, using different lighting characteristics and the performance of key nodes such as light intensity, direction, position, and temperature, and the overall carving Light and shadow, the changes of light and shadow constitute a complete visual system of the film, which reflects the various psychological changes of the characters, thereby expressing the emotion and meaning that the film needs to convey, and accurately expressing the overall artistic atmosphere of the film.

Acknowledgements. This work was supported by the Anhui Province's 2018 University Humanities and Social Science Research Key Project "Innovative Research on Jianghuai Folk Art Intervention in the Packaging Design of Special Agricultural Products in North Anhui" (No.: SK2018A0838); 2019 Anhui Province Quality Engineering Teaching Research Project "Indoor Art Design Professional Teaching Resource Library (No.: 2019zyk32).

References

1. Gronman, C., Le Tourneau, T.: Introduction to Lighting Engineers: Set Lighting Practice, pp. 58–63. Beijing United Publishing Company, Beijing (2017). (United States)
2. Xia Mu, J.: Film Lighting Techniques, pp. 212–215. Shanghai People's Fine Arts Publishing House, Shanghai (2016). (Japan)
3. Brock: Advanced 3DS Max Classic Course-Analysis of CG Lighting Technology, pp. 98–101. Tsinghua University Press, Beijing (2011). (United States)
4. Harry, C.: Box. Lamplighter's Bible, pp. 33–39. Beijing United Publishing Company, Beijing (2017). (United States)
5. Don, G., He, Y.: The Light of Photography: The Technique of Lighting the Center of the Subject, pp. 88–89. People's Posts and Telecommunications Press, Beijing (2013). (America)
6. Barnbaum, B.: The Art of Photography, pp. 198–201. People's Posts and Telecommunications Press, Beijing (2012). (America)
7. Novello, T., da Silva, V., Velho, L.: Global illumination of non-Euclidean spaces. Comput. Graph. 177–178 (2020)
8. Whitted, T., Whitton, M.C., Johnson, C., Kasik, D.: Origins of global illumination. IEEE Comput. Graph. Appl. 63–69 (2020)
9. Desrichard, F., Vanderhaeghe, D., Paulin, M.: Global illumination shadow layers. Comput. Graph. Forum 156–159 (2019)
10. Papaefthymiou, M., Papagiannakis, G.: Real-time rendering under distant illumination with conformal geometric algebra. Math. Methods Appl. Sci. 258–263 (2018)

Computer Media and Communication of Folk Culture in Shaanxi Literary Works Under the Background of Big Data

Shiqing Zhou(✉) and Wengao Sun

Xi'an Traffic Engineering Institute, Xi'an, Shaanxi, China

Abstract. The production and dissemination of literary works have a huge influence and promotion on the inheritance and development of human civilization. The dissemination of literary works is inseparable from the dissemination carrier. New media provides a more novel and convenient way for the dissemination of literary works. The dissemination of literary works based on new media, especially mobile clients, has not only increased the speed of dissemination, but also expanded the dissemination population, allowing people to access reading samples more quickly and effectively. The purpose of this article is to study the dissemination of new computer media and the folk culture of Shaanxi literary works under the background of big data. This article first summarizes the basic theory of big data, and then extends the core technology of big data. Combined with the current status of the spread of Shaanxi literary works and folk culture, the problems and deficiencies in it are analyzed. And combined with the new computer media under the background of big data to conduct research and analysis on the spread of Shaanxi literary works and folk culture. This article systematically expounds the mechanism of new media influencing cultural communication and the inevitability of leading cultural communication. This article uses research methods such as questionnaire surveys, field surveys, observations and experimental research on the communication of new computer media and the folk culture of Shaanxi literary works under the background of big data. The research shows that compared with the traditional ways of dissemination of Shaanxi literary works and folk culture, the dissemination of new computer media and Shaanxi literary works and folk culture under the background of big data is more convenient and diverse.

Keywords: Big data · New media technology · Folk culture communication · Research and analysis

1 Introduction

The socialization of new media is accompanied by the rapid development of our country's "4G+" and "Internet+", and new media has extended to various fields of economy and society [1, 2]. It is foreseen that new media will become the most important carrier of cultural communication in the future [3, 4]. We should look at the important role

B. J. Jansen et al. (Eds.): *International Conference on Cognitive based Information Processing and Applications (CIPA 2021)*, LNDECT 85, pp. 86–94, 2022.
https://doi.org/10.1007/978-981-16-5854-9_11

of new media in cultural communication with a developmental perspective, continue to innovate and explore strategies and routes that are in line with the development of cultural communication in our country [5, 6].

Many experts and scholars have achieved good results in the research on the spread of folklore culture. For example, Zhong Jingwen put forward the specific meaning of folklore in the book "Introduction to Folklore", that is, "folklore, folklore, refers to a country or The life culture created, enjoyed and inherited by the majority of the people in the nation"[7]. This is currently our country's more agreeable view on the concept of and folklore [8, 9]. Zhong Fulan put forward the specific concept of folk culture communication in the book "Folklore Communication", thinking that "the so-called folklore cultural communication [10]. It refers to the use of mass media to disseminate folk culture so that people can understand the culture and folk life of other nations, and at the same time Reflect the changes of the times [11, 12]."

The purpose of this article is to improve the efficiency of cultural dissemination, and to study the dissemination of new computer media and Shaanxi literary works and folk culture under the background of big data. By combining traditional communication methods and new computer media in the context of big data with Shaanxi literary works a comparative analysis of the ways of dissemination of folk culture can be used to judge its feasibility.

2 The Communication and Application of New Computer Media and the Folk Culture of Shaanxi Literary Works Under the Background of Big Data

2.1 Background Connotation of Big Data

In the context of the information age, with the rise of new media such as digital terminals, cloud services, Weibo, and WeChat, information data is also showing an exponential growth trend. At present, there is no unified definition of big data, but the academic circles basically believe that it mainly refers to software and hardware tools, and data related to the perception, collection, processing, management, and service of the assembly. It has the characteristics of low value density, diversity, scale and high speed, can be divided into structured data, semi-structured data and unstructured data.

2.2 Analysis of Shaanxi Traditional Media Culture Communication

The rapid development of new media has made Shaanxi's traditional media and cultural dissemination a major impact, and has been caught in a development dilemma, which is mainly manifested in two aspects.

(1) High-quality content cannot be delivered

Undoubtedly, Shaanxi is a major cultural province, but many of its excellent literary works are still distributed in the form of traditional paper media, which is more authoritative and professional. Traditional media has long been adhering to the concept of "content is king" and has rich experience in providing authentic, high-quality, original, and in-depth content. However, the media ecology has undergone profound changes nowadays. New media are becoming more mobile, socialized, and interactive. Human society has entered an era where "everyone has a microphone". The news production method has been created by mainstream media. The transformation from "mainly, one-way linear communication" to "multiple production subjects and multi-directional interactive communication" of emerging media has resulted in massive amounts of information. Although the content of traditional media is high-quality, the quantity is obviously small, and the news information of traditional media is easily submerged in the information ocean of new media. In addition, traditional media's information products have a long production cycle and relatively lagging dissemination, making it more and more difficult to reach readers directly, facing the dilemma of not being able to distribute high-quality content.

(2) User stickiness is weak

Although Shaanxi traditional media has a stable audience group, and the loyalty of this part of the audience group is relatively high, the traditional media report has a single form, rigid content, and lacks interactivity and participation. It is against this background that traditional media users have plummeted.

2.3 Analysis of the Impact of New Media on Communication

(1) Promote cultural development

The exchange and dissemination of culture can enhance the connotation of culture, inspire more cultural forms, or give birth to new cultures, and inject vitality into the original culture. When new cultures appear, the whole society will change accordingly. The new communication media will also bring different cultural communication effects.

(2) Diversified ways of dissemination

The public's enthusiasm for participation has increased, and their sense of security has also become stronger. In the era of new media, people's thoughts are more diversified, so people break through the limitations of their status, express their opinions and opinions, and expand the scope of communication. It is no longer the communication between acquaintances, but also communication between strangers. Interaction, this way is conducive to cultural communication and communication.

(3) Promote the spread of culture

Under the new media environment, the form of cultural communication carriers has undergone tremendous changes. The new media forms provide new communication methods for cultural communication, break all kinds of communication barriers, and people more truly feel the impact of new media on cultural exchanges and cultural communication. It has also gradually developed digital services in the field of new media, forming a service model with its own industry characteristics.

2.4 Research on the New Computer Media Based on Big Data and the Communication of Shaanxi Literary Works and Folk Culture

(1) Screen touch interactive system

The user can operate the terminal by gently touching the icon or text on the touch screen of the smart terminal device with a finger. The operation is more intuitive, and it can be quickly adapted according to people's cognitive habits without using the manual. Therefore, what is worth learning from Shaanxi cultural boutique communicators is how to make the content-carrying application simple and easy to use is the most important. The complicated design or operation is not intuitive, even if the content is rich and the amount of information is huge, it will not be entered by users. The application design should be as flat as possible to ensure beautiful appearance under the premise of convenient access to information.

(2) Mobile augmented reality technology
Mobile augmented reality can also be used for reference in the development of Shaanxi cultural boutique cultural tourism projects. Shaanxi's tourism projects are mainly humanistic landscapes, and behind this is profound historical and cultural information. When tourists pick up their mobile phones, they can see ancient buildings through their cameras. At the old site, according to the global positioning system and database resources, the scene of the ancient building at that time can be restored on the screen to complete the virtual reconstruction, and relevant introductory text information will appear for visitors to have a deeper understanding.

(3) Voice interaction technology
Voice is the most natural and convenient way of communication for human beings. With the rapid development of the mobile Internet, the wireless bandwidth has been greatly increased and the cloud computing technology system continues to mature, the application of voice interaction technology to mobile smart terminals has gradually become a reality. However, this new technology still has many details that have not been resolved. For example, it still cannot guarantee that all users' voices are 100% recognized by the system. During use, it will still cause dual input of voice input and manual modification, so it is used the dissemination and application of Shaanxi cultural boutiques is not mature at present, and it is too difficult to implement, but it is a future development trend and function development.

2.5 The Proximity Propagation Algorithm for Complex Structure Data

The core idea of the nearest neighbor propagation clustering algorithm is to treat all sample points as potential class representative points, and to continuously transmit two kinds of information, namely, availability and responsibility, to minimize the decision function. Makes the sample similarity in the cluster the largest, and the sample similarity between the clusters is the smallest.

Let X = {x1, x2, ..., xn} be a finite data set of pattern space Rn, where xi(i = 1, 2, ..., N) is a point in a vector space composed of n-dimensional attributes. The similarity s(i, k) between any two samples is measured by a negative Euclidean distance, as shown in formula (1).

$$s(i,k) = -d_{ik} = -||x_i - x_k|| \tag{1}$$

AP algorithm introduces two important information parameters, namely attractiveness r(i, k) and attribution a(i, k). The sample points are updated iteratively through two kinds of news of belonging degree and attractive degree, and k is determined as the final class representative point through the decision matrix E, as shown in formulas (2), (3), (4).

$$r(i,k) \leftarrow s(i,k) - \max\{a(i,k^{'}) + s(i,k^{'})\} \tag{2}$$

$$\mathrm{a(i,k)} \leftarrow \begin{cases} \min\left\{0, \mathrm{r(k,k)} + \sum_{i' \text{ s.t.} i' \in /\{i,k\}} \max\left\{0, \mathrm{r(i',k)}\right\}\right\} \mathrm{i} \neq \mathrm{k} \\ \sum_{i' \text{ s.t.} i' \neq k} \max\left\{0, \mathrm{r(i',k)}\right\} \mathrm{i} = \mathrm{k} \end{cases} \tag{3}$$

$$\mathrm{E(k)} = \mathrm{argmax(a(i,k) + r(i,k))} \tag{4}$$

In order to avoid the algorithm from oscillating in the iterative process, the damping factor $\lambda \in [0, 1]$ is introduced, and the current r(i, k) and a(i, k) values are weighted and summed with the result of the previous iteration to obtain a new one. r(i, k) and a(i, k), this paper chooses λ as 0.5.

$$\mathrm{r}^{(t+1)}(\mathrm{i,k}) \leftarrow (1-\lambda)\mathrm{r}^{(t+1)}(\mathrm{i,k}) + \lambda \mathrm{r}^{(t)}(\mathrm{i,k}) \tag{5}$$

$$\mathrm{a}^{(t+1)}(\mathrm{i,k}) \leftarrow (1-\lambda)\mathrm{a}^{t+1}(\mathrm{i,k}) + \lambda \mathrm{a}^{(t)}(\mathrm{i,k}) \tag{6}$$

Among them, t is the current iteration number. When the class representative point does not change for 10 consecutive iterations or reaches the maximum number of iterations, the nearest neighbor propagation algorithm ends.

3 Experimental Research on the Communication of New Computer Media and Shaanxi Literary Works and Folk Culture Under the Background of Big Data

3.1 Subjects

(1) In order to make this experiment more scientific and effective, this study compares the traditional Shaanxi literary works and folk culture dissemination with the new computer media under the background of big data and the cultural dissemination of Shaanxi literary works. This experiment through the collection of Shaanxi literary works and custom culture in recent years on the Internet, the data analysis is carried out using the analytic hierarchy process.

(2) In order to further experiment and analyze the dissemination of new computer media and Shaanxi literary works in the context of big data, this experiment also conducted an experimental analysis on the dissemination performance of big data methods through interviews with relevant experts. The gender ratio of personnel is equal, which effectively guarantees the scientific nature of the experiment.

3.2 Research Objects

(1) Logic analysis method
This experiment makes the data obtained in this article more rigorous and organized by studying the limitations of traditional media communication and the way, characteristics and algorithms of the new media's dissemination of Shaanxi literary works and folk culture under the background of big data.
(2) Observation method
This experiment observes the dissemination of Shaanxi literary works and folk culture in the context of big data and obtains data. These data not only provide theoretical support for the topic selection of this article, but also provide a reliable reference for the final research results of this article.
(3) Comparative method
This paper compares traditional Shaanxi literary works and folk culture communication channels with the communication methods of new media in the context of big data, in order to judge the excellent performance of new media and folk culture communication based on big data.
(4) Mathematical Statistics
Use relevant software to perform statistics and analysis on experimental results.

4 The Experimental Analysis of New Computer Media and the Communication of Shaanxi Literary Works and Folk Culture Under the Background of Big Data

4.1 Comparative Analysis of Transmission Methods

In order to conduct an experimental analysis on the new computer media and the folk culture of Shaanxi literary works under the background of big data, this experiment will use the traditional way of dissemination of Shaanxi literary works and folk culture and the new computer media and the folk culture of Shaanxi literary works under the background of big data. Comparative analysis and analysis of the progress of the transmission route, and the data obtained is shown in Table 1.

Table 1. Comparative analysis of transmission methods

	Communication quality	User stickiness	Diversity	Others
Big data background	71.5%	64.2%	61.3%	68.2%
Traditional	50.6%	55.6%	52.2%	56.0%

It can be seen from Fig. 1 that compared with the communication methods and channels, it is more diversified and efficient, and especially its communication quality efficiency is about 20% higher, which fully reflects the traditional Shaanxi literary works and folk culture communication methods. Computers under the background of big data new media and the spread of folklore literature in Shaanxi literary works and folklore traditions are backward and need to be resolved urgently.

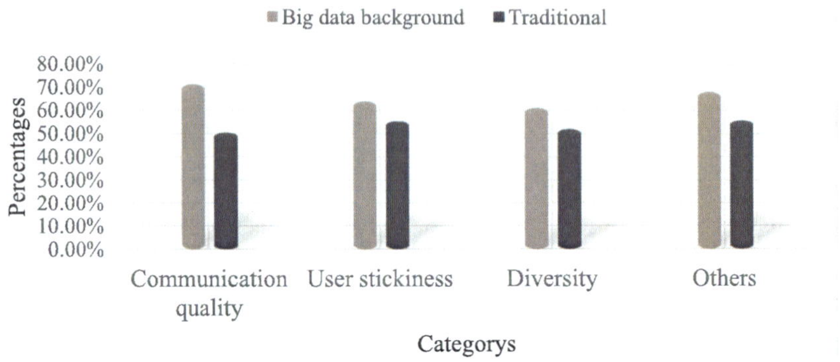

Fig. 1. Comparative analysis of transmission methods

4.2 New Media Communication Analysis in the Context of Big Data

In order to further compare new computer media with Shaanxi literary works and folk culture in the context of big data, this time analyzes the communication of new media in the context of big data. The data obtained is shown in Table 2.

Table 2. New media communication analysis in the context of big data

	Convenience	Diversity	Reliability	Others
Man	34	23	19	15
Woman	29	31	26	11

It can be seen from Fig. 2 that under the background of big data, new computer media and Shaanxi literary works and folk culture communication methods are more excellent, especially its dissemination is more convenient, which fully reflects the new computer media and Shaanxi literary works and folk customs under the background of big data.

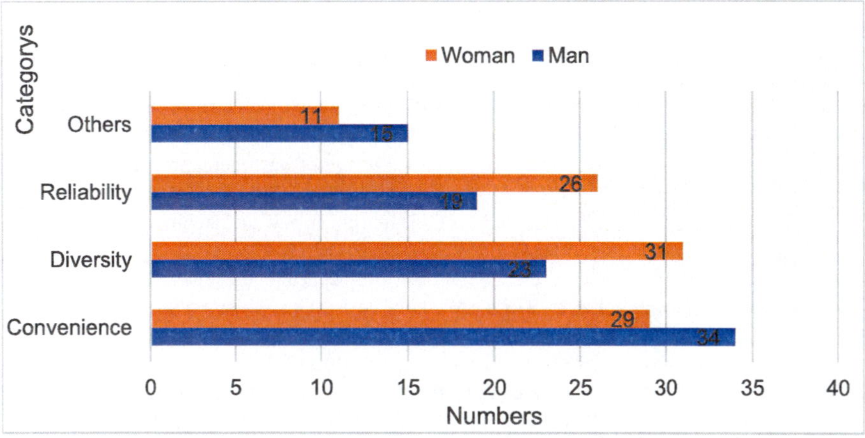

Fig. 2. New media communication analysis in the context of big data

5 Conclusion

The socialization of new media has been accompanied by the rapid development of our country's big data, and new media has been extended to all areas of the economy and society. With the rapid development of digital technology, it is foreseeable that new media will become the most important carrier of cultural communication in the future. We must look at the important role of new media in cultural communication with a developmental perspective, and continue to innovate and explore in line with my country's cultural communication. The development of Internet-related industries is getting deeper and deeper, and new media has developed more broadly. The development of social software WeChat, Weibo, Tencent news client, and other similar media, represented by Tencent, are constantly moving forward. It is foreseeable that new media will become the most important carrier of cultural communication in the future. We must look at the important role of new media in cultural communication with a developmental perspective, continue to innovate and explore strategies and routes that are in line with the development of cultural communication in our country.

Acknowledgements. Scientific Research Program Funded by Xi'an Traffic Engineering Institute: A study of Cognition and Translation Strategies of Culture-loaded Words in Shaanxi Literary Works under the Conceptual Integration Theory.

References

1. Winiecka, E.: Literary internet: online criticism and literary communication. Porównania **27**(2), 289–311 (2020)
2. Zerfass, A., Verčič, D., Wiesenberg, M.: The dawn of a new golden age for media relations? Public Relat. Rev. **42**(4), 499–508 (2016)

3. Basri, W., Siam, M.: Social media and corporate communication antecedents of SME sustainability performance a conceptual framework for SMEs of Arab world. J. Econ. Adm. Sci. **35**(3), 172–182 (2019)
4. Culloty, E., Greene, D., Siapera, E.: Hybrid media and populist currents in Ireland's 2016 general election. Eur. J. Commun. **33**(4), 396–412 (2018)
5. Houston, J.B., Spialek, M.L., Perreault, M.F.: Coverage of posttraumatic stress disorder in the New York Times, 1950–2012. J. Health Commun. **21**(2), 240–248 (2016)
6. Yengin, D., Kınay, Ömür.: Transformation of leisure time in new media: binge watch. Turkish Online J. Design Art Commun. **6**(4), 351–379 (2016)
7. Jingwen, Z.: Cyberbullying, race/ethnicity and mental health outcomes: a review of the literature. Media Commun. **4**(3), 71 (2016)
8. Yudha, R.P., Irwansyah, I.I.: Media baru digital sebagai peretas konteks komunikasi antar pribadi dan kelompok. Islamic Commun. J. **3**(2), 180–183 (2018)
9. Sugimoto, C.R., Work, S., Larivière, V., et al.: Scholarly use of social media and altmetrics: A review of the literature. J. Am. Soc. Inf. Sci. Technol. **68**(9), 2037–2062 (2017)
10. Fulan, Z.: The new public sphere. Adv. Journal. Commun. **04**(4), 149–162 (2016)
11. Borah, P.: Emerging communication technology research: theoretical and methodological variables in the last 16 years and future directions. New Media Soc. **19**(4), 616–636 (2017)
12. Malla, M.W.: Reflections on critique of 'media' in the globalisation debate. J. Media Commun. Stud. **1**(1), 73–85 (2021)

Image Classification Based on Sentinel-1A and Landsat 8 OLI

Tingxi Song[1], Liming Gao[1,2](✉), Lele Zhang[1,2], Yankun Li[1], and Guangrui Zhu[3]

[1] College of Geographical Sciences, Qinghai Normal University, Xining 810008, Qinghai, China
[2] Academy of Plateau Science and Sustainable Development, Xining 810016, Qinghai, China
[3] Xining Institute of Land Survey and Planning, Xining 810008, Qinghai, China

Abstract. In this study, the Sentinel-1A and Landsat 8 OLI images were taken as the basic data in Qarhan Salt Lake area. Combined with the field survey samples, the maximum likelihood classification method and support vector machine method were used to classify the two kinds of images, and the accuracy of the classification results was evaluated. The results show that: (1) The overall accuracy of both maximum likelihood classification and SVM classification is more than 75%. For the same image, the classification accuracy of SVM is higher than that of maximum likelihood classification. (2) The classification accuracy of Landsat 8 OLI data is higher than that of original Sentinel-1A data, but the classification accuracy of Sentinel-1A image based on H/α/Alpha target decomposition is higher than that of Landsat 8 OLI. (3) The classification results of Sentinel-1A images by H/α/Alpha polarization decomposition were verified by actual points. The classification accuracy of salt field and salt lake was 80.00% and 77.78%, respectively. The accuracy of swamp and other unused land was 57.14% and 63.63%, respectively, lower than that of the other four types of land features in the study area.

Keywords: Sentinel-1A · Landsat 8 OLI · H/α/Alpha decomposition · The SVM method · Maximum likelihood method

1 Introduction

Remote sensing image is widely used in the application of optical image. Scholars at home and abroad have done a lot of research, but optical image also has its inevitable shortcomings, such as the difficulty of image interpretation in unstable weather, only imaging in the daytime, etc. [1]. As an active microwave sensor, synthetic aperture radar (SAR) can obtain its covered information through the surface or vegetation, which has the characteristics of all day, all-weather and all-time observation without the limitation of light and climate conditions [2]. As an important application direction of polarimetric SAR data, polarimetric radar image classification has always been a hot research direction [3–6]. Polarization target decomposition was first proposed by Huynen [7] in 1970, and subsequent scholars have done a lot of basic research [8]. Polarization decomposition is widely used in polarimetric SAR image classification because it can well describe

B. J. Jansen et al. (Eds.): *International Conference on Cognitive based Information Processing and Applications (CIPA 2021)*, LNDECT 85, pp. 95–105, 2022.
https://doi.org/10.1007/978-981-16-5854-9_12

the scattering mechanism of targets [9]. Li Kun et al. based on radarsat-2 full polarization data, used Pauli decomposition and decomposition method based on eigenvalue eigenvector to better identify rice [10]. Nigra turson et al. proposed a comprehensive H/α/Alpha decomposition Pauli polarization feature decomposition and support vector machine (SVM) classification strategy are used to supervise classification and accuracy verification of full polarization SAR image [11]. In view of the rare reports on polarimetric radar image classification in this study area, through polarimetric decomposition of radar image, combined with optical remote sensing image, the commonly used SVM classification and maximum likelihood classification methods are selected for image classification and accuracy evaluation.

Qarhan Salt Lake, located in Qaidam Basin, is the main potash fertilizer production base in China. In the past few decades, due to the impact of climate change and human activities, the regional land cover has undergone significant changes, which directly affected the regional ecological security. Therefore, it is of great significance to carry out the monitoring of ground features based on Remote Sensing for mineral resources development and environmental protection in the salt lake region.

2 Research Area and Methods

2.1 Overview of the Study Area

Qarhan Salt Lake is located in the southeast of Qaidam Basin, the inland basin with the highest altitude in China. Its geographical coordinate is 36°20′52″-37°21′42″N, 99° 11′22″-96°01′27″E. It is 20–40 km wide from north to south and 168 km long from east to west, with an average altitude of 2675 m and an area of 20029.37 km^2. The terrain of Qarhan Salt Lake is relatively flat, with an average annual temperature of 5.2 °C, an average annual precipitation of 24.1 mm, and an average annual evaporation of 3549.5 mm, 147.3 times of the precipitation [12]. The annual sunshine duration of the lake basin is 3200–3400 h, and the dominant wind direction is westerly or northwest wind, which is a typical continental arid climate. The salt lake belongs to a comprehensive deposit of potassium, boron and lithium, and is the main potash fertilizer production base in China, covering a number of Salt Lake mines such as potash, potassium magnesium, salt and lithium [13] (As show in Fig. 1).

2.2 Data Preprocessing

The data source. The radar data used in this study are sentinel-1A images, which are downloaded from the official website of ESA (http://www.esa.int/). The satellite is equipped with C-band synthetic aperture radar (SAR). The data types include strip mode SLC, top mode IW SLC, top mode EW SLC and strip mode IW GRD of tops mode and EW GRD of tops mode [14]. Sentinel-1A has different imaging modes with different characteristics, and also has great differences in disaster relief, land monitoring, marine and other applications. In this paper, the interference wide mode slant single look complex data (IW SLC) and the polarization mode VV + VH are selected. As show in Table 1.

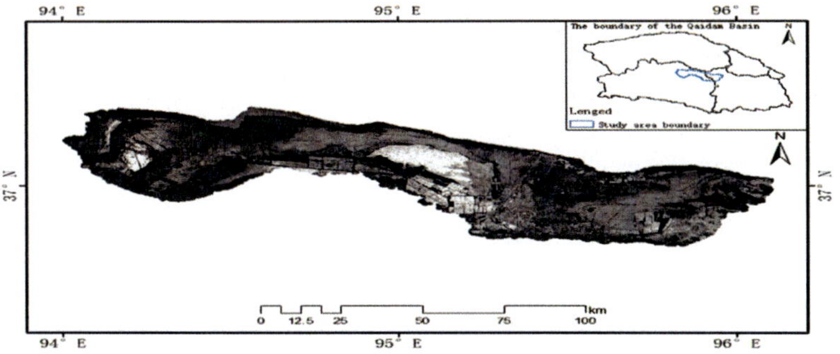

Fig. 1. Location of study area

Table 1. The main parameters of study area Sentinel-1A data

The data format	Working band	Polarization mode	Working mode	Resolution	Channel	Acquisition time
SLC	C band	VV+VH	IW	5*20	Ascending	2020-08-10
SLC	C band	VV+VH	IW	5*20	Ascending	2020-08-15

The optical remote sensing data used in this study are Landsat 8 OLI images. Landsat 8 satellite is equipped with OLI and tirs (thermal infrared sensor) push broom imager. The OLI includes 9 bands, and the spatial resolution of the rest bands is 30 m, except for the panchromatic band with a resolution of 15 m. The optical remote sensing data used in this study is Landsat 8 OLI data. The data set is composed of 7 bands with a spatial resolution of 30m. The study area covers two images, the imaging time is August 28, 2020, and the row and column numbers are 137/034 and 136/034 respectively. Landsat 8 OLI remote sensing data has gradually become the main data source of land cover type monitoring. Data parameters are not described here.

Data preprocessing. In this paper, the radar data processing in the study area adopts ESA snap platform, polsarpro v5.1.3 platform, remote sensing data processing platform ENVI and ArcGIS. Select the time phase on August 10, 2020 and August 15, 2020, select sentinel-1A dual polarization (VV, VH) radar data as the research data. Preprocessing includes precise orbit correction, radiometric calibration, polarization matrix C2 generation, terrain correction, polarization filtering in polsarpro v5.1.3, geocoding and H/a/alpha polarization decomposition of dual polarization data.

Landsat8 OLI image of 28 August 2020 (http://earthexplorer.usgs.gov/) In the 1–8 band, the study area has no cloud coverage, and preprocessed on envi5.3 platform. Firstly, the original image is calibrated by the radiation calibration parameters, and the DN value is converted into the pixel radiation brightness value. FLAASH atmospheric correction converts the radiance value to surface reflectance. Finally, the image is geometric corrected, and the panchromatic band of the corrected image is fused with the 1–7 band image. The resolution of the fused 7 band images is increased to 15 m, which is conducive to the classification of the following research area.

Sentinel-1A filtering methods include gamma map filtering, median filtering, frost filtering, Lee filtering and refined Lee filtering. The above filtering methods are used for C2 polarization matrix after terrain correction, and 5 × 5 filtering window for filtering processing, filtering processing can remove coherent noise, while retaining the image edge information and polarization information, through quantitative analysis of the image mean, standard deviation, equivalent number of views (enl) to select the best filtering algorithm for filtering processing, comparative analysis of five filtering methods [15] (As show in Fig. 2).

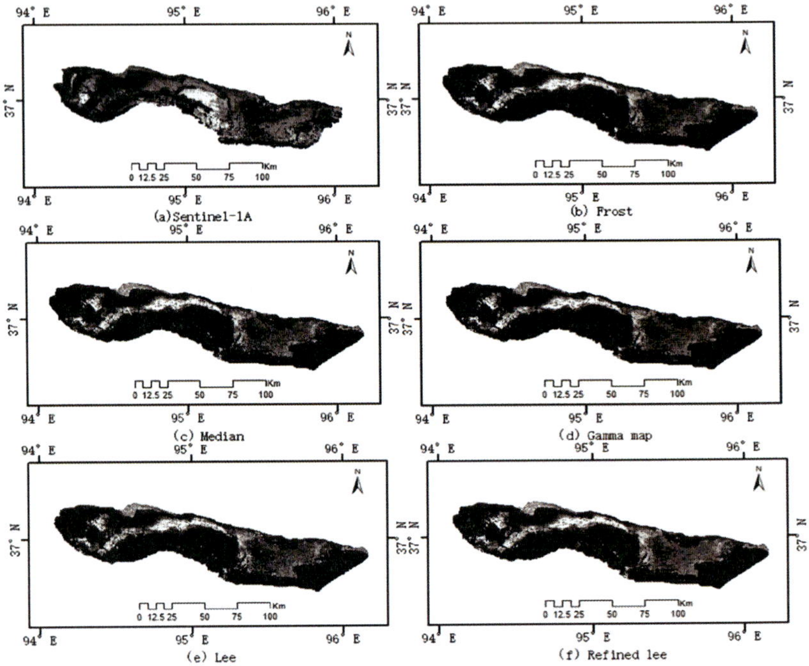

Fig. 2. Sentinel-1A data filtering processing

It can be seen from Table 2 that none of the above five filtering methods can smooth the speckle noise best. Equivalent number of looks (enl) is an index to measure the relative intensity of speckle noise in a SAR image, which reflects the ability of image to distinguish regions with different backscattering characteristics. The larger the equivalent number of looks, the weaker the image speckle, and the better the interpretation [16]. Since polarization decomposition is used in the subsequent processing of this study, Lee filter with the highest edge preserving coefficient after filtering is selected for filtering processing, as shown Fig. 3, C2 matrix image after Lee filtering.

Table 2. Evaluation index of different filtering algorithms

Filtering method	Mean value	Standard deviation	Equivalent number of views
Raw SAR	0.584000	0.197611	2.386209
Frost	0.589000	0.109130	7.958794
Median	−0.074000	0.098669	0.153676
Gamma map	0.645000	0.108939	9.577621
Lee	0.674000	0.108936	10.458802
Refined Lee	0.291000	0.094559	2.587529

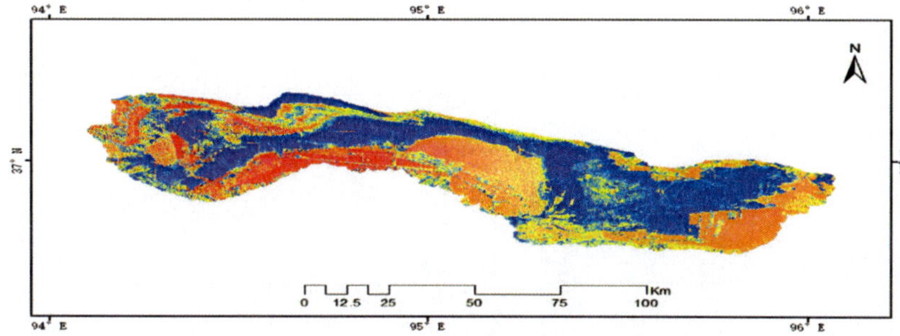

Fig. 3. Lee filter C2 matrix

2.3 Research Method

H/a/Alpha decomposition. Since sentinel-1A is dual polarization radar data, only H/a/Alpha decomposition method can be used. By setting and selecting characteristic parameters, the plane of three characteristic parameters (scattering entropy, average scattering angle and anti entropy) based on H/α/Alpha decomposition can be obtained. As shown in Fig. 4, h from 0 to 1 indicates that the target surface object scatters from isotropic to completely random. When h value is high, the plane of scattering entropy, average scattering angle and anti entropy can be obtained, It indicates that the depolarization effect on the target is strong; When the H value is small, the depolarization effect is weak. There are some defects in the initial classification of targets in H/a space, such as arbitrary boundary setting and overlapping boundaries of similar objects. Therefore, this space classification can not be used as the result of direct classification.

The average scattering angle of sentinel-1A is obtained by polarization decomposition α. Polarization entropy H, anti entropy a and λ1, λ2 eigenvalue 5 parameters, as shown in Fig. 5. It can be seen that the anti entropy a and the average scattering angle α The noise of the image is large, and the physical characteristics of the object are not obvious, which can not reflect the information of the object. Polarization entropy H, λ In this paper, polarization entropy H is used, λ1 eigenvalue λ2 eigenvalue combination for the subsequent study area features classification.

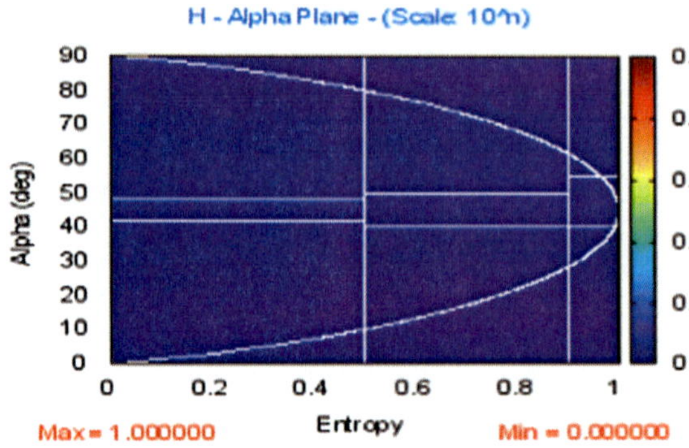

Fig. 4. H/α plane in VV-VH polarization mode

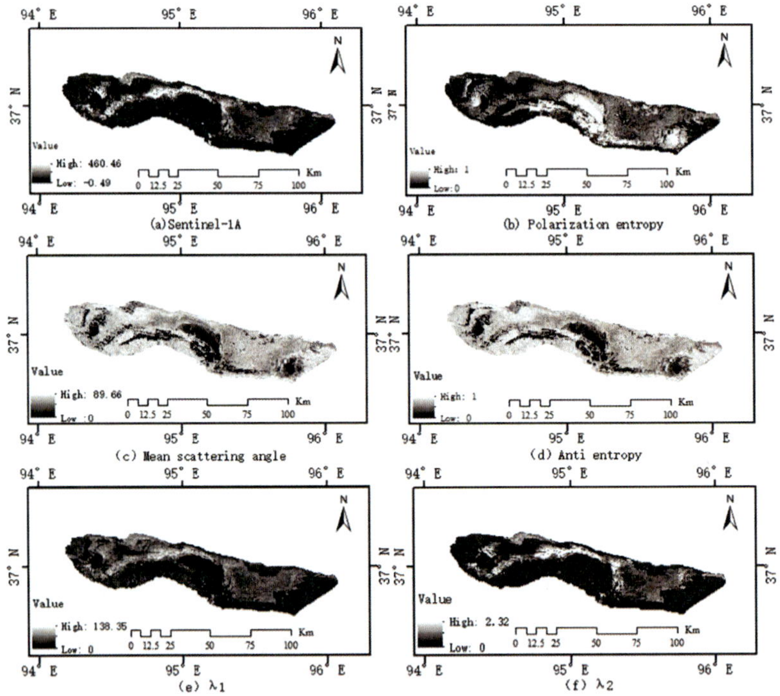

Fig. 5. H/α/Alpha decomposition parameters mean scattering Angle, Polarization entropy, Anti entropy, eigenvalues λ1, λ2 effect

Image classification. Maximum likelihood classification (MLC), also known as Bayesian classification, is the most typical and widely used method in supervised classification, based on Bayesian criterion theory [17]. When judging two or more classes, the maximum likelihood Bayesian decision criterion method is used as the judgment standard. According to the statistical method, the set of nonlinear discriminant functions

Table 3. Image interpretation features and corresponding selected training samples and verification samples

Type	Field photos	Landsat	Sentinel-1A	Interpretation mark	Number of training samples	Number of validation samples
Salt lake				Clear boundary and uniform tone	8	4
Salt field				It has obvious ridge boundary, regular grid distribution and dark green color	24	15
Dry salt flat				Color highlights and large area distribution of flakes	13	5
Unused land				The color is bright and uniform	16	6
Marshland				Dark color and uneven distribution	7	3
Water				The color is dark and has obvious outline	3	2

is compiled. Assuming that each class has a positive distribution function, the training area is selected correctly, and the corresponding attribution probability is calculated for each class to be classified. The advantages of this method are convenience, small classification error and high accuracy, which makes the density distribution function can explain the classification results very effectively [18].

Support vector machine (SVM) can reduce the phenomenon of mixed and missing points caused by nonlinear ground objects, and improve the classification accuracy. The core idea is that for linear separable data, the data can be separated by finding the best partition surface, and the maximum separation distance can be ensured [19]. For the non-linear separable data, based on the idea of linear separability, the original data is mapped into the high-dimensional linear space, and the optimal separation surface is found in the high-dimensional space to maximize the dispersion between the samples. Support vector machine can solve the problems of small sample, nonlinear and high dimension, and has strong generalization ability [20].

Classification results. As show in Table 3. The Landsat 8 OLI image, sentinel-1A original image and polarization decomposed sentinel-1A data in the study area are classified by the maximum likelihood classification and support vector machine classification methods in supervised classification. It can be seen from Fig. 6 and Fig. 7 that no matter the maximum likelihood classification method or the support vector machine classification method is used, when the six types of ground objects in the three images are classified, the maximum likelihood classification method and the support vector machine classification method are used, It can be clearly seen that there is a common problem, that is, the other unused land is wrongly divided into dry salt flats in Landsat 8 OLI image, which can not accurately separate these two types of land features. After repeatedly selecting training samples for experiments, the classification results of dry salt flats and other unused land obtained in Landsat 8 OLI image are consistent, Therefore, the reason of misclassification may be that there is little difference between the reflection values of the two types of ground objects, and there may be the phenomenon of "foreign objects with the same spectrum", which leads to misclassification of classification results. When the original sentinel-1A data and polarization decomposed sentinel-1A data are classified by two kinds of classification methods, good classification results can be obtained and their accuracy can be evaluated.

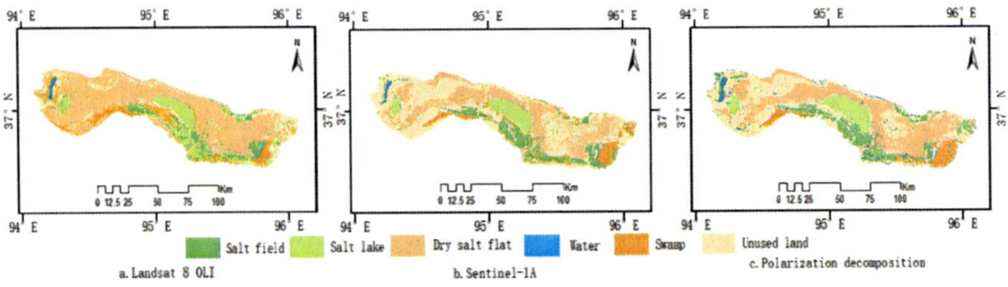

Fig. 6. Maximum likelihood classification results

3 Evaluation and Analysis of Classification Accuracy

3.1 Classification Accuracy Evaluation

In this study, the artificial interpretation method is used to select the ROI validation samples in the study area. The maximum likelihood method and support vector machine

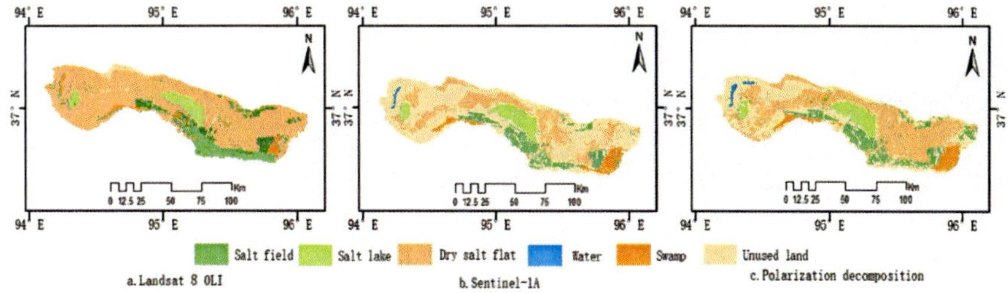

Fig. 7. SVM classification results

classification method are used to classify the objects in the study area. The results of different classification methods for different images are analyzed. The total classification accuracy of six classification models and K coefficient are two indicators reflecting the effect of the classifier, in this paper, the relationship between kappa coefficient and classification quality in Table 4 is selected to show the classification performance. The specific classification accuracy evaluation is shown in Table 5. It can be seen from Table 5 that landsat8 OLI and sentinel-1A images can clearly distinguish six land use types: Salt Lake, salt field, dry salt beach, swamp, other unused land and water body, and the overall interpretation accuracy meets the requirements. Through the comparison and practical verification, the SVM classification results of salt field, salt lake and dry salt flat are the best and the classification accuracy is the highest. The non isotropic degree of the contour and part of the information of salt fields, salt lakes and dry salt flats is low, and the SAR image classification effect is very obvious when showing the trend and shape. However, when using the maximum likelihood method to classify the optical images, the classification accuracy of salt fields and salt lakes is not very high, and it is easy to produce the phenomenon of wrong classification. The overall classification accuracy is only 76.96%, and the kappa coefficient is only 0.67.

Table 4. Relationship between Kappa coefficient and classification quality

Kappa coefficient	Classification quality
<0	It's bad
0–0.20	Bad
0.20–0.40	Commonly
0.40–0.60	Good
0.60–0.80	Very good
0.80–1.00	Excellent

Table 5. Comparison table of different image classification accuracy

Image category	Maximum likelihood method		SVM	
	Overall accuracy/%	Kappa coefficient	Overall accuracy/%	Kappa coefficient
Landsat 8 OLI	76.96	0.67	78.56	0.71
Sentinel-1A	76.32	0.67	77.85	0.70
Polarization decomposition	77.25	0.69	79.03	0.73

4 Conclusion and Discussion

In this paper, sentinel-1A and Landsat 8 OLI images are used as data sources to monitor and classify the research area. The advantages of the two data sources in this research area are as follows:

(1) In this paper, a new method of H/α/alpha decomposition based on sentinel-1A image data is proposed to classify the surface of the Salt Lake area of Qarhan. The maximum likelihood method and support vector machine classification method are selected to classify the original sentinel-1A image and polarization decomposition parameters: polarization entropy H, λ1 characteristic value λ2. The feature values are grouped again and classified. The classification results show that SVM is the best classification precision based on h/a/alpha decomposition, and the overall accuracy and kappa coefficient reach 79.03% and 0.73 respectively. The method can distinguish the similar or similar types of objects, which greatly reduces the phenomenon of wrong and missing points.

(2) For Landsat 8 OLI images, the overall classification accuracy of Landsat 8 OLI images is higher than that of original sentinel-1A images after using maximum likelihood method and support vector machine classification method. The overall accuracy and kappa coefficient are higher than 75%, but the classification accuracy of Salt Lake and salt field is low, and the partial salt fields are classified as salt lake category, It may be caused by the phenomenon of "the same spectrum of foreign matter" in the mixed pixel.

In conclusion, the sentinel-1A image support vector machine classification method based on polarization decomposition of H/α/Alpha is an effective method to obtain land use types in this study area. However, there are still some shortcomings. Through repeated experiments, the parameters of antipentropic A and average scattering angle after polarization decomposition of sentinel-1A are analyzed α The image noise is large, which can not reflect the effective information of the image, and then the two parameters are eliminated and classified; Secondly, the sentinel-1 data used in this research area is only VV and VH polarization methods, which do not fully excavate the effective characteristics of polarization data, fully consider the polarization characteristics of polarizing SAR, and try to use multipolarized radar data to further improve the accuracy of land use classification in this research area by using sentinel data.

Acknowledgments. This study was supported primarily by the Natural Science Foundation of Qinghai Province (2021-ZJ-947Q); 2019QZKK0906 the second comprehensive scientific expedition to the Qinghai Tibet Plateau.

References

1. Xie, H., Luo, X., Xu, X., et al.: Evaluation of landsat 8 OLI imagery for unsupervised inland water extraction. Int. J. Remote Sens. **37**(7–8), 1826–1844 (2016)
2. Greatbatch, I.: Polarimetric radar imaging: from basics to applications. Int. J. Remote Sens. **33**(1–2), 661–662 (2012)
3. Hong, Z.: Polarimetric SAR Theory Method and Application. Science Press, Beijing (2015)
4. Min, C., Zhaoliang, S., Jianyi, Y.: Remote sensing image classification method based on bee colony intelligent algorithm. Acta Surv. Mapp. **42**(5), 745–751 (2013)
5. Chen, B., Wang, H.Z., Xu, X.L., et al.: Automatic classification of medium resolution remote sensing images supported by deep learning google net model. Bull. Surv. Mapp. (006), 29–33, 40 (2019)
6. Zhang, Y.H., Zhang, J.X., Zhang, X.F., et al.: Land cover classification from polarimetric SAR data based on image segmentation and decision trees. Can. J. Remote. Sens. **41**(1), 40–50 (2015)
7. Wang, C., Yu, W.: Comparison and analysis of huyne type goal decomposition methods. J. Univ. Chin. Acad. Sci. **28**(003), 402–409 (2011)
8. Ding, W.: Research on target recognition method based on full polarization radar. Harbin Engineering University, Harbin (2013)
9. He, Y.: Polarimetric SAR image classification based on polarization decomposition. University of Electronic Science and Technology, Chengdu (2013)
10. Kun, L., Yun, S., Fengli, Z.: Rice recognition based on radarsat-2 full polarization data. Remote Sens. Technol. Appl. **027**(001), 86–93 (2012)
11. Based on H/α/Alpha Classification of soil salinization in arid areas by decomposing polarimetric synthetic aperture radar data. Jiangsu Agric. Sci. 047(022), 273–279 (2019)
12. Bao, F.: Development environment and process of Aeolian landform in Qarhan Salt Lake area of Qaidam Basin. Shaanxi Normal University, Xi'an (2016)
13. Yang, X., Xu, X., Xiao, L., et al.: Evolution trend and driving force analysis of Qarhan Salt Lake. Land Resour. Remote Sens. **32**(1), 130–137 (2020)
14. Zhang, S., Jiang, Q., Xi, J., et al.: A extraction of ancient channel information based on decomposition principle: a case study of Western Songnen Plain. J. Jilin Univ. (Geosci. Edn.), **50**(06), 293–304 (2020)
15. Wu, L.: SAR Image Processing and Target Recognition. Aviation Industry Press (2013)
16. Liang, Y., Hu, T.: Research on filtering method of radar image speckle noise. Comput. CD Softw. Appl. (15), 100–101 (2012)
17. He, H., Zhen, K.: Hyperspectral image classification. Sci. Technol. Commun. **011**(001), 105–108 (2019)
18. Tang, B.: Research on rice area detection and yield estimation in Jiangsu Province based on remote sensing and meteorological data. Zhejiang University, Hangzhou (2016)
19. Vapnik, V.N.: An overview of statistical learning theory. IEEE Trans. Neural Netw. **10**(5), 988–999 (1999)
20. Wang, J., Jian-Nong, H.E.: New algorithm of remote sensing image classification based on K-type support vector machine. J. Comput. Appl. **32**(10), 2832–2835 (2012)

Application of Clothing Design and Production Based on Clothing 3D Simulation Technology

Yuwu Xiao(✉)

Beijing Electronic Technology Vocational College, Beijing 100176, China

Abstract. With the continuous development of computer hardware and software technology, people's lives are inseparable from computer networks, especially the virtual simulation technology in the field of computer networks, which has been applied to various fields in people's lives and is also a key research area for researchers. With the rapid development of computer network technology, the clothing industry has increasingly higher requirements for virtual simulation technology. It is hoped that clothing simulation can achieve more realistic effects. However, today's virtual simulation technology cannot meet these requirements, and the realistic effects of clothing simulation are still not ideal. Under this circumstance, this article studies the application of clothing design and production based on clothing 3D simulation technology. This paper first analyzes the application status and potential value of 3D clothing simulation technology in the clothing industry, and studies the status quo of 3D virtual simulation technology in detail; then this paper analyzes clothing fabric simulation methods and collision detection methods; and then analyzes the methods in detail process, design and implement a spring mass point simulation model. Finally, this article carried out a simulation experiment on the model. The experimental results show that the simulation speed of the model is relatively fast, but the number of iterations should be controlled within 200 times as much as possible.

Keywords: 3D simulation · Fashion design · Cloth simulation · Virtual reality

1 Introduction

In today's society, the development of all walks of life is very rapid, and all fields of society are seeking breakthroughs in technological innovation, including the design and production of the clothing industry. The combination of technology and clothing design can not only expand the market scope of the clothing industry, but also improve the economy of the clothing market [1]. This article applies 3D clothing simulation technology to the design and production of clothing. While innovating the design and production of clothing, it increases the market value of the clothing industry and reflects the importance of clothing 3D simulation technology to the clothing industry. In the process of my country's economic development, all areas of society need to develop simultaneously [2]. "Nation, food, housing and transportation" are the basic needs of life, among which clothing is clothing, ranking first. It can be seen that the development

B. J. Jansen et al. (Eds.): *International Conference on Cognitive based Information Processing and Applications (CIPA 2021)*, LNDECT 85, pp. 106–114, 2022.
https://doi.org/10.1007/978-981-16-5854-9_13

of the clothing industry is at the forefront, and the good development of the clothing industry is very important to the growth of my country's economy. In the field of clothing design and production, more and more companies and researchers have begun to apply clothing 3D simulation technology, which is becoming more and more mature in the clothing design industry. Many complex garment structures that cannot be achieved by artificial methods can be perfectly produced through garment 3D simulation technology, and more innovative ideas become possible [3]. For the healthy development of the clothing industry, the clothing industry needs to keep pace with the times and continue to apply emerging technologies similar to clothing 3D simulation technology.

Chinese scholar Chen Jiayu believes that with the rapid construction of our country's economy, many science and technology have also been developed rapidly, including 3D virtual simulation technology. This technology has achieved good application effects in some industries, especially the application in the clothing industry is extremely common, and it provides a huge help in the process of clothing design and production. The application of 3D virtual simulation technology can solve the problems of modeling and system incompatibility, help designers design higher-quality clothing products, and comprehensively improve the effect of clothing design and production [4]. Zhang Jingxuan pointed out that with the rapid development of the clothing industry, people's thinking about clothing has also changed, and clothing design and production are facing new challenges. Through the application of 3D virtual simulation technology, the value potential of 3D virtual simulation technology is analyzed, and some problems existing in the process of transforming clothing design from two-dimensional to three-dimensional are solved [5]. Taking virtual simulation technology as the research object, Bao Rencheng explained the application status of virtual simulation technology in clothing design and production, analyzed the characteristics of virtual simulation technology, combined with clothing three-dimensional structure design methods, and compared virtual simulation technology in the field of clothing design and production. Analysis of the development prospects of China, hoping to provide a certain reference for the application of virtual simulation technology in the field of clothing design [6].

Virtual simulation technology is the core technology of virtual manufacturing. Virtual simulation technology is formed by combining manufacturing technology with simulation technology. This technology is used to provide realistic product simulation support in the manufacturing field. As early as the end of the last century, both domestic and foreign countries have conducted research on virtual simulation technology. The foundation of virtual simulation technology is computer science, probability and statistics. It is an emerging technology that has developed rapidly in recent years [7]. The performance of virtual simulation technology in many fields in our country is very good, and has brought the improvement of economic benefits to the corresponding fields. In this process, the application of virtual simulation technology has become more and more mature. Nowadays, my country is an important base for global clothing processing. With the transformation of the new economic system with the form of scientific and technological innovation as the theme, clothing design and production models also need to be transformed. Although virtual simulation technology has long been used in all walks of life in society, there are still few applications in the field of clothing design and production [8, 9]. Based on this situation, this article studies the application of clothing 3D simulation technology in clothing design and production.

2 Application of Clothing Design and Production Based on Clothing 3D Simulation Technology

2.1 Physically Based Cloth Simulation

High-quality simulation of cloth movement through computer technology has always been a key topic in the field of clothing 3D simulation design. For modeling, the geometric method was used in the past. This method did not study the internal structure and physical properties of the cloth. It only simulated the shape of the cloth. The appearance of the simulated graphics was very stiff, and the realistic effect was not ideal, which could not satisfy the clothing. With the continuous optimization of computer hardware performance and the improvement of computer image algorithms, physical cloth simulation has become the main simulation method now. An important problem in physical cloth simulation is the numerical solution of the dynamics of ordinary differential equations. Although the calculation complexity of this integration method is relatively low, when dealing with differential equations with large stiffness coefficients, only a small integration step length can be used to ensure stability, and the overall performance of the algorithm is not high. However, if the stiffness of the equation is reduced by adjusting the physical parameters of the model, it will lead to the problem of system hypersensitivity [10]. The tacit integration method is always unconditional. Although it is necessary to solve the system of equations, the system can usually take a larger integration step. In terms of simulation accuracy, the indirect integration method introduces many methods, and the resulting excessive damping problem will seriously affect the simulation quality.

2.2 Collision Distance Method

The research on the distance field method includes the following points: The generation of the distance field is usually completed in the preprocessing stage. The issues to be considered involve the time complexity of the distance field calculation and the storage space consumption of the description method, and the distance field is used to detect collisions. The distance field method can eliminate the intersection test at the same time. In practical applications, the update of the distance field is usually the calculation bottleneck. Therefore, this type of method is more suitable for collision detection between rigid objects and deformable objects. Spatial division is a common method to develop the geometric locality of the scene [11]. Used for deformable body collision detection, generally combined with hash index strategy. It has the advantages of simple algorithm implementation and fast processing speed, and can detect self-collision of deformable objects. As long as the conditions for the geometric intersection test are available, the spatial division method also shows good adaptability to the object model composed of non-triangular primitives. In addition, it is not affected by changes in the object's topology. The image space method is mainly to transform the collision problem in the three-dimensional space into the two-dimensional image space through the object projection of different viewpoints, so it is usually necessary to combine graphics hardware. Compared with other collision detection methods, the image space method does not require a time-consuming preprocessing process, which makes it suitable for processing deformable objects. However, the image space inevitably involves the discretization of geometric objects, so the detection accuracy is limited by the discretization error. In addition, the acceleration capability of graphics hardware is not always better than software computing.

2.3 The Meaning of This Article

Although the performance of computer hardware is getting better and better nowadays, due to the huge amount of calculation based on clothing 3D simulation technology, the effect of the system in real-time and interactivity is still very unsatisfactory, especially for situations that require higher realism. In the field of film and television production, the realization of realism is done by offline calculations. For the field of clothing design, timely interaction is a basic requirement, which leads to the inefficient use of clothing 3D simulation technology [12]. Not only that, in terms of technical reliability and dissemination, it is certainly not enough to rely solely on the high performance of the hardware, and users cannot afford expensive and high-quality hardware. For example, in the display of virtual products, merchants cannot ask for it. Therefore, the efficiency of the algorithm and the method used are the directions that should be studied. Therefore, the research significance of this article is to use clothing 3D technology to provide timely interactive support in the field of clothing design.

3 Method and Process of Clothing Design and Production Based on Clothing 3D Simulation Technology

3.1 Particle Simulation

The mass point spring system is the simplest way of modeling cloth particles, where particles are represented as spring mass points, and the basic feature of modeling the interaction between particles as linear springs is. For the mass spring model of the fabric, the model is built on a regular quadrilateral grid. The grid edges that connect adjacent mass points are defined as tension springs, and the diagonal lines are shear springs. The horizontal or vertical spaced mass points are connected to form a bending spring. The elastic forces generated by the above three springs are used to resist the tension of the cloth. Stretching, shearing and bending deformation. This kind of modeling is intuitive and the calculation speed is very fast. It can immediately generate clothing simulations and contact users at any time. However, this method also has certain shortcomings. It can only simulate simple shapes. If the shape is very complex, use this way to simulate is impossible. In order to solve this problem, the mass point spring model can be converted into a triangular mesh. In the triangular grid, the change of the spring is calculated to form a three-variable linear equation system. In short, this mode has done some processing on the triangle mesh, which leads to more accurate clothing modeling. Although the calculation work is twice that of the mass point spring model, the realistic model effect equivalent to the uninterrupted model can be obtained.

3.2 Numerical Solution of Kinetic Equations

The simulation system in this paper adopts the explicit Euler method. The system state parameters at the current time t are known, and the next time point t + h is estimated by constructing a linear function. The integral formula is as follows:

$$\begin{pmatrix} \Delta v^{t+h} \\ \Delta x^{t+h} \end{pmatrix} = h\begin{pmatrix} M^{-1}F^{t} \\ v^{t} + \Delta v^{t+h} \end{pmatrix} \tag{1}$$

The higher-order method improves the accuracy of the solution by combining the information of multiple known data points, and uses the fourth-order method with adaptive time step to obtain more accurate simulation results. In addition, given the upper limit of error, the integration step size can also be appropriately increased, which to a certain extent makes up for the additional computational consumption compared to low-order methods. Choosing the appropriate explicit integration method requires comprehensive consideration of many factors. The first-order Euler method has poor accuracy, but it is more suitable for highly discontinuous situations. Fourth-order is usually a good compromise in cloth simulation based on elastic models. For higher-order methods, they are less stable under nonlinear conditions and cannot effectively handle the mechanical model within the time step. The discontinuity constraint should not be adopted. Reducing the rigidity of the system can improve the stability of the explicit Euler integration method. For this reason, it is necessary to reduce the stubborn coefficient of the spring or increase the mass of the mass point, but it will cause the superelasticity problem in the simulation. The garment part grid is under its own mass point gravity. In this regard, a post-correction strategy is introduced, that is, after each simulation, the over-stretched spring is adjusted by moving the positions of the mass points at its two ends, mainly contracting along the spring line.

3.3 Improved Mass Point Constraint Algorithm

The traditional spring mass point constraint method is simple and convenient to implement, but the mass point position will be offset and corrected during the correction process, which affects the final effect, and also increases the number of iterations and the amount of calculation. The velocity correction method mainly focuses on modifying the velocity of the mass point, setting the velocity component in the direction of over-extension of the spring to 0. The simulation effect of this method is stiff, and the final result presented is unnatural. Therefore, the above two methods have their own shortcomings, and other improved methods are needed to achieve better simulation results. Therefore, the method used in this article combines the above two mass point constraint algorithms. For the points within some conditions, the traditional mass point constraint algorithm is used, and for the points within other conditions, the speed adjustment method can be used. The judgment rule is to use the traditional mass point constraint algorithm for the spring with the largest elongation connected to the mass point, while for other springs that exceed the set value of elongation, the speed adjustment method is used to constrain the mass points on both sides. In the simulation process, the speed adjustment method is adopted in this paper to adjust the speed of the spring to continue to stretch to 0. This method is an improvement on the original method. This article uses a method to set the velocity component along the particle direction to 0, and then only leaves the vertical velocity component, which can make the cloth shrink more to the middle during the simulation process, forming more realistic drape folds, and this method can be effective the simulation of the sagging movement of the cloth makes the simulation of the cloth more realistic and not too rigid. The formulas used in the improvement process are:

$$F_{(i,j)} = -k_{i,j}(\|x_i - x_j\| - l_{i,j}^0)\frac{(x_i - x_j)}{\|x_i - x_j\|} \tag{2}$$

4 Analysis of Simulation Experiment Results

4.1 Analysis of the Simulation Effect of Different Iteration Times

Table 1. Simulation results of different iteration times

Number of iterations	Simulation effect
100	Works well
200	Better effect
300	General effect
400	Bad effect

Cloth wrinkle simulation is an iterative process. The position of the mass point in the previous iteration will affect the initial state of the next iteration. Therefore, the selection of the number of iterations must be appropriate to achieve the real simulation effect. According to Table 1 and Fig. 1, it can be known. As the number of iterations increases, the simulation effect gets worse and worse. In order to get a better simulation effect, the number of iterations should be controlled within 200 times.

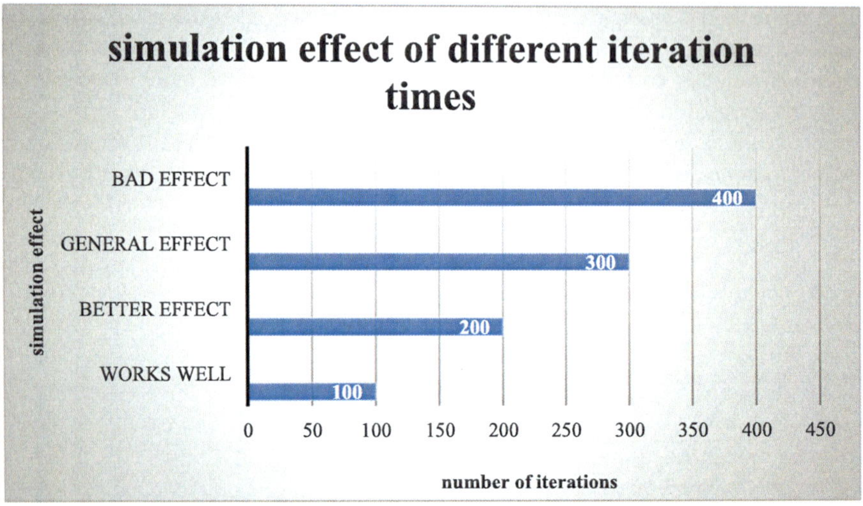

Fig. 1. Simulation results of different iteration times

4.2 System Simulation Speed Test

Table 2. System simulation speed test

Simulation speed	Number of times	Percentage
Very fast	32	64%
Faster speed	13	26%
Average speed	3	6%
Slower speed	2	4%

According to Table 2 and Fig. 2, we can know that 50 simulation experiments were performed, of which 32 simulations were fast, accounting for 64%; 13 simulations were faster, accounting for 26%; there were 3 simulations. Generally, the proportion is 6%; 2 times the simulation speed is slower, and the proportion is 4%. In-depth analysis of the experimental results, the slower simulation speed is mainly caused by network fluctuations. In general, the simulation speed of the system is relatively fast.

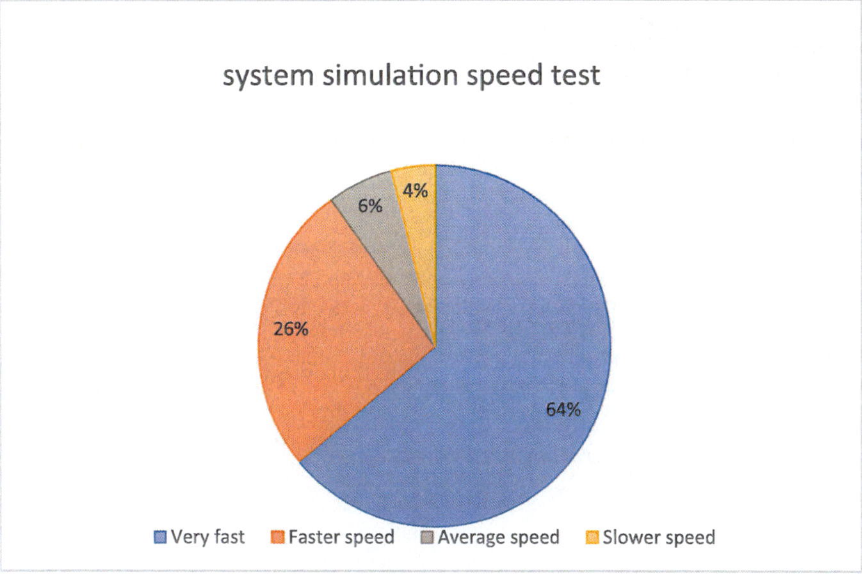

Fig. 2. System simulation speed test

5 Conclusions

Based on the use of clothing 3D simulation technology, this article is oriented to the field of clothing design and production, starting from the needs of manufacturing product design and production, combined with the knowledge of clothing design and production, the virtual simulation technology has been studied in detail, and the design and production a spring mass point model was realized, and simulation experiments were carried out on the model at the same time. Due to the complexity of the clothing design and production system based on virtual simulation technology, the large workload, and the limited research time and conditions, the spring mass point model implemented in this paper still has some shortcomings, and hopes to be further improved in the future work.

References

1. Zhang, J.: Application research of 3D virtual simulation technology in the teaching of fashion design. J. Sci. Technol. Econ. Guide **28**(707, 09), 135–135 (2020)
2. Jiannan, G., Changbin, W.: Application research of 3D virtual simulation technology in clothing design. West China Leat. **041**(021), 77 (2019)
3. Xiaoxu, L.: The application of virtual simulation technology in the teaching of clothing. Text. Cloth. Educ. **034**(004), 363–364 (2019)
4. Chen, J.: Application analysis of clothing design based on 3D virtual simulation technology. Northern Lights, **265**(11), 57–58 (2019)
5. Jingxuan, Z.: Application research of 3D virtual simulation technology in clothing design. Art Technol. **032**(023), 41–42 (2019)

6. Bao, R.: The application and development of virtual reality technology in clothing design. Popul. Color, (003), 51,53 (2019)
7. Jing, Z.: Thoughts on the reform of luggage plate making course based on virtual simulation technology. Mod. Educ. Forum **3**(12), 57–59 (2021)
8. Ji, L., Ming, X.: Design of a clothing aided design system based on three-dimensional images. Modern Electronic Technology, **42**(02), 125–127+131 (2019)
9. He, H., Liu, H.: Application practice of virtual fitting technology in clothing design. Text. Herald, 914(01), 84–87 (2020)
10. Wang, L., Tang, W., Xu, S., et al.: Numerical simulation and improved design of air cushion apron connection structure. China Ship Res. **16**(3), 120–127 (2021)
11. Lu, C.: Application of 3D simulation animation in mechanical design learning. Hubei Agric. Mech. 248(11), 93–94 (2020)
12. Sun, X., Xiao, B., Liu, Z.: Design and implementation of virtual simulation system for group performance activities clothing. J. Syst. Simul. **32**(11), 214–219 (2020)

Image Processing and Restoration of Cultural Relics of Song Dynasty in Sichuan Based on 3D Modeling Technology

Qinbing He[1(✉)], Qiwang Zhao[2], and Zhongyun Wu[3]

[1] School of History and Culture, Sichuan University, Chengdu, Sichuan, China
[2] Shanghai Urban Construction Vocational College, Shanghai, China
[3] Nanjing Museum, Nanjing, Jiangsu, China

Abstract. A large number of Song Dynasty artifacts and decoration images in Song tombs unearthed in Sichuan area are rich in content, unique in shape, and ingenious in composition. They are an important part of the history of ancient Chinese art and an important physical example for the study of ancient Chinese history. This article aims to use modern computer science and technology such as 3D modeling to store unearthed cultural relics and their exquisite images, restore artifacts, realize the matching of plane photos and 3D models, and visually present the overall appearance of the relics, thereby providing more abundant materials to promote archeology for the public and museum exhibitions.

Keywords: Sichuan · Cultural relics of song dynasty · 3D · Image modeling

1 Introduction

In recent years, a large number of Song Dynasty tomb decoration images and various artifacts have appeared in Sichuan. However, due to natural causes such as hot and humid climate, soil, and some unavoidable human factors in Sichuan, a large number of cultural relics were either peeled off in color or broken in utensils when they were unearthed. It is also difficult for well-preserved cultural relics to be displayed 360° to the audience in museum exhibitions. Many precious cultural relics will also be restricted by preservation techniques and exhibition conditions, so they have to choose to seal them for protection instead of displaying them. With the development of computer science and technology and the Internet, 3D technology is used in more and more fields and is widely used in all walks of life. With the continuous deepening of archaeological research and cultural heritage protection, 3D technology has gradually appeared in archaeological work. This article chooses Blender software and uses 3D modeling technology in the restoration of cultural relics, so as to provide a new idea and method for the restoration and preservation of cultural relics.

B. J. Jansen et al. (Eds.): *International Conference on Cognitive based Information Processing and Applications (CIPA 2021)*, LNDECT 85, pp. 115–121, 2022.
https://doi.org/10.1007/978-981-16-5854-9_14

2 The Necessity and Feasibility of Digitalization of Cultural Relic Images

In recent years, a large number of tomb decoration images and various artifacts of Song Dynasty have appeared in Sichuan. The decorative images of the tombs of Song Dynasty mainly point out the various thematic images of the inner walls of the tombs that have been manually carved. They are famous for their large numbers, exquisite carvings, rich content, and diverse themes. Most of the decorative images of Song Dynasty tombs in Sichuan appeared in stone chamber tombs, and a few appeared in brick chamber tombs and cliff tombs. These stone chamber tombs, which are chiseled in the mountains, and built with stones on the spot, decorate the four walls of the tomb with carved images of various wood-like components, figures, flowers, and beasts, and jointly construct the afterlife of the tomb owner. It is a unique type of tombs in Sichuan, and it is also the largest and most widely distributed type of tombs in Sichuan in Song Dynasty. According to statistics, more than 600 Sichuan Song tombs with carved stone images have been published after archaeological excavations and cultural relics surveys. Thousands of tombs are widely distributed in southern Sichuan, eastern Sichuan, northern Sichuan, and Chongqing. The unearthed artifacts include a large number of porcelains, iron, gold and silverware, bronzes, etc., unearthed from the Song Dynasty tombs and cellars, including bowls, plates, stoves, mirrors, etc., with exquisite shapes and rich images.

Affected by natural causes such as hot and humid climate, soil and some unavoidable human factors in Sichuan, a large number of cultural relics were unearthed either in color peeling off or utensils broken. It is also difficult for well-preserved cultural relics to be displayed 360° to the audience in museum exhibitions. Many precious cultural relics will also be restricted by preservation techniques and exhibition conditions, so they have to choose to seal them for protection rather than display them.

With the development of computer science and technology and the Internet, 3D technology is used in more and more fields, and it is widely used in various industries, such as art design, industrial design, medical imaging, etc. In recent years, a large number of emerging technologies have also been applied to archaeological research, museum exhibitions, and cultural relics drawing [1], such as GIS [2], GPS, AutoCAD [3], 3D modeling, 3D printing, 3ds Max technology [4] and so on, so as to establish an effective cultural relic imaging program. "People live in 3D space and are familiar with the way things exist in three-dimensional space. Therefore, the display of 3D information is most in line with people's habits. With the continuous deepening of archaeological research and cultural heritage protection, as well as the rapid development of 3D information analysis and processing methods, the acquisition of all-round spatial information in archaeological work becomes more and more important. [5]" The images and historical relics that were submerged in the long river of history but are of great significance in Chinese culture are presented to the audience in a more diversified manner, which makes up for the past in the presentation of cultural relics and museum exhibition design by simply using artifacts, pictures, and text to interpret. With 3D, archaeology uses diversified scientific and technological means to integrate various artistic expression methods such as sound, light and color to meet the aesthetic needs of the public in modern society and promote the diversification of museum exhibition methods, thereby promoting museum social education and public archaeology development.

3 Software Introduction

Blender is a compact and cross-platform integrated 3D creation suite that can be used to make animated videos and provides a large number of basic tools for editing, including modeling, still images, animation, audio processing, video editing, effect synthesis, etc. Different from other software's single modeling, mapping, and video creation functions, Blender solves the complicated procedures that required multiple software to create collaboratively in the past, and the volume is small. At the same time, Blender is more popular in terms of computer memory and drive performance requirements. It can meet the needs of ordinary 3D modeling, video production, and animation design. It can run under Linux, macOS and Windows systems.

The creation and birth of Blender software was originally completed in NeoGeo, a Dutch animation studio, and the purpose of use was mainly for artistic creation and 3D modeling. In order to solve the problem that the traditional 3D creation kits were too old, complicated and difficult to update, the company independently designed and developed an integrated animation production software. After the software was created, the company continued to improve Blender according to the actual needs of 3D architecture and image production, and established a special company to develop and operate the software, such as rewriting the code, overhauling the built-in rendering performance, adding Cycles features, etc. All these make Blender have more and more unique advantages compared with other software. "Generally speaking, the reversibility of 3D modeling is relatively weak, that is, when I turn a cube into a heterogeneous shape, it is more troublesome to change it back, or to change it into another shape, and there is basically no way out. However, with the continuous development of the modifier design, we have been able to implement quite a wealth of functions that can be adjusted at any time in Blender. For example, to smooth the edges, you don't have to manually smooth them on the model, but programmatically control the range, amplitude, and intensity through the modifier. Simply put, it gives us more control space and sells a dose of 'regret medicine'. [6]" After 2000, under the promotion of Blender Foundation, the Blender open-source project, also known as Blender's liberalization activity, was launched in Europe and quickly gained great support. Blender has more and more registered users worldwide.

It can be seen that Blender is an easy-to-access, efficient and convenient 3D modeling and drawing software, which can not only be widely used in animation design, video production, game development, advertising creation, industrial design and special effects creation, etc., but also has passed a long period of practical testing. Therefore, this article applies Blender to archaeology and museology to serve the digitalization and information construction of archaeology, cultural relics, and museology. It can not only provide scientific aids for archaeology and cultural relics related research, but also provide contemporary audiences with intuitive, vivid, vivid and comprehensive informational images, enrich the audience's sensory experience, meet the aesthetic and spiritual needs of audiences in the new era, and enrich the development path of museum digital construction, museum exhibition design and public archeology.

Thus, applying Blender software to image processing and restoration of ancient cultural relics has very important scientific and social significance.

4 Image Processing and Production

Through the 3D modeling of the artifact itself, it is easy to convert the two-dimensional image into three-dimensional, intuitively see the overall appearance of the artifact, and recombine the broken or broken artifacts through data analysis.

4.1 Image Collation and Data Collection

In the initial stage of collecting cultural relics, experimenters need to obtain the true appearance and accurate information of the cultural relics. Soft and natural lighting should be selected for taking photos. The cultural relics should be shot from multiple angles and omni-directional, to keep the light between the camera and the cultural relics uniform and not blocked [7]. When collecting cultural relic information data, such as height, width, thickness and texture, it is necessary to be precise and careful to organize and restore the true appearance of cultural relics.

The example selected in this article is a bronze mirror of the Song Dynasty unearthed from the Song Tomb in Lu Xian, Sichuan. The bronze mirror was broken when it was unearthed. The overall shape is circular, and the periphery is decorated with double-layer lotus petals with a diameter of 15.2 cm. "The mirror surface is polished, the back is flanged, the nose-shaped button, and the surface is plain. The right side of the back is cast in vertical regular script "Bronze Mirror owned by Nian Ershu in Huzhou", the words are reversed. The middle of the mirror surface is cracked [8]".

4.2 Image Production

The step was processed by Blender. The Blender start interface consists of three parts. The upper bar is the top bar, the wide middle area is the work area, and the bottom bar is the status bar (Fig. 1). There will be a default box behind the main page, a point light source and a camera. In order to facilitate subsequent production, select the default box and point light source on the original page and click delete. Then switch to the orthographic top view, create a reference to import the image of the cultural relics that need to be processed (Fig. 2). Adjust the opacity of the image, and move the image to place the cultural relic image to the center. Create a new column, adjust the value, and adjust the reference diagram to a suitable size. Then enter the edit mode, call up the basic shape of the theme graphics. Then add a mirror modifier for symmetry, and add a surface subdivision modifier to make the surface smooth, and gradually add various details. Next, apply the mirror modifier to the model picture to make one of the diamond pattern ornaments, and use the rotation tool to continue making the remaining diamond pattern ornaments. Here it is needful to repair the edges and remove the extra parts, so that the image looks more regular and smoother. Then merge the adjacent points of the six parts. Move the modifier up in the queue. Press the F key to fill the hole left by the mirror button in the center of the picture, then extrude the filling surface, make the mirror button, and add the size of the bronze mirror (Fig. 3). Then add material to the object, create a new light, and select the surface light option in the light options. And change the view display mode to facilitate the observation of the effect of the light material. Create a new plane as the rendering background, adjust the color, metallicity, highlight, roughness and other attributes of the copper mirror material, adjust the brightness, size, position, quantity, etc. of the light, restore the copper mirror material, and light the copper mirror. Create a new camera, adjust the camera to a suitable position, and rotate the copper mirror to get different angle renderings (Fig. 4). At this point, the 3D modeling image of the bronze mirror has been completed.

Next, continue to complete the image modeling on the back of the bronze mirror and make inscriptions. Select the appropriate edge line on the bronze mirror model image of the finished mirror button, and mark the stitched edge (Fig. 5). Then, expand in the uv view and add texture to the bronze mirror. Import the text image into the blender texture, adjust the uv line, and align the text to the proper position. In the end, complete 3D modeling of the bronze mirror cultural relic image is completed, and realize the cultural relic image processing and online (Fig. 6).

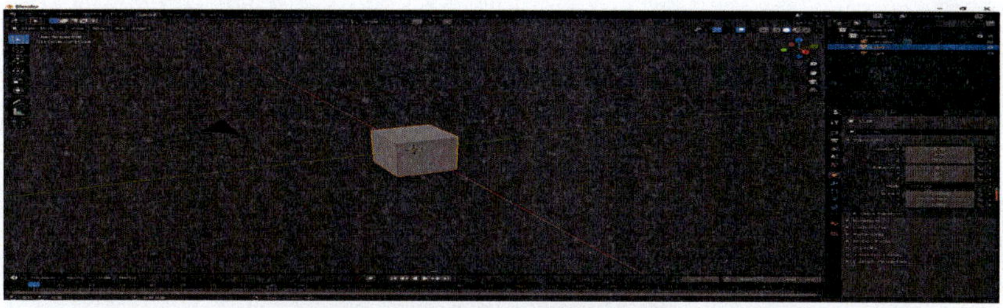

Fig. 1. Start interface of Blender

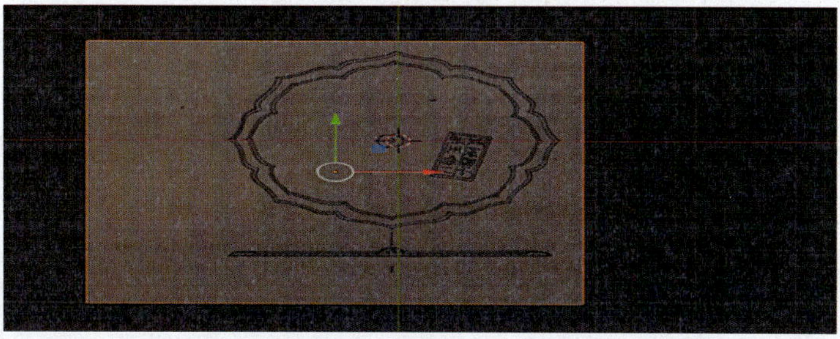

Fig. 2. Embed the picture

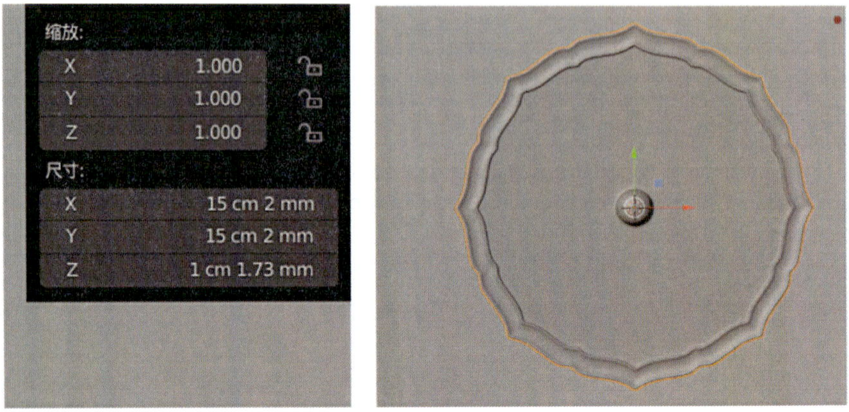

Fig. 3. Embed the size of the bronze mirror

Fig. 4. Restore the material of the bronze mirror and make the surface drawing

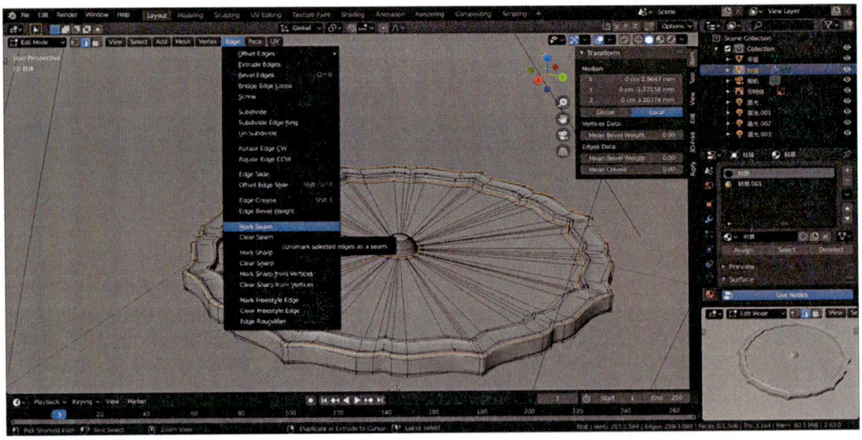

Fig. 5. Mark the edges

Fig. 6. Back view of the bronze mirror

5 Conclusion

As an open-source 3D modeling and all-round 3D animation production software, Blender software has considerable application prospects in archaeology, cultural relics, and museology.

Through the Blender software and the above-mentioned production steps, a staff member who has basically mastered the Blender software program points out the 3D model through a cultural relic picture and a cultural relic line drawing within four to five hours. Compared with the final imaged 3D high-simulation model, although there are minor defects in local positions, the overall effect and image quality are still relatively high. At the same time, 3D model images of cultural relics in different directions can be obtained, which is not only conducive to archeology, cultural relic data storage and related research, but also can be widely used in public archeology and museum exhibitions. If the Blender software related functions are reused in post-production, video animation and audio design are added, the display effect of such cultural relic images will be enriched, thereby playing a certain role in promoting the development of public archeology and museum exhibitions.

References

1. Ma, H.: Archeological Artifact Drawing, p. 99. Peking University Press, Beijing (2008)
2. Zhang, H.: GIS and Archeology Spatial Analysis Practical Tutorial, p. 32. Peking University Press, Beijing (2018)
3. Edited by Phoenix High-tech Education: Chinese version of AutoCAD 2016 Mechanical Drawing Basic Tutorial, p. 211. Peking University Press, Beijing (2016)
4. Era, M. (ed.): 3ds Max 2011 Platinum Manual I, p. 148. Posts & Telecom Press, Beijing (2011)
5. Liu, J.: Multi-view 3D Reconstruction of Archeological Sites, p. 01. China Social Sciences Press, Beijing (2019)
6. Xue, S.: 3D modeling can also be "programmed", Blender shader tutorial. Popul. Comput. Week **04**(26), 98–102 (2021)
7. Zhang, Y.: Introduction to Digital Technology of Cultural Relic Photography, p. 61. Cultural Relics Publishing House, Beijing (2011)
8. Sichuan Provincial Cultural Relics and Archeology Research Institute, Chengdu Institute of Cultural Relics and Archaeology, Luzhou Museum, Lu County Cultural Relics Management Office. Song tombs in Lu County, p. 70. Cultural Relics Publishing House, Beijing (2004)

Chinese Buddhist Literature Database and Image Text Extraction Algorithm

Xuetao Liu(✉)

The College of Literature and Journalism, Sichuan University, Chengdu, Sichuan, China

Abstract. The digitization of Chinese Buddhist literature is the key direction of contemporary Buddhist literature research. Currently, there are three types of Chinese Buddhist literature databases that have been completed: Tibetan scriptures databases, non-Tibetan literature databases, and historical materials databases. Commonly used Buddhist literature database image text extraction algorithms are algorithm based on threshold, algorithm based on edge, algorithm based on area, and algorithm based on deep learning. The current Buddhist literature database still needs improvement in terms of repeated construction, image processing, and text recognition and proofreading.

Keywords: Chinese Buddhist literature · Database · Text extraction · Algorithm

1 Basic Content of Chinese Buddhist Literature

From the day of its birth, Buddhism spread like a vast area outside the place of origin. In China, it was mainly passed through the land route of the Western Regions and the sea route of the South China Sea, and it was introduced to the Central Plains after Han Dynasty. At present, it is known that the earliest Buddhist classic that the Chinese came into contact with was *Buddha Sutra* dictated by Yi Cun, the envoy of the Dayue clan in the first year of the Emperor Ai of Han Dynasty (the first two years) to the disciple Jing Lu [1]. This Buddhist scripture with a world-breaking significance was once widely circulated in the Eastern Han Dynasty, but it has disappeared by the end of the Western Jin Dynasty.

The content of Chinese Buddhist scriptures can be roughly divided into two parts: translated classics and local works [2]. The translation movement of Buddhist scriptures formed three peaks in the Western Jin Dynasty, Southern and Northern Dynasties and Tang Dynasty. There are representatives such as Zhu Fahu, Kumarajiva and Xuanzang. The original Buddhist scriptures on which the translation of Buddhist scriptures in the Middle Ages were based were mostly written in ancient Indian folk languages, Sanskrit, Pali and various Western languages [3]. The content includes not only the Hinayana Buddhism Scripture Tripitaka, which has been sorted out by tribal Buddhism, but also the Mahayana scriptures that have been compiled since the 1st century BC. Local works are Buddhist works written by Chinese monks and lay scholars inspired and influenced

B. J. Jansen et al. (Eds.): *International Conference on Cognitive based Information Processing and Applications (CIPA 2021)*, LNDECT 85, pp. 122–127, 2022.
https://doi.org/10.1007/978-981-16-5854-9_15

by Buddhism. Among them are the classics and annotations compiled by Chinese scholars to explain the Buddhist Tripitaka, the catalogues compiled for the convenience of reading Buddhist scriptures, the sorting of Buddhism historical records of the development process in China, and temple records and geographies that introduce the material achievements of Buddhism at home and abroad. Compared with the system of Buddhist scriptures in other languages such as Pali and Tibetan, the Chinese Buddhist scriptures contain more numbers, richer types, and a longer time span. They are not only indispensable spiritual and material achievements in Chinese culture, but also the most precious cultural relic in the history of civilization. From the perspective of scientific research, Chinese Buddhist classics are also the basic materials for us to study Han Buddhism.

Chinese Buddhist classics are mainly organized in the form of Tibetan scriptures [4]. The compilation history of the Chinese Tripitaka can be traced back to the Southern and Northern Dynasties. Until today, the compilation of various Tibetan scriptures in the Chinese cultural circle is still in progress. Of course, today we should give the Chinese Buddhist database the mission of a general library of documents or a special library of specific documents, rather than compiling a more sophisticated electronic Buddhist scripture.

2 The Current Situation of the Construction of Chinese Buddhist Literature Database

It is not too late for academic circles at home and abroad to start digitizing the cultural heritage of Chinese Buddhism. Material cultural heritage such as pagoda temple gardens, grottoes, sculptures, and remains, as well as intangible cultural heritage such as ritual systems, music and dance recitation, literary legends, and scriptures, have all come out digitally. In terms of the Chinese Buddhist literature discussed in this article, based on the characteristics of the database content, it can be roughly divided into three types: Tibetan scripture database, Tibetan manuscript database and historical material database.

2.1 Tibetan Scripture Database

In the 1990s, the Korean Tripitaka Research Institute had already digitized its collection of "Re-engraving of the Korean Collection of Tripitaka" and published it to the world [5]. In 2012, the Japanese Tripitaka Tekisude-Tabe-Su Research Association also released the SAT Tripitaka text database, the data range is *Taisho Tripitaka* volumes 1–85. The "Hantang Collection" database of Beijing Times Hantang Technology Co., Ltd. contains the "Buddhist Classics Library" that is still under construction, and actually contains *Taisho Newly Revised Tripitaka*, *Pingqiejingshe Tripitaka* and 卍 *Continuous Collection*, and many Supplements to Tibetan Scriptures. Taiwan's Fo Guang Shan Foundation for Buddhist Culture & Education has also launched an electronic version of the *Fo Guang Tripitaka*, which can be read and retrieved online.

At present, the most used and richest Buddhist database of Tibetan scriptures is the "electronic Buddhist scripture integration" built by the Chinese Buddhist Electronic Text Association (Cbeta) [6]. The database integrates the *Taisho Newly Revised Tripitaka*, 卍 *Newly Continuous Collection*, *National Library Rare Book of Buddhist Canon*,

Chinese Translation of Southern Tripitaka, External Tibetan Buddhist Documents and *Orthodox Buddhist Data Collection, The Northern Dynasties Buddhism Stone Inscriptions and Rubbings Hundred Pins, Tripitaka Supplement, Chinese Buddhist Temple Records, The Collection of Buddhist Works of Master Yinshun* and other recent new compilations, CBETA electronic Buddhist scriptures belong to a full-text database. Based on this database, Taiwanese researchers have launched the CBETA online reading system, which is more convenient for reading.

Before the era of printed editions, the Chinese Buddhist Tripitaka used the form of manuscripts. The Buddhist scriptures of the Middle Ages are preserved in the Dunhuang Tibetan Scripture Cave for the most. However, the Buddhist scriptures in the Dunhuang posthumous manuscripts are not the Tibetan scriptures, but there are many single copies. Fortunately, after the Chinese Tripitaka, which was compiled during the Sui and Tang Dynasties, was introduced to Japan, a number of precious Medieval scriptures were produced during the Nara and Heian eras, and they have been preserved in Japan's Shokurain Temple and Amanoyama Jingang Temple, Nagoya Seven Temples, Kyoto Kosho Temple, Myoren Temple and other places. These manuscripts generally have the form and content characteristics of typical Tibetan scripture manuscripts, so they are included in the Tibetan scripture database. The "Japanese Ancient Writings Database" constructed under the auspices of the International University of Buddhism in Japan organizes the ancient scrolls before the Edo period collected by various temples and institutions in Japan, provides two retrieval methods by number and title, and displays part of the color photos of scrolls.

In order to retrieve the Tibetan scripture data more conveniently, some institutions have also specially developed a database of Buddhist scriptures catalogues, such as the "Digital Database of Buddhist scriptures catalogues" produced by Taiwan Chung-Hwa Institute of Buddhist Studies and China Electronic Tripitaka Collection. A total of 24 Tibetan scriptures have been coordinated electronic content. Another example is the "Joint Catalogue of Buddhist Library Collection" of the Taiwan Xiangguang Buddhist College. In addition to the function of searching for collections, it also includes a catalog of Buddhist scriptures from past dynasties, which is more convenient to use.

2.2 Tibetan Manuscripts Database

The database of manuscripts outside Tibet mainly relies on the Buddhist literature in the Dunhuang posthumous writings of China. The Buddhist texts in the Dunhuang posthumous manuscripts are all-encompassing, both scattered and concentrated. A basic fact about the Buddhist texts in the Dunhuang posthumous manuscript is: Although these documents are all-encompassing, their subjects are not well-organized manuscripts of the Tibetan scriptures, but rather complex single manuscripts. Therefore, the Dunhuang posthumous manuscripts are different from the ancient manuscripts of Heian, Nara, Japan mentioned in the previous article.

Currently the most widely used and well-evaluated Dunhuang literature databases include the International Dunhuang Project (IDP), the Dunhuang Academy "Digital Dunhuang" project, the "Dunhuang Posthumous Manuscript Image Database", and the Lanzhou University "Dunhuang Literature Digital Museum", Beijing Erudition Digital Technology Research Center "Dunhuang Archives", etc. Among them, the "Digital

Dunhuang" project launched by the Dunhuang Academy in 2016 provides free high-definition mural images and panoramic roaming of 30 Dunhuang caves to global users for the first time. It can be regarded as a representative of the digitization of Dunhuang cultural relics and documents in recent years.

2.3 Historical Material Database

The Buddhist historical material database mainly includes Buddhist images, rubbings and Buddhist historical books. The representative works of the database of Buddhist historical materials are the sub-sections of the Zhonghua Book Company that have digitalized most of its "Selected Chinese Buddhist Classics" and included them in the "Chinese Classics and Ancient Books Database". Metal and stone rubbings are the most important content of this type of database.

A comprehensive database of metal and stone rubbings often contains Buddhist rubbings. Comprehensive rubbings databases such as "The Essence of Rubbings" by National Library of China, "Special Collection of Jinshi Rubbings" by Peking University, "Compilation of Historical Data of Chinese Moments" by UniHan Digital Technology Co., Ltd, "General Records of Chinese metal and stone" led by Gansu Province Ancient Books Compilation and Translation Center, Taiwan University's "Taiwan Ancient Tablets Database", the Taiwan Institute of History and Language "Liao Jinyuan Rubbings Dictionary Collection", Taiwan's "National Library", "Metal and Stone Rubbings Database", UC Berkeley's "Chinese Rubbings Database" contain various Buddhist rubbing images according to the scope of its collection. In addition, the "Buddhist Stone Inscription Rubbings Database" of the Taiwan Institute of History and Language and the "Chinese Buddhist Stone Scriptures Database" of Heidelberg University are single databases that specifically include Buddhist rubbings. The former is based on the collections of the Fu Sinian Library and has a lot of choicest collections.

3 Commonly Used Database Image Text Extraction Algorithms

In the process of text database construction, image scanning and preprocessing are the pre-steps for extracting image text. The scanned image goes through the steps of grayscale, noise reduction, binarization, and tilt detection and correction to eliminate interference factors in the image as much as possible and improve the clarity of the text, so that the image text can be cut and extracted more accurately and efficiently. There are several common image text extraction algorithms in the current Chinese Buddhist literature databases.

(1) Based on threshold. The method can be divided into global fixed threshold method and local adaptive threshold method. The former is to use a unified threshold for the entire image to perform binarization, and the latter is to set the binarization threshold according to the pixel value distribution of different pixel regions. The local neighborhood block threshold and the Gaussian weighted sum of the local neighborhood block are more commonly used local adaptive thresholds. Generally speaking, threshold-based algorithms are only suitable for grayscale images.

(2) Based on edge. The text is composed of strokes, and the foreground and background in the image can be distinguished by edge detection. The current edge extraction technology of image text is either based on mathematical morphology, or based on gradient operator, or based on transform domain. In view of the complex image conditions of ancient manuscripts, the image can be preprocessed for noise reduction and light and dark enhancement first to filter out the false edges caused by background and pollution, and combine the traditional Canny algorithm for edge extraction. Finally, morphological methods are used to further filter out small-area edges.
(3) Based on area. The regional image algorithm is to measure the consistency of regional features, synthesize neighborhoods with the same features into the same large region, and divide the regions with inconsistent features into multiple subregions. Common area algorithms are the K-means clustering algorithm proposed by Mac Queen, which is suitable for processing continuous data, and the ISODATA algorithm, which is suitable for processing discrete data [7].
(4) Based on deep learning. Before 2012, image text recognition mainly relied on traditional image processing technology and statistical machine learning methods. Since then, with the widespread application of deep learning in the field of computer vision, this algorithm that is more adaptable to complex images and text environments has been introduced into new database construction. A feasible solution is to perform binary preprocessing of the image through the Sauvola algorithm. Then applying the CTPN network neural model for training to obtain model parameters and predicting the position of the text line. Finally performing the text lines' identification based on the DenseNet+ CTC neural network.

At present, the character recognition technology is relatively mature, and the OCR system has been able to undertake most of the character recognition work after segmentation and extraction. Of course, for some Chinese Buddhist scriptures with extremely complicated foreground and background, the application defects of these technologies are quite obvious.

4 Prospects for the Construction of Chinese Buddhist Databases

When planning the digital databases of Chinese Buddhist literature, Mr. Fang Guangchang put forward four basic principles: First, it is the basic work to start from the bottom, and cut all the words and symbols in Buddhist literature; Second, the information should be fully covered, and the database should contain all the words and symbols in Buddhist literature; Third, the process can be traced, the work process data record is undoubtedly, the whole process should be stored for examination; Fourth, the function is expandable, the man-machine interface is extensible and easy to use [8]. Most of the three types of Buddhist literature databases that the author briefly introduced in the previous section have been deeply embedded in the work of contemporary researchers and become essential tools at the desk, facilitating scientific research. However, the dissatisfaction with electronic documents has not disappeared from the minds of any rigorous Buddhist literature worker, because at present, these databases still have the following deficiencies.

The first is the problem of duplicate database construction. Currently, there is no unified plan for the construction of Buddhist literature databases at the high-level. Each presiding unit is limited by its own conditions, and it is inevitable to form an isolated situation in the database construction process. The construction of a unified digital platform for Buddhist literature is extremely difficult, and it can only be realized under coordinated leadership at the national or international level. Currently, all data platforms that have been built generally have the problem of insufficient compatibility.

The second is the problem of document image processing. Except for a very small number of full-color image databases, most Buddhist databases containing images have been processed with black and white binary or grayscale processing when processing the pictures in the engraved copy, which invisibly reduces the identification value of the numerous color marks and symbols in the literature. Problems such as incomplete plates, wrong numbers, and insufficient clarity are not uncommon.

The last is the problem of text recognition and proofreading. This problem is particularly prominent in folk Buddhist scriptures. Many writers in the Middle Ages were not serious when copying Buddhist classics. The calligraphy was clumsy, Chinese characters were added or removed, and there were many misconducts. They were often difficult to distinguish when they were passed on to later generations, which brought great challenges to the proofreading researchers. Even the CBETA electronic Buddhist scripture, which is currently known for its fine examination, still has many textual errors. For example, the word "yu" in "Zhenzhaoquyu" in Sengyou's *Tripitaka Collection* is the same in all editions. This is a variant of "yu" (universe), and CBETA sees it as "yu" (residence) (*Taisho Tripitaka* T55, p0005b21). Another example is "try to pass quicksand at an early age", CBETA mistakenly "sand" for "wading". It can be said that this type of text proofreading error is not uncommon in the Buddhist literature database launched by any institution and unit. Some of the reasons for these errors are the negligence of the staff or the unfamiliarity with the characters. In many cases, the text of the scriptures on which these databases are based -- currently mainly is Tripitaka Collection -- has a lot of text errors, which sometimes mislead users after they are digitized.

References

1. Annotated by Songzhi Pei: Annals of the Three Kingdoms Wei Dynasty, p. 859. Zhonghua Book Company, Beijing (2009)
2. Fang, G.: Briefly talk about the compilation theory of the Chinese Tripitaka and its evolution. Stud. World Religions **01**, 32–41 (2012)
3. Feng, G.: A brief discussion on the necessity to establish the Chinese Buddhism philology. J. Zhejiang Univ. (Human. Soc. Sci.) **39**(06), 125–132 (2009)
4. Zong, S.: Notes on the Chinese Buddhist writings from Karakhoto site. Dunhuang Res. **01**, 82–92 (2001)
5. Bai, H.: On the catalog of Chinese Buddhism books. **45** (01), 142–145 (2008)
6. Fang, G.: Several questions about the national grading standards for Chinese Buddhist ancient books. J. Southwest Univ. National. **08**, 61–65 (2015)
7. Zhou, S., Yin, J.: LBP texture feature based on Haar characteristics. J. Softw. **24**(08), 1909–1926 (2013)
8. Fang, G.: On the construction of the digital database of Chinese Buddhist literature. Stud. World Religions **01**, 10–18 (2016)

A Speedy Brain-Computer Interface Combined with Computer Vision

Kaixuan Liu, Yang Yu(✉), Yadong Liu, and Zongtan Zhou

The College of Intelligence Science and Technology, National University of Defense Technology, Changsha, China

Abstract. Brain-computer interface (BCI) refers to the direct connection created between the human or animal brain and external equipment to realize the exchange of information between the brain and the equipment. This concept has existed for a long time, but it was not until the 1990s that phased results began to appear. the development of BCI is impressive. By decoding subjects' electroencephalogram (EEG) signal, BCI provides a channel for communication between subjects' brain and external devices. However, the interaction speed of current BCIs is relatively insufficient. To this end, we propose CV-BCI, a P300-based BCI combined with computer vision (CV). We recruit 10 participants in this study and ask all participants to finish the same tasks with CV-BCI and a traditional BCI speller. This study analyses the performance of the CV-BCI and BCI speller in accuracy, information transfer rate (ITR), and interaction speed. The experimental results suggest that the proposed CV-BCI surpasses BCI speller in interaction speed (about 6 s per object and 20 s per object respectively) while CV-BCI keeps comparable accuracy and ITR with BCI speller. This study probably proposes a more user-centric and speedy BCI approach, which promotes the development of BCI.

Keywords: Brain-computer interface (BCI) · P300 · Computer vision (CV)

1 Introduction

A brain-computer interface (BCI) is a tool that connects users' brain with external devices. In BCI research, the commonly used EEG components are P300 signal, steady-state visually evoked potentials (SSVEP), and sensorimotor rhythms (SMRs) [1]. Accordingly, there are three main types of BCI: P300-based BCI, SSVEP-based BCI, and SMRs-based BCI, among which the P300-based BCI has a relatively high accuracy, information transfer rate (ITR) [1, 2]. Meanwhile, the P300-based BCI tends not to induce visual fatigue and studies suggest that almost all subjects can use P300-based BCI [3]. For these reasons, we choose to analyse P300-based BCI in this study.

For enriching the application of BCI, numerous interesting BCI systems are proposed. For example, He et al. develops a BCI speller to help users receive and send emails, and browse internet [4]. Yin et al. designs a novel BCI system combined P300 signal with SSVEP signal [5]. Yin et al. recruits 14 participants in their research and the

B. J. Jansen et al. (Eds.): *International Conference on Cognitive based Information Processing and Applications (CIPA 2021)*, LNDECT 85, pp. 128–134, 2022.
https://doi.org/10.1007/978-981-16-5854-9_16

experimental results indicate that their BCI outperforms other BCIs in accuracy and ITR. Yu et al. proposes a self-paced BCI speller that uses SMRs as the switch of the whole system [2]. For solving the adjacency problem in P300-based BCI, Qasem et al. proposes a novel BCI spelling paradigm, i.e., the edges paradigm [6]. The primary purpose of BCI research is to provide human beings convenience, thus it is significant to improve the interaction speed of BCI [3, 7]. Although some studies improve the interaction speed of BCI by optimizing classifiers, e.g., Lenhardt proposes a more speedy and accurate adaptive P300 classification algorithm [8], the interaction speed of BCI speller is still comparatively low.

Studies [9–11] indicate that introducing computer vision (CV) into BCI improves the performance of BCI, which inspires us to design the proposed CV-BCI (a P300-based BCI combined with CV). The major difference of CV-BCI and BCI speller is the interaction method. The interaction method of the BCI speller is to spell out the target, e.g., if a user needs to choose a "bottle" in environment, he/she should spell like "b", "o", "t", "t", "l", "e". In contrast, the CV-BCI detects and decodes the targets in environment. If a user plans to choose a "bottle" in environment with CV-BCI, CV-BCI will detect the whole environment and decodes all objects as options. And then the user can choose the option "bottle" directly in the user interface of the CV-BCI. If the user wants to choose a longer target, such as "refrigerator", the BCI speller needs to spell more times, but the CV-BCI doesn't need to spell. In the CV domain, the 3rd version of You Only Look Once (YOLOv3) algorithm has comparatively high speed, detection accuracy [12], and we thus choose to use YOLOv3 in this study.

We use the accuracy, ITR, and interaction speed as the performance metrics in this study. In order to prove the feasibility of CV-BCI, we compare the CV-BCI with a BCI speller. The remainder of this paper is organized as below. The second section describes the subjects, the structure of the CV-BCI and BCI speller, the experiments, the EEG collection, classification algorithm, and performance metrics. The third and fourth section displays and discusses the experimental results.

2 Materials and Methods

2.1 Subjects

This research recruits 10 participants (ranging from 22 to 32 years of age) who have no BCI experience. Before conducting the experiment, all the subjects are informed of the aim and procedure of this study, and have given the written consent. The experiment is carried out in a normal room with the size and luminance of 5×3 m^2 and 200 lx respectively. There are several books, bottles, chairs, and monitors etc. in the room. The room is closed and only the subjects and researchers are allowed to stay or enter. Subjects sit in a comfortable armchair 60 cm from the computer screen.

2.2 Structure of the CV-BCI and BCI Speller

Figure 1 shows the architecture of the proposed CV-BCI. The CV-BCI includes 3 parts: the computer vision module, the user interface, and the signal decoding module. The computer vision module used in this study is a laptop-run YOLOv3 platform. The CV module detects and decodes targets in environment as options of BCI. The signal decoding module used in this study consists of an amplifier and its Ag/AgCl electrodes (actiCHamp, Brain Products, Germany) and a laptop-run BCI2000 platform. The amplifier and its Ag/AgCl electrodes record and amplify EEG signals from the subject's scalp. The BCI2000 platform extracts the P300 features and decodes subjects' commands. The user interface module receives the options from CV module and then shows them to subjects. Meanwhile, the user interface module also displays the P300 flashes which appear according to the oddball paradigm [13], the decoding results, and the red cue used to remind the subjects the current target.

Figure 2 exhibits the architecture of the BCI speller used in this research. It includes a signal decoding module which is the same as that used in CV-BCI and a user interface module. The user interface used in the BCI speller has the same function as that of the proposed CV-BCI. However, the options of the 2 BCIs are different. The options of the BCI speller are 26 letters, 9 numbers, and the option "DEL" used to delete the current classification result. Figure 3 is a time course depicting the experimental procedure of the CV-BCI and BCI speller.

2.3 Experiments

Each subject attends three experiments: the offline experiment, online experiment, and interaction experiment. In the offline experiment, each subject chooses 20 targets with the CV-BCI and BCI speller respectively. In the online experiment, each subject chooses 25 targets with the CV-BCI and BCI speller respectively. In the interaction experiment, each subject chooses 10 real objects in the environment which are "bottle", "chair", "book", "computer", "refrigerator", "light", "cup", "box", "bowl", and "person" with the CV-BCI and BCI speller respectively.

2.4 EEG Collection

We record EEG signals from five channels according to the international 10–20 standard [1], which are FC1, FC2, CP1, CP2, and Cz. The sampling frequency is 200 Hz. TP9 and TP10 are the reference electrodes and Fpz is the ground electrode. All electrodes are maintained the impedance of below 15 kΩ.

The input of the CV module is the video data captured by a camera linked with the laptop. In environment, there are a bottle, a light, a book, and a computer. After the CV module recognizing and decoding them as options of the CV-BCI, the user interface shows option "Bottle", option "Book", option "Light", and option "Computer" to subjects. The option "Bottle" that marked by a red cue is the current target.

The experimental procedure of the 2 BCIs starts from a cue. Within the 1st second, we use a cue to remind subjects the current target. Then BCIs show flash, which appears 120 ms and disappears 120 ms. Finally, BCIs show the classification results for 1000 ms.

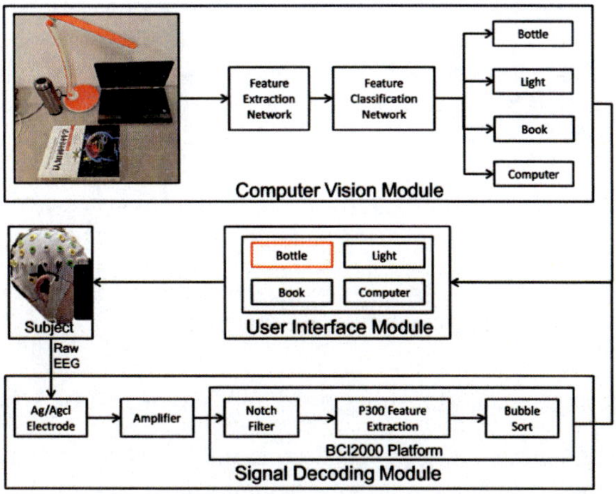

Fig. 1. Structure diagram of the CV-BCI.

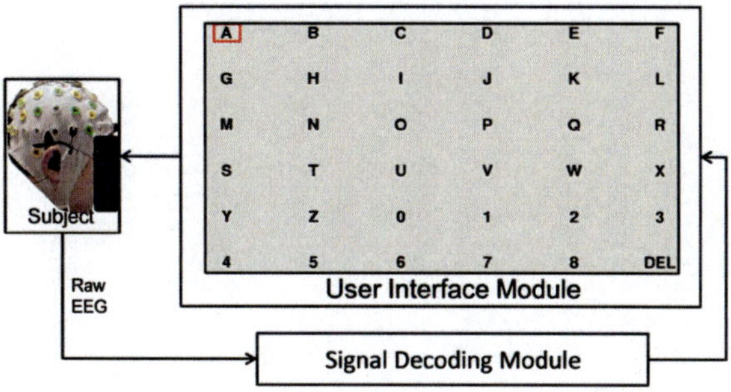

Fig. 2. Structure diagram of the BCI speller used in this study.

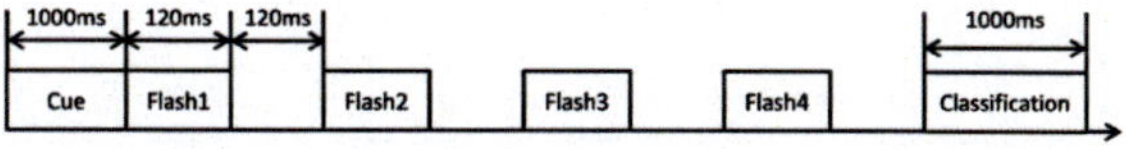

Fig. 3. Time course of the CV-BCI and BCI speller.

2.5 EEG Classification Algorithm

Every time each flash appears, we start to record EEG signal. We record 700 ms for each stimulus. Next, for reducing the size of data, we use a filter with the sampling ratio of one tenth to filter the EEG signals. Then, we utilize step-wise linear discriminant analysis (SWLDA) algorithm which is trained in the offline experiment to extract subjects' P300 feature. In the end, the bubble sort algorithm is used to find the most obvious feature. The option that has the most apparent feature is the classification result.

2.6 Performance Metrics

This study uses accuracy, ITR, and interaction speed as the performance metrics. Formula (1) is used to compute ITR, and P, N, and T are the accuracy, number of options, and the time cost for each choice.

$$\text{ITR} = \left(log_2 N + Plog_2 P + (1-P)log_2\left(\frac{1-P}{N-1}\right)\right)/T \quad (1)$$

3 Results

3.1 The Online Experimental Results with the CV-BCI and BCI Speller

Figure 4 shows the subjects' average signals with the CV-BCI and BCI speller respectively. Intuitively, the two signals are similar, with the peak value of about 1 μV. Meanwhile, the P300 response evoked by CV-BCI and BCI speller both comes at 300 ms. Table 1 shows the accuracies and ITRs of the 2 BCIs. The mean accuracy and ITR of the CV-BCI are 82.0% and 11.7 bits/min respectively, while the mean accuracy and ITR of the BCI speller are 81.6% and 11.6 bits/min respectively. There is no significant difference in the accuracy and ITR between the 2 BCIs.

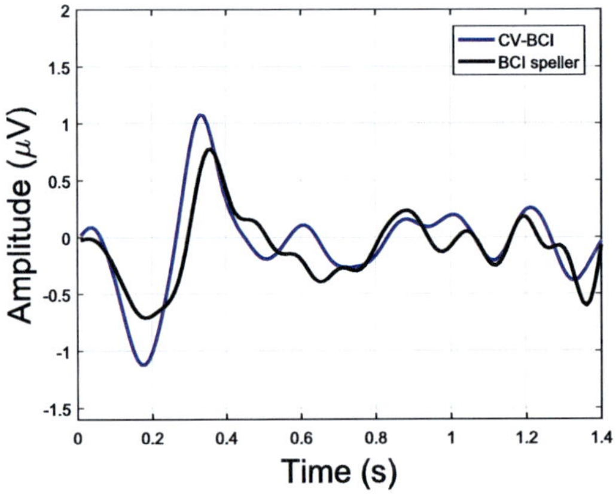

Fig. 4. Subjects' average signals recorded in the online experiment.

3.2 The Interaction Speed with the CV-BCI and BCI Speller

Figure 5 shows subjects' interaction speed with CV-BCI and BCI speller respectively. Subjects' interaction speed with the CV-BCI is about 0.15 objects per second (6 s per object), while the interaction speed with the BCI speller is less than 0.05 objects per second (20 s per object). Consequently, there is a significant difference in the interaction speed between the 2 BCIs.

Table 1. Overview of accuracy and ITR.

Subjects	Age	Accuracy		ITR(bits/min)	
		CV-BCI	BCI Speller	CV-BCI	BCI Speller
S1	26	76%	72%	9.5	8.3
S2	24	84%	84%	12.2	12.2
S3	22	88%	84%	13.7	12.2
S4	32	72%	72%	8.3	8.3
S5	23	76%	76%	9.5	9.5
S6	25	80%	80%	10.8	10.8
S7	24	76%	76%	9.5	9.5
S8	23	84%	84%	12.2	12.2
S9	22	92%	92%	15.4	15.4
S10	25	92%	96%	15.4	17.4
Mean ± SD	24.6 ± 2.9	82.0 ± 7.1%	81.6 ± 8.0%	11.7 ± 2.6	11.6 ± 3.0

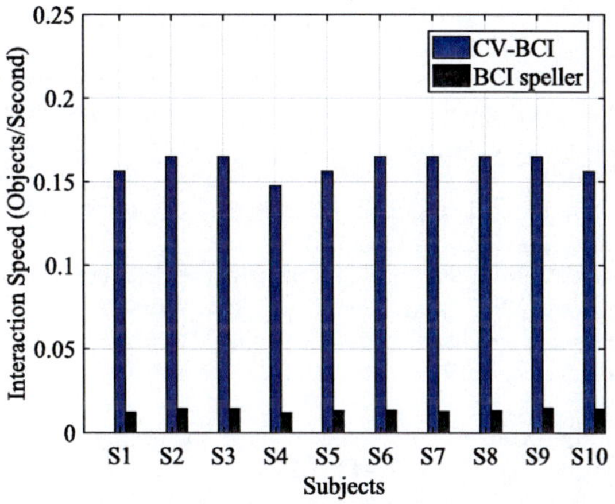

Fig. 5. Illustration for the subjects' interaction speed.

4 Discussion

As shown in the Table 1, the accuracy and ITR of the above 2 BCIs have no significant difference. However, Fig. 5 suggests that the interaction speed of CV-BCI is higher than that of BCI speller. Thereby, CV-BCI outperforms BCI speller in interaction speed while keeps the same accuracy and ITR. The reason is that CV optimizes the interaction processes of BCIs. In other words, the advantage of the proposed CV-BCI is that the users don't need to spell the target because CV automatically detects and decodes the target in the environment. In the traditional BCI speller, if subjects want to choose a target, they should spell out the target. In CV-BCI, CV decodes all objects in environment as options of CV-BCI, and then if subjects want to choose a target, they can directly choose it in the user interface.

The wearability of the proposed CV-BCI should be improved as studies point out that the primary purpose of BCI research is to facilitate people's daily life and make the society develop to a more convenient way [2, 7]. Studies indicate introducing augmented reality (AR) and head-mounted display (HMD) into BCI improves the wearability of BCIs [14]. In future, we intend to combine the proposed CV-BCI with a HMD so that users can wear it.

Acknowledgements. This work is supported by the National Key Research and Development Program (2018YFB1305101), and the National Natural Science Foundation of China (62006239).

References

1. Yin, E., et al.: A speedy hybrid BCI spelling approach combining P300 and SSVEP. IEEE Trans. Biomed. Eng. **61**(2), 473–483 (2014)
2. Yu, Y., et al.: Toward a hybrid BCI: self-paced operation of a P300-based speller by merging a motor imagery-based "brain switch" into a P300 spelling approach. Int. J. Human-Comput. **33**(8), 623–632 (2017)
3. Li, Y., et al.: A hybrid BCI system combining P300 and SSVEP and its application to wheelchair control. IEEE Trans. Biomed. Eng. **60**(11), 3156–3166 (2013)
4. He, S., et al.: EEG- and EOG-based asynchronous hybrid BCI: a system integrating a speller, a web browser, an e-mail client, and a file explorer. IEEE Trans. Neural Syst. Rehabil. Eng. **28**(2), 870–873 (2020)
5. Yin, E., et al.: A novel hybrid BCI speller based on the incorporation of SSVEP into the P300 paradigm. J. Neural Eng. **10**(2), 1–10 (2013)
6. Obeidat, Q.T., et al.: Introducing the edges paradigm: a P300 brain-computer interface for spelling written words. IEEE Trans. Hum.-Mach. Syst. **45**(6), 727–738 (2015)
7. Yu, Y., et al.: Self-paced operation of a wheelchair based on a hybrid brain-computer interface combining motor imagery and P300 potential. IEEE Trans. Neural Syst. Rehabil. Eng. **25**(12), 2516–2526 (2017)
8. Lenhardt, A., et al.: An adaptive P300-based online brain-computer interface. IEEE Trans. Neural Syst. Rehabil. Eng. **16**(2), 121–130 (2008)
9. Mcmullen, D., et al.: Demonstration of a semi-autonomous hybrid brain-machine interface using human intracranial EEG, eye tracking, and computer vision to control a robotic upper limb prosthetic. IEEE Trans. Neural Syst. Rehabil. Eng. **22**(4), 784–796 (2014)
10. Li, W., et al.: A human-vehicle collaborative simulated driving system based on hybrid brain-computer interfaces and computer vision. IEEE Trans. Cogn. Dev. Syst. **10**(3), 810–822 (2018)
11. TylerKabara, E., et al.: 201 brain-machine interface control of a robotic arm for object grasping is improved with computer-vision based shared control. Neurosurgery **62**(1), 233–245 (2015)
12. Redmon, J., et al.: YOLOv3: an incremental improvement. Presented at the IEEE Conference on Computer Vision and Pattern Recognition, Salt Lake City, Utah (2018)
13. Donchin, E., et al.: The mental prosthesis: assessing the speed of a P300-based brain-computer interface. IEEE Trans. Rehabil. Eng. **8**(2), 174–179 (2000)
14. Ke, Y., et al.: An online SSVEP-BCI system in an optical see-through augmented reality environment. J. Neural Eng. **17**(1), 1–15 (2020)

Data Analysis Based on the Processing Method of Displaying Reduced-Layer Dust Handprints

Xinhua Lv[1,2,3](✉) and Shengjiang Wang[4]

[1] Shandong University of Political Science and Law, Jinan, Shandong, China
lxh3002@xueshumail.cn
[2] Key Laboratory of Evidence-Identify in University of Shandong, Jinan, Shandong, China
[3] Forensic Science Academic Innovation Team for Young Teachers of Shandong, University of Political Science and Law, Jinan, Shandong, China
[4] China Maritime Police Academy, Ningbo, Zhejiang, China

Abstract. Because dust has the characteristics of chemical inertness, diffuse reflection, and particle properties, it has always been a problem in the field of handprint display to show reduced-layer dust handprints. This paper selected objects with different background colors and makes thinner and thicker reduced-layer dust handprints with different remaining time. Through experiments, the effects of the "502" glue fumigation method and the conventional display method were compared and analyzed. It is concluded that the "502" glue fumigation method is more effective than the conventional method, and is less affected by the background color of the object. It can obtain satisfactory results for fresh and old thin reduced-layer dust handprints.

Keywords: Diffuse reflection · Reduced-layer dust handprints · 502 glue fumigation method

1 Introduction

Due to the chemical inertness, diffuse reflection and particle nature of dust, the appearance and extraction of reduced-layer dust handprints has always been one of the problems in the field of handprint display. In order to find a better way to visualize the reduced-layer dust handprints, this paper sprayed some water mist on the reduced-layer dust handprints, used the volatility of "502" glue to show them, and achieved good results. At the same time, this visualization method is easy and reliable to operate, and is less affected by background color and handprint remaining time. The visualization result is easy to take pictures and fix, and it has a good promotion prospect.

2 Overview of Reduced-Layer Dust Handprints

2.1 The Concept and Types

The reduced-layer dust handprint refers to the handprints formed when a thin layer of dust on the surface of the object is transferred to the hand when the fingers are in contact

B. J. Jansen et al. (Eds.): *International Conference on Cognitive based Information Processing and Applications (CIPA 2021)*, LNDECT 85, pp. 135–143, 2022.
https://doi.org/10.1007/978-981-16-5854-9_17

with the surface of the object, and a part of the sweat remains on the surface of the object [1].

According to the amount of finger sweat, the reduced-layer dust handprints are mainly divided into the following types:

(1) When the amount of sweat is relatively abundant, while sticking away the dust on the surface of the object, part of the sweat can pass through the thinner dust and stay on the surface of the object, forming a mixture of dust and sweat on the surface of the object.
(2) When the amount of sweat is small, the sweat is imprinted on the ash, and the sweat on the hands will stick a part of the ash and large particles of dust away, causing the handprint lines to break points, resulting in poor continuity of the lines.
(3) When the amount of sweat is very small, the sweat on the mastoid lines of the fingers will be directly left on the dust dirt to form dust-sweat mixed handprints.

This paper mainly studies the first type, which is a thinner layer of dust handprints that can be left on the surface of the object after the dust is transferred to the hands.

2.2 Factors that Affect the Stability of the Reduced-Layer Dust Handprints

When the hand touches the object, the hand exerts a force on the object, and the opponent produces a reaction force [2]. Under these two forces, the surface morphology or attachment of the object changes, forming an impression that reflects the morphology and structure of the lines of the hand, that is, handprint. Therefore, the reduced layer dust handprint is formed under the action of four factors: hand, force, object that bearing the mark and dust, and together affect the quality of the handprint [3, 4].

3 The Principle and Operational Approach

3.1 Shortwave Ultraviolet Reflection Photography

3.1.1 Principle

Ultraviolet light is invisible light in the electromagnetic spectrum. Like visible light, it obeys the laws of linear propagation, reflection and refraction of light, and also obeys the principle of lens imaging. Ultraviolet rays have short wavelengths and large light energy [5]. They are easily absorbed by the atoms or molecules of the substance and converted into the internal energy of the substance. When short-wave ultraviolet rays irradiate the surface of an object with reduced-layer dust handprints, because the handprint lines and the background of the object have a difference in its reflection ability, the brightness distribution of the object in the ultraviolet region is recorded by the camera and film, and the display of reduced-layer dust handprints can be realized.

3.1.2 Operational Approach

Place the tested reduced-layer dust handprint sample in a dark room, fix the camera on the fixed frame above the sample, adjust the camera lens to an appropriate position, and then place the ultraviolet filter in front of the camera lens, adjust the ultraviolet excitation light source, and use the reflection of ultraviolet light to form an image on the camera film. (As is shown in Fig. 1).

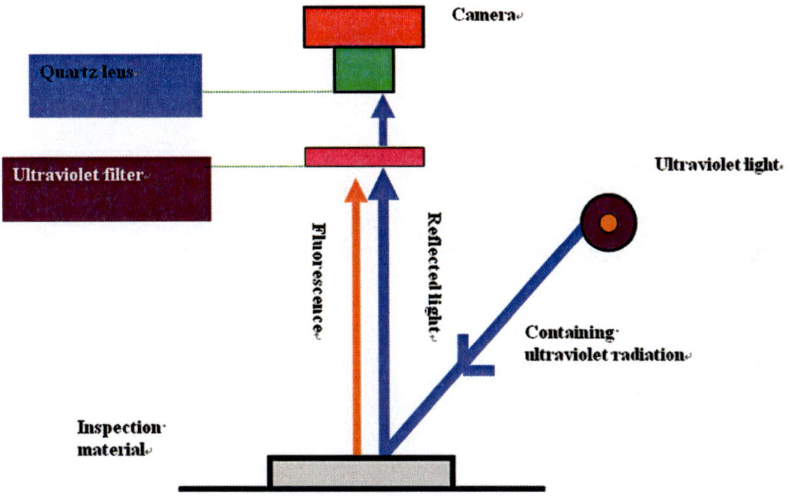

Fig. 1. Shortwave ultraviolet reflection photography

3.2 Photo Paper Transfer Dyeing Method

3.2.1 Principle

Plastic-coated photo paper is mainly composed of gelatin protein, sensitizing agent, and paper base [6]. After the action of water, the photo paper swells and makes the surface sticky. The dust handprint is transferred by the gelatin layer of the photo paper and is fixed by the gelatin protein. Meanwhile, inorganic salt has a charge. After the colored dye is dispersed with a cotton ball, it has a greater surface polymerization performance. During repeated wiping, the surface charge of the dust particles makes the particulate dye quickly adsorb and aggregate on the dust surface to make it color. The paper surface forms a large contrast, which can effectively stick handprints.

3.2.2 Operational Approach

After soaking the cotton ball in water and squeezing out the excess water, wipe the adhesive layer of the photo paper evenly to make the surface protein absorb water swelling and sticky, cover it on the handprint of the reduced layer of dust and press evenly to extract, pay attention to uniform force, avoid displacement, and make the handprint transfer to the photo paper, and wait for the moisture to evaporate for later use. Then crush a small amount of red dye (or carbon powder) on the paper, dipped it with a cotton ball, and wipe it lightly and repeatedly on the reduced-layer dust handprint of the photo paper to attach a small amount of red dye to make the handprint red.

3.3 Cobalt Chloride Fumigation Method

3.3.1 Principle

When heating the lactic acid and ethyl lactate solution containing cobalt chloride, the cobalt chloride crystal loses the crystal water, and the water molecules prevent the process of dehydration of lactic acid to form glue [7]. The ethyl lactate reduces the concentration of lactic acid vapor. When the steam is absorbed by reduced-layer dust handprint, the lactic acid accumulates on the handprint as the ethyl lactate volatilizes. The lactic acid increases the viscosity of the dust particles and the adhesion between the dust particles and the carrier. At the same time, the dust particles have more gaps and the large surface area, which allow it to absorb a large amount of fingerprint powder, thereby showing the dust handprint or the handprint mixed with dust and sweat.

3.3.2 Operational Approach

Add 1 g of cobalt chloride crystals, 2 drops of lactic acid, and 3 drops of ethyl lactate solution to the evaporating dish and mix them thoroughly. Heat the above solution until white smoke appears, and place the object with the fingerprints of reduced layer dust on the evaporating dish. Smoke the object 1cm above for 2 s, remove the object for observation. When there is no effusion on the surface, select fingerprint powder or fluorescent powder that has a large contrast with the object and observe and brush under the ultraviolet light to show the reduced-layer dust fingerprints.

3.4 Bromophenol Blue Method

3.4.1 Principle

The main components in the dust include $CaCO_3$ and $Si0_2$. Bromophenol blue reacts with $CaCO_3$ in the dust, causing the bromophenol blue to change from yellow to blue. After the reaction, the concentration of hydroxide ions increases and the pH value increases to reach the discoloration range of bromophenol blue, thereby revealing the handprint [8].

3.4.2 Operational Approach

Take the light yellow 1.5% bromophenol blue methanol solution, pour it into the sprayer, pinch the rubber ball on the sprayer, and spray the solution evenly on the object carrying the handprint sample to show the reduced-layer dust handprint.

3.5 "502" Glue Fumigation Method

3.5.1 Principle

"502" is a binder with ethyl cyanoacrylate as the main body and a small amount of hydroquinone and sulfur dioxide as a polymerization inhibitor. Due to the presence of strong electron-withdrawing groups, cyano and ester groups, ethyl cyanoacrylate monomers can easily undergo anionic polymerization under the initiation of water or weak bases. Human sweat contains water and amino acids. The water contains a small amount of hydroxide ions. After the ethyl cyanoacrylate monomer is volatilized, it will polymerize under the initiation of water and amino acids. Therefore, after the ethyl cyanoacrylate volatilizes, if it encounters the sweat-stained part of the surface of the object, it will initiate the polymerization of the ethyl cyanoacrylate monomer to form a solid polymer, which can show handprints.

3.5.2 Operational Approach

In order to make the 502 glue have a good display effect on old reduced-layer dust handprints, a special AFC-II type "502" glue display cabinet that can control the environmental humidity to 80c/o for the display of the smoke display was used. The smoke display cabinet is equipped with a special AFC-II "502" plastic smoke display cabinet. The humidity sensor can automatically control the humidity in the cabinet and effectively make the environmental humidity reach the preset target.

The operation steps are as follows: Hang the object to be smoked in the smoke display cabinet, and then drop the "502" glue into the electric heating tank, turn on the power, the "502" glue quickly vaporized into white smoke, after a few minutes, the handprint was appeared. Some handprints took half an hour or even an hour to show up. For example, the old reduced-layer dust handprints should be humidified to make the humidity in the cabinet reach about 80%, and the heating time should be longer.

Using the "502" glue fumigation display cabinet to fumigate the fingerprints is convenient and easy to operate, and the "502" glue fumigation display cabinet has a humidification function, which can significantly increase the success rate of the display, but this method is more empirical, and the amount of "502" glue should be accurately grasped. It should be determined according to the size of the fumigating utensil used,

Table 1. "502" glue dosage reference table

Container volume (cm³) \ "502" glue (g) \ Approach	Heating and fumigation with strong base catalysis	Direct heating and fumigation
15×10×8	0.2~0.4	0.1
20×15×10	0.5~0.8	0.15
30×20×15	0.7~1	0.3
30×30×20	2~2.5	0.5
50×40×30	4~6	1

otherwise it is easy to make the displayed fingerprints too thick. (The reference table for the amount of "502" glue is shown below, As is shown in Table 1).

4 Experiment Results

The shortwave ultraviolet reflection photography method, photo paper transfer dyeing method, cobalt chloride fumigation method, bromophenol blue method, "502" glue fumigation method and other methods are experimented under the same conditions, and their respective effects are shown Table 2, Table 3, Table 4.

Table 2. Comparison of the effects of various display methods on different objects

Object / Effect / Approach	White tile	Black tile	Transparent glass	Paint wooden board	Black rubber
shortwave ultraviolet reflection photography method	+++	++	+++	++	++
photo paper transfer dyeing method	+++	+++	+++	++	+
cobalt chloride fumigation method	+++	+++	+++	++	+
bromophenol blue method	+++	+++	+++	+++	++
"502" glue fumigation method	+++	++++	++++	+++	++++

Note: "++++" means that the lines are the most clear and coherent, the characteristics are reflected the best, and there is a good background contrast; "+++" means that the lines are clear and coherent, and there is a good background contrast; "++" means that the lines are not very clear and relatively coherent, with a certain background contrast; "+" indicates that the lines are unclear, relatively coherent, and have a certain background contrast; "-" indicates that the lines are unclear, inconsistent, and have a certain background contrast.

A comparative analysis of the display effects of the above various methods through experiments shows that the "502" glue fumigation method is more effective than other display methods. The display effect is shown in Fig. 2, Fig. 3, Fig. 4 and Fig. 5:

Table 3. Comparison of the display effects of various display methods on the reduced-layer dust handprints with different dust thicknesses

Effect / Dust thickness / Approach	Thin	Normal	Thick
shortwave ultraviolet reflection photography method	++	++	+++
photo paper transfer dyeing method	++++	+++	++
cobalt chloride fumigation method	+++	+++	—
bromophenol blue method	++++	+++	++
"502" glue fumigation method	++++	++++	+++

Note: Same as Table 1

Table 4. Comparison of the display effects of various display methods on the reduced-layer dust handprints with different remaining time

Effect / Time / Approach	Fresh	Old
shortwave ultraviolet reflection photography method	+++	++
photo paper transfer dyeing method	++++	+
cobalt chloride fumigation method	++++	+
bromophenol blue method	++++	++
"502" glue fumigation method	++++	+++

Note: Same as Table 1

Before After

Fig. 2. The effect of a thin reduced-layer dust handprint on a black tile

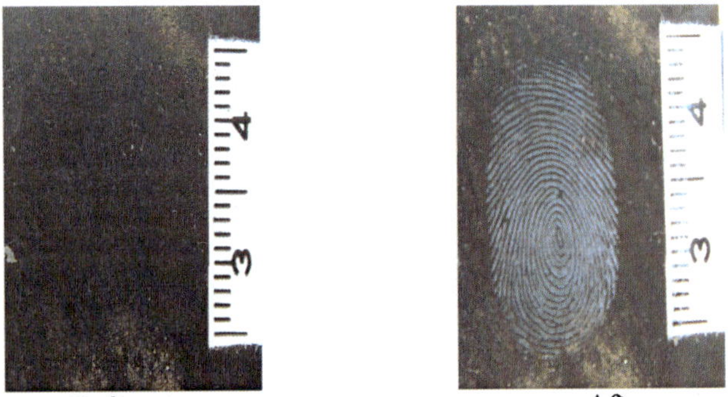

Before After

Fig. 3. The effect of a thin reduced-layer dust handprint on a black rubber

Before After

Fig. 4. The effect of a thick reduced-layer dust handprint on a black tile

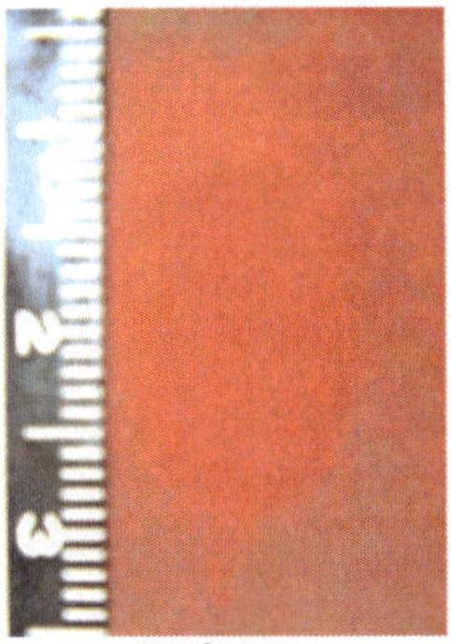

Before

After

Fig. 5. The effect of a thick reduced-layer dust handprint on a red tile

5 Conclusion

Experiments show that the "502" glue fumigation method is suitable for most non-permeable objects, and the appearance effect is less affected by the background color of the object, and the displayed handprint lines are clear and stable. This method can also achieve a satisfactory display effect on old handprints that have been left for a long time. For the two types of reduced-layer dust handprints with thin dust and thick dust, the hand lines appearing on the object with thin dust are clear and the characteristics are obvious. Although the effect on the thick object is not as obvious as the former, the method also has a certain effect comparing with other ways.

References

1. Guo, J., Dong, P.: Talking about the extraction of dust traces on site. Foren. Sci. Technol. **03**, 47–48 (2000)
2. Wang, S.: 502 glue fumigation technology of sweat latent handprint on plastic surface. J. Henan Public Secur. Acad. **20**(01), 124 (2011)
3. Zhu, D., Wen, S., et al.: Research on humidification device with fingerprints on site. Sci. Technol. Innov. **10**, 94–99 (2021)
4. Li, D., Dang, X.: Open up a new field of sweaty handprints display. J. Crim. Invest. Pol. Univ. China **04**, 11–14 (2002)
5. Wang, M., Shen, D., Zhu, Z., et al.: Quantifying contrast of latent handprints developed by fluorescent nanomaterials based on spectral analysis. Talanta **13**, 231 (2021)
6. Wang, M., Shen, D., Zhu, Z., et al.: Dual-mode fluorescent development of latent handprints using NaYbF4: Tm upconversion nanomaterials. Mater. Today Adv. **22**, 87–91 (2020)
7. Zhu, M., Shunping, W.: On the extraction of handprints at the crime scene. Legal. Vis. **04**, 154 (2017)
8. Huang, L.: Research on the extraction method of fingerprints at the crime scene. Legal. Vis. **06**, 110–111 (2015)

Development Status of the Integration of Basic Art Skills and Computer Art Design Under the Background of Information Age

Bei Gong[1] and Ling Fu[2(✉)]

[1] Gongqing College of Nanchang University, Nanchang, Jiangxi, China
[2] Gongqingcheng College of Nanchang University, Nanchang, Jiangxi, China

Abstract. At present, information technology is widely used, which has a strong impact on the traditional art. Before this, many researchers have not systematically studied the operation of computer software, completely ignoring that computer software can be used to implement based on art skills. In particular, the rapid development of computer art design industry, it appears that the basic skills of art in art design is important. Therefore, this paper puts forward the research on the development of the integration of basic art skills and computer art design under the background of the information age. This paper starts from the basic content of computer art design and analyzes the importance of solid basic art skills to computer art design. Based on the background of the information age, this paper combines the basic skills of fine arts and computer art design from the perspective of mutual integration, that is, the traditional artistic consciousness, and sublimates the artistic expression form on the original basis. The results show that the effect of the integration of the two has a qualitative leap compared with the traditional art forms and is deeply loved by the public.

Keywords: Information age · Computer art · Basic art skills · Traditional art · Art design · Integration

1 Introduction

With the rapid development and popularization of computer and mobile Internet technology under the background of information age [1, 2], human beings have rapidly entered the network information age in half a century. With the development of the times, social activities and mode of production have changed, and people's lifestyle and communication mode have also changed quietly. With the rapid development of domestic Internet environment [3], China is in the forefront of this change. With the development of mobile Internet and the emergence of various intelligent communication devices [4], people's communication has broken through the barriers of time and space and is no longer limited by the region and objects. The scope of communication has been expanded to realize all-round information sharing. Museums and exhibitions also opened up a new era of social media, such as micro-blog [5], WeChat official account [6], etc., and publicize

B. J. Jansen et al. (Eds.): *International Conference on Cognitive based Information Processing and Applications (CIPA 2021)*, LNDECT 85, pp. 144–150, 2022.
https://doi.org/10.1007/978-981-16-5854-9_18

exhibitions and activities in various ways. People just need to turn on their mobile phones to learn about the exhibition and related information from various media.

More and more widely used computer art design [7, 8] in advertising, film, television, interior design and other fields, the demand of computer art in society, more and more designers, at this moment, there are more and more profound provisions on the technical ability and education level of human resources in every aspect. However, the actual situation of our country at this stage is that the training level of computer art design talents is lack of basic art skills [9, 10], which is far behind the demand level of talents, and the supply is in short supply. At present, the computer art design industry in technical education schools can be basically divided into two kinds of personnel training methods: the first is the traditional mode, which basically imitates the methods of general higher education and subject knowledge education is systematic, but basically has no characteristics of vocational education; The second is the initial mode of technical school, which emphasizes the computer production skills and ignores the education and cultivation of design concepts and principles. The works design lacks soul, is pale, and has no sustainable development ability and competitiveness.

Based on the above situation, this paper proposes to integrate the basic art skills and computer art design to achieve the perfect presentation of technology. Without a solid foundation of fine arts, excellent works cannot be made. At the same time, with the help of computer software, the color matching, shape design and aesthetic ability of designers' works can be improved. This paper firstly reviews the status quo of traditional art design; secondly, it expounds the art background and computer art design under the background of information age; then designs the overall experimental ideas to verify the feasibility of the integration; finally, the results show that the fusion method can effectively improve the quality of works.

2 Overview of Basic Art Skills and Computer Art Design Under the Background of Information Age

2.1 Development Status of Art Education in the Information Age

The development of information technology and the application of remote communication have become the cornerstone of the development of information society, and to a large extent, promote social change. With the development of information and communication technology, the economic and social fields are more and more closely linked, the boundaries between different countries and regions are broken, and the trend of internationalization and globalization is more and more obvious. The rapid dissemination of information has become a kind of social capital that cannot be ignored. This new type of social capital promotes the social and economic reform, and also has a significant impact on the development of art education. Under the influence of the development of information society, great changes have taken place in the educational concept of art education. Teaching methods are different from those in the past. The roles of teachers and students have changed accordingly. Therefore, the art basic skill teaching course can be carried out smoothly.

2.2 Computer Art Design

Computer art design is to use computer as a tool for art design. It is the product of the crystallization of art and modern computer science and technology. Compared with the traditional design and modeling means, computer art design is more suitable for the needs of modern society. With the rise of computer art and design, great changes have taken place in the means of design modeling, the form of organization and the concept of design and production. Advanced computer design and production system and input and output system have the characteristics of various innovative technologies and high efficiency, fast and convenient. It can not only reduce the transmission deviation and loss of traditional design effect to the minimum, but also can modify the previous steps in the production process at any time, and further extend the designer's creative consciousness based on this. With the help of computer, after the Designer completes the basic idea, he can use the computer software to carry on the entity modeling, so that the designer can edit and modify the graphics in real time according to any characteristics of the virtual object, and the operation is very simple and convenient. More directly reflect the designer's creative intention and bring unprecedented ease of use. At the same time, the digital preassembly is used for comprehensive evaluation.

3 Importance of the Integration of Basic Art Skills and Computer Art Design

3.1 Experimental Ideas

We specifically for the mentioned needs, the local computer art designers to conduct a questionnaire survey, mainly asked about the percentage of various aspects of design. The author also visited the experts, summarized and analyzed the feasibility of the integration of basic art skills and computer art design. We selected 50 students with basic art skills as the art group, 50 students without art background but skilled in using various drawing software as the software group, and 50 students with solid art foundation and proficient in using various drawing software as the comprehensive group. Give them random questions, ask other students to vote and ask experts to rate them. The results show that the expert score and the number of students' votes in the comprehensive group are higher than those in other groups. Verify the current situation and importance of the integration of basic art skills and computer art design in the background of the information age.

3.2 Experimental Design

The birth of computer art design perfectly interprets the advantages of art works, and the pictures divided by color structure make people bright. Starting from the public, we investigate the proportion of people with basic art skills and computer art design ability in the current society and find that the number of people with both basic skills and computer art design ability is relatively small. Shown as Table 1.

Table 1. Survey results

Skill direction	Number of people (%)
Have basic art skills	45
Have computer art design operation ability	53
Namely has the fine arts basic skill and the computer art design operation ability	25

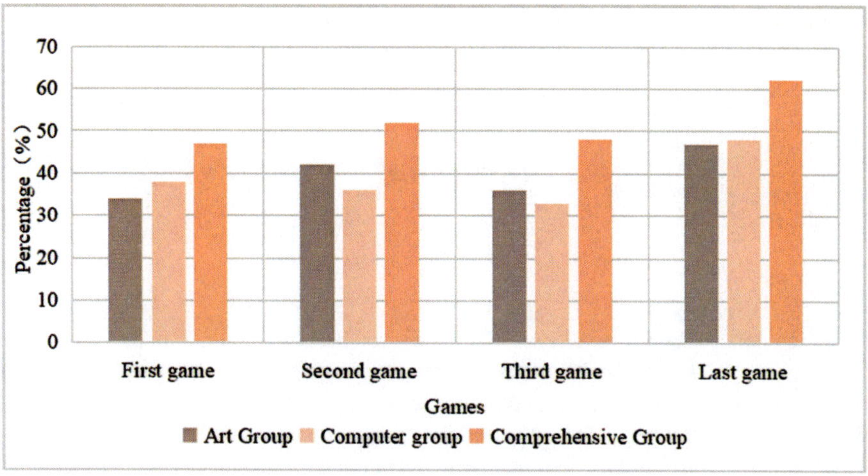

Fig. 1. Number of votes in four matches

4 Discussion

4.1 Number of Votes for Works of Three Groups

Shown as Fig. 1, in the four competitions, the overall number of votes in the comprehensive group was higher than that in the art group and the computer group. Of course, the birth of a work is through the means of art, first conception, hand-painted, and finally the use of computer presentation. Hand drawing and computer shaping are just a design process. That is to say, things are not over, the end is the final shaping. All these require the designer to have good basic skills of art. Even if you have a superb computer software operation ability, it is difficult to achieve artistic creation without profound art skills. On the contrary, the basic skills of solid art in the information age today, if not displayed by computer art, cannot keep up with the trend of the times, cannot meet the needs of the public, the art creation is still incomplete.

4.2 Expert Satisfaction Survey

Shown as Fig. 2, through this competition, we asked professionals to conduct a satisfaction survey on the works of the three groups, and 44% of the experts highly agreed with

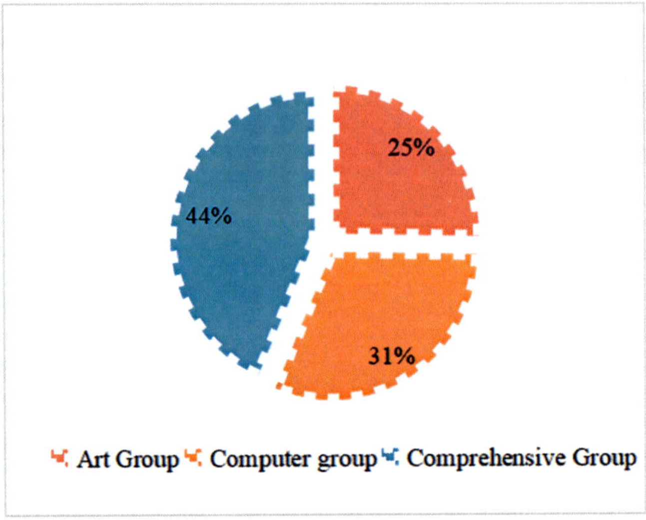

Fig. 2. Expert satisfaction survey results

the design concept and formed works of the comprehensive group. Among them, only 25% of them are satisfied with the works of students with solid art background. And 31% of experts are satisfied with computer art production. Although they have no art foundation, the effect presented by computer software is more vivid than hand drawing. With the development of information age, people prefer works that can keep up with the times. Hand drawing is also an indispensable part of the traditional concept, but in the future file storage, relying on technology is a better choice, and it is easier to pass on the current artistic ideas.

4.3 Wrong Cognition of Computer Art Design

In the era of rapid development of science and technology, it is also an information age. Everyone is experiencing the explosion of knowledge, but also facing a huge amount of information. All kinds of media are full of too many homogeneous advertisements. A lot of low-quality advertising numbs our brains. In this tired situation, people want to see the advertising design with artistic appreciation value. Artistic beauty is needed to buffer the tense psychological state; only in a relaxed and positive state can people have the energy to accept different information. Now there are different levels of designers in society, they also serve different levels of customers. This mainly depends on the designer's value orientation. It's easy to design just to make money. Because it only needs to satisfy the customer, it has nothing to do with it. Many people engaged in design have no artistic foundation or poor foundation. They think that it is too easy and convenient to design with computers, which has formed a social phenomenon.

In recent years, under the impact of market economy, the establishment of art major in Colleges and universities has become a hot spot, and the number of students has increased sharply. In addition to some students with professional knowledge, most of them went to the art department for various reasons. They only take part in short-term training classes in a few holidays before the examination and learn the examination contents such as sketch and color in a surprise way. Easily admitted to art colleges and universities, the enrollment expansion of art major has almost become "indiscriminate", the professional quality of students can be imagined. In learning, although students are very proficient in the operation of computer design software, they often lack new ideas in their homework, and cannot express their own ideas. Maybe they have no ideas and no sense of creativity. There seems to be a lack of motivation in the training of basic theories and basic skills, so it is difficult to achieve results. It is a common fault of students to attach importance to application and ignore foundation.

4.4 Importance of the Integration of Solid Basic Art Skills and Computer Art Design

As an important element in art design, color application plays an important role in the design process of computer art works. If designers ignore the use of color elements in the process of art design, it is likely to lead to the loss of artistic quality of the whole work. Therefore, designers in the design of computer art works, should pay attention to and strengthen the coordinated use of color basis, ensure the use and rationality of color elements in each link, so as to highlight the life and artistic value of aesthetic works of art. At the same time, in order to do a good job in the use of color works, designers put forward higher requirements for the mastery and application of color aesthetic knowledge. The collocation of different color elements is not only a simple combination of color and color, but also an important embodiment of color applied art aesthetics in art works.

As an indispensable part of art design, the completion process of most designers' composition ideas is also the process of the formation and production of art works. For computer art designers need to be able to skillfully use a variety of computer software design space, skilled use of various symbols to complete the composition of art works, in order to avoid unreasonable composition design layout and composition of works, the visual effect of the works of great deviation, thus greatly reducing the overall artistic effect and quality of computer art works. In the actual design of computer art works, designers need to master the basic knowledge and technology of computer composition and master the application of different symbols and images in the design of computer art works. Scientifically and reasonably grasp the overall layout of art works, so as to ensure the balance and coordination of art design composition, and finally realize the perfect combination of plane sense, color sense and three-dimensional sense of art works.

In the design process of computer art works, designers need to show the image features. In the form of mental computer sketch, through keen observation, they can accurately grasp the structure and color of the basic characteristics of each design object, so as to realize the overall aesthetic art works in all aspects. At the same time, in the process of sketch basic literacy training, designers need to pay attention to and strengthen the ability of planting and promoting image perception and spatial imagination thinking, so as to lay a good foundation for basic sketch skills and design better aesthetic art design works.

5 Conclusion

With the rapid development of information technology, computer art design is not only limited to how to skillfully use various software to show, but also needs a solid basic art skill. That is, it does not lose the traditional artistic style, but also shows it in a new form, which makes the viewer feel bright. Through the competition, we found that the public are inclined to the design concept of integrating computer art design and basic art skills. Even experts pay more attention to the use of computer software to achieve artistic creation, which requires solid basic skills. Thus, under the background of information technology, the concept of combining basic art skills with computer art design has a good application prospect.

References

1. Busis, N.A.: Neurology in the electronic information age. Eur. J. Neurol. **6**(4), 385–414 (2015)
2. Antman, E.M.: Saving and improving lives in the information age: presidential address at the american heart association 2014 scientific sessions. Circulation **131**(25), 2238–2242 (2015)
3. Heo, J., Chun, S., Lee, S., Lee, K.H., Kim, J.: Internet use and well-being in older adults. Cyberpsychol. Behav. Soc. Netw. **18**(5), 268–272 (2015)
4. Gong, J.P., Kagata, G., Osada, Y.: Intelligent gel– surface properties and functions of gels–. Macromol. Symposia **159**(1), 215–220 (2015)
5. Wang, C.H., Han, D.: Sentiment analysis of micro-blog integrated on explicit semantic analysis method. Wirel. Pers. Commun. **102**(1079), 1–11 (2018). https://doi.org/10.1007/s11277-017-5144-9
6. Gan, C.: Understanding wechat users' liking behavior: an empirical study in China. Comput. Hum. Behav. **68**(MAR.), 30–39 (2017)
7. Liang, W.: Scene art design based on human-computer interaction and multimedia information system: an interactive perspective. Multimed. Tools Appl. **78**(4), 4767–4785 (2019). https://doi.org/10.1007/s11042-018-7070-6
8. Lenz, M.: Early argentine computer art at the Victoria & Albert museum. J. Design Hist. **31**(2), 154–166 (2018)
9. Marmor, M.: Fine Arts: A Bibliographic Guide to Basic Reference Works, Histories, and Handbooks. Ehresmann Donald l. Englewood, Colorado: libraries unlimited (1990). 3rd ed. xviii, 374 p. ISBN 0-87287-640-3 lc 90-34857 $49.50 ($59.50 outside America). Advances in Applied Clifford Algebras, 24(2), 355–382
10. Can, E., Üzmez, Ö.Ö., Döğeroğlu, T., Gaga, E.O.: Indoor air quality assessment in painting and printmaking department of a fine arts faculty building. Atmos. Pollut. Res. **6**(6), 1035–1045 (2015)

On the Development Status of Student Football Referees' Video Camera Classification in the Development of Campus Football

Jinyu Liu(✉)

Xi'an Physical Education University, Xi'an, Shaanxi, China

Abstract. With the increase in the intensity of modern football, football players need to engage in fierce physical confrontation on the field, which leads to frequent illegal actions in football matches. In this case, it is necessary to use video tape as an auxiliary means to help the referee make accurate and reasonable judgments in the shortest possible time. The purpose of this article is to study the development status of student football referees using video lens classification in the development of campus football, and on this basis, to study the training strategies and development status of student football referees. This article uses a questionnaire survey method to randomly select 10 local colleges and universities to investigate student football referees. At the same time, a large amount of literature was collected, the relevant data was sorted, and the experimental conclusions were finally drawn. Experimental data shows that among the 288 people surveyed, 253 people prefer football referees, accounting for 88% of the total. This is a high percentage, indicating that students have a strong interest in football referees. The experimental results show that 88 people choose to be very interested, accounting for 31% of the total; 165 people choose to be more interested, accounting for 57% of the total; 23 people choose to be uncertain, accounting for 8% of the total, 12 Individuals chose not to be interested, accounting for 4%, and no one chose not to be interested. Therefore, in the future, it is necessary to continuously improve and enhance students' interest and effect in football referee learning, and provide correct guidance to promote them to form a good learning motivation.

Keywords: Campus football development · Student football referees · Video footage classification · Development status

1 Introduction

Football referees are a key part of football. For the football field, the judgment ability of the football referee determines whether the game can be carried out smoothly and orderly, and is the "judge" of the football field. It is a huge and complicated system engineering to improve the judgment ability of football referees. It is necessary to use some auxiliary methods to simplify this process. Colleges and universities are one of the channels for training and exporting football referees. The quality of their training

B. J. Jansen et al. (Eds.): *International Conference on Cognitive based Information Processing and Applications (CIPA 2021)*, LNDECT 85, pp. 151–157, 2022.
https://doi.org/10.1007/978-981-16-5854-9_19

determines the quality of our football referee team to a certain extent [1, 2]. Therefore, the status quo of football refereeing ability training in colleges and universities has been investigated and researched to understand the defects in the training process of football refereeing ability in colleges and universities, and some reference suggestions for the development of football in my country have been put forward, with a view to the development of football in our country. Contribute [3, 4].

C et al. studied how football referees introduced extended training to improve teamwork and increase referees' sense of cooperation. Extended sports is a simple and safe training method. After their research, they found that the team of referees and assistant referees after long-term training, the cooperation is found to be implicitly coordinated [5]. Sauermann and others discussed the combination of each running method and specific punishment. For students majoring in physical education, they need to feel each running method in actual sports training. While pursuing running skills and methods, they will not affect the competition. On the premise, get as close to the ball as possible so that the penalty is more convincing [6]. In order to train grassroots football referees, it is necessary to emphasize the training of the basic skills of referees, cultivate excellent running ability, and master various running skills and methods.

At this stage in our country, due to the obvious deficiencies in actual conditions, theoretical foundations, and school curriculum settings, the overall level of football refereeing for college students in our country is low, which will inevitably affect the level of football refereeing in my country in the future [2, 4]. Therefore, the relevant departments attach great importance to the training and attention of college football referees, provide strong support in terms of people, materials, finances, and spirit, and aim at lifelong education. Therefore, we need to initiate and strengthen the education of college students for the whole society. The training of football referees has contributed to the revitalization of football and the prosperity of the football industry in our country [5, 6].

2 Research on the Current Situation of the Development of Student Football Referees' Video Camera Classification in the Development of Campus Football

2.1 The Focus of Training College Students' Football Refereeing Ability

(1) Accuracy
The smooth progress of a football game depends on the correct judgment and effective control of the referees. The judgment and intensity of football referees directly affect the progress and results of the game. If the judgment of the referee is accurate and conforms to the law enforcement rules, it will contribute to the professional ethics and fighting spirit of the athletes. At the same time, training helps to clean up the atmosphere of the arena, enable athletes to maximize their skills and tactics, and make the game more interesting. In terms of technical operations, college football referees need to accurately distinguish between reasonable actions and foul actions, and control the rhythm of the football game. This is the most basic requirement of referees. What college students lack the most is not to grasp

the accuracy of decision-making in theory, but to lack practical experience in law enforcement against referees in large-scale competitions. From this perspective, students need to participate in large-scale competitions to create more participation opportunities to truly improve their abilities.

(2) Consistency
In order to ensure the continuity and viewing of the game, referees are more likely to use favorable conditions to adjust the actual penalty under the same or similar circumstances. Under similar circumstances, there is no unchanging ruling. However, from the perspective of some experts, referees' consistency with the standards in the game is more important than the accuracy of enforcement.

(3) Fairness
If students want to be real football referees, they need to use the rules as a standard based on facts. First of all, it must be fair and reasonable. On the basis of equality between the two parties, punishment is practical and feasible in the law enforcement process. And apart from athletes and competition regulations, it cannot be disturbed by other factors. Second, the principles of openness and transparency must be reflected in the law enforcement process, so that every whistle of all referees can be checked by fans, players and coaches. This is a stricter standard for football referees on campus.

2.2 Countermeasures for Training College Students' Football Refereeing Ability

(1) Establish a Scientific and Systematic Training System
The current referee training system in colleges and universities is extensive, chaotic, and complex, with few teaching hours and limited opportunities for practical exercises, which greatly hinders colleges and universities from improving their football refereeing capabilities and training their football referee teams. Although some universities focus on developing student refereeing skills from a macro level, they still need to change the shortcomings of college student refereeing ability in the most fundamental way. Relying on the efforts of a few teachers cannot change the overall situation. Therefore, to improve and construct the referee training system, the training system must ensure the depth and breadth of the students' football refereeing theory learning, ensure the quantity and quality of student refereeing practice, set up special refereeing courses and conduct regular assessments as much as possible, and cooperate with the Football Association to strengthen football referees. Transport quality, improve the seriousness of the grade examination, and strictly abide by the declaration and approval system required by relevant regulations.

(2) Increase the importance of football referee learning
Improving the school's emphasis on football referee learning is an important way to improve the quality of referee training. Referee teaching is not only the dissemination of theoretical knowledge, but also the training of practical skills. The improvement of the judging level of any sport depends on for practice, it is far from enough to rely on the learning of theoretical knowledge in the classroom. The cultivation of practical operation ability can only be consolidated and improved in the repeated practice process. Therefore, it is necessary to reasonably arrange the proportion of class hours based on the actual and practical conditions of football

referees, not only to enable students to master the theoretical knowledge system of football referees, but also to give them more practice and opportunities to experience the practice of refereeing in practical classes. Therefore, the school should pay enough attention to the football practice class and even the refereeing class schedule to provide guarantee for the learning of football refereeing practice.

(3) Improve learning interest and learning effect
Interest is the best source of motivation for learning. Cultivating students' interest in learning is an important way to improve the quality of football referee training. A step-by-step learning method from easy to difficult, simple to complex, and shallow to deep is adopted to gradually improve students' football presence Judgment ability, and timely point out the advantages and disadvantages of the judgment process, focus on the analysis of the mistakes that students are more likely to make, find out the crux of the problem, and propose solutions to the problem, so that all classmates and colleagues can gain more experience, And then improve the teaching effect of football referee classroom. Lead students in a reasonable way, let them have a clear goal and motivation for the training of football referees, and guide them to improve their bad learning concepts.

(4) Improving the psychological quality of the students' judgment
Football refereeing has very strict requirements on referees' will and quality and psychological quality. Intense and fierce football games can easily lead to psychological changes of referees, which is a challenge to football referees. In addition, referees have to face pressure from players, coaches, spectators and even the media. Fair law enforcement is the core of the professional ethics of referees in any sports event, and fairness is the fundamental guarantee for football referees to do their job well. In the general environment of the development of football in my country, having a good psychological quality is a prerequisite for a football referee to gain the trust and support of athletes and coaches. On the basis of increasing on-the-spot experience, improve self-control and anti-interference ability and cultivate decisiveness Don't carry huge psychological pressure to judge, otherwise it will cause misjudgment or missed judgment, and then form a continuity of psychological pressure, which will form bad habits in the long run. Therefore, student referees must continue to have targeted exercises. Only in this way can they face the work of football referees with full spirit, high attention and a relaxed attitude.

3 Research on the Current Situation of the Development of Student Football Referees Using Video Lens Classification in the Development of Campus Football

3.1 Questionnaire Survey Method

Randomly conducted a questionnaire survey on teachers from 10 local university campuses. A total of 160 questionnaires were distributed, 16 for each university, and 160 questionnaires were returned, with a recovery rate of 100%, of which 160 were valid questionnaires, with an effective rate of 100%; each school Students from the School of Physical Education randomly selected 80 physical education students (320 in total), issued 320 questionnaires, and returned 300 copies, with a recovery rate of 93.75%, and 288 valid questionnaires, with an efficiency of 90%.

3.2 Mathematical Statistics

Use Excel to sort the collected data, and use SPSS statistical software to perform statistical analysis on the sorted data.

4 Discussion on the Development of Campus Football Development of Student Football Referees Using Video Lens Classification Research and Experimental Analysis

The importance of football referee teaching and training in schools can greatly stimulate teachers' work and students' learning enthusiasm, so as to promote the rapid improvement of teaching quality and even training quality. Through the investigation of the training situation of football referees in Colleges and universities, it is found that most colleges and universities attach importance to the cultivation of football referees, of which 20% do not attach importance to the cultivation of football referees, and 4 schools choose to attach importance to and attach great importance to, accounting for 40% of the surveyed schools. It can be seen that in the process of student training, there is still the phenomenon of ignoring the cultivation of students' referee ability, which seriously restricts the overall training quality of students. Relevant schools or departments should attach great importance to it. Table 1 and Fig. 1 show the importance of football referee training in Colleges and Universities.

Table 1. How much colleges and universities attach importance to the training of football referees

	Take Seriously	Value	More value	Disregard
Number of schools	1	3	4	2
Percentage (%)	10	30	40	20

Sports interest is a psychological tendency for people to actively recognize, explore or participate in sports, and is an important driving force for acquiring sports and health knowledge and promoting physical and mental health. According to a survey of 288 people, 88 people choose to be very interested, accounting for 31% of the total, and 165 people choose to be more interested, accounting for 57% of the total. It can be seen that out of the total of 288 people, 253 people prefer to work as football referees. It accounts for 88% of the total, which is a very high proportion, indicating that the current students have a strong interest in football refereeing. Twenty-three people were uncertain about their choice, accounting for 8% of the total. 12 people chose not to be interested, accounting for 4% of the total. No one chose not to be uninterested. The degree of interest among college students in football referees is shown in Fig. 2.

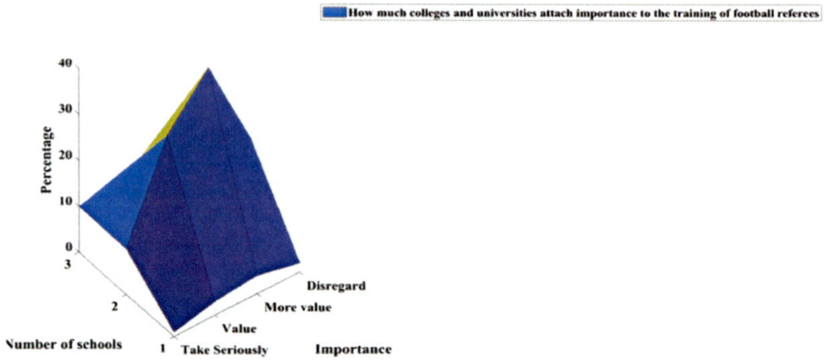

Fig. 1. How much colleges and universities attach importance to the training of football referees

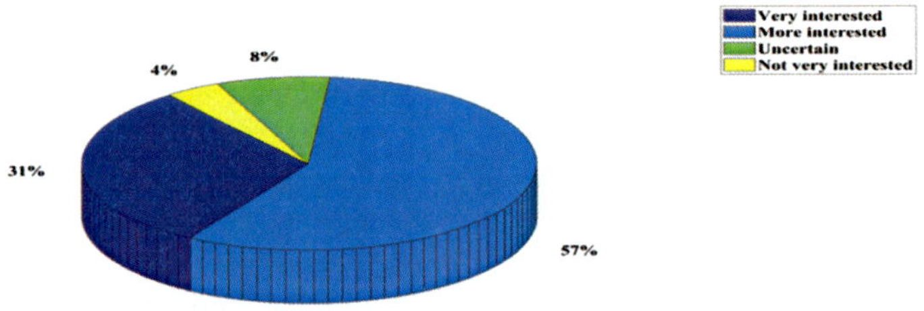

Fig. 2. Degree of interest among college students in football refereeing

5 Conclusion

Football, as one of the most widely developed and most influential sports in the world, is deeply loved by the masses, and its influence has far exceeded the scope of sports. In order to improve the ability of college football referees, the establishment of a scientific and systematic training system is the main problem to be solved urgently. This paper randomly conducts a questionnaire survey on teachers and students of ten local university campuses, collects a large amount of literature and organizes a large amount of relevant data, and finally draws a conclusion of the research. The experimental results show that according to a survey of 288 students' interest in serving as football referees, 88 people choose to be very interested, accounting for 31% of the total; 165 people choose to be more interested, accounting for 57% of the total; there are uncertainties. 23 people, accounting for 8% of the total, 12 of them chose not to be interested, accounting for 4% of the total, no one chose not to be interested.

References

1. Karaçam, A., Pulur, A.: Examining the relationship between referee self-efficacy and general self-efficacy levels of football, basketball and handball referees. Univ. J. Educ. Res. **5**(8), 37 (2017)

2. Gomez-Carmona, C.D., Ortega, J.P.: Kinematic and physiological analysis of the performance of the referee football and its relationship with decision making. J. Human Sport Exerc. **11**(4), 397–414 (2016)
3. Hlasny, V., Kolaric, S.: Catch me if you can: referee-team relationships and disciplinary cautions in football. J. Sports Econ. **18**(6), 560–591 (2017)
4. Oliveira, M.C.D., Reis, L.N., Silva, A.I.D.: Injury incidence in Brazilian football referees. Archivos De Medicina Del Deporte **33**(2), 108–112 (2016)
5. Dell, C., Gervis, M., Rhind, D.J.A.: Factors influencing soccer referee's intentions to quit the game. Soccer Soc. **17**(1), 1–11 (2016)
6. Dohmen, T., Sauermann, J.: Referee bias. J. Econ. Surv. **30**(4), 679–695 (2016)

Development Trend and Research of Computer Drawing Technology in Chinese Farmer's Painting

Lanting Liu(✉)

West Anhui University College of Art, Lu'an, Anhui, China

Abstract. Farmer paintings are an important part of the Chinese cultural system. After decades of development, Farmer Painting has formed its own artistic characteristics and industrial development model. With the development of computer drawing technology, the traditional peasant painting creation mode has been challenged unprecedentedly. In order to better promote the development of Chinese farmer painting industry, this paper studies the development trend of Chinese farmer painting based on computer drawing technology. This article conducts an in-depth study of the two existing creative modes of Chinese farmer paintings. According to the survey results, traditional farmer paintings pay more attention to creativity and artistry. However, after the complete marketization, the quality of the paintings has seriously declined, but the income has increased significantly. In response to this situation, this article analyzes that digitization and industrialization are the main development trends of peasant paintings, but excessive marketization will damage the artistry of peasant paintings and hinder its development. Therefore, in response to these problems and the current situation of computer rendering technology in farmer painting creation and the needs of the development of the digital industry, combined with the current development of the Internet, a set of improvement measures have been proposed. Specifically, from the three aspects of refining the production process, the government's efforts to control development, and the formulation of industry access. The analysis shows that improvement measures can resolve the contradiction between the industrialization and artistry of farmer paintings and establish a long-term management mechanism.

Keywords: Computer rendering technology · Peasant paintings · Art development · Industrialization development

1 Introduction

Peasant painting, also known as Chinese folk painting, is a folk art with Chinese characteristics rooted in agriculture. The earliest active in the 1950s, so far, it has experienced more than 70 years of development, ups and downs, completed the realistic to exaggerated expression of the transformation. In the 21st century, with the vigorous development of cultural industry, farmers' painting has entered a gratifying creative period [1–3]. The

B. J. Jansen et al. (Eds.): *International Conference on Cognitive based Information Processing and Applications (CIPA 2021)*, LNDECT 85, pp. 158–165, 2022.
https://doi.org/10.1007/978-981-16-5854-9_20

theme cultural exhibition of "world feelings and Chinese dream" held at the United Nations headquarters in New York is undoubtedly a great inspiration to the creators of farmer paintings. However, in the face of the impact of the mainstream culture, Chinese peasant painting in the new era is bound to encounter development bottlenecks. The rapid development of modern industrial civilization is disintegrating, supporting the specific environment of the original aesthetic image of farmers' paintings [4, 5].

The report of the 18th National Congress of the Communist Party of China (CPC) proposed to build an excellent traditional culture inheritance system and carry forward the excellent traditional Chinese culture. As an excellent traditional culture, peasant painting should be inherited and carried forward. Farmers have always been good at handicrafts. Carpenters, masons, lacquerers and weavers are famous for their exquisite works and beautiful designs [6–8]. Under the influence of traditional culture and beautiful environment, farmers like painting. In their leisure time, they often put down the hoe and sickle and pick up the brush to daub.Computer-drawn works are also called non-photorealistic rendering and are widely used in engineering technical illustrations, animated cartoons, games and other fields. With the help of computer drawing technology to draw farmers' paintings, the most important and difficult point lies in the color matching rendering, the expression of national culture and regional characteristics. If these problems are solved, coupled with the popularization of mobile Internet, this will be conducive to the dissemination and development of Chinese peasant paintings in the context of big data, and promote the development of local art sources, ecological resources and tourism resources [9, 10].

This paper makes an in-depth investigation and Research on the main creation mode and industrial status of farmers' paintings in China. Through the analysis of the survey results, we can see that at present, there are two opposite trends in the creation mode and industrial development of China's farmers' paintings. Although the traditional art creation mode ensures the quality of creation, it restricts the development of the industry to a large extent. However, the development mode of farmers' paintings after marketization not only improves the income, but also sacrifices the artistry of farmers' paintings. The analysis shows that the two models are not conducive to the development of farmers' painting. Therefore, this article explores the application of computer drawing technology in Chinese peasant paintings. In the research, combining the development status of Chinese peasant paintings and the characteristics of computer drawing technology, proposes the optimization and improvement measures of the development model of Chinese peasant paintings. This measure guides and controls the development of the industry from three aspects: refining the production process, controlling the development efforts of the government, and formulating the industry access. The analysis shows that the optimization measures in this paper can solve the contradiction between industrialization and artistry of peasant paintings while popularizing them.

2 Peasant Painting and Digital Communication

2.1 Peasant Painting

As the name implies, peasant paintings refer to the paintings of peasants. In the period of the great leap forward, peasant painting experienced the wall painting movement. At that time, peasant painting was accompanied by illusory and romantic exaggeration. After the "Cultural Revolution", with the deepening of reform and opening up, the emancipation of the mind, and the change of social environment, farmers' painting took the lead in returning to the traditional folk art of farmers in Jinshan County, explored the emotion and composition of folk art, and created a new painting form. Different from the previous peasant paintings, the "pattern" of Jinshan peasant paintings has become a popular way for various painters to imitate. In the tide of commodity economy, farmers' paintings are gradually market-oriented and popular. Farmer painting is a kind of socialized activity of rural community under the background of national special times, and it is the product of social event selection. Its emergence and development have become an important cultural phenomenon in China, recording and reflecting the changes of the times.

2.2 Concept of Cultural Creation and Digital Communication of Farmer Paintings

Cultural creation is the abbreviation of "cultural creation", which is a new innovation based on the existing art. Digital communication, also known as network communication, is a way of using multimedia devices such as computers and mobile networks to spread information. It is a new way of communication, originated from meeting the needs of most modern users. On the basis of the continuous development of network technology, we can carry out digital dissemination of farmers' paintings. In the new era, new ideas have emerged in the way of communication of farmers' paintings. Modern multimedia technology can be used to change the spread and development of farmers' paintings, and help them achieve a leap in quality.

2.3 Computer Drawing Technology and Artistic Painting

Computer drawing technology mainly relies on the construction of mathematical models and machine algorithms to achieve painting. In the process of generating an artistic painting, it is first necessary to form a mathematical description of the artistic painting. The description includes the direction field model, the color transformation model and the placement of brushes. Model etc. [11, 12].

In the direction field model, it mainly includes three aspects: image layering, main structure analysis, and direction diffusion.

(1) A picture I is superimposed by multiple layers, and how each layer needs to be drawn needs to be determined according to the specific image occlusion. This process often reflects the different personalities of painters through the concept of interaction.

$$W = \pi(I) \tag{1}$$

The π here refers to the user's simple editing based on their own understanding, so as to achieve the required occlusion effect.

(2) In the analysis of the main structure, mathematical formulas need to be used to obtain the main structure of the object. Here we take the Bayesian formula as an example.

$$W' = \arg\max P(I|W')P(W') \tag{2}$$

Here W' represents the main structure of the object, I represents the partial image segmented from the original image, P(W') is the prior probability, and P(I|W') is the likelihood probability.

(3) After the main structure of the object is obtained, the direction of the brush corresponding to any position can be known, but the texture and noise of the object have not been determined, so the object also needs to calculate the direction of the non-main structure of the object.

3 A Comparative Analysis of the Characteristics of Traditional Painting Mode and Market-Oriented Painting Mode

The popularity of mobile Internet has accelerated the marketization of farmers' paintings, which makes the traditional painting mode of farmers begin to change. Adhering to the traditional painting mode and catering to the market demand has become the current two mainstream ways. There are great differences between the two painting modes in terms of creative motivation and creative income. In order to further understand the current development trend of China's farmers' painting, this paper makes an in-depth investigation on the current two painting modes of farmers' painting.

The survey of this paper is mainly based on field visits, which includes 10 farmers' painting artists who are free to create and market-oriented. In the process of investigation, the author makes an investigation and statistics on the motivation, tendency, cycle, interest orientation, competition point and actual income. According to the survey results in Table 1, in the creation of farmers' paintings, the traditional painting mode pursues more originality and artistry, but the disadvantages are that the creation cycle is long, unstable, and the actual income is low; on the contrary, when the market-oriented creation is carried out, the artistry and originality of paintings are greatly reduced, but the quantity and income are relatively stable. The analysis shows that the marketization of farmers' paintings in the era of mobile Internet popularization is justifiable, but it is not advisable to sacrifice the quality of paintings. The creators of farmers' paintings are generally farmers, with low income and unstable income. Income is one of the factors that must be considered when creating. Therefore, this paper believes that it is feasible to carry out marketization under the premise of ensuring artistry. The promotion of farmers' painting products through mobile Internet not only brings benefits to farmers, but also ensures the quality of paintings.

Table 1. Comparative investigation results of creative characteristics between traditional painting mode and market-oriented painting mode

		Traditional painting	Market oriented painting
Creative motivation		The pursuit of originality	Benefit maximization
Creative practice	Creative tendency	Non replicability	Mechanical reproduction
	Creative cycle	Duration	Short time
	Interest oriented	Pursuit of artistic aesthetics	Market interest
	Competitive point	Improvement of art level	Sales volume
Real income		Instable	Stable

4 Discussion

4.1 Research and Analysis on the Development of Farmers' Painting Industry in Huxian County

Mobile Internet has brought new opportunities for the development of farmers' painting. While reforming the traditional painting mode, it is necessary to make market-oriented changes in line with the needs of the development of the times. In view of this, many farmers' painting production areas develop the development route of farmers' painting through the development of mobile app, the promotion of mobile Internet, and the development of cultural and creative products. In this paper, Huxian farmers paint as the object of investigation, its farmers painting industry related investigation. The survey data is selected from the 2014–2018 Huxian government industry report, which is statistically analyzed by this paper.

Since 2014, Huxian has continuously expanded the reputation of farmers' paintings through the "boutique strategy", which has brought good economic benefits to Huxian County by attracting tourists from home and abroad. According to conservative statistics, the sales of farmers' paintings in Huxian County and the added value of related industries driven by farmers' paintings in recent years are shown in Fig. 1. According to the survey results in Fig. 1, from 2014 to 2018, the sales volume of pure farmer paintings in Huxian County was not high, and the annual growth rate was only about 300000. However, due to the influence of farmers' paintings in Huxian County, the added value to the local souvenirs, farmhouse entertainment, farmers' painting performance, accommodation and other tourism related industries is huge.

In addition, in order to further investigate the development of farmers' painting in Huxian County, this paper makes a survey and statistics on the training classes, creation classes, art foundation classes and migrant workers' learning classes. According to the survey and statistics in Fig. 2, it can be found that the training times of various kinds of farmer paintings in Huxian County in 2018 have been greatly increased compared with that in 2014. The analysis shows that this is mainly due to the cooperation of mobile Internet plus excellent marketing strategy. After adopting this measure, Huxian's peasant paintings have been vigorously promoted at home and abroad, forming brand effect. At

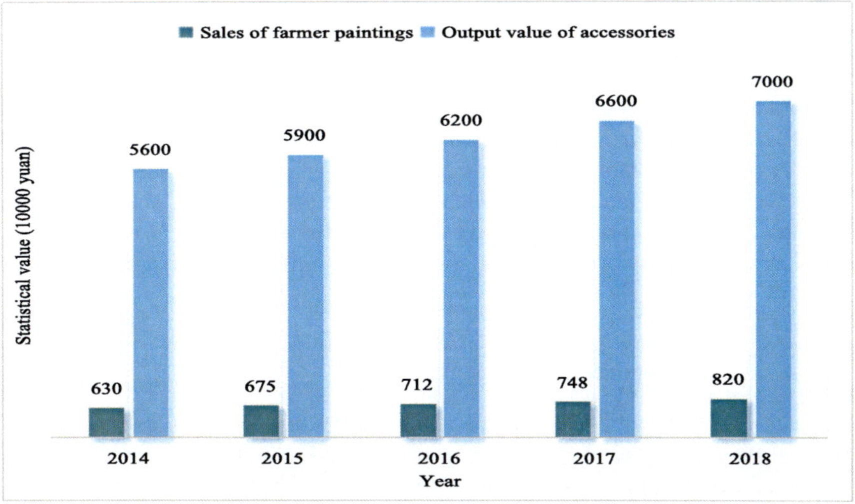

Fig. 1. Investigation and analysis of output value of farmers' paintings in Huxian County

the same time, it improves the industrial environment and establishes the foundation for talent cultivation.

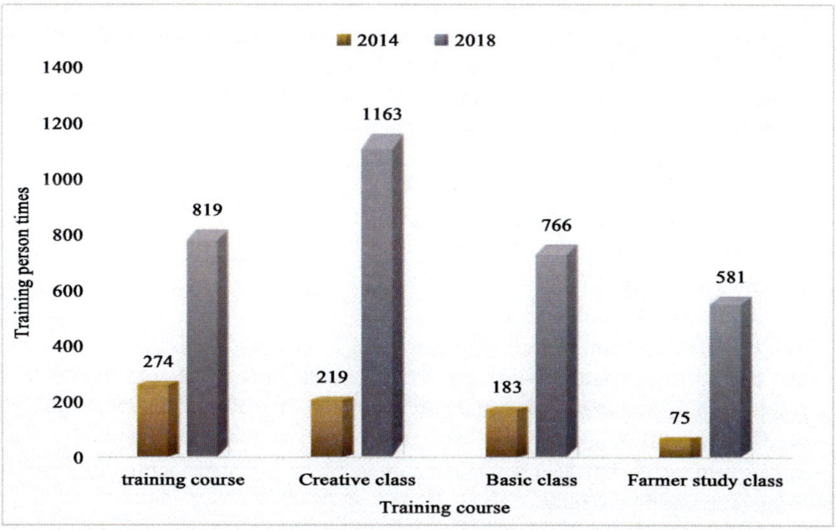

Fig. 2. Comparative analysis of the number of farmer painting training in Huxian County in 2018 and 2014

4.2 Measures to Improve the Contradiction Between Industrialization and Artistry of Peasant Paintings

The contradiction between the industrialization and artistry of peasant painting will be encountered in any cultural industry. The reason why cultural industry is established is to have culture first. If we blindly industrialize culture, we will not be able to make the whole industry into a virtuous circle, which will be difficult to protect the characteristics of art. Therefore, in order to ensure the industrialization development of peasant painting, it can still maintain its own artistic characteristics and ensure its artistic quality, we should do the following:

4.3 Refined Manufacturing Technology

We should set up quality supervision departments related to farmers' paintings, strengthen the supervision of production technology and technology, especially improve the production level of some small workshops and individual production and management households. At the same time, we should strengthen the supervision and management, help the creators to improve the production technology level from the source, and strengthen the overall artistry of farmers, so as to maintain the aesthetic influence ability of their creations.

(1) Government controls development efforts
Although folk cultural resources are rich and colorful, unlike tangible mineral resources, the number of developments is small, but if not protected and controlled, it is easy to fall into the plight of cultural resources depletion. Therefore, local cultural departments should strengthen the guidance of the development of folk cultural resources, especially the control of the development level. When the tide of industrialization of farmers' paintings is about to rise, the government departments begin to strengthen the protection of farmers' paintings, which will bring economic benefits and spiritual enjoyment to the whole industry. At the same time, it can also enhance the repair ability of the industry and enhance the artistry of farmers' paintings after industrialization.

(2) Formulate industry access standards
The original intention of developing cultural industry is to standardize, program and popularize cultural products. Since this is the case, we should formulate the corresponding standardized evaluation standards. The establishment of industry access standards is not only a guarantee for the healthy development of farmers' painting industry, but also a quality management mechanism for industrialized farmers' painting products.

5 Conclusions

Based on the application of computer drawing technology in peasant paintings, this article conducts an in-depth investigation and research on the creative mode and industry status of Chinese peasant paintings. Through the survey results, we have learned about the main issues of the current Chinese farmer painting creation mode and industrial development. In response to these problems and deficiencies, in view of the current development of the farmer painting industry in our country, combined with the characteristics of the mobile Internet background, innovative measures for optimization and improvement are proposed. The main purpose of this improvement is to resolve the contradiction between the industrialization and artistry of peasant paintings. On the basis of not losing the artistic characteristics of peasant paintings, this article combines the current era background and uses peasant paintings as a collection point to effectively promote the industrial clusters. Based on the survey of farmers' paintings in Huxian County, the author found that the farmers' painting industry in Huxian County has been well developed through modern digitalization + excellent industrial development models. It increased the income of farmers' creators, and also promoted the development of related industries, and laid a good foundation for the development of farmers' paintings.

Acknowledgments. Anhui province philosophy and social science planning project "Research on New China Anhui Peasant Painting", No.AHSKY2016D89.

References

1. Mao, Q., Liu, P.: On the attribute of peasant painting: based on the investigation of peasant painting movement in the 1950s and 1960s. J. Baise Univ. **032**(003), 53–57 (2019)
2. Xu, F., Xia, Y.: Research on Liangshan Yi Folk Art Features%. Research on Liangshan yi Folk Art Features%. Guizhou Ethnic Stud. **036**(007), 86–89 (2015)
3. Suopajärvi, T.: Kay, Jon. Folk art and aging: life-story objects and their makers. xii, 131 pp., illus., bibliogr. Bloomington: Indiana Univ. Press, 2016. £22.99 (paper): Book and film reviews. J. R. Anthropol. Inst. **25**(1), 174–175 (2019)
4. Xiong, R., Yi, X.: Books are often stacked like green mountains: on the literary characteristics and dissemination of the folk art Lizhou drum. J. Hunan Univ. Technol. (Soc. Sci. Ed.) **024**(001), 106–112 (2019)
5. Chen, L.: Bringing folk art into full play in cultivating the innovative ability of art students in vocational colleges. Front. Educ. Res.: Chin. Engl. **009**(003), 51–54 (2019)
6. Ping, W., Guangquan, M.: On relations between Zhang Daqian's artistic features and the folk painting. J. Normal Univ. **033**(001), 24–30 (2018)
7. Wan, L., Liu, Y., Dai, H., Feng, W., Zhang, J.: Woodblock image decomposition of Chinese new year paintings. Multimed. Tools Appl. **78**(6), 7621–7641 (2018). https://doi.org/10.1007/s11042-018-6447-x
8. Sarma, A., Chakraborty, S., Nandi, S.: Deciding handover points based on context-aware load balancing in a wifi-wimax heterogeneous network environment. IEEE Trans. Veh. Technol. **65**(1), 348–357 (2016)
9. Yoon, S., Lee, S.S., Kim, S.H.: Robust mutual trust architecture for safety critical service in heterogeneous mobile network environment. Telecommun. Syst. **60**(2), 1–10 (2015)
10. Zhang, Y.: The construction and application of self-study mode based on mobile network environment—taking students of journalism major at Gansu political science and law institute as an example. J. Ningbo Inst. Educ. **019**(005), 15–18 (2017)
11. Ma, C., Cui, X., Chen, F.: Knee arthroscopic navigation using virtual-vision rendering and self-positioning technology. Int. J. Comput. Assist. Radiol. Surg. **15**(3), 467–477 (2020)
12. Wang, Z., Lü, X.: Terrain rendering LOD algorithm based on improved restrictive quadtree segmentation and variation coefficient of elevation. J. Beijing Inst. Technol. **27**(04), 145–150 (2018)

Communication of Advertising Visual Communication Design Art in the Information Network Era

Qiongfang Zhang(✉)

Hainan College of Economics and Business, Haikou, Hainan, China

Abstract. In recent years, visual communication design (VCD) has been used in online advertising, but there is little research on the impact of visual transmission in online advertising. Therefore, this paper proposes the communication research of VCD in the era of information network. This paper discusses the main forms of online advertising and the core concept of visual transmission design. Analysis shows that the use of VCD in online advertising can effectively increase the effect of publicity, thus effectively improving advertising. This article studies the aesthetic needs of online advertising, the characteristics of VCD, and combines the design features of visual communication, combined with the needs of network advertising publicity, and puts forward the design principles of visual communication in network advertising. The main design principles are innovation principle, aesthetic principle, humanization principle and comprehensive principle. The above principles basically meet the application requirements of VCD in online advertising, and also play a role in ensuring the quality of design works. This paper investigates and analyzes the number of Internet users and the scale of online advertising market in China. The survey results show that by the end of 2018, the number of registered users and the total number of statistical data of domestic mobile communication netizens have reached 829 million. At the same time, the scale of network advertising market has also broken through the 600 billion mark, reaching 618.2 billion. The analysis shows that online advertising has great market potential in China, and VCD can help improve the effect of online advertising.

Keywords: Visual communication art · Advertising design · Online advertising · Information dissemination

1 Introduction

Mastering the concept of VCD is very important for the study of how to spread VCD in the new era. The concept of VCD originated in the western countries in the 1960s. It was introduced into Chinese universities in the late 1990s and gradually formed in the promotion of traditional arts and crafts, graphic design and other disciplines [1–3]. In the meanwhile, under the affect of the growth of computer science and Internet, the form of

B. J. Jansen et al. (Eds.): *International Conference on Cognitive based Information Processing and Applications (CIPA 2021)*, LNDECT 85, pp. 166–172, 2022.
https://doi.org/10.1007/978-981-16-5854-9_21

visual culture has changed from printing culture to electronic media culture, from long-distance communication to interactive communication, and the scope and connotation of VCD discipline are rapidly expanding [4, 5].

In the specific design, the design of visual communication is to consider the needs of people's visual psychology. It can make people understand the content to be transmitted more intuitively and concretely through the visual image. Under the influence of the network and multimedia, the effectiveness of information is constantly improved, which has a great impact on vision, and constantly attracts people's attention [6, 7]. Therefore, under the influence of modern science and technology, VCD has developed rapidly. VC designers should fully seize the design connotation of visual communication, follow to the trend of the times, constantly use new technologies to create design concepts, and become the leader of fashion trend [8–10].

This paper deeply analyzes and studies the practical application and development of network visual information communication technology design in China's network advertising communication, and understands that the application of VCD in network advertising can better enrich the propaganda form of network advertising, and play a helpful role in the innovation of content design and element communication of advertising. Due to the fast development of network technology, VCD has been widely used in online advertising. However, the research on the impact of VCD in online advertising is insufficient. Therefore, this article hopes that the communication research of VCD art in the era of information network. I hope that through this study, the application of VCD in online advertising plays a perfect role, and to improve the comprehensive design ability of online advertising in China. This article analyzes the basic elements of visual communication in the design of the network, and puts forward the key elements in the design of advertising. The analysis shows that the arrival of the network era has changed the traditional advertising and publicity mode. Network advertising is one of the most efficient publicity modes at present. In the future development, the growth of enterprises will be inseparable from the application of network advertising.

2 Network Advertisement and Visual Communication Design

2.1 Concept of Network Advertisement

Network advertising information is a kind of advertising information published on the Internet, and it is an inevitable result of the rapid development of modern new media information technology. The basic concept of network advertisement can be roughly divided into two broad sense and narrow sense. Broadly speaking, online advertising information reflects all commercial advertising information published on the Internet, including corporate public welfare activities information, enterprise promotion product information, etc. In a narrow sense, online advertising information refers to a kind of commercial information that can be published and spread on the Internet by Internet advertising paid by the state. It can be viewed, viewed and manipulated by users. Its main business includes various text and hyperlinks, e-mail and other forms of online advertising.

2.2 Concept of Visual Communication Design

Visual information communication art design has been popular all over the world since the middle of 1960, when an International Conference on VCD was held in Tokyo, Japan. Participants gradually realized that in the expanding and developing entity image expression media, vision and entity image can exist together as an independent visual means of expression, so they began to gradually form such a visual design that visual expression can be fully independent, that is, the present VCD. Since the 20th century, with the emergence of digitization, the social environment in which people live has also undergone qualitative changes. The concept of VCD has gone beyond the original category. With the advancement of science and technology, the emergence of new energy development and application materials, the field involved is expanding, and a new design field is gradually formed. The birth of new VCD concept enriches the form of contemporary art design, not only enriches the essence of contemporary art design, but also breaks through the boundaries between traditional design categories.

3 Aesthetic Demand in Network Advertisement Design

In the digital and network era, people's network aesthetic standards and overall psychological quality will change with the development of the Internet, and the former network aesthetic standards have also undergone great changes. The biggest technological change is mainly due to the fact that people increasingly like this dynamic information expression, which makes web browsing more socially participatory. In such a social change, online advertising product design will continue to pursue the best user aesthetic visual effect, closer to the dynamic and flat. Scientific and technological information is an important carrier of network advertising. It can create excellent web advertising effect by using unique network communication mode effect and new network information content, and adding interesting web page visual effect browsing and new aesthetic feeling style at the same time.

4 Discussion

4.1 Analysis of the Current Situation of Network Advertisement

Due to the continuous growth of network technology, the network advertising has gradually developed from the original banner form to the network advertising design works with modern creativity, information dissemination and artistic level.

(1) The number of Internet users in China is soaring

In recent years, the fast growth of China's online advertising market is mainly due to the continuous development and expansion of the network market. Figure 1 shows the number of Internet users in China, and the data is from China's Internet development statistics report. According to the statistics in Fig. 1, there are 829 million Internet users, 817 million mobile Internet users and 222 million rural Internet users in China. The mainstream media of the Internet play an important role. In order to seize their own piece of cake, major advertisers have launched online advertising. Network advertising has also broken through the past simple visual expression and paid more attention to creative expression.

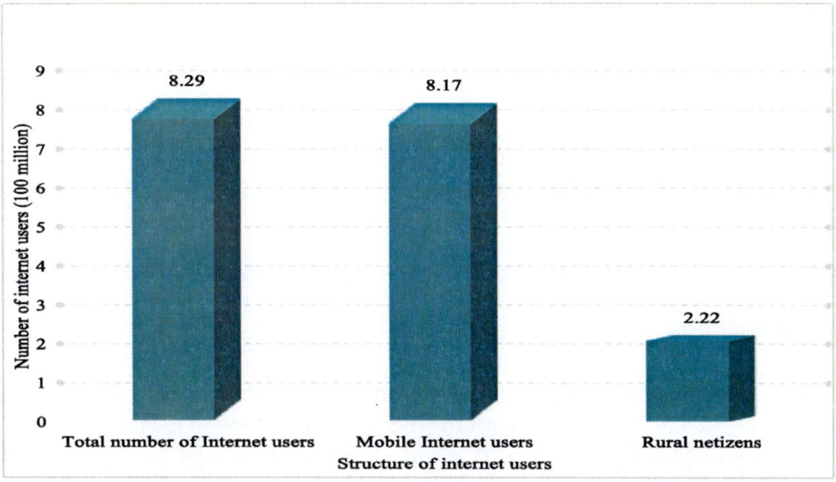

Fig. 1. Statistics on the number and structure of Internet users in China by 2018

(2) Market scale of online advertising industry

The progress of technology promotes the rapid development of network advertising market. With the development of technology and the acceleration of transmission speed, the content and form of consumers' access to information platform have changed, and the network advertising market has also developed rapidly. This paper analyzes the scale of China's online advertising market from 2013 to 2019. The details are shown in Table 1 and Fig. 2. The data in Table 1 and Fig. 2 come from the collation of public information. The results show that the scale of China's online advertising market has reached 618.2 billion in 2019, indicating that the scale of China's online advertising market is huge.

Table 1. Statistics of China's online advertising market size from 2013 to 2019

	2013	2014	2015	2016	2017	2018	2019
Market size (100 million)	1105.2	1546	2184.5	2884.9	3750.1	4884	6182
Growth rate (%)	43.0	39.9	41.3	32.1	30.0	30.2	26.6

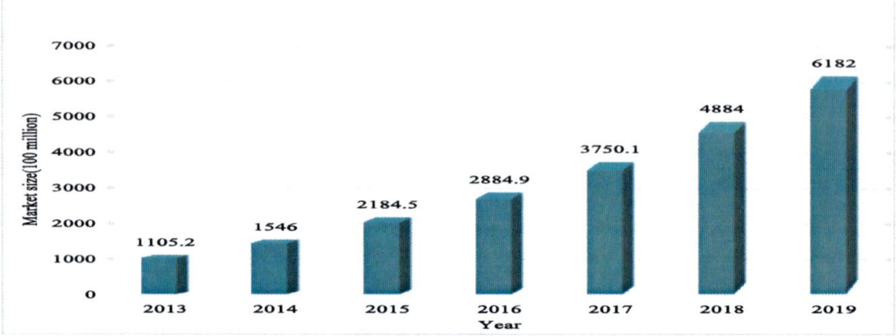

Fig. 2. Statistical analysis of China's online advertising market size from 2013 to 2019

4.2 Composition Characteristics of Visual Communication Design

(1) Intrinsic characteristics

As one of the main characteristics of VCD art, ethnic characteristics make the design of online advertising more unique and attractive. Integrating the connotation and cultural characteristics of different ethnic groups into advertising can make the visual communication effect of online advertising better. In addition, the rendering of national culture will make advertising more approachable, reduce the boredom of netizens, and have a good resonance with netizens. In addition, regional culture is also one of the inherent characteristics of VCD art. According to the cultural elements of different regions, the advertising effect can be better expressed in the advertising design, so that people can easily accept the content of the advertisement.

(2) External characteristic network

Advertising visual communication depends on the external characteristics of image features, color performance, because different people have different concerns. Some people pay attention to the use of color, it is easy to be attracted by bright colors. Some people pay attention to the use of graphics and are easily attracted by the graphic design in advertisements. Therefore, the VCD of network advertisement should pay attention to different contents, such as paying attention to the diversity and regionality of graphic design, paying attention to the rendering and deployment of colors, and making the advertisement more attractive through balanced collocation.

4.3 Constituent Elements

In terms of advertising, the ultimate goal of advertising is to persuade the audience to buy goods or services voluntarily. Visual information transmission is also an important element in the design of advertising art, greatly enhancing the visual vividness and visual impact of text advertising. Among them, character design, as the most important essence embodying the wisdom of human civilization, is the most direct information carrier of human information expression and transmission. However, its application space in the design of text design should not be greater than that. In practical choice, it should be concise and accurate, and assist the expression of the connotation of works. At the same time, color design, as an indispensable part of visual information transmission form design in modern network advertisement, directly affects the consumer psychology of advertising audience.

4.4 Analysis of the Design Principles of Visual Communication in Network Advertisement

(1) Innovation principle

In this Internet era, this advocates the innovative spirit of communication, the design of various forms of visual information transmission of online advertising needs to pay more attention to technological innovation. Therefore, the designer industry should always adhere to the design principle of continuous innovation, constantly carry out technical innovation on the basis of inheriting the original traditional design expression method and technical concept, and constantly seek for new design expression ways.

(2) Humanization principle

Network advertising is to let people see the content and information of products. Because the traditional network advertising design is people-oriented, designers in the specific process of design work, that is, they can pay attention to cater to people's various emotional activities. In the network advertising, some emotional factors can be added to cause all kinds of emotional resonance of advertising people, so as to achieve a direct impact on the advertising audience.

(3) Aesthetic principles

With the continuous development of higher education and the continuous improvement of knowledge and culture level, people's cultural aesthetic concept is constantly changing, and the ability of aesthetic feeling is also constantly improved. Therefore, designers in the process of designing advertising need to focus on the cultural and artistic application value of art advertising, and improve their own different aesthetic and cultural ability.

(4) Comprehensive principle

In the overall design and production process of network advertising, we should try to present different advertising design scheme styles at the same time to make it conform to the current network culture design trend. Under the trend of diversified development of modern culture and art, designers need to make full use of the different characteristics of various advertising elements to organically integrate them, boldly design, and create a network advertising brand that can quickly be widely welcomed by the public.

5 Conclusions

This article makes an in-depth study on the practical application of VCD in online advertising. Analysis shows that in the traditional network advertising design, the integration of VCD concept, for the design effect of network advertising will play a qualitative improvement. There are many deficiencies in the traditional network advertising design, which leads to the unsatisfactory publicity effect. According to the publicity needs of online advertising, combined with the characteristics of VCD, this paper makes relevant optimization design. By establishing the design principles of visual communication in online advertising, the overall design ability of online advertising is further improved. Analysis shows that the information age has brought new opportunities to online advertising, but also brought new challenges. The application of VCD in online advertising is the future development trend, and it is also the necessary skills that network advertising designers need to master.

References

1. Guan-Dong, S.: The image creation of science and technology from the perspective of visual communication design. Packag. Eng. **99**(5), 2584–2601 (2015)
2. Sun, Y.: On the application of traditional tang costumes in visual communication design. J. Heihe Univ. **010**(006), 178–180 (2019)
3. Baek, S.-Y.: A study on visual communication design for effective trial presentation. J. Korea Contents Assoc. **16**(5), 601–611 (2016)
4. Cao, L., Yu, Z.: On the aesthetic tendency of visual communication design in modern Japan: taking Alan Chan's works as an example. Des. Art Res. **008**(003), 25–31 (2018)
5. Zhang, Y., Wu, W., Zhang, L.: Development trend of visual communication design combining low carbon environmental concept. Environ. Sci. Manag. **043**(006), 53–56 (2018)
6. Tang, J., Zhang, P., Wu, P.F.: Categorizing consumer behavioral responses and artifact design features: the case of online advertising. Inf. Syst. Front. **17**(3), 513–532 (2015)
7. Wu, C.H., Kao, S.C., Wu, C.C., Huang, S.: Location-aware service applied to mobile short message advertising: design, development and evaluation. Inf. Process. Manage. **51**(5), 625–642 (2015)
8. Pei, Y.: "C Drive" creative advertising design. Macromolecules **26**(7), 1767–1773 (2015)
9. Liang, L.: Research on the teaching of advertisement creative and print advertising design experiment course. Heilongjiang Sci. **009**(007), 66–67 (2018)
10. Kim, J., Kim, J.: Consumer interest areas and attention to bluetooth product advertising design. J. Digit. Des. **17**(2), 93–103 (2017)

The Influencing Factors of Business English Intercultural Communication Based on Data Mining

Xiaohong Zhang(✉)

Foreign Language Department, Ganzhou Teachers College, Ganzhou 341000, Guangdong, China

Abstract. With the deepening of economic globalization and the closer communication between countries, English, as a global language, is particularly important in cultural exchanges. Business English is an independent major in our college, so we should pay attention to the training mode. This paper focuses on the influencing factors of business English cross-cultural communication based on data mining clustering algorithm, and innovates the mode of business English teaching and training in Colleges and universities, so as to improve the deficiencies in cross-cultural communication. This paper analyzes the current research situation of intercultural communication ability in China by searching relevant keywords through literature retrieval method, summarizes previous theories, and puts forward the importance of cultivating business cultural communication ability, and then uses questionnaire survey to preliminarily understand the main problems existing in the cultivation of intercultural communication ability, then, based on data mining clustering algorithm, the influencing factors of cross-cultural communication are obtained. Finally, combined with situational teaching method and cross-cultural communication, this paper explores a classroom teaching model to cultivate the cross-cultural interactivity and communicative competence of business English Majors in our university. Through three levels of classroom teaching, a large number of students' cross-cultural interactivity and communicative competence in Business English industry are cultivated. After the experiment, it is found that after the situational teaching training, students' test scores have been significantly improved, from 56.54 to 66.32, which proves that the appropriate situational teaching training mode is helpful to improve students' intercultural communication ability.

Keywords: Data mining · Business English · Intercultural communication · Teaching English

1 Introduction

Since the 21st century, human science and technology has developed rapidly. At the same time, with the advocacy of the concept of world integration such as economic globalization and cultural globalization, countries have close exchanges in politics, trade and other aspects [1, 2]. More and more people pay attention to intercultural communication

B. J. Jansen et al. (Eds.): *International Conference on Cognitive based Information Processing and Applications (CIPA 2021)*, LNDECT 85, pp. 173–181, 2022.
https://doi.org/10.1007/978-981-16-5854-9_22

[3, 4]. It is worth noting that, due to different cultural concepts, in the process of cross-cultural communication, mistakes, contradictions and conflicts are gradually emerging. What are the key factors causing these problems and how to solve them are the issues that scientists and scholars pay close attention to [5, 6].

Many scholars have made researches on the factors influencing transnational cultural exchanges [7, 8]. For example, some scholars believe that due to different thinking modes and values of people in different countries and nationalities, pragmatic failure often occurs in communication, which makes cross-cultural communication unable to proceed smoothly [9]. Some scholars believe that a lot of new Internet words have emerged in the Internet era. Because English and Chinese are two different languages, it is difficult to accurately express the meaning of Chinese in English, which often leads to the misinterpretation of the meaning of words and cannot communicate smoothly [10].

This paper analyzes the current research situation of intercultural communication ability in China by searching relevant keywords through literature retrieval method, summarizes previous theories, and puts forward the importance of cultivating business cultural communication ability, and then uses questionnaire survey to preliminarily understand the main problems existing in the cultivation of intercultural communication ability, then, based on data mining clustering algorithm, the influencing factors of cross-cultural communication are deeply obtained [11]. Finally, the situational classroom teaching method and cross-cultural interactive communication classroom teaching are combined to explore a kind of classroom teaching model to cultivate the cross-cultural interaction and communication ability of business English Majors in our university. Through three levels of classroom teaching, a large number of students' cross-cultural interaction and communication ability in Business English industry are cultivated [12].

2 Business English Intercultural Communication Research

2.1 Current Situation of Intercultural Communication Competence

Through the selection of master's and doctoral dissertations in China excellent thesis database, this paper finds that the Cultivation Strategies of intercultural communication competence account for the vast majority, nearly 60%. Therefore, in the research of cross-cultural communication in China, the research on the cultivation strategies and models of students' cross-cultural communicative competence in English teaching is the main one, followed by the investigation on the current situation of English learners' cross-cultural communicative competence in China. If "business English cross-cultural communication + influencing factors" is used as the search item, the result shows that it is 0, which shows that there is little research on the influencing factors of business English cross-cultural communication in China.

2.2 The Importance of Cultivating Business English Intercultural Communicative Competence

In the specific communication practice, business English majors always express their ideas from the perspective of their own culture, without taking into account the cultural differences between different languages, which can easily lead to misunderstanding and conflict, thus leading to the failure of business activities. Therefore, it can be said that intercultural communicative competence is very important in foreign communication, and it is an important part of business English Majors' communicative competence.

2.3 Data Mining Technology

Data mining can find hidden and meaningful knowledge from the database, and has the functions of forecasting future trends, clustering and deviation detection. In this paper, we need to use data mining technology to analyze the current situation of cross-cultural ability research. Association rules can find potential associations in massive data, and the most important parameters are support and confidence. Support rule is used to set the frequency of data set, count the occurrence frequency of items, and eliminate the items with small occurrence frequency. The confidence rule is used to determine the frequency of itemset y in transactions containing itemset X. The support calculation formula in association rules is as follows:

$$\sup(X) = \frac{\|\{d \in D | X \sqsubseteq d\}\|}{\|D\|} \tag{1}$$

The calculation formula of confidence is as follows:

$$\mathrm{conf}(\mathrm{X} \rightarrow \mathrm{Y}) = \frac{\sup(X \cup Y)}{\sup(X)} \tag{2}$$

The strength of association rules is generally evaluated by support and confidence.

2.4 Main Problems in the Cultivation of Intercultural Communication Ability

The Cross-Cultural Quality of Teachers Themselves Needs to Be Improved. At present, the main English teachers in China are the traditional exam oriented education, they have good English language skills, but they don't know the western culture deeply and comprehensively, have not strong cross-cultural awareness and lack of initiative and enthusiasm in cultural teaching.

Lack of Corresponding Evaluation Methods for Intercultural Communication Ability. The English cross-cultural teaching takes the learning achievement as the main evaluation standard of students' English learning, and lacks or even does not have the content and standard of cultural evaluation. Therefore, although some English teachers can plan to integrate the content of cultural teaching and try to improve their cross-cultural literacy, the teaching effect is not satisfactory.

Lack of Cross-Cultural Content in English Textbooks. English textbooks have gradually integrated some cross-cultural communication contents, but the cultural knowledge involved in the level of narrow and superficial, cultural information system is not strong, cultural content source is single, students' contact surface for cross-cultural communication is relatively narrow, and the combination of English language knowledge and cultural content is not closely, which is not in line with the requirements of English as an international communication language.

Cultural Knowledge is the Key to the Success of Cross-Cultural Communication. Cross cultural communication is the process of communication between the two sides. If a participant only considers this issue from the perspective of their own culture, does not fully understand and understand the other party's culture, regardless of their feelings and their feelings, and does not fully respect their cultural preferences, it may directly cause their unhappiness with each other, and even lead to contradictions in serious cases, thus, they will directly lead to the failure of cross-cultural interaction and communication. Therefore, it is very important to have a certain international cross-cultural knowledge and the ability of cross-cultural communication at home and abroad, whether in daily life or in actual work.

The Influence of Conceptual Metaphor on Intercultural Communication. Firstly, conceptual metaphor is realized in cultural practice. Conceptual metaphor has a strong material existence in a specific culture. When the realization of metaphor changes, culture will also change. Second, language is an important part of culture and the expression of conceptual metaphor. Sapir Whorf hypothesis once believed that language and culture are interrelated. Therefore, metaphor marked by language is also related to culture. Thirdly, the metaphor system is relatively stable in keeping with the specific culture. Due to the relative time stability of material reality, the group's understanding of the world and the tools for observing culture also remain stable. Fourth, according to the definition, culture can be summarized as a series of common understandings of the world, and metaphor plays a role of metaphorical understanding in culture. Concrete and abstract objects and events constitute people's understanding of the world, and image thinking plays an important role in abstract objects and events. This is because abstract thinking is understood through the mapping of specific domain and experience. In other words, cultural patterns in abstract and complex fields, such as business, emotion or diplomacy, are all metaphorical. Fifthly, conceptual metaphor is an invisible cognitive process, which is expressed in metaphorical language in discourse. These language expressions have distinct social and cultural functions. Through the above analysis, we can draw a conclusion: conceptual metaphor is not only a part of culture, but also a cognitive process to shape culture. Because of the consistency between conceptual metaphor and cultural reality, it can be traced back to the presentation of specific culture.

2.5 ICC Framework of Foreign Language Teaching for Chinese College Students

The relationship among the three levels sensitivity, tolerance and flexibility in foreign language teaching is hierarchical. The lowest level is based on sensibility, the middle tolerance and the highest flexibility. Although the three components are graded, the upper component cannot be regarded as the natural product of the lower component. For example, a person with cross-cultural sensitivity and tolerance to cultural differences cannot guarantee flexible solutions to problems in cross-cultural communication. However, this hierarchical structure of intercultural competence means that the cultivation of intercultural competence in foreign language teaching should start with the enhancement of sensitivity and intercultural awareness. When students can understand and respect cultural differences, the focus can be shifted to enhancing flexibility through skill training. In the undergraduate stage, the cultivation of intercultural communicative competence should be carried out in a spiral cycle, rather than divided into three linear stages in different semesters or grades.

3 Business English Situational Teaching Model Experiment

This paper designs a situational teaching model experiment to study the effect of teaching mode on cross-cultural communication of business English.

3.1 Experimental Objects

This study selects the students of business English major of the school of foreign languages and foreign trade of our city as the research sample. The experiment time is the first semester of grade 2, which starts from April 2020 and lasts for 10 months from January 2021.

3.2 Experimental Arrangement

In the 10 month experiment, the business class is arranged into three stages.

The first stage is the "Western food etiquette" stage, with 4 classes arranged. The differences between Chinese and Western food etiquette are studied. The correct arrangement of Western food tableware is recognized and learned, the correct use method of Western tableware, napkin and wine cup is mastered, and the arrangement principle and order of serving the formal western food are understood. The purpose of this paper is to help students form a state of view of respecting cultural differences and stimulate their love of life.

The second stage is the "Western idiom" stage, which is arranged with 4 classes. The skills to learn are to feel the strong cultural connotation of Western idioms, master the correct meaning of the Western idioms related to color, animals and numbers, correctly use Western idioms related to color, and correctly use Western idioms related to animals and numbers to communicate. The purpose of this paper is to improve the students' awareness of using English language and to cultivate their sense of language beauty in the context of western culture.

The third stage is the "western festival" stage, which also has 4 classes. The skills to learn are to understand the background knowledge and related celebrations of major western festivals such as Christmas and Thanksgiving, aiming to feel the charm of western traditional culture and improve the interest in learning western culture.

The above teaching activities are arranged according to the teaching arrangement, and the teaching tasks of each unit are divided into two classes. The first time, students will explore the situation, study independently, display the results of the classroom learning, and carry out the learning evaluation and summary; teachers set up the task of expanding activities. The second class students report the development of learning results, and the teachers will summarize and evaluate.

Curriculum Plan of Western Food Etiquette. The teaching content and arrangement are shown in Table 1.

Table 1. Course schedule of Western food etiquette

Schedule	Teaching activities
Week 4 to week 5	A questionnaire survey of Western food etiquette was conducted among the students
	Classroom teaching of Western food etiquette with situational teaching mode
	Students report classroom learning outcomes
	Classroom teaching evaluation
Week 6	Classroom teaching evaluation teachers arrange and expand learning tasks according to learning themes
	Students work in groups to complete the tasks assigned by teachers in extra-curricular time
Week 7	Students show group learning results, teachers organize classroom evaluation, and summarize the learning activities of this topic
	A questionnaire survey was conducted among the students
	Reflect on the problems in this round of action research and find solutions

Curriculum Arrangement of Western Idioms. The teaching content and arrangement are shown in Table 2.

Table 2. Western idiom curriculum arrangement

Schedule	Teaching activities
Week 9 to week 10	A questionnaire survey of Western idioms was conducted among the students
	A questionnaire survey was conducted among the students, and the situational teaching model was adopted in the classroom teaching of Western idioms
	A questionnaire survey was conducted on the students in the Western idiom pre-test. The situational teaching mode was adopted to teach the Western idioms in the classroom. The students reported the classroom learning results and evaluated the classroom teaching
Week 11	Teachers arrange and expand learning tasks according to learning themes
	Students work in groups to complete the tasks assigned by teachers in extra-curricular time
Week 12	Students show group learning results, teachers organize classroom evaluation, and summarize the learning activities of this topic
	A posttest survey of Western idioms was conducted among the students
	This paper conducts a post-test questionnaire survey on Western idioms to reflect on the problems in this round of action research and find solutions

Western Festival Culture Curriculum Arrangement Plan. The time of the western festival culture curriculum is the 14th week to the 17th week. The experimental arrangement is similar to the previous one. The 14th week to the 15th week is a stage, the 16th week is a stage, and the 17th week is a stage.

4 Analysis of the Application Effect of the Situation Teaching Mode

4.1 Comparative Analysis of Pre and Post-test of Questionnaire

The author conducted three rounds of action research before and after the issue of 276 pre and post-test questionnaires, of which 276 effective papers were effective, with an effective efficiency of 100%. As shown in Table 3, the t-test results of pre-school and post-test of intercultural communication ability training based on situational teaching mode are presented.

Table 3. T-test results of three pre and post tests

Manners book	Sig. (bilateral) = .000 < 0.01	The difference is very significant
Western idioms	Sig. (bilateral) = .000 < 0.01	The difference is very significant
Western festivals	Sig. (bilateral) = .000 < 0.01	The difference is very significant

By analyzing the statistical results of the above experimental data, we can know that the significant probability of t-test is $p < 0.01$, and there are significant changes in Business English before and after the situational teaching, which shows that the situational teaching mode constructed in this paper has a significant effect in the teaching of cultivating the cross-cultural communication ability of business English Majors in secondary vocational schools, and proves the effectiveness and feasibility of the situational teaching mode.

4.2 Comparative Analysis of Mid-term and Final Examination Results

In this paper, we set up mid-term and final examinations for business class students. We use Excel and SPSS to do paired sample t test.

As can be seen from Fig. 1, the final average score is 66.32, which is higher than the mid-term score, and the difference is significant, indicating that students have made significant progress in English learning. In recent years, with the deepening and accumulation of cross-cultural basic knowledge in Colleges and universities at all levels in China, students' communicative expression level and ability in English are gradually improved, and the communicative expression level and ability of business English is like an educational model to cultivate students' intercultural communication level in business and other regions, it has effectively improved the cross-cultural communication ability of teachers of business English and other types of professional at all levels.

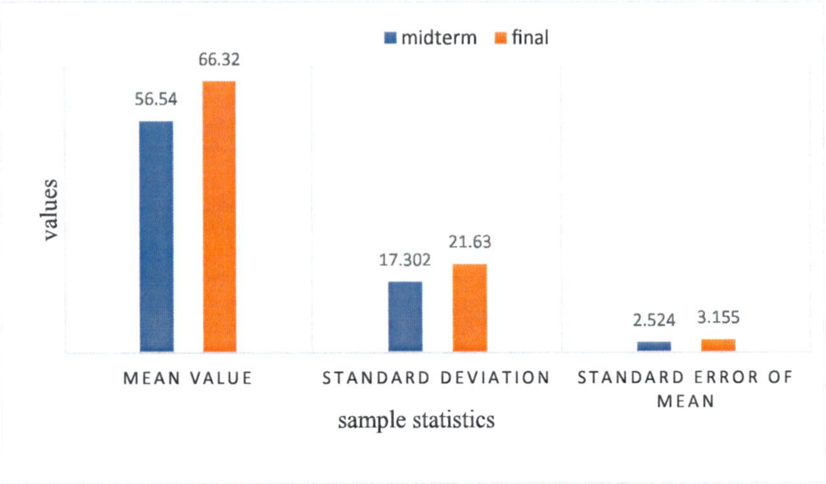

Fig. 1. Paired sample statistics

4.3 Student Interview Analysis Based on Situational Teaching Mode

This paper focuses on: 1) the interest in situational teaching mode; 2) The role of situational teaching mode; 3) The satisfaction degree of the teaching effect of situational teaching mode is three questions. The interview results are shown in Fig. 2.

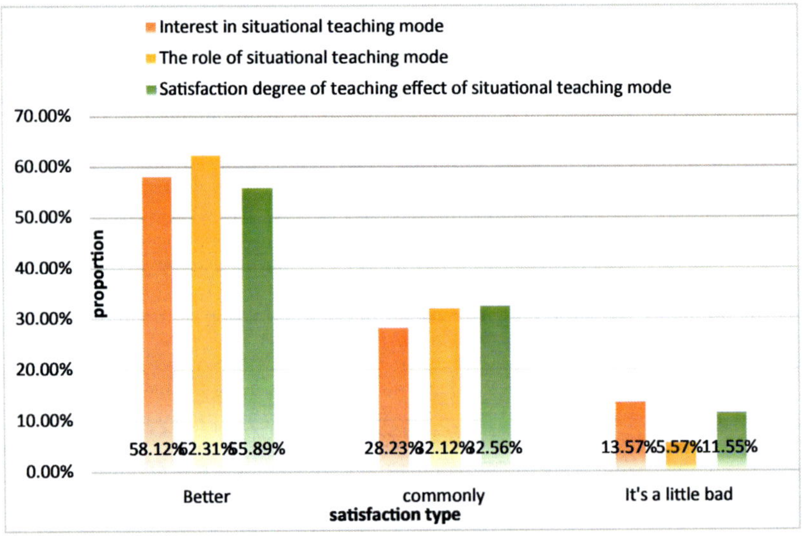

Fig. 2. Interview results

It can be seen from Fig. 2 that most students are satisfied with the situational teaching mode and think it can greatly promote cross-cultural communication.

5 Conclusion

This paper discusses the factors influencing business English cross-cultural communication, and conducts a situational teaching experiment to verify the influence of business English training mode on cross-cultural communication ability. In this paper, situational teaching method is applied to the teaching of cross-cultural communication ability training. By combining the two methods, a specific and feasible teaching mode is explored, which has certain innovation, enriches the content of English teaching and research, and has a certain reference for English teaching reform.

References

1. Luan, Y.: The PAD class teaching of business English translation. Overseas Engl. **342**(02), 120–121 (2017)
2. Wang, P.: The innovation of business English teaching and the cultivation of students' ability in vocational schools. Overseas Engl. **417**(05), 279–280 (2020)
3. Zinchenko, O., Ogrenich, M., Shepel, M., et al.: Future economists cultural competence and business English speech etiquette formation. Revista Romaneasca pentru Educatie Multidimensionala **12**(2), 290–310 (2020)
4. Wen, B.: On teachers' professional qualities structure and development mechanisms in the business English discipline. Chin. J. Appl. Linguist. **42**(2), 218–235 (2019)
5. Zhang, J.: Investigation and survey of graduates employment of business English major - in the case of employees. Overseas Engl. **000**(011), 243–244 (2017)
6. Paternotte, E., et al.: Intercultural communication through the eyes of patients: experiences and preferences. Int. J. Med. Educ. **8**, 170–175 (2017)
7. Ramírez-Cruz, H.: No Manches, Güey! service encounters in a hispanic American intercultural communication setting. J. Pragmat. **108**, 28–47 (2017)
8. Liu, J., Fang, F.G.: Perceptions, awareness and perceived effects of home culture on intercultural communication: perspectives of university students in China. System **67**, 25–37 (2017)
9. China, Y.: One belt, one road, "and construction", "and the communication and intercultural, communication **35**(002), 31–38 (2018)
10. King, C.S.T., Bailey, K.S.: Intercultural communication and US higher education: how US students and faculty can improve. Int. J. Intercult. Relat. **82**, 278–287 (2021)
11. R'Boul, H.: Re-imagining intercultural communication dynamics in TESOL: culture/interculturality. J. Multicult. Educ. **14**(2), 177–188 (2020)
12. Shen, J., Sheng, Y., Zhou, Y.: Learning for adaptation: an edusemiotic perspective on intercultural communication competence. Chin. Semiot. Stud. **16**(4), 603–625 (2020)

Effect of Remote Sensing Image Pixel Decomposition on the Spectral Response of Soil Heavy Metals

Jiabin Huang[1], Qiong Zhang[1](✉), and Yunkai Guo[2]

[1] GuangZhou City Construction College, Guangzhou, Guangdong, China
[2] Changsha University of Technology, Changsha, Hunan, China

Abstract. In view of the disadvantages of low spectral resolution of remote sensing images, weak correlation between the original spectrum and heavy metal content, this paper starts from different data sources to study the response relationship of soil spectral reflectance to soil heavy metals and images under pixel unmixing. Remote sensing image has high spatial resolution, wide data source and fast update, which is conducive to the establishment of large-scale inversion of heavy metal content in soil spatial dynamic analysis. In this paper, we take Gangkou Town of Yueyang County and Dapu Town of Hengdong County as the research area, using Landsat8 data and Sentinel-2 image data as the data sources, and using the pixel linear unmixing model to extract the pure soil spectra in the image mixing end members, respectively Perform different spectral transformations on the two data sources, and compare the correlation between the original image spectrum and the decomposed soil spectrum and heavy metals. The research results show that whether it is for Landsat8 image data or Sentinel-2 image data, the number of bands that pass the sig < 0.05 two-sided correlation test after spectral linear decomposition is greater than before unmixing, and pass sig < 0.01 two-sided correlation The number of bands for sex testing has also increased, indicating that the spectral response of the soil spectrum and heavy metal Cr extracted through pixel decomposition has been effectively improved, and the spectral information has been enhanced.

Keywords: Pixel linear decomposition · Spectral response · Landsat8 remote sensing image · Sentinel-2 remote sensing image · Correlation test

1 Introduction

With the rapid development of China's industry and agriculture, the hot issues caused by environmental pollution follow. In recent years, especially in the cultivated land crop system, the "three wastes" emitted by industry and the drugs sprayed by agriculture have caused serious pollution to the cultivated soil in China, which caused the sharp accumulation of heavy metal elements in cultivated land crops to exceed the standard, and caused many cases of excess content of heavy metals in the soil. The traditional

B. J. Jansen et al. (Eds.): *International Conference on Cognitive based Information Processing and Applications (CIPA 2021)*, LNDECT 85, pp. 182–193, 2022.
https://doi.org/10.1007/978-981-16-5854-9_23

heavy metal content pollution monitoring and evaluation is not only time-consuming and labor-intensive, but also difficult to detect for large-scale soil heavy metal pollution. How to effectively prevent and control heavy metal pollution at present is premised on how to quickly grasp the spatial distribution trend of heavy metal content in the research area.

With the progress of remote sensing science and technology, and the continuous development of new remote sensing devices, remote sensing images have been widely used in recent years. Remote sensing images provide rich spectral information for different space-time scales with their macro, time continuity and data comprehensiveness. At present, the main data source of soil heavy metal remote sensing monitoring is geostransspectral data, hyperspectral can often provide hundreds of thousands of spectral information, identify different types of soil properties and reflect certain soil physical and chemical properties, for soil composition quantitative information extraction provides a way of thinking. Kooristra L and others built a PLSR model based on the significant correlation between soil visible light band reflection spectra and heavy metals, successfully estimated the content of heavy metals Zn and Cd in soil along the Rhine River, and pointed out the feasibility of inverting cadmium and zinc in soil heavy metals by PLSR [1]. Wang et al. compared the effects of the general Partial Least Squares Regression (PLSR) and Partial Least Squares Regression Combined with Genetic Algorithm (GA-PLSR) on inverting low-polluting concentrations of heavy metals Zn, As, Pb and Cu in soil, and found that the GA-PLSR model had better predictive power [2]. Compared with hyperspectral data, remote sensing image multispectral has the advantages of easy access to data, fast time update and high image quality, which can provide strong support for the rapid dynamic monitoring of heavy metals. Murakami and others obtain the required ground data through SPOT image data, and explore the spatial distribution of crops in the study area [3]. Remote sensing images for soil heavy metal inversion work, because the image of cells is not an independent embodiment of a certain kind of land, but a mixed expression, resulting in the image reflected in the spectral information is not simple, directly used in cultivated soil heavy metals inversion may lead to poor estimation effect. The simulation band of Landsat TM image and ASTER image multispectral data used by Lu Wang and others to estimate the heavy metal content in the study area shows the feasibility of using remote sensing imagery to monitor the content of heavy gold elements in soil [4]. Zheng et al. simulate sentinel-2 sensor multispectral red-edge band data from hyperspectral data and build new vegetation indices using simulated red-edge band data to monitor wheat strip rust [5]. Tianxiang Cui and others used Landsat TM remote sensing images to estimate vegetation coverage in the study area using linear decommissioning models. The results show that the end metamodel based on linear deconscing can fully reflect the spectral differences in the study area and further improve the inversion accuracy of the model prediction [6]. Zuowei Huang, etc. use the optimized end-element judgment method and the end-element extraction method, put forward a new hybrid cell decomposition method, and get accurate end-of-earth element and abundance, the results of the mixing provide an effective reference information for the classification of urban land [7]. Yachun Zhang and other nature reserves as the research object, using remote sensing image data, based on object-oriented classification combined with spectral demixing method to invert the vegetation coverage of the study area. The results show that the accuracy of inversion results is improved by using this method compared with traditional multi-ended metaspectral disassembling [8, 9].

Based on the research results of heavy metal remote sensing monitoring at China and abroad, this paper will explore the linear decomposition of remote sensing cells using the bi-binary decomposition model of cells, extract the spectral reflection information of soil spectroscopic images of remote sensing images, and perform spectral enhancement transformation processing on the soil reflectivity and the original image reflectivity of the extracted images and compare the effect of the reflectivity of remote sensing images before and after spectral disassembling on the response of heavy metal Cr spectra. This provides data support for subsequent model construction, which improves the accuracy of soil heavy metal estimation model.

2 Data Acquisition and Processing

2.1 Overview of the Research Area

The research areas selected in this paper are located in Gangkou Town of Yueyang County and Dapu Town of Hengdong County, both of which are important grain producing areas. Gangkou Town is located in the northeast of Yueyang County, is the largest town, but also the surrounding area of the transportation hub channels and economic exchange center. The town is located at 29.1° east longitude and 113.2° north latitude, and the climate in the study area is humid, with an average annual precipitation of 1350 mm. The total area of Gangkou Town is approximately 139.38 km^2. Among them, the arable land crop area is about 47 km^2, more than one-third of the town area, is an important local grain producing area. With the accelerated development of various industries in the research area, the impact of waste from industry and agriculture on cultivated crops is gradually increasing. Dapu town in the research area is located in the southwest direction of Hengdong County, the geographical location is 112°45′ east longitude, 26°59′ north latitude, the study area climate is mild and humid, and the annual rainfall is abundant. Dapu Town's convenient transportation, several highways through the research area, geographical advantages are obvious. Dapu Town area of 115.05 km^2, of which the mountain area of about 5766 acre, arable land area of up to 5898 acre. In recent years, the research area has vigorously developed the industrial and agricultural industries, and also built industrial parks with chemical and heavy metal industries as the main factors, which have caused some heavy metals exceeding the standard, which has greatly affected the agricultural ecological environment of Dapu Town.

2.2 Data Acquisition and Processing

In order to improve the accuracy of geometric correction of remote sensing images, the location of the control point of the research area should be laid in advance before field experiments can be carried out. Control point laying should cover the whole research area, the position of control point should have obvious characteristics. In the actual collection, the road intersection and the road corner point are the majority. Secondly, according to the shape and size of the study area, the ground sample point of 15 m ×

15 m is selected in the study area, and the center of the sample point is used as the center point of the point, and the geographical coordinates of the center point are measured. Gangkou Town collected a total of 15 control points, 30 sample square points. Dapu Town collected a total of 21 control points, 60 sample square points.

2.3 Soil Sample Collection and Physical and Chemical Properties Determination

During field operations, soil sample points are distributed in the form of "S". When collected, the sampling depth of soil sample points was approximately the same and the depth was not more than 20 cm. Mix the collected soil samples evenly and pack them into a black plastic bag and mark the dot. At the same time, the form is filled with sampling date, climatic conditions, health status of cultivated land and other basic information. Bring the soil back to the laboratory for natural air drying, screening for debris in soil samples, and filtering through 100 nylon sieves. The filtered fine soil is mixed evenly and loaded into the sealed bag with a marking number and placed in the dryer for backup. The determination of heavy metal content in soil is strictly carried out according to the standard, and the laboratory specific determination work is carried out by the professional testing unit according to the national standard.

3 Remote Sensing Image Acquisition and Processing

3.1 Remote Sensing Image Pre-processing

Remote sensing images are susceptible to the interaction between electromagnetic waves and atmosphere, satellite changes, random noise and other factors in the imaging process, resulting in the actual acquisition of image grayscale values cannot fully reflect the size of electromagnetic wave energy of their subjects. Hence, image distortion and geometric distortion are caused. In order to eliminate the influence of the environment, improve the quality of the image and obtain the real radiation information of the earth, it is necessary to pre-treat the remote sensing image.

In this paper, the data source selected by Gangkou Town is Landsat8 satellite imagery, which is synchronized with the data collected by field experiments. In this paper, Landsat8 pre-processing process includes: (1) radiation correction: radiation correction includes radiation calibration and atmospheric correction, the purpose is to eliminate the sensor itself, atmosphere, terrain and other factors, to obtain the real ground reflectivity; (2) geometric correction: the remote sensing image imaging process by different factors led to the image of geometric deformation, through the ground measured control point to correct the deformation, can get the correct ground features; (3) image cropping: a single remote sensing image contains a wide range of areas, and the experimental research area is only part of it, through image cropping, to obtain only the study area of the image; (4) classification of image features: different types of earth objects have different characteristics of different geographical differences, according to this difference, all pixels in the image according to their own nature into several different categories of the process. According to the actual situation of the research area, the land is divided into forest land, cultivated land, construction land, water body and five other categories, which is convenient for follow-up treatment.

With the continuous development of remote sensing images, the spatial resolution of remote sensing images is getting higher and higher, and the quality of images has been improved. The remote sensing image data source selected in Dapu Town is sentinel-2 series of images. Sentinel-2 images contain different levels of data, commonly L1C-class multispectral data, Sentinel-2 L1C data is an orthodontic and geometric precision-corrected atmospheric apparent reflectance products and is not atmospherically corrected. According to the time of soil acquisition in the field, remote sensing images similar to the experimental time were selected, and the image level was S2-L2A, which is the data containing atmospherically corrected atmospheric underlying reflectivity. Based on this, the S2-L2A data processing in this paper mainly includes: (1) resampling: sentinel-2 image data contains 13 bands, each with three different resolutions, the data process is not unified resolution, the subsequent process will be difficult to continue. In this paper, the SNAP software processing tool is used to resample and format the data by the closest method; (2) band synthesis and cropping: the data obtained by resampling above is a single-band full-color image, using ENVI for band synthesis, output as a complete image, and then through cropping to obtain the study area image data; (3) image classification: The same type of land is divided into the same post-crop images, and the data of different land types are saved for easy subsequent processing.

3.2 Cell Linear Decomposition

Remote sensing image cells usually contain vegetation and soil information, and the cells corresponding to GPS quadrat points in the field are often the common expression of the spectrum of various land types. In order to explore the response relationship between heavy metals and spectral characteristics of remote sensing soil, mixed cells were decomposed, pure cells containing only soil spectral information were extracted, and the effects of remote sensing image spectra on soil heavy metal spectra before and after the disassembling were compared. In this paper, both research areas are grain-producing areas, mainly soil and vegetation, so the use of cell ditomy model for linear decomposition. The basic idea of the cell dilysis model is to assume that a cell contains only two parts of vegetation spectrum and soil spectrum, with its area ratio in the cell as its weight factor. Commonly used dilyst models are expressed as:

$$T = T_{vb} * f_v + T_{sb} * f_s \tag{1}$$

In the formula, T represents a cell spectroscopic information of the image, Tvb, Tsb represents vegetation information and soil information in the various bands of sentinel II, and fv and fs are the proportion of vegetation information and soil information in the cell. Fv indicates the vegetation cover of the study area, expressed as follows:

$$\mathrm{f_v} = (\mathrm{NDVI} - \mathrm{NDVI_{min}})/(\mathrm{NDVI_{max}} - \mathrm{NDVI_{min}}) \tag{2}$$

$$f_s = 1 - f_v \tag{3}$$

In the formula, NDVImin and NDVImax are represented as NDVI thresholds in the study area. According to the above three formulas, it can be concluded that the soil spectral information expression under the cell dioritation model is:

$$Tsb = (T - Tvb^{*}fv)/(1 - fv) \tag{4}$$

The two research areas in this paper use the same division model and processing process, the following is the basic processing process of the two-part model:

The vegetation index of the study area was calculated. In order to estimate vegetation cover fv in the study area, the vegetation index must be selected as the index factor. There are many common vegetation indices, of which vegetation index NDVI is recognized as the optimal index factor for vegetation development and vegetation cover area. In this paper, the commonly used NDVI vegetation index is used to estimate vegetation cover fv, and the calculation of NDVI is completed by ENVI software band processing.

Land category masking file production. To remove background cells from the study area, unused information is eliminated through mask files. According to the results of the classification of the land, the species land category mask file is saved for the calculation of the maximum minimum value of the subsequent NDVI.

Table 1. Landsat8 NDVI thresholds for all types of lands

The type of land cover	$NDVI_{min}(NDVI_{soil})$	$NDVI_{max}(NDVI_{veg})$
Woodland	0.041409	0.858117
Cultivated land	0.04247	0.864183
Construction land	0.04134	0.84935
Body of water	0	0
Other	0.041919	0.839189

Table 2. Sentinel-2 NDVI threshold for all types of land

The type of land cover	$NDVI_{min}(NDVI_{soil})$	$NDVI_{max}(NDVI_{veg})$
Woodland	0.5132	0.8659
Cultivated land	0.1304	0.8408
Construction land	0.0359	0.7118
Body of water	0	0
Other	0.0685	0.6561

NDVI threshold and parameter file acquisition. According to the results of remote sensing image classification, the cells of five land object categories are counted. In this paper, with 95% confidence, the $NDVI_{max}$ and $NDVI_{min}$ of the land cover category are counted respectively. As shown in the Table 1 and Table 2, water bodies are generally considered not to cover vegetation, so the vegetation coverage in the water body area is 0. $NDVI_{max}$ and $NDVI_{min}$ cover the threshold for all kinds of land objects, using the ENVI band calculation function, taking the NDVI threshold of the various object categories as the coefficient, taking the land category mask file as the input parameters, and finally calculating the $NDVI_{soil}$ and $NDVI_{veg}$.

Soil spectral information extraction. According to the obtained threshold $NDVI_{max}$ and $NDVI_{min}$ sub-formula (2), the vegetation coverage fv of the study area can be obtained. According to the statistics of NDVI, the vegetation spectrum of the study area screened out several cell spectroscopic features with NDVI greater than 0.8 as the result. By combining the above data results, according to the formula (4), the reflectivity information containing only the spectral characteristics of the soil can be obtained.

3.3 Image Spectrum Preprocessing

Due to the influence of satellite platform, measurement conditions, instruments and equipment and other random factors, there are different degrees of error in the spectral data obtained. The scattering direction, different particle size distribution and size of light all affect the acquisition of spectra. In order to weaken the irrelevant information and noise of spectral aggregation and improve the spectral characteristic response ability of heavy metal elements, a series of pre-processing of the original reflectivity of the image is required.

Commonly used pre-processing methods of spectral data include differential transformation, penultimate log spectral transformation, anti-sine square root transformation, band combination transformation and other different processing methods. Spectral differential transformation can eliminate the influence of baseline drift, noise and other factors on the spectrum. The processing of the reciprocal logarithm can enhance the spectral differences in the visible light band. The spectrum is close to the normal distribution under the transformation of the square root of the antisine, which brings the variance closer. Band combinations highlight band information and reduce interference with background noise.

In this paper, first-order differential transformation, second-order differential transformation, countdown conversion and other methods are used to process the soil reflectivity of the image after the dissolving and unmissable image reflectivity, and to study the spectral response relationship between two different spectral reflectivity and heavy metals.

4 Results and Analysis

Correlation analysis is an indispensable step in the study of remote sensing quantitative inversion. Correlation analysis is the analysis of the relationship between two and more variables to determine the degree of correlation between variable factors. In this paper, using SPSS data processing software, the correlation analysis between Landsat8 image data and Sentinel-2 image data before and after demixing spectrum and heavy metal element Cr is analyzed, and the response relationship between image spectrum before and after disassembling to heavy metal Cr is explored.

4.1 Analysis of Landsat8 Image Data Results

Table 3. Landsat8 spectral reflectivity and heavy metals correlation (unmixed)

Center wavelength/nm	Raw	Reciprocal logarithm	First-order differentials	Second-order differentials
440	−0.367	0.347		
480	−0.373	0.353	0.394	
560	−0.443	0.442	0.333	0.384
655	−0.400	0.380	0.363	0.049
865	0.300	−0.308	0.085	0.355
1 610	−0.208	0.180	0.403	
2 200	−0.333	0.293		

**Significant correlation at 0.01 level (both sides).
*Significant correlation at 0.05 level (both sides).

Table 4. Sentinel-2 NDVI threshold for all types of land

Center wavelength/nm	Raw	Reciprocal logarithm	First-order differentials	Second-order differentials
440	−0.474*	0.465*		
480	−0.468*	0.455*	0.021	
560	−0.489*	0.497*	0.302	0.071
655	−0.465*	0.460*	0.426	0.559*
865	−0.556*	0.583**	−0.778**	0.361
1610	−0.468*	0.462*	0.302	
2200	−0.457*	0.447*		

**Significant correlation at 0.01 level (both sides).
*Significant correlation at 0.05 level (both sides).

The results of the Landsat8 image data are shown in Table 3 and 4. Where sig < 0.05 represents a significant correlation between spectral reflectivity and heavy metal Cr at the level of 0.05, sig < 0.01 represents a significant correlation between spectral reflectivity and heavy metal Cr at the level of 0.01; that is, through correlation test, it is shown that the spectral reflectivity and heavy metal Cr spectroscopic response are significant, there is a good linear relationship. According to the data results of Table 3, Landsat8 images are not obvious in spectral linear decomposition, the spectral response to heavy metal Cr is weak, and neither band has passed the correlation test. After the spectral transformation of the original spectrum of the image, such as reciprocal logarithm, first-order differential and second-order differential, the spectral characteristics can not be improved and the weak information of the spectrum has not passed the correlation test. According to Table 4, Landsat8 image obtains the image soil spectrum after the bipolar decomposition of the cell, and the spectral response of each band to heavy metals is significantly improved without spectral transformation, and the correlation between each band and heavy metal Cr is tested by 0.05. After the reciprocal logarithm to the transformation, each band passed 0.05 two-sided significant test, and at 865 nm, the correlation coefficient is up to 0.583, indicating that the band after spectral transformation of the heavy metal Cr's spectral response is most sensitive. After the first-order differential transformation, the other band spectroscopic information was not enhanced, only reached 0.01 two-sided significant correlation at 865 nm, the maximum correlation coefficient is −0.778. At this time, the band spectrum reacts strongly with the heavy metal Cr. After the second-order differential transformation, the spectral information was not effectively improved. The data of the two tables can be combined to predict that the linear decomposition of landsat8 image spectra has effectively improved the weak information of the image spectrum and enhanced the spectral response of soil reflectivity and heavy metal Cr.

4.2 Analysis of Sentinel-2 Image Data Results

The statistical results of Sentinel-2 image data are shown in Table 5 and 6. Where sig < 0.05 represents a significant correlation between spectral reflectivity and heavy metal Cr at the level of 0.05, sig < 0.01 represents a significant correlation between spectral reflectivity and heavy metal Cr at the level of 0.01; that is, through correlation test, it is shown that the spectral reflectivity and heavy metal Cr spectroscopic response are

Table 5. Sentinel-2 spectral reflectivity and heavy metals correlation (unmixed)

Center wavelength/nm	Raw	Reciprocal logarithm	First-order differentials	Second-order differentials
443	−.318*	.363**		
490	−.274*	.296*	0.062	
560	−0.15	0.172	−0.192	−0.03
665	−.259*	.276*	0.028	.316*
705	−0.061	0.047	.313*	.330*
740	0.233	−0.231	.355**	−0.11
783	.368**	−.364**	.364**	−.275*
842	.406**	−.415**	.381**	−0.256
865	.410**	−.406**	−0.102	−.460**
945	.270*	−.280*	−.397**	0.033
1610	−0.132	0.129	−.292*	
2190	−0.172	0.192		

**Significant correlation at 0.01 level (both sides).
*Significant correlation at 0.05 level (both sides).

Table 6. Sentinel-2 spectral reflectivity and heavy metals correlation (after disassembling)

Center wavelength/nm	Raw	Reciprocal logarithm	First-order differentials	Second-order differentials
443	0.224	−0.234		
490	0.254	−0.241	.323*	
560	.341**	−.313*	−0.25	0.077
665	0.074	−0.018	.314*	.372**
705	.345**	−.344**	.373**	.372**
740	.367**	−.378**	.377**	−.336**
783	.379**	−.398**	.381**	−.357**
842	.378**	−.401**	.432**	−.296*
865	.389**	−.411**	−0.027	−.424**
945	.377**	−.394**	−.397**	−0.145
1610	.271*	−.314*	−.379**	
2190	0.126	−0.152		

**Significant correlation at 0.01 level (both sides).
*Significant correlation at 0.05 level (both sides).

significant, there is a good linear relationship. As can be seen from Table 5, sentinel-2 images have a good spectral response between the original spectral reflectivity and the heavy metal Cr, and more than half of the bands pass the 0.05 two-sided significance test and the 0.01 two-sided significance test. The original spectrum before unmissable was transformed by a reciprocal logarithm, and the band number of the two-sided significance test increased by 0.01, and the correlation increased overall. After the first-order differential transformation, the number of bands passed the 0.01 two-sided significance test increased. After the second-order differential transformation, at 865 nm, the correlation coefficient reaches a maximum value of −0.460. According to Table 6, Sentinel-2 images obtain the image soil spectrum after pixel binary linear decomposition. After unmixing, the soil spectrum has passed the 0.01 two-sided significance test for more than half of the band without spectral transformation, indicating that these bands have a better spectral response with heavy metal Cr. After unmixing, the original spectrum undergoes reciprocal logarithmic and first-order differential transformation, and the overall correlation is improved, indicating that after spectral transformation, the spectral difference is improved, and the weak information of the spectrum is enhanced. After the second-order differential transformation, the spectral characteristics have not been significantly improved, only reaching 0.01 two-sided significant correlation at 865nm, and the maximum correlation coefficient is −0.424. From Table 4.3 and Table 4.4, it can be seen that after unmixing Sentinel-2 image pixels, the number of bands that pass the 0.05 two-sided significance test and the 0.01 two-sided significance test are improved, and the correlation between the image spectrum and the heavy metal Cr is overall Improved. It shows that the Sentinel-2 image is decomposed by pixels, which effectively enhances the weak information of the image spectrum, and enhances the soil reflectivity and the spectral response of heavy metal Cr.

4.3 Results and Discussions

Based on the above results, it can be seen that the response of the image spectrum and heavy metals before unmixing shows that the correlation between Landsat image data and heavy metals is numerically greater than the correlation between Sentinel-2 image data and heavy metals. However, none of them passed the significant correlation test. The main reason was that the number of samples collected in the field of Gangkou Town was relatively small. In statistics, a small number of samples can improve the correlation coefficient, but it does not mean that the two have a good linear relationship. Moreover, after spectral transformation, the overall spectral response between the spectrum of each band and heavy metal based on Landsat images has not been greatly improved, but the maximum correlation can be achieved at a certain band. For Sentinel-2 image data, after spectral transformation, except for the second-order differential that has not been improved, the reciprocal logarithmic and first-order differential transformation have improved the overall correlation. All in all, whether it is based on Landsat image data or Sentinel-2 image data, after the pixel binary linear unmixing, the spectral information and the spectral response of the heavy metal Cr are improved, indicating that the soil spectral information extracted after the pixel binary model decomposition is better. The original spectral information of the image is more sensitive to the response of heavy metals.

5 Monitoring Data Analysis and Processing

In this paper, Landsat8 image multi-spectrum and Sentinel-2 image multi-spectrum are used as data sources, and the pixel dichotomy model is used to linearly decompose remote sensing image pixels, so as to explore the characteristics of the influence of remote sensing image pixel decomposition on the spectral response of heavy metals. From the research results, the soil reflectance spectrum and the spectral response of heavy metal Cr obtained after the decomposition of the above image data sources have been improved, indicating that this method can effectively enhance the weak information of the spectrum and improve the sensitivity between the two. The analysis of the correlation between spectral reflectance and heavy metals is a key step in the quantitative inversion of remote sensing images. Only when the spectrum and heavy metals have a good spectral response can a remote sensing inversion model with high accuracy be constructed. The current research on quantitative inversion of heavy metals using remote sensing images is mainly focused on model construction. This paper attempts to start from the response of spectra and heavy metals to study the relationship between the original spectrum of the image and the response of pure endmember spectra to heavy metals. This is a new research idea. In follow-up research, how to further improve the reaction between weak spectral information and heavy metals is a key issue for remote sensing directed inversion.

Acknowledgments. This paper is funded by the National Self-Fund Project (41671468) and the Guangdong Provincial Focus Area Special (2020ZDZX1066).

References

1. Kooistra, L., Wehrens, R., Leuven, R.S.E.W., et al.: Possibilities of visible-near-infrared spectroscopy for the assessment of soil contamination in river floodplains. J. Anal. Chimica Acta **446**(1), 97–105 (2001)
2. Wang, J., Cui, L., Gao, W., et al.: Prediction of low heavy metal concentrations in agricultural soils using visible and near-infrared reflectance spectroscopy. J. Geoderma **216**(4), 1–9 (2014)
3. Mkhabela, M.S., Bullock, P., Raj, S., et al.: Crop yield forecasting on the Canadian Prairies using MODIS NDVI data. J. Agric. Forest Meteorol. **151**(3), 385–393 (2011)
4. Wang, L., Lin, Q.D., Jia, D., et al.: Research on predicting the content of soil heavy metal elements based on reflectance spectroscopy. J. Remote Sens. **11**(6), 906–913 (2007)
5. Zhong, Q., Huang, W.J., Cui, X.M., et al.: New spectral index for detecting wheat yellow rust using Sentinel-2 multispectral imagery. J. Sens. **18**(3), 868–887 (2018)
6. Cui, T.X., Gong, Z.N., Zhao, W.J., et al.: Methods for extracting wetland vegetation coverage under different end-member models-taking Beijing Yeyahu Wetland Nature Reserve as an example. J. Acta Ecologica **33**(04), 1160–1171 (2013)
7. Huang, Z.W., Hu, G.W., Xie, S.X.: Research on urban ground object classification based on spectral unmixing. J. Agric. Mach. **49**(10), 205–211 (2018)
8. Zhang, C.Y., Na, X.D., Zang, S.Y.: Improvement of multi-terminal spectral unmixing model and inversion of vegetation coverage. J. Northeast Forestry Univ. 46(01), 68–71+79 (2018)
9. Zhang, Q., Guo, Y.K., Qian, J., Jiang, M., Ding, M.Q.: Remote sensing inversion of soil heavy metal Cr based on pixel binary model. J. Surv. Mapp. Sci. **45**(04), 58–63 (2020)

Computer Artificial Vision Image Processing System Based on FPGA

Xiuzhi Su(✉) and Guocai Zuo

Hunan Software Vocational Institute, Xiangtan, Hunan, China

Abstract. With the continuous development of signal processing technology, digital image processing technology has also made a breakthrough, the application field is also expanding. This paper summarizes the concept of FPGA, introduces the advantages of FPGA, summarizes the concept of artificial vision image, and puts forward the requirements of artificial vision image. The results show that the number of FPGA related literatures increases with the year. In 2021, 255 papers were published, and in 2021, 201 papers were published on HowNet.

Keywords: FPGA · Computer vision technology · Artificial vision image · Salient object detection algorithm

1 Introduction

With the rapid development of artificial intelligence technology, the function of vision sensor chip is more and more powerful, and the image processing based on vision sensor is more and more intelligent. As an important part of artificial intelligence, visual image information has received unprecedented attention.

With the continuous development of science and technology, many experts have studied the visual image. For example, some domestic teams have studied the global saliency detection algorithm of multiple visual features, and proposed an advanced image processing software environment based on FPGA, whose purpose is to hide hardware details from users as much as possible. This paper presents a very advanced image processing coprocessor (IPC), whose core instruction set is based on image algebra. The environment includes a generator that generates an optimized architecture for specific user-defined operations. An active vision system architecture based on FPGA is proposed. Active vision is another way to solve the problem of artificial vision. Its core idea is to consider the perception of visual tasks, which is inspired by the biological visual system. Therefore, a method based on programmable chip system is proposed. The system is implemented by FPGA and connected with CMOS imager and inertial device. The system is based on programmable device structure and has strong universality, which allows parallel image processing algorithm. In an image analysis system, 1 million reconfigurable hardware and two CMOS cameras are used. The system is a part of the robot sensor system, which can transmit the position information of the robot and the relative distance information of other objects in real time. This paper presents

B. J. Jansen et al. (Eds.): *International Conference on Cognitive based Information Processing and Applications (CIPA 2021)*, LNDECT 85, pp. 194–201, 2022.
https://doi.org/10.1007/978-981-16-5854-9_24

a system that integrates retinoplasty chips into a platform. The platform includes a circuit board with reconfigurable hardware (FPGA) and a traditional computer to evaluate image processing schemes, such as front-end motion detection. Using an artificial retina, the light intensity is converted into spikes, and they are sent using event driven protocol. In order to build the development platform, FPGA is used to build the driver for the circuit board. As the interface between retina and PC, FPGA stores the captured peak signal [1]. Some experts have studied the image processing algorithm in the real-time 3D information acquisition system, and proposed a multi-core architecture based on ASIP's Viola and Jones algorithm. A new scanning strategy is adopted to make data access more regular. At the same time, through the multi-level optimization of detection processing and application-oriented ASIP, the design can meet the overall requirements. This paper compares GPU and FPGA, two main hardware indexes of computer vision algorithm real-time optimization. VHDL language is used to transplant the algorithm to FPGA, and OpenCL language is used to transplant the algorithm to heterogeneous GPU/CPU structure. In addition, the problems in the process of code migration are introduced in detail. The measurement from the direction sensor can be used by comparing at least one of the yaw angle, pitch angle and/or roll angle derived from the relative rotation matrix of one or more solutions with at least one of the yaw angle, pitch angle and/or roll angle derived from the measurement of the direction sensor. Its main purpose is to help drivers by providing important information about traffic environment and vehicle behavior. This design is based on the embedded computing platform. Due to the limitations of physical, memory and computing power, the existing image processing algorithms based on PC platform are not suitable for this kind of application. On the contrary, simple and efficient image processing algorithm is most suitable for our embedded Architecture [2]. Some experts have studied the defogging algorithm of stereo vision camera based on FPGA, and proposed a probability based method to design and implement the simulation engine of energy-saving signal processing system. Firstly, in the framework of probability based processing, this paper introduces how to decompose a given DSP application into different functional modules. In addition, it will show how these individual functional modules can be easily mapped to the limited analog block selection in commercial FPAA devices. This paper presents a hardware implementation method of dense recovery in stereo vision 3D measurement. The traditional three-dimensional stereo vision system imposes the largest volume of stereo communication, which brings great limitations to the artificial vision algorithm. The system has good performance and efficiency, and has a scalable structure. It can be used in many different situations to solve the real-time processing problem of stereo image stream. Parametric SOC is implemented by using double buffering technology, pipeline processing and reconfigurable hardware, which provides designers with the opportunity to determine its correct size and characteristics. Since the processing is completed when the image stream arrives, the structure does not need any external memory. While maintaining the accuracy of 3D data, we can get higher processing speed by using the least resources. Configurable parameters can be controlled through the late/parallel phases of visualization algorithms executed on embedded processors. Considering the 100 MHz hardware FPGA clock and two mpix images, we can obtain more than 30000 depth point dense stereograms of 50 frames per second with minimum initial delay. The implementation of computer vision algorithm on reconfigurable hardware and explicit underlying processing open up a broad prospect for the application of computer vision algorithm in autonomous system. They can be used as coprocessors to reconstruct 3D images with high density information in real time [3]. Although the research of visual image is fruitful, the research of computer artificial vision image processing system based on FPGA is not enough.

In this paper, in order to study the computer artificial vision image processing system based on FPGA, through the research of FPGA, computer vision technology and artificial vision image, we find the prominent target detection algorithm based on FPGA. The results show that FPGA and computer technology are beneficial to the image processing of artificial vision.

2 Method

2.1 FPGA and Computer Vision Technology

FPGA Overview. FPGA is a new high performance programmable logic device developed on the basis of CPLD, and even integrates CPU (central processing unit) and DSP (digital processor) on the new generation FPGA [4]. Therefore, compared with CPLD, FPGA is more suitable for the design of high-speed and high-density digital logic circuits. The basic components of FPGA include not only programmable I/O unit and basic programmable logic unit, but also embedded RAM, rich wiring resources and embedded special hard disk [5].

Advantages of FPGA. As a programmable logic device, FPGA is developed from pal, gal and CPLD [6]. At present, the FPGA chip has reached tens of millions of gates, and the internal integration also includes rich soft core resources such as IPCore, arm, DSP, etc. FPGA has the ability of parallel processing, which is very suitable for machine vision image acquisition system. Different logic functions can be realized by loading different configuration data into FPGA. This reconfigurability will bring great convenience to the design of digital system [7].

Computer Vision Technology. Computer vision, also known as machine vision, uses image sensors instead of human eyes to obtain the target image, converts the image into a digital image, and uses the computer to simulate human discrimination criteria to understand and recognize the image, so as to analyze and draw conclusions on the image [8]. Computer vision is the application of digital image processing and analysis technology based on computer. It uses optical non-contact sensitive technology to automatically collect and analyze the image of the actual scene, so as to obtain useful information and control the operation or production process of the machine. Computer vision is a comprehensive technology involving sensitive technology and computer technology [9].

2.2 Artificial Vision Image

The Concept of Artificial Vision Image. In artificial vision systems, images are captured or converted into signals [10]. The signal is transmitted to the retina through the implant. The implant includes a polymer substrate made of compliance materials such as poly (dimethylsiloxane) or PDMS. The shape of polymer matrix is consistent with that of retina. Electrodes and conducting leads are embedded in polymer substrates. Wires and electrodes transmit signals representing images to retinal cells. Signals representing images stimulate retinal cells [11]. The size control of artificial vision is becoming a standard tool for industrialists interested in this remote non-contact measurement method. The expected accuracy of these systems depends on the resolution of the camera. High precision requires very expensive charge coupled device sensors and frame capture devices.

Requirements of Artificial Visual Image. The requirements of high performance image processing are: (1) the processing speed is fast, the data is large and the calculation is large; (2) The system should adapt to the change of application environment; (3) The system requires small volume, light weight and low power consumption. The above three requirements pose new challenges to the design of modern image processing system. There are two choices in traditional image processing: one is based on special hardware system, the other is based on general processor, and the basic algorithm is realized by software programming system.

2.3 Salient Target Detection Algorithm

Super pixel refers to a group of pixels with similar texture or color features, uniform size and good boundary preservation. Super pixels have more feature information and less quantity. SLIC super pixel segmentation algorithm is a simple and effective segmentation algorithm. The distance between pixels I and j is defined as Eq. (1):

$$d_p = \sqrt{(x_i - x_j)^2 + (y_i - y_j)^2} \tag{1}$$

For each super pixel block, the minimum appearance distance from the surrounding nodes is taken. Then the average of the minimum distances of all pixel blocks is selected as the "meaningless" threshold. If any distance is less than the threshold, the distance is considered unimportant and clipped to 0. This inner edge weight calculation is very effective, as shown in Eq. (2):

$$w_{ij} = |c_i - c_j| \tag{2}$$

In the case of only given image boundary blocks, the saliency value of each boundary node is calculated as the weight of the edge. This step makes the boundary prior and geodesic significance more robust. The difference between adjacent nodes is expressed by the following Eq. (3):

$$\mathrm{T} = \frac{1}{\mathrm{N_i}} \sum\nolimits_{\mathrm{j}=1}^{\mathrm{N}} |\mathrm{c_i} - \mathrm{c_j}| \tag{3}$$

Adaptive threshold segmentation algorithm can find the appropriate segmentation threshold adaptively according to the saliency state of the image. The threshold value of salient image segmentation is twice of the average value of salient image pixels, as shown in Eq. (4):

$$T_a = \frac{2}{W \times H} \tag{4}$$

3 Experience

3.1 Extraction of Experimental Objects

After generating key descriptor, SIFT feature extraction algorithm is completed. The key point with descriptor is called image feature point. The matching of image features can be realized by using these feature points. In FPGA, the exhaustive method is used to match the descriptors of two images, and the parallel characteristics of FPGA are used to speed up the matching. Since two sets of descriptor data need to be matched and the description of the module is input in an image sequence, the descriptor of the first image needs to be saved in advance and matching can only be started after the description of the second image is saved. The design takes advanced image as the first image and backward image as the second image. The order of input images is identified by signal symbols. After the matching of two images is completed, the next group of two images is matched.

3.2 Experimental Analysis

Firstly, a layer of 10% waterproof coating is laid on the ground, and then the machine vision image acquisition system installed in the center of the car bottom detects the edge information of the image between the paved ground and the unpaved ground. The current position of the car with waterproof coating is determined by the edge information. Finally, the position information is transmitted to the vehicle control platform through the system to control the vehicle to drive along the edge straight line. This paper mainly designs a machine vision image acquisition system based on FPGA, which mainly collects and processes the simulated scene image, obtains its edge information, calculates the average value of the pixel coordinates of the edge points through the edge information, and then detects the frame rate of the image collected by the system, and then designs a VGA display window to get the image. The average value of pixel coordinates of characters and edge points, and the frame rate are displayed on the screen. Through the display content to verify whether the system can achieve the design purpose, whether it can realize the accurate detection of the car position. The main work flow of the system is as follows: FPGA is used as the core control chip, CMOS image sensor is configured to output video image in YUV422 format, and then the image is collected and processed by image acquisition module and preprocessing module to detect the frame rate of video image, and then the video image is buffered by SDRAM to read SDRAM data, calculate the average value of the pixel coordinates of the image edge points, and finally carry out VGA display.

4 Discussion

4.1 The Pruning Rate of FPGA

In the process of pruning FPGA, we need to consider the pruning rate, the size of the network model after pruning, the accuracy of the network after pruning and the running speed of the network after pruning. For the convenience of later description, please note that these four aspects are a, B, C and D respectively. As shown in Table 1.

Table 1. Experimental data of different pruning rates

Type	A(%)	B(M)	C(%)	D(s)
Original network	10%	32.1	20%	24
Maximum pruning rate	15%	23.5	25%	12

It can be seen from the above that the pruning rate of the original network is 10%, the size of the pruned network model is 32.1 m, the accuracy of the pruned network is 20%, and the operation speed of the pruned network is 24 s; the pruning rate of the maximum pruning rate is 15%, the size of the network model after pruning is 23.5 m, the accuracy of the network after pruning is 25%, and the operation speed of the network after pruning is 12 s. The results are shown in Fig. 1.

It can be seen from the above that the operation speed of the network is the average of the time required for recognition and classification of multiple test images. It can be seen from Table 1 that the larger the pruning rate is, the smaller the network model is, the smaller the network precision is, and the faster the network operation speed is.

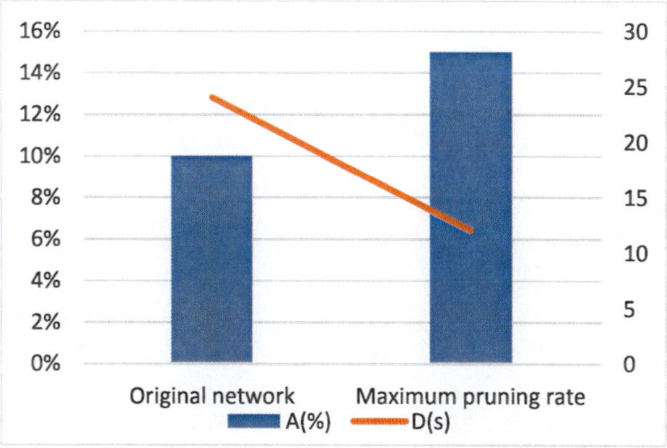

Fig. 1. Experimental data of different pruning rates

4.2 Statistical Analysis of Research Status

In CNKI resource database, 638 related records are retrieved by fuzzy search with "FPGA" as the title keyword; Using "artificial vision image" as the key word, 603 related records were searched by fuzzy search; The statistics of relevant papers from 2017 to the end of 2020 are shown in Table 2.

Table 2. Statistics of keyword search results in 2018–2021

	FPGA	Artificial vision image
2018	94	108
2019	124	127
2020	165	167
2021	255	201

It can be seen from the above that in 2018, there were 94 articles about "FPGA" as keyword retrieval, 124 articles about "FPGA" as keyword retrieval in 2019, 165 articles about "FPGA" as keyword retrieval in 2020, and 255 articles about "FPGA" as keyword retrieval in 2021. In 2018, there were 108 articles about "artificial visual image" as keyword retrieval, 127 articles about "artificial visual image" as keyword retrieval in 2019, 167 articles about "artificial visual image" as keyword retrieval in 2020, and 201 articles about "artificial visual image" as keyword retrieval in 2021. The results are shown in Fig. 2.

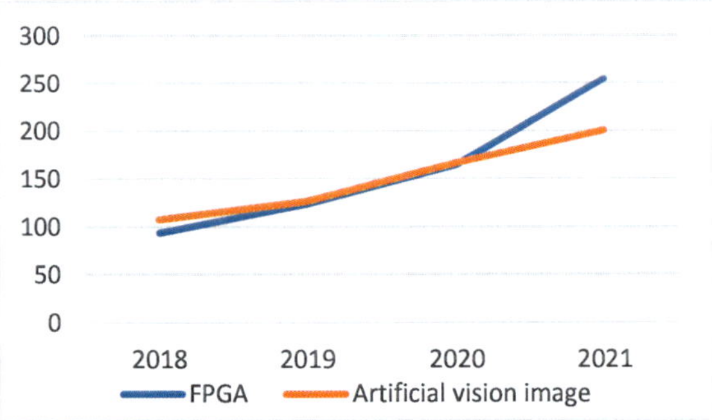

Fig. 2. Statistics of keyword search results in 2018–2021

It can be seen from the above that the number of relevant literatures published on "FPGA" increases with the growth of years. In 2021, the number of relevant literatures published is up to 255, and in 2021, the number of relevant literatures published on HowNet is up to 201.

5 Conclusion

Face detection system is a system that extracts face information from images or video frames. It plays an increasingly important role in computer vision, image processing, artificial intelligence and other fields. An effective design method of sigmoid function of neural network based on FPGA is proposed. The mixed CORDIC algorithm is used to describe sigmoid function in VHDL language at register transfer layer. In order to improve the efficiency and accuracy of Altera FPGA, pipeline technology and look-up table technology are used. By comparing the simulation results of EDA tool with the direct calculation results of MATLAB, we can see that the designed model is accurate and effective.

Acknowledgement. Research projects supported by Hunan Provincial Department of Education (No. 20C1043); Natural Science Foundation of Hunan Province (No. 2020JJ7007).

References

1. Moralistically, R., Linares-Barranco, A., Jimenez-Fernandez, A., et al.: Neuromorphic LIF row-by-row multiconvolution processor for FPGA. IEEE Trans. Biomed. Circuits Syst. **13**(1), 159–169 (2019)
2. Chai, Z., Liu, W., Wu, Q., et al.: FPGA virtualization mechanism based on heterogeneous Zynq platforms. J. Circuits Syst. Comput. **28**(12):1950199.1–1950199.24 (2019)
3. Rui, L., Liu, X., Wang, X., et al.: The design of FPGA-based digital image processing system and research on algorithms. Int. J. Future Gener. Commun. Netw. **10**(2), 41–54 (2017)
4. Esquembri, S., Nieto, J., Ruiz, M., et al.: Methodology for the implementation of real-time image processing systems using FPGAs and GPUs and their integration in EPICS using Nominal Device Support. Fusion Eng. Des. **130**(MAY), 26–31 (2018)
5. Craciun, S., Kirchgessner, R., George, A.D., Lam, H., Principe, J.C.: A real-time, power-efficient architecture for mean-shift image segmentation. J. Real-Time Image Proc. **14**(2), 379–394 (2014). https://doi.org/10.1007/s11554-014-0459-1
6. Li, L.: Research and implementation of fast image processing algorithm based on FPG. Revista de la Facultad de Ingenieria **32**(3), 749–756 (2017)
7. Wang, X., Gu, L., Wang, Z.: Computer medical image segmentation based on neural network. IEEE Access **PP**(99), 1 (2020)
8. Rao, P.S., Yedukondalu, K.: Hardware implementation of digital image skeletonization algorithm using FPGA for computer vision applications. J. Visual Commun. Image Represent. **59**(Feb), 140–149 (2019)
9. Kyrkou, C., Bouganis, C.S., Theocharides, T., et al.: Embedded hardware-efficient real-time classification with cascade support vector machines. IEEE Trans. Neural Netw. Learn. Syst. **27**(1), 99–112 (2017)
10. Jee Ba Ratnam, N., Nayak, S.S., Patra, N., et al.: FPGA design for implementation of the 2D discrete cosine transformation for image processing. Test Eng. Manag. **83**(March–April), 17220–17224 (2020)
11. Peng, C.W., Yang, Z.: Image feature extraction and object recognition based on vision neural mechanism. Int. J. Pattern Recognit. Artif. Intell. **34**(06), 3340–3342 (2020)

Analysis and Application of Computer Graphics Processing and Recognition Technology

Xiuxian Li and Yurong Zhu(✉)

Department of Artificial Intelligence, Shanghai Normal University Tianhua College, Shanghai, China
lxx4353@sthu.edu.cn

Abstract. The ultimate goal of computer graphics processing is to replace people with computers to recognize images and be able to discover what people are interested in images. It is the first step in computer recognition. At this stage, it mainly simulates human vision and hearing in real life. Graphic recognition refers to the use of a machine to process and recognize graphic information (such as text, images and landscapes) to solve an important process of direct communication between the computer and reality. This article analyzes the analysis and application of computer graphics processing and recognition technology. After consulting relevant literature, it summarizes the related technology of computer graphics processing and the algorithm of computer graphics recognition. After experimenting with graphics processing technology and recognition algorithms, through experiments the accuracy of the detection algorithm provides a basis for the application of pattern recognition. The experimental results show that the recognition rate of LeNet-5 in the MNIST image library is only 93 at the beginning, and the latter two models are respectively 95 and 96 at the beginning. CNN1 only achieves 30 times the stability, while the improved CNN2 and CNN3 models achieve a relatively stable recognition rate of nearly 20 times, thereby increasing the relative learning speed. In the end, the recognition rate of the CNN1 model reached 97.8, and the improved CNN1 model achieved a maximum recognition rate of 97.82. The model can reach 98.8 and 99.5, and the improved model algorithm significantly improves the image recognition rate. And in the case of the same number of levels and different node numbers, the recognition rate of the picture first increases, then reaches a peak, and then decreases. The overall rate of return is a trend that first rises and then declines. It can be seen that when the data is 500, that is, the number of nodes is set to half of the input data, and the artificial neural network has the highest recognition rate, the data has the best performance.

Keywords: Graphics processing · Recognition technology · Recognition algorithm · Artificial neural network

1 Introductions

The 21st century has reached an emerging era full of information [1, 2]. As an important visual basis and foundation for modern humans' perception of the world, image art has

B. J. Jansen et al. (Eds.): *International Conference on Cognitive based Information Processing and Applications (CIPA 2021)*, LNDECT 85, pp. 202–212, 2022.
https://doi.org/10.1007/978-981-16-5854-9_25

become the main tool for information transmission and communication [3, 4]. Because of the various useful information that images can bring to us [5, 6], image recognition and recognition technology has developed rapidly with the advancement of modern computer technology, multimedia science and other technologies [7, 8].

In view of the analysis and application research of computer graphics processing and recognition technology, some researchers have conducted research on traditional image recognition technology, and proposed that many image recognition technologies are based on large-scale computers for calculation, and in these calculations and calculation accuracy. There are also very incompatible conflicts between [9]. Due to the large-scale distributed information storage and the customization of artificial neural network technology and the large-scale custom parallel processing, it fully meets the requirements for real-time processing of big data target images, and its high fault tolerance makes the target image. A large number of backgrounds in the video become more blurred [10]. Recognition technology has been successfully widely applied to relatively simple (or strict constraints in the recognition environment) and is basically a two-dimensional space [11]. The main problem that the current image recognition needs to face is: the designer usually needs many different processing procedures to complete the image recognition, and the image recognition is the result of the synthesis of these processing procedures. However, there are no general principles to guide how to organize and coordinate these processes to perform specific tasks. Secondly, various existing image recognition algorithms are more or less restricted. Even with a certain degree of flexibility, the algorithm with better effect still has the disadvantages of large amount of calculation and difficulty in real-time application [12–15].

This article analyzes the analysis and application of computer graphics processing and recognition technology. First, it summarizes the graphics processing technology through the method of literature research, expounds the graphics recognition algorithm, and then conducts computer graphics processing and algorithm recognition experiments. According to the experiment the results provide some relevant basis for the application of computer recognition technology.

2 Research on Computer Graphics Processing and Recognition Technology

2.1 Computer Graphics Processing

Threshold Segmentation. Threshold segmentation is a regional segmentation technique. When the contrast between the object and the background is strong, the segmentation effect is also very good. This calculation is very simple, and the area that does not overlap can be represented by a boundary defined as closed or interconnected. When we perform the gray value segmentation of the target and the background, the segmentation of the target and the background is based on comparing the gray value of the pixel at its designated position with the designated limit value. Pixels whose gray value is not less than a certain threshold are regarded as the target, and pixels whose gray value is less than this threshold are regarded as the target background. The most common threshold processing method is to perform image copy processing, that is, to select a threshold and separate the image target and background into a binary image with only 0 and 1 grayscale representation, that is, to convert into black and white that is usually widely used in image segmentation. Binary image.

Regional Growth. Area development is a method of grouping pixels based on the similarity between the gray pixel values in a specific area and a specific target or a fixed background in the image. Assuming that the area and position of a single pixel in the region are determined, start from a known point, and then intersect the pixel around that point. If the neighboring pixel point is similar to the known point, then the neighboring point will be included in the area. According to the similarity criterion, the area will continue to expand the scope of the area. The similarity criterion can be the characteristics of the image itself, such as gray-scale pixel values, pixel organization, or other characteristics. The image limit can be used as a method to determine the similarity measurement. The similarity measurement method starts from the point that meets the detection criteria and expands the area in all directions, that is, it intersects with surrounding neighboring pixels and meets the detection at its neighboring points. The standard time is merged into a small area. The area is merged into a new point that meets the conditions, and the new area is used to detect whether the surrounding points meet the similarity standard, and the above process is repeated until all acceptable points are added to the area, thus completing the area Segmentation process.

Edge Detection. Marginality is one of the most fundamental basic features of images, and the marginality is also the focus of image analysis and processing. The image edge refers to a pixel group with a step or upper limit change in the gray pixel plane around the target image of the object or the meaningful area. Edge refers to the existence between the target and the background in an image, between the target and the background, and between the original features in the image and the original background. These boundary characteristics make them suitable for image segmentation.

2.2 Pattern Recognition Mode

(1) Pattern recognition includes two interrelated stages, namely the learning stage and the realization stage. The former is to select the characteristics of samples and find the rules of classification, and the latter is to classify and identify unknown sample sets according to the classification rules. Generalized pattern recognition belongs to the research category of intelligent simulation in computer science. The content is very extensive, including voice and language recognition, character recognition, fingerprint recognition, sonar signal and seismic signal analysis, photo and image analysis, chemical pattern recognition and so on. Computer pattern recognition has realized part of the mental work automation.
(2) Image pattern recognition is the use of machines to process and recognize pattern information such as text, images, pictures, and scenery to solve the important problem of direct communication between the computer and the external environment. Its purpose is to develop and use a certain instrument or equipment to automatically process certain information, replace people to complete the task of classification and

identification, and be able to quickly and accurately identify graphics. Generally speaking, an image recognition consists of three main parts: image preprocessing, image feature extraction and image pattern classification. Pre-processing generally refers to operations such as smoothing, enhancing, restoring, edge detection, and segmentation of the image, and its purpose is to simplify the input image into a segmented mode. Feature extraction refers to extracting the main features of the image under the condition of meeting the requirements of classification and recognition accuracy, and selecting as many features as possible for correct classification and recognition according to certain criteria, so that the classification and recognition task can be completed with fewer features. Image pattern classification is the most important part. It maps the feature vector space of the previous part to the type space based on the extracted features, and assigns the corresponding image to a known type of pattern.

2.3 Graphic Detection Method

Image Preprocessing. In the actual production environment, the collected color print image occasionally contains noise and defects, and image preprocessing is required: convert the color print image to grayscale image, reduce the amount of calculation, reduce the calculation time, and improve the efficiency of scanning time.; Filter the image to remove non-target noise and stains, and at the same time ensure that the original details of the image will not be destroyed in order to prepare for further image processing.

Image Grayscale. The collected color print images not only take up more storage space, but also increase the amount of calculation in the next edit. The real-time processing of color images is time-consuming and laborious, and the real-time performance is low. Therefore, before processing the image, first convert the color image to a grayscale image. Without affecting the processing results, the processing speed of the algorithm can be effectively improved, and the real-time performance can be improved.

Image Denoising. The main purpose of image denoising is to remove image distortion caused by the environment and imaging equipment, such as image noise caused by unstable lighting, camera shake and other factors. The presence of these noises reduces image quality, even covers important image information, and seriously affects Increased the image resolution. Therefore, image preprocessing requires denoising.

Image Matching and Positioning. Image mapping and placement is to superimpose two or more printed images taken at different times, different perspectives and different image acquisition methods, using the standard image as a reference, and geometrically separating, editing and processing the captured image and the standard image. Image mapping and placement must first determine the mapping relationship between the standard image and the subsequent acquired images, correct the distorted image according to the geometric mathematical transformation between the images, and finally calculate the difference between the two images. The matching accuracy will seriously affect the subsequent detection results and detection accuracy, so the matching and positioning of the image is a very important step in the entire printed product defect detection process.

Image Geometric Transformation. During the image printing process, the position of the printed product image to be inspected relative to the printed product standard image may undergo rotation or translation changes. The difference should be adjusted to make the relative positions of the two printed products the same for subsequent detection.

Image Feature Extraction. Feature extraction is a key step in image recognition. It is very important to extract features with good resolution and high reliability from complex images. The effectiveness of features is directly related to the image recognition rate. Gray-scale image features mainly include shape and texture features. Among them, the texture capability can show the uniformity of pixels in the image without relying on color difference and brightness changes. It can measure the relationship between pixels in a local area of an image. Texture can be described as the gray dots of pixels in a specific pattern, expressing the correlation of the gray space between adjacent pixels, or expressing the understanding and response to human vision based on the change of gray space. Image texture features represent the offset and combination of image pixels and the relationship between the image and its surrounding environment. Texture mode is the most basic feature of an image and one of the main features used in image recognition.

2.4 Limitations of Detection Algorithms

(1) The online detection algorithm is insufficient in acquiring the semantics of some complex self-non-self-spaces. The binary encoding method and its corresponding matching rules limit its representation of the complex detection space.
(2) The binary coded detector and the continuous r-bit matching rule lack actual physical meaning, and its knowledge expression ability is also very poor. In the actual print defect detection process, the positioning matching cannot better reflect the change of print quality and the type of defect.
(3) It is very difficult to achieve data fusion and cannot be combined with other artificial intelligence methods. Because other artificial immune algorithms and advanced intelligent algorithms are mostly represented by high-dimensional data, it is difficult to combine an online detection algorithm represented by low-dimensional data with other intelligent algorithms represented by high-dimensional data.
(4) The binary coded online detection algorithm is particularly sensitive to non-self data. It can detect all subtle anomalies except normal data, it cannot set thresholds, and its anti-noise ability is very weak.
(5) There is no adaptive learning ability.

Due to the shortcomings of the online detection algorithm, it is difficult to directly apply the algorithm to print defect detection, but its unique way of dealing with the problem-the use of a limited number of antibody detectors to detect an infinite variety of abnormalities, is worthy of in-depth study.

2.5 Computer Graphics Recognition Algorithm

Image Recognition Based on Artificial Neural Network. BP neural network is a multi-level feedforward neural network. The entire network is divided into three levels: input level, hidden level and output level. It is also a multi-level learning network.

When the system has inputs, the input data will be transferred from the input layer unit to the hidden layer unit, and then after processing the hidden layer, they will be passed to the output level, thereby creating output data. This process is a layer-by-layer status update, which is called relay forwarding. If the output data does not match the expected data, the error will be transmitted backwards, and each connection weight will be corrected at each layer. For a given set of training data, the network must be continuously trained in each training mode, and forwarding and errors must be repeated continuously. After learning, the neural network can start to work. When there is a data input set, the BP neural network will identify and judge unknown samples based on the conventional recognition and calculations obtained in the learning process. This is also the embodiment of the self-adaptation and self-learning ability of BP neural network.

1) Initialize the network. The weight of each connection is assigned a random number in the interval (−1, 1), the error function is set to e, and the value of calculation accuracy e is given and the number of times of maximum learning time is m.
2) Randomly select the input sample and the corresponding expected output $d_0(k) = (d_1(k), d_2(k), ..., dq(k))$.
3) Calculate the input hi_hk of each neuron in the hidden layer, and then use hi_hk the sum activation function to calculate the output ho_hkho_hk of each neuron in the hidden layer:

$$hi_hk = \sum_{i=1}^{n} w_{ih}x_i(k) - b_h \tag{1}$$

4) Using the expected output $d_0(k) = (d_1(k), d_2(k), ..., dq(k))$ of the network and the actual output $yo_o(k) = f(yi_o(k))$ o = 1, 2, … q, calculate the partial derivative of the error function to each neuron in the output layer:

$$\delta_o(k) = (d_o(k) - yo_o(k))yo_o(k)(1 - yo_o(k) \tag{2}$$

5) Use the output $\delta_o(k)$ of each neuron in the output ho_hk layer and the output of each neuron in the hidden layer to modify the connection weight $w_{ho}(k)$ s and thresholds $b_o(k)$:
6) Use the input $\delta_o(k)$ of each neuron in the hidden layer and the input of each neuron in the input $x(k)$ layer to modify the connection weight and threshold:
7) Calculate the global error E.

3 Computer Graphics Processing and Recognition Technology Experiment

3.1 The Purpose of the Experiment

This paper conducts experiments on the image extraction accuracy of different image recognition and analysis technologies, and analyzes the results of the experiments that the image recognition and analysis technology has a higher accuracy rate. Provide a basis for the application of image recognition technology.

3.2 Experimental Design

In order to verify the effectiveness of related experiments, the experiment used the classic CIFAR-10 image library. There are 8 different categories in the image library, namely numbers, flowers, etc. There are a total of 60,000 different images, each type contains 6,000 images, the image format is 28 × 28 color images, of which 40,000 images are used as training data, and the remaining 8,000 Two images are used as test data, and then related experiments are carried out through MATLAB.

3.3 Data Enhancement

In machine learning and deep learning, in order to avoid excessive placement, a large amount of data is usually needed to train a data model. In fact, collecting large amounts of data is a bit difficult, so increasing the amount of data is a common method. Image data enhancement is the geometric transformation of the original data and the expansion of a person's data set to the assumption that the image attributes remain unchanged. Currently, commonly used data enhancement methods mainly include: image rotation transformation, inversion transformation, zoom transformation, translation transformation, noise interference, contrast transformation, color transformation, random image cropping, etc.

3.4 Data Normalization

In machine learning, data normalization is a necessary step in data processing. After the image attributes are extracted, due to the existence of multiple attribute vectors and the inconsistent size of the attribute values, there are often large differences in the price range, and the distribution is irregular. Inputting these data directly into the sorting algorithm will cause the model to converge slowly. And the accuracy is reduced. Therefore, in order to solve this problem, it is necessary to standardize the data, especially for data with different characteristics, and convert the data dimension to non-dimensional data.

3.5 Image Recognition Process

The basic functions of the process are described as follows: First, initialize the capture card, and then perform CCD capture and capture. Firstly, the image information is preprocessed, including filtering, enhancement and image segmentation. Then the key feature points of the image are identified and classified. Identify the point information, and store the position coordinates of the point information, and then sharpen the point information, and then identify whether the line information is based on the point information. The main method is the edge detection method, and then the line information position is recorded. Since there are many pixels in a frame, a sliding table is used to drive the CCD.

3.6 Data Analysis

1) The main meaning of a correlation relationship in the objective correlation analysis method is to generally refer to a certain relationship between various objective phenomena, but they do not strictly correspond to each other in quantity. There are two main forms of determining the relevant properties of objective phenomena here: qualitative analysis and quantitative analysis. The main purpose of qualitative analysis is to rely on the scientific theoretical knowledge and practical experience of the researcher to accurately judge whether there are correlations between various objective phenomena. Or what kind of factor, the subjectivity of this analysis method is relatively strong. Among them, the commonly used calculation formula is expressed as:

$$r = \frac{S^2\,xy}{Sx\ Sy} = \frac{\sum (x-\overline{x})(y-\overline{y})/n}{\sqrt{\sum (x-\overline{x})^2/n}\sqrt{\sum (y-\overline{y})^2/n}} \tag{3}$$

$$r = \frac{n\sum xy - \sum x \sum y}{\sqrt{n\sum x^2 - (\sum \overline{x})^2}\sqrt{(n\sum y^2 - (\sum \overline{y})^2)}} \tag{4}$$

4 Analysis of Experimental Results

4.1 Image Recognition Based on Artificial Neural Network

In order to verify the influence of the number of nodes and the number of levels on the network structure on the extraction performance of the artificial neural network model, experiments were designed for verification. The ANN1 artificial neural network model uses a three-level network model, the input level is 780 dimensions, the number of hidden level nodes is designed to be 200, 300, 400, 500, 600, 700, 800, and the output level dimension is 10, in order to make the network converge faster, Set the learning rate to 0.2, set the mini-batch of the tilt algorithm to 40, and repeat the experiment 40 times. The ANN2 artificial neural network model uses a 3-level network model, the input level is 780 dimensions, the first hidden level is fixed, the number of nodes is 500, and then the number of nodes at the second hidden level is designed to be 200, 300, 400, 500, 600, 700, 800, and the dimension of the output level is 10. To speed up network convergence, set the learning rate to 0.2. The small batch size results of the tilted cathode algorithm are shown in Table 1.

Table 1. Accuracy of models with different numbers of nodes and layers (%)

	200	300	400	500	600	700	800
ANN1	94.44%	94.56%	95.75%	96.64%	95.78%	94.81%	94.01%
ANN2	94.81%	94.45%	95.92%	97.37%	96.36%	95.38%	94.45%

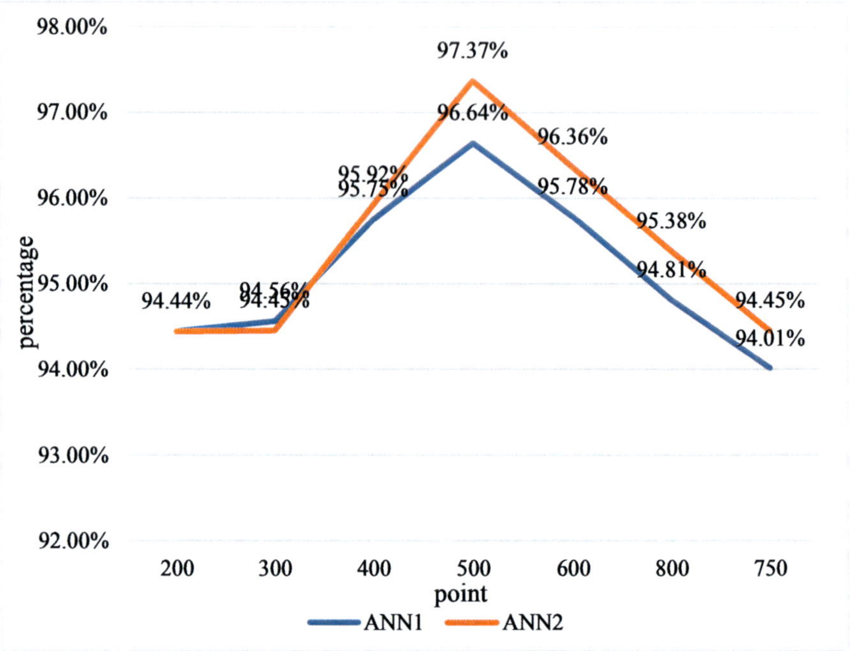

Fig. 1. Accuracy of models with different numbers of nodes and layers (%)

Figure 1 shows the small batch size results of the tilted cathode algorithm in the form of Line chart. By comparing the experimental data, it can be seen that in the case of the same number of levels and different node numbers, the recognition rate of the picture increases first, then reaches the peak, and then decreases. The overall rate of return is a trend that first rises and then declines. It can be seen that when the data is 500, that is, the number of nodes is set to half of the input data, and the artificial neural network has the highest recognition rate, the data has the best performance.

4.2 Image Recognition Based on Convolutional Neural Network

In order to verify the effectiveness of different improved methods of convolutional neural network, a confirmatory experiment was carried out in the MNIST image library. For the convolutional neural network models CNN1, CNN2 and CNN3, in order to make the network converge faster, the learning rate is set to 0.1, the mini-batch size of the gradient descent algorithm is set to 50, and the experiment is iterated 55 times. Statistics are 3 times and 8 times respectively. 113 times, 23 times, 33 times, 43 times, 53 iterations of image recognition accuracy, the results are shown in Table 2, Fig. 2.

It can be seen from the statistical results that the recognition rate of LeNet-5 in the MNIST image library is only 93 at the beginning, while the latter two models are 95 and 96 respectively at the beginning. CNN1 only achieves 30 times the stability, while the improved CNN2 and CNN3 models achieve a relatively stable recognition rate of nearly 20 times, thereby increasing the relative learning speed. In the end, the recognition rate

Table 2. Accuracy rate of different models in MNIST (%)

	3	8	13	23	33	43	53
CNN1	93%	94%	95%	96%	97%	97.7%	97.8%
CNN2	95%	96%	97%	98%	98.6%	98.7%	98.8%
CNN3	96%	97%	98%	99%	99.3%	99.4%	99.5%

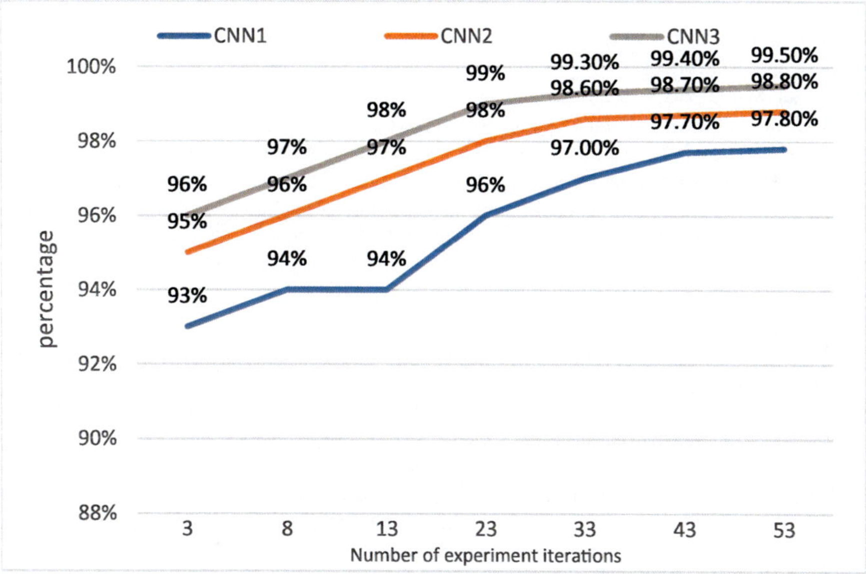

Fig. 2. Accuracy rate of different models in MNIST (%)

of the CNN1 model reached 97.8, and the improved CNN1 model achieved a maximum recognition rate of 97.82. The model can reach 98.8 and 99.5, and the improved model algorithm significantly improves the image recognition rate.

5 Conclusions

Image recognition technology is one of the key points of modern artificial intelligence. It plays an extremely important role in modern automatic control technology and the fifth generation of computers. It was originally established in the late 1950s and began to undergo rapid development in the early 1960s. In just over two decades, it has received extensive attention from many scientific research industries and has been applied to various scientific research and industries produce. Its basic meaning mainly refers to the use of computer-processed images to obtain specific expected effects and to collect and extract useful information from these images in order to facilitate the analysis, description, judgment and identification of things or phenomena.

References

1. Gu, D.X., Patane, G.: Special issue on "Heat diffusion equation and optimal transport in geometry processing and computer graphics". Comput. Aided Geom. Des. **67**(Dec), 96 (2018)
2. Mizera-Pietraszko, J.: A sampler of useful computational tools for applied geometry, computer graphics, and image processing. Comput. Rev. **57**(5), 285–286 (2016)
3. Fan, M., Li, Y.: The application of computer graphics processing in visual communication design. J. Intell. Fuzzy Syst. **39**(8), 1–9 (2020)
4. Sahasrabudhe, S.C., Kulkarni, A.D.: Shift variant image degradation and restoration using SVD. Comput. Graph. Image Process. **9**(3), 203–212 (2016)
5. Tanimoto, S.L.: Pictorial feature distortion in a pyramid. Comput. Graph. Image Process. **5**(3), 333–352 (2016)
6. Livesu, M., Ellero, S., Martínez, J., et al.: From 3D models to 3D prints: an overview of the processing pipeline. Comput. Graph. Forum **36**(2), 537–564 (2017)
7. Park, Y., Hong, Q.Y., Kim, M.S.: Geometric processing for freeform surfaces based on high-precision torus patch approximation. J. Korea Comput. Graph. Soc. **25**(3), 93–103 (2019)
8. Markus, S.: On dynamic scheduling for the GPU and its applications in computer graphics and beyond. IEEE Comput. Graph. Appl. **38**(3), 119–130 (2018)
9. Ozdemir, D., Uur, M.E.: A model proposal on the determination of student attendance in distance education with face recognition technology. Turk. Online J. Dist. Educ. **22**(1), 19–32 (2021)
10. Sun, C., Wang, L., Wang, N., et al.: Image recognition technology in texture identification of marine sediment sonar image. Complexity **2021**(2), 1–8 (2021)
11. Park, G.H., Su, J.K., Cho, Y.S.: Development of a voiding diary using urination recognition technology in mobile environment. J. Exerc. Rehabil. **16**(6), 529–533 (2020)
12. Huai, X., Kitada, S., Choi, D., et al.: Heart sound recognition technology based on convolutional neural network. Inform. Health Soc. Care **2**, 1–13 (2021)
13. Qin, J., Zhong, M., Liu, Y., et al.: Covert attack detection based on Hi/H^{∞} optimization for cyber-physical systems. IFAC-PapersOnLine **53**(2), 4487–4492 (2020)
14. Ou, Z., Wei, X.: Online magnetic flux leakage detection system for sucker rod defects based on LabVIEW programming. Comput. Mater. Continua **58**(2), 529–544 (2019)
15. Li, C., Wei, Z., Jing, X.: Online inspection system for the automatic detection of bolt defects on a freight train. Proc. Inst. Mech. Eng. Part F J. Rail Rapid Transit **230**(4), 271–288 (2016)

Effective Application of Sports Video Tutorial Mode Teaching Based on Information Processing Technology

Bin Qian(✉)

Jiangxi University of Engineering, Xinyu, Jiangxi, China

Abstract. With the development of information technology and the extensive use of multimedia equipment in actual teaching, sports video tutorials have gradually been popularized and used in teaching. Educational institutions are knowledge-intensive organizations. The task of the physical education major is to train qualified graduates who meet the needs of modern education, and play an important role in improving my country's quality education and cultivating qualified personnel who meet the needs of modern development. Physical education in primary and secondary schools is a very important link in our education system. A healthy student requires physical, mental and social adaptation to meet certain requirements, among which physical health is the basis for other aspects to meet the requirements. Therefore, the education department in my country has put forward higher requirements for the physical health of middle school students, especially from the high school entrance examination. It can be seen from the whole process of the implementation of physical education and the proportion of the total score that in order to improve the physical quality of middle school students and promote their physical and mental health, higher requirements are put forward for middle school physical education teachers, and physical education teachers are required to develop in all aspects. The purpose of this article is to explore the effective application of sports video tutorial mode teaching based on IPT. Studies have shown that the gymnastics skills of the students in the experimental class and the control class for one semester have improved significantly, from a few grades at the beginning to the current score of more than 8 points. The experimental class and the control class adopted different teaching modes. After a semester of practice, the results of the two classes of students with the same foundation were compared. After the t test, $p < 0.05$, which proved the performance of the two classes. There is a significant difference in the final scores of the students, that is, the scores of the students in the experimental class are significantly better than those of the control class.

Keywords: Information processing · Sports videos · Video tutorials · Effective use

B. J. Jansen et al. (Eds.): *International Conference on Cognitive based Information Processing and Applications (CIPA 2021)*, LNDECT 85, pp. 213–221, 2022.
https://doi.org/10.1007/978-981-16-5854-9_26

1 Introduction

After the introduction of sports video tutorial mode teaching in our country, various tentative experiments and researches have been caused in different subject areas in our country, and we strive to explore a new teaching method that promotes the overall development of students [1, 2]. The sports video teaching mode uses multimedia players to play standard movement techniques, allowing students to observe the movements they want to practice, analyze their movement processes and principles, so as to master basic gymnastics skills and cultivate students' self-study, practical and research abilities, Update the gymnastics learning model [3, 4].

The use of micro-video assisted teaching can not only improve students' self-learning awareness, but also improve self-solving problems and personal skills [5, 6]. Students can see their own problems in the process of practicing, and make self-correction and practice. The micro-video can be watched repeatedly, which can not only strengthen the students' impression of the important and difficult points and practice by themselves; it can also enable parents and friends to understand the content learned in the classroom by watching the video with the students, and provide relative guidance and supervision to the students. Conducive to improving the relationship between students and parents [7, 8]. Replying or sharing in WeChat groups after practice can also cause healthy competition among students and stimulate their willingness to learn. At the same time, increase the communication between teachers and students, so that teachers can better understand students' learning progress and students' specific needs. Compared with traditional monotonous theoretical teaching, video teaching is more interesting and vivid. In the process of playing the video, teachers can use the video to concisely and effectively connect the theoretical knowledge that needs to be explained with the actual situation, which can better stimulate students to learn Interest [9, 10].

This article conducts an experimental study on the teaching mode of student observation video learning. This teaching mode is under the guidance of the teacher, through the observation of the video teaching model, the students master the essentials of technical movements, learn its teaching and protection methods, and through the students' self-study and self-training. By discussing the feasibility and effectiveness of the teaching mode of observation video model learning in gymnastics general courses, it is hoped that a new method and theoretical basis will be proposed for further improving the teaching research of gymnastics class [11, 12].

2 Effective Application of Sports Video Tutorial Mode Teaching Based on IPT

2.1 Characteristics of Sports Video Tutorial Mode Teaching Based on IPT

The video teaching mode allows students to understand the actions more vividly, intuitively and perceptually. Introduce the video teaching mode into the classroom, and better combine practical teaching with theoretical teaching. Make full use of modern teaching methods and network resources, broadcast standard movement techniques, for students to observe the movements they want to practice, analyze their movement processes and principles, so as to master the basic skills of gymnastics, cultivate students' self-study ability, practice ability and research ability, update Gymnastics learning mode. By watching the video case to comprehend the essentials of the action and understand the difference between the correct action and the wrong action, deepen the students' understanding of the correct action, and effectively prevent the appearance of the wrong action in the video case. During the playback of the video case, students can watch and disassemble repeatedly Watch, and you can ask questions about the video case at any time. The teacher explains the scoring standard of the decomposition action for the case for the students to understand.

After class, make full use of the network platform to establish a video recording network platform through the QQ group, WeChat group and WeChat official account, etc., to collect actual materials related to the theoretical content in the teaching process, so that students can use and freely grasp during their spare time. Study time to make up for the lack of time in class.

2.2 Sports Video Tutorial Mode Teaching Based on IPT

This article reforms the teaching mode of physical education from the perspective of physical education science, teaching methods and practical teaching concepts. Design a suitable teaching environment, maximize the use of teaching resources inside and outside the school, and organically combine practical problems in sports with theoretical teaching, so that students can have a more intuitive and in-depth understanding of the standards of sports in sports, and further improve the quality and effect of teaching.

(1) Data collection
Make full use of the network resource platform to extensively collect video cases, action essentials videos, etc. Look for construction-related forums, if necessary, download them through Zhulong, Taoke Baba, Taobao, etc.
(2) Establish a video recording network platform
Construct a network platform by using the collected video recording resources. Through the QQ group, WeChat group and WeChat official account, collect audio-visual materials related to the theoretical content in the teaching process. Enable students to use their spare time as they learn, freely master the learning time, to make up for the shortcomings of insufficient time in class, enhance the understanding and understanding of teaching content, and improve learning efficiency.
(3) Video discussion to deepen understanding
Make full use of modern teaching methods and network resources, broadcast standard movement techniques, for students to observe the movements they want to practice, analyze their movement processes and principles, so as to master the basic skills of gymnastics, cultivate students' self-study ability, practice ability and research ability, update Gymnastics learning mode. By watching the video case to comprehend the essentials of the action and understand the difference between the correct action and the wrong action, deepen the students' understanding of the correct action, and effectively prevent the appearance of the wrong action in the video case. During the playback of the video case, students can watch and disassemble repeatedly watch, and you can ask questions about the video case at any time. The

teacher explains the scoring standard of the decomposition action for the case for the students to understand.

3 Effective Application Experiment of Sports Video Tutorial Mode Teaching Based on IPT

3.1 Investigation Method

Questionnaire survey method is a qualitative research method, commonly used in sociology and psychology research, its characteristic is to understand some characteristics of the whole by analyzing selected samples. In addition to the questionnaire survey method, this paper also uses the comparative experiment method. Through the comparison of the sports video teaching experiment based on IPT, we observe that sports video teaching based on IPT has some advantages in mastering technical movements, physical fitness, and sports learning attitudes. Different influences have more significant effects with traditional teaching models and methods.

3.2 Respondents

A student in a general gymnastics course majoring in physical education in a college in this province. This article takes the students in the general gymnastics course of a physical education major in a university in this province as the survey object. Before the test, the students in the two classes are analyzed. Two classes are randomly selected as experimental classes, and two classes are controlled classes. There are 28 boys in each class. The students in each class have their own special subjects. The college entrance examination special subjects include track and field, football, basketball, volleyball, and martial arts. Before the experiment, the two classes of students' physical learning attitude, basic gymnastics skills, and special gymnastics physical fitness were tested. The special physical fitness tests included parallel bars arm flexion and extension, push-ups, lying on both ends, and standing against the wall. The parallel difference test showed that there was no significant difference between the control class and the experimental class.

3.3 Statistics

This article uses SPSS 22.0 software to count and analyze the results of the questionnaire, and conduct a t test. The t-test formula used in this article is as follows:

$$t = \frac{\overline{X} - \mu}{\frac{\sigma X}{\sqrt{n}}} \tag{1}$$

$$t = \frac{\overline{X_1} - \overline{X_2}}{\sqrt{\frac{(n_1-1)S_1^2+(n_2-1)S_2^2}{n_1+n_2-2}\left(\frac{1}{n_1} + \frac{1}{n_2}\right)}} \tag{2}$$

$$t = \frac{\overline{d} - \mu_0}{s_d / \sqrt{n}} \tag{3}$$

Re-testing the reliability of the questionnaire this article uses the retest method for testing. When the questionnaire is first issued, each student's questionnaire is marked according to his or her student number. Two weeks after the first questionnaire was issued, a second questionnaire survey was conducted on some of the students, and the subjects of the survey were kept consistent according to their student number. The consistency index of the two measurements R1 = 0.915 indicates that the reliability of the first survey is high, and it meets the reliability standards required by surveying, indicating that the questionnaire method has high feasibility.

4 Experimental Analysis of Effective Application of Sports Video Tutorial Mode Teaching Based on IPT

4.1 Test of the Physical Learning Attitude of the Two Groups of Students Before the Experiment

Students' learning attitudes in gymnastics classes can reflect a student's enthusiasm for gymnastics learning. Before the two rounds of experiments, the students' attitudes towards gymnastics skills learning in the two classes were tested, and the students' attitudes towards physical education learning were classified as good (80–100 points), good (61–80 points), average (41–60 points), poor (21–40 points), and poor (10–20 points). The details are shown in Table 1.

Table 1. Investigation and analysis of sports attitudes of the two classes before the experiment

	Learning attitude	80–100 points	61–80 points	41–60 points	21–40 points	10–20 points
Experimental class	First round	5	6	10	5	2
	Second round	4	6	12	3	3
	Total	9	12	22	8	5
Control class	First round	4	6	11	4	3
	Second round	3	7	12	3	3
	Total	7	13	23	7	6

It can be seen from Fig. 1 that before the two rounds of experiments, the physical learning attitudes of the experimental class and the control class are generally higher, each class has about 10 students, and the attitudes are good, good, poor, and very poor students. The number of people is small, and there is no difference in sports learning

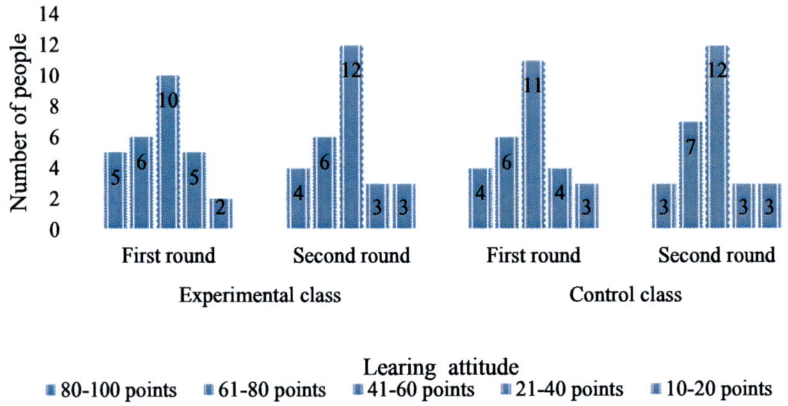

Fig. 1. Investigation and analysis of sports attitudes in two classes before the experiment

attitudes between the experimental class and the control class, which ensures that the students have the same foundation before the start of the experiment and the accuracy of the experimental results.

4.2 Influence of the New Teaching Model Applied to the Teaching Model of the General Gymnastics Course on the Student's Performance

The application of the new teaching model in the general gymnastics course has played an important role in improving students' physical performance. After a semester of study, most of the ten movements required by the mat exercise can be completed by students, and there are also individual difficulty coefficients. Larger movements such as headstand and other movements cannot be completed well by individual students. Many students still have certain problems in the connection of the whole set of movement techniques. The final score test is to mix the students in the experimental class and the control class. Teachers with extensive gymnastics experience combine the exercise assessment standards on the gymnastics mat to score the students' movements, and finally do statistical analysis. After two rounds of experiments, statistics were made on the number of students in the experimental class and the students in the control class, and the general situation of the students' scores was obtained. The details are shown in Table 2.

Table 2. After the experiment, the comparison of the technical performance evaluation of the experimental class and the control class

		Number of people	Grades	Average value	T value	P
First round	Experimental class	28	8.65 ± 0.25	0.62	1.24	p < 0.05
	Control class	28	8.03 ± 0.14			
Second round	Experimental class	28	8.91 ± 0.16	0.75	2.13	p < 0.05
	Control class	8	8.17 ± 0.22			

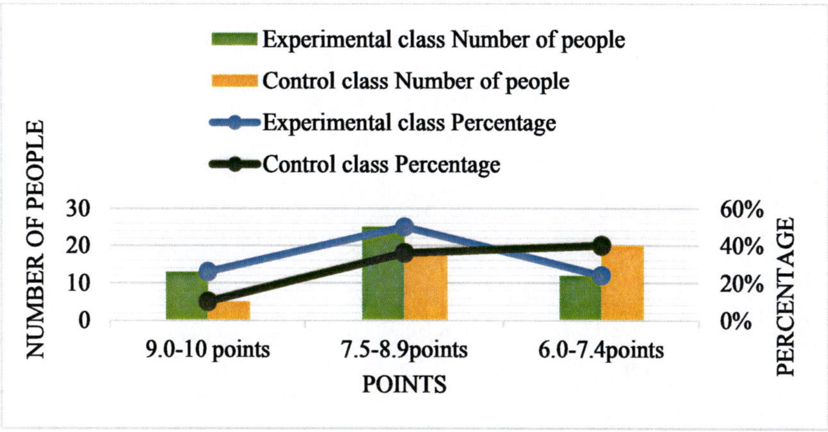

Fig. 2. Statistics of the number of students at each stage

From Table 2 and Fig. 2, it can be seen that the gymnastics skill level of the students in the experimental class and the control class of a semester has been greatly improved, from the initial level of about a few points to the current level of more than 8 points. The experimental class and the control class adopted different teaching modes. After a semester of practice, the results of the two classes of students with the same foundation were compared. After the t test, $p < 0.05$, which proved the performance of the two classes. There is a significant difference in the final scores of the students, that is, the scores of the students in the experimental class are significantly better than those of the control class. Experiments have proved that the teaching of sports video tutorial mode based on IPT is effective for students to improve their performance.

5 Conclusions

Nowadays, the education mode is changing from the traditional transmission mode to the experience mode. The advancement of information technology and the widespread use of multimedia equipment in teaching are also changing the teaching mode of teachers and the learning mode of students. The video teaching mode better integrates practical teaching and theoretical teaching, and realizes the characteristics of applying what has been learned. Based on the sports video tutorial mode teaching under the IPT, the prepared video cases are played for students to watch, so that students can understand the essentials of the actions and understand the difference between correct and wrong actions by watching the video cases, and deepen the students' understanding of correct actions, and effectively prevent the occurrence of wrong actions in the video case. During the playback of the video case, students can watch repeatedly, disassemble and watch, and can ask questions about the video case at any time. The teacher explains the scoring standard of the disassembled action for the case for students to understand. After the student practice, replay the videos of the students' practice in real time and analyze them one by one, focusing on correcting wrong actions and giving specific analysis scores. The practice videos are played back and analyzed in real time, so that students can more vividly and concretely find their own actions and correct them in time, so as to deepen their grasp of actions, and learn from each other and sum up their experiences by watching other students' practice videos. The real-time playback of the practice video helps students find mistakes in summarizing and correct them in time, or prevent problems before they happen, and inspire students to learn from each other and make progress together.

References

1. Li, C.: Research on an enhanced web IPT based on AIS text mining. Recent Adv. Electr. Electron. Eng. (Former. Recent Pat. Electr. Electron. Eng.) **14**(1), 29–36 (2021)
2. Tarn, C., Zeng, W.F.: pDeep3: toward more accurate spectrum prediction with fast few-shot learning. Anal. Chem. **93**(14), 5815–5822 (2021)
3. Gao, Y., Sun, J., Li, X.: Research on platform environment simulation platform based on station status monitoring and IPT. J. Phys. Conf. Ser. **1802**(4), 042106 (7 pp) (2021)
4. Liu, T., Cheng, L.: Research on key IPT of environmental law enforcement based on big data. J. Phys. Conf. Ser. **1648**(3), 032055 (6 pp) (2020)
5. Sobko, I.M., Chucha, Y., Podmaryova, I.A., et al.: Application of the video-tutorial "Challenge for Referees" in sports training of young basketball referees for the game season. Health Sport Rehabil. **7**(1), 42–53 (2021)
6. Meyerhoff, H.S., Papenmeier, F., Huff, M.: Studying visual attention using the multiple object tracking paradigm: a tutorial review. Atten. Percept. Psychophys. **79**(5), 1255–1274 (2017). https://doi.org/10.3758/s13414-017-1338-1
7. van den Berg, M., et al.: Video and computer-based interactive exercises are safe and improve task-specific balance in geriatric and neurological rehabilitation: a randomised trial. J. Physiother. **62**(1), 20–28 (2016)
8. Coreno, A., Ciccia, A.H.: Supporting students with TBI: a clinically focused tutorial for speech-language pathologists. Semin. Speech Lang. **41**(02), 161–169.v (2020)
9. Graziani, S.: Video tutorial program. IEEE Instrum. Meas. Mag. **24**(1), 1–1 (2021)
10. Marlina, M., Tresna, P., Karmila, M.: Tutorial video multimedia development learning of white garden. Innov. Voc. Technol. Educ. **15**(1), 24 (2019)

11. Okajima, H., Hashitsume, S., Oishi, R., et al.: Reconstruction of sports competition video using fixed camera based on receding horizon strategy. Trans. Soc. Instrum. Control Eng. **55**(2), 135–143 (2019)
12. Ramadhan, R.F., Kustandi, C., Nursetyo, K.I.: Pengembangan video tutorial membuat halaman website. J. Pembelajaran Inovatif **2**(1), 55–61 (2019)

Exploration and Research of Computer Image Processing Technology in Clothing Design Patterns

Li Wang(✉)

School of Fashion, Dalian Polytechnic University, Dalian 116000, Liaoning, China

Abstract. With the reform and opening up and the improvement of the national economy, globalization and internationalization are affecting us step by step. Especially in terms of fashion design, our local Chinese designers use Western elements and traditional Chinese symbols in the fashion design patterns to design trendy clothing. However, various complicated elements and symbols need to be sorted and selected to be better used. Therefore, this article studies a method of using computer image processing technology to solve this historical problem for fashion designers. According to the survey report, women buy three times as many clothes as men each quarter; more than 50% of people like cute animal images such as cats and dogs, and small fresh patterns such as daisies are only favored by young women. With the use of computer image processing technology explored in this article, fashion designers can design exquisite clothing more easily.

Keywords: Computer image processing technology · Clothing design · Pattern design · Visual communication

1 Introduction

International trade has become popular since the time of the ancient Silk Road. With the convenience of land, sea, and air transportation, friendly exchanges between countries have become more frequent, with both cultural output and cultural input. In the 21st century of internationalization, Chinese fashion designers smell the taste of fashion and combine traditional Chinese culture with Western classical culture to design patterns on clothing. Fashions designed by Chinese fashion designers are sweeping the world, and they can be seen on various large-scale fashion runways. When designing clothes, designers will encounter the problem that it is difficult to choose a combination of fashion elements in pattern design, such as English letters and Chinese characters, animals and plants, or foreign characters and traditional characters. Therefore, in order to solve the above problems, this article attempts to explore the application of computer image processing technology in clothing pattern design.

When reviewing relevant research results, we found that many experts and scholars have explored this field. Al ZNE elaborated on the application of computer-aided design in modern clothing design teaching. He took typical clothing product-based enterprises

B. J. Jansen et al. (Eds.): *International Conference on Cognitive based Information Processing and Applications (CIPA 2021)*, LNDECT 85, pp. 222–230, 2022.
https://doi.org/10.1007/978-981-16-5854-9_27

as his teaching content, took the actual tasks of enterprises as his own classroom teaching method, and paid attention to students' learning consistency with practical work, and strive to meet the requirements of practical teaching, exploration classrooms, and enterprise integration technology innovation [1]. On the basis of visual communication, Liu K et al. introduced the characteristics of web pages and visual communication information. He pointed out that the content of web visual communication is designed from the perspective of web design computer image processing technology. From the aspects of corporate website color requirements, color characteristics, coordination rules, design principles and methods, it focuses on how to apply the concept of visual communication to web design. Starting from the simplicity, recognition and communication of Internet analysis, they used design, psychology, communication, aesthetics and other methods in artistic factors [2]. Liu Jingde et al. showed that clothing patterns are an important manifestation of regional culture and national characteristics. Through digital image processing technology, clothing pattern recognition can be objectively and accurately realized. They compared and analyzed the extraction techniques of various pattern elements, and then proceeded This paper summarized the extraction methods of clothing pattern color, contour and fabric texture, and constructed the core technology chain model of clothing pattern extraction. Through the construction and testing of the model, they summarized the research status, core technology and development trend of the pattern element extraction technology based on two-dimensional images, which provided a reference and reference for the subsequent research on the pattern of clothing and the upgrading of the textile and apparel industry [3].

Since ancient times, human beings have always been persevering in their pursuit of beauty, and the development of modern technology has opened up new paths for human research and innovation in clothing trends. Looking at the research results of predecessors, it is not difficult to find that the integration of computer technology and clothing design has already collided with fierce sparks. This research is only a further improvement on the basis of previous research results. It is expected that computer image processing technology will be applied to clothing pattern design and contribute to the innovation and development of the clothing industry.

2 Proposed Method

2.1 Computer Image Processing Technology

The Emergence of Computer Image Processing Technology. As the visual basis for humans to perceive the world, images are an important means for humans to obtain information, express information and transmit information. With the development of computer network information, human beings have gradually entered the era of data, and digital images were born [4]. Because image processing is cumbersome and complicated, and the emphasis on craftsmanship and technical requirements are relatively high, if you use computer machines to collect, analyze, and process them through specific programs and code systems, the "one-stop" service will be greatly improved. Reducing labor input will also effectively improve the efficiency of image processing. This is the origin of computer image processing technology. This technology originally originated in the 1920s, and later became perfect with the further development of science and technology, until the 21st century gradually entered the public's field of vision and was widely used [5].

Common Methods of Computer Image Processing Technology

(1) Image transformation
Image transformation is actually the preprocessing of the image. Since the original image collected is often a large array, if it is directly processed in the spatial domain, the amount of calculation will be very large. Therefore, it is necessary to transform the image first to make it conform to Processing requirements in the space domain. Commonly used transform processing techniques include Fourier transform, wavelet transform, etc. [6, 7].
(2) Image coding and compression
The purpose of encoding and compressing images is to reduce the amount of image data, reduce the number of bits of the image, so as to increase the speed of image transmission and reduce the amount of memory it occupies [8].
(3) Image enhancement and restoration
In the process of image acquisition, transformation and compression, certain damage and impurities may be generated more or less. Therefore, it is necessary to restore and denoise the image to enhance the quality and clarity of the image. At the same time, enhancement processing is needed to highlight the local features in the image to strengthen the high-frequency components of the image and highlight the contours and details of the local features.
(4) Image segmentation
If the meaningful part of the image needs to be extracted separately, it is necessary to apply the image segmentation technology, which generally includes the edge and local area of the image [9]. Through segmentation, the image area of interest can be identified and analyzed more directly.
(5) Image description
The prerequisite for image recognition and understanding is to describe the characteristics of the image. According to the different characteristics of the object, it can be divided into shape description, volume description, surface description and texture description.
(6) Image classification
The purpose of image classification is to facilitate effective image recognition. After the image is preprocessed, it is segmented and feature extracted, and then classified according to different characteristics. Commonly used classification methods include pattern recognition methods, statistical pattern classification and syntactic pattern classification.

Computer Image Processing Technology Algorithm. Image processing technology algorithms mainly include matching algorithms and retrieval algorithms. Matching algorithms include feature-based matching algorithms and image gray-scale matching algorithms; retrieval algorithms include content-based retrieval algorithms and text-based retrieval algorithms [10]. Generally, feature-based matching algorithms are the most common and are the most commonly used technical methods in image processing. They can be divided into texture features and shape features.

(1) Matching algorithm based on texture features
The texture feature is the most direct and accurate information of the object image, and the corresponding object can be quickly matched according to the texture feature of the image. Gray-level co-occurrence matrix method is the most commonly used image texture feature description method, and its expression is shown in formula:

$$M(i, j|d, \theta) = \frac{m(i, j|d, \theta)}{\sum_i \sum_j m(i, j|d, \theta)} \tag{1}$$

(2) Matching algorithm based on shape features
Commonly used shape feature description methods include geometric moment invariant descriptors and Fourier descriptors. The geometric invariant moment is a feature description of the shape of the image area. Suppose the image objective function has center distance and order moment expressions (as shown in Eqs. (2) and (3)), where is the center coordinates of the shape area.

$$U_{m,n} = \sum_x \sum_y x^m y^n f(x, y) \tag{2}$$

$$W_{m,n} = \sum_x \sum_y \left(x - x^l\right)^m \left(y - y^l\right)^n f(x, y) \tag{3}$$

The Fourier descriptor is transformed from the geometric invariant moment symbol, namely:

$$Z(l) = x(x) + jy(l)l = 0, 1, \ldots\ldots, n - 1 \tag{4}$$

$$Z(l) = \sum_{k=0}^{n-1} Z(k) \exp(-j2\pi kl/L)l = 0, 1, \ldots\ldots, n - 1 \tag{5}$$

2.2 Clothing Pattern

In the primitive society, human beings began to have the consciousness of dressing in order to hide their shame and avoid the cold. With the development and progress of society, clothing has gradually become a symbol of identity, status and taste, and the display of this feature is indispensable to the decoration of clothing [11]. It can be said that patterns have a very close relationship with clothing. In ancient times, people used traditional crafts such as embroidery, printing and dyeing to transfer various vivid patterns to clothing. However, with the advancement of technology in modern society, people seem to be more comfortable with the application of patterns in clothing design.

There are various types of clothing patterns, which can be divided into different categories from different angles, mainly as follows:

From the perspective of space, clothing patterns can be divided into plane patterns and three-dimensional patterns [12]. The so-called flat pattern refers to the pattern and other patterns on the fabric used in clothing and its accessories, showing a two-dimensional plane effect; three-dimensional patterns are fabrics made of folds, rows of folds, bows and other forms, presenting a three-dimensional effect of three-dimensional space.

In terms of pattern modeling, it can be divided into concrete and abstract patterns, and concrete is divided into realism and freehand brushwork. The so-called realism is to pay attention to the truthful description of the prototype features of the pattern itself, while the freehand brushwork emphasizes the design interest of the creator himself.

In terms of materials, it can be divided into character patterns, animal patterns, landscape patterns, flower patterns, and geometric patterns.

In terms of process characteristics, it can be divided into traditional printing and dyeing or tie-dyeing patterns, hand-painted patterns, embroidery patterns, knitting patterns, and jacquard patterns.

2.3 Application of Computer Image Processing Technology in Clothing Pattern Design

The integrated development environment for Windows platform applications-Visual Studio 2014, which is based on NET Framework 4.5.2. Not only does it include most of the tools needed in the entire software life cycle, but the written object code is applicable to all platforms supported by Microsoft. Therefore, this paper chooses Visual Studio 2014 as the software development platform, and uses the gray-level co-occurrence matrix method and Fourier descriptor mentioned above to realize the application of computer image processing technology in clothing pattern design.

Summarize the small patterns covering various fashion elements to the computer storage system, and combine them into fashion patterns inlaid with small patterns through various algorithm processing. This technology provides the independent input function of the parameters of some small patterns, through the small changes of the parameters, generate completely different mosaic graphics; and design a Chinese interface that is convenient for fashion designers to operate, and save the generated fractal graphics in jpg format for subsequent modification and processing of clothing patterns. The various patterns generated by the computer are carefully selected by the designers and then processed on the clothing, so more colors are added to our country's clothing.

3 Experiments

3.1 The Purpose of the Experiment

In order to collect and classify the image data used by the computer image processing system, it is convenient for Chinese fashion designers to choose and adopt, and to design more and more fashionable images into various fashions, so as to give more to Chinese citizens and foreign friends who love Chinese clothes. This experiment uses questionnaires to collect how men and women of different ages like different clothing patterns and how often they buy clothes each quarter.

3.2 Experimental Method

This experiment chooses to hold a questionnaire in a shopping mall with a large traffic volume, and uses a small gift to attract shopping crowds to fill in and submit the questionnaire on the spot. A total of 213 questionnaires were distributed, with a recovery rate of 100%, but only 200 valid questionnaires. Among the valid survey questionnaires, there are 100 males, and the number of respondents who are 16–25, 26–35, 36–45 and 46–55 are 25 respectively. The total number of women and the total population distribution of all ages are the same as those of men.

3.3 Experimental Results

The variance formula used in the data analysis of this experiment is as follows:

$$\frac{(x_1 - M)^2 + (x_2 - M)^2 + (x_3 - M)^2 + \cdots + (x_n - M)^2}{n} \tag{6}$$

The data analysis chart of this experiment will be displayed, discussed and analyzed in the fourth part, so I won't elaborate on it here.

4 Discussion

4.1 How Often Do People Buy Clothes Every Quarter

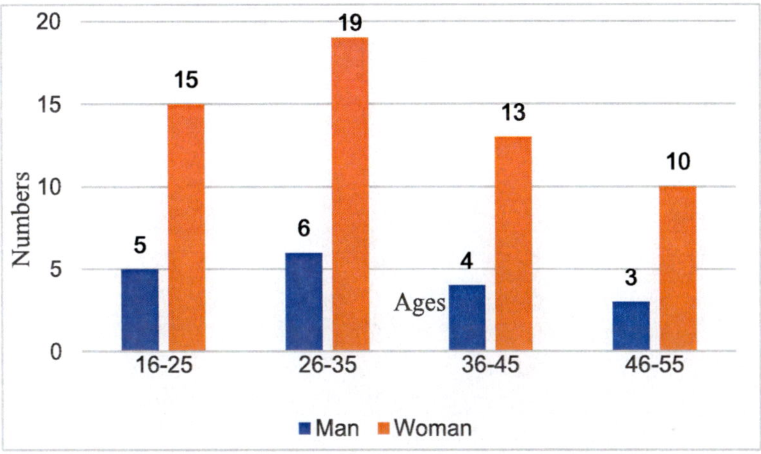

Fig. 1. Buy clothes masses quarterly frequency

It can be seen from Fig. 1 that the people aged 26–35 buy the most clothes every quarter, including 6 for men and 19 for women; the number of clothes purchased by people aged 46–55 is the least in each quarter, and the number of clothes purchased by men and women is 3 and 10 respectively; people of 16–25 and 36–45 age groups buy almost the same amount of clothes every quarter. Interestingly, it can be clearly seen from the figure that no matter what age group, women buy about three times as many clothes as men every quarter. Therefore, it can be estimated that the purchasing power of women is far greater than that of men in China's clothing market.

4.2 Clothing Patterns Preferred by Men and Women of Different Ages

Table 1. Clothing patterns that women of different ages like

	Lotus	Peony	Daisy	Cat	Dog
16–25	2	3	6	8	7
26–35	3	2	8	6	7
36–45	5	5	2	7	6
46–55	7	8	1	4	5

Table 2. Clothing patterns preferred by men of different ages

	Lotus	Peony	Daisy	Cat	Dog
16–25	1	0	2	12	10
26–35	2	2	5	10	6
36–45	6	5	2	7	5
46–55	7	8	1	4	5

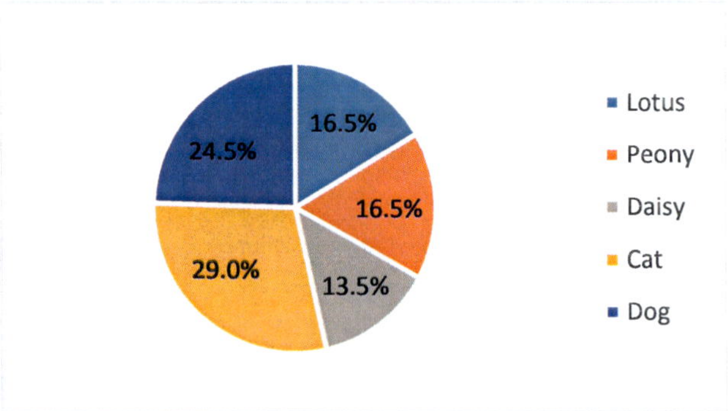

Fig. 2. The number of people who like different patterns

From the survey data in Table 1 and Table 2, it can be clearly seen that men and women of different ages have great differences in their choice of clothes with lotus flowers, peony flowers, daisies, cats and dogs. The data show that both men and women in the age group of 16–25 prefer cats and dogs, and more like cats than dogs; There are 8 women who like cats and 7 women who like dogs. There are 12 men who like cats and 10 men who like dogs. It is worth noting that no man likes daisies at this age in the survey report. Among the people aged 26–35, they still like cats and dogs, and more people like cats than dogs. But the number of people who like daisies is second only to dogs, among which 8 are women and 5 are men. Among the people aged 36–45, the number of people who like lotus and peony is similar to that of cats and dogs; however, the number of people who like daisies is relatively small, with two men and two women. Finally, people aged 46–55 years old. Surprisingly, at this age, men and women like the same pattern on clothing, and they all prefer lotus and peony. The number of people who like lotus, peony, daisies, cats and dogs are 7, 8, 1, 4 and 5 respectively.

It can be seen from Fig. 2 that more than 50% people like cats and dogs, 29% like cats and 24.5% like dogs. The total number of people who like daisies is the least, accounting for only 13.5%; The number of people who like lotus and peony is the same, accounting for 16.5% of the total number.

To sum up, people of any age are more acceptable to add cute animal patterns such as cats and dogs to their clothes; relatively speaking, small and fresh patterns such as daisies are more popular with young women; middle-aged people prefer elegant flowers such as lotus and peony on their clothes.

5 Conclusions

Everyone can't live without clothes. No matter people in primitive society or us now, clothes can not only cover up our shame and keep warm, but also beautifully designed clothes can enhance our aesthetic feeling. As the old saying goes, "People depend on clothes, while Buddhas rely on gold." Although clothing can bring us many benefits, fashion designers often encounter material selection problems when designing clothing patterns. Therefore, this paper studies the application of computer image processing technology in fashion design patterns. Firstly, we investigate the number of clothes people of different ages and genders buy in each quarter and their tendency towards clothing patterns. The results show that women buy about three times as many clothes as men; in terms of clothing patterns, people generally like cute animal images such as cats and dogs, and older people like elegant patterns such as lotus flowers and peony. I hope that computer image processing technology can bring convenience to the majority of fashion designers, and also hope that China's clothing will be popular.

References

1. Ninga, Z.: Research on the application of computer aided design in clothing design teaching in higher vocational colleges. Turk. J. Comput. Math. Educ. (TURCOMAT) **12**(3), 4817–4821 (2021)
2. Liu, K.: Research on computer image processing technology in web design based on visual communication. Revista de la Facultad de Ingenieria **32**(11), 90–95 (2017)
3. Liu, J., Zhuang, M., Ai, R., et al.: Analysis on the application of image processing technology in clothing pattern recognition. J. Donghua Univ. Engl. Ed. **3**, 224–231 (2020)

4. Anifah, L., Harimurti, R., Qoiriah, A.: Beef weight prediction system based on image processing in order to increase livestock productivity. Adv. Sci. Lett. **23**(12), 11892–11896 (2017)
5. Jamsranjav, E., Kuge, K., Ito, A., et al.: Evaluation of noise limits to improve image processing in soft X-ray projection microscopy. J. Xray Sci. Technol. **25**(4), 653–665 (2017)
6. Erell, E., Pearlmutter, D., Boneh, D., Kutiel, P.B.: Effect of high-albedo materials on pedestrian heat stress in urban street canyons. Urban Clim, **10**, 367–386 (2020)
7. Kuzmichev, V., Moskvin, A., Surzhenko, E., et al.: Computer reconstruction of 19th century trousers. Int. J. Cloth. Sci. Technol. **29**(4), 594–606 (2017)
8. Jankoska, M.: Application CAD methods in 3D clothing design. Tekstilna Industrija **68**(4), 31–37 (2020)
9. Sheng, S.: Implementation and test of new media technology in clothing design system. IPPTA: Q. J. Indian Pulp Paper Tech. Assoc. **30**(8), 443–447 (2018)
10. Yan, J., Li, H., Chen, X.: Application of Yangzhou freehand embroidery flower-and-bird subject in Chinese woollen clothing design. Wool Text. J. **45**(6), 48–52 (2017)
11. Qian, M., Wu, Y.: Study on the application of vegetation printing and dyeing in the green clothing design. Shandong Text. Sci. Technol. **058**(003), 50–53 (2017)
12. Chen, X., Zhu, Y., Yu, R.: On the application of VR technology in the digital display of Chinese han clothing. Des. Art Res. **009**(001), 85–91 (2019)

Research on Computer Intelligent Image Recognition Algorithm in Chinese Local Opera Clothing Education Teaching

LuLu Jia(✉) and Tao Gao

Jiangxi Normal University Science and Technology College, Gongqing 332700, Jiangxi, China

Abstract. Image recognition is one of the key technologies in the information age. The use of image recognition technology can complete the tasks that traditional sensor technology can not achieve. At the same time, the fusion of image recognition and multi-source information can make the monitoring effect more accurate. With the continuous development of computer technology, computer computing power is further improved, and the processing ability of information has been greatly improved. This paper analyzes the characteristics of costume color in opera and the factors influencing its evolution, and expounds the important role of costume color in conveying the status and psychological temperament of characters, as well as customs and stage effect in opera performance, which can provide certain reference for the design and research of opera costume color.

Keywords: Intelligent image · Local opera costume · Algorithm and technology

1 Introduction

With the continuous progress of computer technology, more advanced algorithms are applied in image recognition technology, such as neural network, genetic algorithm, wavelet algorithm, etc. the realization and application of these algorithms and technologies play an important role in improving the accuracy and speed of image recognition. The modern term "opera" was put forward by Wang Guowei, a master of modern literature. It is the general name of Chinese traditional opera culture. Chinese opera is a combination of farce, folk song and dance and rap. It is a combination of singing, reading, doing and playing. It enjoys a high reputation in the world and is known as Chinese opera. The performance form of opera is changeable, the singing is flexible, the clothing is various, and it contains profound cultural heritage. It not only conveys the personality characteristics of the characters, but also lays a good foundation for the rendering of stage atmosphere and the expression of aesthetic feeling. In this paper, through the analysis of opera costume color, it can provide some reference for the design and research of opera costume color.

B. J. Jansen et al. (Eds.): *International Conference on Cognitive based Information Processing and Applications (CIPA 2021)*, LNDECT 85, pp. 231–236, 2022.
https://doi.org/10.1007/978-981-16-5854-9_28

2 Principle and Development of Image Recognition Technology

2.1 Principle of Image Recognition Technology

Image recognition technology is a very popular technology in the field of research, through image recognition can solve the problems that were difficult to solve in the past. The computer can use the collected image information to analyze and process the image, and then realize the image recognition [1]. Computer image recognition technology is to use the relevant features in the image to judge, through these features can analyze the image, and the computer itself stores a large number of image information features, through the computer recognition and database information comparison, can realize the image recognition.The image recognition process is shown in Fig. 1. This kind of recognition technology is similar to human image recognition, which realizes the recognition of new images through previous learning.

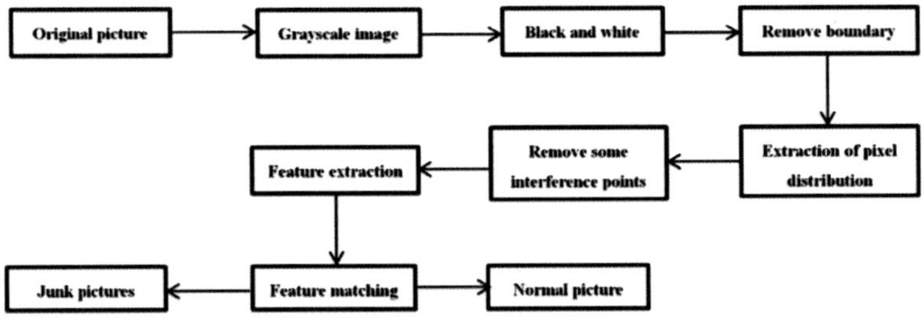

Fig. 1. Process chart of image recognition technology

2.2 The Development of Image Recognition Technology

The development of image recognition technology has changed from text image to digital image. From the research of text recognition in the middle of the 20th century to the research of digital image processing in the middle and later period, this development not only improves the storage quality, but also improves the efficiency of information analysis. The field development provides abundant research results. Compared with the traditional analog image recognition methods, the application of digital image recognition technology can greatly ensure the authenticity and integrity of digital information, and provide practical image processing methods for employees in various fields, so as to better show the digital information content contained in image information [2]. But now image recognition is also facing great challenges. With the application of image recognition technology in all walks of life, the requirements for its recognition accuracy and recognition speed are also higher and higher, and the image recognition process itself needs to process a lot of data, so it will be the development trend of intelligent image processing technology to do well in image feature extraction and eliminate the influence of redundant information.

3 Realization of Computer Intelligent Image Recognition Technology

Intelligent image recognition technology can identify image information more accurately and quickly. In order to realize these functions, it is necessary to establish the relevant image model in advance, and to optimize the model by using learning algorithm, so that the image features contained in the model are more comprehensive. In this process, in order to reduce the pressure of image recognition equipment, the training process of the model can be carried out on the cloud system. The cloud has a large storage capacity, which can meet the storage of image data [3]. At the same time, with the training ability of cloud, it can use intelligent algorithms such as deep learning for model training. With the continuous enrichment of database, the model can be more accurate.

3.1 Computer Intelligent Image Recognition Algorithm

For digital images, in the discrete state, for the images whose gray distribution is f (x, y), the $(p + q)$ order ordinary moments of f (x, y) are defined:

Origin moment:

$$m_{qp} = \sum\nolimits_{x=1}^{M} \sum\nolimits_{y=1}^{N} x^p y^q = f(x, y), (q, p = 0, 1, 2, L) \tag{1}$$

At the same time, we need to define the $(p + q)$ order central moment.

Center distance:

$$\mu_{qp} = \sum\nolimits_{x=1}^{M} \sum\nolimits_{y=1}^{N} (x - x_0)^p (y - y_0)^q f(x, y)(q, p = 0, 1, 2, L) \tag{2}$$

4 The Principles of Reference for Traditional Chinese Opera Costumes

The reference principle of the color evolution of opera costumes refers to the continuous absorption of external factors in the process of inheritance and development of opera costumes to change the existing performance form and connotation. Sometimes, although the performance forms of some operas still exist, the connotation of the original opera has changed; while the form and connotation of some operas have changed, forming different styles and inheritance. The color of opera costumes is also gradually changing in relative stability to adapt to the national characteristics and customs of the spreading area. The aesthetic view of opera costume color is formed after the interaction, precipitation and integration of historical, cultural, religious, regional and other factors (such as learning from folk costumes, calligraphy, sculpture, national painting, etc.). The use of freehand brushwork and expansion of artistic expression techniques to design clothing color, so that it has a more distinctive artistic and national characteristics, so as to form a unique regional or national characteristics of the drama.

5 Research on Local Opera Costumes Based on Recognition Algorithm

5.1 The Characteristics of Opera Costumes

The color of opera costume contains three basic aspects of color psychological function, namely color expression, color association and color symbol. The psychological function of color can be different from culture, region, class, race, religion and times. Even if the real color can not be described concretely, it can also show meaningful truth. The same people have different associations and psychological functions. For example, orange means patience and magnificence for old people who have gone through life training. For young people in love, they represent warmth and enthusiasm. For children who are naive and lively, they represent ripe and edible fruits.

For operas with special performance forms, the use of color in opera costumes can be said to reach the extreme. According to the needs of the stage performance, opera costumes often show the beauty of "inlaid colors and inlaid gold, full of carved feedings" through special craftsmanship, giving people a pleasant feeling. The colors of opera costumes are mainly concentrated in 10 colors, i.e. upper five colors and lower five colors [4]. The upper five colors refer to black (neutral color), white (neutral color), red (big red), yellow (bright yellow) and green (old green), while the lower five colors refer to Lake (lake blue), pink (pink), purple (purple red), blue (dark blue), and autumn fragrance (similar to olive green).

5.2 The Evolution of Opera Costume

Opera actors in different times, regions and backgrounds have different understandings of the characters in the drama. This subjective understanding is not only reflected in the performance form of the opera, but also in the understanding of the costume color. The color of opera costume is not unchangeable, but changes with time, region, culture, belief and understanding of characters. As mentioned above, the color of Guan Yu's opera costumes changes with the understanding of the characters, which shows that the color of opera costumes has certain stability, also has certain changes, and gradually changes in stability.

The evolution of the color of opera costume is inseparable from the development of costume materials and the progress of dyeing technology. Human beings have gone through a long process from covering their shame with plant leaves and animal fur to the application of various textile fibers and the development and utilization of dyeing and finishing processes. The increase of clothing color selection also promotes the progress of colorology in the application of clothing design. The diversity of costume color on the stage of opera can not be separated from the support of textile and clothing science and color science. The evolution and change of costume color is closely related to the science and technology of an era.

It's too fuckin 'bullshit. There's no shortcut to understanding human nature. It needs a lot of life experience. Sometimes I really doubt whether I've ever done sales or not. It's hearsay. Without rich life experience and rich experience, it is impossible to understand human nature. It is useless to talk on paper. According to years of sales practice, we find that the relative shortcut to understand human nature is not only to summarize our own experience and be good at reflection, but also to study more history, psychology, independent thinking ability and reading volume. Understand human nature, there is no shortcut, no shortcut, no shortcut, important things to say three times!

There happened to be a very handsome man there who answered this question. Then I saw a lot of people in the comment area. Many people are jealous of him. And then I published some. I am very enthusiastic about some of the comments made by the respondents. And guess what?

There are all kinds of answers to this comment. I like this comment. And then analyze it according to this sentence. What happened to him? What is he like? Why can he say such a thing? Then click on his avatar to see what the main color of his avatar is, and then see how he answers other questions and what the color of its cover is. Then I'll come back to see his comment, guess what kind of experience he had, and what kind of people he was. Then I found a lot of Qiu Shuai. The answer is a bit messy. But maybe I really understand. I think it's quite interesting.

The essence of human nature is to pursue existence and strengthen existence. There are some other views, such as nature is good, nature is evil, seeking advantages and avoiding disadvantages, pursuing value and so on, which are all for the purpose of pursuing one's own existence and strengthening it.

The complexity of human nature comes from the cognitive mechanism of human beings to pursue existence and strengthen existence. For example, how do humans confirm their existence? It needs cognition. How to recognize it? You need feedback. It's better to look in the mirror. Human beings must look in the mirror to confirm their existence. If you look in the mirror, there will be deviation, and cognitive feedback will magnify the deviation and complicate human nature. For example, herding, the cultural revolution, and the Stockholm syndrome. These complicated cases of human nature are even more complicated if they are reflected in the mirror of human culture, morality and law.

In fact, human nature has never been complicated. What is complicated is how human beings look at themselves in the mirror, and how the world inside and outside the mirror interact with each other, cause and effect, and how the circle strengthens.

6 Conclusion

Opera is a treasure of Chinese culture and art, and plays an important role in Chinese culture and art. By analyzing the characteristics and evolution of costume color in traditional Chinese opera, this paper expounds the important application of costume color in opera performance and the various factors influencing the development of costume color in opera. It is helpful to promote the Chinese opera civilization to play its unique artistic charm on the world cultural stage.

Acknowledgements. Jiangxi Social Science Foundation Project "Art and Multicultural Value of Gan Opera Dress from the Perspective of Chinese Aesthetics", Project Number: 20YS34.

References

1. Zhang, L., Huang, J., Zhang, T., et al.: Human image intelligent analysis and algorithm optimization in video investigation. Inf. Netw. Secur. **5**, 88–93 (2020)
2. Yang, L.: Research on the color of ancient opera. Henan University, Kaifeng (2013)
3. Deng, X.: Theoretical breakthrough of computer intelligent image recognition technology. Hubei Agric. Mech. **10**, 109–110 (2020)
4. Zhou, P.: On the color implication of Chinese opera costume. J. Henan Univ. Sci. Technol. **3**, 75–78 (2010)

Analysis of the Non-invasive Identification of Sensitive Load Based on Voltage Sag Monitoring Data Based on BP Neural Network

Ben Wang(✉), Qing Zhao, Bo Li, Xiaosong Jing, and Qi Diao

Beijing China Power Feihua Communication Co., Ltd., Beijing, China
wangben@sgitg.sgcc.com.cn

Abstract. Residential users are important consumers of strong smart grid, and also an important part of microgrid. The scientific and reasonable use of electricity by residents plays a vital role in promoting the energy conservation and emission reduction of the whole society and alleviating the energy crisis. An effective household energy management system is the basis of energy saving work. It can effectively help residents to know the power consumption of electrical equipment in different periods of time, improve the awareness of users to participate in demand response, guide users to choose energy-saving equipment reasonably, and make scientific energy-saving plan, so as to reduce household consumption and electricity expenditure. Therefore, it is of great significance to monitor the power consumption details of residents for improving the energy utilization efficiency, realizing the sustainable development of energy, alleviating the current energy pressure and building an energy-saving society. Based on the summary of previous studies, an intelligent nilm monitoring scheme is proposed, and the accuracy and effectiveness of the proposed scheme are analyzed in detail.

Keywords: Household energy management system · Demand response · Neural network pattern recognition · Non intrusive load monitoring · Energy consumption optimization

1 Related Concepts of Personalized Recommendation System

In the face of the pressure of energy supply, in order to meet the needs of economic and social development, it can be achieved mainly through two aspects: open source and saving. Among them, open source refers to the way to expand the power system and increase power supply; throttling refers to improving the use of electricity, saving electricity and more reasonable and scientific electricity consumption, such as power demand side management. The establishment of new energy power plant is a very effective open source measure, and has achieved certain results [1]. At present, our country has been vigorously developing nuclear power, wind power, solar power and other new energy systems in the expansion of power system, in order to gradually change the situation that thermal power and other resource consumption power supply dominated in the

B. J. Jansen et al. (Eds.): *International Conference on Cognitive based Information Processing and Applications (CIPA 2021)*, LNDECT 85, pp. 237–241, 2022.
https://doi.org/10.1007/978-981-16-5854-9_29

past. For example, the bird's nest, the main stadium of Beijing Olympic games, is one of the buildings that use new energy to generate the most power in the world. However, the vast majority of power sources are still fossil fuels, which means that long-term and large-scale government investment is needed to change the energy structure through open source. In addition, a large number of new energy power plants will destroy the local natural environment and affect the local ecology. Therefore, more attention has been paid to the energy saving scheme of throttling, that is, DSM. This measure enables all parties to work together to improve energy utilization rate, change energy consumption mode, and achieve the purpose of reducing power energy consumption and saving power energy resources.

2 Neural Network Pattern Recognition Model

2.1 Current Situation of Neural Network Pattern Recognition

Pattern recognition refers to the analysis and processing of various information reflecting things or phenomena, so as to realize the description, theory is closely related to the development of computer science. With the rapid development of microelectronic technology, although pattern recognition technology has made some achievements, it still can not directly perceive the external information such as voice, text, pictures and so on. For this reason, in order to solve this problem, it is urgent to take Ann and artificial intelligence as the core technology to open up new research fields [2].

Ann is a nonlinear adaptive information processing system based on a large number of processing units. Based on the research results of contemporary neuroscience, it attempts to process information by simulating the principle of human brain neuron memory and processing information. So far, in the field of pattern recognition, almost all the basic models of neural network have been successfully applied. The improvement and progress of neural network theory can greatly promote the development of pattern recognition technology; at the same time, a large number of in-depth research on pattern recognition theory can greatly promote the progress of neural network theory. The relationship between them is mutual influence and penetration.

2.2 Structure of Artificial Neural Network

Artificial neural network (ANN) is a complex network system which consists of a large number of simple basic elements neurons connected with each other, and carries out information parallel processing and nonlinear transformation by simulating the way of human brain neural processing information. The advantage of neural network is multi input and multi output, which realizes the parallel processing of data and self-learning ability. Back propagation (BP) network and radial basis function (RBF) network are the most mature and widely used networks. There are many kinds of neural network structures, the most common of which is feedforward neural network structure, as shown in Fig. 1.

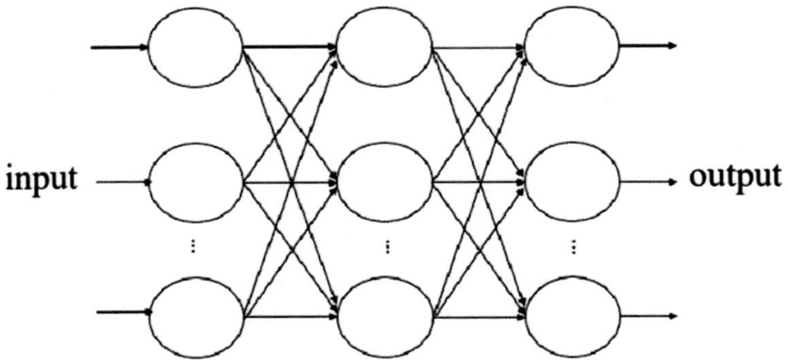

Fig. 1. Neural network structure

2.3 Basic Principles of Neural Network

The data show that more than 80% of the neural network models use BP network or its deformation structure. BP network shows the most critical and perfect content of neural network. The theory of BP network is complete, it started earlier, and it is widely used. Therefore, the structure model of neural network in this paper is BP network [3].

The neurons in the same layer are not connected with each other. The hidden layer and the output layer can process the data transmitted from the upper layer of neurons. In the process of processing, they will multiply the corresponding weights and add a threshold of appropriate size, and then take the processed data or information as the input value of the layer of neurons. When the input neuron node obtains the training data samples, the output value of the current node is transferred from the input layer through each hidden layer to the output layer, and then the input information of the network is obtained according to the information of the output node. In the direction of reducing the error between the output value of the network and the output value of the actual sample, the output value of the current node passes through each hidden layer from the output layer and then returns to the input layer, We call this algorithm error back propagation algorithm, that is BP algorithm. In the whole mathematical theory of BP algorithm, the most critical one is undoubtedly the theory of "negative gradient descent", which can be described as: the adjustment direction of BP network is always in the direction of the fastest error descent. The weight and threshold of general three-layer BP neural network are adjusted by the following formula:

$$w_{ij}(t+1) = -\eta \frac{\partial E}{\partial w_{ij}} + w_{ij}(t) \tag{1}$$

$$B_{jk}(t+1) = -\eta \frac{\partial E}{\partial B_{jk}} + B_{jk}(t) \tag{2}$$

In the above formula, E is the sum of square errors between the network output value and the actual sample output value; η is the learning rate of the network, that is, the weight adjustment range.

3 Non Invasive Load Monitoring Method Based on NNPR

3.1 Nilm Model Based on NNPR

NNPR is a new research direction in the field of pattern recognition in recent years. It has many advantages, such as robustness, fault tolerance, strong self-learning and so on. Different from traditional pattern recognition, neural network pattern recognition is better at extracting hidden association between patterns. We know that the switching states of household appliances may affect each other, which means that nnpr model can get better results than traditional methods. Generally speaking.

3.2 Input Layer of NNPR Model

For non-invasive load monitoring, the selection of input layer parameters of nnpr model is very important, which can directly affect the accuracy of load identification. Different from other nllm technologies, this paper not only considers the total load current (I) and total active power (P) data from each sampling point, but also makes full use of the corresponding time information of each sampling point. In order to prove that time information is closely related to the working condition of household appliances, this paper puts forward the concept of time density curve of household appliances for the first time based on the historical data obtained from monitoring. The abscissa and ordinate of the curve are explained as follows: abscissa is the time coefficient, and each point corresponds to a specific time of the day; ordinate is the current amplitude obtained by monitoring at each time [4].

3.3 How to Determine the Working Mode of Electrical Appliances

The number of neurons in the output layer of NNR model is closely related to the types of load working modes. Therefore, it is very important to determine the types of load working modes according to historical monitoring data. Previous studies have shown that: because the current is not affected by the fluctuation of supply voltage, compared with the active power, the monitoring current data can more accurately determine the working state of electrical appliances. In order to determine the working mode of the load to be identified, we make the current density curve of each electrical appliance based on the historical monitoring data. The abscissa and ordinate of the curve are explained as follows: 1) abscissa refers to all current values monitored by electrical appliances in a long monitoring period; 2) ordinate refers to the number of times of each monitored current value.

3.4 Summary of this Chapter

In this chapter, we build a neural network pattern recognition model for load identification. Two kinds of density curves are proposed innovatively: time density curve and current density curve. Through the former, it shows that the working state of household appliances is closely related to time, and according to this, it gives a method to convert the time parameters of each monitoring point of smart meter into integer form, so as to determine the content of the neural network input layer; through the latter, we determine the type of working mode of the load to be identified, and then determine the number of nodes of the neural network output layer. In addition, this chapter also establishes the construction method of nnpr model based on MATLAB, and introduces in detail the main contents of each layer of nnpr model under the condition of practical application problems, that is, the input layer is the total load information of electrical appliances, the output layer is the specific working state of each electrical appliance, and the hidden layer is the sigmoid excitation function. Finally, we also give the detailed algorithm formula of each layer transformation, including the excitation function of neuron, the mathematical expression of training algorithm and so on. It is difficult to detect the switching events of load at low sampling frequency.

4 Conclusion

With the rapid development of smart grid, global energy Internet and other technologies, the energy-saving awareness of power users has been greatly improved. It is more and more meaningful for the scientific use of electrical appliances and the application of demand response policy to refine the total energy consumption of residents' homes to each or each type of electrical appliances level, that is, residential power consumption detail monitoring technology or non-invasive load monitoring technology. It can effectively help residents to know the consumption of electrical equipment in different periods of the family, improve the awareness of users to participate in demand response, guide users to choose energy-saving equipment reasonably, formulate scientific energy-saving plan, and then achieve the purpose of reducing household consumption and electricity expenditure. This paper mainly studies the non-invasive load monitoring technology based on neural network pattern recognition, and puts forward a new scheme of residential power load recognition, which solves the problem that the existing load decomposition methods require high sampling frequency of equipment, and can not effectively deal with the use of multi-mode electrical appliances. In addition, the proposed method considers the time characteristics of load for the first time, which further improves the accuracy of load identification.

References

1. Wang, Z.: Research on key technology of low voltage side power consumption management based on load characteristic identification. Wuhan University (2011)
2. Wang, B., Li, Y.: DSM planning and implementation mechanism for smart grid. Power Autom. Equip. **12**, 19–24 (2010)
3. Huang, Y., Hou, F., Zhou, Q., Fu, B., Guo, X.: A method of user load pattern combination analysis for DSM. Power Syst. Prot. Control **13**, 20–25 (2013)
4. Li, P.: Non invasive power load decomposition and monitoring. Tianjin University (2009)

Study on Population Prediction by Two Different Neural Network Algorithms

Jiangnan Wang(✉)

Beijing 21St Century International School, Beijing 102208, China

Abstract. Population problem is the premise of sustainable development of a country or region, especially in today's increasingly prominent population problem, whether the population scale is reasonable or not is more critical to the coordination and sustainable development of its economy, society and ecological environment. In this paper, RBF neural network and BP neural network are established by using MATLAB programming software. Taking the total male population of China population Yearbook in 1999–2003 as the sample, RBF neural network and BP neural network are trained to predict the total male population in the next five years. The results show that the two methods are feasible with good effect and small error.

Keywords: RBF neural network · BP neural network · Population forecast · Economic forecast

1 Introduction

There are many methods for the study of population forecast. The general methods are: growth curve model, grey system model, autoregressive model, social and economic model, etc. in actual forecasting work, autoregressive model can analyze the periodicity and randomness of population, A comprehensive model has been established by combining the effects of trend fluctuation, periodicity and randomness. Based on the data collected according to time series, some curves such as polynomial curve and exponential curve are used to fit the data. This method is relatively simple, However, the accuracy of grey system model is not high. The main purpose of grey system model is to analyze the fluctuation of population growth trend. 1. The number of population is a very complex nonlinear system, and the artificial neural network has the nonlinearity between neurons, so it can be used to express the nonlinear system of population number to improve the accuracy of the model [1].

In this paper, two different neural network algorithms, RBF neural neural network, used train forecast network respectively, and compare and analyze the actual male population.

B. J. Jansen et al. (Eds.): *International Conference on Cognitive based Information Processing and Applications (CIPA 2021)*, LNDECT 85, pp. 242–247, 2022.
https://doi.org/10.1007/978-981-16-5854-9_30

2 Construction of Population Forecast Model

2.1 Establishment of RBF Network Model

With input function layer 1 and radial basis function layer 2 (RBFNN), RBF network is shown in Fig. 1.

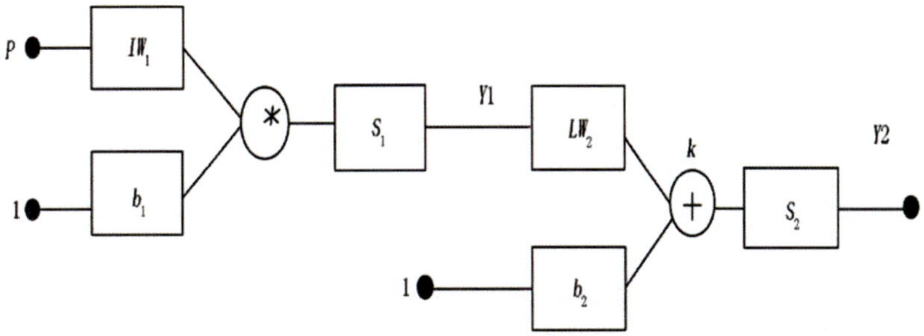

Fig. 1. Schematic diagram of RBF artificial network structure

Let the learning center be the clustering center, where $S_i(i = 1, 2, \cdots, i)$ is the center of the nth iteration time base function determined by experience [2]. Randomly select l different samples from the training sample set as the initial center $S_i(0)(i = 1, 2, \cdots, i)$. Let $S_i(n)$ be the first center of the nth iteration time base function.

$$i(X_k) = \arg\min\|X_k - S_i(n)\|, i = 1, 2, \ldots I \tag{1}$$

Find the center nearest to X_k and use:

$$\begin{cases} S_i(n+1) = \{S_i(n) + \eta[X_k(n) - S_i(n)], i = i(X_k)\} \\ s_i(n) \end{cases} \tag{2}$$

To adjust the center of the base function until the center of the base function is found.

2.2 Establishment of BP Neural Network Model

The basic idea of BP network is to attribute the errors which are inconsistent with the facts to the "fault" of the connection weights and threshold values among the nodes in the connection layer. The error of the output layer node is propagated to the input layer layer layer by layer to "share" each connection layer. Thus, the reference error of each connection point can be calculated and the connection weight can be adjusted accordingly, Mapping to adapt networks to requirements The structure is shown in Fig. 2.

BP network learning steps: 1) give the initial value of network parameters; 2) give the input and target output vectors and initialize; 3) calculate the hidden layer, the output of each unit in the output layer; 4) calculate the error e between the target value and the output of network calculation; 5) judge whether e reaches the specified error; 6) calculate the unit error and error gradient of hidden layer, constantly modify the weight, and continue training until the error accuracy requirements are met [3].

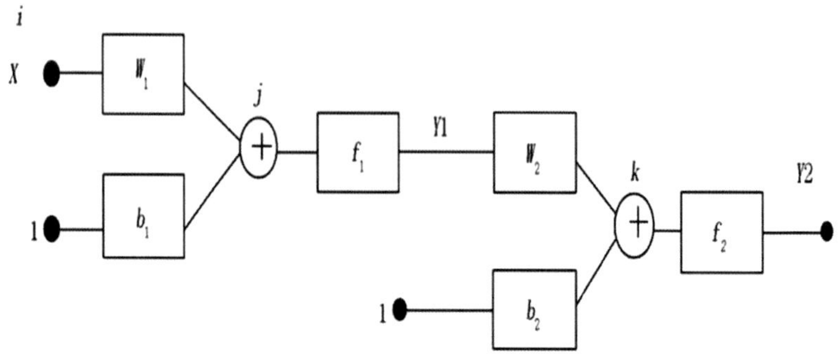

Fig. 2. Schematic diagram of BP artificial neural network

In this paper, nine input nodes and one output node are used to predict the population size. The hidden layer uses S-type activation function, and the output layer uses linear function. At the same time, the network uses momentum method and traingda. M function with adaptive learning rate to further reduce the training time.

3 Prediction Results

After training RBF network and BP network for many times, the forecast results of population in 2004–2008 are shown in Table 1.

Table 1. The actual number of male population and the predicted population of 10000 from 2004 to 2008

Particular year	Actual population	BP forecasts population	Population forecast by RBF
2004	66976	67363	67689
2005	67375	67048	66316
2006	67728	66896	68555
2007	68048	66896	63281
2008	68357	68265	67102

4 Population Forecast Model

4.1 Input Output Vector Design

The establishment of neural network prediction model based on time series takes the population data of the first three years as the input variable of the neural network, and the total population data of the fourth year as the output variable. That is, the input is 3 data and the output is 1 data.

In the sample set, 80% of the total amount is used sample is used as the test sample set, and then the accuracy of the model is verified by comparing the consistency between the test data and the real data. Verified by comparing is shown in Fig. 3.

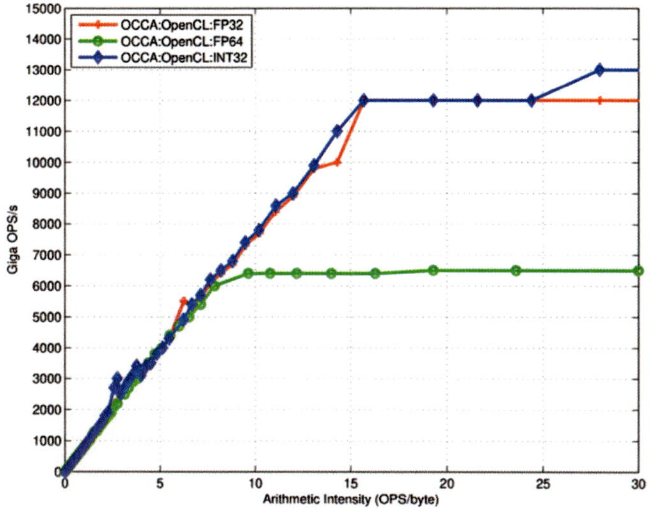

Fig. 3. Verified by comparing

4.2 Network Model

According to the principle of BP network, the general prediction problem can be realized by a single hidden layer, because in theory, approach any nonlinear function. This study constructs three layers of After the network layer structure is determined, the number of neurons in each layer and neuron is set as three, and the output layer neuron can be defined as one. For the selection of the number of neurons in the hidden layer, there is no unified and accurate method to determine the number of neurons. Empirical formula or trial and error method is often used. The empirical formula method is used in this paper.

The methods of testing formula are mostly derived from general induction. To determine the number of hidden layer neurons, the following formula can be used to determine the number of hidden layer neurons:

$$n_1 = \sqrt{n+m} + a \tag{3}$$

Among them, n is the number of neurons in the input layer, M is the empirical formula for the number of neurons in the output layer, and a is between 1 and 100.

$Sqr(3+1)+3=5$. After comprehensive test and comparison, it is concluded that the hidden layer neurons can be determined as five transfer functions, also known as excitation functions, which are an important part of BP network. BP network usually selects S-type logarithm or tangent and linear function as transfer function. In the design of predictive neural network model in this study, sigmoid logarithmic function is used as input layer transfer function.

Logsig is an S-type logarithmic function, which can map the input (range of the whole real number set) of neurons to (0,1), and it is also small. Therefore, it is very suitable to use the neural network hidden layer to output layer transfer function of BP algorithm using linear function. As shown in Fig. 4. Purelin is a linear transfer function, and a linear transfer function is more suitable for the output layer transfer function in the prediction model [4].

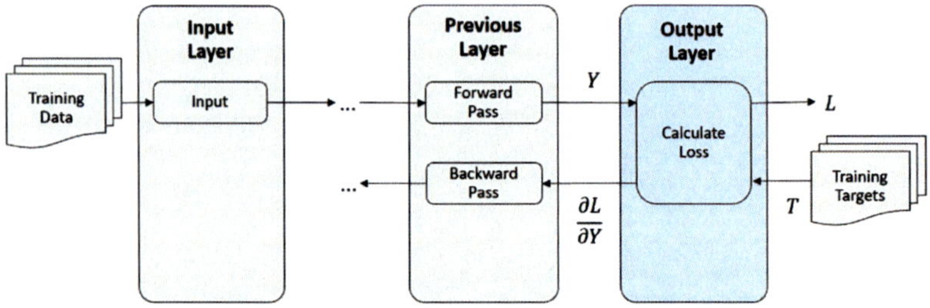

Fig. 4. Layer to output layer transfer function

Scientific prediction of the total population can provide comprehensive and objective reference for population planning and future policy. At the same time, the more effective is that the forecast result data will have immeasurable basic significance for the formulation and Realization of the national economic development plan. At present, the main prediction methods are: Logistic equation method, ordinary differential equation method and dynamic prediction method. However, compared with the traditional neural network model, the distributed neural network provides a flexible and effective way of data storage and prediction. Specifically, each neuron in the neural network has a self-organizing compound pattern and reflects the nonlinear characteristics, even if the neural network can reconstruct any nonlinear continuous function. Through learning, we can use the network to obtain the internal law of the sequence, and then we can predict the change of the sequence. Neural network can be applied to prediction in two ways: neural network prediction based on regression analysis and neural network prediction based on time series.

5 Conclusion

After running for many times, the BP forecast population and RBF forecast population are compared with the actual population. The results show that the two methods are feasible and the error is very small.

References

1. Huang, R.: Reflections on population prediction. Popul. Res. **28**(1), 88–90 (2004)

2. Yang, L., Yang, G., Yuan, S.: Application of mathematical model in population prediction: a case study of Jiangsu Province. Resour. Environ. Yangtze River Basin **15**(3), 287–291 (2006)
3. Cong, S.: Theory and Application of Neural Network for Matlab Toolbox. China University of Science and Technology Press, Hefei (1998)
4. Tian, B.: Artificial neural network and its application in system theory. Beijing Normal University, Beijing (2001)

Development and Algorithm Research of Deep Learning Training Cloud Platform

Li Li(✉) and Jian Xiao

Wuhan Technology and Business University, Wuhan 430065, China

Abstract. In recent years, colleges and universities pay special attention to the cultivation of students' practical ability, which requires efficient and practical cloud platform training system. Due to the rapid increase in the number of students, the shortage of teachers and the shortage of special venues, the centralized traditional experimental training mode can not meet the actual needs. This paper designs a training system based on cloud platform technology: the development board and experimental box used by students can be connected to the cloud platform anytime and anywhere, and the experimental process and results can be accurately recorded in the cloud database; Students don't have to carry out experiments at fixed time and place, which solves the problem of shortage of experimental sites and stimulates students' autonomous learning ability; Teachers can track the progress of students' experiments online and guide students interactively, which makes full use of limited teachers' resources; Managers can analyze equipment data and faults online, evaluate students' experiments and teachers' guidance accurately, and solve management problems.

Keywords: Training system · Cloud platform · System development

1 Introduction

The Internet of things technology has developed into a full swing stage, which has a great impact on the teaching of colleges and universities. The Laboratory Laboratory of Internet of things in Colleges and universities has more and more functions. The management system of laboratory is also intelligent and scientific based on the Internet of things technology. But the corresponding experimental teaching mechanism is not perfect, and students have not a deep understanding of the application of Internet of things. Although some companies focusing on laboratory teaching are also developing various Internet of things training platforms, they can be seen in the meeting hall of relevant electronic technology seminars in schools. However, the current teaching mode or experimental mode is still the former experimental mode of electronic information specialty, which belongs to centralized duck filling teaching, which is difficult to make each student explore independently and deeply, and can not stimulate the interest of students.

The rapid development of Internet of things promotes the next generation innovation with user as the core, namely, the formation of Innovation 2.0. Innovation 2.0

B. J. Jansen et al. (Eds.): *International Conference on Cognitive based Information Processing and Applications (CIPA 2021)*, LNDECT 85, pp. 248–254, 2022.
https://doi.org/10.1007/978-981-16-5854-9_31

is people-centered, emphasizing that anyone can participate in innovation at any time and place. Based on this, under the background that Internet of things has gradually become the mainstay of digital economy, the real educational innovation should be to fully respect personality, adopt open teaching concept, implement subjectivity education, enrich knowledge and activate thinking in practice, combine theory and practice deeply, and give full play to students' initiative.

Therefore, the Internet of things cloud platform training system needs to have the function of stimulating students' interest and giving full play to their own learning consciousness [1]. The cloud platform training system designed in this paper makes full use of the wireless boundless characteristics of the Internet of things, and gives students the development board and networking module in hardware; Provide the students with experimental manuals, burning firmware, debugging tools, etc. on the infrastructure; Provide the operation interface at the application level. The development board can carry on the various sensors to collect the data; The LAN can be built freely by using communication module; According to the open API interface and website development guide manual, we can learn the technology of website development, and provide a variety of development boards and supporting platforms to further enrich the application scenarios.

2 Analysis of Cloud Platform Technology of Training System

The main protocol of Internet of things equipment access to the Internet is mqtt protocol. The cloud platform of each Internet of things training system supports the cluster of equipment servers of mqtt protocol.

In the server performance level of the long connection application, the problem of C10K has developed to the c10m problem, that is, the possibility of realizing millions of connections by a single machine. Some mature high-performance server design methods are produced, such as IO multiplexing, multi-core parallel processing, reactor reactive design.

The system products need to have better disaster tolerance ability, so the equipment server must be clustered. Clustering means to route the device, and the whole server cluster becomes a stateful cluster. However, under the limitation of cap theorem, the intra cluster needs to be balanced between availability and consistency. A series of distributed data consistency algorithms, such as raft algorithm and Zab protocol, are generated. In contrast, it is the open source implementation of a series of mature consistency algorithms, such as etcd and zookeeeper.

In the application diversification today, traditional relational database is no longer the preferred storage scheme in some scenarios. Many non relational databases have been born to complement each other with relational databases, such as mongodb, a document database and redis database, which is commonly used as cache.

With the rise of cloud prototypes, container becomes the mainstream of development. Because of its simple deployment, flexible combination and low resource occupation, the majority of developers pay attention to the container technology, which is represented by docker'container. But when the number of containers increases, the correlation between containers, the management and operation of containers bring more complex problems.

Kubernetes, which supports container choreography, is undoubtedly the leader in this field. The appearance of kubernetes has greatly affected the pass manufacturers and will have a profound impact on the future application development mode.

3 Overall Design of Cloud Platform

3.1 Overall Function of the System

The panorama of the hierarchical structure of the cloud platform training system designed in this paper is shown in Fig. 1:

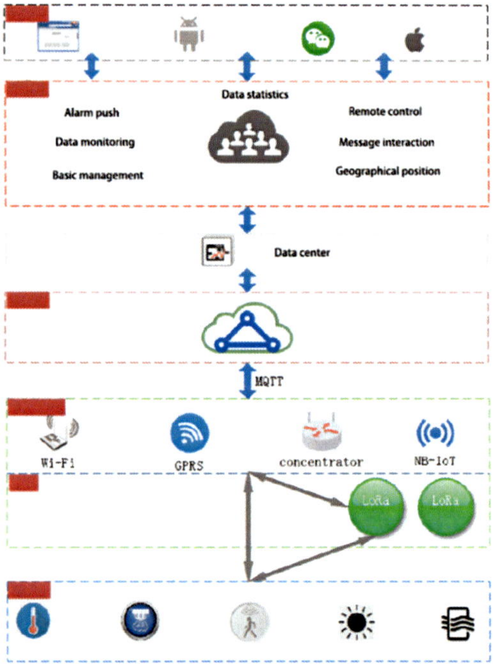

Fig. 1. Overall hierarchical structure of training system

The bottom layer is the access of all kinds of sensors, and then the inner network Lora node in the edge network can collect the information of each sensor, and then report it to the external network devices in the edge network, such as the GPRS concentrator. Because the GPRS concentrator has built-in Lora module and GPRS module, it can act as the gateway in the star network topology. Or the sensor data can be collected directly by the external network devices in the edge network. The external network devices report the collected data to the device cloud platform through mqtt protocol, and the device cloud platform stores it in the data center. Business cloud platform provides a reliable interface for the upper application, and it is the business enabling layer of the application. Users can access the rich functions provided by the system through a variety of applications [2].

In the edge network, the system provides a variety of functional development board. The external development board includes Wi Fi, GPRS, Nb IOT, and GPRS Lora concentrator. The internal development board is Lora node. Through the interconnection of development boards, a variety of collection networks can be realized.

Device cloud supports external network devices to access cluster and client through mqtt protocol for interaction and control, which plays a connecting role.

In the data center, the main storage is equipment data and business data.

In the business cloud, it provides interface calls to the upper layer, and also opens part of the application programming interface (API) to developers, so that they can use the API to realize some web page functions.

At the application level, various kinds of applications have been developed: students use wechat to control web pages, general industrial gateway monitoring background, lorawan management background, teacher management background and manager background, etc.

3.2 Server Architecture Design

Software system is a process from simple to complex. In order to solve the problems of high complexity of software, it is necessary to analyze the complexity before architecture design.

In the server cluster, the users are mainly students, teachers and administrators of training related majors. The average number of visits per user per day is not very much, so it can be estimated that the performance requirements of the server cluster are not high.

At present, the presentation layer still has a lot to improve, so as a business server providing services for the presentation layer, it needs high scalability, easy iterative modification, easy rapid deployment and testing. At the level of software development, it is necessary to choose the development framework and separate the interface and implementation; At the operation and maintenance level, docker deployment is needed to achieve the purpose of rapid deployment and simplified operation and maintenance. However, as the core architecture and facilities are unchanged, for business server cluster, scalability only refers to how to expand when the current server configuration is not enough. From the perspective of cost, we don't need special cloud load balancing, we just need to use nginx software to do seven layers of load balancing, and reverse proxy the back-end business server. The back-end program is actually deployed on the Tomcat server software.

There is no strict requirement for the high availability of business system, even if the short-term downtime has little impact on the whole, just need to make a good alarm mechanism. But the business system has requirements for high reliability of storage. If all the relevant data are lost, it will be very troublesome to repair. Therefore, the complexity of this system mainly focuses on the high reliability of storage [3].

In terms of security, since it does not involve financial scenarios such as transactions, three things need to be done well: configuring nginx to do ACL verification, user account password authentication and database access control.

To sum up, the overall architecture of the server cluster is shown in Fig. 2:

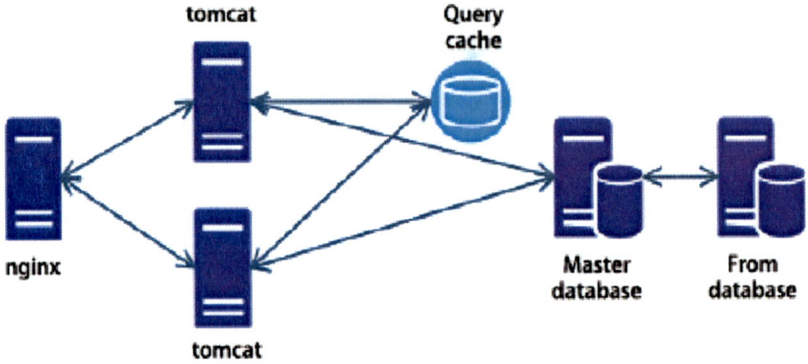

Fig. 2. Overall architecture of server cluster

4 Module Development of Training System

The production software writes the randomly generated deviceld and the encrypted generated auth into the flash of the development board, and then calls the production data interface provided by the business server. The main function of the interface is to store the production data of the device. The production data table serves as the inventory list and provides data for the statistics of the management system. Then it goes to mongodb to create the corresponding storage set of the device.

After the device is successfully added, you can view the monitoring points under it. Here, the monitoring point does not involve the manual addition operation. As long as the device reports a message with the subject of # data by using the mqtt protocol's publish message, the mqtt server software will write the data of the message to mongodb database. In addition, the mqtt server node will also write a message to rabbitmq. The queue name of the message is pointid, and the message format is as follows:

{"deviceid": "device ID,",
"Pointid": "monitoring point ID"
}

4.1 Quick Data Search

The real-time data reported by the device is stored in mongodb database. Since mongodb is a document database, the document format of each monitoring point designed in this paper is as follows:

```
{_ id: ObjectId(4f442120eb03305789000000'),
data: 88,
time: ISODate("2018-10-10T20:55:36Z"),
pointId:1
}
```

Each set corresponds to a device, and each document corresponds to a piece of data of the device at a certain time point. Among them, time represents the time point of reporting the data, and data represents the corresponding value of the time point. Because of the self description of JSON format, there is no restriction on the data type. Because the system is based on the time period to query data, so the index of time. And because the data to be queried is the data of a monitoring point for a period of time, the pointid parameter needs to be added to each document.

4.2 Message Interaction Module

Students can leave messages and teachers can reply to students. The message system designed in this paper, for students, does not need to know who the teacher is, only focus on the problem. As mentioned in the data table domain model above, students are associated with classes, schools are also associated with classes, and teachers are associated with schools. These associations are bound in the administrator background, so teachers can see the list of students in the class after logging in, and also can see the students' messages in the class.

Mongodb database is selected for the storage implementation of message module. Due to the need to connect with different schools, each school corresponds to a collection, and each document in each collection corresponds to a message from a student to a teacher. The teacher's reply to the student is also directly written in the corresponding document.

```
{
"_ id":ObjectId,
"StudentID": "student number,",
"Scontent": "message content":
"Stime": "message time",
"Tcontent": "reply content,",
"Ttime": "reply time"
}
```

When students leave a message, the business server will insert a new document into mongodb, in which 'tcontent' and 'ttime' are set to null, and other parameters are filled in according to the actual content_ The ID is generated automatically by mongodb. When students view their message list, they will go to mongodb to query message information according to their studentld and school ID, and then combine these message information and return it to the front end. For the teacher did not reply to the message, the front only shows the student's message [4].

5 Conclusion

This paper designs the cloud platform in the training system and studies the key technologies. The overall architecture design from the reliable storage layer to the service support layer and then to the application enabling layer is realized, and the server cluster and device server cluster are developed.

References

1. Fan, T.: Research and design of Internet of things embedded development platform based on cloud service. Shanghai Normal University (2018)
2. Liu, Q.: Research on fog resource scheduling technology for intelligent manufacturing service. University of Electronic Science and technology (2018)
3. Shao, X.: Design and implementation of intelligent training room management system based on Internet of things. Netw. Secur. Technol. Appl. (12), 96+116 (2018)
4. Wei, Z.: Analysis on the construction of IOT training room Ding. Occupation (05), 101–102 (2018)

Research on Optimization Algorithm of Digital Supply Chain Finance in Science and Technology Enterprises

Xiaojie Ren and Xiaoyi Wang(✉)

Eastern College of Shandong University of Finance and Economics, Shandong, China

Abstract. This paper studies the application scenarios of digital supply chain finance in science and technology enterprises under the optimization algorithm. Digital supply chain finance has the endowment characteristics of precision management services, network-based credit transmission and intelligent risk control, which can meet the requirements of high-quality development of science and technology enterprises for financial innovation. Based on the perspective of digital technology penetration, this paper discusses the operation process and critical control point design of supply chain financial innovation mode in the application of technology-based enterprises under the optimization algorithm. In view of the possible constraints of risk control, technology application and legal system in the process of digital supply chain financial services for technology-based enterprises at this stage, this paper proposes that we should control the risk source, avoid the risk, and improve the quality of service. We should speed up the digital technology enabling ecological innovation and improve the legal system to standardize the development.

Keywords: Digital supply chain finance · Optimization algorithm · Innovation · Enterprise application

1 Introduction

Supply chain finance is that banks provide financing and other settlement and financial services to customers (core enterprises of supply chain), and provide loans or other financial services to suppliers of these customers, or provide prepayment and inventory financing services to distributors. When the supply chain finance issues loans with raw materials, work in process, finished products and so on as collateral, the introduction of logistics supervision enterprises participate in the whole process of supply chain finance collateral supervision. In this sense, supply chain finance is an integrated development mode of Finance and logistics industry, which can, to a certain extent, cope with the great challenges brought by China's current economic transformation to the development of Finance and logistics industry.

The core goal of supply chain finance is to provide financing solutions for SMEs through the innovation of service mode. Logistics supervision enterprises are established

B. J. Jansen et al. (Eds.): *International Conference on Cognitive based Information Processing and Applications (CIPA 2021)*, LNDECT 85, pp. 255–261, 2022.
https://doi.org/10.1007/978-981-16-5854-9_32

in accordance with the law, with professional qualifications in transportation, distribution, warehousing and supervision, and provide third-party goods supervision services for bank supply chain financial business through bank credit. This paper assumes that logistics supervision enterprises not only provide pledge supervision services, but also provide warehousing, transportation and distribution business. Their professional management can improve the management level of pledge, reduce the risk of supply chain finance, and promote the smooth development of new business. Logistics supervision enterprises can participate in supply chain finance to expand logistics value-added business and realize logistics value-added benefits. Logistics supervision enterprises mainly undertake two responsibilities in supply chain finance. One is to ensure the quantity and quality of the pledge during the pledge period through warehousing services, so that the value of the pledge can meet the minimum requirements of bank lending; The second is to provide transportation and distribution services for the financial pledge of the supply chain to meet the logistics needs of all links in the supply chain and ensure the smooth flow of the goods.

This paper discusses the optimization of the distribution path of the pledge, that is, by choosing a reasonable distribution path, the pledge can be quickly and economically delivered to the customers in all links of the supply chain, so as to improve the customer satisfaction and enhance their profitability. Pledge distribution route optimization is one of combinatorial optimization problems, and it is also a traveling salesman problem in the type of uncertain problems [1]. It is difficult to find the optimal solution of this kind of NP complete problem by using the global search method, so it is necessary to explore other suitable and effective algorithms to obtain the optimal solution or approximate optimal solution. The common and effective algorithms used to solve the distribution route optimization problem mainly include artificial neural network algorithm, ant colony algorithm, artificial immune system algorithm, hybrid genetic algorithm and so on.

2 The Endowment Characteristics of Digital Supply Chain Finance

The concept of rural supply chain finance in academic circles has not been unified. Some researchers have suggested that rural supply chain finance refers to innovative financial business which combines the products and individuals in the supply chain of agricultural and superior agricultural products and the interests of farmers or consumers of small and medium-sized enterprises in the upstream and downstream, takes banks as the leading and enterprises as participants, and designs financial products based on the supervision science of third-party logistics to meet the financing needs of all links in the agricultural supply chain. Rural supply chain finance can reduce the risk of capital flow for enterprises through credit and financial services provided by banks. Some researchers believe that rural supply chain finance is market-oriented, through which the bank financial institutions integrate the interests of upstream and downstream small and medium-sized enterprises or consumers as one, and regard the supply chain as the main body of credit, and provide financing and financial services. This kind of financial business avoids the shortage of funds through the financing integration of upstream and downstream enterprises, The internal turnover rate of the enterprise has been improved. As shown in Fig. 1. Some foreign researchers believe that rural supply chain finance refers to the financial business that the trade groups composed of enterprises in the supply chain provide transaction financing for the supply chain portfolio through financial institutions. In this combination, the resources and services provided by the enterprises in the supply chain can achieve value.

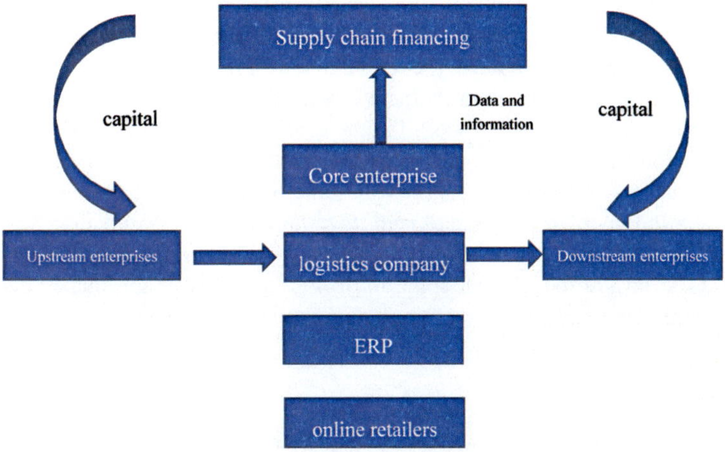

Fig. 1. Network structure of digital supply chain finance

3 Construction of Digital Supply Chain Financial Model Based on Optimization Algorithm

The premise of building the model is to simplify the research problem according to the actual situation, so as to make the problem more simple [2]. The biggest difference between the distribution of pledged goods and the distribution of ordinary goods is that the distribution of pledged goods can only be carried out after the supervision is removed. That is to say, the customer must return a certain amount of loan according to the contract signed with banks, core enterprises and logistics supervision enterprises before the supervision of corresponding value goods can be removed. At this time, the pledged goods can enter the distribution process. In this paper, the optimization of the distribution path of the pledge is simplified as follows: from the pledge warehouse of the logistics supervision enterprise, multiple vehicles are used to deliver goods to the customers at the upstream and downstream nodes of the supply chain, the location of each customer is unchanged, the input of the pledge logistics is basically determined, and the carrying capacity of the vehicles that distribute the pledge is determined, At the same time, it is constrained by the following three conditions.

1) The collateral of all customers on each route is the same product that is deregulated, and the sum of the distribution volume does not exceed the rated load capacity of the vehicle;

2) The total length of each distribution route is less than or equal to the maximum mileage of a single operation of distribution vehicles;
3) All customers' collateral must be delivered in place and can only be delivered by one delivery vehicle.

By simplifying the objectives and constraints of the pledge distribution path optimization problem, we know that this kind of optimization problem is a typical traveling salesman problem. In order to make the research more simple and clear, this paper only discusses the optimization problem of single vehicle distribution route selection. If more than two distribution vehicles are needed in the actual work, the route selection of multiple distribution vehicles can be decomposed into the optimization of single vehicle distribution route. The optimized distribution route of each vehicle can be obtained by decomposing the distribution demand, and then the complete optimization scheme of multi vehicle distribution can be obtained by integrating according to certain rules.

In order to ensure the optimal solution of digital supply chain finance under the support of digital technology, the minimum cost tree is solved by using plam algorithm. As shown in Fig. 2.

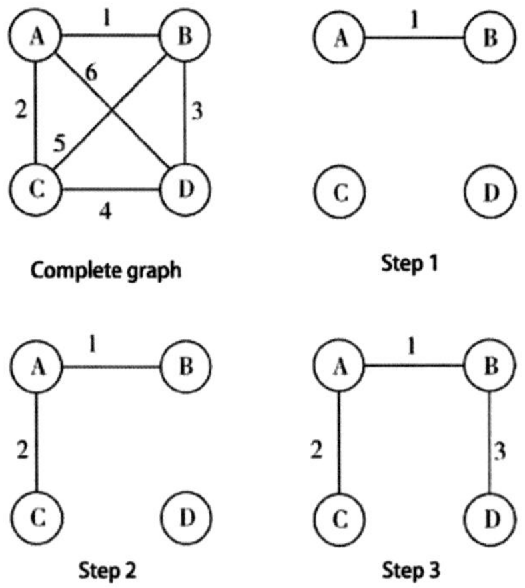

Fig. 2. Step diagram of minimum cost tree generation

The questionnaire of this study was conducted in five months from August to December in 2019. The respondents were mainly employees of logistics enterprises. A large amount of questionnaire information was collected from employees of different levels in different enterprises, and different questionnaire schemes were designed for management and ordinary employees, The different identities of employees can reflect the development of enterprises and avoid the research bias caused by a single survey group. In order to ensure the quality of the questionnaire, this paper uses offline and online questionnaires to collect data. s; According to the division of enterprise turnover, the number of enterprises with turnover between 3 million and 5 million is the largest, accounting for 38%, those with turnover less than 1 million account for 25%, those with turnover between 1 million and 3 million account for 14%, and those with turnover more than 5 million account for 25%. The above results show that the distribution of enterprises with different nature is more uniform.

1) Let a be the distribution center and the starting point of the distribution path, then the set $U = \{A\}$ searches for the adjacent edge with the smallest weight (distance) of A. in this case, the edge (a, b) is added to the set u, and the edge (a, b) from the distribution center a to the customer B is added to the set E, that is, $U = \{A, B\}, E = \{A, B\}$. By adding the set u, all the customers who need to distribute the pledge have got the goods, and the calculation ends [3].

The inventory pledge financing mode of digital supply chain usually occurs in the production process of technology-based enterprises with inventory pledge. It requires dynamic monitoring and value evaluation of movable property pledge of technology-based enterprises through technologies such as Internet of things and cloud computing, and technology-based enterprises apply for loans from financial institutions with inventory as pledge, And the capital inflow of its future inventory sales is used as the source of repayment.

4 Optimization Algorithm Digital Supply Chain Finance in the Innovation of Science and Technology Enterprises

The application of optimization algorithm in digital supply chain finance accelerates the speed of optimization and improves the quality of understanding space. Generally speaking, it optimizes the search efficiency and effect of the distribution path of pledged goods, reduces the total distance of distribution, accelerates the distribution speed of pledged goods, thus can improve the customer satisfaction of logistics supervision enterprises and increase the financial income of supply chain.

There are various ways for rural logistics enterprises to participate in supply chain finance, mainly including the mode of pledge of goods rights, the mode of joint pledge of goods, the mode of financing through warehouse and the mode of financing through confirmed warehouse. First, the mode of pledge of goods rights refers to the mode that logistics enterprises as media provide credit guarantee for dealers, and the banks that have obtained the credit guarantee issue letters of credit for dealers, Then purchase the goods from the supplier through the letter of credit, and pledge the right of the remaining goods. Second, warehouse receipt pledge financing mode means that logistics enterprises have the pledge right of products and issue warehouse receipts. Banks provide loan financing business for enterprises by measuring the value of goods in the warehouse. In this process, logistics enterprises make full use of their own advantages and use their strong storage capacity as a medium between manufacturers and dealers, And improve their own business performance. Third, the financing mode of rongtongcang refers to that the rural logistics enterprises use their asset advantages to apply for the credit line from the bank [3]. After the bank applies, the logistics enterprises provide the financial services of pledge loan to the financing enterprises. The logistics enterprises realize the financing purpose by using their own loan line through the credit guarantee to the financing enterprises. Fourth, in the confirmed warehouse financing mode, the dealers send a certain amount of production goods to the logistics enterprises through the acceptance bill issued by the bank, and the logistics enterprises become the credit intermediary again.

Through the empirical analysis of the structural equation model, we can see that the hypotheses proposed in this paper have passed the significance test and got the theoretical support. The above results show that each dimension of rural supply chain finance has a significant impact on the economic performance of rural logistics enterprises. Among them, assuming that the path coefficient of H1 is the highest, it means that the rural supply chain finance has the most significant impact on the process turnover optimization of the internal business of rural logistics enterprises, that is, under the same supply chain finance mode, the process turnover of the internal business of logistics enterprises is the most affected, and the improvement effect is the most obvious [4]. Assuming that the value of path relationship among the elements of H4 is the smallest, it indicates that the influence between supply chain finance and logistics financial cost is not high, and the influence between other elements is higher than that of financial cost on supply chain finance. It is assumed that the path coefficients of H2 and H3 elements are almost the same, and their sizes are in the middle, which indicates that in the rural supply chain financial system, the creativity of logistics enterprises has the same impact on customer value. Rural logistics enterprises can optimize the internal organizational structure of enterprises, improve the internal turnover rate, strengthen the degree of technological innovation, enhance the competitiveness of enterprises, and enhance the competitiveness of enterprises by participating in the rural supply chain financial system Reduce financial costs.

5 Conclusion

In order to simplify the calculation, the objective function of the mathematical model only considers the distribution distance of pledged goods. In the actual operation of supply chain finance, logistics regulatory companies need to consider the value, distribution volume and traffic congestion of pledged goods when distributing the pledged goods. Digital supply chain finance has obvious cross-domain and cross temporal characteristics, which requires coordination and cooperation of regional regulatory agencies. And increase the punishment of illegal acts, to provide a good legal environment for the technology-based enterprises to use digital supply chain finance.

Acknowledgements. Research on Financial Supply Support for Micro, Small and Micro Enterprises Impacted by the Epidemic in the Big Data Era.

References

1. Li, B., Yu, B., Yang, X.: Optimal path selection of emergency logistics based on fuzzy neural network. Logist. Technol. **28**(12), 162–163, 174 (2009)
2. Dou, Y., Zheng, M.: Research on the theoretical framework, constraints and governance strategies of B2B supply chain finance. J. Nanjing Inst. Technol. (Soc. Sci. Edn.) **19**(2), 46–51 (2019)

3. Gao, H.: Thinking on digital economy and financial innovation. Sci. Technol. China (8), 97–100 (2017)
4. Song, H., Yang, X.: Sources of supply chain financial risk and systematic management: an integrated framework. J. Renmin Univ. China (4), 119–128 (2018)

A Corpus-Based Multi-dimensional Study of Tourism English Register Features

Jiawen Sun(✉)

Maanshan Teachers College, Maanshan 243000, China
Sunjiawen555@aliyun.com

Abstract. In this paper, the corpus based research method is used to study the register characteristics of tourism English, and the multi-dimensional analysis and corpus research method are used to carry out qualitative analysis and research on the quantitative data. Corpus linguistics mainly by providing real corpus of multidimensional character description and analysis on the tourism English text research, introducing corpus to tourism English text and the translation study, can in the past research on tourism text words, chapter case analysis, on the basis of multidimensional, multi-angle on the depth of tourism English text features of research and analysis of characteristics of the language. Therefore, this paper mainly combined with some tourist attractions in Anhui Province, based on the webpage text of these tourist attractions, and then used CAWS4 software to annotate the tourism English text corpus with part of speech, on this basis, combined with Wordsmith and Antconc corpus tools to retrieve and analyze these data. Then it analyzes whether there are differences in tourism English text features among different corpora.

Keywords: Corpus · Tourism English · Multidimensional

1 Travel English

So-called tourism English refers to the specialized study of English and tour-ism a basic theoretical knowledge of subject, fluent in listening, speaking, read-ing, writing and translating skills and development of China's tourism industry guidelines, policies and regulations, understand the tourism economy law of de-velopment and the strategy of market economy, familiar with master guide indus-try related business, to be able to English as a tool to engage in tourism related work, Tourism English has strong requirements for communicative competence in English. For example, tourism English needs to skillfully master English ex-pression skills, and be able to explain in detail the local conditions and practices, historical and cultural background, human geography, and relevant characters and deeds involved in a historical scenic spot and cultural landscape [1, 2].

Under normal circumstances, the tourism English contains many aspects, such as comprehensive English, hear English, read and write English, english-chinese translation, English country profiles, English tour guide, tour guide prac-tice, laws and regulations and professional ethics of tourism, tourism geography, hotel English, hotel

B. J. Jansen et al. (Eds.): *International Conference on Cognitive based Information Processing and Applications (CIPA 2021)*, LNDECT 85, pp. 262–268, 2022.
https://doi.org/10.1007/978-981-16-5854-9_33

management, and other related fields, but the current situa-tion of the development of, especially for domestic most of the people, When it comes to tourism English, often encountered in the process of tourism from sev-eral aspects, such as tourist traffic comes to English, room and board comes to English, shopping involves English, these contents are basically practical English dialogue, no more complex English vocabulary, sentence pattern is relatively simple, clearer dialogue scenes, so in the process of tourism can be directly ap-plied [3].

For example, travel English related to transportation. When people travel, they first need to buy a traffic ticket. For long-distance travel, they need to choose trains, planes, high-speed trains, etc., while those with a distance relative to the traffic police will choose buses, motorcycles, etc. Therefore, for tourism English in this field, we first need to understand the professional terms and phrase structures related to transportation. On this basis, we also need to know the site, price and time after buying a ticket, and also need to know how to ex-press in English in case of emergency, so as to ask for help [4, 5].

Another example is travel English related to accommodation and food. Travel accommodation, food and drink is essential, with accommodation as an example. Is out accommodation way, there is a lot of, have high star hotel and ordinary hotel, general folk, even some people will choose the hostel long journey or the backpacker hotels, etc., can say that there are so many ways to travel accommodation, can according to own hobby, travel budget situation, such as a specific choice, but we need to solve problems when we check in or have temporary accommodation. These basic words need to be learned. For another example, when traveling, basic food problems require us to remember the names of some foods we have seen in daily life, such as alcoholic beverages, vegetables, fruits, meat and other major food English expressions, as shown in Fig. 1:

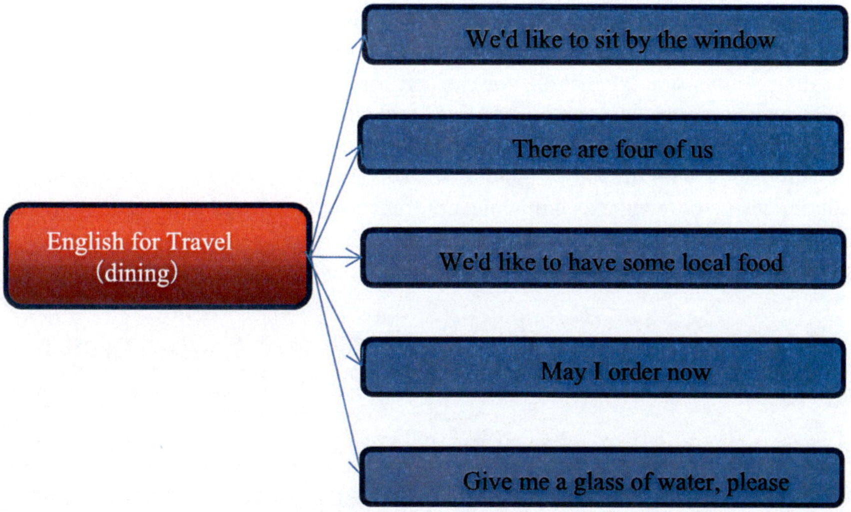

Fig. 1. English for travel (dining)

Tourism English is often compared with Business English. The biggest difference between the two lies in the vocabulary range. Both English and Tourism English need to master basic spoken English. Also need to use in the process of their own expression has strong requirements. Tourism English involves many aspects, especially in the process of travel, food and clothing live line a lot of problems hard to avoid all need to solve in tourism English, tourism English as a branch in the structure of English, has its own uniqueness, the particularity lies in the emergence of high frequency words and scenario, these characteristics make people to master in a short time, After all, tourism English does not need too much grammar and word memorization, emphasizing more practicality.

2 Multi-dimensional Analysis

Multidimensional analysis is a high-level statistical analysis method. Multidimensional analysis is to put a product or a market phenomenon in a two-dimensional spatial coordinate for feature analysis. For example, a three-dimensional coordinate market portfolio model requires one axis to represent market share, one axis to represent the growth rate of market demand, and one axis to represent the profit rate of enterprises. If we use multidimensional analysis to study the sales status of a product in the market, it will be more intuitive.

MDA (Multidimensional Analysis) is actually developed based on corpus. Multidimensional Analysis is based on empirical and quantitative methods to identify potential co-occurrence characteristics of discourse. This method was first used by Biber and later widely used in the comparative study of register. In the multi-dimensional analysis method, the analysis of language co-occurrence features is mainly realized with the help of factor analysis. The results are obtained through factor analysis and quantitative statistical analysis, and then some language features with high frequency in the text are identified, namely, "dimensions". Different dimensions in the text, and each register after the average dimension calculation, you can compare the any kind of register corresponding language features, and can be explained from the perspective of language function accordingly, then based on the different language domain in a dimension of similarities and differences, with the aid of qualitative analysis to explain that is associated with each dimension of language communication function, Compared with the comparative analysis of a single feature, the final result of such analysis method can more clearly reflect the characteristics of different texts.

3 Multi-dimensional Comparative Analysis of Tourism English Texts

3.1 Overall Statistical Features of English Text

This paper first carries out cross-corpus comparative analysis of domestic tourism English texts and the original tourism English text type, category, category/category ratio and standardized category/category ratio. The corpus comparison is as follows: Anhui Tourism English Web Page Text Corpus and American Tourism English Web Page Text Corpus. The results are shown in Table 1:

Table 1. Comparative analysis of the two tourism English texts

Compare the item	Anhui tourism English web page text corpus	American travel English web page text corpus
Token	110293	104529
Type	10729	10765
TTR	9.73	10.27
STTR	41.15	44.85

It can be seen from the above table that the proportion of class symbols/type characters in Anhui Tourism English webpage text corpus is 9.73, which is lower than that in American Tourism English webpage text corpus, which is 10.27. Of course, due to the differences in the data capacity of the two corpora, the ratio of standardized types needs to be further compared and analyzed. The normalized type ratio of the latter is 44.85, which is still higher than the former's 41.15. Thus it can be seen that the English introduction of relevant scenic spots in domestic tourism websites lacks diversity from the perspective of vocabulary usage. Of course, this result is greatly related to the lack of professionals in the English translation of most domestic websites.

On the other hand, through to the average word length in the two corpora comparative analysis it can be seen that used in the text of the English group tourism English words compared with domestic tourism English text is longer, in combination with the class operator ratio comparison results in Table 1, further clarified the domestic tourism website English text words with dimension simplify has obvious characteristics, as shown in Table 2:

Table 2. Comparative analysis of average word length and average sentence length in two tourism English texts

Compare the item	Anhui tourism English web page text corpus	American travel English web page text corpus
Average word length	4.57	4.85
Average sentence length	19.57	17.32

Can be seen from the above two kinds of comparison and analysis results, the average word length of the former obviously lower than the latter, and the average sentence length on the basis of the latter than the former, thus it can be seen, from the perspective of language expression, the author or translator is mostly use simple statements of the former, the complexity of the tourism English can't compare with this group of speakers.

3.2 Analysis of Lexical Density Features

The lexical density in English text reflects the proportion of content words in a text. The higher the proportion of content words, the greater the amount of information reflected in the text and the higher the difficulty of the corresponding text will be. Here, we also make a comparative analysis of the lexical density of the two corpora, as shown in Table 3:

Table 3. Comparative analysis of lexical density in the two tourism English texts

Compare the item	Anhui tourism English web page text corpus	American travel English web page text corpus
Total text vocabulary	110293	104529
Number of word frequencies of content words	64985	54702
Lexical density (%)	58.92	52.27

Can be seen from the comparison and analysis of the above results, the latter of lexical density is lower than the former, the result showed that the amount of information contained in the tourism English translation text more abundant, also can be seen at the same time, the text of the translator or author in expression a things piled up more accustomed to information, so can let the reader feel content overload, it's easy to have a reading disorder, Here, for example, the following example is Anhui Taibai Tower English text introduction:

The total vocabulary of Taibai Tower in Anhui introduced above is 45, among which 36 are content words, and the lexical density is 80%. For foreign tourists, seeing such cumbersome and overloaded vocabulary will not only feel strange, but also increase the burden of reading, which may eventually lead to the failure of the cross-cultural communication process of tourism resources.

3.3 Grammatical Feature Analysis

The analysis of grammatical features is divided into tense and voice.

(1) Characteristics of tense.
Tense analysis involves five aspects, namely the present simple tense, the past simple tense, the ongoing tense, the future tense and the perfect tense, as shown in Table 4:

Table 4. Contrastive analysis of the temporal features of the two tourism English texts

Compare the item	Anhui tourism English web page text corpus	American travel English web page text corpus
Present simple tense	856	2150
Past simple	528	622
During	62	82
Future tense	203	212
Perfect tense	241	177

Can be seen from the table above, two kinds of "present tense" the most frequently used in the English text, followed by "the past tense", based on the temporal characteristics of these two kinds of text analysis and comparison, the latter of temporal variety of applications, because "present tense", "the past tense", "on" and "future" are higher than the former, The application of perfect tense is more than that of the latter, which indicates that the former pays more attention to the continuity and influence on the current situation in the description of past actions, especially the introduction of some historical scenic spots, which pays more attention to reflect the influence of historical scenic spots on the region and even the whole country.

(2) Voice characteristics.
There are two kinds of voice in English text: active voice and passive voice. The frequency of passive voice and passive voice in tourism English web pages in China is higher than that in foreign countries. This characteristic indicates that a large number of active sentences are translated into passive forms in the translation of tourism texts in China. It also indicates that tourism translated texts are more objective than original texts. In fact, a large number of passive voices will be used when introducing some historical tourist attractions, which will appear too rigid. However, a less passive voice will make the introduction of tourism texts more vivid and stimulate the interest and enthusiasm of sightseeing in class.

4 Conclusion

In this paper, we study theme from the Angle of the corpus to analyze the multidimensional characteristics of tourism English, for this place in Anhui tourism English web page text corpus and the tourism English text corpus as the research object, this paper compares and analyzes the comprehensive analysis of both the overall features, lexical density, grammatical characteristics of three aspects, through three aspects of the analysis and comparison, There are great differences in lexical richness, word length, sentence length, lexical density, tense, voice and so on. For the tourism English web page text corpus to benchmarking analysis can be found that the domestic travel website English text in language use and stylistic dimensions such as the increasing of normative remains to be perfect, from this point, in the future tourism English text compilation or is in the process of translation, the translator should be more reference to some native English text, After all, domestic tourism English mainly serves foreign tourists, so the translation of tourism English text should be more concise and interactive, so as to realize the purpose of cross-cultural communication of tourism resources.

Acknowledgement. The excellent talents support program of Anhui Province in 2020 (gxyq20240). "English writing" (2020kfkc542), an excellent offline open course in Anhui Province in 2020. Anhui teaching team "marketing teaching team" in 2020 (2020jxtd258). The key project of visiting and studying at home and abroad for outstanding young and middle-aged backbone talents in Colleges and universities of Anhui Province (gxfxzd2016354). Key project of Humanities and social sciences of Anhui Provincial Department of Education (SK2016A0908). The Ministry of education's higher vocational education innovation and development action plan project in 2016 (XM-2-326).

References

1. Bruyèl-Olmedo, A., Juan-Garau, M.: Coexisting varieties of English in the linguistic landscape of tourism: the Bay of Palma. J. Multilingual Multicult. Dev. **41**(2), 157–174 (2020)
2. Terauchi, H., Maswana, S., Yamada, M.: A genre-based study of tourism language in English textbooks. J. Glob. Tour. Res. **2**(2), 115–122 (2020)
3. Zhang, W., Zhai, C.: On stylistic features of tourism English & its translation strategies. In: Proceedings of 2016 3rd International Conference on Economic, Business Management and Education Innovation (EBMEI 2016 V55). Singapore Management and Sports Science Institute (SMSSI), Singapore, Academic Conference Institute, USA, p. 5 (2016)
4. Li, R.: Research on functions of tourism English in ecological tourism development in Qiannanyu area of Xingtai county. In: Proceedings of 2015 2nd International Conference on Education, Management and Computing Technology (ICEMCT 2015). Computer Science and Electronic Technology International Society, p. 5 (2015)
5. Gandin, S.: Translating the language of tourism. A corpus based study on the translational tourism English corpus (T-TourEC). Proc. Soc. Behav. Sci. **95**, 325–335 (2013)

Basketball Players' Behavior Prediction Method Based on Video

Yu Long[1](✉) and Jiaojiao Long[2]

[1] Yangtze Normal University, Chongqing 408000, China
[2] Tongliang Experimental Middle School, Chongqing 402560, China

Abstract. Basketball is a very competitive sport, in which the behavior of basketball players is the key factor to win. If we can fully predict the behavior of basketball players, then we can benefit more easily (successful defense or successful score). In this case, many basketball teams have set up a professional analysis team. The main work of these teams is to make statistics on the behavior of basketball players according to the basketball game video. According to the statistical results, we can predict the behavior of basketball players. However, simple statistics can not cover the complexity of basketball players' behavior on the court, so although the statistical results have predictive utility, there are defects in the utility level. At this time, we need to use more advanced behavior prediction methods to predict basketball players' behavior according to the basketball game video, this paper will take the improved RBF network player behavior prediction method as the research object, discuss the basic concept of RBF method, and then put forward the application scheme of the improved RBF network player behavior prediction method. Finally, taking the basketball game video as an example, the behavior of the players in the game is predicted, and then compared with the results, the effectiveness of this method is verified by the comparison results.

Keywords: Basketball video · Basketball players · Sports behavior prediction

1 Introduction

In the basketball game, the behavior of basketball players is very complex, which involves the arrangement of game tactics, the personal ideas of the players to deal with defense or attack, the positioning of the players in the game (center, forward, guard), the position of the players in the current time of the game, the time of the basketball game, whether the basketball players hold the ball, the current score situation and other factors, therefore, relying on artificial statistics of each basketball player's behavior data can only be judged from the probability, and there are differences in the players' ideas in each game, which indicates that relying on probability to judge is difficult to guarantee the correctness of the prediction results, and once the error is made, it may lead to the loss of the game. In this case, in order to replace the artificial statistical method, the research community proposed an improved RBF network player behavior prediction method. This method includes the factors that have an impact on the behavior of basketball players, and integrates various

B. J. Jansen et al. (Eds.): *International Conference on Cognitive based Information Processing and Applications (CIPA 2021)*, LNDECT 85, pp. 269–275, 2022.
https://doi.org/10.1007/978-981-16-5854-9_34

factors to form a variety of influence relationships. At this time, further analysis can get more accurate and complete prediction results, The results are as follows: in the situation of delaying the attack time according to the tactical requirements, the players should deal with the defense, the players are guards, the players are in their own bottom line position, the basketball game time is coming to an end, the basketball players hold the ball, the score is leading but may be overturned, the basketball players generally choose to slow forward to achieve the purpose of delaying the time, and pay great attention to the protection of the ball. According to this result, the opposite team members can set the tactics of grabbing the ball, and their own team members can set the tactics of passing and delaying time, so as to win the most. However, it is difficult to improve the application of RBF network team member behavior prediction method, so it is necessary to carry out relevant research [1].

2 Improve the Basic Concept of Behavior Prediction Method of RBF Network Members

The core of improving the behavior prediction method of the members of RBF network is RBF radial neural network, which is a very efficient feedforward three-layer neural network. Compared with other neural networks, it has two advantages, i.e. excellent performance and global optimization. Therefore, it is very fast in training speed, and the structure of RBF radial neural network is simple, so it is easy to operate. RBF radial neural network is mainly composed of input layer, hidden layer and output layer, in which input layer is signal source node, and the number of nodes in this layer is the specific number of actual entry and exit variables. The hidden layer mainly includes implicit elements. Each hidden unit represents an influencing factor on input node, so it can be used to describe the complexity of the problem, the output layer is the layer that outputs the change of a node in the input layer affected by the hidden layer. The output information is the final result. In this process, the information transformation between the input layer and the hidden layer is nonlinear, while the information transformation between the hidden layer and the input layer is linear. Figure 1 is a topological map of the model for improving the behavior prediction method of the members in RBF network.

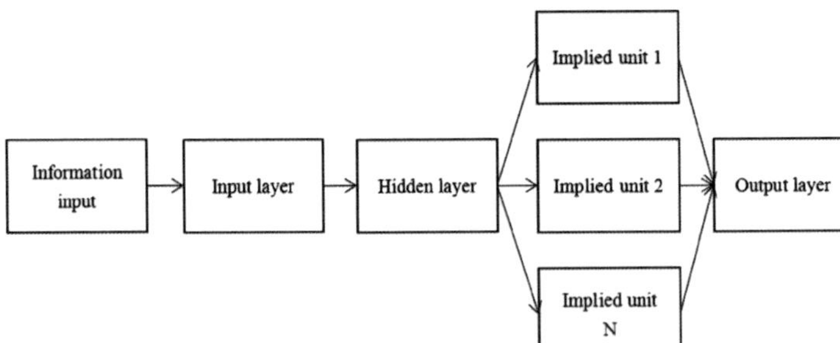

Fig. 1. The model topology of the improved RBF network member behavior prediction method.

3 Application Scheme of Improved RBF Network Member Behavior Prediction Method

3.1 Prediction Process Design

The prediction process of the improved RBF network player behavior prediction method can be roughly divided into four steps: first, data normalization, that is, all the basketball video player behavior related data are normalized in the range of [0, 1], which can make the subsequent prediction operation more convenient. The specific method of data normalization is shown in formula (1); Second, network structure setting, that is, in order to predict accurately, it is necessary to set the network structure of RBF before forecasting. In the process of setting, it is necessary to confirm the number of input layer nodes, network sliding window value, and the number of hidden units in hidden layer based on the actual demand (the specific number of hidden units cannot be directly confirmed, but in general, it is only necessary to input the main factors, If the actual requirements are higher, the more the number of hidden units, the better. It is suggested to use MATLAB software to set parameters. The setting method is shown in formula (2); Thirdly, training and testing, that is, using the simulation and testing function of MATLAB software to simulate and test the RBF network which has set parameters, according to the test results to adjust the distribution density value of radial basis function until the distribution density value is optimized; The third error calculation and check is to use RBF neural network to make a single prediction, and compare the prediction results with the actual results to know the difference between the two. If the difference is large, it means that the error of the prediction results is large. It is necessary to adjust the distribution density of parameters and radial basis function again, so that the error can be checked, When the error does not exceed the maximum allowable value, the prediction can be formally started [2].

$$X = \frac{x - x_{\min}}{x_{\max} - x_{\min}} \tag{1}$$

Where $x_{\min}$ is the minimum value in the experimental data, $x_{\max}$ is the maximum value in the experimental data, and X is the data value after normalization.

$$net = newrbe(P, I, spread) \tag{2}$$

Where P is the input vector matrix, I is the training output vector, and read is the distribution density value of radial basis function, and the default value is 1.0.

3.2 Prediction Model

Firstly, the learning algorithm of RBF neural network prediction model is selected. Generally, this algorithm is divided into two types, namely, online algorithm and offline algorithm. Online algorithm can handle more factors and has good controllability. The disadvantage is that the convergence speed is slow, while the advantages and disadvantages of offline algorithm are opposite to that of in-line algorithm. According to the performance of both, the online algorithm will be selected for the prediction model in this paper, the reason is that there are many factors affecting basketball players' behavior, but offline algorithm does not meet this requirement. Therefore, only online algorithm is selected, and the shortcomings of online algorithm will not affect the prediction operation. It can be ignored, which shows that online algorithm is the most suitable for this prediction model. Secondly, around the online algorithm, considering the shortcomings of traditional RBF neural network can not guarantee the accuracy of prediction, it needs to be improved. The improvement methods are: introducing new hidden layer hidden elements, adjusting network parameters, deleting the original hidden layer hidden elements. Formula (3) is a prediction model for improving the behavior prediction method of RBF network players, and the behavior of basketball players can be known through the model [4].

$$r_{kj}^{i} = \frac{\left\| O_{k,j}^{i} \right\|}{O_{j,\max}^{i}}, (k = 1, \ldots K) \tag{3}$$

Where r is the result of the behavior of the players, k_{j} and i are the factors affecting the behavior of the players, O is the input information, K is the hidden unit to be deleted, and K is the number aof k.

4 Behavior Prediction of Basketball Players

4.1 Case Selection

This paper takes one player from each side in a CBA game video as the object of prediction (hereinafter referred to as player a and player B), and the prediction conditions are as follows: (1) player a is in the tactical requirement to delay the attack time, the player should deal with the defense, the player is a guard, the player is in his own bottom line position, the basketball game time is coming to an end, the basketball player holds the ball, the basketball player is playing basketball, the player is playing basketball, the player is playing basketball the situation that the score is leading but may be overturned; (2) B player is in the situation that the tactics require the whole court to press and steal, the player should deal with the delaying attack, the player is the guard, the player is near the center line of the opposite court, the basketball game time is coming to an end, the basketball player is the defender, the score is behind but may turn over. According to the situation of team members a and B, the above method is used for prediction.

4.2 Prediction Logic Design

A. Team B may make a variety of behavior choices in their own situation, so to predict the behavior of two team members, we must first confirm their possible behavior. Then, according to the situation of the two players, their possible behaviors are shown in Table 1 and 2.

Table 1 and 2 show the behaviors that the A and B players may make under their own circumstances. When these behaviors are input into the improved RBF network member behavior prediction system, corresponding prediction logic will be generated.

Table 1. A possible behaviors of team members

Team member	Behavior	Possible explanation
A	Slow progress	Slow progress can avoid its own mistakes, and play a role in delaying time, which is conducive to our party's keeping ahead
	Pass the ball	It is possible to encounter the other side's pressing steals. In order to avoid losing the right to the ball, they will choose to pass the ball
	Stop and defend	In the face of the other side's pressing steals, and they don't have the right chance to pass, they may stop the ball and defend, the purpose is the same as the pass

Table 2. B possible behaviors of team members

Team member	Behavior	Possible explanation
B	Targeted press defense	Strive to press defense to force the other side to show things or flaws, so as to snatch as soon as possible to win the counterattack
	Perimeter defense	Considering that the opponent's players will choose to pass the ball when they encounter press defense, they may defend on the side of the ball, which is convenient for intercepting the ball at the first time and then counterattack
	Give up	He thought it was impossible to win, so he chose to give up and wait for the end of the game

With reference to logic and the analysis of influence relationship between the hidden elements of the hidden layer, the probability of various behavior choices can be obtained for members a and B, the behavior with probability over 80% is the prediction result. In addition, in order to ensure the accuracy of prediction results, two new units are added in the implicit layer unit, which are historical records of two players under the same circumstances and conventional logic of thinking of choice (that is, when different players face multiple behavior choices, they will definitely judge the degree of benefit of various behaviors to our party, and finally choose the behavior that is most beneficial to our party, the judgment behavior is beneficial to our side involves the thinking logic of the players. According to the historical records and interviews with myself, this paper confirms the normal selection logic of a and B players).

4.3 Prediction and Analysis Process

Based on the process of improving the behavior prediction method of the members of RBF network, the author forecasts the behaviors of members a and B under their own conditions. The prediction process is as follows: (1) input the situation of team members a and B, analyze the selection orientation and achievement probability of the two players according to the historical records. For example, team a will choose to push slowly whenever they encounter similar situations in the history record, if the player encounters the pressing defense, he will stop the ball and die. Only one pass has been selected in 13 similar choices. At this time, the selection orientation of player a is selected. In the behavioral strategy of 12 slow push + stop, the benefit rate of the player reaches 100%, indicating that the player has successfully won every time in this case, which will affect the choice of team a at the time of the game, therefore, it is necessary to include video prediction reference. (2) According to the regular thinking logic of team a and B, it is judged whether there is blind choice phenomenon between the two sides when facing the choice. That is, taking team a as an example, although he will choose the behavior strategy of slow push + stop and death, it does not mean that the choice is made consciously by team a, and it may be the subconscious choice of team a, which is blind, it is not necessary to cope with all situations. If this phenomenon exists, it will be shown that team a will not judge the degree of benefit of various behaviors to our party before the selection. On the contrary, if team a is conscious of this behavior, then the selection of team a may be different. According to the historical records and interview results, a B players will consciously choose the most favorable behavior strategies, not blind choice, which is their normal choice thinking logic (3) forecast and output results.

4.4 Comparison Between Predicted Results and Actual Results

After the above analysis process, the results are obtained by improving the behavior prediction method of the members of RBF network, as shown in Table 3.

Table 3. A, B players basketball behavior choice results

Team member	Select results	Probability	Reason
A	Slow push + pass	93%	Because the defender is more flexible and good at seizing, a player does not have enough self-confidence to defend the ball, so this time he chose slow forward + pass
B	Targeted press defense	99%	Because the strength and flexibility of the attacking side are weaker than themselves, and there is a high probability that they can steal by using the targeted press defense, so player B chooses the targeted press defense

Then this paper plays the video to watch the actual behavior of players a and B. The results show that the actual results are consistent with the predicted results. The final winner is the team of player a, because the defense of player B is difficult to deal with the fast passing ball.

5 Conclusion

To sum up, the improved RBF network player behavior prediction method can predict the player behavior according to the basketball video. This paper discusses the basic framework and application method of the method. At the same time, a case is used to test the accuracy of the method, and the results show that the method is accurate.

References

1. Chen, L., Wang, W.: Analysis of technical features in basketball video based on deep learning algorithm. Signal Process. Image Commun. **83**, 115786 (2020)
2. Azzam, M.G., Throckmorton, T.W., Smith, R.A., et al.: The functional movement screen as a predictor of injury in professional basketball players. Curr. Orthopaedic Pract. **26**(6), 619–623 (2015)
3. Glickman, M.E., Sonas, J.: Introduction to the NCAA men's basketball prediction methods issue. J. Quant. Anal. Sports **11**(1), 1–3 (2015)
4. Rexhepi, A.M., Brestovci, B., Mucolli, A.: Prediction of explosive power of basketball players based on their morpho-functional characteristics. Int. J. Morphol. **30**(2), 719–722 (2012)

Evaluation of Enterprise Network Marketing Performance Based on Improved BP Neural Network

Huojun Zhu, Liangting Wang(✉), and Zhishan Zheng

Jiangxi University of Engineering, Xinyu 338000, China

Abstract. In the Internet age, many online marketing companies have appeared on the market, and most of the companies that used to do business have also developed their own online influence channels. Therefore, the success of online marketing is linked to the economic benefits of the company. In this case, companies naturally want to ensure their own economic output capacity for network marketing, and need to perform performance evaluation on network impact. At this time, many companies have introduced BP neural network for evaluation. However, the traditional BP neural network has the problems of local smallness and slow convergence speed in the evaluation, and it is difficult to guarantee the quality and efficiency of the evaluation. Therefore, this article will propose an improved method of the traditional BP neural network and use the improved BP neural network to the performance evaluation of enterprise network marketing.

Keywords: BP neural network · Neural network improvement · Corporate network marketing performance evaluation

1 Introduction

In fact, many companies have long introduced the BP neural network to evaluate the performance of their own network marketing results, and the introduction of the BP neural network has indeed played a certain evaluation effect, making the evaluation results more objective, scientific, reliable and complete. However, with the popularization of traditional BP neural network performance evaluation methods, the problems of local smallness and slow convergence speed in traditional BP neural network have been exposed, which also makes enterprises put forward higher requirements for evaluation quality and efficiency. In this case, some scholars have proposed a traditional BP neural network improvement method based on the LM algorithm and the adaptive adjustment strategy of the learning rate. The improved BP neural network successfully eliminates the problems of the traditional BP neural network and satisfies the enterprise network marketing Performance evaluation needs, so companies should have an understanding of this method, and it is necessary to carry out relevant research.

B. J. Jansen et al. (Eds.): *International Conference on Cognitive based Information Processing and Applications (CIPA 2021)*, LNDECT 85, pp. 276–283, 2022.
https://doi.org/10.1007/978-981-16-5854-9_35

2 Defects and Improvements of Traditional BP Neural Network

2.1 Defects

The BP neural network consists of three parts: input layer, hidden layer, and output layer. Among them, there is a relationship between the hidden layer and the input layer, and between the output layer and the hidden layer, which are connected one by one according to the transmission strength. The form is similar to the human nerve. The connection structure between the nerve cells in the network. Under this structure, the well-known BP algorithm was born, which can train the BP neural network and promote the BP neural network to complete self-learning based on information. The entire learning process can be divided into two links: signal forward propagation and error back propagation. The first link starts from the input layer, from which the information is input to the BP neural network, the information will then be received by the hidden layer and processed by the hidden units of the hidden layer, and then enter the output layer after completion, and the latter step is Starting from the output layer, it is mainly to compare the output result of the previous link with the expected value. If the output result is found to meet the expected value, it will be back propagated, and the propagation route is completely opposite to the previous link. In the entire learning process, if it enters back propagation, after the propagation ends, the neuron contact weights in the entire BP neural network will be adjusted to reduce the error, and enter the forward propagation link again, repeating this until the forward direction the output results of the propagation meet the expected value requirements. It can be seen that the BP neural network has the function of self-iteration to improve accuracy, so high-precision evaluation can be achieved in the evaluation of network marketing performance. However, the traditional BP neural network has the problems of small locality and slow convergence in this learning process. The reason is that the traditional BP neural network is prone to oscillating and has defects in redundant connections or nodes. At the same time, the specific number of hidden layer nodes of the traditional BP neural network has not yet been unified [1, 2].

2.2 Improvement

In view of the shortcomings of the traditional BP neural network, modern scholars have proposed a corresponding improvement method. The method is mainly implemented by the LM algorithm, and the adaptive adjustment strategy of the learning rate is mainly used to operate the improved BP neural network. See below for specific improvement methods and strategies.

(1) LM algorithm improvement
From the perspective of the LM algorithm, the self-learning process of the BP neural network is a nonlinear objective function optimization process. For this process, the LM algorithm can effectively use the first-order derivative information and the second-order derivative information in the learning, which is different from the traditional BP neural network algorithms are quite different (traditional methods can only use the first derivative). The iterative expression of LM in the improved BP neural network is shown in formula (1). In this case, the improved BP neural

network with the help of the LM algorithm can dynamically adjust the iterative convergence direction according to the iteration results, which can ensure that the error of each iteration is reduced, and the convergence speed can also be improved [3].

$$W_{k+1} = W_k - \left(J^T J + uI\right) J^T e \quad (1)$$

Where e is the error vector; J is the matrix of the derivative of the network error to the weight; I is the identity matrix; u is a standard value.

(2) Adaptive adjustment strategy of learning rate
The traditional BP neural network does not have the function of adaptive adjustment of the learning rate. The reason is that the learning parameters η in the measurement are determined by the manager's experience. Therefore, the learning parameters η are prone to be too large or too small. When the learning parameters η are too large, It will cause the error function to oscillate, the learning parameter is too small, and it will cause the convergence speed to slow down (representing a slower learning rate). In this case, the selection of learning parameters η has become a difficult problem in the improvement of BP neural network, because no matter how to choose, the learning parameters η are always a fixed value, and the environment in actual operation will continue to change, so the fixed value will inevitably be too large. Or too small. However, with the deepening of research, the research community has proposed a method that can realize the adaptive adjustment of learning rate, see formula (2) for details. In this method, the learning parameter self-adjustment algorithm is mainly used to select learning parameters η. The basic principle is When the weight of the connection between element i and j is far away from the target point, the learning parameter takes a high value, otherwise it takes a low value [4].

$$\eta(t+1) = \eta(t) \bullet E(t)E(t-1) \quad (2)$$

Where E is the global average error; t is the connection weight between i and j of the neuron.

3 Enterprise Network Marketing Performance Evaluation Analysis

3.1 Evaluation System Construction

The construction of the corporate network marketing performance evaluation system is divided into two steps, namely indicator construction and standard construction. The specific content of each step is shown below.

(1) Indicator construction
With reference to the network influence activities of general enterprises, this article mainly constructs the enterprise network marketing performance evaluation index composed of three primary indicators and corresponding secondary indicators. See Table 1 for details.

Table 1. Performance evaluation index of enterprise network marketing

First-level index	Secondary indicators	Introduction (only for secondary indicators)
Website benefits	Website design	Website design includes web page aesthetics and service functions. If the design is good, it is beneficial to the consumer experience, so it can bring higher benefits, and vice versa
	Website promotion	Website promotion refers to how many consumers know the website, which is related to the number of consumers. The higher the number of consumers, the higher the benefit, and vice versa
	Website traffic	Website traffic refers to the activity of website users. The higher the activity, the higher the efficiency, and vice versa
Business efficiency	Financial benefit	Financial benefits refer to the indirect benefits generated under financial management, such as cost reduction
	Publicity benefits	Publicity benefits refer to the benefits generated under the construction of a corporate brand. For example, the higher the brand awareness, the higher the benefits, and vice versa
	Competitive benefit	Competitive benefits refer to the benefits an enterprise obtains in competition, and it represents that the company has improved its competitiveness with the help of network marketing, thereby generating benefits
Consumer benefit	Consumption benefit	The benefits that consumers get in the consumption process, such as the reduction of shopping costs
	Satisfaction	Satisfaction refers to the degree of consumer satisfaction with corporate network marketing. The higher the degree, the higher the benefit, and vice versa
	Loyalty	Loyalty refers to the trust or favor of consumers on the influence of the corporate network. The higher the degree, the higher the benefit, and vice versa

(2) Standard construction

With reference to the classification of network marketing performance by most companies, the performance level can be divided into four grades: best, good, fair, and poor. In order to evaluate better, these four grades need to be assigned, so as to form a complete evaluation standard. The evaluation criteria constructed in this paper are shown in Table 2.

Table 2. Evaluation criteria for enterprise network marketing

Grade	Points
The best	≥0.9
Good	≥0.7
General	≥0.5
Difference	≥0.3

3.2 Evaluation Implementation

Refer to the indicators and standards of the corporate network marketing evaluation system, take a certain company as the evaluation object, and use the BP neural network improved by the LM algorithm for evaluation. The evaluation plan is divided into three steps, see below for details.

(1) Network structure design
Following the basic operation logic of the BP neural network, the network structure in the evaluation can be divided into two parts: ① Indicator input and standardization processing, that is, statistics corresponding to the data information according to the secondary indicators, and then the statistical information is input into the network as the input vector, synchronously accept standardized processing. The standardization process is to filter each data information. The filtering master adopts the matching degree strategy, and analyzes the matching degree based on the data information of the same secondary index. If the data information matching degree exceeds 95%, it means that the data information is duplicate information. Perform a preliminary screening, and then manually perform a secondary matching analysis on the data information screened under the preliminary screening (the amount of data in this screening is relatively small, so it can be done manually to ensure the integrity of the data information). In addition, if unreadable data information is encountered in the filtering process, it will be screened out and judged manually; ② Forward/reverse cycle, that is, forward output for all standardized processed data information, with the help of learning the rate-adaptive adjustment strategy dynamically adjusts the number of learning parameters to avoid slow learning rate or oscillations. Synchronously rely on the LM algorithm to iterate and loop continuously to obtain the evaluation result with small error and the maximum allowable error value between the result and the expected value. It is 0. 1. Before the direct error between the result and the expected value does not meet this value, the BP neural network will continue to perform a forward/reverse cycle.
(2) Network training
Because the performance of network marketing is high or low depends on the subjective cognition of the company, but companies generally measure the level according to the common standards of similar companies, so in order to ensure that the performance evaluation results of the BP neural network are objective and consistent with the actual situation In the evaluation process, the general standards of other

similar enterprises should be selected as references, and a performance evaluation standard that meets the requirements of the enterprise and is more objective can be obtained through the independent training of the neural network. In the evaluation of this article, we mainly refer to Tables 1 and 2 to select 20 companies with the same industry, similar assets, and small differences in scale as the network training samples. The final results are recognized by the evaluation objects, so they can be used for evaluation. Implement.

(3) Fuzzy comprehensive evaluation
Because there are too many factors involved in performance evaluation, the analysis and evaluation of all factors will greatly reduce the evaluation efficiency, and it is not necessary to do so. Therefore, this article mainly uses the fuzzy comprehensive evaluation strategy to evaluate the performance level of each secondary indicator with the help of the LM algorithm. Obtain the performance level score of a secondary index, and then refer to Table 2 to make a final judgment. The fuzzy comprehensive evaluation strategy can be roughly divided into five steps: ① Confirm the evaluation domain, that is, select a number of factors with the highest weight level among the many performance evaluation factors as the reference object, and the other factors do not need to be considered, and the selected factors constitute the evaluation domain; ② Establish the fuzzy relationship matrix of each factor; ③ Confirm the vector of each factor; ④ Normalize; ⑤ Result output.

3.3 Evaluation Results and Discussion

See Table 3 for the evaluation results of this article.

Table 3. Evaluation results

First-level index	Secondary indicators	Result	Evaluation	Overview
Website benefits	Website design	0.6	General	Good
	Website promotion	0.8	Good	
	Website traffic	0.8	Good	
Business efficiency	Financial benefit	0.5	General	
	Publicity benefits	0.7	Good	
	Competitive benefit	0.7	Good	
Consumer benefit	Consumption benefit	0.9	The best	
	Satisfaction	0.8	Good	
	Loyalty	1.0	The best	

According to the results in Table 3, the overall performance level of the company's network marketing is good, of which the highest performance contribution level is consumer benefit, and the performance contribution levels of the other two indicators are equivalent. Under this circumstance, if an enterprise wants to further improve its network marketing, it is necessary to strengthen the work of both website efficiency and corporate efficiency. That is, for website efficiency, it should focus on optimizing website design, and for corporate efficiency, it should focus on improving financial efficiency. Promotion, website traffic and publicity benefits, and competition benefits should also be further optimized. It is worth noting that on the basis of the improvement of website efficiency, the three secondary indicators in consumer efficiency will also benefit. Take satisfaction and loyalty in consumer efficiency as examples. When the website design is optimized, consumers will you can get a good consumer experience, so consumer satisfaction will increase, or under the effect of website promotion, consumers will trust the website more and actively consume through the website, which can effectively maintain consumer loyalty. In addition, there is a circular relationship among the three first level indicators, that is, the website benefit can enhance the enterprise benefit, the enterprise benefit can enhance the consumer benefit, and the consumer effect can enhance the website benefit. Figure 1 shows the circular relationship among the three indicators of the enterprise.

Fig. 1. Shows the circular relationship of the three indicators of the enterprise

4 Conclusion

In summary, this article uses the LM algorithm to improve the traditional BP neural network, and uses the improved BP neural network to evaluate the corporate network marketing performance. The evaluation process is very smooth, and there are no problems in the traditional BP neural network. Using the three steps of evaluation implementation in the article, the performance evaluation of the evaluation object was successfully carried out, and the defects and optimization direction of the evaluation object in network marketing were pointed out. Therefore, the improvement method of BP neural network in this paper is effective and the evaluation result is good.

References

1. Li, X., Wang, G.: Research on food network marketing performance evaluation based on improved BP algorithm. Adv. J. Food Sci. Technol. **9**(8), 637–642 (2015)
2. Jie, Y., Qu, H., Liang, Z.: Research on the evaluation methods of bid of construction project based on improved BP neural network. WSEAS Trans. Comput. **9**(1), 93–102 (2010)
3. Cao, Q.K., Li, J.: Reliability analysis and evaluation on member enterprise of manufacturing supply chain based on BP neural network, pp. 474–479. IEEE (2008)
4. Zhang, H., Yuan, W., Hua, J.: Performance evaluation on human resource management of China's commercial banks based on improved BP neural networks. Int. J. Adv. Comput. Technol. **4**(11), 304–310 (2012)

Check for updates

Data Mining Method of Tennis Technique and Tactics Association Analysis

Mengran Niu(✉)

Chengdu Polytechnic, Chengdu 610041, China

Abstract. Tennis competition is not only a competition of personal skills, but also a competition of skills and tactics. Therefore, to a certain extent, skills and tactics play a decisive role in the result of tennis competition. Therefore, in order to win the competition, relevant personnel need to analyze the characteristics and laws of competitors' skills and tactics through relevant means, so as to set targeted coping strategies. In the past, the analysis method of tennis technique and tactics is generally data statistics method. Although this method has certain practicability, it has many defects, such as long event of technique and tactics data statistics, large amount of data involved (it is difficult to deal with large amount of data manually), and subjective analysis. Therefore, the results of previous technique and tactics analysis may be wrong, and the effect of the results is difficult to give full play. In the face of this situation, in order to give full play to the role of technical and tactical analysis results and provide scientific basis for strategy setting in tennis matches, this paper will analyze the data mining method of tennis technical and tactical association analysis, introduce the basic concept of the method, then discuss the method application scheme, and finally verify the application effect of the method with a case.

Keywords: Tennis · Technology and tactics · Association analysis data mining

1 Introduction

In the early days of the concept of tennis technology and tactics, some related personnel thought to analyze the technical and tactical characteristics and laws of the other party by means of data statistics, which can provide a good direction for their own technology and tactics, and also analyze the characteristics and rules of their own skills and tactics, and clarify what techniques and tactics are more conducive to their special features and effectively respond to each other's skills and tactics, In this way, it can improve the probability of winning in tennis. However, with the continuous development of technology and tactics, the technology and tactics adopted in modern tennis match are more complex. In addition, due to the changes of different athletes' techniques and tactics over the years, the technicians and tactics analysts need to know the relevant data of the athletes in the past years, which makes the analysis of the characteristics and laws of technology and tactics appear a large data body, and the relationship within the

B. J. Jansen et al. (Eds.): *International Conference on Cognitive based Information Processing and Applications (CIPA 2021)*, LNDECT 85, pp. 284–290, 2022.
https://doi.org/10.1007/978-981-16-5854-9_36

data body is complex, and the artificial capacity is limited, It is not objective enough, not only difficult to deal with large and complex data bodies, but also difficult to give reliable objective results. These problems are very prominent in the analysis of modern tennis technology and tactics. At this time, with the development of big data and other related technologies, the industry has put forward the data mining method of association analysis, which can be applied to the analysis of various sports techniques and tactics, tennis is no exception. therefore, it is necessary to replace the previous methods with association analysis data mining method, so as to better analyze tennis technology and tactics, and it is necessary to carry out relevant research.

2 Basic Concept of Data Mining Method of Association Analysis

In essence, association analysis data mining is a kind of big data application technology, that is, big data itself is a huge data body, which contains valuable application information. But to obtain this information and put it into practical application, it is necessary to analyze big data. However, the large data volume and complex data relationship make it difficult to analyze it by ordinary means, at this time, the concept of big data application technology was born, and the method of association analysis data mining is such a technology. The association analysis data mining method was known by people in the 1990s. At that time, the definition of this method was "A method that can dig out knowledge in a large amount of data". In operation, it can be found that there are valuable implementation and rules between data, which is different from the similar method of transmission. The so-called traditional similar method generally refers to a statistical method used to verify whether the hypothesis is established. That is first, it puts forward various hypotheses, and then confirms whether the hypothesis exists through data statistics. If it exists, it is true. On the contrary, it will not be true. After continuous circulation and elimination, all kinds of hypotheses are verified and the hypothesis confirmed can be combined to obtain the final result, The association analysis data mining method is to directly mine new information or rules from big data. This process is called knowledge discovery in professional field, which is different from traditional method knowledge verification.

Combined with the operation principle and basic mode of association analysis data mining method, this method can be applied to data analysis in any field, and tennis technology and tactics analysis is no exception. According to the actual performance of this method in tennis technology and tactics analysis, it can be confirmed that the traditional methods can also obtain and can not get the information and laws of technical and tactical characteristics, it shows that the results of this method are more complete and in-depth, and the corresponding techniques and tactics can be set up with the analysis results of the method. It is proved that the analysis results of association analysis data mining method are more reliable. Therefore, in tennis technology and tactics analysis, the data mining method of joint analysis is more advantageous than the traditional method. In addition, the association analysis data mining method is realized by the algorithm, and the algorithm is not fixed. The commonly used algorithms are apriori algorithm, apriori II algorithm, FP growth algorithm, etc. the applicable conditions of various algorithms are different. In tennis technology and tactics analysis, apriori algorithm is generally recommended. Generally speaking, association analysis data mining is a big data application technology, which has strong universality and can be applied to any data for association analysis. So since tennis technology and tactics can generate data, it is shown that the association analysis data mining method can be used for analysis [1].

3 The Application Scheme of Association Analysis Data Mining Method in Tennis Technical and Tactical Analysis

3.1 Basic Framework Setting

Based on apriori algorithm, the basic framework of the application of association analysis data mining method in tennis technical and tactical analysis is as follows: (1) Set the set of all analysis items in tennis technical and tactical problems as $I = (i_1, i_2, \cdots, i_n)$ and the set of all things (excluding those that may exist) is $T = (t_1, t_2, \cdots, t_n)$, and anything t_i are subsets of itemset m, and the expression is $T \subseteq I$; (2) Based on the setting of the first step, it can be seen as association analysis. In the process, the number of things containing a specific itemset represents the support of the itemset, and the support of each itemset can be counted as $\sigma(A)$. In addition, the association rules in the application of association analysis data mining method are A $\rightarrow$ B, and the strength of rules can be measured by support s and confidence C. See formulas (1) and (2) for their definitions, where n is the total number of things. According to the minimum support threshold and the minimum confidence threshold, the analysis is carried out. If the result shows that the relevant support of the item set meets the minimum support threshold standard, this indicates that the itemset is a frequent itemset. At the same time, if the threshold of a given thing in the result is greater than or equal to the minimum support threshold, and the confidence exceeds the minimum confidence threshold, it indicates that there are association rules in the thing, and the analysis should be carried out according to these rules [2]. Table 1 is the main goal of association analysis data mining, and Table 2 is the case of frequent itemsets that may be generated after algorithm calculation.

$$S(A \rightarrow B) = \frac{\sigma(A \cup B)}{N} \tag{1}$$

$$C(A \rightarrow B) = \frac{\sigma(A \cup B)}{\sigma(A)} \tag{2}$$

Table 1. Main targets of association analysis data mining

Target ranking	Concrete content
1	Find all frequent itemsets
2	Generating strong association rules from frequent itemsets

Table 2. Possible frequent itemset cases after algorithm calculation

Serial number	Itemset
1	a, b, c
2	a, b
3	a, c

3.2 Scheme Application and Mining Process

According to the basic framework, the application process of association analysis data mining method in tennis technical and tactical analysis is shown in Fig. 1.

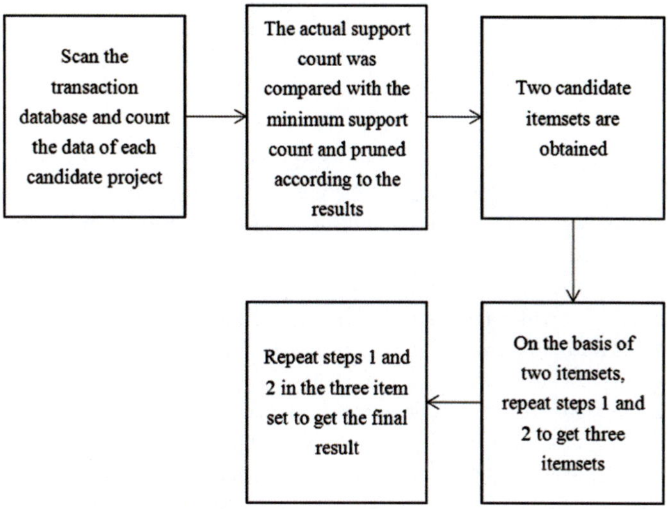

Fig. 1. Application flow of association analysis data mining method

According to Fig. 1, the mining process of the method is as follows: (1) because the association analysis can only be carried out according to the association between a certain technique index and the corresponding batting result in the tennis technical and tactical analysis, before the analysis, the relationship data between the hitting point and the gain and loss points should be obtained according to the technical and tactical decision-making needs of the tennis match; (2) in order to analyze the relationship between hitting point and score, weka platform will be used for data mining and preprocessing. The preprocessing is mainly to remove duplicate data and expired data (i. e. technical and tactical features or rules that are no longer used), and then formally enter the analysis stage. Figure 2 shows the data mining process of association analysis data mining method [3, 4].

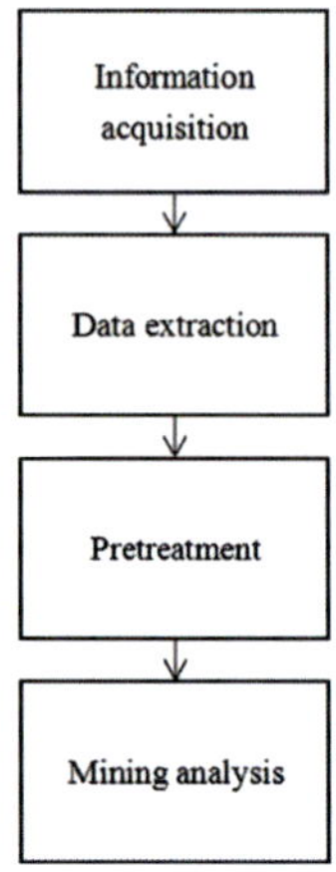

Fig. 2. Data mining process of association analysis data mining method

3.3 Detailed Explanation of Mining Process

First of all, the relevant software is used to collect information on the Weka platform. The collection path is the game video of the players in the tennis competition. The collection target includes the hitting point, hitting position, hitting technique, hitting racket position,etc. according to the collection results, the corresponding technical and tactical database is obtained, and the technical and tactical big data body is produced, association rules are used for pattern mining of technical and tactical information. Secondly, we extract the data from the mining data. According to the needs of our own technical and tactical decision-making content, we use database interface program to extract the data sequences and incentives that meet the needs from the big data body, and then preprocess them. In the preprocessing process, we transform all the extracted data into format, and propose the data that is consistent with the original data format and type, exclude data that is no longer applicable. Finally, the data mining analysis is carried out, and the rules of technical and tactical information can be obtained by using the mining tools of the platform, and the mining results can be reasonably explained, and the corresponding results can be output [5].

According to the characteristics of association rules, apriori algorithm combined with association analysis data mining method can produce frequent itemsets through combination, but this method is not consistent with the order requirements of tennis shot down decision-making, at the same time, it can not link the hit point with the hit effect, therefore, after the frequent itemsets are generated, we need to mine the characteristics of the hitting points in tennis techniques and tactics, such as the hitting points of serve and receive, and learn the characteristics of different hitting points and the situation before the hitting score, so as to increase the correlation strength and integrity of the analysis process, and ensure the quality of the analysis results [6].

4 Case Verification

4.1 Case Overview

According to the application requirements of apriori algorithm of association analysis data mining method, the whole data analysis must be aimed at a certain tennis player (it can be one's own contestant or the other's contestant, but it must be targeted). Therefore, in terms of case selection, this paper takes a domestic well-known female tennis player as an example, the video of the athlete's last four games is obtained through the network, and the information of the video is collected through the data mining method of association analysis, and the correlation relationship is established. The basic parameters of association rule data mining are: minimum support 10%, maximum support 100%, minimum confidence 70% [7].

4.2 Analysis Results

According to the tennis technical and tactical data after processing, this paper analyzes the relationship between the impact point and score of the last two strokes of each game, and the impact point and score of the first three strokes of the serve round. First of all, according to the association decision rule, the last two rackets of each game belong to catching the ball, and the opponent's ball path is mostly in the middle, and the player often swings along the middle. The result shows that the player's score in the last two rackets is not good, the confidence of losing points is as high as 96%, and the support is only 13.9%, which indicates that the player's losing points rate of hitting the ball back to the middle is very high, rough statistics more than 80%, and then analyze the player will return the ball to the left and right sides of the last two rackets, the results show that the left and right sides of the last two rackets scoring rate is more than 70%. It can be seen from here that the player's ability in the last two rackets is on the left and right sides, so the tennis technique and tactics should focus on returning the ball to the left or right side, and avoid the middle return as far as possible. Secondly, in the first three shots of the serve round, the player's landing point is generally near the center of the right side of the opponent's court and near the center of the back court. According to the scoring situation, when the player serves to the middle of the right side of the opponent's court, the opponent will return the ball to the middle of his own back court. At this time, the player's loss confidence is as high as 99%, when the player catches the opponent's return stroke to his own backcourt center, the impact point of the return stroke is generally the center of the opponent's backcourt. When the opponent returns stroke again, the player's confidence is still as high as 97%. On the contrary, when the player serves to the left and right sides of the opponent's right court, the scoring rate of service is more than 67%, and the scoring rate of return service is more than 53%. Therefore, the player's tennis skills and tactics should focus on the left and right sides, it is best to move left and right, try not to give the other side the chance to hit back in the middle, so that the player's winning probability will be maximized.

5 Conclusion

To sum up, in order to better give tennis players technical and tactical decision-making, adjustment support, this paper analyzes the tennis technical and tactical association analysis data mining method. Through the comparison, it can be confirmed that the data mining method of tennis technology and tactics association analysis is more reliable than the traditional method, and has obvious advantages, so this method is worth promoting. The scheme in this paper can give full play to the role of this method. Combined with the case verification results, it can be seen that this method can find out the technical and tactical weaknesses and strengths of tennis players. In this way, athletes will have a clear direction in the technical and tactical setting, at least be able to set the most suitable technical and tactical.

References

1. Mai, F., Tomoya, H., Kouichi, H., et al.: Association between statin use and cancer: data mining of a spontaneous reporting database and a claims database. Int. J. Med. Sci. **12**(3), 223–233 (2015)
2. Weng, J., Zhu, J.Z., Yan, X., et al.: Investigation of work zone crash casualty patterns using association rules. Accid. Anal. Prevent. **92**(7), 43–52 (2016)
3. Jeon, J., Sohn, S.Y.: Product failure pattern analysis from warranty data using association rule and Weibull regression analysis: a case study. Reliab. Eng. Syst. Saf. **133**(6), 176–183 (2015)
4. Das, A., Ahmed, M.M., Ghasemzadeh, A.: Using trajectory-level SHRP2 naturalistic driving data for investigating driver lane-keeping ability in fog: An association rules mining approach. Accid. Anal. Prev. **129**(8), 250–262 (2019)
5. Nagao, M., Seki, H.: An FCA approach to mining quantitative association rules from multi-relational data. Int. J. Comput. I **6**(4), 366–383 (2017)
6. Gonfiotti, A., Santini, P.F., Campanacci, D., et al.: Malignant primary chest-wall tumours: techniques of reconstruction and survival. Eur. J. Cardio-Thorac. Surg.: Off. J. Eur. Assoc. Cardio-Thorac. Surg. **38**(1), 39–45 (2010)
7. Guo, H.T.: Research on application of IT in techniques and tactics analysis of basketball. Adv. Mater. Res. **926–930**, 2586–2589 (2014)

Security Issue in Cognitive-Inspired Computing

Implementation of Application Asset Analysis and Management Tool Based on Zowe

Xiaosong Lv(✉) and Biyue Fan

School of Computer Science, Beijing University of Posts and Telecommunications, Beijing, China
lvxiaosong@bupt.edu.cn

Abstract. With the development and progress of the society, information changes at an amazing speed and customers leave more and more program codes on the mainframe. As society's demands on information technology continue to grow and mainframe customers' work items change, some program assets are shelved, deleted, hidden, or even forgotten. When customers need them again to build new projects, they can't find where the code is, which obviously makes code reuse difficult. Based on the needs of users for the reuse of project assets, this paper designs and develops a web graphics tool based on Zowe to help users explore project assets deeply, find the location and value of program code, and make old code get better use in new projects, so as to effectively solve the related pain points of users.

Keywords: Zowe · Application assets · Management

1 Introduction

IBM's z/OS is known for its compatibility, allowing decades-old compilers to run continuously. At present, these valuable program assets accumulated over many years in the customer environment are difficult to understand the original program due to the absence of corresponding reference documents and even the original code, and more and more problems are encountered in the management.

At present, the value mining of application assets by z/OS users is a realistic problem that needs to be solved urgently. However, there are very few studies based on Zowe, let alone analysis of application assets based on Zowe. The lack of documentation and source code for program assets makes this problem more difficult to solve.

The simple and easy-to-use program asset analysis and management tool based on Zowe proposed in this paper reduces the pressure of the server through the development mode of front-end and back-end separation [1]. The front-end uses React framework to build pages, and the back-end provides RESTful interface based on Spring Boot framework to interact with the front-end. It can excavate the value from the accumulated program assets and effectively solve the related pain points of users, which has great social research value and practical significance.

B. J. Jansen et al. (Eds.): *International Conference on Cognitive based Information Processing and Applications (CIPA 2021)*, LNDECT 85, pp. 293–303, 2022.
https://doi.org/10.1007/978-981-16-5854-9_37

2 Relevant Platforms

2.1 Mainframe

The mainframe is the top product in the computer. Because of its security, high reliability and powerful data processing ability, it is widely used in finance and other industries. As mainframe hardware technology improved, so did IBM mainframe operating systems [2]. Currently, IBM mainframes support multiple operating systems, including VSE, MVS, OS/390, VM, AIX, z/OS, and more.

2.2 Z/OS

Z/OS operating system is developed on the basis of OS/390 operating system. It is a 64-bit operating system developed with z/Architecture sequence mainframe. It is the latest mainframe operating system of IBM and the mainstream operating system of mainframes.

The advantages are as follows:

(1) Trusted Infrastructure: IBM z/OS provides an infrastructure that is highly secure, easy to scale, and fully capable of disaster recovery.
(2) Continuous delivery: New function are introduced every quarter without compromising stability or increasing complexity or cost.
(3) Better Hybrid Cloud: IBM z/OS brings new enterprise capabilities to the hybrid cloud, increasing agility and trust.

2.3 Zowe

Zowe is the first Open source and extended framework in the history of z/OS. It is both an open source project under the Open Mainframe Project (OMP) and a collaborative Project under the Linux Foundation [3]. Zowe reduces the learning threshold of z/OS and improves production efficiency. At the same time, it attracts more people into the mainframe ecosystem. Zowe's goal was to build a large console ecosystem that could be used to solve business problems at low cost.

Zowe framework is shown in Fig. 1:

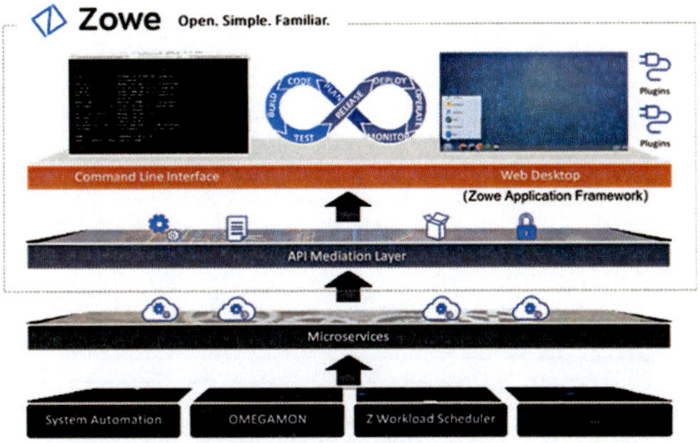

Fig. 1. Zowe framework

Zowe connects the user's system management and development tools to the mainframe, so that users can use the most familiar cloud platform to deal with z/OS [4].

(1) CLI allows users to interact with z/OS from a variety of other platforms.
(2) Web Desktop is a browser-based application container and virtual Desktop system, including the Zowe Desktop and various application plug-ins.
(3) The API Mediation Layer provides a single access node to the host REST API.
(4) The z/OS service provides REST APIs for accessing z/OS resources.

3 Design and Implementation of the System

3.1 Overall Architecture of the System

The system can be divided into front-end and back-end [5]. The overall framework of the system is shown in Fig. 2. The front-end of the system is divided into application layer and component layer, and the back-end is divided into service layer and resource layer.

Application layer: the application layer provides browsing files, parsing program structure, browsing program structure and other functions.

Component layer: Provides common front-end components, such as folder components, file components, tree components, etc.

Service layer: According to the requirements of the system, it encapsulates a variety of basic services into interfaces to provide services for the front-end. Such as file browsing, determine whether the program is executable, parsing programs and other services.

Storage layer: The storage layer mainly consists of databases and file systems, which are used to store user data and documents of parsed files.

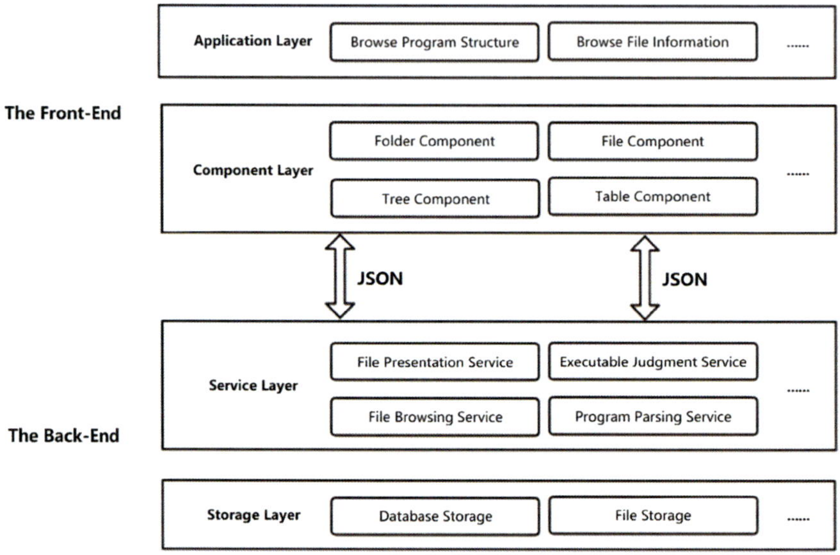

Fig. 2. System architecture diagram

3.2 Front-End Design and Implementation

The front-end of this system is component-based development based on the React framework and uses Redux to manage the state of components [6]. The front-end architecture is shown in Fig. 3. Material-UI is an open-source React component library that implements the Material Design specification of Google [7]. The components in the library cover all kinds of UI components necessary for single-page applications. Part of the UI components and icons in the front-end of this system are designed and developed by using the components in the Material-UI library.

Front-end framework—react. React is one of the most popular front-end frameworks today. It defines JavaScript control mechanisms for Web pages to respond quickly. It renders the UI by changing the state and dynamically controls the rendering of various areas of the Web page. React's strengths lie in its declarative design, componentized development and virtual DOM manipulation [8]. Declarative programming is easier to implement, read, and maintain. Componentized development divides the UI into separate, reusable parts, and each component maintains its own internal state independently, which facilitates code reuse. The biggest difference between React and other JavaScript frameworks is the virtual DOM technology that controls page rendering. The traditional JavaScript framework directly controls the DOM of the page. When the data changes, it will frequently go to the browser to layout and draw the page, which consumes a lot of computing resources. While, React maintains a virtual DOM tree in memory. When the state of a component changes, React will use the diff algorithm to figure out what parts of the virtual DOM really need to change, and only carries out native DOM operation for this part, rather than rendering the whole page, which makes the page more responsive [9, 10].

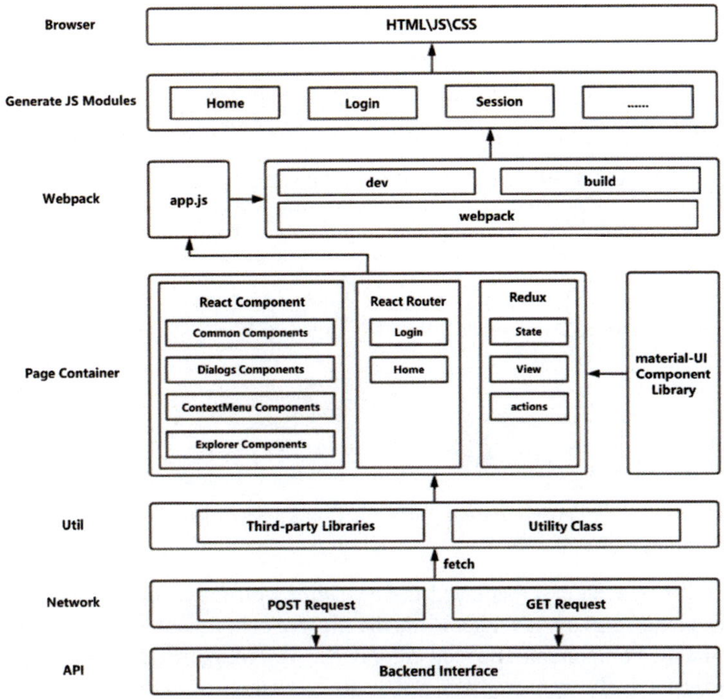

Fig. 3. Front-end architecture

React-redux state management. In the React development, it is often encountered that components pass arrays and shared arrays. However, some data is global in nature, and it needs to be nested layer by layer if it needs to be passed to the innermost component. To solve this problem, the front-end uses Redux to manage the common data [11]. The process of data state management by Redux is shown in Fig. 4.

The initial public State is stored in the Store, and the Views the user is exposed to depend on the State. When an interaction event is generated between the user and the Views, the Actions corresponding to the event will request data to the RESTful API via Middlewares, and the data returned asynchronously by the RESTful API will be dispatched to the corresponding Reducer function with the current State. The Reducer function changes the State to make changes to the Views.

Introduction of main front-end modules.

(1) File browsing module
The page for the file browsing tool is shown in Fig. 5. After the user successfully login we can browse the file on USS (UNIX Subsystem) through the file browsing tool. We can also browse the z/OS specific MVS (Multiple Virtual Storage) DataSet, including the sequential DataSet, PDS, PDSE, VSAM and other datasets. Click the switch button on the page to realize the free switch between the two types of file browsing functions. After switching, the current search path and the corresponding

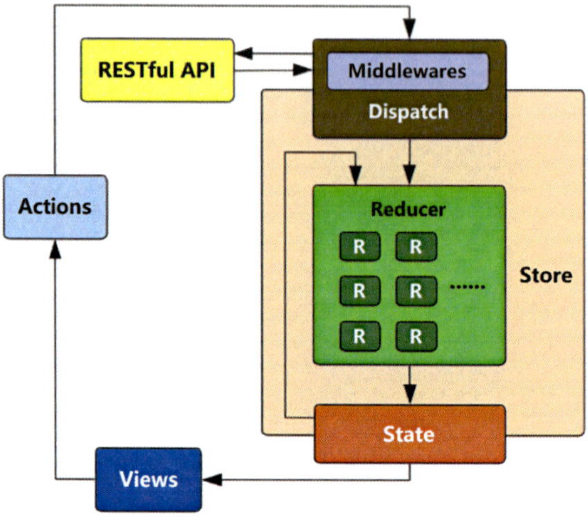

Fig. 4. Redux data state management process

file will be saved and displayed the next time you return to the page. The main components used by this module are the file tree component, the folder component and the file component. The icon color of the file component is shown in black and gray respectively according to whether the file is an executable file. For executable files, right-click to show the parsed program structure.

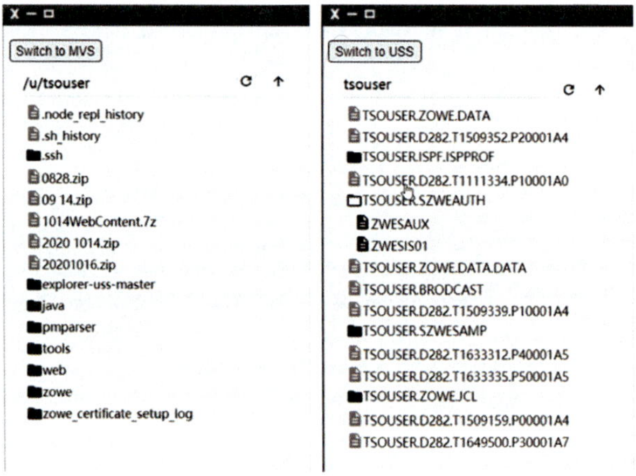

Fig. 5. File viewing page

(2) Program structure browsing module
The page for the program structure browsing tool is shown in Fig. 6. This page appears when you select an executable and choose to show its program structure functionality. Through this page, we can understand the basic properties of the

program, segment information and compilation unit information. The main components used in this module are tree component, file component, folder component and Table component. Among them, the folder and file components are used to visualize the internal structure of the program, and the Table component is used to visualize the specific information of each part. When the mouse is placed on a line in the Table, it will also display a comment to the contents of that line in the form of.

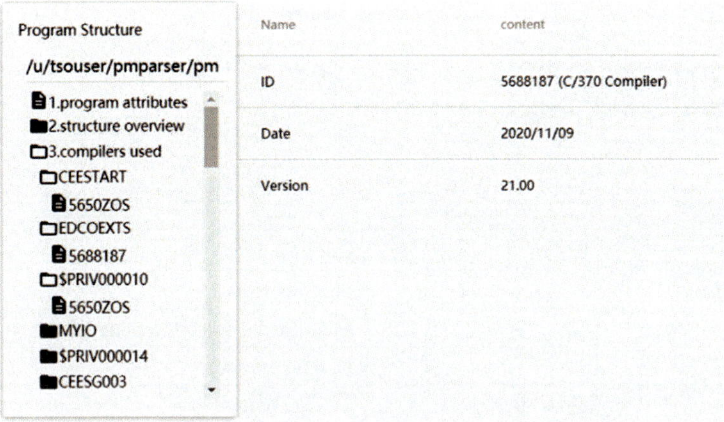

Fig. 6. Program structure browsing page

3.3 Back-End Design and Implementation

The back-end of the system provides a RESTful style API for the front-end through Spring Boot. Spring Boot is a module of the Spring framework.It is used to create stand-alone, production-level base applications [12]. Spring Boot has built-in Servlet container, which can be run independently in the form of JAR package, greatly improving the efficiency of development and deployment [13, 14]. At the same time, Spring Boot provides the function of application monitoring and management at runtime. We can use HTTP, SSH protocol to monitor the runtime of the service. The main APIs provided in Spring Boot of this system include user login verification, browsing files, judging whether the program is executable, parsing the internal structure of the program, etc. The API Mediation Layer provides the API for user login verification and browsing files. Next, I'll look at two APIs, one for determining whether a program in a folder is executable, and the other for parsing the internal structure of a program.

Determine whether the program in the folder is executable. The API receives parameters of path and type. The type parameter refers to the type of the file, and the path parameter is the path to the file. The API first looks for all the files in the folder, and then judges them one by one. By looking at the first seven characters of each file to determine whether the file is executable, the result of the judgment is stored in the JSON data along with the file name as a key-value pair.

Parses the internal structure of the program. The receiving parameters of this API are path and type. The type of file is determined by the type parameter, and the path of the file to be parsed is determined by the path parameter. If the corresponding file cannot be found according to the path parameter, the content of the JSON data returned to the front-end is false. If the corresponding file is found according to the path parameter, the internal structure of the program will be analyzed, the report of the text file will be generated, and the content will be converted into JSON format and returned to the front-end.

The basic properties of the text file are as follows:

(1) Program Attributes: Includes Addressing Mode (AMODE), Residency Mode (RMODE), Link Time, etc. As shown in Table 1:

Table 1. Program attributes

Program	Bin/at	Unix file
PMAR_LFMT	Program object 2.0	Program format
PMAR_MAMD	31	Main entry point addressing mode
RMODE	31	Residence mode
PMARL_1STOR	68160 bytes	Storage required for loading code segment
PMAR_EPM	0	Main entry point offset
PMAR_RENT	YES	Reenterable
PMAR_1BLK	NO	LM has only one block of text
PMARL_XPL	NO	Module requires XPLINK
Binder ID	5695PMB01 (02.04)	Binder version
…	…	…

(2) Structure Overview: Includes class, offset information, and length information. As shown in Table 2 and Table 3:
(3) Compilers Used: Includes information about the number of compiled modules, compiler version, and compilation time. As shown in Table 4:

The back-end needs to return JSON data to the front-end. The JSON data can be divided into the following two types according to whether there is an array of Children:

(1) {"Type": "file", "name": "name1", "contents": "contents1",…}.
(2) {"Type": "directory", "name2": "segments", and "contents2": "contents1", "children": [],…}.

The first kind of data is file data, and the second kind of data is folder data. The corresponding value of children in folder data is an array composed of several file data and folder data.

Table 2. Structure overview

Segment ID	RMOOD	Offset	Length	Name
1	ANY	0	10A40	B_TEXT
1	24	0	0	B_MAP
2	24	0	014D	B_RLD
3	24	0	0357	B_ESD
4	24	0	0087	B_IDRL
…	…	…	…	…

Table 3. Structure overview

Name	Type	Offset	Length
B_TEXT	C	0000	10A40
CEESTART	S	0000	007C
@@PPA2	S	0080	002C
@ST00001	S	00B0	00A8
@ST00002	S	0158	0389
…	…	…	…

Table 4. Compilers used

ID	Version	Data	Section
5650ZOS	BM XL C/C++ Compiler	02.03,2019/11/12	CEESTART
5650ZOS	BM XL C/C++ Compiler	02.03,2019/11/21	@ST00001
5688187	C/370 Compiler	21.00,2019/11/21	EDCOEXTS
5688187	C/370 Compiler	21.00,2019/11/21	EDCOEXTL
569623400	HIGH-LEVEL ASSEMBLER	01.06,2017/03/07	CEESG003
…	…	…	…

We need to convert the table contents to the specified JSON format. The Program Attribute and Compilers Used are folder data and they only have file data in their children array, so you only need to pair the table content with the table header content as the elements in their children array. Since the contents of Structure Overview are stored in two tables respectively and there is a parent-child relationship between nodes in the two tables. When traverse Table 3, search for its parent node in Table 2, store the child node into the children array of the parent node, and complete the JSON format conversion after the final traversal.

3.4 Project Deployment

This project uses Webpack to package the front-end React project and deploy the packaged content to the Zowe desktop environment. Meanwhile, the SpringBoot program is deployed to the Zowe API Mediation Layer. For cross-domain problems existing after project deployment, this project is solved by CORS protocol in SpringBoot, and CORS can be enabled by annotating the entire controller with @crossOrigin attribute [15]. The two parameters in @crossLogin are Origins (the list of fields that are allowed to access) and MaxAge (the maximum time the cache will last before preparing a response).

4 Conclusion and Future Work

This paper is based on the idea of front-end and back-end separation. The front-end uses the React framework for componentized development, and the back-end uses the RESTful style service interface provided by Spring Boot. This project designed and developed a Zowe Plugin, and developed a web graphical interface in Zowe Desktop to display the internal information of the executable program, such as compilation time, compiler version, link time, symbol table of each segment, etc. This project will help customers extract more value from their accumulated program assets. For mainframe customers, the implementation of this project will provide them with convenient graphic retrieval tools.

There are still some scientific challenges to be addressed in the future:

(1) To achieve more program information analysis: support analysis of the object file, display the compiler version of the object file, compilation time and other information, as well as the symbols contained in each section, code and data offset length and other information.
(2) Realize the analysis of different types of files: support the analysis of Arhive files, Arhive files can contain multiple target files and binaries.
(3) Determine whether the program has been optimized or patched: analyze whether the program has been processed by an optimization tool, such as IBM COBOL Automatic Binary Optimizer, and which patches have been installed on the executable program.

References

1. Gong, Y., Gu, F., Chen, K., Wang, F.: The architecture of micro-services and the separation of frond-end and back-end applied in a campus information system. In: 2020 IEEE International Conference on Advances in Electrical Engineering and Computer Applications (AEECA), Dalian, China, pp. 321–324 (2020). https://doi.org/10.1109/AEECA49918.2020.9213662

2. Berry, C., Wolpert, D., Vezrytzis, C., et al.: IBM z14: processor characterization and power management for high-reliability mainframe systems. IEEE J. Solid-State Circ. 1–12 (2018). https://doi.org/10.1109/JSSC.2018.2873582
3. Open Mainframe Project; Open Mainframe Project Announces the Launch of Zowe - an Open Source Framework that Strengthens Integration with Modern Enterprise Applications. Computer Weekly News (2018)
4. Baker, B.: Open source project Zowe: fast, simple, familiar z/OS development [EB/OL]. https://www.ibm.com/blogs/systems/open-source-project-zowe-fast-simple-familiar-zos-development/. Accessed 15 Aug 2018
5. Guan, S., Hu, W., Zhou, H. : Front-end and back-end separation-react based framework for networked remote control laboratory. In: Proceedings of the 37th China Control Conference, pp. 1200–1205 (2018)
6. Kaderka, J., et al.: The behavior, constraint, and scenario (BeCoS) tool: a web-based software application for modeling behaviors and scenarios. In: Aiaa Aerospace Sciences Meeting (2018)
7. Google. Material-UI: A popular React UI framework [EB / OL]. https://material-ui.com/. Accessed 10 Oct 2020
8. Javeed, A.: Performance optimization techniques for ReactJS. In: 2019 IEEE International Conference on Electrical, Computer and Communication Technologies (ICECCT). IEEE (2019)
9. Voutilainen, J.P., Mikkonen, T., Systä, K.: Synchronizing application state using virtual DOM Trees. In: Casteleyn, S., Dolog, P., Pautasso, C. (eds.) Current Trends in Web Engineering. ICWE 2016. Lecture Notes in Computer Science, vol. 9881, pp. 142–154. Springer, Cham (2016). https://doi.org/10.1007/978-3-319-46963-8_12
10. Chęć, D., Nowak, Z.: The performance analysis of web applications based on virtual DOM and reactive user interfaces. In: Kosiuczenko, P., Zieliński, Z. (eds.) KKIO 2018. AISC, vol. 830, pp. 119–134. Springer, Cham (2019). https://doi.org/10.1007/978-3-319-99617-2_8
11. Badapanda, R.P.: Web application development using a web component framework. US20190324729 (2019)
12. Gajewski, M., Zabierowski, W.: Analysis and comparison of the spring framework and play framework performance, used to create web applications in Java. In: 2019 IEEE XVth International Conference on the Perspective Technologies and Methods in MEMS Design (MEMSTECH). IEEE (2019)
13. Yi, Y., Li, Z.: Design and implementation of music web application based on Vue and Spring Boot (2020)
14. Guntupally, K., Devarakonda, R., Kehoe, K.: Spring boot based REST API to improve data quality report generation for big scientific data: ARM data center example. In: 2018 IEEE International Conference on Big Data (Big Data). IEEE (2018)
15. Kolam, H., Kennedy, B.: Method to enable cross-origin resource sharing from a webpage inside a private network. US9697188 (2017)

Negative Transfer of L1 Thinking on L2 Writing Based on Iwrite Evaluation System

Dan Shen[1,2](✉) and Dan Huang[2]

[1] Universiti Utara Malaysia, Bukit Kayu Hitam, Kedah, Malaysia
[2] Guangzhou Institute of Science and Technology, Guangzhou, China

Abstract. In the process of learning a second or a foreign language, learners are often disturbed by the negative transfer of their native language thinking, which is especially evident in the process of foreign language writing. This thesis analyses the common errors caused by the negative transfer of native language in college students' English writing from three aspects of lexicon, syntax and grammar by taking students' writing texts on the iwrite evaluation system as the source of corpus analysis and explores to reveal the influence of the negative transfer of native language thinking on college students' English writing.

Keywords: Text corpus · Iwrite evaluation system · Negative transfer

1 Introduction

As one of the basic English skills, writing is an integrative reflection of students' English learning effect and application ability, and also an important indicator to measure the level and ability of English teaching in higher education. There are many factors resulting in the inability to write well, such as students' insufficient vocabulary, poor knowledge of grammar and writing strategies, language transfer and so on. Among all these factors, language transfer is the main cause affecting second language learners' writing [1]. The study conducted by Guo Chunjie and Liu Fang concluded that native language thinking also has an impact during the output of the second language [2]. On this basis, this thesis attempts to analyze the negative transfer of native language thinking on college students' English writing by analyzing their writing texts and relevant data based on the iwrite evaluation system, which aims to guide students to diminish the impact of this negative transfer correspondingly and improve writing effect.

"Language transfer has always been a central issue in applied linguistics, second language acquisition and language teaching" [3]. Robert Lado mentioned in his book *Linguistics Across Cultures* that "in a second language learning situation, learners rely extensively on their native language that they tend to transfer the forms and meanings of their native language and culture to the foreign language". And when learners communicate in a new language, they often try to express their ideas with the help of previously acquired sounds, meanings, structural rules or cultural habits. This phenomenon is called

B. J. Jansen et al. (Eds.): *International Conference on Cognitive based Information Processing and Applications (CIPA 2021)*, LNDECT 85, pp. 304–312, 2022.
https://doi.org/10.1007/978-981-16-5854-9_38

mother tongue transfer or native language transfer [4], which is very common in foreign language learning, especially in the beginning stages of foreign language learning.

Ellis defined transfer as "a hypothesis concerns the influence of the previous task A to task B" [5]. Odlin proposed a widely accepted definition that "transfer is the influence resulting from the similarities and differences between the target language and any other language that has been previously acquired (and perhaps imperfectly acquired)" [3]. As to the effect of language transfer, it is mainly manifested in two types: positive transfer and negative transfer. Positive transfer means that the similarities between the native language and the target language facilitate the learning of the second language, helping learners to comprehend and master certain language items in the foreign language. While negative transfer, or the mother tongue interference, occurs when foreign language learners inappropriately substitute the expression or comprehension of the foreign language with that of the native language.

2 The Negative Transfer of Native Language on College Students' English Writing Based on the Iwrite Evaluation System

Iwrite evaluation system, combining machine and human assessment, is designed to help teachers improve the effectiveness of writing instruction and thus truly improve students' writing skills based on an in-depth research on the teaching of college students' English writing. Thus, analyzing student' writing texts on the iwrite system provides an insight into the negative transfer of English writing that occurs during the writing process.

For a successful writing text, language and thoughts are indispensable. However, most of the college students usually write in their native language by conceiving the essay in their brain at first and then converting it to the appropriate language, which does not conform to the language norms of the foreign language [6] and make many serious grammatical errors. In this section, errors caused by negative transfer will be analyzed by using examples of college students' English compositions from the iwrite evaluation system.

2.1 Negative Transfer in Lexicon

Lexical errors account for a large proportion of college students' native language transfer errors. Moreover, Mei Hua stated in her study that in the writing process, college students are seriously influenced by their mother tongue in word choice, word usage and word collocation [7]. The most typical errors are the following three major categories.

(1) Improper use of part of speech
Part of speech in Chinese have very few variations compared to that in English. Therefore, Chinese college students seldom consider the problem of misuse of part of speech in their writing. The errors of part of speech mentioned here mainly focus on verbs, nouns, adjectives and adverbs. There are example 1 "The little girl in the photo is looking at a pile of unfinished homework. She feels very upset and worry." and example 2 "Especially among the younger generation, such social network has gained enormously popularity." The word "worry" in example 1 is a verb rather than

an adjective. Although it shares the same meaning in Chinese, it cannot work as a predicative in English sentence, thus it should be replaced by "worried". While in example 2, the word "popularity" is a noun that should be modified by an adjective "enormous" rather than the adverb "enormously". The above sentences present similar mistakes of ignoring the lexical differences between English and Chinese part of speech, which reflects that students are easily influenced by Chinese thinking in writing.

(2) Improper use of words
Many students hold a system of Chinese-English vocabulary correspondence in their minds. However, the vocabulary between Chinese and English does not correspond to each other exactly. When students translate them literally, they ignore the connotative meanings of the words and make transfer errors instead. The word "body" in the sentence "Drinking will do harm to our body" only means "the whole physical structure of a human or an animal" in English, while in Chinese, it has another meaning "health condition". Since this sentence clearly tends to indicate the latter, hence, it would be more logical to change the word "body" to "health". Another example "A harmonious neighborhood life needs everyone to protect", the replacement with "maintain" is more consistent with its English connotation in this context. Usually, when students write in English, they only pay attention to the superficial meaning of the words but ignore the differences in the use of the word meaning, which inevitably leads to mistakes in the use of vocabulary.

(3) Improper use of collocations
English words do not have exact equivalents in Chinese and need to be written in different ways in different contexts. It is precisely because of the differences in expression between Chinese and English that the interpretation of one language is often outside the range of the other, resulting in some inconsistencies in expression. For example, there are two phrases "raise people's level of knowledge" and "learn more knowledge" extracted from students' composition. "Raise" is collocated with "improve" or "advance" instead of "level" In English; the English expressions that go with the word "knowledge" are usually "acquire/accumulate/obtain/gain/absorb knowledge". The collocation habits and the range of collocations vary between the English and Chinese languages due to the influence of their respective linguistic features and cultural factors. Thus it is important for college students to understand the range of collocations in their English studies so that they can use them flexibly in English writing.

2.2 Negative Transfer in Sentence

The syntactic structure between English and Chinese is very different. Chinese attaches great importance to parataxis without morphological change in words, and the syntactic structure is more flexible. Yet English, by contrast, is hypotaxis-prominent, with a variety of morphological changes among words, such as tense, voice, and person, and emphasizes the integrity and rigor of sentence form and structure. Consequently, the rigor of English syntactic structure is very likely to cause syntactic errors for learners whose native language is more flexible in syntactic structure.

(1) Improper word order
Due to the differences in syntactic structure between English and Chinese, the two languages differ in the word order of sentence expressions, especially the word order of attributes, adverbial modifiers, negations, etc. College students are frequently influenced by their native language thinking and make syntactic errors in word order. For example: Nowadays, China is a prosperous, socialist, modern powerful country. In English, when there are multiple attributes modifying a noun, the more essential the attribute is, the closer it is to the noun it modifies, but the opposite is true in Chinese. Therefore, the correct order of sentence should be "Nowadays, China is a modern, prosperous, powerful socialist country". Besides, in Chinese the order of words denoting time, place and manner is arranged so sequentially, whereas in English the order is usually as "verb, manner, place and time". It is seen that when expressing ideas, the English way of thinking is used to state the main points in a straightforward manner, but the Chinese tends to state the main points from the side. This difference in thinking style also has a negative transfer effect on students' English writing to some extent.

(2) Improper use of conjunctions
English is connected by strong logical relationships between linguistic symbols and grammar, and conjunctions are used more frequently in compound sentences. Chinese sentences, on the other hand, are connected by the meaning of words, and conjunctions are less often used. Therefore, college students often miss or misuse conjunctions in the process of English writing. For example: Although there are many long-lived and memorable stories abroad, but I prefer the ancient Chinese stories. Pairs of conjunctions appear more frequently in Chinese, but in English, if both "although" and "but" are used together, it becomes a mixed and contradictory structure. Due to the influence of negative transfer of Chinese, students often use "although…but…" wrongly to express the adversative relation. In terms of morphological structure, English has a very high use of conjunctions and the logical relationships between sentences are very clear. College students therefore overlook these differences between English and Chinese conjunctions in the process of writing in English, which can easily lead to the omission or misuse of conjunctions.

(3) Improper use of special sentence patterns
The negative influence of Chinese thinking on college students' English writing is also manifested in the use of special sentence patterns in English. "There be" sentence pattern in English is a very typical one, which can't be randomly replaced with "there have" as Chinese thinking pattern. Apart from that, students often make mistakes when writing negative sentences in English. An example like "Anybody who habitually procrastinates cannot achieve real success", in English the predicate verb cannot be used in the negative form when a compound word formed with "any" or a word modified by "any" is the subject. Therefore, the expression "any…not…" is not in accordance with the English expressions, and this sentence can be changed to "Nobody can achieve…". The negation in Chinese is usually placed before the verb, while it has to be placed in front of the verb in English negative sentences, and if the negation is in a complex sentence, the negation is placed in front of the verb of the main clause. This kind of negative order, which is contrary to Chinese, often makes college students write sentences with transfer errors in English writing.

2.3 Negative Transfer in Grammar

English grammar is a set of linguistic rules that are systematically summarized and generalized. From the analysis of college students' English writing texts, the number of errors in grammar accounts for the largest part, mainly covering three aspects: tense, voice and subject-predicate agreement.

(1) Improper use of tense
Tense is the form of verb that expresses the behavior, action and state in various conditions of time. In Chinese, tense is usually reflected by the use of time adverbs as complements. In English, however, there are many variations of verbs that indicate tense, so students often use the wrong tense or forget to reflect it when they use it in a specific way. There is a sentence excerpted from a student's composition "It was common that people use WeChat and many other media platforms to create and maintain his or her social relationship". In the sentence, the adjective "common" indicates that the content following "that" clause is a frequent, recurring behavior, which can be expressed in simple present tense. In English writing, different contexts require corresponding tenses to promote the overall logic of the statement. However, in the actual writing process, students often use the wrong tense in the syntax of English writing due to their native language's habitual way of thinking.

(2) Improper use of voice
Voice is divided into active and passive voices. Both voices are used at a higher frequency in English, while the voice in Chinese is usually the active voice. Therefore, students are apt to make writing mistakes in the voice influenced by Chinese language. For example: Great changes have been taken place in China since the reform and opening up. It should be noted that in English, only transitive verbs have a passive structure, and intransitive verbs and their phrases cannot be used in the passive voice. In the example, the phrase "take place" is an intransitive verb phrase that cannot be used in the passive voice. Therefore, the correct expression of sentence B should be "Great changes have taken place in China…". This poses a potential difficulty for college students' English writing due to the differences in the expression of the passive voice between Chinese and English.

(3) Negligence of the subject-predicate agreement
Another common error in English writing is shown in the subject-predicate agreement of sentences, that is, the inconsistency between the person and number of the subject and the corresponding form of the predicate. The typical one is that those collective nouns are not followed by the plural form of the predicate verb accordingly. In the sentence "A number of students is unwilling to do the homework ahead of time when they are allowed more time to finish it", when "a number of" is followed by a plural noun, its predicate verb should also be in plural form, so "is" in this sentence should be changed to "are". Due to the lack of subject-predicate agreement in Chinese, many college students do not know the rules of subject-predicate agreement in English thoroughly enough, which makes it more difficult for them to grasp this grammar phenomenon.

3 Data Analysis of Negative Transfer of Lq Thinking in L2 Writing Based on the Iwrite Evaluation System

In this section, the impact of negative transfer of native language thinking on students' English writing is further explored through an analysis of data on errors in writing texts on the iwrite evaluation system. Table 1 shows the distribution of students' errors in the iwrite system, which is based on the text analysis of sophomore students' writing on the topic of *Endless Delay*. There were 208 college correspondents participating in the writing with the average score 74.2 (out of 100), among whom 156 students with score 70 or more (including 70) were selected for this study. The chosen students are generally at an intermediate level and therefore the data analysis on this basis is representative. The data are demonstrated as follows:

Table 1. Error types of sophomore students' works on Endless Delay in the iwrite evaluation system

Error types		Number of errors	Percentage
Lexical errors	Errors of words	21	3.72
	Errors of parts of speech	61	10.8
	Errors of spelling	59	10.44
	Errors of collocation	66	11.68
Syntactic errors	Misuse or missing of conjunction	47	8.32
	Improper word order	13	2.3
	Improper use of special sentence pattern	11	1.95
Grammatical errors	Errors of tense	46	8.14
	Errors of voice	18	3.19
	Errors of subject-predicate agreement	67	11.86
	Missing of predicate	24	4.25
	Misuse of preposition	56	9.91
	Misuse of singular and plural form	69	12.21
Idiomatic errors		7	1.24

The data in the table show the number and percentages of errors in lexicon, syntax, and grammar in compositions. Besides, a qualitative analysis is integrated through a questionnaire survey conducted on 122 native Chinese college students on their errors in vocabulary, grammar, and sentences.

3.1 Lexical Transfer Errors

English vocabulary differs greatly from Chinese vocabulary in terms of conceptual meaning, connotative meaning, and collocation ability, etc. College students who do not understand these differences tend to apply these words with the laws of thinking in their native language, and then this will inevitably lead to the phenomenon of negative transfer of native language [8]. From the above table, the frequency of lexical errors among college students is second only to that of grammatical errors, with a total of 207 errors, accounting for 36.63% of all errors, of which only 10.44% are purely intralingual errors, that is, spelling errors caused by incomplete learning of the target language items. While the rest of errors are directly or indirectly related to the influence of the native language, with a total of 148 errors, accounting for 71.5% of all lexical errors. In other words, more than 70% of the lexical errors in the English writing of native Chinese speakers are related to their native language knowledge. Not only that, among all the lexical errors related to the influence of native language, 61.35% of them are related to lexical collocation and the transfer of parts of speech. There are huge differences in the meaning and usage of words in English and Chinese languages, therefore, the connotation and extension of English and Chinese vocabulary are not exactly equivalent [9]. As the data of the questionnaire survey shows, at the level of vocabulary, college students have great problems in the mastery and application of word collocation, transformation of parts of speech and word meaning, and the error rate is over 95%. Among them, college students are the weakest in mastering and using parts of speech of words, which appear the most serious errors, the error rate is as high as 99.17%. Besides, 95.04% and 96.69% of college students make mistakes in word collocation and use of vocabulary respectively. It is the influence of Chinese thinking that leads to the result that the words expressed often do not conform to the English language norms.

3.2 Grammatical Transfer Errors

From the perspective of errors, college students made the most serious mistakes in grammar, with 280 mistakes in total, accounting for 49.56% of all errors, which shows that college students have the most difficulties in grammar mastery and application. From the data, it can be seen that subject-predicate inconsistency and singular-plural misuse are both the most frequent grammatical errors in English writing, with 67 and 69 errors respectively, accounting for 23.93% and 24.64% of the grammatical errors, followed by preposition misuse and tense error, with 56 and 46 errors respectively, accounting for 20% and 16.43% of the grammatical errors respectively. In addition, due to the strict subject-predicate structure of English sentences, the subject is indispensable and the predicate verb is the center of the sentence, both are coordinated. In English, the subject and the predicate verb go together to form the core of the sentence. Chinese sentence structure, however, is relatively flexible, which therefore leads college students to often omitting the predicate, resulting in incomplete sentence structure. As shown in the table, the number of errors in the missing of the predicate and the errors in the voice is relatively small, which account for 8.57% and 6.43% of the grammatical errors. It still cannot be ignored, because the questionnaire survey finds that 98.44% of college students use Chinese thinking to conceive English compositions, and almost all of them use Chinese translation to check whether the English sentence expressions are complete, which makes it difficult to avoid the phenomenon of missing predicates. The writing texts of native Chinese students reflect that they prefer to use the active voice in English writing, resulting in a monotonous sentence style and lack of variation. It is because of the interference and influence of the negative transfer of native language thinking that they produce such Chinese expressions in English writing that do not conform to the rules of Western language and cultural habits.

3.3 Idiomatic Transfer Errors

Idioms are the result of long-term language use and are closely related to the environment in which people live. As the phrase "Tomorrow and tomorrow, how many tomorrow" in the table above, this kind of Chinglish expression probably leads readers to confusion because of its literal translation from the Chinese idiom. Because of the unique phonetic characteristics of Chinese, the phrase is repeated in Chinese with a stretched effect, but loses its rhythm when translated directly into English. Another case of idiomatic errors is the phenomenon that they cannot find the English counterparts of these slang, idioms or proverbs, and use some synonyms indiscriminately to replace them, which leads to unintelligible sentence meaning. The data show that among the 156 students' composition, there are 7 idiomatic errors, accounting for 1.24% of all errors. Although the proportion is relatively small compared with other errors because of the less absolute number of their use in the writing, almost all college students use Chinese thinking to conceive when expressing idioms, and even 56.2% of them often use Chinese thinking to express idioms, which means that in the process of English writing, more than half of the college students cannot express English idioms accurately with English thinking.

4 Conclusion

Since negative transfer of native language is an inevitable phenomenon in the writing process of English learners, college students should shift their mindset and face up to its existence, develop good writing habits in writing and try to communicate in English. In conclusion, overcoming negative transfer of native language thinking cannot be done in a hurry, but should be improved continuously to lay the foundation for the formation of English thinking.

Acknowledgements. This work is supported by Guangdong provincial project "The Study on Language Innovation in College Students' English Acquisition Process Based on Big Data Analysis" (No. 2020WTSCX156), and funded by The 10th batch of China Foreign Language Education Foundation Project (No. ZGWYJYJJ10B026).

References

1. He, X., Niao, L.: A probe into the negative writing transfer of Chinese college students. English Lang. Teach. **8**, 21–29 (2015)
2. Guo, C., Liu, F.: A dynamic research into L1 influence on L2 writing. Mod. Foreign Lang. **4**, 32–40 (1997)
3. Odlin, T.: Language Transfer: Cross-linguistic Influence in Language Learning. Shanghai Foreign Education Press, Shanghai (2001)

4. Lado, R.: Linguistics Across Cultures: Applied Linguistics for Language Teachers. University of Michigan Press, Ann Arbor (1957)
5. Ellis, H.: The Transfer of Learning. Macmillan, New York (1965)
6. Kobayashi, H., Rinnert, C.: Effects of first language on second language writing: translation versus direct composition. Lang. Learn. 183–215 (1992)
7. Mei, H.: Negative transfer of mongolian on english writing and counterplan research. J. Inner Mongolia Univ. (2008)
8. Li, C.: Negative transfer of native language in the second language acquisition and countermeasures. Continue Educ. Res. 116–119 (2008)
9. Wang, T.: The influence of negative transfer of native language on English writing in intercultural context. Exam Week 74–75 (2013)

The Civil Law Protection of Citizens' Personal Information in the Context of Big Data

Chao Jin(✉)

Law School of Zhengzhou Sias University, Zhengzhou 451150, Henan, China

Abstract. The 21st century is an era of big data. Massive information is produced in society at all times. In today's era of rapid information development, citizens' personal information has also become an asset, forming an information trading market. Our country has certain legal regulations on the protection of personal information, but there are still many problems and many gaps and deficiencies. The purpose of this article is to research and analyze the civil law protection of citizens' personal information based on big data. This article first conducts research on the current status of personal information leakage of citizens in our country, puts forward suggestions for improvement and solutions, and starts with defining personal information in the network environment, analyzes the legal characteristics, and proposes the current status of imperfect legal protection system, and proposes its own View. Then, in response to issues such as personal information protection in the current network environment, it is recommended to improve the upper-level legislative construction, and to further improve the guarantee system by strengthening industry supervision, increasing industry self-discipline and self-protection awareness. Research shows that in the context of big data, it is difficult to protect the personal information of citizens, and my country's current laws need to improve the personal information protection system.

Keywords: Big data background · Citizen information · Information protection · Civil law research

1 Introduction

In the current era of explosive growth of data, people have data everywhere in their lives, ranging from national security, crime prevention and control, to clothing, food, shelter, and transportation. The Internet has completely penetrated our lives [1, 2]. However, while enjoying the convenience brought by these technologies, we neglected to tap into our personal privacy step by step [3, 4]. The arrival of big data means that a large amount of personal information will be collected and analyzed. The collected personal information will form an "information person" in the database to match us in reality. Our privacy will be exposed and our actions will be exposed. Will be predicted, information has become a huge treasure of this era [5, 6].

My country has not yet formulated specific legislation for the protection of personal information on the Internet, but it is covered in some scattered legal provisions: Article

B. J. Jansen et al. (Eds.): *International Conference on Cognitive based Information Processing and Applications (CIPA 2021)*, LNDECT 85, pp. 313–320, 2022.
https://doi.org/10.1007/978-981-16-5854-9_39

13, Article 32, and Article 38 of the constitution; General Provisions of the Civil Law "Article 100 and Article 101; Article 252 of the Criminal Law [7, 8]. The above regulations have an important guiding role in our country's research on the lack of personal information protection legislation and systems under the network environment. Foreign experts and scholars have relatively few researches on personal information protection under big data [9, 10]. The United States is a developed country with a very complete legal system. The United States believes that people's privacy rights should include personal information. Personal information should be protected through privacy laws. This is also very reasonable [11, 12].

This article analyzes the current situation of citizen information leakage in our country, extends the direct source of citizen information leakage, and uses the existing law to protect personal information to expand the research on the civil law protection of citizens' personal information in the context of big data.

2 Research on the Civil Law Protection of Citizens' Personal Information in the Context of Big Data

2.1 Personal Information Crisis in the Big Data Environment

(1) The economic benefits of big data are becoming more and more obvious. In the past society, the meaning of the existence of information is to recognize the information subject represented by the information. The commercial value and use value of personal information have not been developed and valued. Therefore, there is no collective or individual collection of personal information on a large scale. In the era of big data, data, as a basic resource that is increasingly valued by Internet companies and even companies in the traditional secondary industry, plays an increasingly important role in decision-making and analysis.

(2) In the era of big data, the popularity of the Internet and the rapid development of the 4G era are about to enter the 5G era. The Internet gradually occupies a large part of individuals regardless of work, life and leisure. The Internet is in this way for everyone. The circumstances are closely related. When individuals are engaged in network activities, all kinds of personal information are generated all the time, which reduces the difficulty and cost of collecting information to a certain extent.

(3) Because of the highly developed technology, the collection of information will not be easily known by others. For example, using the user's cookie to collect information, the user's shopping on Taobao and other websites leaking transaction information, etc. What's even more frightening is that unless the information stolen by this kind of high-tech technology is discovered by the store of the stored information and published, otherwise it will be difficult for ordinary people to know that their personal information has been used by other people.

(4) In the era of big data, because enterprises in the entire market have a great demand for personal information, the risk of information leakage is greatly increased, and because of the concealment of information collection, compared with the information collection in the traditional era, the information subject does not. If you don't know that personal information is collected by others, so you don't care

about the protection of personal information, so you won't have a strong sense of self-protection.

2.2 My Country's Problems in the Personal Information Protection Law

(1) There is no clear personal information right
At present, from the perspective of specific judicial practice, the protection of citizens' personal information rights is basically based on the legal provisions of privacy rights. This kind of protection is basically indirect protection, while distinguishing personal information rights. There are no clear regulations regarding privacy. At the same time, the benefits of personal information have not yet risen to the level of personal information rights. At the same time, there have not been clear regulations on the attributes and content of personal information properties, which may easily cause difficulties in rights protection.

(2) There is no special legislation for personal information
My country's existing laws and regulations also include the protection of personal information, but these regulations are not direct and are scattered in some laws and regulations. There is no systematic special legislation for the protection of personal information. Generally speaking, we can only find a basis for the protection of human rights and human dignity, and there is no special mention of the protection of personal information on the Internet, and the effectiveness of the existing regulations is low. All laws exist in terms of legislative purposes and legislative techniques. Differences, there are still legal conflicts between each other, and there is a mismatch with the important status of personal information, and because of this problem, it is impossible to effectively protect personal information.

(3) Lack of a dedicated department in charge of personal information protection
From a global perspective, many countries, regions or international organizations have become specialized protection agencies in order to further enhance the protection of personal information. When infringement of citizens' personal information occurs, these institutions are responsible for handling. For example, in Germany, a personal data protection committee is set up to supervise the handling of personal information by public agencies. At present, our country has not established a dedicated personal information protection department, and these deficiencies have more or less impact on the implementation of personal information protection measures.

(4) Difficulties in legal relief
At present, a large amount of personal information is embodied in the form of electronic, and the storage, transmission, and processing are basically carried out through the network platform. The corresponding electronic evidence and network evidence have relatively difficult problems. In addition, the infringement of personal information itself involves professional and technical knowledge, as well as the problem of information asymmetry between the personal information subject and the information manager, which leads to an increase in the difficulty of proof, which will bring greater protection to the information subject's rights burden.

(5) Lack of special legislation and related systems
The protection of personal information in our country only has legislative protection in individual fields, and there is a lack of unified and specialized legislation to

protect and regulate personal information. Specific to the network field, it is also a mess. In terms of legislation, personal information protection in the network field shows problems such as scattered legislation, lack of systems, and lack of necessary coordination and cooperation between various departments. There are no specific corresponding explanations and regulations on fundamental issues such as the scope of use of network information, the field of use, whether the use process is reasonable, whether it is infringing, and how to investigate after improper use, only limited protection is provided.

2.3 Suggestions on Improving the Information Security Protection Law

(1) Establish the legal status of personal information rights
The right of personal information is a basic civil right, derived from the dignity of the personality. Its essential attribute is the right of personality, which is different from the ordinary right of personality. It is a new type of right of personality. The "Personal Information Protection Law" should determine the legal status of personal information rights. From the perspective of the information subject, it has the right to require the person in charge to stop infringement, and it also has the right to demand compensation for its own losses. Losses also include compensation for mental damage.
(2) Improve legal awareness and knowledge of netizens
Strengthen legal propaganda and guidance, strengthen citizens' personal information rights awareness and legal awareness, and let citizens know their information rights and how to protect personal information, do not fill in or leak personal information at will, and protect personal account information and passwords on websites, software, etc. When discovering that your personal information has been leaked, you must maintain your rights in accordance with the law in a timely manner, take remedial measures, and seek relief in a timely manner. When encountering a criminal act of infringing personal information, it is necessary to report and expose the crime to form a good atmosphere for combating acts of infringing on citizens' personal information.

2.4 Regression Portfolio Estimator

Using the positive correlation between the auxiliary variables and the research variables in the previous two surveys, a binary regression model is established, and the following binary regression estimators are obtained:

$$\overline{y'_{2m}} = \overline{y_{2m}} + \beta_{1,2}(\overline{y_1} - \overline{y_{1m}}) + \beta_{2yx}(\overline{X_2} - \overline{X_{2m}}) \quad (1)$$

The regression portfolio estimator about the unknown population mean θ_2 can be constructed as.

$$\widehat{\theta_2} = \varphi_2\overline{y'_{2m}} + (1 - \varphi_2)\overline{y'_{2m}} \quad (2)$$

3 Experimental Research on Civil Law Protection of Citizens' Personal Information in the Context of Big Data

3.1 Subjects

(1) To investigate the leakage of citizen information. This time, two age groups of middle-aged and young people in a certain place were investigated to ensure the scientific validity of the experimental data. This time, a questionnaire survey method was adopted, and a total of 50 people were surveyed, of which 25 people were equally young and middle-aged. This questionnaire survey uses a semi-closed form, the purpose of which is to ensure the accuracy of the experimental data.
(2) In order to investigate the protection of citizens' personal information by existing laws in my country, this paper conducted face-to-face interviews with experts and scholars through field surveys, and used regression estimators to analyze the results to ensure the source of experimental data Reliability.

3.2 Research Methods

(1) Document method
This article collects cutting-edge industry data by analyzing the research results of relevant experts and scholars and reading a large number of previous research documents. These data not only provide strong support for the topic selection of this article, but also provide a reference for the experimental results of this article.
(2) Questionnaire survey method
This article uses a semi-closed form to obtain more accurate experimental data by issuing targeted questionnaires to social personnel.
(3) Field research method
Through field surveys, face-to-face interviews with relevant scholars and experts, this paper sorts out and analyzes the collected data, which provides a reliable reference for the experimental results of this article.
(4) Mathematical Statistics
Use relevant software to sort out the experimental data.

4 Experimental Analysis of Civil Law Protection of Citizens' Personal Information in the Context of Big Data

(1) Analysis of Citizen Information Leakage
In this era of explosive growth of data, people's behavior will be recorded as information, and its economic benefits are becoming more and more obvious. In order to investigate the leakage of personal information of Chinese citizens, this experiment conducted a questionnaire survey on young and middle-aged people, and the data obtained are shown in Table 1:

Table 1. Analysis of citizen information leakage

	Merchant service	Personal network	GPS positioning	Others
Youth	7	9	6	3
Middle aged	10	6	7	2

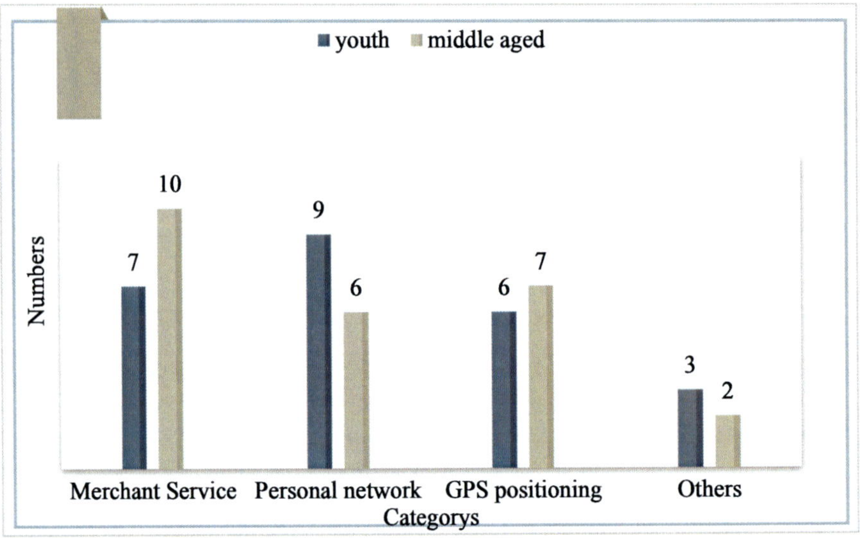

Fig. 1. Analysis of citizen information leakage

It can be seen from Fig. 1 that in the context of big data, people's personal information is being leaked in multiple ways. Among them, young people suffered more information leakage due to personal networks, and middle-aged people suffered more information leakage due to business services. This shows that it is difficult for citizens' information to be protected by law.

(2) Analysis of the protection of citizens' information by existing laws
Although our country's current laws will also have provisions on the protection of citizens' information, these regulations are not direct. They are scattered in the laws and regulations. Legislation on the protection of citizens' information is urgent. This article passes an interview with experts and scholars, and use the ten-point system. The collected data is shown in Table 2:

It can be seen from Fig. 2 that experts and scholars agree that the current laws on the protection of citizens' identity information are not perfect, especially the lack of targeted individual legislative research.

Table 2. Analysis of legal protection of citizen information

	Unclear personal information rights	There is no legislation on the right to personal information	Difficulty in legal aid	Lack of related systems
Man	7	8	7	9
Woman	6	9	8	7

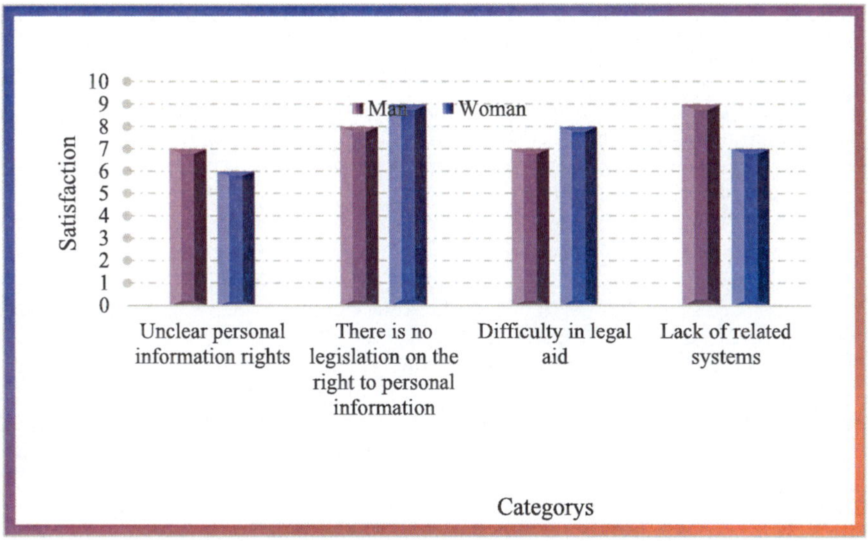

Fig. 2. Analysis of legal protection of citizen information

5 Conclusion

In the context of big data, the huge value of personal information has gradually emerged, and it has a positive impact on economic development, technological progress, and the construction of a harmonious society. If the civil law protection of personal information still adopts a strict protection attitude, it will not be able to meet the needs of the development of the times, and it will not be possible to well tap and use the value contained in personal information. Therefore, it is necessary to adopt unified legislation to recognize the reasonable use of personal information, and to guide it to avoid the risk of possible infringement of personal information rights, so as to promote the use of big data technology to rationally use personal information, which will bring us convenient life and an increasingly humane environment. At the same time, attention should be paid to the improvement of relief methods for victims whose personal information rights have been violated, so as to balance a proper balance between the rational use and protection of personal information.

References

1. Wang, T., Duong, T.D., Chen, C.C.: Intention to disclose personal information via mobile applications: a privacy calculus perspective. Int. J. Inform. Manage. **36**(4), 531–542 (2016)
2. Zhou, Z., Zare, R.N.: Personal information from latent fingerprints using desorption electrospray ionization mass spectrometry and machine learning. Anal. Chem. **89**(2), 1369 (2017)
3. Brown, S.A., Jouni, H., Marroush, T.S., et al.: Disclosing genetic risk for coronary heart disease: attitudes toward personal information in health records. Am. J. Prev. Med. **52**(4), 499–506 (2017)
4. Lin, P.C., Lin, P.Y.: Unintentional and involuntary personal information leakage on facebook from user interactions. KSII Trans. Internet Inf. Syst. **10**(7), 3301–3318 (2016)
5. Peer, E., Acquisti, A.: The impact of reversibility on the decision to disclose personal information. J. Consum. Mark. **33**(6), 428–436 (2016)
6. Allmang, S., Jou, J., Gadoth, A., et al.: Legislative protection from discrimination in access to employer-provided training. Int. J. Train. Dev. **23**(4), 276–290 (2019)
7. Forere, M.A.: The effect of section 43 of the BCEA on employment contracts and legislative protection of minors. De Jure **49**(1), 23–40 (2016)
8. Schroer, S., Huggins, B.J., Azam, C., et al.: Working with inadequate tools: legislative shortcomings in protection against ecological effects of artificial light at night. Sustainability **12**(6), 2551 (2020)
9. Zaharia, M., Xin, R.S., Wendell, P., et al.: Apache spark: a unified engine for big data processing. Commun. ACM **59**(11), 56–65 (2016)
10. Obermeyer, Z., Emanuel, E.J.: Predicting the future - big data, machine learning, and clinical medicine. N. Engl. J. Med. **375**(13), 1216–1219 (2016)
11. Akter, S., Wamba, S.F.: Big data analytics in E-commerce: a systematic review and agenda for future research. Electron. Mark. **26**(2), 173–194 (2016). https://doi.org/10.1007/s12525-016-0219-0
12. Hashem, I., Chang, V., Anuar, N.B., et al.: The role of big data in smart city. Int. J. Inf. Manage. **36**(5), 748–758 (2016)

Development of Network Security Based on Information Technology

Renxian Zeng(✉)

School of Information Engineering, Nanchang Institute of Technology, Nanchang 330099, Jiangxi, China

Abstract. With the development of science and technology, information technology has become more and more developed, and network security has become a new challenge. The times are advancing, the scale of the network is constantly expanding, and the number of network users is also uneven. With this, there are endless network security incidents. The development of network security has attracted more and more attention. At the same time, the accompanying network security issues have brought certain risks to individuals and even countries. While accepting that the Internet brings convenience to life, it also makes it a huge hidden danger. The purpose of this article is to study the development of network security based on information technology. Based on information technology, this article focuses on the development status of network security and the challenges it faces, and then studies the current network information problems, and puts forward effective suggestions. Mainly from the information encryption technology to the main functions of the network information security protection barrier firewall, and the functions that the new generation firewall should have. Then there is the method of maintaining network information security. First, it introduces the principles of maintaining network information security, and secondly, it describes the main methods of maintaining network information security. Finally, some techniques commonly used in hacker attacks are studied, and some available preventive measures are proposed for these attacks. Ensure the normal development of society, maintain network order, and resist the adverse effects of network security issues. The experimental research results show that the number of times that users A, B, and C use Trojan removal tools for protection in the first year are 13, 16, and 15, respectively, and the number of times that users use Trojan removal tools for protection in the third year are 34, 41, and 39, respectively. In summary, the number of protections is increasing year by year, and the development of network security protection technology is also improving year by year. At the same time, network viruses are becoming more and more pervasive, which also indicates that there is still a lot of room for improvement in the development of network security.

Keywords: Internet · Information technology · Network security · Development

1 Introduction

With the development of science and technology, information technology has become more and more developed, and network security has become a new challenge [1, 2].

B. J. Jansen et al. (Eds.): *International Conference on Cognitive based Information Processing and Applications (CIPA 2021)*, LNDECT 85, pp. 321–328, 2022.
https://doi.org/10.1007/978-981-16-5854-9_40

The times are advancing, the scale of the network is also expanding, and the network users are also uneven, followed by endless network security incidents [3, 4]. The development of network security has attracted more and more attention. At the same time, the accompanying network security issues have brought certain risks to individuals and even countries [5, 6]. While accepting that the network brings convenience to life, it also makes it a huge hidden danger [7, 8].

In the research on the development of network security based on information technology, many scholars at home and abroad have conducted research on it, and have achieved good results. Langer pointed out that in the non-traditional security environment, rights are all in terms of system, technology, and norms [9]. Only by improving comprehensive rights can the country's network security be effectively protected [10, 11]. Zhang Y indicated that it is necessary for our country to fully understand and understand the latest international development trend of network security under the overall structure of the central network security and informationization leading group, continuously improve the system and mechanism, and build a network security strategic framework with Chinese characteristics [12].

Based on information technology, this article focuses on the development status of network security and the challenges it faces, and then studies the current network information problems, and puts forward effective suggestions. Mainly from the information encryption technology to the main functions of the network information security protection barrier firewall, and the functions that the new generation firewall should have. Then there is the method of maintaining network information security. First, it introduces the principles of maintaining network information security, and secondly, it describes the main methods of maintaining network information security. Finally, some techniques commonly used in hacker attacks are studied, and some available preventive measures are proposed for these attacks. Ensure the normal development of society, maintain network order, and resist the adverse effects of network security issues.

2 Research on the Development of Network Security Based on Information Technology

2.1 The Status Quo of Network Security Development

(1) Current status of network security
Since the development of network security, network security problems have become more and more serious day by day. In order to meet the needs of users, relevant companies must update information security technologies in real time. Our country's network security research is still at a slightly mature stage. At present, various firewall technologies have been developed to resist various computer viruses. To ensure network security and promote social progress, we still need to work hard to explore and develop it to gradually mature.

(2) Challenges
The first is technical issues. Our country's Internet started late, and technology has developed rapidly. In the process of continuous technological improvement, network security problems are also increasing. It has become a hidden danger that is

difficult to solve in the field of network security and has brought huge challenges to my country's network security. The second is the lack of relevant technical personnel. Network security technical personnel are the guarantee of national network security and also determine the future development prospects of the country's network security.

2.2 Principles of Maintaining Network Information Security

(1) The principle of initiative
Actively construct the network and occupy the propaganda position in the network, and cannot wait to make up for it. We must take the initiative to build our line of defense, but also take the initiative to publicize our ideological theories. In response to the use of the Internet to spread rumors, take the initiative to attack, both offensive and defensive is the basic principle of our work.

(2) The principle of openness
Openness is the basic principle of our country's current socialist ideology endorsed by the times. With the reform and opening up of the political, cultural, and economic global exchanges and cross-border development determining our country's socialist ideology, the world is inherently pluralistic. It is closed and unchanging, but is moving forward and advancing with the times. Maintaining the security of our country's ideology is not blindly exclusive or narrow-minded national prejudice, but the protection of the core and content of our country's ideology from damage on the premise of absorbing and drawing on all advanced ideas and outstanding achievements. All the outstanding national cultural traditions, the historical and practical achievements of civilizations from all countries in the world, including developed capitalist countries, are worthy of our reference. In the information age, no country can hide information in front of the people. First, this approach cannot be done in the pervasive cyberspace. Second, doing so will undoubtedly damage the image and image of a country in front of the people and other countries. Reputation hinders the formation of a country's economic competitiveness in the era of globalization.

(3) Respect the laws of the network
Only when you are familiar with the basic laws of the Internet and respect the rules of the game, can you effectively use the Internet to respect the laws of the Internet. First of all, we must cultivate network technical talents, let the Internet use and serve me. Only scientific and technological talents who have mastered advanced network technology can guarantee the security of our country's network to the greatest extent. Second, it is necessary to cultivate leaders with network awareness. In the Internet age, leaders are required to actively embrace the Internet and build an online ideological position with a positive attitude. How can our current work in the ideological field be truly and effectively integrated with the Internet? To build an online ideological position into a home where people are happy, happy to see, happy to browse, and convinced, requires leaders and ideological construction teams to understand the laws of the Internet and features.

(4) Combination of network construction and offline construction
In the Internet age, although we want to seize the position of the Internet, at the same time we also have to combine with the work offline, and neither should be neglected. Strengthen our country's network laying and connection, increase network technology and network usage frequency, principles, directions and methods, and shorten the "digital divide". And to make use of a set of valuable experiences and practices that have been formed in our country's ideological work for a long time, and combine our previous offline work experience with the new online situation in the information age, so as to formulate a method of maintaining ideology that is in line with our country's characteristics.

2.3 Hacking Techniques

(1) Trojan horse
Use the client to operate the control side to lure the target object to run the server-side program. After the intrusion is successful, a program is hidden in the target object's server. When the target object server starts, the hidden program will also be started, thereby achieving control of the target object the purpose of the computer. It can modify the target object's files, log in to its registry, steal passwords, etc. to achieve remote control. The characteristics of this technology are concealment and special functionality.
(2) Web deception
The technology has a realistic but completely wrong copy, with the same web pages and links. It can monitor and record various operations of users on the webpage. This operation does not need to store the content of the entire web page, only a copy of the site needs to be created.

2.4 Common Prevention Techniques

(1) Prevention of Trojan Horses
First, do not run programs or software from unknown sources. Stop them from the source. After downloading the software, you must first kill the virus to make sure there is no problem in using it. Second, use the detection function of the firewall to find the Trojan horse and delete it. The third is to use the Trojan removal tool, which can find the Trojan horse program, terminate the operation, and then remove it further.
(2) Prevention of Web fraud
First, prohibiting the JavaScript function in the browser will expose all kinds of rewriting information; secondly, confirming the browser's connection status in real time. The above is a short-term solution. If you want to remove it for a long time, you must change the browser so that the browser can truly reflect the URL information.

2.5 Population and Sample

Suppose a finite population contains N population units, and the k-th population unit is identified by k, where the finite population can be expressed as:

$$U = \{1, \cdots, k, \cdots N\} \tag{1}$$

Let y denote a research variable, which is the value of the research variable of the k-th overall unit in the population. Usually before the investigation, the value is unknown. And what we need to estimate is the total value of the research variable y

$$t = \Sigma_{u} y_{k} \tag{2}$$

Or study the overall mean of variable y

$$\overline{y}_{u} = t/N = \Sigma_{u} y_{k}/N \tag{3}$$

3 Experimental Research on the Development of Network Security Based on Information Technology

3.1 Experimental Subjects and Methods

This experiment conducts research from two aspects of network security intrusion and protection. With network users as the research object, the number of intrusions and protections of users is investigated through the method of questionnaire survey. Analyze the actual situation of the development of network security.

3.2 Data Collection

Investigation task decomposition makes a detailed breakdown of the total amount of investigation activities, assigns investigation tasks to investigators or investigation teams, and assigns tasks equally by default. The survey task is broken down into specific task assignments, such as how many answers each investigator or survey team needs to collect, the time range for collecting the answer sheets, and the location of the survey. The survey tasks can also be manually assigned. If the system automatically divides them into an inappropriate way, you can modify the assigned tasks.

4 Based on the Experimental Research Analysis of the Development of Network Security Under the Information Technology

4.1 Analysis of Network Security Intrusion Technology

This experiment takes network security intrusion technology as the main research object, analyzes the type and number of virus intrusions on users' computers in the past three years, and provides a research basis for the development of network security. The experimental results are shown in Table 1:

Table 1. Computer virus intrusion analysis

		User A	User B	User C
The first year	Trojan Horse	15	25	19
	Web deception	24	23	26
The second year	Trojan Horse	12	20	18
	Web deception	21	16	16
The third year	Trojan Horse	14	17	9
	Web deception	16	13	11

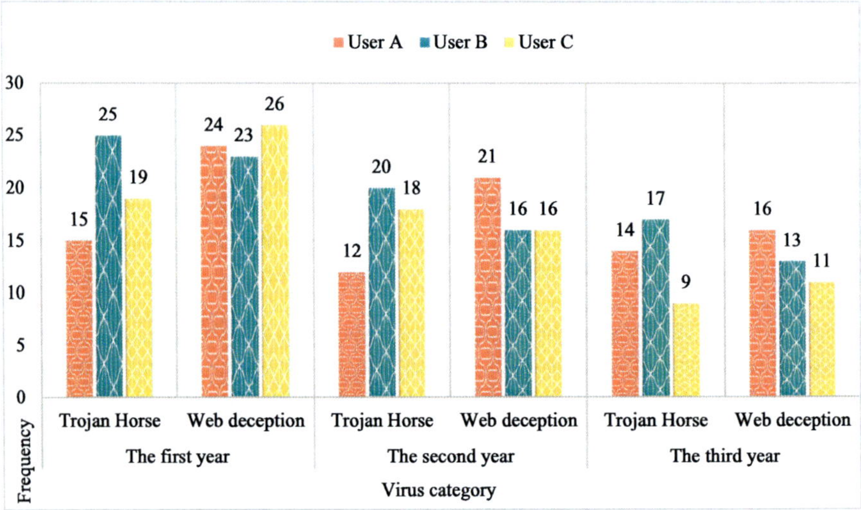

Fig. 1. Computer virus intrusion analysis

As shown in Fig. 1, the general trend shows that virus intrusions are declining year by year, especially in Web deception intrusion technology. User A, User B, and User C were 24 times, 23 times, and 26 times in the first year, respectively, and in the second year. They were 21 times, 16 times, and 16 times. In the third year, they were 16 times, 13 times, and 11 times.

4.2 Analysis of Network Security Protection Technology

This experiment takes network security protection technology as the main research object, and analyzes the number of times the user's computer has been protected against virus intrusion in the past three years. The experimental research results are shown in Table 2:

Table 2. Analysis of computer protection times

		User A	User B	User C
The first year	Firewall	27	25	31
	Trojan removal tool	13	16	15
The second year	Firewall	33	37	41
	Trojan removal tool	25	28	30
The third year	Firewall	43	39	47
	Trojan removal tool	34	41	39

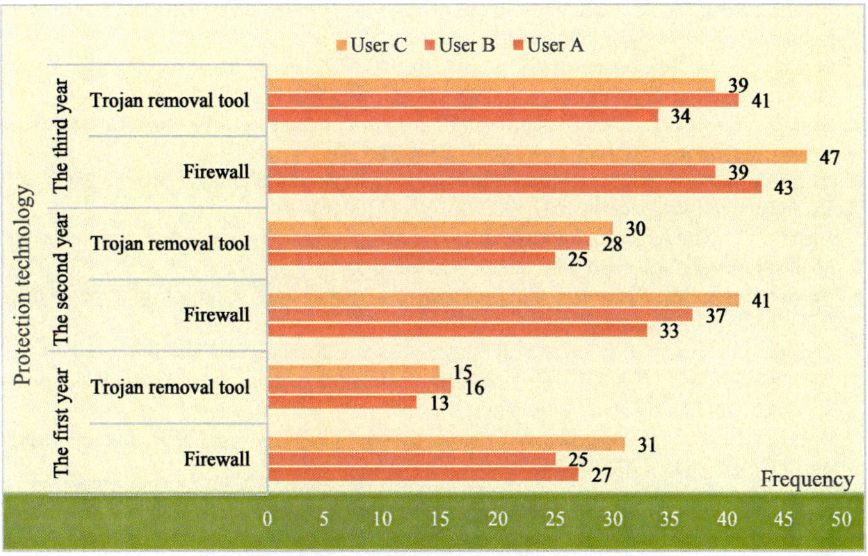

Fig. 2. Analysis of computer protection times

As shown in Fig. 2, the number of times users A, B, and C used Trojan removal tools for protection in the first year were 13, 16, and 15, respectively, and the number of times users A, B, and C used Trojan removal tools for protection in the third year were 34, 41, and 41, respectively. 39. In summary, the number of protections is increasing year by year, and the development of network security protection technology is also improving year by year. At the same time, network viruses are becoming more and more pervasive, which also indicates that there is still a lot of room for improvement in the development of network security.

5 Conclusion

With the expansion of computer network services, the network structure is becoming more and more complex, and the control and management of the network has become a focus of research. Conduct research on network security issues, put forward countermeasures for network security issues, effectively analyze the development status of network security, and better implement optimal network management. While the information network era brings opportunities to society, it also brings huge challenges. Strengthen the protection of network security from all aspects, start with security awareness, technology research and development, laws and regulations, etc., to establish a complete network security guarantee mechanism.

References

1. Liao, X., Song, Y.: Research on furniture design system based on big data and information technology. J. Phys. Conf. Ser. **1744**(3), 032025 (2021)
2. Mentsiev, A.U., Magomadov, V.S., Ashakhanova, M.Z., et al.: How the development of blockchain affected cybersecurity. J. Phys.: Conf. Ser. **1399**(3), 033048 (2019). (4pp)
3. Kilji, A.: Cybersecurity and remote working: Croatia's (non-)response to increased cyber threats. Int. Cybersecur. Law Rev. **1**(1–2), 51–61 (2020)
4. Horowitz, B.M., Lucero, D.S.: System-aware cyber security: a systems engineering approach for enhancing cyber security. Insight **19**(2), 39–42 (2016)
5. Haass, J.C., Sampigethaya, K., Capezzuto, V.: Aviation and cybersecurity: opportunities for applied research. Tr News (304), 39–43, 45 (2016)
6. Chronopoulos, M., Panaousis, E., Grossklags, J.: An options approach to cybersecurity investment. IEEE Access **PP**(99), 1 (2017)
7. Ferrara, A.: Cybersecurity in medical imaging. Radiol. Technol. **90**(6), 563–575 (2019)
8. Zare, H., Zare, M.J., Azadi, M.: Cybersecurity vulnerabilities assessment (a systematic review approach). In: Latifi, S. (ed.) Information Technology - New Generations. Advances in Intelligent Systems and Computing, vol. 738, pp. 61–68. Springer, Cham (2018). https://doi.org/10.1007/978-3-319-77028-4_10
9. Langer, S.G.: Cyber-security issues in healthcare information technology. J. Digit. Imaging: Off. J. Soc. Comput. Appl. Radiol. **30**(1), 117–125 (2017)
10. Wang, X.L.: Criminal law protection of cybersecurity considering AI-based cybercrime. J. Phys. Conf. Ser. **1533**(3), 032014 (2020). (5pp)
11. Brilingaite, A., Bukauskas, L., Juozapavicus, A.: A framework for competence development and assessment in hybrid cybersecurity exercises. Comput. Secur. **88**(Jan.), 101607.1–101607.13 (2020)
12. Zhang, Y.: Research on the security protection of E-commerce information under the background of mobile consumption. Int. J. Netw. Secur. **20**(6), 1215–1220 (2018)

Status Quo of TCM Tumor Diagnosis and Treatment Based on Literature Analysis

Liang Han[1,2](✉)

[1] Henan Province Hospital of Traditional Chinese Medicine, Zhengzhou 450002, Henan, China
[2] The Second Affiliated Hospital of Henan University of Traditional Chinese Medicine, Zhengzhou 450002, Henan, China

Abstract. The document analysis method mainly refers to the method of collecting, identifying, and collating documents, and forming a scientific understanding of facts through the analysis of the documents. Traditional Chinese medicine understands and classifies the cause of tumors under the guidance of the overall concept. Although modern medicine is the main treatment method for tumors, it still has many shortcomings. With the deepening of TCM's understanding of tumors, it was discovered that TCM decoctions combined with treatment can increase efficacy and reduce toxicity. The ingredients in them play an important role in improving the body's immune function and improving the diagnosis and treatment of TCM tumors. Based on this, this article has launched an analysis and research on the status quo of diagnosis and treatment of TCM tumors based on literature analysis methods. In this paper, we systematically searched the related literatures of TCM treatment of tumors, used TCM literature analysis methods, used prescriptions as clues, explored the law of TCM diagnosis and treatment of tumors, and summarized the similarities and differences between ancient and modern diagnosis and treatment of tumors. This article mainly uses the literature analysis method to summarize and analyze the domestic literature on the diagnosis and treatment of TCM tumors, to understand the development trends of Chinese scholars on TCM, and to have a clearer understanding of the status quo and development of TCM diagnosis and treatment of tumors in my country. Problems in the process, and propose corresponding solutions based on the problems found. In this paper, by consulting the literature, we will sort out the tumor's pathogenic factors, pathogenesis, and the current status of Chinese medicine treatment, and systematically summarize and summarize the modern experimental basic research to provide a basis for the diagnosis and treatment of tumors. The results of literature analysis show that the total frequency of medication is 1021 times, the relative frequency is 1021 88%, indicating that these 8 types of traditional Chinese medicine occupies a dominant position in ancient prescriptions. It can be seen that tonic drugs are the core drugs used in modern Chinese medicine in the treatment of tumors.

Keywords: Literature analysis method · Traditional Chinese medicine · Diagnosis of Tumor · Treatment of tumor

B. J. Jansen et al. (Eds.): *International Conference on Cognitive based Information Processing and Applications (CIPA 2021)*, LNDECT 85, pp. 329–336, 2022.
https://doi.org/10.1007/978-981-16-5854-9_41

1 Introduction

Tumors are new organisms formed by the long-term synergy of internal and external tumor-causing factors of the body's cells that lead to mutations at the gene level and abnormal function regulation, which causes the cells to continue to proliferate and lead to transformation [1, 2]. Tumors are classified into benign and malignant. Cancer is the general term for malignant tumors [3, 4]. Although modern medicine is the main treatment method for tumors, it still has many shortcomings. For example, studies have shown that chemotherapy can induce diabetes, cause nausea and vomiting in the gastrointestinal tract [5, 6]. With the continuous deepening of Chinese medicine's understanding of tumors, it has been found that Chinese herbal decoctions combined with treatment can increase efficacy and reduce toxicity, and its components play an important role in improving the body's immune function, improving microcirculation, and inducing tumor cell differentiation and apoptosis [7, 8].

Many experts and scholars have done in-depth research on the diagnosis and treatment status of Chinese medicine tumors based on literature analysis methods, and they have achieved good practical results. For example, the Yang WP study found that in the process of treating tumors, a new Chinese medicine treatment system was established to provide free treatment services. However, as residents' requirements for medical services and the price of medicines increase, the government's financial support is insufficient and Problems such as low efficiency [9]. Han L pointed out that although a tumor treatment system with corresponding specifications has been established in China, the lack of incentive measures such as salary incentives and performance evaluation has greatly affected the improvement of the level of tumor treatment technology and led to an outflow of high-quality medical talents [10].

In order to better carry out the research, this article firstly provides a brief overview of the current status of TCM tumor diagnosis and treatment based on literature analysis methods, and compares and analyzes the use of TCM for different types of tumors based on literature. Secondly, through the analysis of literature based on the use of prescription drugs for tumors, we can understand the current status of the diagnosis and treatment of TCM tumors and the actual problems faced in the application. For this reason, in order to verify the effectiveness of the measures in this article, we have investigated the data was analyzed in depth.

2 Analysis and Research on the Status Quo of Diagnosis and Treatment of TCM Tumors Based on Literature Analysis Methods

2.1 Comparative Analysis of the Use of Traditional Chinese Medicine for Different Types of Tumors Based on Literature

(1) Frequency analysis of traditional Chinese medicine. Licorice, ginseng, tuckahoe, tangerine peel, pinellia, atractylodes, astragalus, and angelica rank the top ten in the frequency of use of traditional Chinese medicine for tumors. Comparing different types of tumors, the frequency of use of these drugs is still in the top 10, reflecting the

treatment of traditional Chinese medicine. The thought of syndrome differentiation and treatment also embodies the thought of treating different diseases in traditional Chinese medicine at the same time.

(2) Complex network analysis. Combining the results of the statistical analysis of syndrome types, the composition of the core drugs illustrates the combination of syndrome differentiation and disease treatment in TCM. According to the clinical symptoms, if the common clinical symptoms of lung cancer are cough or blood in sputum, add phlegm and cough medicine; colorectal cancer is common with constipation, drenching turbidity, etc., add diuresis and dampness medicine; breast cancer is common with liver qi, uncomfortable for lumps, add Bupleurum and other Qi regulating drugs.

(3) Analysis of Chinese patent medicines. Among the injections used by patients with lung cancer and colorectal cancer, Kangai injection is used the most frequently, and among the injections used by breast cancer patients, Shenqi Fuzheng injection is used the most frequently. The most frequently used oral preparations for patients with lung cancer, colorectal cancer, and breast cancer are Yincaoqingfei Oral Liquid, Kangfuxin Liquid, and Diyushengbai Tablets. Diyu Shengbai tablets are mainly used for leukopenia syndrome, suggesting that leukopenia is more common.

2.2 Analysis of the Usage of Chinese Medicine Prescriptions Based on Literature Analysis

In the literature collected in this article on the prescription of traditional Chinese medicine for oncology, each prescription uses at least 4 traditional Chinese medicines. It can be seen that there are many prescriptions for cancer treatment, and the amount of medicine used is also large. Because cancer patients have more complicated diseases, it is necessary to use drugs with concurrent certificates and minor concurrent certificates in the prescription, so the medicines have more flavors and the dosage is also larger.

(1) Analysis of the frequency of traditional Chinese medicine: the total frequency of use reached 50,100 times, and the frequency of use of 31 medicines exceeded 400 times. The top 3 traditional Chinese medicines are: licorice, ginseng and atractylodes. Licorice is the most frequently used, and it is a product that harmonizes various medicines. Poria has the effect on the lung meridian, diuresis and dampness, and invigorates the spleen. The extracts of Pinellia ternata dry dampness, relieve phlegm, relieve vomiting in the stomach, and its extracts can effectively fight tumors in terms of cytotoxicity, reversal of drug resistance, and tumor cell apoptosis.

(2) Classification analysis of traditional Chinese medicine. The top 3 types of drugs are: tonic drugs, heat-clearing drugs, and phlegm-relieving cough and asthma-relieving drugs. Among them, the tonic drugs are much higher than the yin, yang and blood tonic drugs. The frequency of use of antipyretic drugs ranks second. It is suggested that in the treatment of tumors, the main therapies are to strengthen the body and invigorate qi, clear away heat and detoxify. Combined with the analysis of syndrome types, the syndromes of deficiency of qi and blood and mutual congestion of phlegm and blood stasis are more common, indicating that it is more reasonable.

(3) Analysis of Chinese patent medicines. The composition of Kanglaite injection is coix seed oil for injection. Coix seed is a diuretic and dampening medicine, which is beneficial to water and swelling, invigorating the spleen and dehumidification. Pharmacological experiments show that coix seed oil can enhance immune function, and the anti-tumor mechanism is mainly constant. Tumor cell metastasis and proliferation, induce tumor cell apoptosis, etc.

2.3 Document Analysis Algorithm Formula

The main idea to realize the vector expression of the object to be studied is to use the relationship between the MESH article and the gene article to establish the relationship between the MESH gene. In this way, the vector expression of the gene g by the MESH subject word can be realized. According to the described process, the weight of each feature item is calculated in turn, and the vector expression result of the entity object is obtained. The mathematical calculation process is as follows: For genes, the weight corresponding to the m-th MESH feature item is expressed as shown in formula (1):

$$w_{gm} = T_{gm} \times IDF_m = T_{gm} \times \log\left(\frac{N}{n_m}\right) \tag{1}$$

The formula for calculating the degree of correlation is shown in the following formula (2):

$$T_{gm} = \frac{\sum_{i=1}^{k} d_i}{\sum_{j}^{c} d_j \times n_{gi}} (k \le c) \tag{2}$$

From the previous step, the vector expression results of each MESH feature item corresponding to different entities can be obtained, and each entity corresponds to a vector. The calculation formula is shown in the following formula (3):

$$association = cos\left[W_g, W_t\right] = \frac{\sum_{i=1}^{M} w_{ti} w_{gi}}{\sqrt{\sum_{i=1}^{M} w_{ti}^2 \times \sum_{i=1}^{M} w_{gi}^2}} \tag{3}$$

3 Experimental Research on the Diagnosis and Treatment Status of TCM Tumors Based on Literature Analysis Methods

3.1 Research Materials and Experimental Design

This article included 26 articles on lung cancer, 44 articles on colorectal cancer, and 15 articles on breast cancer. The papers included in this study all mention random methods. Four papers in lung cancer describe the specific random methods used, and 21 papers in colorectal cancer describe the random methods used. None of the included literatures described allocation concealment. The lung cancer, colorectal cancer, and breast cancer literatures described the follow-up in 2 articles, 3 articles, and 2 articles respectively. The frequency of using self-made prescriptions of Chinese medicine decoctions in the control group of the three disease types is the highest.

3.2 Analysis Method and Evaluation Content

Use spss21.0 to analyze the results of chi-square and rank sum surveys on personal gender, age, country, national history and other related data. For the preparations of Chinese patent medicines, the types of Chinese medicine decoctions, the frequent use and the use status of clear differentiation of symptoms, a specific analysis was made. The actual application of traditional Chinese medicines with relatively high frequency of use was analyzed, and various single medicines and the frequency of use of their functions were classified, and some high-frequency medicines were obtained. A complicated network analysis was carried out using MATLAB. Gephi 0.8.2 constructed a visual network diagram.

4 Research and Analysis of the Diagnosis and Treatment Status of TCM Tumors Based on Literature Analysis Methods

4.1 Comparative Study and Analysis of Ancient and Modern Drug Use Rules Based on Literature Analysis Methods

This article uses the literature analysis method to investigate the medication frequency of various drugs of Gujin Recipe. The comparison data of the medication frequency of various drugs of Gujin Recipe is shown in Table 1 below:

Table 1. Comparison of the frequency of various drugs used in ancient and modern prescriptions

Drug	Previous prescription (%)	Current prescription (%)
Gentle veins, digestion	7.5	2.3
Strengthen the body	18	32.1
Desilting	47.8	29.7
Activating blood to remove blood stasis	10.4	12.1
Fight with poison	3	4.8
Detoxification	13.3	19

Through the comparative analysis of various medicines of ancient and modern prescriptions in Fig. 1, it is found that ancient and modern physicians have both similarities and differences in the treatment of tumor prescriptions. The ancient and modern medicines for the treatment of tumors are mainly medicines for removing phlegm and relieving congestion, strengthening the body tonic, and clearing away heat and detoxification. The proportions of the three kinds of medicines are about 80%, the ancient prescription is 79.21%, and the current prescription is 80.82%. The ancient prescription focused on removing phlegm and dispelling masses, but the status of strengthening and replenishing deficiencies in the prescriptions has been significantly improved, which reflects the treatment thinking of both removing phlegm and dispelling masses and reinvigorating deficiencies.

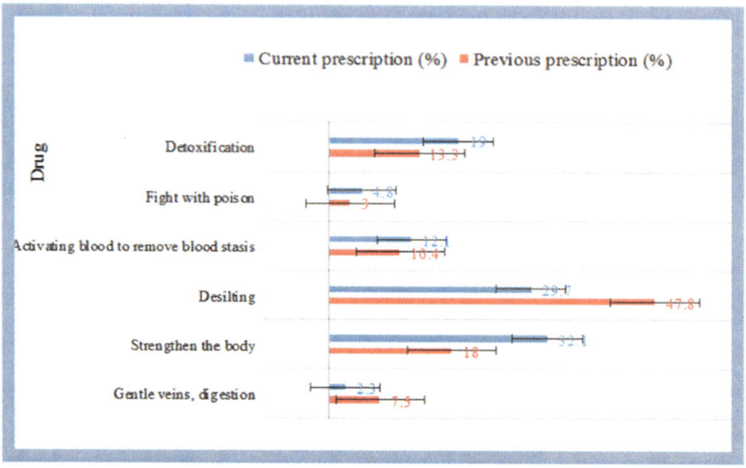

Fig. 1. Comparison of the frequency of various drugs used in ancient and modern prescriptions

4.2 Comparative Analysis of the Relative Frequency of the Classification of Tonic Drugs in Ancient and Modern Prescriptions Based on Literature Analysis Methods

This article uses the literature analysis method to investigate the relative frequency of the classification of the tonic drugs in the ancient and modern prescriptions and makes a statistical survey. The data results of the relative frequency of the classification of the tonic drugs in the ancient and modern prescriptions are shown in Table 2:

Table 2. Comparison of the relative frequency of the classification of tonic drugs in ancient and modern prescriptions

Drug	Previous prescription (%)	Current prescription (%)
Qi Replenishing Medicine	41.5	67.7
Blood tonic	43.4	13.1
Yin tonic	6.8	9.8
Tonic	8.3	12.4

According to Fig. 2 above, it can be seen that the frequency of modern physicians using Qi-tonifying drugs is much greater than that of ancient physicians, and in the application of blood-tonifying drugs, it is less than that of ancient physicians; the application frequency of Qi-tonifying drugs in ancient prescriptions is not much different from that of blood-tonifying drugs., Which means that the ancients paid attention to the double tonic of qi and blood in the treatment of colorectal cancer; in the present prescription, the frequency of application of qi tonics in tonics is as high as 64.71%, while blood

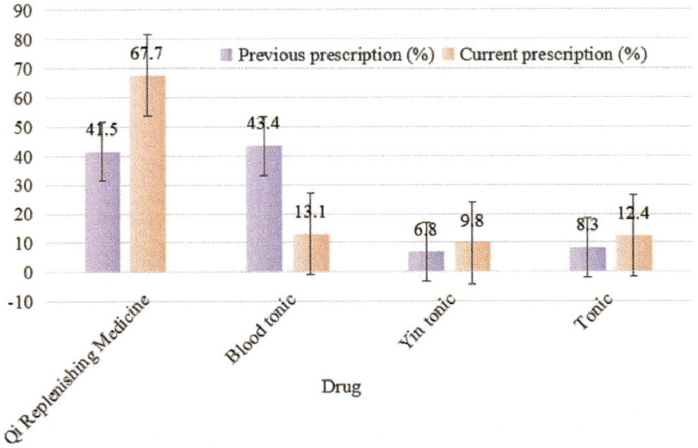

Fig. 2. Comparison of the relative frequency of the classification of tonic drugs in ancient and modern prescriptions

tonics is only 13.17%, and they are used with yin and yang drugs. There is little difference in frequency, which shows that when treating this disease, people nowadays mainly replenish qi, while replenishing blood, replenishing yin and replenishing yang. Taken together, the trend of ancient and modern physicians in applying tonic medicines has gradually shifted from qi and blood tonic to qi, and to supplement blood, yin, and yang at the same time.

5 Conclusions

Traditional Chinese medicine is more and more widely used in the treatment of tumors, but there are also some problems. At present, in the study of pharmacological effects and mechanism, there are more studies on the pharmacological effects and mechanisms of single medicines, and fewer studies on compound Chinese medicines. Traditional Chinese medicine treatment pays attention to the holistic concept. The decoction is about monarch, minister, adjuvant, and enlightenment. It is not a certain medicine that works. The decoction acts on multiple parts and multiple targets to fully regulate the body and restore the balance of yin and yang. Research on adverse reactions of traditional Chinese medicines is also rare, and basic research needs to be strengthened. In addition to decoctions and proprietary Chinese medicine preparations, there are few studies on the use of acupuncture and moxibustion to treat tumors in traditional Chinese medicine. In the future, with the improvement of scientific research and clinical level, Chinese medicine can play a more important role in the diagnosis and treatment of tumors.

References

1. Tang, Y., Li, Z.: Research of insomnia on traditional Chinese medicine diagnosis and treatment based on machine learning. Chin. Med. **16**(1), 46–47 (2021)

2. Ye, H.L., Fu, F.Y.: Study on TCM syndrome types and syndromes of knee osteoarthritis based on modern literature. J. Hainan Med. Univ. (Engl. Edit.) **000**(001), 45–51 (2020)
3. Jiang, Y.: Analysis on status quo and influence factors of key staff's job burnout in TCM hospital. Chin. Hosp. **023**(007), 42–43 (2019)
4. Li, H.M., Liao, X.: Analysis of status quo of RCT outcome indexes in treatment of acute ischemic stroke with traditional Chinese medicine. China J. Chin. Mater. Med. **45**(9), 2210–2220 (2020)
5. Ferlay, J., Soerjomataram, I., Dikshit, R.: Cancer incidence and mortality worldwide: sources, methods and major patterns in GLOBOCAN2012. Int. J. Cancer **136**(5), 359–386 (2015)
6. Fishchenko, I.V., Kravchuk, L.D., Perepechay, O.A.: Lumbar spinal stenosis: symptoms, diagnosis and treatment (meta-analysis of literature data). Pain Med. **3**(1), 18–32 (2018)
7. Zhang, R.: Research on the status quo of Chinese academic evaluation based on co-word analysis. J. Qiqihar Univ. (Philos. Soc. Sci. Edit.) **000**(005), 144–147 (2017)
8. Ha, Y., Huang, P.: A systematic review and meta-analysis on a disease in TCM: astragalus injection for gathering qi depression. Evid. Based Complement. Altern. Med. **2020**(3), 1–10 (2020)
9. Yang, W.P.: Analysis of ethnic minority education policy research of present situation and its problemsr—based on literature metrology and knowledge map. Contemp. Educ. Theory Pract. **010**(004), 79–87 (2018)
10. Han, L.: Status quo of clinical nurse specialists and inspiration for specialization in TCM nursing. J. Nurs. **032**(011), 93–97 (2017)

Use the TCM Diagnosis and Treatment Information Mining Platform to Carry Out the Diagnosis and Treatment of TCM Liver, Gallbladder, Spleen and Stomach Diseases

Qianqian Su[1,2](✉)

[1] Henan Province Hospital of Traditional Chinese Medicine, Zhengzhou 450002, Henan, China
[2] The Second Affiliated Hospital of Henan University of Traditional Chinese Medicine, Zhengzhou 450002, Henan, China

Abstract. With the acceleration of people's life rhythms, people are under greater and greater pressure, making people more and more prevalence of liver, gallbladder, spleen and stomach, which seriously threatens people's health, and in the treatment of liver, gallbladder, spleen and stomach diseases, most people choose TCM treatment because the side effects of TCM treatment of diseases are small, while in the treatment of Western medicine, the side effects of Western medicine are very large. With the development of information technology, information platform data mining technology is applied to the diagnosis and treatment of Chinese medicine. This article is based on the diagnosis and treatment information platform of Chinese medicine, focusing on data mining, information technology, statistics, evidence-based medicine, and integrated methods of disease and syndrome. Complex diseases and other related analyses provide scientific basis for the prevention and clinical treatment of liver, gallbladder, spleen and stomach diseases. The results show that the chance of dry mouth in traditional Chinese medicine is gastritis. 67% of the patients are gastritis, and 42% of the patients with dull gastric pain are gastritis. Therefore, in the dialectic of gastritis, it is not necessary to ask about other diseases when the dull gastric pain occurs. The medicinal material of Muxiang is mainly used to treat gastritis. Among them, 80% of patients with gastritis join Muxiang for treatment, and about 70% of the other two medicinal materials are added, which may be related to other medicinal properties.

Keywords: Chinese medicine diagnosis and treatment · Information mining · Liver · Gallbladder · Spleen and Stomach · Diagnosis and Treatment

1 Introduction

Today is the era of information explosion, and a large amount of information is flooded [1, 2]. The rapid growth of the amount of data has become a problem that cannot be ignored. What is completely different from 20 years ago is that the data could not be found in the past because the amount of data was too small [3, 4]. Today's era is no

B. J. Jansen et al. (Eds.): *International Conference on Cognitive based Information Processing and Applications (CIPA 2021)*, LNDECT 85, pp. 337–344, 2022.
https://doi.org/10.1007/978-981-16-5854-9_42

longer the lack of data, but the need to selectively extract valuable data from a large amount of data and make decisions and important information [5, 6]. The application of data mining technology in the diagnosis and treatment of liver, gallbladder, spleen and stomach in Chinese medicine can solve many intractable diseases [7, 8].

In the diagnosis and treatment of TCM diagnosis and treatment of liver, gallbladder, spleen and stomach diseases using the TCM diagnosis and treatment information mining platform, many scholars have conducted research on them, and achieved good results. For example, Ahmed Z aimed at the ambiguity of pulse status and the determination of pulse status by traditional Chinese medicine. Thinking method, the artificial neural network method is applied to the recognition of the pulse state of the human body. Using the fuzzy processing of the input samples, the self-customized algorithm that adds momentum to the algorithm is improved [9]. Laval combines data mining and online Chinese medicine formula data analysis technology, through the Internet processing system, using fast-developing network communication technology to perform higher-level data processing of information [10].

This article is based on the TCM diagnosis and treatment information platform, which mainly integrates data mining, information technology, statistics, evidence-based medicine and the combination of disease and syndrome methods. Based on the diagnosis and treatment data of a large number of patients in a large general Chinese hospital, it mines the diagnosis and treatment information of patients with liver, gallbladder and spleen, correlation analysis of combined diseases provides a scientific basis for the prevention and clinical treatment of liver, gallbladder, spleen and stomach diseases.

2 Research on Diagnosis and Treatment of TCM Diagnosis and Treatment Information and TCM Diagnosis and Treatment of Liver, Gallbladder, Spleen and Stomach Diseases

2.1 Research Method

(1) Literature research method
The main purpose of the literature research method is to broadly refer to that scholars can obtain research data through various methods such as various books, newspapers, magazines, and electronic books that they need to consult, and generate scientific research ideas and inspirations. Its greatest advantage is that we can understand the historical changes and development history of the object to be studied from our own sources, and understand the historical changes of the object to be studied. By comparing with related materials, we can prompt us to have a more comprehensive understanding of the scope and objects of the research.
(2) Quantitative analysis
Qualitative analysis is related to quantitative analysis. Quantitative analysis refers to the analysis of mathematical hypothesis determination, data collection, analysis and testing, and it is also quantitative and qualitative.

Qualitative analysis refers to the qualitative analysis of the research object. It refers to the process of conducting research based on subjective understanding and qualitative analysis, through research and bibliographic analysis.

2.2 Main Methods of TCM Diagnosis and Treatment Information Mining

(1) Association rules
If the values of two or more data elements are most likely to be repeated, then there is a specific correlation, and the correlation rules for these data elements can be determined. The purpose of correlation analysis is to discover hidden networks in the database. Usually, you use the two closed values "support" and "reliability" to eliminate these useless connection rules.

(2) Decision tree method
The decision tree is represented in the form of a tree, similar to a flowchart. Generally, a decision tree consists of a root node, a set of internal nodes and some branch nodes. Each internal node represents the test of a specific attribute, each branch represents the test output, each node represents a class, and sometimes different nodes of the leaf can represent the same class.

(3) Neural network
An artificial neural network is a network that is widely connected to a large number of simple processing units. It is used in an intelligent computer system that simulates biological neural networks. Usually a neural network is a group of processing units similar to neurons, and each processing unit is related to body weight. In the learning phase, by adjusting these weights, it is possible to predict the correct class order to enter the labels. Because of the connections between units, learning neural networks is also called connection learning. Neural networks require a large number of parameters, which usually depend mainly on experience, such as network topology.

2.3 Utilizing the Advantages of Chinese Medicine in the Diagnosis and Treatment of Liver, Gallbladder, Spleen and Stomach Diseases

(1) According to the dynamic changes of human health and diseases, TCM treatment should focus on the overall concept. When people feel a little uncomfortable, they can prescribe and keep the body "seeing, smelling, asking and pulse". Dynamic balance eliminates the lesions, effectively achieving the purpose of maintaining health and preventing disease development. However, the treatment of western medicine has been delayed. When the disease has progressed to a certain extent, it is necessary to use some biochemical and medical imaging examinations of western medicine to judge the situation.

(2) TCM treatment focuses on the treatment of the root cause and uses TCM to regulate it. It relies on the normal balance of the individual to mobilize the body's own immune function, so as to achieve the goal of "people-centered" Yin and Yang balance and internal organ harmony in medicine and health care. However, treatment with western medicine focuses on the disease, ignores the cause of the disease, and often treats the symptoms rather than the root cause.

(3) In addition to the prevention and treatment of diseases through Chinese herbal medicine, Chinese medicine also uses acupuncture, massage, and other diagnostic and therapeutic techniques and methods to focus on regulating the overall functions of the body. It has reliable effects and low toxicity and side effects.

2.4 The Necessity of TCM Diagnosis and Treatment Information Mining for the Diagnosis and Treatment of TCM Liver, Gallbladder, Spleen and Stomach Diseases

(1) TCM methodology has significant legacy features. After thousands of years of theoretical research and clinical practice, Chinese medicine has accumulated a lot of information. In the past, the theory and experience of traditional Chinese medicine were mainly maintained in two ways. One was to use literature and cultural relics as the carrier; the other was to use the teacher's continuous guidance as the carrier. The bibliography represented by ancient medical books is an important way for future generations to learn from the experience of famous doctors. Due to age, it is difficult to collect these documents. Although they have been passed down through generations, they are only simple text summaries that can reveal the medication experience and rules of famous and senior Chinese doctors, and the amount of information we can use is still very small. Although the traditional "teacher and apprentice" method has played an important role in the inheritance of the famous and ancient Chinese medicine experience, the quantity alone is far from meeting the needs of the society, and it is often limited to the experience and skills of the famous and traditional Chinese medicine. Ancient Chinese medicine is difficult to reflect the overall technical and disciplinary level, and there are limitations.

(2) With the idea of data mining proposed in this century, its economic value has gradually emerged and has been gradually realized in many fields. Data mining technology can handle a large amount of chaotic data, find patterns and laws in a large amount of data, summarize default logical relationships, find equivalence rules, and conduct predictive research to develop data analysis and research topics. Our country has not yet established a complete data mining system for traditional Chinese medicine. Existing databases can only provide conventional services such as retrieval and statistics, and a large amount of indirect knowledge has not been excavated and used. Data mining technology and ancient Chinese medicine have a lot in common in the ideas and methods of identifying things. Data mining technology can discover hidden previously unknown but potentially useful information from a large amount of information accumulated in the development of traditional Chinese medicine, clinical syndromes and treatments, prescriptions, drugs and related knowledge. The further inheritance and promotion combined with the experience of TCM theoretical experts has contributed to the further search for innovation in the research of TCM characteristics.

2.5 Calculation Algorithm of Chinese Medicine Diagnosis and Treatment Information Mining

(1) PEP algorithm

The top-down method is used for the entire decision tree, and the number of unsorted samples is compared before and after pruning to determine whether pruning is necessary. Since the algorithm does not require independent pruning, the training samples will be used to create a decision tree and pruning. The pruning results are biased and optimal pruning cannot be achieved. Therefore, a continuity correction based on the binomial distribution is introduced to reduce the difference and increase the reliability and accuracy of future case predictions. That is, when the following formula is satisfied and the subtree T with t as the root node is pruned and replaced by a leaf node, pruning is performed.

$$e'(t) \leq e'(T_t) + S_e(e'(T_t)) \tag{1}$$

Among them

$$e'(\mathrm{t}) = [\mathrm{e}(\mathrm{t}) + \frac{1}{2}] \tag{2}$$

$$e'(T_t) = \sum e(i) + \frac{N_t}{2} \tag{3}$$

3 Investigation on the Diagnosis and Treatment of Liver, Gallbladder, Spleen and Stomach Diseases Based on the TCM Diagnosis and Treatment Information Mining Platform

3.1 Purpose of the Investigation

Based on the diagnosis and treatment data of a large number of patients in large-scale general hospitals, mining information about patients with liver, gallbladder and spleen and stomach, syndrome differentiation, comorbid diseases, etc., provide scientific basis for the prevention and clinical treatment of liver, gallbladder, spleen and stomach diseases.

3.2 Database Construction

The medical record data of patients with liver, gallbladder, spleen and stomach were extracted and diagnosed through the medical record system of the Provincial Hospital of Traditional Chinese Medicine, and the database was constructed using Excel2010.

3.3 Standardization of Data

We first standardize the symptoms, syndromes and prescriptions in the medical records according to the "Standard Disease List", "Chinese Medicine Clinical Diagnosis and Treatment Terminology" and "Chinese Medicine List". The rules here mainly refer to the rules that have the same meaning, show different symptoms and distinguish the syndrome.

3.4 Experimental Design

(1) As far as traditional Chinese medicine is concerned, stagnation of qi and blood stasis is a common syndrome of chronic atrophic gastritis. We examine all medical records by distinguishing the "qi stagnation syndrome in gastric insufficiency" from the history of gastritis to the history of traditional Chinese medicine, and focus on the relationship between the disease and symptoms. Here, we rely on the symptoms and syndrome differentiation in the TCM gastritis medical records to establish data.

(2) According to the medicinal materials corresponding to the syndrome differentiation of "deficiency and qi stagnation" in the TCM gastritis medical record, the data from the frequent item set to the largest frequent item set reflecting the relationship between syndrome differentiation and medicinal materials are obtained.

4 Data Analysis

(1) Diagnosis of liver, gallbladder, spleen and stomach diseases in TCM

This article analyzes some of the symptoms collected in the TCM database to see if the doctor's diagnosis is a gastritis disease. The results are shown in Table 1.

Table 1. Is it a gastritis disease?

	Thin moss White	Stomach prolapse	Dry mouth	Pulse string
belong	20%	42%	67%	26%
not belong	80%	58%	33%	74%

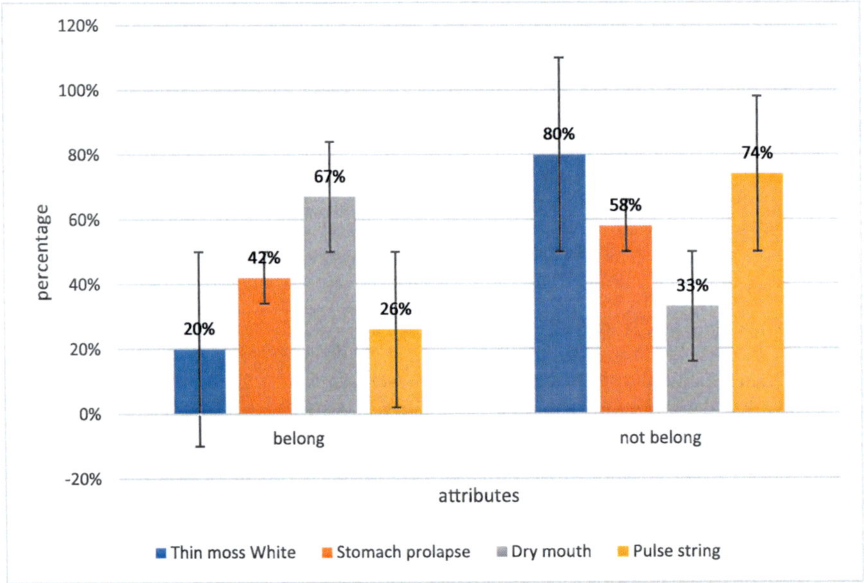

Fig. 1. Is it a gastritis disease?

It can be seen from Fig. 1 that there is a greater chance of gastritis in the case of dry mouth in traditional Chinese medicine. 67% of patients are due to gastritis, and 42% of patients with dull gastric pain are due to gastritis. Therefore, in the dialectic of gastritis, it is not necessary to ask about other diseases when the dull gastric pain occurs.

(1) TCM treatment of liver, gallbladder, spleen and stomach diseases
This article looks at the medicinal materials that appear in the traditional Chinese medicine formulas collected in the TCM Receipt Bank to see if they are the treatment medicinal materials for gastritis. The results are shown in Table 2.

Table 2. Is it a treatment medicinal material for gastritis?

	Taizishen	Fried atractylodes	Wood incense
deficiency and stagnation of qi	67%	70%	80%
other	33%	30%	20%

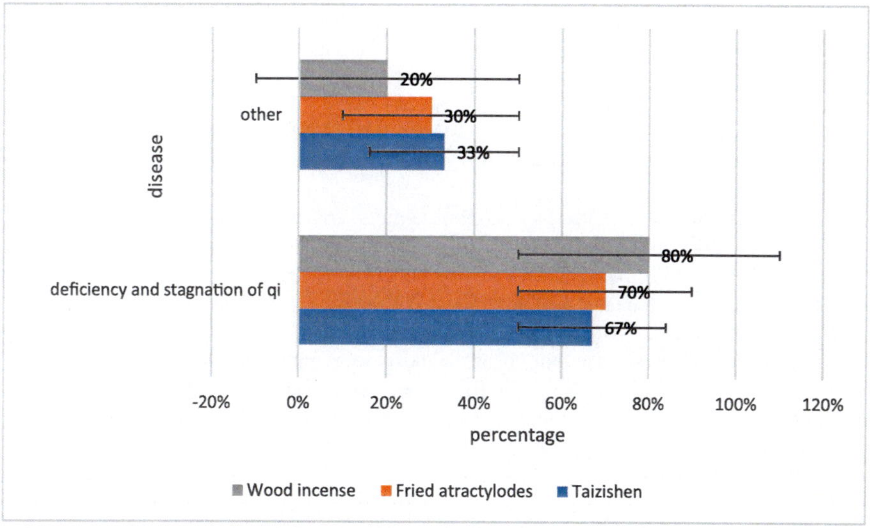

Fig. 2. Is it a treatment medicinal material for gastritis?

It can be seen from Fig. 2 that Muxiang is a medicinal material mainly for the treatment of gastritis. Among them, 80% of patients with gastritis join Muxiang for treatment, and about 70% of the other two medicinal materials are added, which may be related to other medicinal properties.

5 Conclusion

Data mining of Chinese medicine is a very promising but difficult research direction. I believe that anyone who hopes that TCM will continue to develop will support the application of modern science and technology in TCM research, thereby accelerating the information and innovation of TCM knowledge and promoting the development of TCM.

The process of industrialization and internationalization of traditional Chinese medicine. From the diagnosis of traditional Chinese medicine, the identification of syndromes in prescriptions, the process described in words and words accounts for a large proportion, which makes the quality content relatively high. In addition, in the diagnosis process of Chinese medicine, the neglect of many links leads to the lack of certain information, which also leads to changes in the compatibility of prescription drugs, and to a certain extent, causes the uncertainty and incompleteness of Chinese medicine information.

References

1. Chen, W., He, S.: Application of data mining technology in TCM diagnosis and treatment. In: IEEE International Conference on Bioinformatics & Biomedicine. IEEE Computer Society, no. 1, pp. 1350–1353 (2017)
2. Tian, X.P., Zeng, X.F.: Enhance treat-to-target strategy by standardize the diagnosis and treatment of rheumatoid arthritis. Zhonghua nei ke za zhi Chin. J. Intern. Med. **57**(4), 240–241 (2018)
3. Liu, Y., Yu, Z., Yang, Y.: Diabetes risk data mining method based on electronic medical record analysis. J. Healthcare Eng. **2021**(6), 1–11 (2021)
4. Wang, H., et al.: Feature-level attention-guided multitask CNN for fault diagnosis and working conditions identification of rolling bearing. IEEE Trans. Neural Netw. Learn. Syst. **PP**(99), 1–13 (2021). (2)
5. Ip, H.K., Chandran, A., Cranshaw, I., et al.: Multidisciplinary pigmented lesion clinic at Auckland district health board: impacts on melanoma diagnosis and treatment outcomes. N. Z. Med. J. **134**(1530), 30–37 (2021)
6. Yevtushenko, D.O., Taraban, I.A., Minukhin, D.V., et al.: Current questions of diagnosis and treatment of patients with peritoneal adhesive disease. Kharkiv Surg. Sch. **5–6**, 22–25 (2020)
7. Nassereddine, S., et al.: The impact of diagnosis to treatment interval on the outcome of patients with acute myeloid leukemia. Blood **134**(Supplement_1), 2212–2212 (2019)
8. Zhang, Q.: Using TCM to treat gynaecological disorders in accordance with the tradition and methods of Zhang Zhongjing. Chin. Med. Chin. Med. **34**(1), 30–45 (2019)
9. Ahmed, Z., Bujanda, D., Befeler, A.: A case of gastric variceal bleeding: diagnosis and treatment of a rare anatomical variant. J. Gastroenterol. Hepatol. Res. **8**(5), 2990–2993 (2019)
10. Laval University, Canada, et al.: A case of disseminated infection caused by streptococcus equisubspecieszooepidemicus. Can. J. Infect. Dis. Med. Microbiol. **20**(2), 59–61(2016)

Water Quality Security Guarantee and Monitoring System Based on Cloud Technology

Li Wei, Hui Xie(✉), and Hong Lv

School of Software and Communication Engineering, Xiangnan University, Chenzhou 423000, Hunan, China
xiehui@xnu.edu.cn

Abstract. In recent years, with the rapid development of social economy, water quality and safety issues are very common, and they are also a topic of our daily concern. Water resources are an important foundation for social development, and the quality of water sources will have an extremely important impact on human production, life, and economic and social development. China is a severely water-scarce country, and the quality of water will directly affect the development of society and the health and safety of the human body. Therefore, efficient technology is needed to detect water quality to ensure safety and improve the water quality from the source. In addition, it is necessary to use advanced information technology to develop early warning and water quality monitoring systems to analyze and predict water quality changes and early warning of accidents. Based on the network Cloud technology, real-time monitoring is carried out through the water quality security monitoring system, and data sharing is completed through information transmission. This paper analyzes the water quality security monitoring based on the development and application of Cloud technology.

Keywords: Cloud technology · Water quality security · Monitoring system · Efficient work

1 Introduction

Water resources are one of the important basic resources for social development. Whether it is human daily life, daily production or the supply of water resources for economic and social development, water quality issues are dependent on water quantity. Therefore, ensuring the safety of water quality monitoring is the top priority. This technology involves physics, chemistry, biology, information science, hydraulics, etc., and requires interdisciplinary application of multiple disciplines. [1] In recent years, with the rapid development of cloud computing technology, this technology has become one of the important technologies widely used in all walks of life. Water quality safety assurance and monitoring, a work involving multi-disciplinary issues, is indispensable through Cloud technology calculation and analysis. With the widespread application of this technology, it is necessary to study how it affects people's monitoring and protection of water quality and safety issues and industrial development and upgrading.

B. J. Jansen et al. (Eds.): *International Conference on Cognitive based Information Processing and Applications (CIPA 2021)*, LNDECT 85, pp. 345–350, 2022.
https://doi.org/10.1007/978-981-16-5854-9_43

2 The Impact of Cloud Technology on Water Quality Security Monitoring System

With the advancement of social science and technology, people are in-depth research and supervision on the safety of water quality, risks and the impact on human health. [2] The application of cloud technology calculations strictly controls the data management and analysis of each indicator. The application of this technology is also affecting the way relevant companies monitor water quality safety.

2.1 People's Satisfaction with the System Monitoring Water Quality Security Based on Cloud Technology

In order to understand the degree of satisfaction of related industries with the construction of the cloud platform's water quality monitoring system, we conducted a detailed survey of relevant personnel in aquatic and environmental enterprises through questionnaire surveys. Among them, there are 300 valid questionnaires, and the specific survey data are shown in Table 1.

Table 1. Survey of satisfaction with production informatization by fishery and environmental companies

Satisfaction with Informationization of Aquatic and Environmental Companies				
Very satisfied	Relatively satisfied	General	Not so satisfied	Unsatisfied
51%	24%	13%	9%	3%

It can be seen from the above table that related aquatic and environmental companies based on the Cloud technology platform are generally satisfied with the water quality safety monitoring system, more than half are very satisfied, and dissatisfied and dissatisfied only account for 12% of the total number of surveys. It can be seen from the data that most aquatic and environmental companies are using cloud technology platform for development and relying on the construction and application of Cloud technology to promote the efficient and sound development of enterprises.

2.2 Cloud Technology Brings Advantages to Water Quality Safety Guarantee and Monitoring System

The rapid development of cloud computing technology and its scope of application are becoming wider and wider. In addition to related enterprises, relevant people in the government and academia in the industry are constantly paying attention to the development of this technology and vigorously supporting its development. [3] Based on the water quality safety assurance monitoring system developed by Cloud technology, through the design of shared water quality assurance infrastructure, the overall water quality safety assurance data analysis is controlled, and relevant staff can also use cloud technology to monitor the water quality assurance situation at any time. Data analysis will also be accurately uploaded to the cloud, and be analyzed and saved in real time, saving time and space costs. This is also realizing informatization, automation, intelligence, and enabling the system to adapt to ecological changes, and conduct timely and efficient detection of microorganisms and trace elements in the water.

3 The Necessity of Developing Cloud Technology for Water Quality Monitoring

In water quality safety, the basis for ensuring safety monitoring risks is water quality standards and benchmarks, but people cannot conduct comprehensive inspections on every cube. However, the application of Cloud technology can continuously improve the detection efficiency and simultaneously perform multiple water quality detection tasks. It can be said that efficiency is the key to the development of modern society. [4] We use the understanding and application of Cloud technology by relevant companies to continuously adjust and upgrade the water quality safety monitoring system and promote the upgrading of industrial models to illustrate the necessity of the combined development of the two.

3.1 Time for Related Companies to Apply Water Quality Control System Based on Cloud Technology

In addition to the survey of satisfaction with Cloud technology, we also understand how long it should take for related industries to use cloud platform technology to monitor water quality activities. We conducted a detailed survey of relevant personnel in aquatic and environmental enterprises in the form of questionnaires, including 300 valid questionnaires. The specific survey data are shown in Table 2.

Table 2. The length of time that aquatic and environmental companies use cloud technology for water quality testing

The Length of Time that Aquatic and Environmental Companies Use Cloud Technology for Water Quality Testing				
More than 10 years	8–10 years	5–8 years	Within 5 years	Unused
12%	24%	29%	35%	0%

As can be seen from the above table, the number of related aquatic and environmental companies using Cloud technology platform construction is increasing. Only 12% have used it for more than ten years, and the number of users has accounted for more than half of the total in the past 5–10 years. In the past five years, the convenience of the Internet and the number of popular users increased rapidly. Among the surveyed people, there is no system that has not used Cloud technology to ensure and detect water quality safety.

The collected data shows that most aquatic and environmental companies not only make good use of the water quality security monitoring system based on the development of Cloud technology. And after years of continuous upgrading and development of the system, these companies can save more time and labor and complete the work of water quality safety supervision more quickly. In addition to the people who received the questionnaire survey, many companies are also constantly accepting the development of innovative models brought to the industry by Cloud technology.

3.2 Industrial Structure Upgrading

In the context of the rapid development of science and technology, the contemporary water quality security monitoring system is relying on scientific and technological means to continuously increase efficiency. Cloud technology has quantified the water quality standards with unified indicators. The economic growth achieved in the process of efficient and intelligent development is essentially the process of completing the optimization and upgrading of the industrial structure. [5] The main factor that promotes the upgrading of the industrial structure is the progress of science and technology. The water quality security monitoring must adhere to high standards and seriously deal with existing and potential water quality safety issues. With the advancement of science and technology, the water quality assurance system is also constantly updated and upgraded with the new technology-Cloud technology, and the corresponding testing plan and systematic supervision methods are formulated to ensure that the water quality meets the standards of safety and sanitation. [6] Through the development of Cloud technology, the operation of the monitoring system for water quality safety guarantee is promoted, thereby promoting the upgrading and development of the industrial structure.

4 Cloud Technology Improves the Efficiency of Water Quality Security Monitoring System

At present, both domestic and foreign water quality testing is a very important task. For example, the detection technology of aquatic organisms includes microbial sensor water quality monitoring support system, water quality safety online biological early warning system, and various detection systems. [7] We use various technologies to ensure the safety of water quality, integrate data through Cloud technology calculations, and conduct centralized and effective analysis and processing.

From a practical point of view, China is a water-scarce country. As we all know, water resources have become one of the important factors restricting the sustainable development of a country's society and economy. [8] The water quality safety guarantee and monitoring system based on Cloud technology can not only effectively monitor the water resources that can be directly managed and used, but also treat rainwater or sewage for reuse of water resources [9, 10].

Protecting the water environment relies on monitoring water quality information. The traditional methods of monitoring water quality information include manually collecting water samples and sending them back to the laboratory for experimental analysis to obtain appropriate water quality information. However, completing this series of tests requires a lot of manpower and material resources, and will cost a lot of costs relative to Cloud technology support. And the sampling area may be in an area where human sampling is difficult. Similarly, bad weather will also bring a lot of inconvenience to the collection, but the most important thing is that these monitoring methods are not performed in real time. It takes time to send water back to the laboratory, and there may be errors between the water quality information obtained through laboratory analysis and the real-time water quality information of the water samples, resulting in inaccurate monitoring of water quality information. Therefore, automatic monitoring of the aquatic environment has been adopted, and good results have been achieved. However, the existing automatic monitoring method only collects water quality information through sensors, and then sends it back to the monitoring center for monitoring. At the same time, using existing monitoring methods, information is concentrated in the monitoring center. When managing the aquatic environment, employees often need to communicate with the monitoring center employees to understand the water quality information after treatment. And it cannot be displayed directly, and information about water quality will reduce work efficiency. However, the water quality information monitoring system based on cloud servers includes cloud servers, water quality detection nodes located in different geographical locations, and multiple terminal devices that can perform mobile monitoring. Each water quality monitoring node can be connected to the cloud server through the network to complete data sharing, without frequent movement of personnel, and reducing consumption in all aspects.

Each water quality monitoring node is equipped with a positioning module, and the water quality monitoring information and positioning information of each node are uploaded to the cloud server at the same time. In this way, it is convenient for relevant personnel to query water quality data at any time, and real-time water quality monitoring information through mobile monitoring terminals can be used to understand the current state of the water environment and improve work efficiency.

5 Conclusion

All in all, with the advancement of science and technology, the water quality safety assurance supervision system has been continuously and efficiently developed through the support of Cloud technology, bringing people safer and more secure water resources. The support of Cloud technology has also brought development and innovation to the industrial structure, and it will increasingly become more convenient for us with advanced technology in the future. There will also be more breakthroughs in the water quality safety guarantee and monitoring system based on the continuous development of cloud technology.

Acknowledgments. This work was financially supported by:

1. Scientific Research Project of XiangNan University in 2019, Monitoring and Control System for Water Quality Security in Longhu Cave Qater Source Area Based on Cloud Technology

2. Innovation and Entrepreneurship Base of Hunan Province in 2020, Network Security Innovation and Entrepreneurship Education Base (NO. 65)

3. XiangNan University Research Learning and Innovative Learning Experiment Program for College Students in 2020, Internet + Cloud Service System for Health Management of Villagers in Remote Mountainous Areas (NO. 57)

4. Scientific Research Fund of Hunan Provincial Education Department, Research on the Review system of wireless Self-organizing Network conference (NO.18C1033).

References

1. Dapeng, L.: Water quality safety guarantee in urban water society cycle. J. Environ. Sci. **29**(01), 50–53 (2009)
2. Dapeng, L., Jiuhui, Q.: Research and technology development trend of drinking water quality safety assurance: II. Water purification technology based on micro-interface process. Environ. Eng. J. **4**(09), 1921–1925 (2010)
3. Lin, L.: The development and application trend of cloud technology. Inf. Technol. Inf. **11**, 175–176 (2015)
4. Jiuhui, Q.: Water quality safety guarantee strategy based on synergy between standards and effects. Environ. Life **06**, 72–73 (2017)
5. Co-construction of smart water services to promote the development of the industry-congratulations on the successful convening of the "Sixth National Building Secondary Water Supply and Water Quality Safety Assurance Technical Experience Exchange Conference". China Water & Wastewater, vol. 31, no. 21, pp. 16–19 (2015)
6. Bi, G., Xiong, J., Luona, Z.H.: Research on modern urban swimming pool water quality safety assurance and management methods: comment on urban water safety and water industry regulatory capability research. Chin. Saf. Sci. J. **31**(01), 199–200 (2021)
7. Yun, S., Dongping, Q., Changfei, L.: Groundwater dynamic monitoring system based on intelligent network sensing technology. Trans. Chin. Soc. Agric. Eng. **S2**, 68–71 (2008)
8. Jianzhi, X., Yubao, Z., Ying, Q., Na, L., Yuewu, Z., Shan, Z.: Water quality safety assurance technology for water sources and its application in large water sources in Northern China. China Water Resour. **01**, 11–13 (2015)
9. Hu Hongying, W., Qianyuan, H.J., Huang, H., Xin, Z.: Analysis of the safety guarantee system and technical demand for Urban sewage recycling. China Constr. Inf. (Water Ind. Mark.) **08**, 8–12 (2010)
10. Hong, G.: Research on Biological Monitoring and Early Warning of Water Source Areas in Shijiazhuang City and Capacity Building of Water Quality Safety Guarantee. Hebei Province, Shijiazhuang City Environmental Monitoring Center, 19 December 2015

Performance Test of Corrosion System Based on Simulation Control Technology

Sen Cai(✉)

Wuhan University of Science and Technology, Wuhan, Hubei, China

Abstract. Materials are used in various environments, so they are easily corroded and damaged. The economic loss caused by material corrosion damage is huge every year. Stress corrosion damage occupies a high proportion of corrosion damage and its process is complicated. A large number of stress corrosion tests are required to determine the stress corrosion failure mechanism of materials. Due to its own characteristics, the slow strain rate test method has been widely used to explore the stress corrosion mechanism. This paper aims to study the performance test of corrosion system based on simulation control technology. Based on the analysis of the characteristics of stress corrosion, the theoretical basis of the slow strain rate test method, and the evaluation method of the slow strain rate test results, a full strain rate stress corrosion test device was designed, and the performance of the device was tested. The test results show that the sample in the repeatability test comes from a 1 mm thick sheet, so the yield strength and elongation are relatively large, which is in line with the actual situation.

Keywords: Simulation control technology · Corrosion system · Stress corrosion · Performance test

1 Introduction

Materials are one of the pillars of the development of contemporary civilization. When materials are used in various environments, various forms of damage will occur. Every year, various forms of damage to materials will cause major life and property accidents. Many of these accidents are caused by ignorance of the destruction mechanism of materials [1, 2]. The most common failure forms of engineering materials are fracture, corrosion, and abrasion. Corrosion accounts for up to 42% of the failure failure of engineering materials [3, 4].

In the case of material corrosion damage, stress corrosion accounts for more than 40% [5, 6]. Many accidents occur because the mechanism of stress corrosion failure of materials is not clear enough. Since the study of stress corrosion began in the late 19th century, it has been more than 100 years. There have been more than a dozen models for the mechanism of stress corrosion damage, but there are certain ones. There is no unified theory at present. The reason is that the stress corrosion process is complex and there are many influencing factors. Various factors such as stress conditions, material composition

B. J. Jansen et al. (Eds.): *International Conference on Cognitive based Information Processing and Applications (CIPA 2021)*, LNDECT 85, pp. 351–358, 2022.
https://doi.org/10.1007/978-981-16-5854-9_44

properties, and corrosion environment will affect the stress corrosion process and results [7, 8].

Based on the analysis of the characteristics of stress corrosion, the theoretical basis of the slow strain rate test method and the evaluation method of the slow strain rate test results, this paper designs a full strain rate stress corrosion test device, and conducts a performance test on the device. The results show that the sample in the repeatability test comes from a 1 mm thick plate, so the yield strength and elongation are relatively large, which is in line with the actual situation.

2 Research on Performance Test of Corrosion System Based on Simulation Control Technology

2.1 Characteristics of Stress Corrosion

Stress corrosion refers to the phenomenon of material damage caused by the combined action of stress and corrosive media. In this damage phenomenon, the phenomenon of delayed cracking or even delayed fracture of the material is called stress corrosion cracking [9, 10]. In the phenomenon of stress corrosion, the damage caused by stress and corrosion to the material is not a simple superposition, but a synergistic effect, which promotes each other and increases the degree of damage. Under the common effect of stress and corrosion, cracking and fracture will occur when the stress of the material is far less than the tensile strength. The occurrence of stress corrosion requires three characteristic conditions: a certain stress level, a specific corrosive medium environment, and a specific material composition structure. Only when these three conditions exist at the same time, the material will be destroyed [11, 12].

Stress corrosion has the following characteristics:

(1) The stress corrosion fracture has time lag. The less the material is stressed, the longer the fracture time t. After the stress is less than the critical value σscc, t tends to infinity, and σscc is called the critical stress of stress corrosion.
(2) Generally speaking, stress corrosion only occurs under specific stress conditions. The stress can be externally applied stress, residual stress left by machining, surface treatment, etc., or the wedging effect of corrosion products Introduced expansion stress.
(3) Different metal materials have different sensitivity to different corrosive medium environments. Each metal material can only undergo stress corrosion in a specific corrosive medium environment.
(4) The composition of alloy materials has a significant effect on SCC. According to existing research experience, alloys are more prone to SCC than pure metals. The reason is that the electrochemical stability and uniformity of the material are affected by alloying elements, which will induce selective corrosion.
(5) From a macro point of view, SCC is a brittle fracture. Even for materials with good plasticity, the brittle fracture characteristics are very obvious when SCC occurs, the fracture is straight and there is no obvious necking.

2.2 Theoretical Basis of the Slow Strain Rate Test Method

Slow strain rate test, also known as constant strain rate test, is to apply a constant and slow loading speed to a sample placed in a corrosive medium to make it strain until the sample breaks, and to evaluate the stress corrosion cracking sensitivity of the material with a certain evaluation index Test method.

SSRT method, constant load method, and constant deformation method are three basic loading methods for studying stress corrosion. The stress corrosion that occurs in daily life is slow, and the constant load method and the constant deformation method are test methods that simulate the state of the material in daily life. Therefore, the constant load method and the constant deformation method have the characteristics of time-consuming, which often takes as long as several months or even years. In order to speed up the test, these two methods often need to change the state of the corrosive medium (such as the concentration and temperature of the medium) to speed up the test. The results obtained by this method are quite different from the actual situation.

The slow strain rate test method is an accelerated test method designed according to the mechanism of stress corrosion, and it usually only takes 0 to 3 days to get the test results. The current representative stress corrosion mechanism anodic dissolution theory believes that stress corrosion must go through the three processes of passive film rupture-surface alloy dissolution-crack propagation.

Usually, the metal surface is covered with a passivation film with relatively stable chemical properties. When the material is deformed by force, the surface passivation film is broken. Due to the stress concentration at the tip of the crack, the passive film on the surface is more likely to rupture, and at the same time, chemically active points appear at the tip of the crack due to the local yield of the stress concentration, which reduces the activation energy required for dissolution and forms an active dissolution pathway. Crack tip cyclic film rupture-the process of dissolving, forming a new protective film and then rupturing the film, the cracks continue to grow forward, the effective force area of the material continues to decrease, and the material breaks under the action of mechanical force.

In the slow strain rate test, the strain rate is constant, and the sliding steps are always formed at a constant speed, which will continuously cause the rupture of the passivation film to accelerate the crack propagation and shorten the test time. In the constant deformation method and constant load method, there is no way for continuous strain generation, and no active dissolution pathway can be formed. The test often takes a long time before the sample breaks. The SSRT method captures the essence of anodic dissolution of stress corrosion without changing the nature of the test. It is the best way to accelerate the test.

2.3 Evaluation Method of Slow Strain Rate Test Results

There are many methods for judging the sensitivity of the material-corrosive medium system to SCC in the slow strain rate test, and the common ones are the following two types:

(1) Fracture analysis. (A) After the sample is broken, observe with the naked eye or a low-power microscope whether there are secondary cracks near the main fracture surface of the sample. If there is a secondary crack, it means that it is sensitive to SCC. (B) Looking at the morphological characteristics of the fracture, if it shows obvious brittle fracture (the fracture is flat and bright, perpendicular to the normal stress, without obvious necking), it is sensitive to SCC.
(2) Comparative analysis. The results of the slow strain rate test of the same sample exposed to corrosive media and exposed to inert media (vacuum, dry air, silicone oil, etc.) are compared, and the stress corrosion sensitivity index Iscc is obtained. The larger the Iscc (between 0 and 1), the more significant the stress corrosion sensitivity. It is generally believed that the material-corrosive medium system has stress corrosion sensitivity when Iscc >5%.

$$I_{scc} = 1 - \frac{A_{\text{corrosion}}}{A_{\text{Inertia}}} \times 100\% \tag{1}$$

Where is the parameter obtained by the slow strain rate test of the sample in a corrosive medium, and is the comparative parameter obtained by the same test of the same sample in an inert medium.

Integrating the various data obtained by SSRT can obtain the comprehensive stress corrosion sensitivity index $I_{\text{sccComprehensive}}$. $I_{\text{sccComprehensive}}$ can better respond to stress corrosion sensitivity than I_{scc}.

$$I_{SCCComprehensive} = 1 - \frac{\sigma_{corrosion} \cdot (1 + \delta_{corrosion})}{\sigma_{Inertia} \cdot (1 + \delta_{Inertia})} \times 100\% \tag{2}$$

In addition, the average rate of stress corrosion cracking in the slow strain rate method can be determined by dividing the longest crack length measured on the specimen by the cracking time. The crack is assumed to be initiated at the beginning of the experiment. In fact, there may be an incubation period, which means that the time of cracking is not accurate, but the measured result is in agreement with the value measured by a more accurate method.

3 Experiment

3.1 Design of the Test Device

The mechanical loading test device is the core component of the slow strain rate stress corrosion in-situ test system. Unlike traditional commercial instruments, this instrument needs to be compatible with commonly used stand-type optical microscopes to realize in-situ monitoring of the test process. Record the deformation, cracking, fracture and other phenomena of the specimen during the whole test. In order to achieve compatibility with commonly used microscopes, the device is designed as a miniaturized instrument with a horizontal structure. The miniaturization of the instrument will lead to the weakening of the rigidity and strength of the mechanical structure. This paper has carried out a design check on the key components of the test device.

3.2 Simulation Analysis of the Test Device

3.2.1 Dynamic Analysis of the Whole Machine

The test device is assembled from various parts. Although the rigidity of the parts is intentionally strengthened during the design process to increase the reliability of the assembly, there are inevitably some weak links, such as the displacement sensor of the slender rod, the thin-walled medium pool and the Cantilever mounted motors, etc., these are places that easily cause vibration. Moreover, due to the existence of a DC motor in the test device, the high-speed rotation of the DC motor may excite the entire test device and cause the instrument to resonate. In order to prevent vibration from damaging the test device, this paper uses ANSYS finite element analysis software to perform a modal analysis of the whole machine.

The analysis results show that the vibration mainly occurs in the two places of the motor and the heating rod installed on the cantilever, and the stiffness in other places is relatively large, and there is no resonance phenomenon. The resonance fundamental frequency of the test device is 413 Hz, the sixth-order vibration frequency is 670 Hz, and the maximum motor speed of the test device is 4500 r/min, which means that the maximum operating frequency is 75 Hz, which is much smaller than the resonance fundamental frequency. Therefore, the slow strain rate stress corrosion in-situ extensometer designed in this paper will not resonate.

3.2.2 Statics Analysis of the Whole Machine

To ensure the accuracy of the test, the whole machine must have greater rigidity and strength. In the design process, wear-resistant, high-strength abrasive steel SKD-11 is selected for the key fixtures. The hardness after quenching Reaching above 40 HB, corresponding to the tensile strength of 1400 MPa, the yield strength ratio is as high as 0.9, that is; the yield strength can reach 1250 MPa. This section carries out a static simulation of the whole machine to verify the strength and rigidity of the test device. In order to facilitate the simulation calculation, the medium pool that is not stressed (that is, the simulation result is not affected) is removed, and a pulling force of 2500 N, which is slightly larger than the device range (2000 N), is applied to the two fixtures in the opposite direction. The simulation results show that the maximum stress is 97.5 MPa at the end face of the fixture, which is much smaller than the tensile strength of the material. The maximum deformation appears on the compression block, which is 0.02 mm, the deformation is very small, and the impact on the loading accuracy is negligible. The whole machine has a reasonable structure and meets the design requirements.

4 Performance Test

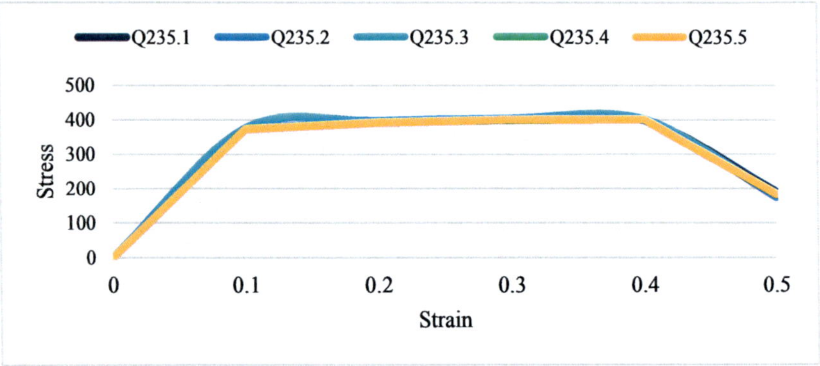

Fig. 1. Q235 tensile repeatability test

The repeatability of the testing machine refers to the ability to reproduce the test results under the same test conditions, and good repeatability is the basic requirement of the test device. This article uses a typical material Q235 steel to perform multiple tests to test the repeatability of the test device. The material in this test is derived from the same rolled plate and processed by a slow-moving wire cutting machine. The axis of the sample is consistent with the rolling direction. The test was repeated five times, and the tensile rate was 0.1 um/s. The results are shown in Fig. 1. The stress-strain curves obtained from the five sets of tests have good repeatability.

From the stress-strain curve obtained from the test, four mechanical properties data of Q235, namely elastic modulus E, yield strength σs, tensile strength σb, and elongation δ, are sorted out, as shown in Table 1 and Fig. 2. The relative deviation between the various indicators does not exceed 7%, indicating that the instrument is reliable. The mechanical properties of domestic Q235 are related to the thickness of the rolled sheet. The yield strength of 16 mm thick Q235 is 235 MPa. The thinner the rolled sheet, the higher the yield strength. In the repeatability test, the sample comes from a 1 mm thick sheet, so the yield strength and elongation are relatively large, which is in line with the actual situation.

Table 1. Q235 repeatability test mechanical performance index

Test number	E (GPa)	σ_s (MPa)	σ_b (MPa)	δ (%)
Q235.1	197	260	412	52.4
Q235.2	194	259	403	51.3
Q235.3	202	270	408	50.8
Q235.4	198	267	407	50.2
Q235.5	204	264	411	48.8

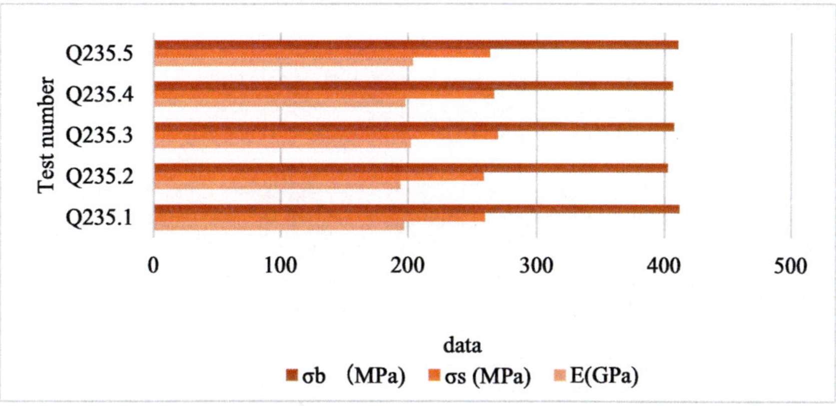

Fig. 2. Q235 repeatability test mechanical performance index

5 Conclusions

Due to the many influencing factors and complex processes of stress corrosion, there is no unified theory for the internal mechanism of stress corrosion. Therefore, the life and reliability of materials under the combined action of corrosive environment and stress cannot be simply and effectively evaluated. The traditional slow strain rate method focuses on the study of experimental phenomena such as the stress-strain curve and fracture morphology of materials, but lacks in-depth research on the process of stress corrosion cracking crack initiation and propagation. In-situ testing technology can monitor the detailed process of material microstructure change and damage in real time during the test, which is of great significance to the study of stress corrosion mechanism.

References

1. Qiang, Y., Dong, C.F., Liang, J.X., et al.: Stress corrosion cracking behavior of PH13-8Mo stainless steel in Cl^- solutions. J. Iron. Steel Res. Int. **24**(003), 282–289 (2017)

2. Venugopal, A., Narayanan, P.R., Sharma, S.C.: Evolution of microstructure and stress corrosion cracking behavior of AA2219 plate to ring weld joints in 3.5WtPct NaCl solution. Metall. Mater. Trans. A **47**(4), 1–14 (2016)
3. Zhong, Y., Zhou, C., Chen, S., et al.: Effects of temperature and pressure on stress corrosion cracking behavior of 310S stainless steel in chloride solution. Chin. J. Mech. Eng. **30**(1), 200–206 (2017)
4. Lee, H., Kang, S.J., Choi, J.B., et al.: An extended finite element method-based representative model for primary water stress corrosion cracking of a control rod driving mechanism penetration nozzle. Fatigue Fract. Eng. Mater. Struct. **41**(1), 138–145 (2017)
5. Wan, H., et al.: The effect of hydrogen on stress corrosion behavior of X65 steel welded joint in simulated deep sea environment. Ocean Eng. **114**(Mar.1), 216–223 (2016)
6. Du, D., et al.: Effects of chloride and oxygen on stress corrosion cracking of cold worked 316/316L austenitic stainless steel in high temperature water. Corros. Sci. **110**(Sep.), 134–142 (2016)
7. Meo, D.D., Diyaroglu, C., Zhu, N., et al.: Modelling of stress-corrosion cracking by using peridynamics. Int. J. Hydrog. Energy **41**(15), 6593–6609 (2016)
8. Cui, Z., et al.: Effect of plastic deformation on the electrochemical and stress corrosion cracking behavior of X70 steel in near-neutral pH environment. Mater. Sci. Eng. A **677**(nov.20), 259–273 (2016)
9. Ming, H., et al.: Microstructure, local mechanical properties and stress corrosion cracking susceptibility of an SA508-52M-316LN safe-end dissimilar metal weld joint by GTAW. Mater. Sci. Eng. A **669**(jul.4), 279–290 (2016)
10. Scatigno, G.G., et al.: The effect of prior cold work on the chloride stress corrosion cracking of 304L austenitic stainless steel under atmospheric conditions. Mater. Sci. Eng. A **668**(jun.21), 20–29 (2016)
11. Zhong, X., Bali, S.C., Shoji, T.: Accelerated test for evaluation of intergranular stress corrosion cracking initiation characteristics of non-sensitized 316 austenitic stainless steel in simulated pressure water reactor environment. Corros. Sci. **115**(FEB.), 106–117 (2017)
12. Nakamura, N., et al.: Inspection of stress corrosion cracking in welded stainless steel pipe using point-focusing electromagnetic-acoustic transducer. NDT & E Int. **83**(oct.), 88–93 (2016)

Application of Computer Virtual Simulation Technology in Building Structure Design and Construction

Zhenhua Liu(✉) and Yuechun Feng

Ningxia Institute of Technology, Shizuishan City, Ningxia, China

Abstract. The application of virtual simulation related technology in architectural design will make the architect's creative process start from three dimensions. It not only provides a new means for architectural design, but also brings about a change in creative methods and concepts. This article attempts to study the application of computer virtual simulation technology in building structure design and construction from the perspective of architects. This article first gives a quick introduction to the characteristics and research status of computer virtual simulation technology, then analyzes the application range of virtual simulation technology in the field of architectural design and the new design process, and finally designs a computer virtual simulation system using related technologies. In addition, simulation experiments are carried out on the convergence speed of the system during collision and non-collision. The experimental results show that the convergence speed of the virtual simulation system is more reasonable. The convergence speed of the system in the non-collision process is faster than that in the collision process.

Keywords: Virtual reality · Architectural design · Simulation system · Simulation experiment

1 Introduction

The application of computer simulation technology, computer graphics and image software technology created a 3D simulation model of the virtual simulation system, and was modeled in the created 3D model through a series of experimental testing techniques. Three-dimensional simulation technology has the advantages of high efficiency, safety, and strong anti-interference ability. Computer three-dimensional simulation is interactive, conceptual, and in-depth. In-depth refers to the real degree of the user in the virtual environment [1]. Interactivity refers to the degree to which the user can manipulate the object model in a 3D virtual environment and the natural degree of feedback received from the environment. For example, the user can directly control an object in the virtual environment through which the user can move freely, and when he encounters a tree or wall in the virtual scene, the user will be blocked by the object and stop moving [2]. Conceptual z refers to the virtual space that will make users feel immersive, and even forget the real world. This is to use the user's feeling and cognition to comprehensively

B. J. Jansen et al. (Eds.): *International Conference on Cognitive based Information Processing and Applications (CIPA 2021)*, LNDECT 85, pp. 359–367, 2022.
https://doi.org/10.1007/978-981-16-5854-9_45

improve the effect of capturing the space, give full play to subjective initiative and create new feeling [3].

Chinese scholar Wang Junyan and others pointed out that my country's economic development has led to the rapid development of social science and technology, and the construction industry has also developed rapidly. People's demand for buildings has increased, and architectural design has become more and more difficult. In addition to extensive management of buildings, there is an urgent need to change existing construction methods and architectural designs. First, use virtual simulation technology for parameter modeling and analysis, then use virtual simulation technology to program the information model and construction management system, and finally, through the creation of a virtual simulation technology platform for theoretical guidance of building construction [4]. Mao Zhenhua believes that virtual simulation technology is applied to various cutting-edge fields, and the application of virtual simulation technology in the field of construction is also expanding. Designers use virtual simulation technology to improve the effectiveness of architectural design, so that the designer's design can be visually, audio-visual and tactile. On the one hand, it is generated in a highly simulated digital environment and uses various science and technology to complete architectural design work in buildings. Virtual simulation technology allows designers to display various external conditions more intuitively and three-dimensionally when designing buildings [5]. Zhang Songkuo believes that the generation and application of virtual simulation technology can better solve the problem of information faults in the life cycle of construction projects. The use of virtual simulation technology can complete the comprehensive operation of building construction, so that the construction of various steps of the construction project can cooperate with each other. Not only has the quality of the project been guaranteed, but the efficiency of the construction performance has also been improved [6].

Virtual simulation technology is a very key technology in the field of architectural design. Many problems between abstract thinking and creating entities can be overcome by virtual simulation technology [7]. In the 1990s, the field of physical modeling became more and more inclined to use virtual simulation technology, and more and more designers began to understand virtual simulation technology. Virtual simulation technology was recognized by many designers in the field of architectural design. The architectural design supported by virtual simulation technology is still in the initial stage of implementation, that is, the realization of the design function; although it is applied in the conceptual stage of architectural design, its potential has not yet been fully utilized. The emergence of this situation is of course related to the software itself, but it is also related to the manufacturer's traditional concept [8]. Architects have no in-depth and systematic understanding of the functions provided by related design software. They believe that virtual simulation design will hinder the creative thinking of architects. In a sense, they did not use virtual simulation to help initiate the concept of architectural design.

2 Application of Computer Virtual Simulation Technology in Building Structure Design and Construction

2.1 Application of Virtual Simulation Technology in the Field of Architectural Design

Before entering the digital age, virtual simulation has been associated with architecture. Since the renaissance, architects are happy to create virtual spaces in the four murals of buildings, hoping to make people in the building feel that the area of the house they are in is more wide [9]. After entering the 21st century, virtual simulation technology is mainly used for the research or repair of buildings. Many architects have similar experiences. Although these buildings look perfect in design and have a good understanding of the model, the actual results after completion are unexpected. The reason is simple, there is a gap between the fantasy of the image and the real scene. Although there are various two-dimensional images and models as media, they still cannot bridge the gap between the above differences. The virtual simulation system can not only simulate the building itself, but also replicate and realistically predict the environment in which the building is located [10]. With the help of virtual simulation, it is possible to evaluate and reasonably evaluate and finally obtain a form with excellent physical and aesthetic characteristics. When the wind slides along the curved surface of the building, a certain wind pressure will be formed around the surface of the building, thereby improving the efficiency of natural ventilation inside the tower.

2.2 Application Range of Virtual Simulation Technology in Architectural Design

Through virtual simulation, the functions of the building can be fully integrated in the architectural creation stage, so that the space experience in the architectural design is more interactive and flexible, expressing the spirit of space as much as possible, and significantly improving the authenticity. In the virtual simulation system, you can really feel the proportion of the space layout through custom roaming and custom design elements in the building and design. Different from the previous design methods, designers can use their imagination in the virtual simulation system to their fullest, and are not restricted by geographical space. The creativity and inspiration of designers have been greatly improved. The virtual simulation system can provide architects with free multi-dimensional, multi-dimensional human-computer interaction program experience [11]. Using virtual simulation technology, it can also be used to research and review other aspects of the design. For example, by simulating the wind environment of the lower part of a high-rise building, the shape and appearance design of the building are adjusted to improve the wind environment around the high-rise building. In addition, the safety of fire rooms, smoke control and evacuation can be verified. In short, today, the virtual simulation system can support architectural design from all aspects, integrate different disciplines and conduct a comprehensive inspection, so that the design of the building can be more complete.

2.3 New Design Process

First of all, in the drawing media stage, from the initial sketch to the completion of the final design plan, we will need many different types of drawings to express. Therefore, the continuity of these drawings cannot be guaranteed. Under the condition of virtual simulation-aided design, the generation of graphics is usually a process of continuous modification and accumulation from the initial plan. For example, it is more common that the perspective drawing is separated from the design process, and the drawing of the construction drawing is independent of the plan design. This separation of design and expression also leads to the separation of the design process itself. What virtual simulation shows is that the design can be used directly based on the results of the previous stage. The entire design process is a process of continuous adjustment and improvement of the design object [12]. To be sure, as far as the current domestic auxiliary design is concerned, these integration processes are not uncommon. This will be a new problem facing our architects. Secondly, virtual simulation breaks the relatively independent design thinking in the architectural design process. It can always use the three-dimensional space form as the main way of design thinking in the architect's design process, and really think about the content of the design from space. Another major change of virtual simulation to the original design method is to move from the traditional phased design process to synchronization, so that the pre-planning, design conception, construction and building management after completion of the building become a unified whole, and the architectural design can be directly transformed into actual construction and construction.

3 Modeling of Appendages in Simulation Platform

3.1 Modeling of Various Objects

The modeling of trees, flowers and grass is a difficult problem in a 3D simulation system, because the huge model resources cannot create detailed models of branches and leaves like 3D animations. This is not only inefficient, but also the amount of data does not meet the visual simulation. The three-dimensional simulation system in this article uses transparent channel mapping. The model creates intersecting surfaces, and uses the center point as the axis to process the photos of complex objects into channel maps by graphics software, which is used as the transparent map channel to make it in three-dimensional only trees, flowers and grass are displayed in the scene, and the model can create a cross-shaped patch in the view, and attach its texture map to the cross-shaped patch. In the real-time interactive dynamic display of complex scenes in 3D simulation simulation, trees, flowers and grass can be displayed truly, and the texture channel is processed into the transparency effect of the map channel. Therefore, these complex models are a form no matter where they are viewed. According to the degree of fineness of the model, the creation of a street lamp can first perform detailed modeling of the various details of the street lamp, and then associate the corresponding texture map and self-luminous map, or create a double-sided display face sheet, and set it can be visualized, assign the processed channel map to the created street lamp and determine the texture of the channel, so that the street lamp becomes a double-sided display and supports real-time interaction. The formulas used in the modeling process are:

$$\mathrm{Mind} = \left\| \sum_{i=1}^{m} \varphi_i x_i - \sum_{j=1}^{n} \sigma_j y_j \right\| \tag{1}$$

3.2 Optimization of Computer Simulation System

The number of faces and texture calculations of the developed campus display system are very large. In order not to affect the operating efficiency of the system, a series of real-time optimization of the computer simulation system is implemented to optimize the model library and the texture resource library. The 3D model resource library divides the models into high-precision models, medium-precision models and low-precision models to meet the different needs of close range, medium range and wide range in 3D simulation scenes. The camera travels along the path and is limited to projection. The displayed model is a low-precision model, which can greatly save the calculation of the model in the scene and meet the requirements of real-time update of the interactive frequency. The camera follows a path limited to projecting close-up views, and the displayed model is a high-precision model that ensures real-time interaction. The optimization of the system is mainly to appropriately divide the tetrahedron on the 3D model into triangles by applying the modification tool of polygon modeling. Adjust the size and format of the texture map and adjust the map mode, thus greatly improving the running speed of the model library resources in the system and the fluency of real-time interaction. The formulas used in the system optimization process are:

$$x(j) = a(j) + u_0(j) * (b(j) - a(j)) \tag{2}$$

3.3 Application Setting in Computer Simulation System

The trees of the distant view model are matched with the transparent channel map, and the cross-shaped face sheet is used to simulate the effect of real trees through the transparent map channel. This function is widely used to create computer simulation scenes, because for simulation systems that need to be viewed in real time, this is the key to reduce the amount of calculation generated by the system as much as possible in order to perform 3D modeling of surface texture. The geometric object adopts texture texture. The production technology performs meticulous processing, and seamless texture processing can reduce the amount of calculation of the resource system through iteration, thereby saving a lot of storage space. Traditional computer graphics models and designs have large circles. Although offline during the modeling process, creating accurate 3D geometric patterns is time-consuming and laborious, which has always been an obstacle to graphics applications. On the basis of the perfect texture database and model, the virtual building construction time texture library, the virtual building model library construction time, the construction of the texture library and the model library are inseparable. As for the process of changing the typical texture characteristics of the building and the surface characteristics of the changing three-dimensional physical characteristics, any building will inevitably lead to changes in the texture database and model library. Therefore, the creation of the texture database and the 3D model database is a gradual update process. To complete a simulation platform system applied to the construction industry, we must first create a texture resource database and model resource database that can update resources, and then create a virtual three-dimensional environment, and test and update the texture data database and model database in the practical environment.

4 Analysis of Simulation Experiment Results

4.1 System Simulation Experiment on Collision Detection

Table 1. System collision detection simulation experiment results

Total group number	System convergence time (seconds)
500	3.755
1000	11.563
1500	18.427
2000	26.718

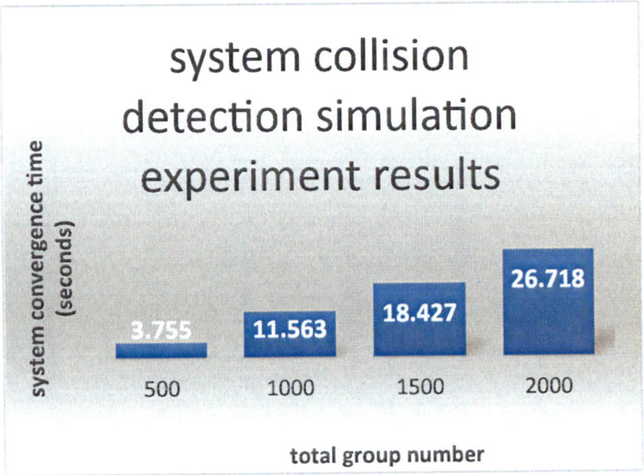

Fig. 1. System collision detection simulation experiment results

According to Table 1 and Fig. 1, when the population size is 500, 1000, 1500, 2000, the convergence speed of the virtual simulation system is 3.755 s, 11.563 s, 18.427 s, and 26.718 s, respectively. From 1500 populations to 2000 populations, the convergence rate of the system changes the most, followed by 500 populations to 1000 populations, and finally from 1000 populations to 1500 populations. This shows that in the collision detection process of the virtual simulation system, when the population size changes, the convergence speed of the virtual simulation system is not linear and has no regularity, but the convergence speed is still relatively reasonable.

4.2 System Simulation Experiment on Non-collision Detection

Table 2. System non-collision detection simulation experiment results

Total group number	System convergence time (seconds)
500	3.267
1000	10.418
1500	16.273
2000	22.576

According to Table 2 and Fig. 2, when the population number is 500, 1000, 1500, 2000, the convergence speed of the virtual simulation system is 3.267 s, 10.418 s, 16.273 s, and 22.576 s, respectively. From 500 populations to 1000 populations, the convergence rate of the system changes the most, followed by 1500 populations to 2000 populations, and finally from 1000 populations to 1500 populations. This shows that in the non-collision detection process of the virtual simulation system, when the population size changes, the convergence speed of the virtual simulation system is not linear and has no regularity, but the convergence speed is still relatively reasonable. In general, the convergence speed of the virtual simulation system in the non-collision detection process is faster than the convergence speed in the collision detection process.

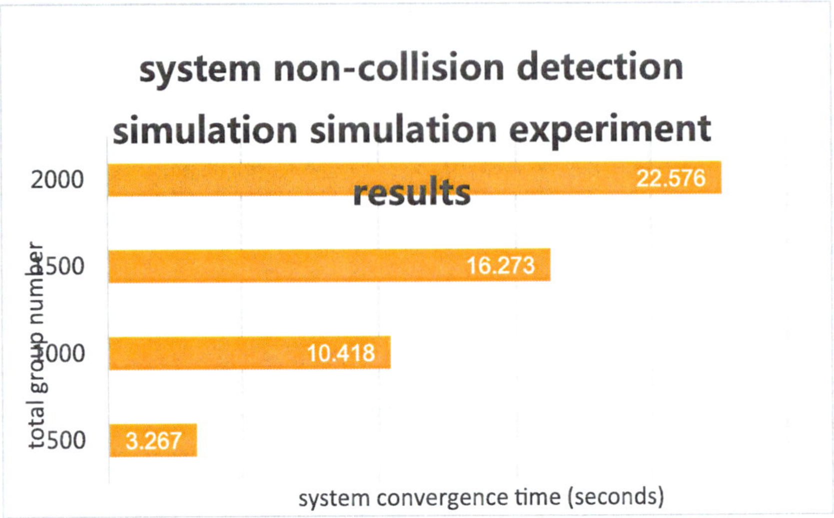

Fig. 2. System non-collision detection simulation experiment results

5 Conclusions

The application of computer simulation technology in the construction industry will have a wide range of development, and at the same time, it will also greatly expand the design art. The requirements for environmental design in the virtual environment of building roaming have brought new challenges. The creation of various shapes and delicate colors in the virtual environment, as well as the combination of the particle dynamic effects of the three-dimensional environment and the virtual model of dynamic objects, can create a more exquisite visual experience. In this paper, the application of computer virtual simulation technology in building structure design and construction is researched. The performance of virtual modeling graphics and image software used by computer virtual simulation is analyzed in depth, and the system is simulated in terms of convergence speed.

References

1. Bin, C., Fei, Z.: Research on talent cultivation of building prefabricated construction based on virtual simulation. J. Wuhan Inst. Eng. Technol. **031**(001), 102–105 (2019)
2. Jiao, W.: Research on green construction technology of building engineering based on virtual simulation technology. Build. Technol. Dev. **410**(08), 151–152 (2019). v.46
3. Chen, S.: Construction of virtual simulation training platform for construction engineering. China Educ. Technol. Equip. **480**(06), 62–63+68 (2020)
4. Wang, J., Liu, Y., Ma, X.: Building and construction design management platform design based on BIM and virtual simulation simulation technology. Doors Windows **176**(20), 140–141 (2019)
5. Mao, Z.: The application of virtual reality technology in architectural design. Beauty Times (Urban Edit.) **863**(09), 23–24 (2020)

6. Zhang, S.: The development and application of BIM technology in construction engineering. Shaanxi Archit. **302**(08), 21–23 (2020)
7. Wang, W., Teng, J.: Curriculum reform and practice of garden engineering construction organization design based on virtual simulation. Agric. Technol. **358**(17), 180–182 (2020). v.40
8. Hao, W.: Application of virtual simulation technology in road engineering practice teaching. J. Chifeng Univ. (Nat. Edit.) **035**(004), 160–162 (2019)
9. Xiao, Y.: The practice of BIM virtual simulation technology in the teaching of construction technology. Eng. Technol. Manag. (Engl.) **004**(002), 207–208 (2020)
10. Pei, W.: Architectural engineering design optimization based on BIM and virtual reality technology. Microcomput. Appl. **35**(05), 89–91 (2019)
11. Chongyang, Y., Xiaoyan, W., Yanmei, D., et al.: Evaluation of the effect of digital virtual simulation training system for pre-clinical teaching of medullary cavity crown preparation. Chin. J. Stomatol. **56**(05), 479–484 (2021)
12. Ma, D., Hu, Y., Zhang, X.: Road construction virtual simulation experiment teaching construction. Lab. Res. Explor. **288**(02), 142–146 (2020). v.39

Financial Technology Risk Management and Control in the Big Data Era

Xiuwen Wang[1(✉)], Zhen Wu[1], and Shuangcheng Shen[2]

[1] National Computer Network Emergency Response Technical Team/Coordination Center of China, Beijing, China
[2] Golaxy Data Technology, INC., Beijing, China

Abstract. With the advent of the era of big data, new technologies such as cloud computing and artificial intelligence are changing with each passing day. More and more science and technology are applied to the financial field, which greatly promotes a new round of prosperity and development of China's financial market. However, because big data has the characteristics of fast spreading speed, wide penetration area, strong concealment, and difficulty in supervision, the integration of technology and finance in the era of big data will also generate some new risk problems. Based on this, this article starts with the connotation and characteristics of financial technology, summarizes the risk performance of financial technology, and proposes corresponding countermeasures, aiming to promote the healthy and stable development of China's financial technology.

Keywords: Big data · Finance · Technology · Financial risk · Management and Control

1 Introduction

Data has already penetrated into every industry and business functional area. Data is closely related to us. We are dealing with all kinds of data every day, entering the era of "big data". The emergence, rise and development of big data technology relying on modern Internet technology [1]. The mass storage characteristics of network resources provide the necessary conditions for big data storage. The analysis and storage characteristics of big data technology help to collect more users. With the rapid development of the business of financial institutions and the expansion of the service population, traditional risk management and control have exposed related problems, such as high labor costs, lengthy review procedures, etc., which cannot meet the real-time analysis of massive data generated in the current business and the identification of diversified risks. In order to better develop business and serve customers, financial institutions need to strengthen the use of big data [1].

B. J. Jansen et al. (Eds.): *International Conference on Cognitive based Information Processing and Applications (CIPA 2021)*, LNDECT 85, pp. 368–374, 2022.
https://doi.org/10.1007/978-981-16-5854-9_46

2 Big Data and Finance

The characteristics of big data have experienced the evolution of 3 V, 4 V to 5 V. The University of Amsterdam proposed the 5 V characteristics of the big data architecture framework, which are Volume, Velocity, Variety, Value and Reality. The first is the huge volume of data, which is reflected in the huge amount of data collection, storage, and calculation [2]. The storage unit ranges from GB to TB in the past, to PB, EB, and ZB. The second is the variety of data types (Variety), big data includes multiple formats and different types of data, divided into structured, semi-structured and unstructured data. The third is the low value density (Value). The value density is inversely proportional to the total amount of data. It is necessary to dig out valuable information from the massive data. The fourth is Velocity, which is the most significant feature that distinguishes big data from traditional data mining. In terms of data processing speed, there is the "1 s law", that is, if the data does not provide analysis results within seconds, the data will lose its value [2]. The fifth is the veracity of data. Big data happens in the real world, and the accuracy and reliability of the data must be ensured.

The financial industry has the characteristics of "massive users and big data" due to its large number of customers, complex business types and high degree of informationization. In terms of scale, financial institutions have accumulated massive amounts of data, including customer information, product information, transaction information, market information, etc., and are collecting and processing data all the time [3]. In terms of speed, in order to meet the requirements of customers, products and transactions, financial institutions attach importance to the development of big data technology, improve the processing speed of business, and continuously optimize the data collection, storage and analysis platform. In terms of type, the data generated by financial institutions not only includes structured data, but more of them are semi-structured and unstructured data.

3 Fintech Overview

The integration of technology and finance has spawned a new financial format, namely financial technology. The definition of financial technology is: non-traditional companies use the power of technology to turn their target market into the financial sector, seize the market with the convenience and efficiency of technology, and improve the efficiency of financial services while enhancing the prediction and judgment of financial risks [3]. In China, the most people-friendly manifestation of financial technology is people's use of wealth management apps such as Alipay, WeChat, Ping An Pocket Bank, and CITIC Payroll. Financial technology has penetrated into all aspects of people's daily consumption. It has the following main characteristics:

First, it exists based on big data. The operation of fintech needs to be based on big data architecture, and analyze financial data through artificial or intelligent modeling, so as to provide users with reliable financial information more efficiently, as shown in Fig. 1;

Second, the degree of artificial intelligence is relatively high. Artificial intelligence is a main direction of the development of financial technology [4]. By introducing artificial intelligence, the data that the human brain needs to collect and analyze are transferred to the computer system for processing, which can greatly improve the efficiency of financial services.

Third, focus on user experience. Through the optimization and upgrading of system function modules, use procedures, etc., users can better obtain knowledge and benefits from technology finance while feeling the charm of science and technology [4].

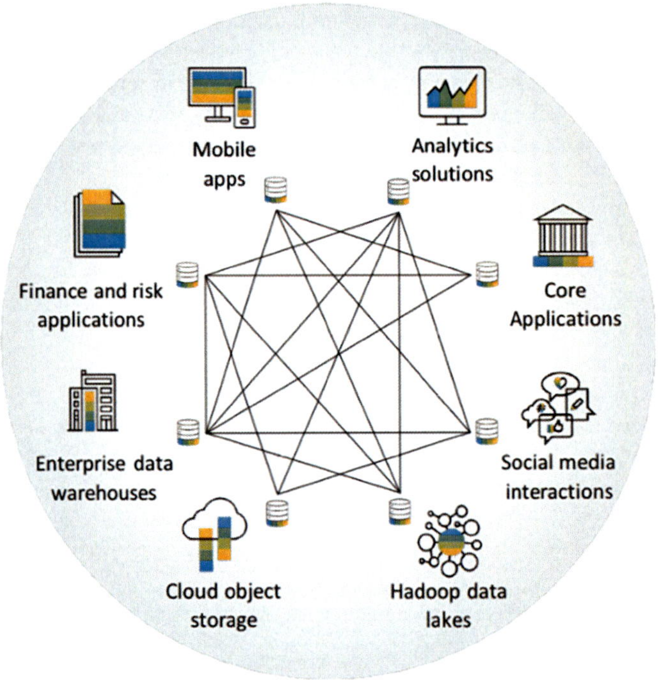

Fig. 1. Fintech under the big data environment

4 Fintech Risks and Performance

The development of information technology has promoted the prosperity of financial technology, and the development of financial technology has further promoted continuous innovation in the financial field and injected more momentum for the development of China's financial market. However, the development of financial technology is still in its infancy [5]. In addition to its own shortcomings, there are many behaviors that wander on the edge of law and regulation, and fail to be discovered and controlled in time, which can easily evolve into various financial risks.

4.1 Credit Risk

As far as the field of financial technology is concerned, credit risk is the possibility that one party to a transaction is unwilling or unable to perform the established contract or agreement due to some reason, causing the counterparty to suffer losses [5]. The occurrence of general credit risk is directly related to the lack of timely information disclosure or lack of disclosure. In the financial industry, information is the basic element of financial asset prices, and the symmetry and traceability of information are much more important than statistical data. However, in China, financial consumers and financial institutions have unequal status in transactions [5]. Consumers are often in a disadvantaged position, and their awareness and ability in information acquisition and information application are relatively poor.

4.2 Data Security Risks

In the era of big data, technology and finance are more closely connected, and the financial industry uses technological means more frequently. This transfers some inherent security risks of information technology to the financial field, the most prominent manifestation in data security risks aspect. Due to the explosive growth rate of data in the big data environment, the amount of data generated and transmitted every day can be said to be massive and unpredictable. Regulators cannot audit all financial data one by one, which leads to data fraud and data leakage [6]. The emergence of security issues provides objective conditions. In recent years, it has become common for all industries to rely on financial technology tools for transactions. The phenomenon of criminal behavior is also increasing day by day, such as financial fraud, telecommunications fraud, database attacks, cloud data theft, etc., which pose a great threat to the data security of financial technology.

4.3 Regulatory Risk

In the financial technology environment, big data, block chain and other technologies have become the technical support framework for the development of financial technology. On the one hand, it has changed the past financial transaction habits and trading methods, making it difficult for traditional financial legislation to adapt to the new environment of financial market development [7]. On the other hand, in the existing legal framework, how to regulate, guide and restrict some new financial formats has not formed complete and accurate legal regulations, leading to a dilemma in financial supervision, which can easily lead to a series of regulatory risks. For example, in the financial market, the application of smart contracts has gradually become popular, while the applicability of smart contracts in China's contract law does not make clear provisions [7].

5 Fintech Risk Control Measures

Objectively speaking, in the financial market, development and risk always coexist. Therefore, while promoting the development of financial technology, we should actively explore ways and means to prevent and control various financial risks.

5.1 Improve Information Disclosure and Risk Notification

In the current financial technology environment, the unequal relationship between Chinese financial consumers and financial institutions needs to be corrected urgently. In this regard, it is necessary to further enhance the awareness and ability of financial consumers to rationally choose financial institutions, rational investment, and rational consumption [8]. In the system design of financial laws and regulations, perfect the principle of appropriateness of financial institutions, clarify the obligations of financial institutions, and especially emphasize that financial institutions must perform information disclosure obligations, and improve the symmetry and traceability of information. When consumers choose a certain financial product or service, the relevant information should be clearly communicated to consumers so that consumers can understand the necessary information such as the whereabouts, uses, and risk levels of financial funds [8]. At the same time, the functions of financial industry associations and consumer associations should be brought into full play to effectively safeguard and protect the legitimate rights and interests of financial consumers.

5.2 Strengthen Information Protection and Data Security

For the data security issues of financial technology in the big data era, financial institutions must first strengthen the awareness of information protection and data security, strengthen the basic management of financial information, and improve the level of refinement of information management; improve the information security system, and implement and implement safe operations [9]. Security monitoring, security early warning, emergency response and other aspects of the work requirements, even if there are hidden risks, they can be prevented, controlled and resolved in a timely manner in accordance with the prescribed system and procedures; secondly, financial institutions must introduce new security prevention and control technologies, as shown in Fig. 2. For the security risks that arise in the era of big data, it is also necessary to use big data-related technologies to deal with it [9]. Financial institutions should be aware of the advantages of big data technology in enhancing information and data security capabilities, and actively research the application of big data, cloud computing, artificial intelligence and other technologies in risk prediction, risk monitoring, risk management, risk resolution, etc., to promote the financial market.

5.3 Innovating Regulatory Ideas and Methods

Since the introduction of financial technology in China, regulatory technology has also emerged. The definition of regulatory technology is very similar to that of financial technology. It is the integration of supervision and technology [10]. Its role is to use information technology to improve the supervision methods and procedures of the supervisory department, so that the operation of financial technology is legal and compliant. At present, more and more government departments, financial institutions, economists, etc. have proposed a penetrating supervision based on the financial technology environment. The concept of penetrating supervision has been recognized by the policy, and the concept of penetrating supervision has been mentioned many times in China's Internet

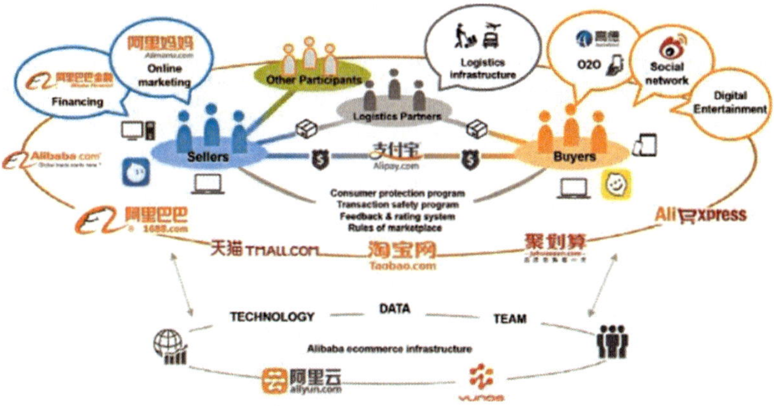

Fig. 2. Alibaba financial data protection system

special rectification plan, thus establishing the concept of penetrating supervision [10]. Under this regulatory model, the supervision of financial technology no longer emphasizes the identity and type of the subject of supervision, but starts from the nature of the business of financial institutions, and conducts a coherent and penetrating approach to the source, operation process, and final flow of their funds (Fig. 3).

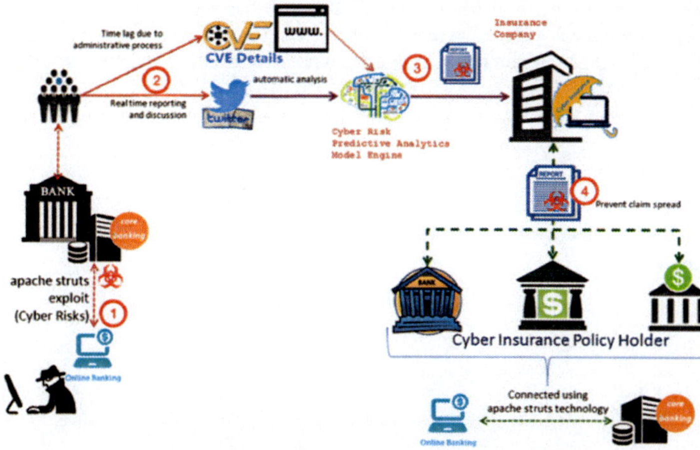

Fig. 3. Regulatory innovation of big data fintech

6 Conclusion

Fintech is not a simple combination of finance and technology, but an organic integration of finance and technology. Based on the research of financial technology development related issues, it is not difficult to find that financial technology under the big data environment has prominent risk prevention problems during the development process, and the lack of financial supervision has led to a series of financial risks. On the basis of the organization's information symmetry work, strengthen data security protection work, and continuously improve and optimize the supervision model. It can be predicted that in the future, big data and related science and technology will have more intersections with various industries, the degree of mutual integration will continue to deepen, and the subsequent risk issues will become more complex and diverse. The government, financial institutions, investors and all parties of society need to cooperate and cooperate closely to jointly explore a path suitable for the development of financial technology.

References

1. Zhong, H.A.: Research on the development of financial technology and risk prevention. Financ. Dev. Res. **13**(01), 57–59 (2017). (in Chinese)
2. Zhang, H.W.: Opportunities and risk prevention of financial technology innovation and development. Northern Econ. Trade **11**(02), 56–58 (2018). (in Chinese)
3. Wang, Q.J.: On the rise of financial technology. Mod. Bus. **14**(9), 132–133 (2017). (in Chinese)
4. Hu, P.F.: The application of financial technology in the field of risk prevention in the Internet finance industry. Big Data **11**(01), 41–42 (2018). (in Chinese)
5. Yang, L.: Steadily promote financial technology innovation to prevent systemic financial risks. Banker **11**, 75–77 (2017). (in Chinese)
6. Zhao, P.: Talk about the challenges and countermeasures facing the financial industry in the era of big data finance. China Electron. Commer. **12**, 83–86 (2018). (in Chinese)
7. Feng, P.F.: A brief discussion on Internet financial risk prevention and control methods in the era of big data. Mod. Mark. (late issue) **12**(07), 158–162 (2016). (in Chinese)
8. He, C.B.: Research on the application model and value of big data based on Internet finance. China Circ. Econ. **12**, 24–26 (2016). (in Chinese)
9. Liu, Y., Luo, M.X.: Thoughts on internet finance model and risk supervision. Chin. Mark. **10**(43), 29–32 (2013). (in Chinese)
10. Li, S.Y.: Internet finance research based on big data. Consum. Guide **9**(01), 131–132 (2017). (in Chinese)

Module Design of Network Security Course Based on Attack and Defense

Guofang Zhang(✉)

Hainan College of Software Technology, Qionghai, Hainan, China
Guofang_Zhang@haoxueshu.com

Abstract. Network security includes hardware, software, human error and unpredictable objective factors. As a profession skill computer professional courses in higher vocational colleges, how to set up the teaching module of network security course, bring students into the scene classroom, can experience and understand the principle of network security in the present situation is the key point of this course. From the perspective of attack and defense of network security incidents, this paper constructs the curriculum structure module according to the idea of red blue confrontation, and designs the learning scene according to the cognitive psychology and cognitive rules of higher vocational students, so that students can be personally on the scene in the classroom, exercise and improve their skills in practice. It is found that through the group learning of attack and defense module, students' interest in classroom learning and exploration is stimulated, their professional skills are constantly proficient, and the new problems found in practice are constantly enriched into the curriculum construction, which has certain reference significance for the classroom reform of colleges and universities.

Keywords: Network security · Network scanning · Red-Blue confrontation · Attack host · Target aircraft

Computer network is a complex composed of hardware and hardware equipment of multiple manufactures and different software developed in different languages. The security of computer network involves many problems such as physics, mathematical logic, software vulnerabilities, hardware failures, human errors, etc. Is caused by natural disasters. As a professional technical subject of computer network technology course is how to design course teaching content, how to organize student training, improve students' sense of situational experience in the classroom practice. The focus of more attention is also the hot-spots of teaching reform in higher vocational colleges. Start from the characteristics of the network security technology course, this paper studies the knowledge points and scenarios of network security technology in the restoration of network security incidents, and designs the teaching module of the network security technology course. Experimental teaching proves that this new course teaching mode is more suitable. College classroom teaching mode, students can get a sense of professional experience in advance under this teaching mode, and lay a professional foundation for future careers [1].

B. J. Jansen et al. (Eds.): *International Conference on Cognitive based Information Processing and Applications (CIPA 2021)*, LNDECT 85, pp. 375–381, 2022.
https://doi.org/10.1007/978-981-16-5854-9_47

1 Problems Faced by Course Teaching

Network hacking and defence technology course is a special course in higher vocational colleges. It requires rigorous theory and realizable technical operation. The teacher explained the transmission process of data packets in the network protocol in class. Then choose the right platform and tools to build a specific experimental environment, in the experimental environment to simulate the hacker‘s attack to achieve the desired effect. The problem is that the choice of tools is difficult. The reason for this problem is that some security tools will be killed as viruses by the anti-virus software with continuously upgraded virus library. As a result, many experiments can not be carried out on new computers, often using the platform before Windows XP, and shutting down or even uninstalling antivirus software. And this is not consistent with the training objectives of this course [2].

1.1 Wide Range of Knowledge and Scattered Tools

Computer security events involve a wide range of knowledge, both communication protocol vulnerabilities, and logical vulnerabilities. The existing security tools for these different vulnerabilities have their own standards,can not solve all the problems. As a result, the knowledge of this course is messy, the tools are piled up, and the theories are empty. Teachers have no way to start, and they can only grasp it at random. Students can't build a systematic knowledge system when they are learning. So students lose interest and motivation in learning, and teachers also lose interest in teaching. Therefore, the first problem of this paper is to reconstruct the knowledge system of this course according to the cognitive law of the higher vocational students. According to the requirements of program modular design, the knowledge module of the course is reconstructed, the network security events are restored in the laboratory network environment, the scene of network security events is designed, the key knowledge points are highlighted, the students can be personally on the scene, and the key to the course design is to grasp the excitement of the students. Based on the principle of " easy to difficult", this paper takes the basic knowledge of network protocol as the starting point, and takes the difference and cognitive psychology of students into full consideration around the center of network security events, and constructs knowledge modules from shallow to deep, so that students can drill in with all their heart [3–5].

1.2 Design Traps in Knowledge Module

According to the professional job requirements of network security engineers, network security engineer should have the ability to deal with all kinds of common problems. In order to let students experience common problems in professional work in advance and accumulate experience in solving problems, we should set traps in the modules of each chapter, and let students fall into traps without knowing it, so that they can experience the feeling of being in deep trouble. Under the guidance of teachers, students are rescued from traps. The knowledge point of setting traps should be extracted from the basic knowledge of the chapter module, and the key point of this chapter is to rescue them from the traps. The effect of doing so is that the knowledge of this chapter is not separated from the chapter knowledge, and students' impression of the chapter knowledge points is deepened. Therefore, it is a key point to design knowledge trap around each chapter knowledge point [6–8].

1.3 Selection of Experimental Platform and Security Tools

At present, there are various operating system platforms in the network. Windows is a popular personal desktop platform for users. Different versions of windows have different characteristics. There are many network security tools on each operating system platform. Different security tools may be created by different standards, and tools based on different standards may produce different experimental results, this brings some confusion to learners. Choosing the right platform and tools is another key of this study. Considering the characteristics of various platforms and tools, combined with learners' own foundation, the paper makes a trade-off between platforms and security tools according to the principle of sufficiency in different environments. Experimental platform and security tools are set up for modules in different situations, so that the selected tools can be used in different situations, so that learners have a sense of acquisition and achievement, and achieve the goal of modular learning. The platform and security tools of unity and ease of use is often a pair of contradictions, this paper uses a compromise method, sometimes sacrifice ease of use for unity, sometimes sacrifice unity for ease of use [9–12].

2 Experimental Environment and Equipment

It is an important feature of this course to stimulate network security events in the laboratory environment. Because there is no regularity in the occurrence of security events, the network equipment involved in different experiments is also different. If you buy the required experimental equipment, it will increase the construction cost of the laboratory and increase the difficulty of laboratory management. fortunately, the development of virtual technology injects new vitality into network technology. Virtual technology can be used to complete network experiments that require multiple devices on a single host, which is flexible and can simulate various complex experiments. On the one hand, the use of virtual software technology reduces the cost of network experiment, on the other hand, it avoids a lot of time and work for the maintenance of network equipment.

At present, the mainstream network virtual software has a variety of products, including GNS3 and PT simulators. Using these simulation software, we can simulate a variety of network equipment in the laboratory environment. There are VMware and virtual box to simulate computer, among which virtual box is an open source virtual software. Using these two kinds of software, we can simulate and build all kinds of target aircraft. A key problem of network security experiment is that both sides need their own attack platforms. We can use the host built by virtual software to install all kinds of required operating system platforms. Considering the friendly interface of Windows platform and the friendly interface of security tools, we can choose to do the preliminary experiment attack platform, and Linux is the open source software, the running environment has low configuration requirements and high running efficiency, but it is not used to be a pair of contradictions with open source tools. In order to master ht security of learners, this paper uses open source Linux system as the attack platform. The mainstream Kali Linux and Back Track Linux are adopted. These two platforms integrate most of the network security tools. All the tools are classified according to their functions and are consistent and effective. This ensures that this course is an experimental link and saves the risk of being killed due to the upgrade of the virus library. Use the virtual software to simulate the experimental environment of multiple target aircraft, as shown in the Fig. 1.

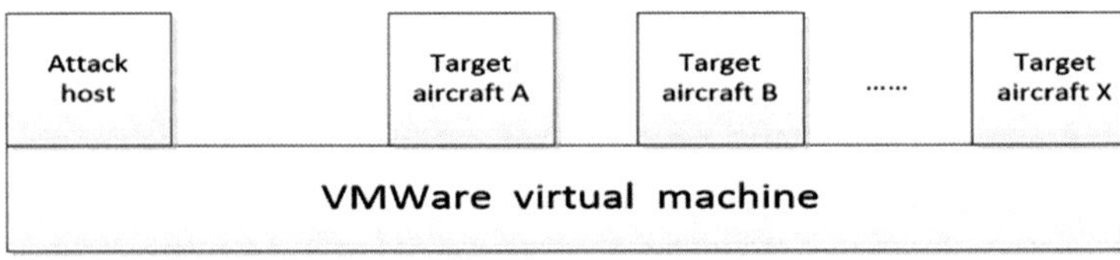

Fig. 1. Virtual attack and defense environment

3 Design of Red-Blue Confrontation Course Module

Using a virtual network to simulate an actual network requires simulating the deployment of the actual network and the setting of network security policies. According to the actual situation in a network security incident, it is often the network security incident that occurred before, and the network security measures are remedial measures taken after the network security incident. To prevent the recurrence of similar security incidents, security measure lag behind security incidents. Therefore, the network security course are designed from the two directions of attack and defense [11, 12].

3.1 The Design of the Red-Blue Counterattack Course Module

From the perspective of the network attacker, the target of an infiltration attack needs to know oneself and the enemy. Based on the infiltration process of the network attacker, this article combines the cognitive laws and cognitive processes of higher vocational students. From the periphery of the network to the inside of the network, seven course teaching modules are designed, including information collection, host discovery, service identification, operating system identification, vulnerability scanning, penetration targets, and data packet capture. Each teaching module is under set up the sub-module again, as shown in Fig. 2.

The design of the information collection module is an indispensable link before the peripheral penetration of the target network. According to the information collection method of the target network, two sub-modules are set up under this module: the passive information collection module and the active information collection module. The host discovery modules is designed to detect the specific target host after mastering the target network information. The module is divided into two sub-modules: the second-level discovery module and the third-level discovery module. The service identification

information gathering	passive information collection
	active information collectionn
host discovery	second layer discovery
	third layer discovery
service identifacation	port scan
	service identifacation
operating system identifacation	Windows operating
	Linux operating
vunlerability scanning	service version information
	open vunlerability library
permeating target	Web permeation
	SQL injection
packet hijacking	Wireshark
	TCPdump

Fig. 2. Course attack module

module is to detect the service running on the target host after confirming that the target host is online. The module is divided into a port scanning module and a service identification module. The operating system identification module design is to detect the operating system platform of the target host. The vulnerability scanning module is designed to scan and detect the vulnerability of the service software based on the service version information that has been detected on the target host and the vulnerability database of the service software that has been disclosed. The design of the penetration target module is a specific penetration of the target host. According to the penetration method, the module is divided into a web penetration sub-module and a SQL injection sub-module. According to the needs of network penetration, a data packet capture module is designed, which is divided into the Wireshark sub-module and the TCPDUMP sub-module to adapt to the actual network host running in the GUI environment and the NO-GUI environment.

3.2 The Design of the Course Module of Red and Blue Against Defenders

As the defender of the red and blue confrontation, according to the cause of the network security incident and the network intrusion path, the targeted access security strategy for network equipment is implemented. According to the network scope, the network equipment is classified into security strategies, and four teaching modules are designed: host reinforcement module, firewall deployment module, log audit module, and virus protection module. The host hardening module is mainly for the host, and is divided into a system account and password hardening sub-module and a file directory permission sub-module. The firewall deployment module mainly sets the firewall setting strategy, which is divided into the system firewall setting sub-module and the network firewall setting sub-module. The log audit module mainly audits the traffic in and out of the network. According to the traffic direction, this module is divided into an internal network access external network log audit sub-module and an external network access internal network log audit sub-module. The anti-virus module is the design of the virus database of network antivirus tools, including the update setting sub-module of the virus database and the regular anti-virus sub-module. as shown in Fig. 3.

machine host reinforcement	system account password reinforcement
	file direction permission setting
firewall deployment	system firewall setting
	network firewall setting
access policy setting	internet access policy setting
	intranet access policy setting
antivirus policy setting	update strategy of virus database
	system anti-virus strategy

Fig. 3. Course reinforcement module

4 Rules of Network of Confrontation

How to conduct practice teaching based on different roles of network red-blue confrontation is a key of this research. If effective practice rules and experimental strategies cannot be set up, classroom teaching will not be able to grasp the key points, and the students are also confused. The design of detailed experimental rules and steps of network countermeasure can not only ensure the integrity of the experiment, but also the basis for learners to benefit from it, indirectly improve the classroom learning effect and achieve the teaching gaol. In this paper, the rules of network countermeasures are set as follows.

(1) network construction. The first stage is network construction, which is the first stage of each chapter module of network countermeasure. The tasks in this stage include sub-net division, switch and router configuration, building target aircraft, network service configuration on target aircraft, firewall configuration and network connectivity test. The adversary and opponent have their own target. The target machine opens the service required by the module in this chapter as the target of the later attack.
(2) Network reinforcement. The second stage of network confrontation is network reinforcement. In this stage, the main task is to reinforce their own network, especially their own target service. The rule of reinforcement is to ensure that the target service of one's own team can be accessed by the other party in third stage and cannot be shutdown. The reinforcement time is exact and cannot be overtime.
(3) Network attack and defense. The third stage is network attack and defense. In this stage, the attack and defense use tools to scan each other's target on their own attack platform. Once a flag of each other's target is detected, the attack is considered successful and the corresponding flag score is obtained. Get different flag score to accumulate, and finally get the winner with high score.
(4) Making experimental documents. The last stage of network confrontation is to make documents. The documents produced in this stage are also the final results of network confrontation experiment. The documents include the contents of each part of the network construction in the first stage, the content of the second stage network reinforcement and the flags that the third stage attacks the other party and obtains. The final document is also apart of the evaluation of students' achievements.

5 Conclusion

Network security confrontation is a course with strong practical operability, but the complexity of practice and the indirectness of result confirmation are often out of proportion. The reasonable allocation of the total time and the progress of the experiment in the practice of the course is an evaluation index of the whole practice. The quantification of the evaluation index should be explored repeatedly in practice, it take time to set.

Acknowledgement. This paper is supported by "Research project of higher education and teaching reform of Hainan Provincial Department of Education: Hnjg2020-144".

References

1. Zhang G., Zhou R.: Research on the breakthrough of defense system from the perspective of red team. In: Xu, Z., Parizi, R.M., Loyola-González, O., Zhang, X. (eds.) Cyber Security Intelligence and Analytics. CSIA 2021. Advances in Intelligent Systems and Computing, vol. 1342 (2021). Springer, Cham. https://doi.org/10.1007/978-3-030-70042-3_31
2. Wang, D.: Teaching reform of computer network security course. Chinese and Foreign Entrepreneurs, No. 6, pp. 35–36 (2019)
3. Zhang, G., Wu, H.: The web penetration based SQL injection. In: Hung, J., Yen, N., Chang, J.W. (eds.) Frontier Computing. FC 2019. Lecture Notes in Electrical Engineering, vol. 551 (2020). Springer, Singapore. https://doi.org/10.1007/978-981-15-3250-4_89
4. Zhang, Q.: Exploration on the integration of red blue confrontation and safety supervision. Electron. Des. Eng. **11**(8), 36–37 (2015)
5. Zhang, G., Li, W.: A security reinforcement scheme for the server. WOP Educ. Soc. Sci. Psychol. **14**, 92–92 (2018). 7
6. Weidman, G.: Penetration test. China, people's post and telecommunications press, pp. 124–125 (2019). 5
7. https://baijiahao.baidu.com/s?id=1631597347538496274&wfr=spider&for=pc
8. Red blue confrontation. https://he1m4n6a.github.io/
9. Red blue confrontation. https://www.secpulse.com/archives/107194.html
10. Common network attack technology. https://blog.csdn.net/qq_39507276/article/details/88883421
11. https://baike.baidu.com/item/2018%E4%B8%AD%E5%9B%BD%E7%BD%91%E7%BB%9C%E5%AE%89%E5%85%A8%E6%8A%80%E6%9C%AF%E5%AF%B9%E6%8A%97%E8%B5%9B/23380041?fr=aladdin
12. https://blog.csdn.net/weixin_34342992/article/details/92304115

Design and Implementation of an Information Security Platform for the IoT Based on Blockchain

Chao Duan(✉)

School of Engineering, Guangzhou College of Technology and Business, Guangzhou, Guangdong, China

Abstract. As in the blockchain system, with the continuous access of various devices, there will be information security risks. Based on the security vulnerabilities of the IoT, this article combines blockchain technology to design a distributed Internet of Things information Security management system. The system can access devices through identity authentication strategies, and control the authority of various devices. At the same time, the security arbitration mechanism is integrated into the smart contract. In terms of information management, the data is classified and stored to ensure data security. This article is based on the design and implementation of a blockchain-based IoT information security platform.

Keywords: Blockchain · Internet of Things · Information security · Platform design

1 Introduction

As an important technology of informatization, the Internet of Things (IoT) has been widely used in all walks of life. For example, there are applications of IoT technology in power systems, water supply companies, transportation departments, and smart life [1]. However, as a large number of devices are connected to the network, this may cause external intrusions or virus attacks, and information security issues have become more and more prominent. According to surveys in foreign markets, on average, nearly tens of billions of terminal devices are connected to the IoT every year. These devices are not only computer terminals, but also wireless access devices such as mobile phones and tablets. More devices will be connected in the future. In this way, the information security of the IoT will also become a hot issue in the information industry [1].

Research on blockchain technology has received extensive attention at home and abroad. Initially, blockchain technology was mainly used in Bitcoin, but now it is mainly used in policy departments, financial institutions, securities and futures, corporate production, medical and health, and information Industries, etc., have their own research projects. These projects are paying attention to the global academic community, and they are constantly technologically updated and developed. Blockchain technology will be more widely used in the future [1].

B. J. Jansen et al. (Eds.): *International Conference on Cognitive based Information Processing and Applications (CIPA 2021)*, LNDECT 85, pp. 382–389, 2022.
https://doi.org/10.1007/978-981-16-5854-9_48

2 Correlation Technique

2.1 Blockchain Related Technologies

The blockchain is a data storage system based on a distributed network. The ledger is the core mechanism of the blockchain, which mainly includes data blocks and links. The blockchain technology has the characteristics of decentralization [2]. The entire network can be regarded as a distributed ledger, and the ledger of each node in the blockchain network is backed up by information; the blockchain has a strict authentication mechanism, In the process, the data generated by all transactions must be confirmed by more than half of the nodes before it can take effect. This is the authentication mechanism that all nodes comply with; the blockchain data has the characteristics of non-tampering, and all nodes in the blockchain network are stored If there is data information, if you want to modify the data information, you must be able to verify the hash of the previous node, and at least more than half of the nodes must be verified [2].

2.2 Authentication Mechanism

Before an IoT device is connected to the network, the identity of the device must be authenticated according to the authentication mechanism, to ensure that the system users are legal and effective. In addition, users who access the network can be authenticated and filed for future system use certificate. The authentication mechanism of the blockchain mainly includes two processes of identification and authentication, which are an important part of entering the IoT system. For illegal users, the authentication mechanism will prohibit the user from entering the network, while for legitimate users; the system will authenticate and allow the system to connect. Enter the system to prevent the network from being invaded by the outside world [2]. At the same time, for the equipment terminals connected to the network, when communicating with each other, identity authentication must be carried out first, that is, the user's authentication certificate is available for inspection at any time, and the system performs data on the authentication process. Backup, which can ensure the security of the network.

2.3 Permission Control Strategy

For each user of the IoT, they have their own usage rights, and access the corresponding network resources according to the rights. Users will not be able to access those outside the rights. Rights control is an important technology for IoT security. Through the rights control strategy, Real-time monitoring of all users, preventing illegal users from accessing unauthorized network resources, so as to ensure the security of network resources [3]. Permission control generally uses role control. In this mode, the user and access permissions are still in a one-to-one relationship, but roles are replaced in the two, that is, the user may appear in other roles, and this role is granted accordingly the scope of the permission is mainly determined by the specific content of the work. When the permission is removed, the role will also be cancelled, which can ensure the speed of authorization [3].

Smart contract is a digital contract method that uses blockchain technology. Generally, program code is written into the blockchain, and the transaction is guaranteed through a specific operating mechanism. The operation of the contract is not interfered by the outside world. The smart contract method first negotiates through the content of the contract between the two parties. If the two parties reach an agreement, the system will publish the content of the contract in the system through the program code according to the contract logic. Once the contract is established and written into the system, the contract will automatically take effect, and the outside world will not be able to modify it [4].

2.4 Intrusion Detection Mechanism

In the IoT, external intrusions pose a serious threat to network security, and the system must have a corresponding mechanism to deal with this security threat. The intrusion detection mechanism is a protection mechanism for the system to deal with external intrusions [4]. The mechanism is mainly based on the operator's behavior, operation logs, system audit results and other data information for effective monitoring. According to these data information, the system will analyze the big data to determine whether the user's operation behavior is an intrusion, and the system will immediately take an early warning measures, using a forcible mechanism to prohibit the user's behavior and prevent the intrusion from occurring in one step [4].

3 System Design Analysis

3.1 System Requirements Analysis

The system requirements analysis is mainly divided into functions such as equipment access management, equipment operation management, equipment data transmission, data and resource acquisition and other functions for analysis and authentication. For device access management, it is mainly that the terminal device must be registered and authenticated before accessing the IoT. After the authentication is passed, the registration is successful, and then you can access the IoT. Use the platform to perform various operations [5]. The authentication process is mainly for The legality of the ID card of the device is mainly realized by the contract function of the blockchain, and every node on the blockchain is certified as much as possible, which can ensure the authenticity and effectiveness of the certification process. At the same time, the system will verify the certification after passing the certification. Information is automatically backed up to prevent tampering; equipment operation management mainly requires unified management of all equipment after users access the IoT. Generally speaking, the operating status of IoT equipment is mainly divided into two types: online and offline [6].

3.2 System Use Case Diagram

According to the analysis of the system function, all the devices involved in the system are those connected to the IoT. They must be authenticated before entering the system.

The authentication process is automatically completed by the contract program. During the authentication process, the data information will be stored in the ledger and log in to the system. After the equipment operation, the system will automatically read the registered equipment registration form of the ledger for comparison and authentication. When the equipment is authenticated, data transmission and information acquisition within the network will also pass the equipment authority authentication. In order to carry out data transmission, the acquisition of data resources [5]. The system also needs to be authenticated before proceeding.

3.3 System Architecture Design

This system mainly deals with the data exchange between devices in the IoT. The exchange of information requires effective management of authentication mechanisms and permissions, as well as intrusion detection mechanisms [6]. Each node in the system is mainly used for information processing and processing, packaging services, system management, etc., and each node of the blockchain is mainly used for data storage and authentication management, shown as in Fig. 1.

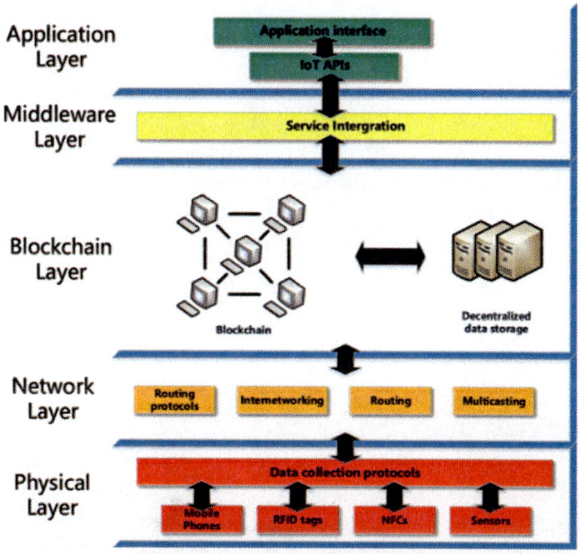

Fig. 1. Blockchain architecture

The bottom layer of the system is mainly a network formed by each node of the blockchain. Each node is the Geth client. The main function is to verify the information during the transaction process between the two parties, the backup of transaction information, and the return of contract data. In addition, each node authenticates the request of the upper layer before passing it, and it must pass after unanimously passed [7].

3.4 System Function Module Design

According to the demand analysis of the entire IoT system and the assumption of the overall architecture, this paper divides the functional modules of the system into four major functions: authentication management module, operation management module, security detection module, and information management module [7], shown as Fig. 2. Each device terminal of the IoT must perform identity authentication before accessing the network, and the system authentication request is sent to the access management module, which converts the format of the request and sends the converted request information to the node of the blockchain. The block chain system will call the contract program for identity authentication, return the authentication result to the module, and store the authentication information at the same time.

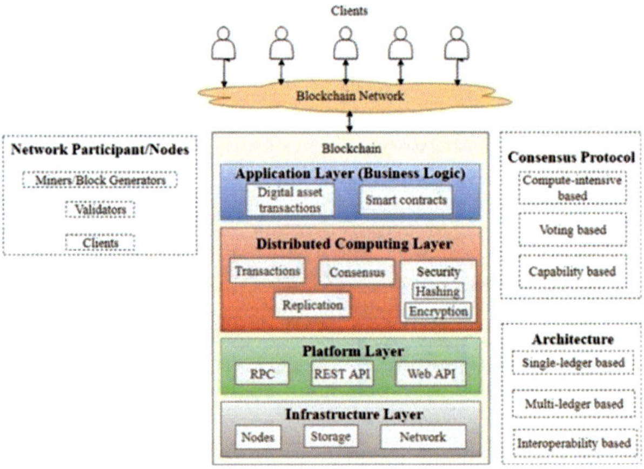

Fig. 2. Blockchain function module design

4 Blockchain Network Construction

Since the blockchain runs in a distributed environment, the client is mainly installed on multiple computers, and each computer is used as the node of the blockchain. Deploy nodes in the computer separately. The main task is to set up the genesis block of each computer. Create the first one, and then add new blocks to other computers in order, so that each node can be connected to a simulated blockchain. In the network, the same Genesis block is set by each computer terminal in order to maintain the ledger together [8]. After each node is deployed, run the Geth client to link each other, and at the same time, add each node to a list so that each node can transmit data, thus forming a simple blockchain network, shown as Fig. 3.

5 The Function Realization of the System

5.1 The Realization of Authentication Management Module

In the current IoT system, there are three main modes for device identity authentication, namely, static password authentication, dynamic password authentication and biometric authentication [8]. The static password authentication is mainly through the user's preset password for authentication. The password is static. If the user does not modify it, the password will always be valid; the dynamic password is mainly based on the built-in password chip to calculate the dynamic password, and the password authentication must be dynamic. The password and the built-in password chip are completed together; the biometric authentication is mainly based on a certain feature of the human body, such as face recognition, fingerprint recognition, expression recognition, etc. [9]. The structure of the certification table mainly includes the equipment name, identification code, equipment characteristics and remarks.

5.2 Implementation of Operation Management Module

When information is sent and information is obtained between devices of the IoT, the interaction module of the system receives the information of the device and transmits the sending request to the data transmission module, and the data transmission module sends a permission verification request to the blockchain system. The contract program will automatically verify the authority of the device and return the verification result to the data transmission module [9]. If the verification is passed, the data transmission module will transmit the information to the interactive module for information transmission; otherwise, the interactive module will not be notified.

5.3 Realization of Safety Detection Module

The safety detection function is mainly realized through the CIDF program. The CIDF program mainly includes three subroutines, namely the event generation program, the event analysis program and the data writing program. The system first monitors the running status of the sent information, and then the test results It is sent to the event analysis program, and the event analysis program compares the message behavior with the malicious event database [10]. If the comparison is successful, it means that the information is a malicious message, and the message is returned to the system to restrict the authority of the device.

5.4 The Realization of the Information Management Module

The data processing function is mainly realized by calling the JSON-PRC interface program [10]. The program packs and encapsulates the information according to the type of request through the call form, and then must be sent to the corresponding address; the message management is mainly realized through the data processing function, JSON -The PRC interface program will call the message manager of the system to classify and save the information [10].

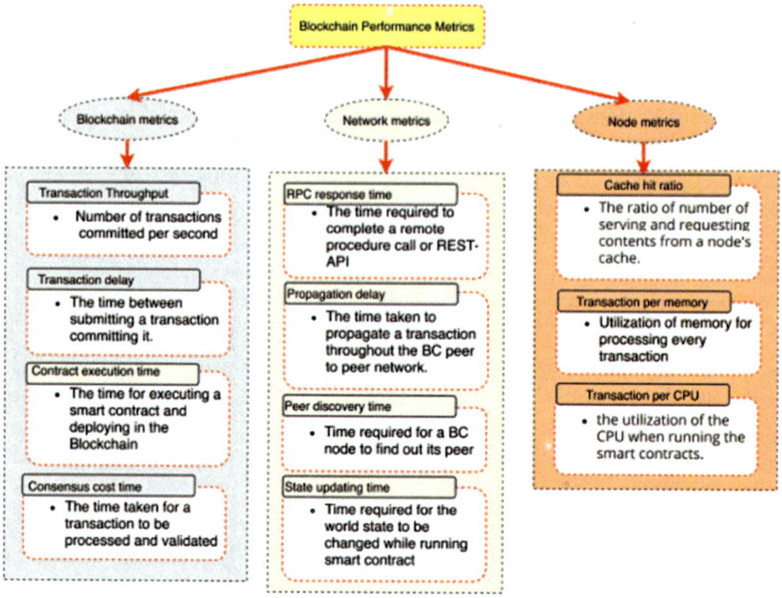

Fig. 3. Blockchain safety detection module

6 Conclusion

As the IoT has been widely used in various industries, a large number of terminal devices are connected to the IoT, which has greatly promoted work and life. However, the information security problem of the IoT has become increasingly prominent. Especially the network management is very tricky. For this reason, this paper proposes a design plan for an information security platform system based on blockchain technology to solve this technical problem and achieve certain scientific research results, but the system still has some shortcomings. For security detection, the detection range of malicious messages needs to be broadened, and the detection strategy needs to be improved to make the detection range more comprehensive and the detection mechanism more complete. In addition, for the integration of blockchain technology, there may sometimes be incompatibility with the system.

Acknowledgement. Fund: 2019 Guangdong Province Ordinary University Young Innovative Talent Project (Natural Science): 2019KQNCX232.

References

1. Chu, R., Zhou, X.N., Li, S.G.: Research and application of data governance collaboration method based on blockchain. Comput. Appl. **7**(11), 962–964 (2016)
2. Dang, J.Y.: Discussion on the application of blockchain technology in the medical field. China Med. Educ. Technol. **11**(12), 1080–1083 (2017)

3. He, Y.Y.: Research on service composition strategy in cloud computing environment based on blockchain mechanism. Comput. Appl. Res. **11**(06), 256–257 (2017)
4. Wang, B.N., Jiang, P.H., Qing, X.X.: Research on the characteristics and laws of network public opinion information dissemination under the blockchain environment. Intell. Mag. **8**(08), 4–7 (2014)
5. Huang, K.: Research on library scientific data storage model based on blockchain technology. J. Libr. Sci. **6**(18), 164–165 (2016)
6. Wu, J.: Data in the key technology of blockchain. Comput. Technol. Dev. **10**(05), 16–19 (2015)
7. He, W.Y., Yang, X.D.: The application advantages and risk response of blockchain technology in the financial industry. Contemp. Econ. Manag. **9**(01), 38–43 (2018)
8. Liu, D.M., Zhang, R.N.: Digital currency evolution based on blockchain. Comput. Appl. Res. **10**(8), 10–13 (2018)
9. Xu, S., Miao, Y.F.: Research on the application of blockchain technology in the field of investigation. J. Inf. Resour. Manag. **11**(05), 32–37 (2018)
10. Jin, Z.: Mobile terminal authentication management method based on blockchain technology. Electromech. Eng. Technol. **12**, 22–24 (2017)

Check for updates

Improvement and Software Realization of Residential Area Landscape Design and Protection System Based on the New Coronavirus Epidemic

Yang Wang[1](✉), Yiting Wu[1], and Xiaohe Chen[2]

[1] Shanghai China Overseas Development Group Co., Ltd., (East China), Shanghai, China
[2] Shanghai China Overseas Development Group Co., Ltd., (Suzhou), Jiangsu, China

Abstract. The novel coronavirus pneumonia was discovered in Wuhan at the end of 2019. The outbreak of the new coronavirus has disrupted normal social life. The novel coronavirus pneumonia epidemic is a public health event sweeping the world and has a profound impact on all aspects of people's lives. The purpose of this article is to study the improvement and software realization of the landscape design and protection system of residential areas under the influence of the new coronavirus epidemic. Based on protection measures and national strategies for responding to major public safety incidents, this article aims to improve the protection system of residential areas and change the landscape design, and design a software system. This article collects, organizes and analyzes relevant information by using literature methods. Conducted on-site investigation and sandbox simulation, designed and implemented a software system based on the residential area landscape design and protection system, and finally conducted experiments on the system. The experimental results show that the system can improve the protection system of residential areas, redesign the landscape, and improve the protection ability in response to major security incidents.

Keywords: Novel Coronavirus · Sandbox simulation · Landscape design · Software system

1 Introduction

Generally speaking, public health emergencies usually seriously endanger people's production and life. In this special social situation, any negative information about the event will aggravate public anxiety and panic. Rumors may have serious negative effects [1]. The impact of novel coronavirus epidemic on the public at the peak of the epidemic is inevitable [2]. As the protagonist of major public health emergencies, new coronavirus poses a serious threat to people's health and causes social panic because of its sudden, infectious and wide spread characteristics. Therefore, it is necessary to improve the protection system of residential areas [3].

B. J. Jansen et al. (Eds.): *International Conference on Cognitive based Information Processing and Applications (CIPA 2021)*, LNDECT 85, pp. 390–396, 2022.
https://doi.org/10.1007/978-981-16-5854-9_49

The intelligent system of residential area is the organic combination of modern building technology, computer technology and network control technology, security defense, information service and management, home office and intelligent system integration to provide services for residential areas [4]. Only by providing high-tech and intelligent auxiliary management means and improving the service level of property management in residential areas can we realize fast and efficient service and provide residents with safer, more convenient and more comfortable home environment territory [5]. However, since major public safety incidents can cause fever, infrared thermometry should be increased in residential areas to reasonably screen the people in and out of the residential areas [6].

Residential green space is the main place for residents' leisure and entertainment [7]. 67% of China's land exists in the form of mountains and rivers. Mountain city landscape is the bottleneck and difficulty of urban development. At present, most of the theoretical studies on the environmental assessment of residential areas are conducted from the perspective of landscape aesthetic objects, which brings more landscape lines, recreation space and plant community levels to residents. With the promulgation of the national new urbanization plan (2014–2020), the urban environment began to change to humanism [8, 9]. From the perspective of residents' subjective aesthetics, the environmental evaluation factors suitable for the construction of Mountainous Residential areas are selected. The landscape environment of residential areas in mountainous cities is particularly important [10].

2 Method

2.1 Intelligent Introduction of Wireless Sensor Network

Wireless sensor network (WSN) is composed of a large number of different rules. The self-organizing method is used to construct a small-scale wireless network with sensing, computing and communication capabilities of grid and functional micro sensor nodes. A large number of sensor nodes can understand, monitor and collect the information of the monitored objects or the surrounding environment in the distribution area through real-time perception.

2.2 Introduction of Intelligent Wireless Sensor Networks

Because of the independence of intelligent protection in residential areas, it is difficult to realize the integration and timely sharing of information. In the context of information technology, wireless sensor network is used to establish intelligent protection system to redesign the traditional protection system, promote the integration of residential protection system and national public security system, and realize the full sharing of information, which is conducive to strengthening protection and promoting the innovation of overall management of residential areas. The specific implementation is as follows: the whole network coverage is realized in the residential area. Once someone has strong fluctuation of vital signs under monitoring, his medical records will be immediately transferred out and reported to the residential area administrator. If the residential area administrator is not in, the system will immediately contact the medical and public security systems for rescue. Moreover, the intelligent residential area based on wireless sensor network system can also be analyzed and compared to select the most appropriate set of landscape design schemes, which can be combined with modern technology to set up virtual intelligent landscape.

2.3 Application of Modern Safety Protection Theory to Establish Intelligent Residential Area

On the basis of modern security protection theory, we first put forward the idea of reorganizing all the traditional protection standards. Secondly, we should formulate a unified information standard, complete the basic work, and finally improve the public basic network and ideas. Further strengthen network security management. To sum up, although the theory of wireless sensor network system has entered our country for a few years, until recently, the theory of wireless sensor network has not been applied to residential areas and the research on protection and monitoring in domestic and foreign literatures, which lacks certain systematicness. Therefore, we need to explore and try our best to build real intelligent residential areas.

3 Experiment

3.1 Sandbox Simulation

Because of the complexity of the experiment, because it involves the majority of residents in several residential areas, we first use sandbox simulation to get the general details, and analyze and discuss the repeated experiments to draw general conclusions and countermeasures, and then conduct field experiments.

3.2 1800 Households in 15 Residential Areas with the Same Conditions were Selected as the Survey Objects

1800 households in 15 residential areas with the same conditions were randomly selected to conduct field simulation experiment and subsequent questionnaire survey. Two sets of questionnaires are used. The first is about the use of old residential protection systems and landscape design, and the second is about the use of new intelligent residential systems. For these 15 residential areas, our questionnaire is mainly divided into two parts, that is, asking the residents' attitudes towards the residential areas before and after the change and their views on the future prospects.

4 Results

4.1 Results of the Questionnaire

The results of previous questionnaires are shown in Table 1 and Fig. 1. Comparing the results of Table 1 and Fig. 1, it is not difficult to find that the optimized intelligent

Table 1. Evaluation of residents on the management of new and old residential areas

Evaluation attitude Number of copies/servings	New Residential Management System	Old Residential Management System	Total number of copies/servings
Good	1600	300	1900
General	120	1200	1320
Not good	80	300	380
Total number of copies/servings	1800	1800	3600

residential area management is more popular than the old residential area management before.

According to the data in Table 1, through the intuitive data, we can see that nearly 90% of the residents express good attitude after experiencing the new residential area management, while in the old residential area management, only less than 20% of the residents express a good attitude, and most of them follow the trend before experiencing, They think it is general, and express the welcome to the intelligent residential area quickly after the experience, so it shows the benefits of the new intelligent residential area.

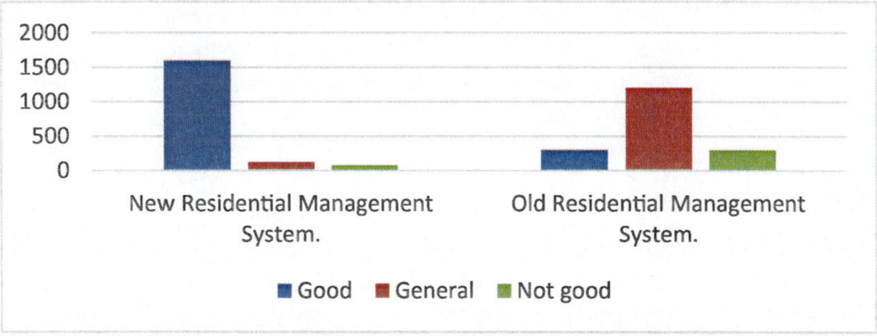

Fig. 1. Residents' Evaluation on the management of new and old residential areas

Figure 1 is a column chart based on Table 1. Figure 1 expresses the evaluation of residents on the management of new and old residential areas more intuitively, highlighting the advantages of new residential area management.

4.2 Preparation of Requirements for Intelligent Residential Areas

Wireless sensor network is composed of data acquisition network, data distribution network and control management center. The main components of the system are integrated sensor node, data processing unit and communication module. Each node forms a distribution. The network optimizes the collected data and transmits it to the information processing center through radio waves. It has the characteristics of fast construction, convenient deployment and not easy to be restricted by the target environment. It is especially suitable for environment monitoring, military reconnaissance, medical monitoring, intelligent community, warehouse management and other fields where personnel are difficult to reach or hard to reach. It has broad application prospects. So we finally chose it as the cornerstone of intelligent residential areas.

Housing security in residential areas involves the needs of safety and fire protection. Such as the installation of smoke sensors, temperature sensors, special gas sensors, infrared sensors, vibration sensors, wireless microwave ovens, etc., to prevent indoor fire and excessive harmful gases. The alarm device through the monitoring of the community environment and public facilities, to prevent the invasion of thieves, timely elimination of hidden dangers. This involves multidisciplinary data collection, while wireless sensor networks involve multidisciplinary. At present, one of the research hotspots in the field of industrial technology is known as one of the four high-tech in the 21st century.

We can define smart home as a process or system. In other words, the use of advanced computer technology, network communication technology, generic cabling technology, all kinds of equipment and resources related to home life, unified management, so that family life more comfortable, safe and efficient. Smart home and ordinary residence not only have the traditional residential function, but also can provide comfortable and safe family living space, and combine the original single static living space into an intelligent and active environment. It provides a full range of information exchange function for people's life, meets the requirements of free exchange of information between family and the outside world, greatly improves people's life style and strengthens people's life style. In order to meet the people's pursuit of fast-paced work in the information society of the 21st century, the safety and comfort of family life can be improved. Style and completely open living environment.

4.3 Safety Guarantee for Its Normal Operation

However, in order to ensure its normal implementation, we need certain conditions. The first is a secure physical environment. The protection of safe physical environment includes physical site selection, physical access control, anti-theft and anti-damage, lightning protection and fire prevention, waterproof and moisture-proof, anti-static, temperature and humidity control, etc. The protection of secure communication network includes network architecture, communication transmission and credibility verification. Access authentication is carried out for each layer of the network architecture. Finally, safe operation and maintenance management. It can continuously monitor the abnormal, loopholes and risks, malicious code, performance failure and other behaviors of important systems, networks, personnel assets, business and other services, deeply analyze the causes and formulate emergency plans, so that the system can recover quickly and the failure can effectively prevent losses.

4.4 Requirements of Landscape Design

The landscape is the nature that people yearn for, the landscape is the human habitat, the landscape is the artificial handicraft, the landscape is the material system which needs the scientific analysis to be understood, the landscape is the problem to be solved, the landscape is the resource that can bring wealth, the landscape is the ideology reflecting the social ethics, morality and values, the landscape is the history, the landscape is the beauty. As the object of landscape design, landscape refers to the complex composed of land and space and objects on the land.

The survey results show that, in terms of visual beauty, the vegetation on the hillside > the residential area at the top of the mountain > the residential area in the valley. The aesthetic value of plant landscape in valley type residential area is the lowest, and most areas are negatively correlated. Valley type residential areas are mostly located in depressions, and plant landscape tends to gather to the center, resulting in residential chaos. Plant landscape recreation satisfaction decreased. For the study of plant landscape planning and design in Mountainous Residential Area, although the coefficient of the model reflects the influence of various factors on the landscape score, it cannot reflect the influence of each factor on the landscape score. In order to coordinate the collocation of plant landscape elements, we should pay attention to the color matching of plant landscape. When planning the plant landscape of residential area, we should decorate the color leafed plants properly and pay attention to the coordination of plant landscape color in the regional space.

According to the survey, evergreen broad-leaved forest is the main natural vegetation type in the city, the seasonal change of plant color is not obvious, and the satisfaction degree of regional landscape effect is improved. When planning the plant landscape of residential area, the plant landscape color should be properly decorated. The higher the importance of vegetation in residential areas, the better the stability of plant communities, the richer the plant landscape levels in the field of vision, the more likely residents are to feel close to nature, so they are easily loved by residents.

According to the above discussion, we should better set up an intelligent residential area. Sometimes we can't verify it from one aspect, just like the cross of students and subjects. Many applications will lead to some problems in the application. We can only restrict and develop each other. In order to normal development, we will make more efforts and hope to make every one of them in the future Residential areas can truly achieve intelligent protection and landscape design.

5 Conclusion

Novel coronavirus pneumonia is the most rapid public health event since the founding of new China. Its spread is the fastest and the most widely spread. Control and prevention are the most difficult. China's novel coronavirus pneumonia prevention and control has quick and effective mobilization and organization capabilities, coordinated and orderly deployment capabilities and professional effective emergency treatment capacity, and has won high praise from WHO and the international community. At the same time, epidemic prevention practice has also exposed the deficiencies in the field of public health emergency treatment in China, which has sounded the alarm siren, the alarm bell and the alarm bell. Therefore, the purpose of this paper is to study how to protect the smallest unit of residential area to avoid more serious situation in the future. We hope that in the future, we can get the information we need in time and face it in time. Therefore, we will continue to revise and make progress towards perfection.

References

1. Pan, H., Yu, X.: Study on the development trend of tourism in the cloud under the background of new crown epidemic. Open Access Libr. J. **07**(8), 1–6 (2020)
2. Garber, K.: The new liver epidemic. Nat. Biotechnol. **37**(3), 209–214 (2019)
3. Nakhaee, S., Mehrpour, O.: Letter to the editor: opium addiction as new source of lead poisoning: an emerging epidemic in IRAN. EXCLI J. **17**, 513–515 (2018)
4. Cao, B., Zhao, J., Yang, P., et al.: 3D Multi-objective deployment of an industrial wireless sensor network for maritime applications utilizing a distributed parallel algorithm. IEEE Trans. Ind. Inf. **14**(12), 5487–5495 (2018)
5. Osamy, W., Khedr, A.M.: An algorithm for enhancing coverage and network lifetime in cluster-based wireless sensor networks. Int. J. Commun. Netw. Inf. Secur. **10**(1), 1–9 (2018)
6. Zhao, M., Tian, Z., Chow, T.W.S.: Fault diagnosis on wireless sensor network using the neighborhood kernel density estimation. Neural Comput. Appl. **31**(8), 4019–4030 (2018). https://doi.org/10.1007/s00521-018-3342-3
7. Kobayashi, S., Yamada, H.: The proposal of housing external lighting incorporating light into wall lattice. Aij J. Technol. Des. **24**(56), 243–246 (2018)
8. Okada, N.: Adaptive process for SMART community governance under persistent disruptive risks. Int. J. Disaster Risk Sci. **9**(4), 454–463 (2018). https://doi.org/10.1007/s13753-018-0204-7
9. Farooqi, N., Gutub, A., Khozium, O.: Smart community challenges: enabling IoT/M2M technology case study. Life Sci. J. **16**(7), 11–17 (2019)
10. Yang, Y.: Smart community security monitoring based on artificial intelligence and improved machine learning algorithm. J. Intell. Fuzzy Syst. **38**(1), 1–13 (2020)

Optimizing the Configuration of Financial Supervision Equipment by Using Key Privacy Protection Technologies of Blockchain

Jianfei Ke[1(✉)] and Jianxiong Lin[2]

[1] Fujian University of Technology, Fuzhou, China
kejianfei@fjut.edu.cn
[2] Fujian Jiangxia University, Fuzhou, China

Abstract. The use of blockchain technology in the financial field can effectively break information barriers; track and audit the entire funding cycle, and help improve the transparency of fund use. It can simplify the transaction process, help reduce transaction costs and improve efficiency. However, compared with Western developed countries, the application of blockchain technology in China is still in its infancy, and its application in the financial field faces challenges such as the limitations of traditional concepts, low technology levels, and technology shortages. Research and application talents, laws and regulations are not perfect. It is necessary to innovate ideas, strengthen technology research and development, vigorously cultivate technology research and application talents, and improve relevant laws and regulations to promote the wide application of blockchain technology in the financial field. Based on this, this article studies the use of blockchain key privacy protection technologies to optimize the configuration of financial supervision equipment. This article mainly focuses on blockchain technology and its application in the financial industry, combined with an overview of blockchain technology as the main basis, and then analyzes the application of blockchain technology in the promotion of the financial industry, and makes recommendations. Use blockchain technology to better promote the development of the financial industry. Research shows that blockchain is a new application model of computers. Blockchain is second only to network technology. Mainframes and personal computers have the greatest potential and widest influence, and are essential for triggering industrial change power. Therefore, blockchain technology has great potential in the financial governance system.

Keywords: Blockchain technology · Financial industry · Financial field · Privacy protection

1 Introduction

With the continuous progress of China's social economy, the financial industry [1–3] has been well developed. In order to make the financial industry move forward faster, we need to constantly innovate and improve financial technology. The emergence of

B. J. Jansen et al. (Eds.): *International Conference on Cognitive based Information Processing and Applications (CIPA 2021)*, LNDECT 85, pp. 397–403, 2022.
https://doi.org/10.1007/978-981-16-5854-9_50

blockchain technology has laid a good foundation for the long-term development of the financial industry. Blockchain technology [4–6] is a new technology, which contains information about consensus building, decentralization and non-infringement. Using these functions can effectively solve some problems in the financial industry. In this paper, the blockchain technology and its application in the financial industry are deeply studied, so the blockchain technology can be better applied in the financial industry.

Blockchain technology is mainly an electronic financial system based on point-to-point information [7]. Department researchers believe that bitcoin is based on blockchain technology. Especially in the global scope, the first application of blockchain technology is the emergence of bitcoin. In bitcoin's e-money system, the storage unit is mainly in the form of blocks, and then blocks are connected by encryption algorithm to create a blockchain. In fact, blockchain technology is not a new technology, but a combination of existing technologies. At present, there is no specific definition of blockchain technology in the world, but in essence, it belongs to a distributed ledger [8, 9] and has the characteristics of decentralization [10]. It can also share data in the Internet, belonging to a database. At present, blockchain technology is divided into three types: private blockchain, public blockchain and joint blockchain. Among them, the main purpose of public blockchain is to ensure the security of the system through the central distributed operation and fair mechanism.

This article studies the application of blockchain technology in multi-level supply chain finance. The composition and functions of blockchain technology were explained, and several characteristics of supply chain financial services were summarized, and a blockchain technology compatible with supply chain financial services was found. On this basis, the role of blockchain technology in improving supply chain finance is analyzed. Finally, in the field of supply chain finance, the application prospects of blockchain technology are prospected in order to provide references for current practice and future research and exploration.

2 Opportunities and Challenges of Computer Technology in the Era of Big Data

2.1 Connotation and Characteristics of Blockchain Technology

(1) Decentralization
Blockchain technology is to establish a point-to-point self-organizing network, on the basis of cryptography, to create a trust system that does not rely on third parties, regulators or non-centralized hardware. In each node, all the data of the blockchain is stored, and the obligations and rights of each node are the same. When a node is damaged, it will not have any impact on the operation of the system, so as to achieve mutual constraint.

(2) Transparency
All content in the system is open, sustainable, and non-modifiable. Therefore, no node can cheat other nodes within the time and rules specified by the system. Upgrading or deleting nodes does not affect the operation of the entire system.

(3) Security
The system adopts the distributed network, which can make each node have a complete database. The node cannot modify the database, nor can it have any impact on the data content of other nodes. The more nodes we have, the stronger the computing ability and the stronger the fault tolerance function. We can transmit data stably and safely without central server and supervision.

2.2 Blockchain Technology and Its Composition

In the rapid development of blockchain, its activities are not limited to the electronic currency that can be used in the first stage of development, but gradually enriched, including smart contracts and infrastructure that can be implemented. The overall model of blockchain system consists of four main parts: block, tag, consensus and smart contract. Each chain collects the information generated by each participant in the chain into a "block" at a fixed time interval, and the data blocks form a sequential data block chain. This paper summarizes three minimalist formulas of blockchain application.

Token economy:

$$\text{Dapp} = \text{wallet} + \text{Smart contract} + \text{oracle} \tag{1}$$

Payment scenario:

$$Dapp = \text{wallet} + \text{browser} \tag{2}$$

The unidirectionality of *Hash*() function in cryptography is difficult to find x through y:

$$Hash(x) = y \tag{3}$$

Payment status and signal economy may be closely related to wallets. Wallet content is not used to store digital assets, but to store personal keys. Digital assets are linked together. Only the private key in the wallet can guarantee that these resources or rights belong to users. The identity of the participant is derived from the asymmetric key of the account, which is used to record the latest information. Each participant has the same information chain and cannot be tampered with unilaterally. Any information modification can only be carried out with the consent of the client, and only new information can be added, while old information cannot be modified.

3 Experimental Thinking and Design

3.1 Experimental Ideas

This paper studies the application of blockchain technology in supply chain finance by stages, obtains the score data through questionnaire survey, and compares the financial business characteristics of traditional payment and supply chain payment. This paper analyzes the composition and functions of the blockchain technology, and finds the blockchain technology compatible with the supply chain financial services based on the characteristics of the supply chain financial services. This paper also analyzes the application of financial analysis in the whole supply chain.

3.2 Experimental Design

Compared with the traditional paper money, digital currency has lower cost, stronger convenience and higher security in the process of circulation, which meets the needs of rapid development of e-commerce and e-finance. At present, digital currency has been widely recognized by many countries and regions.

As the whole society becomes more and more digital, the application conditions of blockchain will be more abundant. Digitization is not only related to the application of blockchain technology, but also related to the country's economic and social transformation and the desire for improvement in the new era. In the process of digitization, we need to focus on improving social credit management and blockchain technology application scenarios as digital strategic objectives and thinking mode. We should effectively accelerate the rapid development of information, break the ownership of information departments, promote the development of digital resource sharing strategy, and realize the sharing of digital resources in the whole community.

4 Discussion

4.1 Making Financial Payment and Settlement More Convenient, Faster and Cheaper

Table 1. Comparison between traditional payment and supply chain payment

	Traditional payment	Supply chain payment
Trading hours	More than one day	Seconds or hours
Transaction security	High risk	Reduced transaction risk
Currency circulation cost	Higher	Greatly reduced
Authenticity of bill transaction	There is a risk of illegal operation	No intermediary of bill delivery

The comparison between traditional payment and supply chain payment is shown in Table 1. Compared with the previous payment measures, the blockchain payment is faster, more secure and costs less. In terms of transaction time, the processing time of previous payment methods is usually more than one day. If cross-border settlement or payment is required, it will take longer; the blockchain payment process takes about several seconds to complete the transaction, and the length of transaction time may have a significant impact on specific transactions. In the case of cross-border settlement, it usually takes two to three days to process the transaction of the previous payment method, but this transaction has been in effect for some time. In the frozen two or three days, the investment cost is very large, and the blockchain technology needs less time to process the business. Therefore, the total cost is lower; in addition, rapid processing of business can greatly reduce the risk probability caused by delaying transaction time, greatly increase the competitiveness of banks in cross-border payment, and form positive external effects. In the future, when handling international business, the financial industry can directly use the method of point-to-point payment with the help of blockchain technology, and no longer need the intermediate step of other three-party financial institutions. The blockchain technology can not only provide 24-h round the clock service, but also realize real-time payment. At the same time, the total cost will be greatly reduced, which can also meet the needs of cross-border e-commerce for more convenient, faster and timelier settlement. At present, blockchain technology has been widely used in the cross-border settlement and payment business of financial technology companies in many countries.

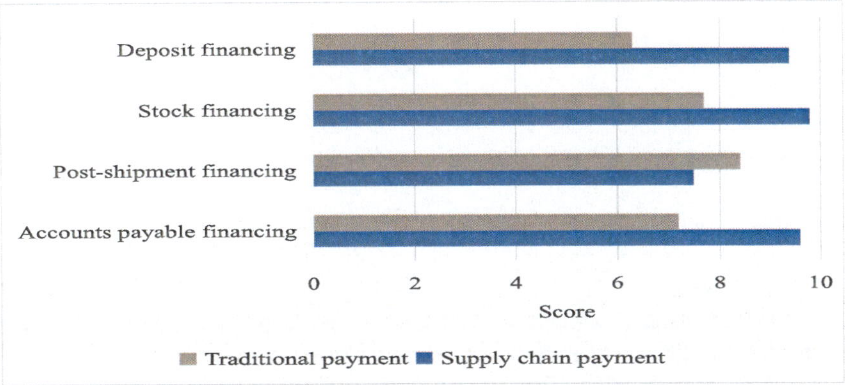

Fig. 1. Comparison of financial business between traditional payment and supply chain payment

It can be seen from Fig. 1 that in the comparison of financial business between traditional payment and supply chain payment, the accounts receivable, post freight financing, inventory financing of various events in the actual supply chain and other invoices and processes related to the supply chain will generate financial demand before and after shipment. The financing policy provides services for the needs of these links. From the above analysis, it is obvious that the function of blockchain technology is closely related to the characteristics of supply chain finance. The supply chain financing of blockchain technology can solve the following problems: promote the growth of supply chain financing, improve financial convenience, enhance the security of transactions and reduce financial costs.

As can be seen from Fig. 2, blockchain has the functions of disintermediation, unforgeability, consensus mechanism, openness, anonymity, cross platform, etc., and can play multi-level functions in a variety of situations. Generally, blockchain can understand the functions of three dimensions: trust model construction, information infrastructure and economic management mechanism. It provides a consistent and decentralized data storage system for data blocks, making it an excellent carrier for information and data capture. Blockchain uses the principle of confidentiality to store and link information, and associate each block with the chain structure, so that the stored information can be verified mutually, so as to ensure that the information will not be tampered with after it is on the chain.

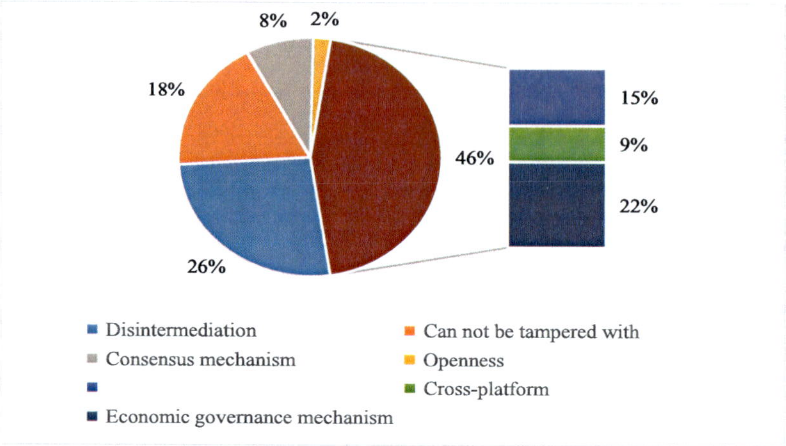

Fig. 2. Functions of blockchain

4.2 Application of Blockchain Technology in Financial Industry

(1) Digital currency
In the sustainable development of society, the transaction between people can be divided into three stages: barter transaction, physical currency purchase and digital currency transaction. The low circulation and issuing costs of digital currency and high transaction transparency greatly promote the convenience of economic activities and keep in line with the current e-commerce era. If the blockchain loan method can be widely used in the world, paper currency will be replaced by digital currency, and will become a general payment method.

(2) Cross border payment and settlement
Compared with traditional payment, the application of blockchain technology in payment has greater advantages in transaction cost, time and security. In terms of transaction cost, the traditional payment needs to charge a certain fee; however, the service charge for payment by using blockchain technology can be ignored. In terms of transaction time, the traditional payment method needs a longer time. In cross-border payment and settlement, due to the different degree of settlement in each country, the procedures are cumbersome and the efficiency is low, thus affecting the investment opportunity cost; however, the application of blockchain technology for payment only takes a few seconds.

(3) Application in bill
Nowadays, bank accountants are deeply involved in accounting, and there may be many risks and violations in accounting. At present, the third-party central bank is usually used to ensure the security of electronic transactions. In the accounting industry, the implementation of blockchain technology to simplify and control billing transactions can greatly reduce the impact of human factors, improve the security and efficiency of transactions, and reduce the cost of bills.

5 Conclusions

In the process of studying the application of blockchain technology in boosting the financial industry, this paper analyzes and explores the application of blockchain technology in supply chain finance. This paper analyzes the composition and function of blockchain technology, combs the characteristics of supply chain finance business, finds that blockchain technology and supply chain financial business have matching, and takes factoring as a typical representative of supply chain finance to carry out the whole process application analysis. Research shows that blockchain technology has been widely used in the financial industry, which is the key to promote the development and operation of digital currency, and has great advantages compared with traditional transactions. It has a good development direction in China's cross-border credit reporting and settlement technology, and has a good development direction in the cross-border financial industry and the application of digital payment technology in China.

Acknowledgments. Research on the path and strategic choice of China's industrial upgrading base on the perspective of trade imbalance and trade friction (GY-S18101).

References

1. Cull, R., Gine, X., Harten, S., Heitmann, S., Rusu, A.B.: Agent banking in a highly underdeveloped financial sector: evidence from the democratic republic of congo. Soc. Sci. Electron. Publ. **107**, 54–74 (2018)
2. Pradhan, R.P., Arvin, M.B., Mahendhiran, N., Bennett, S.E., Hall, J.H.: The dynamics between energy consumption patterns, financial sector development and economic growth in financial action task force (FATF) countries. Energy **159**(sep.15), 42–53 (2018)
3. Schrimpf, A., Drechsler, A., Dagianis, K.: Assessing identity and access management process maturity: first insights from the german financial sector. Inf. Syst. Manag. **1**, 1–22 (2020)
4. Dai, M., Zhang, S., Wang, H., Jin, S.: A low storage room requirement framework for distributed ledger in blockchain. IEEE Access **6**, 22970–22975 (2018)
5. Brewer, T.L.: Regulating black carbon emissions in international maritime shipping: can distributed ledger technologies help? The Magazine for Environmental Managers (APR.), pp. 21–24 (2019)
6. Rahmadika, S., Ramdania, D.R., Harika, M.: Security analysis on the decentralized energy trading system using blockchain technology. J. Online Inf. **3**(1), 44 (2018)
7. Saputra, D.E., Sutikno, S., Supangkat, S.H.: General model for secure electronic cash scheme. Int. J. Netw. Secur. **21**(3), 501–510 (2019)
8. Wang, S., Wang, Y., Zhang, Y.: Blockchain-based fair payment protocol for deduplication cloud storage system. IEEE Access **7**, 127652–127668 (2019). https://doi.org/10.1109/ACCESS.2019.2939492
9. Barman, R., Singha, P., Zaman, S.: Electronic attendance management system. IOSR J. Electron. Commun. Eng. **14**(3), 31–35 (2019)
10. Minh, A.T.T.: Challenges in implementing decentralization of foreign direct investment management in viet nam – case study of the hung nghiep formosa ha tinh steel project in ha tinh province. Asia-Pacific Sustain. Dev. J. **26** (2019)

Analysis of Key Technologies of Cloud Computing Security Based on Trust Model

Fangfang Dang(✉), Weijian Zhang, Shuai Li, and Han Liu

State Grid Henan Information and Telecommunication Company, Zhengzhou, Henan, China

Abstract. With the development of cloud computing, cloud computing provides more and more computing services. In today's fiercely competitive environment, the service flexibility and choice provided by this highly scalable technology are becoming more and more attractive to enterprises. However, behind the rapid development of cloud computing, its security and trust issues have become increasingly unignorable. The purpose of this paper is to analyze the key technologies of cloud computing security based on the trust model (TM). This paper first proposes a subjective and objective TM based on the ideal point method for the trust problem in the cloud environment, and verifies the performance of the model through experiments. This paper proposes a decentralized TM based on blockchain, and verifies the effectiveness of the model through experiments. This paper starts from the current problems of cloud computing, and on the basis of understanding the shortcomings of traditional security technologies, proposes a key technology of cloud computing security based on the TM. The solution proposed in this paper has a minimum average time consumption of 0.12 ms in the copy + label generation stage, while the minimum average time consumption of a traditional PDP is 16.3 ms. It is obvious that this solution has great advantages.

Keywords: Trust model · Cloud computing · Protection and security · Data mining · Task Scheduling

1 Introduction

With the rapid development of society and science, trust is playing an increasingly important role in various fields [1, 2]. In terms of its function, trust reflects the subject's objective evaluation of the object's comprehensive ability, and usually this evaluation can assist the subject in making decisions [3, 4]. The TM is a trust evaluation system that uses a specific trust calculation method. It is mainly used to evaluate the trust of nodes in a large-scale system or network, and to find malicious nodes and undesirable collusion in the network [5, 6]. At the same time, in the field of cloud computing security, data integrity audit technology is a hot topic in data security research. Although there are various methods to ensure the security of cloud computing, they still cannot solve the security loopholes [7, 8]. Research on the key technologies of cloud computing security based on the TM is of great significance to ensure the security of cloud servers.

B. J. Jansen et al. (Eds.): *International Conference on Cognitive based Information Processing and Applications (CIPA 2021)*, LNDECT 85, pp. 404–412, 2022.
https://doi.org/10.1007/978-981-16-5854-9_51

In the related fields of TM and cloud computing security key technologies, more and more researchers at home and abroad have paid attention to it. At present, the field is developing very rapidly. For example, Chen J combines sociological theory with trust and proposes a method based on Sociological theory of trust relationship prediction model [9]; Benziker SA proposed a TM based on accurate recommendation trust, which can accurately detect dishonest recommenders with switch attack mode [10]; Huszti A proposed a cloud environment Under the adaptive TM, the model uses IOWA operator and rough set theory to calculate the trust value [11].

The purpose of this paper is to analyze the key technologies of cloud computing security based on the TM. This paper first puts forward a subjective and objective TM based on the ideal point method for the trust problem in the cloud environment, and verifies the performance of the model through experiments. Then it is proposed that the solution to cloud verification is the PDP (Provable Data Possession) method with an error location mechanism. After verification, the PDP method with an error location mechanism proposed in this paper is better than the general traditional PDP method after possibility verification and parameter optimization. Finally, starting from the current problems of cloud computing, and on the basis of understanding the shortcomings of traditional security technologies, the key technology of cloud computing security based on the TM is proposed.

2 Analysis of Key Technologies of Cloud Computing Security Based on TM

2.1 Establishment of a Trust Model in the Cloud Environment

2.1.1 System Architecture

The system is mainly composed of two major parts, namely Trust service Provider (TSP) and Cloud Service Provider (CSP). CSP represents the service provider and provides services to users. It needs to interact with users and TSP. TSP stands for Trusted Service Provider, and it is the core part of the system [12]. The function of each component of TSP is described below.

Data Storage Module: The data storage module is the basic module of the TSP, responsible for storing the real-time data monitored by the SLA Monitor Module, the data feedback from the user, recording which CSPs are registered in the TSP, and recording the trust value of the CSP. When the CSP registers with the TSP, the Service Registration Module will write the basic information of the CSP into the module. Other modules, such as the Service Recommend and negotiation Module, can read data from this module when recommending trusted CSPs to users.

Service Registration Module: The service registration module is responsible for the registration of the CSP. When a new CSP requests to join the system, the service registration module will write the basic information of the CSP such as ID, CPU capacity, memory size, external memory size, etc. into the system. Only CSPs that have been registered in the system can be used by users.

SLA Monitor Module: The service level monitoring module is responsible for monitoring the service quality of the CSP and is the core part of measuring the objective trust value. After reaching the SLA agreement with the CSP, the user starts to use the services provided by the CSP. The SLA Monitor Module will monitor the quality of the various services provided by the CSP in real time according to the SLA agreement and enter it into the Data Storage Module.

User Feedback Module: The user feedback module is used to collect user feedback data and is the core part of measuring subjective trust value. After completing the transaction with the CSP, the user needs to evaluate the service provided by the CSP according to the previously reached SLA agreement, and the evaluated data will be entered into the Data Storage Module by the User Feedback Module.

Trustalue Computing Module: The trust value calculation module is the core part of the model, which is used to calculate the trust value of the CSP in the system. The trust value is divided into two parts, namely the objective trust value and the subjective trust value. The two parts of trust value are collected by SLA Monitor Module and User Feedback Module respectively. The trust calculation module retrieves the data from the Data Storage Module, calculates the objective and subjective trust values of the CSP according to our model, and finally combines these two trust values to obtain the final trust value.

Service Recommend and Negotiation Module: The Service Recommend and Negotiation Module is responsible for recommending the appropriate CSP to the user and negotiating the final CSP to be used with the user. When the user selects the CSP in the TSP, the module will recommend the CSP that meets the user's needs to the user, and the user selects the appropriate CSP from these recommended CSPs. After the user selects a suitable CSP, he needs to negotiate the SLA with the CSP. This module will play the role of an intermediate agent between the user and the CSP, and is responsible for coordinating the negotiation process between the user and the CSP.

2.2 Cloud Computing Security Key Technologies Based on Trust Model

(1) *PDP (Provable Data Possession) verification method with error location*
The administrator first runs the Key algorithm to obtain the public/private key pairs of CSPs and users. Then, the client uses Tag to generate tags for outsourced data resources. At any time, this Proof protocol is executed by a five-step interactive proof protocol, and it is carried out between a verifier and multiple CSPs. In this, CSPs do not need to be valid for each of the verification processes, but an organizer needs to be used to organize and manage all CSPs.
This protocol is described as follows:

   (1) The organizer initiates the agreement and sends a promise to the verifier;
   (2) The verifier returns a challenge (a set of random index coefficients to Q) to the organizer;
   (3) The organizer relays each data block to each Pi in P according to the exact location of each data block;
   (4) Each Pi returns its challenge answer to the organizer;
   (5) The organizer receives the data sent by each Pi from the server, does some calculations, and then sends it to the verifier.

(2) *Efficient integrity of data under cloud computing*
The system model of this solution mainly includes three entities: data owner (Actor), data manager (IM), and third-party audit (TPA).

Actor: In AMI-based systems, Actors are generally considered to be various information processing devices, including smart devices, user terminals, embedded devices, etc. Its main function is to collect user and environmental context information, and send the collected information To IM.

IM (Information Management): It is the information manager. Usually, it is an agent running on a cloud server or workstation managed by a cloud service provider, which is mainly used to store the received data information, analyze and manage the collected information, make corresponding decisions and send it to the Actor, etc. Update the environment context.

TPA (Third Party Auditor): It is a third party audit. Due to the large difference in computing power between various actors, TPA has audit experience and higher computing power that Actors do not have. It can replace Actors in performing audit tasks on the data stored in IM to monitor whether IM is performed correctly. The storage specification reduces the computational burden of the Actor in the verification phase.

The Actor will collect the context information of the environment as a data file, generate tags for the data file, and upload the data and tags to IM for storage. The Actor itself does not need to save relevant data. IM stores the data after verifying the validity of the data sent by the Actor. TPA sends a data integrity verification request to IM, IM replies to the corresponding storage certificate according to the verification request, and TPA judges the integrity of the data based on the obtained storage certificate and sends the verification result to the Actor.

2.3 Cloud Computing Platform Task Scheduling Algorithm

The insufficiency of the IBA algorithm only considers the number of PEs requested by the task and does not consider the execution time of each independent task, which results in a great task starvation, which eventually results in a slower task execution time and a larger average delay. The execution time T (i) of the independent task i is as follows:

$$T(i) = \frac{MI_i}{PE_i \times MIPS_j} \tag{1}$$

MI_i is the length of task i, and PE_i is the number of processors requested by task i. $MIPS_j$ is the processing capacity of virtual machine j. The processing power of $MIPS_j$ is unchanged. Therefore, we define T_{ratio} as the time coefficient, so the formula (2) can be obtained:

$$\begin{cases} T(i) = T_{ratio} \times \frac{1}{MIPS_j} \\ T_{ratio} = MI_i / PE_i \end{cases} \tag{2}$$

In the case of virtual machine selection, the execution time of the independent task depends entirely on T_{ratio}. The T_{ratio} tasks scheduled in the previous part are not much different. Once the size difference of T_{ratio} increases, the previous algorithm shows its disadvantage, and the virtual opportunity causes a great idle starvation state. The task with a small number of applied PEs is called a narrow task, and vice versa is called a wide task; a task with a larger Tratio is called a long task, and vice versa is a short task.

3 Experimental Research on Key Technologies of Cloud Computing Security Based on TM

3.1 Experimental Environment

The experimental development environment is Visual Studio, which uses the open source software gmp for large number operations, openssl for cryptographic operations, and pbc for bilinear pairing operations. The network environment of the experiment is a local area network with a bandwidth of 1.0 gBPS.

3.2 Parameter Setting

In order to ensure fairness, the large number calculation methods, basic cryptographic algorithms and parameters, and bilinear pairing parameters used in the two schemes in the experiment are completely the same, and the network environment is also completely the same. Among them, symmetric encryption and decryption uses AES with 192-bit key CFB mode, the hash method is shal one-way hash, and the bilinear pair uses standard D159 parameters in pbc. The experiment selected 5 files for testing. For each file, the block sizes of 10K, 100K, 1M, 10M, and 100M were tested.

3.3 Experimental Settings of Trust Model Accuracy and Recall

(1) Data source
This experiment uses the attribute set of Weibo users as the experimental data. First, 200 initial users are selected, including normal users and untrusted users, and then the Weibo crawler is used to iteratively grab the information of the current user's adjacent user nodes according to the relationship of concern. Get a social network with 6954 user nodes.

In order to meet the needs of subsequent experiments, this research manually added some special node data, such as disguised malicious nodes with oscillating trust values, pretending to be normal nodes in the early stage, and starting to publish fraudulent information and false information when the trust value accumulates to a certain value. The malicious node finally obtained an unlabeled data set containing 40,000 records.

4 Data Analysis of Cloud Computing Security Key Technologies Based on TM

4.1 Cloud Computing Data Dynamic Security

The experiment mainly tests the total time consumption of the copy generation and label generation phase and the total time consumption of the proof phase. Table 1 lists the final experimental results. It is easy to see from the table that the scheme proposed in this article is based on copy + label generation. The average time consumption of the stage is at least 0.12 ms, while the average time consumption of traditional PDP is at least 16.3 ms. It is obvious that this solution has great advantages.

Table 1. Statistics table of average time consumption per KB at each stage (unit: milliseconds)

	Average time per KB in copy + label generation phase		The proof phase is more time-consuming than the average time per KB	
	Article	Traditional PDP	Article	Traditional PDP
10K	2.4	120	1.6	290
100K	1.46	50	0.38	25
1M	0.50	44.7	0.16	7
10M	0.65	45.2	0.1	0.24
100M	0.12	16.3	0.05	0.13

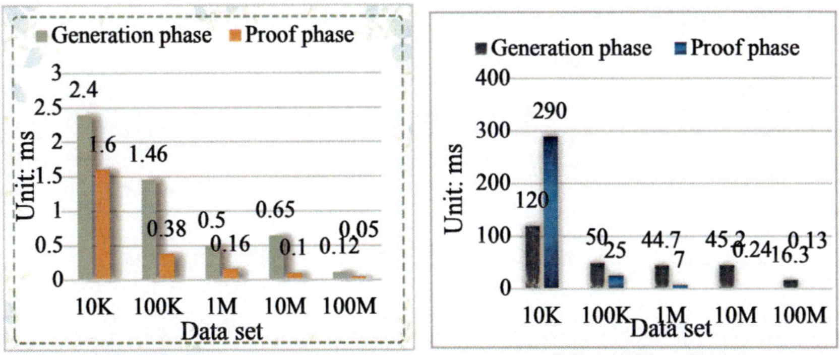

(a) Article (b) Traditional PDP

Fig. 1. Statistics table of average time consumption per KB at each stage (unit: milliseconds)

It can be seen from Fig. 1 that with the increase of file size, the average time per KB of the two schemes has rapidly become smaller, but the time consumption of the scheme in this paper is within 1.6 ms, while in the traditional PDP solution, the average time per KB can be as high as about 280 ms when the block size is 10K. In fact, the time cost of the proof phase is mainly affected by the number of challenged blocks. The smaller the number of challenged blocks, the lower the time cost. Finally, it is obvious that the number of copies has a linear relationship with the time consumption of the copy + label generation phase and the proof phase.

4.2 Accuracy and Recall Rate of Trust Model

Accuracy (Precision) and recall (Recall) are commonly used machine learning evaluation criteria. Among them, the accuracy is for the prediction result, indicating how many of the samples whose predictions are positive are real positive samples, and the recall is for the original sample, it indicates how many positive examples in the sample have been correctly predicted. For further comparison, the experiment takes a window length of 20 and a data set size of 40,000. At the same time, the TM proposed in this article and the TM based on sociological theory, the TM based on the division of knowledge domains, and the trust based on the interaction model and reputation model and the results are shown in Table 2:

Table 2. Accuracy and recall rate of TM

	Accuracy (%)	Recall rate (%)
Research	0.42	0.44
TM based on sociological theory	0.36	0.37
TM based on the division of knowledge domain	0.35	0.365
TM based on interaction model and reputation model	0.36	0.375

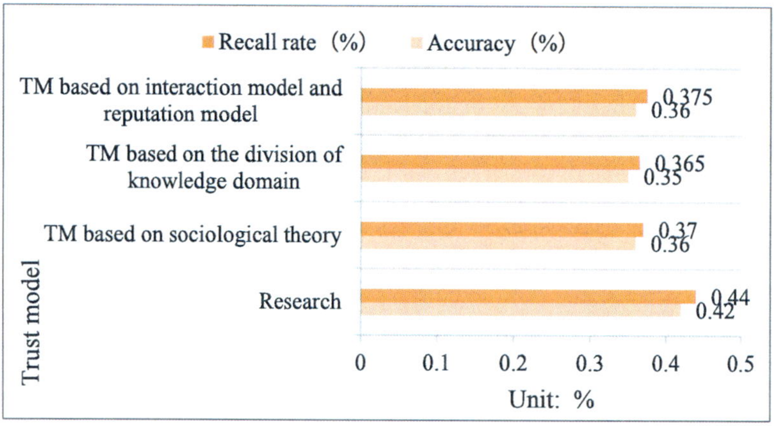

Fig. 2. Comparison of accuracy rate and recall rate of TM

It can be seen from Fig. 2 that the accuracy and recall rate of the TM proposed in this paper are higher than the other two TMs. This is because the TM proposed in this paper effectively utilizes the global optimization capability of the PSO algorithm and the powerful LSTM neural network. Different from the traditional TM that only focuses on calculating the trust value of the current node, the learning ability and presentation ability of, can grasp the characteristics of all nodes as a whole, find the most essential difference between nodes with different trust levels, and calculate more accurate trust rating.

5 Conclusion

Cloud computing provides users with powerful data computing and storage services, and has brought revolutionary changes to all areas of society. With the rapid development of Internet technology, big data has inevitably become the trend of the times. The TM in the big data environment is an extension of the traditional TM. How to use massive data with high dimensionality, high redundancy, and unknown credibility, how to preprocess large data sets, how to build with high accuracy and wide versatility, which are the challenges faced by traditional TMs. This article aims to improve the accuracy and versatility of the TM, analyzes the key technologies of cloud computing security based on the TM, and innovatively solves the problems of the traditional TM. Some problems that can be solved. A subjective and objective TM based on ideal point method in cloud environment is proposed. The subjective and objective trust are combined through the feature matrix, and the ideal point method is used to calculate the final trust value. Considering the temporality of trust, recurrent neural network is used as the classifier of trust. The solution proposed in this paper is the PDP (Provable Data Possession) method with an error location mechanism, which enables the client to reduce the burden of its local storage and can exchange information with the cloud service provider to ensure that the storage is correct. The PDP method with error location mechanism includes PDP index hashing layer, homomorphic verification, BDH, collision resistance hashing, signature encryption and other process methods to ensure that the user and the cloud server are in the process of mutual transmission and storage, and the data will not be opponents. Attack, package and encrypt the data to be transmitted to make the transmitted data more secure.

References

1. Mahmud, M., et al.: A brain-inspired trust management model to assure security in a cloud based IoT framework for neuroscience applications. Cogn. Comput. **10**(5), 864–873 (2018). https://doi.org/10.1007/s12559-018-9543-3
2. Jelenc, D., Tamargo, L.H., GoTtIfredi, S., et al.: Credibility Dynamics: A belief-revision-based TM with pairwise comparisons. Artif. Intell. **293**(2), 103450 (2021)
3. Meng, X.: speedTrust: a super peer-guaranteed TM in hybrid P2P networks. J. Supercomput. **74**(6), 2553–2580 (2018)
4. Pinsker, J.E., Laguna, S., Bok, L.J., et al.: Evaluation of an artificial pancreas with enhanced model predictive control and a glucose prediction trust index with unannounced exercise. Diabetes Technol. Ther. **20**(7), 455–464 (2018)
5. Atinkut, H.B., Yan, T., Zhang, F., et al.: Cognition of agriculture waste and payments for a circular agriculture model in Central China. Sci. Rep. **10**(1), 10826 (2020)
6. Al-Rakhami, M.S., Al-Mashari, M.: A blockchain-based trust model for the internet of things supply chain management. Sensors **21**(5), 15 (2021)
7. Hamzelou, N., Ashtiani, M., Sadeghi, R.: A propagation TM in social networks based on the A* algorithm and multi-criteria decision making. Computing (2), 1–41 (2021)
8. Yang, H.: A study on improving secure routing performance using trust model in MANET. Mob. Inf. Syst. **2020**(1), 1–17 (2020)
9. Chen, J., Gong, B., Wang, Y., et al.: Construction of Internet of things trusted group based on multidimensional attribute TM. Int. J. Distrib. Sens. Netw. **17**(1), 155014772198988 (2021)

10. Benziker, S.A., Kb, N.: Design and Implementation of novel security approach designed for cloud computing with load balancing. Solid State Technology **64**(2), 1877–1889 (2021)
11. Huszti, A., Oláh, N.: Security analysis of a cloud authentication protocol using applied pi calculus. Int. J. Internet Protoc. Technol. **12**(1), 16–25 (2019)
12. Xia, H., Yang, W.: Security access solution of cloud services for trusted mobile terminals based on TrustZone. Int. J. Netw. Secur. **22**(2), 201–211 (2020)

Innovating the Management Method of Students' Growth Files by Using Computer Devices

Lin Hou(✉)

Liaoning Jianzhu Vocational College, Liaoyang, Liaoning, China

Abstract. With the development of information technology, computer networks have gradually penetrated into information management. However, the current college education in our country, whether it is the evaluation of talent training, the overall quality development of students, or the evaluation of school education and teaching, cannot meet the increasing needs of students. Introducing computer devices into file management can effectively solve file storage, sharing, analysis and other issues. The purpose of this article is to use computer devices to innovate the management methods of student growth files. This article first summarizes the basic theory of computing network, and then extends its core technology, and combines the management status of modern student growth files in our country, analyzes the problems and shortcomings, and combines computer devices to analyze the growth files of college students. This paper systematically elaborates the functional requirements, module design and system function design of the use of computer to solemnly design student growth archives management, and uses questionnaire surveys, field surveys and other research methods to carry out research. Experimental research shows that it is relative to traditional student growth archives management methods. The use of computer devices to innovate student growth file management system has a higher feasibility.

Keywords: Computer network technology · File management · Method research · System design

1 Introduction

The current archives of college students are mostly based on basic student information, semester grades, and semester comments [1, 2]. There is a lack of relevant materials reflecting the colorful growth history of college students, and most of them are based on paper records, which are not convenient for long-term storage and reference, and it is also not convenient for various departments in the school, students themselves, parents of students, and corporate employers to inquire at any time[3, 4].

In the research on the innovation management of student growth archives, many scholars and experts have achieved impressive results [5, 6]. For example, in "History of Archives Work in Colleges and Universities", Chen Huaxin discussed that since the

B. J. Jansen et al. (Eds.): *International Conference on Cognitive based Information Processing and Applications (CIPA 2021)*, LNDECT 85, pp. 413–421, 2022.
https://doi.org/10.1007/978-981-16-5854-9_52

founding of the People's Republic of China, my country's archives work has gone through four development stages: "self-care and decentralization, initiation of management, centralized management, and in-depth development." [7]. It also pointed out that the current shortcomings of archives work in colleges and universities are mainly manifested in the "weak awareness of archives, the relatively frequent changes in the leadership of the library, the team structure needs to be adjusted, the school funding is insufficient, and the development of archive information resources is insufficient" [8, 9]. Zhang Yijun pointed out in the "Study on the Status and Development Trends of Archives Work in Colleges and Universities" that the current status of archives work in colleges and universities in the emerging stage is mainly "the north-south difference is large, and the development is unbalanced, which is mainly manifested in the small scale of college archives and the oversimplification of the institution setting. There is a serious shortage of personnel [10]. The educational background of personnel needs to be improved. The modernization of archives management is seriously lagging behind. Archives awareness is weak, the effect of archive utilization is not good, and the characteristics are not obvious" [11, 12].

The purpose of this article is to improve the management ability of students' growth files, and to use computer devices to innovate the management methods of students' growth files for the purpose of research. By comparing the traditional management methods of student growth files with computer-based management methods of students' growth files, analyze it to judge the feasibility of the computer device innovating the management method of student growth files.

2 Using Computer Devices to Innovate the Application Research of the Management Methods of Students' Growth Files

2.1 Computer Related Technology

(1) PHP technology
PHP provides object-oriented support. When choosing PHP for project development, you can choose to use process-oriented or object-oriented development, or a mixture of the two to develop. PHP is an open source software with many extension libraries, such as mnoGoSearch search engine function, IRC gateway function, multiple compression tools (gzip, bz2, zip), calendar conversion, translation, etc. It is very easy for any programmer to extend additional functions for PHP.

(2) B/S structure model
B/S is a browser and server structure model. It is a three-tier structure system composed of a client browser, a Web server, and a database server. This mode stores the PHP program in the Web server and the data in the database server, which is convenient for the administrator to manage, upgrade and maintain the system. For users, there is no need to download client software, just open the browser and enter the URL to directly access the system, which is convenient for users.

2.2 Functional Module Design of Computer Network System

(1) The system is mainly divided into identity authentication module, student/parent module, enterprise user module, teacher module (class teacher, subject), administrative staff module (school leader, professional department) and system administrator module.

1) Identity authentication module
After the user logs in through the account and password verification, the system automatically enters the corresponding module according to the user's role, and obtains the permissions of the corresponding role.
2) Student parent module
The online archives platform of secondary vocational students is the link between parents and schools, and the function of student and parent module is the core function provided by the platform. Through this module, students and parents can inquire about student status information, grade information, conduct records, evaluations, student honors, skill works, practice records, semester testimonials, parents' messages, etc. in the student growth bag. Students can manage their own personal honors, Skills works, practice records and semester testimonials can participate in peer reviews, and parents can post messages to their children. In order to facilitate the management and use, both parents and students use the student ID as the user name when logging in, and use different login passwords, which can be modified after logging in.
3) Enterprise user module
This module provides companies with a window to understand the performance of students at school. Business users can query student growth profile information based on student ID and name. Companies use a unified account and password when logging in, and they need to add the company name when evaluating students. Address, contact person and telephone information, and enterprise evaluation information shall be reviewed by the head teacher.
4) Teacher module
In the teacher function module, the head teacher can manage the student's school status, grades, conduct records, semester comments, various evaluations, student practice records, etc. of the teaching class. The subject teacher can query the student's growth file information and import subject exams. Achievements and records of student conduct.
5) Administrative staff module
The administrative staff module includes school-level administrative staff and professional department staff. Users can view student growth files, perform statistics on student file information (such as the completeness of school status, completion of various evaluations, student social practice, etc.), and file appraisal.

6) Administrator module
The system administrator module can set the basic parameters of the system (such as: semester, grade, class, course name, class teacher settings), can manage teacher users and enterprise users, can back up the database, and can register student change information (such as transfer Class, transfer, drop out, suspension, etc.).

(2) Database design
Database design is the core and foundation of the entire student growth electronic archive management platform. The quality of the database structure design directly affects the efficiency of data storage and data integrity, and directly affects the stability and operating efficiency of the application system. When designing a database, it is necessary to ensure data consistency and integrity, and minimize data redundancy. Table 1 is a table of user roles.

Table 1. tb_role

Field name	Types	Is it empty	Annotation
Roel_id	Int (8)	Not null	Primary key, increment
Role_name	Char (20)	Not null	Role name
Role_value	Tinyint (2)	Not null	Role identification
Role_lock	Char (4)	Not null	Login lock control

(3) Analysis of system non-functional requirements
In addition to specific functional requirements, the file management system for student growth has the following non-functional multimedia requirements. The requirements include:

1) The software system can be upgraded
Each module file management system with standard interfaces is used for student growth, and the system provides online upgrades, and upgrades do not need to install the system.
2) A certain degree of fluency is required for the network
The transmission of student files and the transmission of multimedia data should be less delayed. Multimedia data, stable quality, matching bandwidth, continuous video image, audio and video synchronization.
3) Provide users with easy-to-understand authority management
Student file management can let teachers, parents, and other users managed by different situations, groups or authorities to students, let students be allowed by real name and password settings.
4) System pressure bearing

Since the student growth file management system is for all teaching units in this city, there are a huge number of teachers, parents, and student users involved. In general, the maximum pressure of the system requires 500 people to be online at the same time.

(4) The application of ALOHA algorithm in the anti-collision of student growth file management label

In the file management system, each layer of filing cabinets has dozens or more electronic tags, and each layer has only one reader antenna for data collection. Therefore, collisions between tags are inevitable. The label collision problem is one of the key tasks of this paper.

Assuming that the number of time slots in a frame is N and the number of tags to be identified in the effective area is n, the probability of k tags selecting the same time slot obeys the binomial distribution, namely:

$$P(X = k) + C_n^k(\frac{1}{N})^k(1 - \frac{1}{N})^{n-k} \tag{1}$$

When there is only one electronic tag in a certain time slot, the time slot is called a successful time slot. Therefore, the probability of a successful time slot, that is, the value of k is 1, and k = 1 is inserted into Eq. (1) to obtain:

$$P = (X = 1) = C_n^1(\frac{1}{N})^1(1 - \frac{1}{N})^{n-1} \tag{2}$$

The expected value of the number of successful time slots is:

$$E(X = 1) = NC_n^1(\frac{1}{N})^1(1 - \frac{1}{N})^{n-1} \tag{3}$$

From this, the throughput rate S of the system can be obtained as:

$$S = \frac{E(X = 1)}{N} = \frac{n}{N}(1 - \frac{1}{N})^{n-1} \tag{4}$$

When the throughput rate S reaches the maximum value, the number of tags to be identified in the system is basically the same as the frame length. When the frame length is small and the number of electronic tags is relatively large, the throughput rate of the system becomes particularly low, and the conflict between tags is also more serious. When the number of electronic tags is relatively small, there are more idle time slots in the system, resulting in a waste of time slots.

3 Experimental Research on the Management of Student Growth Files Using Computer Devices

3.1 Subjects

(1) In order to make the experiment more scientific and effective, this experiment compares and analyzes the traditional management methods of student growth

files and the use of computer devices to innovate the management methods of student growth files, and analyzes the obtained data using the analytic hierarchy process. This research conducted a survey by going to a local university and using a questionnaire survey method.

(2) In order to further conduct experimental research on the computer network-based student growth file management system, this experiment conducted face-to-face interviews with teachers. A total of 20 teachers were interviewed in this experiment, among which the gender ratio was equal to ensure the validity of the experimental data.

3.2 Research Methods

(1) Questionnaire survey method
This article sets up a targeted questionnaire by going deep inside a university in a certain place, asking relevant experts, and using this closed form to investigate and research students, the purpose of which is to promote students to fill in correctly.
(2) Field research method
This article goes deep into the school, investigates and studies the management methods of student growth files and collects data. These data provide a reliable reference for the final research results of this article.
(3) Interview method
By conducting face-to-face interviews with teachers and recording the results of the interviews, this paper organizes and analyzes the recorded data. Do these data not only provide theoretical support for the topic selection of this article, but also provide data support for the research results of this article.
(4) Mathematical Statistics
Use related software to perform statistical analysis on the final research data.

4 Experimental Analysis of the Management Method of Students' Growth Files Using Computer Devices

4.1 Comparative Analysis of the Management Methods of Student Growth Files

In order to make the experiment more scientific and effective, this experiment compares and analyzes the traditional student growth file management method and the computer network-based student growth file management system. The data obtained is shown in Table 2.

Table 2. Comparative analysis of the management methods of student growth files

	Authentication	Information management	Course management	Others
Computer device	65.2%	69.7%	63.2%	52.9%
Traditional	55.2%	50.3%	54.1%	46.3%

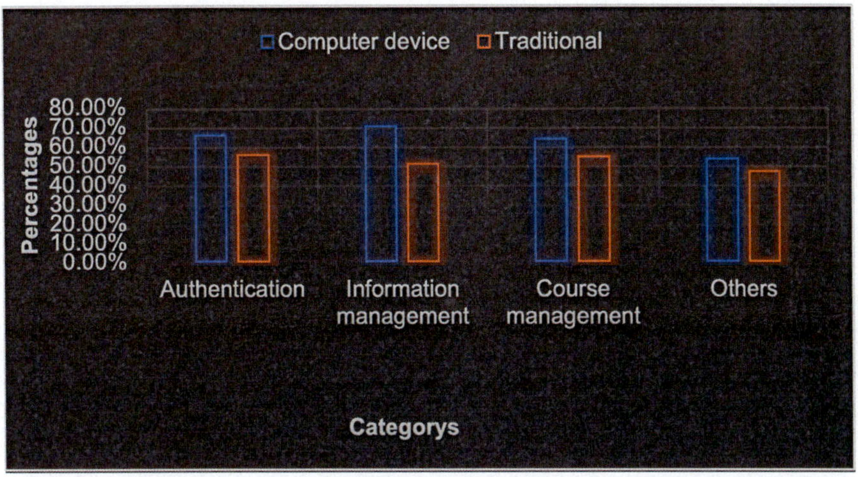

Fig. 1. Comparative analysis of the management methods of student growth files

It can be seen from Fig. 1 that compared to the traditional management methods of student growth files, the management methods of student growth files using computer devices are superior in many aspects of management, especially in the management of student information, which is nearly 20% higher. It reflects the excellent performance of the student growth file system based on the computer network and the traditional management methods of student growth files urgently need to be innovated.

4.2 Performance Test of Student Growth File Management Based on Computer Network

(1) In order to further conduct an experimental analysis on the management methods of using computer devices to innovate student growth files, this experiment was conducted by concurrently accessing the client to 20 users. This time, a ten-point scoring system was adopted, of which 1 means disapproval. 10 indicates agreement, and the data obtained will be analyzed using mathematical statistics. The data obtained are shown in Table 3.

It can be seen from Fig. 2 that most people agree with the management system of student growth files based on computer devices. Especially in terms of its convenience, it has the highest proportion, which fully reflects the feasibility of using computer devices to innovate the management system of student growth files.

Table 3. Performance test of student growth archives management based on Computer network

	Extensibility	Convenience	Robustness	Safety
User 1	5	8	7	6
User 2	7	9	7	6
User 3	7	10	5	6
User 4	6	7	6	7
User 5	5	7	8	8
User 6	7	8	7	8
User 7	8	6	5	9
User8	8	8	5	7
User 9	6	7	6	8
…	6	9	7	5
User 20	6	7	8	5

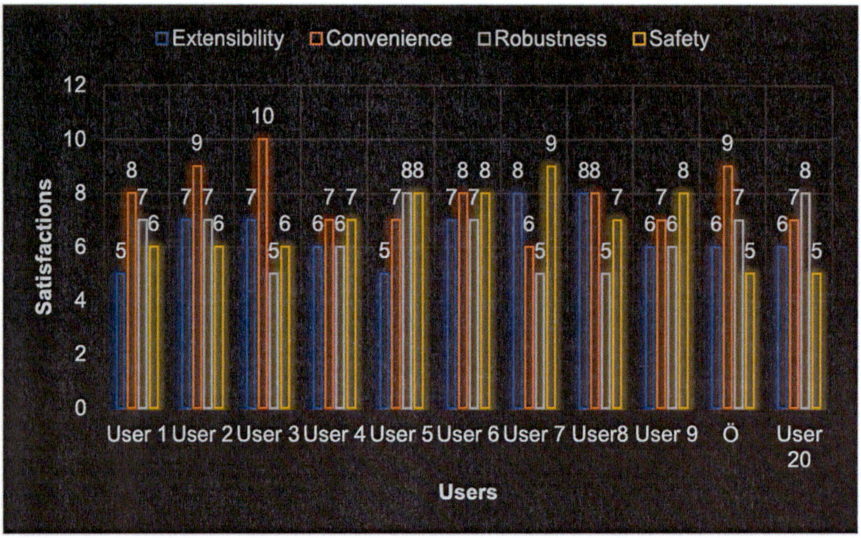

Fig. 2. Performance test of student growth archives management based on computer network

5 Conclusion

This article combines China's basic education reform and comprehensive quality evaluation, design and implementation of the development direction of the student record management system. What the system realizes through computer technology, communication technology, and database technology is that it is not helpful for portfolio management and will record the development of students one by one. The attitude of students to study and life, together with the evaluation of others, shows the growth process of students. In order to show the students of the growth process from multiple perspectives, multi-level, comprehensive evaluation, provide daily life records, the comprehensive quality of evaluation and developmental evaluation for actual materials and objective foundations.

References

1. Hope, J.: Focus on student needs to achieve enrollment growth. Enroll. Manag. Rep. **22**(11), 9 (2019)
2. Alsaif, S., Li, A.S., Soh, B.: A snapshot survey of Saudi higher education students' perceptions of SNSs and Web 2.0 as a support learning tool based on the proposed social learning hub model. Int. J. Serv. Econ. Manag. **8**(4), 287 (2017)
3. Zakharova, I.G.: Machine learning methods of providing informational management support for students' professional development. Educ. Sci. J. **20**(9), 91–114 (2018)
4. Sauvage, T., Schmidt, W.E., Suda, S.: A metabarcoding framework for facilitated survey of endolithic phototrophs with TufA. BMC Ecol. **16**(1), 8 (2016)
5. Filho, I., Litschig, S.: Long-run Impacts of Intergovernmental Transfers. Work. Pap. **46**(5), 1045–1060 (2016)
6. Wynne, K., Filante, R.: Investor behavior in a university student managed portfolio. Invest. Manag. Financial Innov. **11**(3), 28–33 (2017)
7. Chen, H.: The application of FP-growth algorithm in student performance analysis. Inf. Comput. (Theoret. Ed.) **359**(13), 109–110 (2016)
8. Han, R., Lei, L., Ding, Y.: Research on student performance analysis system based on FP-growth algorithm. J. Changchun Univ. Sci. Technol. (Nat. Sci. Ed.) **41**(05), 127–131 (2018)
9. Aithal, P.S.: student centric curriculum design and implementation – challenges & opportunities in business management & IT education. Mpra Paper **4**(3), 423 (2016)
10. Zhang, Y.: The application of association rule mining algorithm FP-growth in the analysis of the performance of network technology students in higher vocational colleges. J. Hechi Univ. **36**(158(02)), 67–72 (2016)
11. Lewis, L.S., Williams, C.A., Dawson, S.D.: Growth mindset training and effective learning strategies in community college registered nursing students – science direct. Teach. Learn. Nurs. **15**(2), 123–127 (2020)
12. Sta, C.C., Hiotti, S., Astolfi, A., Gronchi, A.: Evolution of Dermatofibrosarcoma Protuberans (DFSP) to DFSP-derived Fibrosarcoma (FS-DFSP): an event marked by epithelial mesenchymal transition-like process and 22q loss. Mol. Cancer Res. **14**(9), 820–829 (2016)

Analysis and Research of Civil Engineering Construction Application Based on Computer BIM Technology

Jie Wang(✉)

School of Road Bridge and Architecture, Chongqing Vocational College of Transportation, Chongqing, China
wyji@bjtu.edu.cn

Abstract. At present, with the continuous development of my country's information technology, there are related information technology applications in the construction industry, which greatly improves the development space of my country's construction industry. The combination of computer technology and various technologies allows us to combine modular computer room prefabrication with more new design methods and manual technologies. The application of BIM Technology in civil engineering projects greatly improves the construction quality of the project, which can not only save costs, but also effectively shorten the construction period. On this basis, the application of BIM Technology in civil engineering is analyzed in detail. This article proposes the main high-end technical support for the related research system by turning the computer-related simulation research technology, data control technology into technology and computer-aided design three major technologies. The test results show that the wide application of BIM Technology in civil engineering construction can effectively control the project cost, progress and quality, and ensure the completion of civil engineering projects with the lowest cost, the fastest speed and the best quality.

Keywords: BIM technology · Civil engineering · Construction application · Analysis and research

1 Introduction

With the development of BIM technology, the promotion of BIM technology in the construction industry has led to the use of this technology in more and more projects. At present, the civil engineering around us is using this technology. In the early stage of the project, BIM technology will be used in the implementation of the project, through the computer virtual construction method, the construction of various buildings and equipment of three-dimensional components, Realizing engineering three-dimensional simulation technology, as well as BIM technology, three-dimensional technology, etc. occupy a place in the engineering field. Compared with the traditional work model, the existence of these emerging technologies can to a certain extent achieve work results

B. J. Jansen et al. (Eds.): *International Conference on Cognitive based Information Processing and Applications (CIPA 2021)*, LNDECT 85, pp. 422–428, 2022.
https://doi.org/10.1007/978-981-16-5854-9_53

that the traditional model cannot achieve, and can effectively reduce work errors and work risks. It is the main tool for the development of various industries in our country.

The United States is the birthplace of BIM technology and is relatively advanced in development. The application and development of information technology is very much in line with the rapid development of information technology. BIM technology came into being [1–3]. The birth of BIM Technology has brought reform and development opportunities for the development of the construction industry. It can promote the development of the information technology industry and construction industry. Therefore, at this stage, many countries are conducting in-depth research on BIM technology in construction. At the beginning of the 21st century, the United States took the lead in implementing a series of measures to relevant parties and formulating relevant norms and standards. Immediately afterwards, other countries in the world have also formulated relevant norms and standards in this regard. Our country is relatively late in formulating standards in this regard. It only put forward relevant planning requirements in 2011. After 5 years, it only promulgated the main specifications and standards in 2016. The BIM technology in "Building Information Model Application" was promulgated [4–6]. "Uniform Standard" and "Building Information Model Application Standard" are defined as a common name. "The core content is to use digital and information technology to simulate all the information of the construction project in the plan, so that the construction personnel can discover the problems in the construction plan in advance, so as to correct errors in time, reduce costs, shorten the construction period, and reduce coordination work. Realize resource sharing. From the perspective of civil engineering construction, through the simulation of BIM technology [7–10], construction personnel can clearly understand all content and errors in the construction process, and even understand the quality, function and quality of each component and component in the engineering structure in detail. Quantity. At present, BIM technology is being implemented vigorously in Asia, South Korea, Britain and other construction industry countries.

The application advantages of BIM technology mainly involve the following aspects: First, it is the visualization of the model. The visualization of the model is reflected in the fact that the project can be passed through the similar model, and the scale is reduced to show the complete project. At present, some civil engineering construction projects in China are still limited to traditional buildings. Construction can be carried out according to design drawings. However, traditional drawings have great limitations It is to establish and simulate the same scheme as the actual civil engineering construction work to simulate the construction effect. Engineers can build it by developing architectural model software. When there are potential problems in construction, they can be specialized. The second is the simulation technology BIM Technology can simulate all aspects of civil engineering construction. According to the design plan, it can simulate the content of the construction in progress compared with traditional civil engineering construction. Even if the designer chooses the best construction decision, the design effect cannot be predicted in advance, and the problems that need to be changed in the later design cannot be avoided. BIM technology can be used to construct a three-dimensional entity through comprehensive simulation to show all possible construction conditions. And even 4D, 5D simulation.

2 Algorithm Establishment and Optimization

2.1 Adaboost Algorithm

On the basis of this algorithm, the first thing we need to do is to use probability distribution. Classify and learn from the given data to obtain basic data, and select classifiers based on the data:

$$G_a(x) : X \rightarrow \{-1, +1\} \tag{1}$$

The next step is to determine the correctness of the data, and calculate the correct and incorrect objects of the calculated data given by the classifier:

$$e_a = P(G_a(xi) \neq yi) = \sum\nolimits_{i=1}^{N} wmiexp\left(-a_a yiG_a(xi)\right) \tag{2}$$

The coefficient in the calculation formula:

$$b_a = \frac{1}{2}\log\frac{1 - e_a}{e_a} \tag{3}$$

Update the data set during the calculation process, and correctly calculate the probability distribution of the data:

$$D_{m+1} = (W_{m+1}, 1, \ldots, W_{m+1}, I, \ldots, W_{m+1}, N) \tag{4}$$

$$W_{m+1,I} = \frac{W_{mi}}{Z_m}\exp(-a_m y_i F_n(x_i)), (I = 1, 2, \ldots, N) \tag{5}$$

According to the method mentioned above, a modified linear combination of classification is established:

$$f(x) = \sum\nolimits_{m=1}^{M} a_m F_n(x) \tag{6}$$

Finally, according to our above experimental data, the final result can be obtained, and the classification data can be obtained

$$F(x) = seg(f(x)) = seg\left[\sum\nolimits_{m=1}^{M} a_m f_n(x)\right] \tag{7}$$

3 Modeling Method

3.1 Predictive Model

In the process of establishing a predictive model, the first thing we need to do is to analyze and classify the information, conduct a comprehensive analysis of the collected information, and model the collected data according to a certain method. The original neural network at the beginning will be terminated. It is expected that the error in the output will gradually decrease until the requirements are met. As a result, the neural network model will be changed, and will be more mature compared with the previous model, which can meet the requirements of higher data information processing. After the experimental model is successfully constructed, you can input the data for prediction. According to the price prediction demand, use the input value as the price to be predicted, and train and educate the established model to achieve future price trends and sometime in the future price.

The objective function of the prediction model is shown in Eq. (8):

$$o(t) = \sum_{i=1}^{n} L(y_i, \hat{y}^{t-1} + f_i(x_i)) + \Omega(f_i) + c \tag{8}$$

Among them, formula (8) is a loss function and a regular term, and c represents a constant. Combining the knowledge of Taylor formula, perform Taylor expansion, and the expanded formula is used as the loss function in formula (9)

$$f(x + \Delta x) \approx f(x) + f'(x) + \frac{1}{2} f''(x) \Delta x^2 \tag{9}$$

After performing Taylor expansion, the objective function becomes Eq. (10):

$$o(t) \approx \sum_{i=1}^{n} \{L(y_i, \hat{y}^{t-1}) + g_i f_i(x_i) + \frac{1}{2} h_i f_i^2(x_i)\} + \Omega(f_i) + c \tag{10}$$

In formula (10):

$$g_i = f'(x) = \partial L\left(y_i, \hat{y}^{t-1}\right) / \partial \hat{y}^{t-1} \tag{11}$$

$$h_i = f''(x) = \partial^2 L\left(y_i, \hat{y}^{t-1}\right) / \partial \hat{y}^{t-1} \tag{12}$$

The positive term in formula (10) is:

$$\Omega(f_i) = \gamma T + \frac{1}{2} \lambda \sum_{i=1}^{T} w_j^2 \tag{13}$$

In the above formula (13), the parameter represented by T is the number of leaf nodes in a tree, and the parameter represented by W is the weight of a leaf node. Both of these two parameters are it is a hyperparameter, and the following formula can be obtained by simplifying the objective function:

$$o(t) \approx \sum_{i=1}^{n} \{a_i b_i(x_i) + \frac{1}{2} c_i b_i^2(x_i)\} + \Omega(f_i) + c \tag{14}$$

The new formula (15) is derived from the previous formula (13) and formula (14), and (13) is incorporated into our objective function (14). After calculation, the following formula can be obtained:

$$o(t) \approx \sum_{i=1}^{n} \{a_i b_i(x_i) + \frac{1}{2} c_i b_i^2(x_i)\} + \gamma T + \frac{1}{2} \lambda \sum_{i=1}^{T} w_j^2 + c \tag{15}$$

Through the above method, the dynamic query of engineering quantity and the price prediction of man, material, machine, etc. are realized. Multiply the two to get the dynamic prediction of direct cost described in this article.

Table 1. Benefits of BIM technology application

BIM application value	Investment budget error	Save investment estimate time	Reduce construction time	Reduce project investment
Survey results	5%within	60%	15%	10%

4 Data Results

The data in Table 1 is obtained from our investigation on a certain construction site. Through the study of these data, we have reached the conclusion that BIM technology can bring great convenience to the project in civil engineering projects, and its existence is obvious. The advantages. Using BIM technology in an investment budget error can lead to an enviable calculation of the enviable investment budget and control the budget error by 5%. Can be controlled at a higher level. Small scale. And it can save a lot of time given the estimated investment time. Using BIM technology can reduce the estimated investment time by almost 60%. As for cost control, the benefits of BIM technology are also very obvious. Accurate calculation using BIM technology can reduce costs by about 10%. The benefits of BIM technology in civil engineering projects are not only those shown in the table, only there are also great advantages in other aspects. BIM technology is an indispensable part of civil engineering projects (Fig. 1).

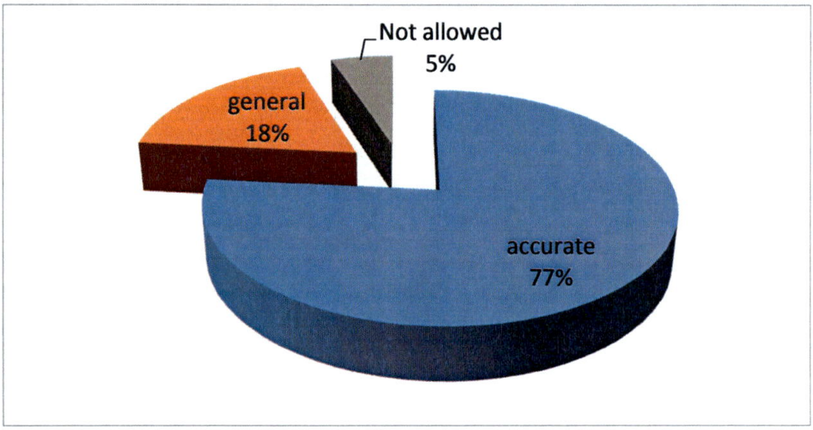

Fig. 1. Forecast model accuracy

The spread result of this test is an analysis made on the data of multiple engineering projects, which has a high degree of credibility. On the basis of the experimental data, the experimental data is analyzed according to the budget model proposed in this article, and the project quantity and other aspects are dynamically predicted. The results are shown in the figure. Judging from the budget results of the budget model, 77% of the calculated budget results are very accurate, and there is no deviation from reality. Among them, 18% of the budget data results show that the budget results are average compared with reality. These results The reference value is not great. Finally, 5% of the budget data results are quite different from the actual results. In summary, it can be seen that the model can still accurately budget some data in civil engineering projects. These data can bring great reference value to project construction. Data on the dynamic prediction results before project construction Analyzing, being able to avoid the risks in some projects is still very useful to the project.

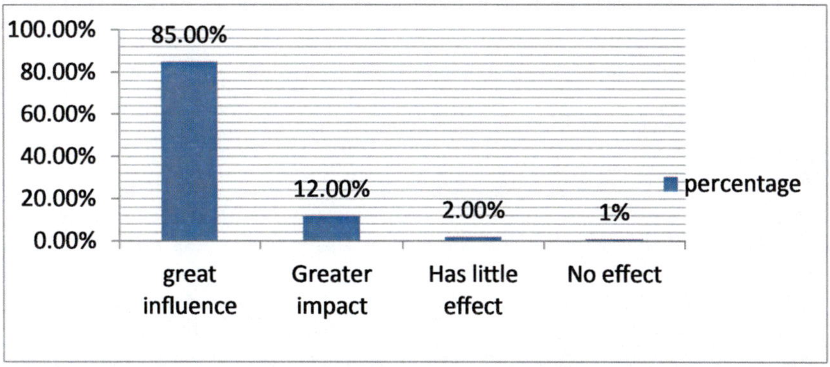

Fig. 2. The impact of BIM technology on all aspects of the project process

Together with the data in Fig. 2, we examined the status of BIM technology application on a particular construction site during the test. After a thorough analysis of the data along the way, we can draw the following conclusions: The application of this technology on the construction site is very extensive. "This is very important, because and very much promote the success of the project. It can bring project convenience in many aspects. For example, forecasting and risk prevention, cost and cost reduction. The implementation of BIM technology in civil engineering projects has a major impact on all aspects of the project. Two percent of the results show that the implementation of BIM technology in civil engineering projects does not have a major impact on all aspects of the project and 1% - the application of the technology in civil engineering projects does not have any impact on all aspects of the project. According to our survey results from this site, the implementation of BIM technology can indeed bring convenience to the project to some extent and supports BIM technology, so that staff on site have reduced a lot of workload during the project. [BIM technology].Especially after the continuous development of BIM technology and some breakthrough development, he gradually began to get the attention of workers on the construction site.BIM technology has led to major changes in the implementation of construction projects very important.

5 Conclusion

In conclusion, in the further development of civil engineering works, the use of BIM technology to perform related tasks can effectively reduce the workload of construction workers and technical management, as well as effectively improve the overall quality of work and strengthen scientific and complete engineering work. save space, reduce disadvantages, promote stable project development, reduce project costs and increase project benefits.BIM technology is widely used in civil engineering, which can effectively control construction costs, schedule and quality, and ensure the completion of civil engineering projects at the lowest cost, fastest speed and best quality. The relevant departments and units must attach great importance to the research and implementation of BIM technology. It covers a wide range of disciplines. Therefore, this article has some shortcomings and limitations and still needs to be done. Perfect. BIM technology has applications to advance the implementation of our civil engineering projects so that BIM technology can better serve society and the public.

References

1. Akinade, O.O., Oyedele, L.O., Ajayi, S.O., et al.: Designing out construction waste using BIM technology: stakeholders' expectations for industry deployment. J. Clean. Prod. **180**(APR.10), 375–385 (2018)
2. Ji, Q., Chen, W.: The application of BIM technology in the cost management of the whole process of construction projects. J. Phys.: Conf. Ser. **1648**(3), 032016 (2020). (6pp)
3. Moravej, H., Chan, T.H.T., Nguyen, K.D., et al.: Vibration-based Bayesian model updating of civil engineering structures applying Gaussian process metamodel. Adv. Struct. Eng. **22**(16), 3487–3502 (2019)
4. Thyagaraj, T.: Ground Improvement Techniques and Geosynthetics Volume 14 (IGC 2016 Volume 2). Influence of Granulated Blast Furnace Slag Contents on California Bearing Ratio Value of Clay GBFS Mixture. Lecture Notes in Civil Engineering, Chap. 31, pp. 277–284. Springer, Heidelberg (2019). https://doi.org/10.1007/978-981-13-0559-7
5. Bonelli, S., Jommi, C., Sterpi, D.: Internal Erosion in Earthdams, Dikes and Levees Volume 17 (Proceedings of EWG-IE 26th Annual Meeting 2018. Simplified Estimation of Some Main Characteristics of Pores and Constrictions in Granular Materials. Lecture Notes in Civil Engineering, Chap. 18, pp. 189–199. Springer, Heidelberg (2019). https://doi.org/10.1007/978-3-319-99423-9
6. Scott, R., Vidakovic, M., Chikermane, S., et al.: Encapsulation of fiber optic sensors in 3D printed packages for use in civil engineering applications: a preliminary study. Sensors **19**(7), 1689 (2019)
7. Xie, M., Meng, F., Feng, W., et al.: Education reform of theoretical mechanics for civil engineering in big data era. J. Phys.: Conf. Ser. **1624**(3), 032037 (2020). (5pp)
8. Abdurrahman, M.A., Latief, R.U., Mustari, A.S., et al.: Green construction application of the vida view apartment project in Makassar. In: IOP Conference Series: Earth and Environmental Science, vol. 419, no. 1, p. 012145 (2020). (6pp)
9. Hahn, L., Rittner, S., Nuss, D., et al.: Development of methods to improve the mechanical performance of coated grid-like non-crimp fabrics for construction applications. Fibres Text. East. Eur. **27**(1), 51–58 (2019)
10. Gao, L., Gong, Y., Zhou, H., et al.: Construction and application of a PCC energy pile model test system. In: IOP Conference Series: Earth and Environmental science, vol. 570, no. 3, p. 032065 (2020). (12pp)

Digital Protection of Dance of Intangible Cultural Heritage by Motion Capture Technology

Lebei Zhang(✉)

Cai Jikun Conservatory of Music, Minjiang University, Fuzhou, Fujian, China

Abstract. Motion capture is an application field involving electronic communication, human-computer interaction, motion biomechanics, navigation and so on. With the widespread popularization of computer technology and artificial intelligence, motion capture technology has broad prospects in the fields of medical rehabilitation, film and television animation production, virtual interactive games, sports training, intelligent monitoring and other fields. With the motion capture technology, the realization of resource sharing will become an important part of Fujian intangible cultural heritage dance industry development. The traditional dance data platform is established by using motion capture technology, and the virtual space of traditional dance information dissemination is constructed. The new resource-sharing mode based on shared platform can break the closed state of information and resources and avoid the consumption cost of repeated construction of resources or data platforms. This also realized the effective integration and reasonable allocation of resources. Actively cultivate modern dance talents of Fujian intangible cultural heritage. For the dance industry of Fujian intangible cultural heritage, it is necessary to reasonably combine online promotion with offline courses and services, so as to better develop the industry. "Under the influence of motion capture technology, the transmission rate of intangible cultural heritage dance in Fujian has reached 80%. This study has positive significance for promoting the inheritance and development of Fujian intangible cultural heritage dance.

Keywords: Fujian intangible cultural heritage dance · Motion capture · Dance inheritance and development · Resource sharing

1 Introduction

The continuous development of science and technology not only optimizes the efficiency of people's lives, but also saves offline costs and improves the speed of social and economic development. The rapid development of the sharing field cannot be separated from the rapid development of network technology. In the past era, ownership and use rights were not separated, but as a state of property rights. With the development of the sharing economy, the transition from products to services is accelerating the separation

B. J. Jansen et al. (Eds.): *International Conference on Cognitive based Information Processing and Applications (CIPA 2021)*, LNDECT 85, pp. 429–436, 2022.
https://doi.org/10.1007/978-981-16-5854-9_54

of these two rights. The advent of the digital age is first to separate the ownership of information products into possession and use rights. With the development, this situation does not stay online, but quickly extends to offline.

Therefore, the circulation of art needs the support of the market [1, 2]. Of course, business does not fully benefit art. We need to implement the inheritance basis of traditional culture and the quality of art to develop and create diversified industries [3, 4]. As an imported product, intangible cultural heritage dance has developed rapidly since it was introduced to my country, but its popularity is not as good as other sports [5, 6]. The Internet has changed all aspects of human life, making every individual an inseparable part of "Internet plus", so the paper also provides theoretical reference for other projects in theory [7, 8]. Therefore, the development of Fujian's intangible cultural heritage dance cultural industry chain requires the integration of different talents, which requires more development paths for the traditional cultural industry. Their construction is of great significance to the development of dance education [9, 10].

Facing the current economic gap between Fujian Province and the first-tier provinces and cities and the backward intangible cultural heritage dance business, the research believes that: reasonable resource allocation, serving employment and economic development as the purpose, adhere to the market-oriented, to improve Fujian Province The overall level of cultural heritage dance is the core, and the popular sharing concept is used to establish a "lessor" based on the particularity of the intangible cultural heritage dance industry. Change the backwardness of the Fujian region, open up the market in the Fujian region, promote economic development, and tap the huge potential of the unopened market. So as to improve the overall level of intangible cultural heritage dance in Fujian Province.

2 Inheritance and Development Path of Fujian Intangible Heritage Dance

2.1 Motion Capture Technique

The algorithm structure of the whole body motion capture system consists of five functional modules: sensor error correction, SINS solution, skeleton animation rendering, zero speed correction and multi-sensor calibration. Firstly, the sensor error correction module is used to correct the three sensors of gyroscope, accelerometer and magnetometer to improve the measurement accuracy of the sensor unit. Secondly, the data fusion of the three sensors is carried out by the strapdown inertial navigation solution module to obtain accurate and stable attitude motion data. Thirdly, the 3D virtual character model is established by the bone animation rendering module, so that the virtual character can interact with the real human movement, and the problems such as jumping and frame dropping in the animation are corrected by the zero-speed correction module. Finally, the multi-sensor calibration module connects the whole body node sensors to ensure the trajectory of the whole body movement.

2.2 Combination of Dance and the Internet

Entering the Internet age, we can always feel the changes in our lives brought by the Internet. Everyone can get information from the outside world just by opening the Internet without leaving home.

Assuming that the total number of samples in X is total, then:

$$P(c_i) = n_j/total\, j = 1, 2, \ldots m \tag{1}$$

For the data set X, the expected information is calculated as:

$$I(n_1, n_2, \ldots, n_m) = -\sum_{j=1}^{m} P(c_j)\log_2(P(c_j)) \tag{2}$$

The entropy obtained by dividing the data set Af by the description attribute X is:

$$E(A_f) = \sum_{x=1}^{m} \frac{n_{1x} + n_{2x} + \ldots + n_{mx}}{total}(n_{1x}, n_{2x}, \ldots, n_{mx}) \tag{3}$$

The change in audience needs is of course a higher demand for Internet-related functions. The audience in the dance industry is also naturally affected by the general environment. More and more people are accustomed to obtaining information by paying attention to public accounts, searching for videos through websites, displaying and promoting themselves through self-media, and conducting dance online courses in remote places. Study out of time and expand the sales of dance supplies through Internet technology. Because of the popularization of Internet technology and various demands, the continuous development of "dance + Internet" is bound to exist in the future.

3 Inheritance and Development Path Exploration Experiment of Fujian Intangible Cultural Heritage Dance

3.1 Relying on Network Technology

Use the network platform to manage resources, and provide scientific and effective technical support in a reasonable system. In the network platform environment, the realization of resource sharing will become an important part of the development of Fujian intangible cultural heritage dance industry. Use metadata suitable for dance to establish a traditional dance data platform to build a virtual space for the dissemination of traditional dance information. Taking into account the storage of the underlying data of the platform, the establishment of the database can prevent and manage information resources, and achieve effective control of information resources.

3.2 Enhancing the Awareness of Shared Resources

The new resource sharing model based on the sharing platform can break the closed state of information and resources and avoid the consumption cost of repeated construction of resources or data platforms. This also realizes the effective integration and reasonable allocation of resources. Actively build modern talents for Fujian intangible cultural heritage dance. For the Fujian intangible cultural heritage dance industry, online promotion and offline courses and services need to be reasonably combined in order to develop this industry better. Table 1 shows the average cost per 100 people for expanding art audiences in the network resource library.

Table 1. The average cost per 100 people of the network resource library to expand the art audience

Number of media contacts	Actual expansion	Total cost	Expansion of audience average cost per hundred people
50	42	8650	2.060
100	84	8750	1.042
150	126	8810	0.699
200	168	8840	0.526
250	210	8860	0.422

4 Inheritance and Development Path of Fujian Intangible Heritage Dance

4.1 Recruitment Channels of Intangible Cultural Heritage Dance Internet Companies

Table 2 shows the analysis results of recruitment channels of intangible cultural heritage dance Internet companies. Through performance or academic achievements, Fujian intangible cultural heritage dance can achieve certain economic benefits. Through publicity, the traditional folk dance performance market, tourism market and commodity market will be actively explored, the protection and inheritance of intangible cultural heritage will be developed by means suitable for current trends, and the local traditional folk dance will be brought out of the mountain by commercial methods. With the development of the Internet era, some social platforms have been widely used. Dance Internet companies are increasingly using social platforms such as We Chat, QQ, and Wei bo for recruitment. According to a survey, dance Internet companies use We Chat to recruit 75.56%. 42.22% of QQ recruits. The demand information is sent out through social platforms such as We Chat and QQ. Recruitment information can be quickly spread. There are usually hundreds of people in the teacher group, dance examination group, and substitute class group, which can also attract potential outstanding people. Recruiting through these social platforms can not only improve the efficiency of recruitment, but also promote the culture of dance Internet companies and enhance their brand image. In the Internet age, data can be used for analysis in recruitment work, to make statistics on the entire recruitment process, accurately record the number of interviewers, the number of applicants, the number of hires, and the required costs, etc., do a good job of recruitment issues and experience summaries, and timely Recruitment work is well adjusted.

Table 2. Analysis results of recruitment channels of intangible cultural heritage dance Internet companies

Recruitment channels	Quantity	Percentage
Five to eight city	25	55.56%
Gan ji	18	40%
We Chat	34	75.56%
QQ	19	42.22%
Employee recommendation	18	40%
Campus recruiting	6	13.33%
Flyer recruitment	21	46.67%
Headhunting	3	6.67%
Other	17	37.78%

4.2 College Dance Teaching

Figure 1 shows the analysis results of college dance teaching. In the questionnaire survey of college students, only 29.65% of the students think that the school attaches great importance to Internet teaching, 58.99% of the students think that the school attaches great importance to it, and a small number of students think that the school does not take it seriously. Insufficient attention from the school will affect the enthusiasm of dance teachers to use the Internet to assist teaching, and thus also affect the students' awareness of learning dance through the Internet. Interview items with a recognition rate of more than 80% include: Internet dance teaching promotes innovation in dance teaching models, and Internet dance teaching can enrich teachers' teaching methods: Interview items with a recognition rate of 70% to 80% include: Internet dance teaching has increased The cultural heritage of dance promotes the diversity of dance types, increases the opportunities for students to communicate with famous teachers, and improves the teachers' own teaching ability. Interview items with an approval rate of 60%–70% include: Internet dance teaching can help students improve their self-confidence, can increase the accuracy of student dance movements through video recording, can help students continue to solve problems until they find a solution, and can help students Self-development and improvement. The interview items with an approval rate of less than 60% include: Internet dance teaching can teach students the strategies needed for effective learning. It can be seen that most teachers have a high evaluation of Internet-assisted dance teaching, and at the same time believe that Internet dance teaching can not only improve teachers' own ability, but also improve students' learning level. Combined with the relevant conclusions of this questionnaire survey, we should pay full attention to the effective role of online teaching videos in basic teaching. The "Internet plus" mode is a low-threshold entry method that can maximize the acquisition of knowledge for students in terms of time and space. The frequency of students using Internet dance learning is not high. A large part of the reason is that the school's investment in Internet education is insufficient, and the network teaching equipment is backward, especially in economically backward areas. This situation is particularly obvious. The teacher's Internet use level is not high. Causing some teachers to reduce or not use online media for teaching, which will have a great impact on students' cognition of the Internet.

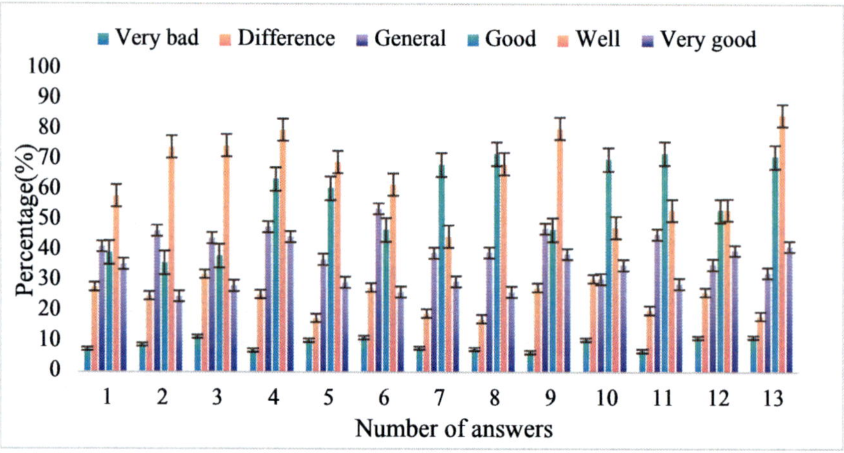

Fig. 1. Analysis results of college dance teaching

4.3 Digital Inheritance of Intangible Cultural Heritage Dance

Figure 2 shows the analysis results of digital inheritance of intangible cultural heritage dance. Digital communication makes full use of digital technologies such as augmented reality, virtual reality, and computer animation to make the display and presentation of traditional dance more three-dimensional and diversified. The application of somatosensory technology and human-computer interaction technology has enhanced the interactivity of dance display. The use of network technology breaks the barriers of communication and realizes the spread of traditional dance across time and space. All in all, digital communication not only breaks through the one-way nature of traditional display and communication methods, but also has advantages such as spanning time and space and saving costs. With the influence of Internet plus, the spread rate of intangible cultural heritage dance in Fujian reached 80%. Regarding the sharing and dissemination of traditional dances, we should also see that the introduction of new technologies will gradually become an important means of supplementary protection and inheritance. The development of digital new media not only helps people from different regions to communicate without restrictions, but also does not suffer from regional restrictions in the display of intangible cultural heritage, and the local culture and characteristics can also be felt in unrealistic spaces. The most basic function of establishing a digital database is to protect and manage the intangible cultural heritage dances and related cultural and creative products derived from them in different regions. Secondly, in this way, people of different nationalities can feel the power of culture, and at the same time, they can never understand the region. People can learn more about and appreciate the national culture of other regions, and it is also convenient for scholars from all over the world to exchange and learn dance, which ensures the inheritance of intangible cultural heritage dance and enhances national identity and pride. Finally, it is possible to sell unique cultural and creative designs related to intangible cultural heritage dance to interested audiences to promote the development of cultural and creative economic aspects of intangible cultural heritage dance. These three advantages are enough to show that the feasibility of establishing digital dance is very obvious and worthy of promotion.

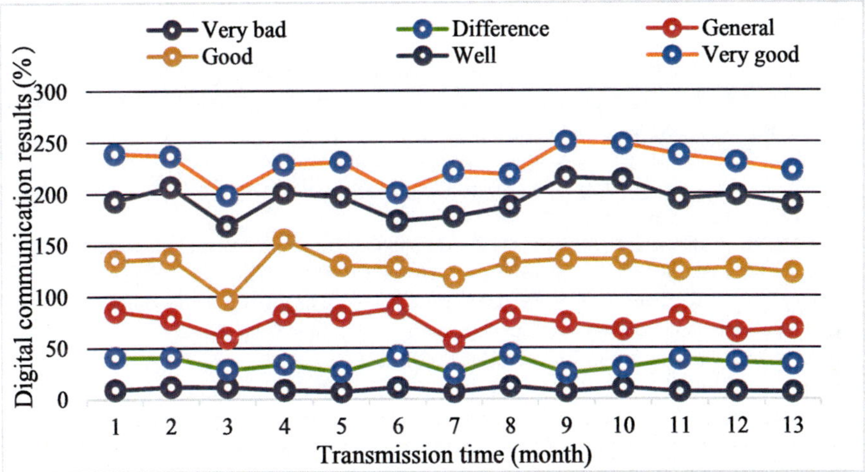

Fig. 2. Analysis results of digital inheritance of intangible cultural heritage dance

5 Conclusion

Under the impact of multiple cultures, Fujian intangible cultural heritage dance needs to seek more inheritance methods to better resist the storm of art and fashion, and under the general trend of cultural industry, Fujian intangible cultural heritage dance must seek new formats from the market, With the help of business models for dissemination and development. Art and capital should not be opposed, because people often create spiritual culture on a rich material basis, and its circulation and development will inevitably flow through the capital market.

While promoting market development, optimize the path of inheritance and development. After analyzing the possibility and necessity of the development of Fujian's intangible cultural heritage dance cultural industry, creating a "diversified" cultural product brand, and improving the capabilities of practitioners, these aspects will definitely be conducive to the marketing of intangible cultural heritage dance in Fujian. And development, the energy development and conversion of modern resources, so that more people can understand and inherit it. Witness the modern style of Fujian intangible cultural heritage dance.

Acknowledgments. Foundation project: 2018 Fujian Provincial Education Hall youth project JASI80348 "one belt and one road" cultural disseminat perspective Fujian heritage dance inheritance and Development Research.

References

1. Matthias, J., Kanagasundaram, T., Kopka, K., et al.: Synthesis of a dihalogenated pyridinyl silicon rhodamine for mitochondrial imaging by a halogen dance rearrangement. Beilstein J. Org. Chem. **15**(1), 2333–2343 (2019)
2. Maedza, P.: SOLD!: restaging dance, death, and disability. TDR **63**(4), 181–188 (2019)
3. Phuchomsri, S., Wongkhamchan, K., Wongpongkham, N.: Creativity in Isaan local dance. J. Eng. Appl. Sci. **15**(5), 1140–1147 (2019)
4. Belmar, S.: Singularities: dance in the age of performance. TDR **62**(4), 167–169 (2018)
5. Park, Y., Kimyoungjae. Exploring the factors of Drop out of arts high school dance major. Korean J. Dance **18**(3), 47–56 (2018)
6. Montanari, F., Scapolan, A., Mizzau, L.: Embeddedness and locational choices: a study of creative workers in a dance organisation. Urban Stud. **55**(5), 1121–1138 (2018)
7. Gerasimov, Y.V., Smirnov, A.K., Kodukhova, Y.V.: Assessment of possible causes of changes in abundance and sexual structure in populations of Prussian Carp (Carassius auratus gibelio Bloch. 1783). Inland Water Biol. **11**(1), 72–80 (2018)
8. Wenn, B., Mulholland, R., Timmons, W., et al.: Towards a developing construct in dance education-exploring the relation of emotional intelligence to teacher's sense of efficacy and teaching experience among dance education student teachers in the United Kingdom. Res. Dance Educ. **19**(1), 14–38 (2018)
9. Dominguez, J., Moral, M.D., Chio, J., et al.: Improving cognition through dance in older filipinos with mild cognitive impairment. Curr. Alzheimer Res. **15**(12), 1136–1141 (2018)
10. Nishizaki, T.: PKCε increases extracellular elastin and Fibulin-5/DANCE in dermal fibroblasts. Cell. Physiol. Biochem. **46**(1), 291–302 (2018)

Analysis and Research of Students' Mental Health Status Based on K-Means Clustering Under the Background of Big Data

Yao Li(✉), Ruijie Liu, Yu Bo, and Huimei Wei

School of Digital Technology, Dalian Institute of Science and Technology, Dalian, Liaoning, China
liyao@dlust.edu.cn

Abstract. With the continuous improvement of the quality of education in our country, mental health is one of the important basic conditions for students to become talents, and the management of students' mental health has received more and more attention from the society. Traditional student mental health management generally still stays at the establishment of questionnaires, and the application of data only stays at the stages of addition, deletion, modification and investigation. With the development of big data and artificial intelligence, We should elevate data analysis and application to an objective, visual, and predictable level, and take precautions and early warnings for student mental health management. This article uses data analysis technology to develop a student mental health data analysis platform based on the R-Shiny framework. The platform model uses the K-Means clustering algorithm to make full use of student data, scientifically and efficiently analyze the mental health of students, which is helpful for early warning and targeted scientific mental health counseling and management for students.

Keywords: Data analysis · K-means clustering algorithm · Student mental health analysis · Data visualization

1 Background Analysis

With the rapid development of our country's economy, the learning environment, living environment and ideological concepts of contemporary students have undergone tremendous changes. Facing problems in graduation, employment, and personal development, many students have negative emotions such as anxiety, low self-esteem, fear, depression, and tension. Mental health problems are becoming increasingly severe, which has seriously affected the overall development of students. Therefore, the mental health of students is an issue that deserves great attention from society and educational institutions. In most schools, the schools have set up mental health consultation offices, opened student mental health courses, and carried out student psychological tracking and evaluation work, and the mental health data of many students will be stored in the corresponding student psychological database. The basic operation of the database has

B. J. Jansen et al. (Eds.): *International Conference on Cognitive based Information Processing and Applications (CIPA 2021)*, LNDECT 85, pp. 437–444, 2022.
https://doi.org/10.1007/978-981-16-5854-9_55

also achieved certain results in solving mental health problems, but these data have not really been fully analyzed and utilized, and they have shortcomings such as lagging and one-sidedness [1–3].

2 Research Goals

This article proposes a method of mental health analysis based on K-Means clustering, using the ideas of data analysis and data mining on the basic functions of traditional systems to make secondary use of student psychological data, through the iterative process optimization of the K-Means algorithm, a large number of valuable parts of student psychological data deposited in the database are excavated, a data model is established, and a real-time visualized student mental health analysis platform is built based on the R-Shiny framework. Provide decision-making guidance to parents, schools, and students' mental health managers.

3 K-Means Clustering Algorithm

3.1 Definition of K-Means Clustering Algorithm

K-Means Clustering Algorithm (K-Means Clustering Algorithm) is an iterative solution of clustering analysis algorithm. Its steps are mainly to select K objects as initial cluster centers, and then calculate the distance between each object and each seed cluster center, Assign each object to the nearest cluster center. The cluster centers and the objects assigned to them represent a cluster [5]. Each time a sample is allocated, the cluster center of the cluster will be recalculated based on the existing objects in the cluster. This process is repeated until a certain end condition is met. The end condition can be that no object is reassigned to different clusters, no cluster center changes again, and the error square sum is locally minimum.

3.2 The Main Steps of K-Means Clustering Algorithm [9]

Input: sample set; number of clusters.

Steps:

1:Randomly select samples from D as the initial mean vector $\{u_1, u_2 \ldots u_k\}$;

2:**Repeat**

3: Set $C_i = \emptyset (1 \le i \le k)$

4: **for** $j = 1,2 \ldots m$ **do**

5: Calculate the distance between the sample and each mean vector (1≤i≤k):

$d_{ji} = \| x_i - u_i \| 2$;

6: Determined x_j 's cluster label according to the nearest mean vector: $\lambda_j \in = \arg\min \in \{1,2,..k\} d_{ji}$;

7: Divide samples into corresponding clusters: $C\lambda_j = C\lambda_j \cup \{x_j\}$;

8: **end for**

9: **for** $i = 1,2, \ldots k$ **do**

10: Calculate the new mean vector: $u'_i = \frac{1}{ci} \sum x \in ciX$

11: **if** $u'_i \neq u_i$ **then**

12: Update the current mean u_i to u'_i

13: **else**

14: Keep the current mean vector unchanged

15: **end if**

16: **end for**

17:**until** None of the current mean vectors are updated

Output: Cluster partition $C = \{C_1, C_2 \ldots C_k\}$

Note: The first line initializes the mean vector;

In lines 4–8 and 9–16, the current cluster division and mean vector are updated iteratively. If the clustering result remains unchanged after the iterative update, the current cluster division result is returned in line 18.

4 Analysis of Student Mental Health Based on K-Means Clustering

Using big data technology to objectively analyze and predict the various psychological data generated by college students on campus can effectively evaluate the psychological state of students [4].

Based on the R-shiny framework, this paper calls the K-Means clustering analysis algorithm to classify and visualize a large amount of data, and build a student mental health analysis platform. The psychological data of students is classified and visualized according to important characteristics, and then the characteristics of each type of students are analyzed. Potential mental health problems of such students can be discovered in advance, and a data analysis and early warning mechanism for proactive defense against mental health problems is proposed.

4.1 Data Sources

This paper did a questionnaire survey on sleep, social skills, stress, classroom, exercise and other categories of students in two classes (60 students). Each category has one or more characteristic quantities, which are finally randomized in a ten-week survey the 72-h feature value was extracted as the data source of this research.

The data types and characteristic quantities are shown in Table 1:

Table 1. Examples of data sources

Class	Class2	Exercise	Mood2	Sleep	Social	Stress
Experience	Challenge	Have	How	Hour	Number	Level
Hour	Effort	Schedule		Rate		
Due	Grade	Exercise		Social		

4.2 Selection of Feature Quantities

Each category in the data source has one or more feature quantities. The selection of feature quantities is closely related to the mental health of research students. Choosing the best feature quantity means that the final clustering result has a higher degree of reliability [6, 7]. By referring to the visual dimensions of the domestic student mental health standards and other standards-making systems, we continue to analyze and compare them, and finally we use "Sleep_hour", "Exercise_exercise", "Class_hour", "Stress_level", and "Social_number" as the five best Characteristic quantity, among which "Stress_level" is a characteristic quantity that best reflects the mental health of students.

4.3 Data Preprocessing

1. In order to ensure the accuracy of the data analyzed by the K-Means algorithm, the data needs to be preprocessed. the purpose of data preprocessing is as follows:

   (1) Ensure the validity of the data: the collected data should be related to the content of the research to avoid data loss;
   (2) Eliminate isolated data: Eliminate obviously inaccurate data;
   (3) Unified data scale: quantify data to facilitate the calculation of clustering algorithms;

2. The data processing steps are as follows [8]:

   (1) Perform average processing on the data of the characteristic quantity of each student sample within the planned time;
   (2) According to the corresponding data conversion table, convert the mean value of the feature quantity into the mean initial vector required by the K-Means clustering algorithm;

3. After data preprocessing, as shown in Table 2:

Table 2. Examples of some data after reprocessing

Class_hour	Exercise_exercise	Sleep_hour	Social_number	Stress_level
−1.5455	−1.1146	−0.18	0.2142	−0.2937
−0.9104	3.7491	−0.18	1.4492	−0.2937
−0.2752	−1.1146	−2.5805	−1.0709	−0.2937
−1.5455	−1.1146	1.0202	−2.3559	−2.2517
2.9005	0.1013	−0.18	−2.3559	−0.2937

4.4 K-Means Cluster Analysis Results

Based on R-shiny, a student mental health analysis platform was built. The number of clusters, namely the K value, can be selected arbitrarily. Each time a K value is selected, the sample data will be divided into K groups. The selection of the best K value is of great significance to the accuracy of the analysis of students' mental health. Comparing ground analysis and clustering, clustering into 5 categories has the best effect.

(1) The K-Means cluster centers are shown in Table 3:
(2) The student mental health analysis platform is shown in Fig. 1.

Table 3. K-Means cluster center table

Category	Class hours	Exercise	Sleep hour	Social number	Stress level
C1	−0.3006	−0.2391	−0.132	0.4712	−0.2937
C2	−0.9898	−0.5066	0.7201	−1.3921	−0.5385
C3	1.457	−0.1197	0.6929	−0.4868	−0.2937
C4	−0.2752	2.012	−1.2088	−0.153	−0.014
C5	0.1482	−0.304	−0.18	0.6425	1.6643

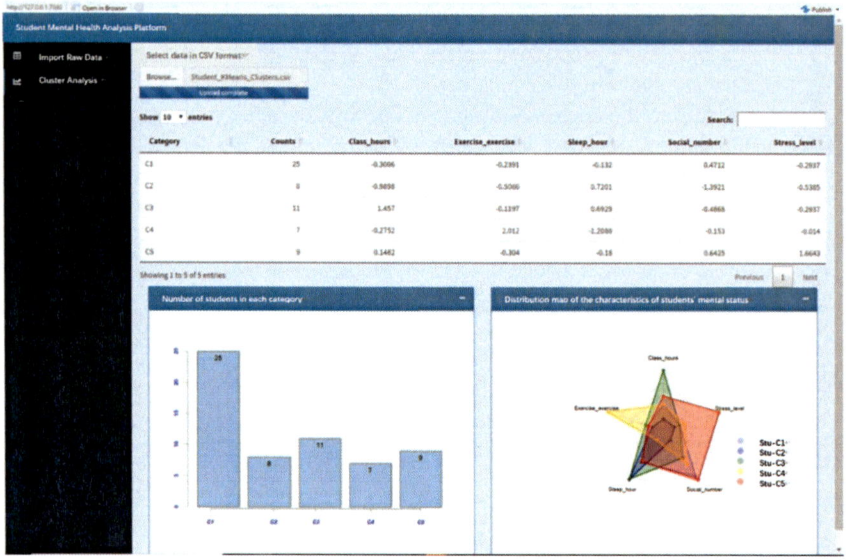

Fig. 1. Student mental health analysis platform

5 Analysis and Early Warning Based on K-Means Clustering Results

The number of students in C1 is 25, accounting for 42%. Most of these students have high EQ, are good at communication, lively and active, and their mentality is quite optimistic. Their college life is rich and colorful, and they can be regarded as happy children. They are willing to spend more. If they spend more time doing things they like, the chances of their mental health are quite small, but teachers or parents should encourage such students to spend more time on their studies and output outstanding talents to the society.

The number of student group C2 is 8, accounting for 13%. This type of student should be a group worthy of attention by teachers or parents. The phenomenon they reflect is that the direction is not clear, and they cannot find their own goal in the university, and they encounter a little setback. It is extremely easy to degenerate, and emotions are also easy to lose control, and the mental health problems of such students are difficult to find in the early stage, so teachers or parents should promptly provide psychological counseling and direction for such students.

The number of students in C3 is 7, accounting for 18%. Most of these students have excellent academic performance. They are often active members of the class, class leaders or student union leaders. For this kind of students, they should be well cultivated, and their positive, optimistic and cheerful mental state should be brought to every student in the class. This group is the main group that transmits positive energy in the class. Teachers or parents should treat them Guide them to improve their learning and morals.

There are 9 students in the C4 group, accounting for 12%. This kind of students are naughty and playful. They like to play games and also love and exercise. They are so-so about learning, but they have a strong self-esteem. When they are continuously hit by self-esteem, they are likely to overreact. If they have mental health problems, it is easy to detect them in time. While protecting their self-esteem, they should also point out their mistakes appropriately.

The number of C5 students is 11, accounting for 15%. This kind of students are prone to mental health problems. Their inner pressure is too great. Under the accumulation of small emotions, it is easy to have a mental breakdown. Most of them are the source Because their actual actions cannot catch up with their inner desires, they have only three minutes of enthusiasm for most things, and they are easily stumped by new knowledge. It is obviously impossible to achieve high results in this area. Teachers or parents should encourage such students to do things down-to-earth, and freezing three feet is not a cold day [10].

6 Conclusion

The methods of data analysis and data mining not only help school counselors and student administrators provide more and better mental health services to students, but also provide a certain reference value for student mental health educators and improve relevant The work efficiency of managers makes up for the limitations of traditional analysis methods, so as to achieve the purpose of scientifically, reasonably and quickly reflecting the psychological state of students.

References

1. Meyer, D.: Big data study for coping with stress. In: Scientific Stream at Big Data in Health Analytics, Sydney, NSW, Australia (2015)
2. Surma, J.: The privacy problem in big data applications: an empirical study on Facebook. In: 2013 ASE/IEEE International Conference on Social Computing, Washington, DC, USA (2013)
3. Larson, R.R.: Integrating data mining and data management technologies for scholarly inquiry. In: 2014 IEEE International Conference on Big Data (2014)
4. Zhao, G.: Advances in emotion recognition based on physiological big data. Jisuanji Yanjiu yu Fazhan **1**, 80–92 (2016)

5. Wu, T.: Research on the psychological management system of college students based on k-means clustering algorithm. Hubei University of Technology (2017)
6. Li, F., Jun, L.: Big data to mental health education: functional boundary and implementation orientation. Ideol. Theor. Educ. (2016)
7. Wu, J.: Intelligent era: big data and intelligent revolution redefine the future (2016)
8. Xia, Q.: Analysis of student campus activity management based on k-means algorithm. Soochow University, Suzhou
9. Zhou, Z.: Machine Learning, 1st edn. Tsinghua University Press (2016)
10. Wencong, L., Beilei, L.: Common psychological problems of college students and intervention strategies. Gener. Nurs. **16**(24), 2982–2983 (2018)

Financial Crisis Early Warning Model of Small and Medium Sized Enterprises Based on Big Data

Tao Yu(✉)

Aba Normal University, Aba, Sichuan, China

Abstract. Reform and opening up is the rapid growth of China's enterprises. At present, most of China's enterprises are mainly small and medium-sized enterprises, but in recent years, the problems encountered in the development of small and medium-sized enterprises are more and more obvious. The development of things always has a gradual process, so does the financial crisis. The purpose of this paper is to study an early warning model, which can collect, analyze and sort out the evidence, identify the precursor of financial crisis, and strangle it in the cradle. The research work of this paper is to establish an early warning model of financial crisis by means of data mining. Firstly, this paper selects 10 financial and non-financial indicators that can describe eight aspects of enterprise risk level, development ability, market value, relative value and corporate governance. Firstly, Mann Whitney method was used to test the indicators, and six indicators were obtained to identify the two types of enterprises. Then PCA Cox and pls Cox models were established by combining 7 factors extracted from PCA and 3 factors extracted from pls with Cox model Financial crisis early warning model, and through the fitting and prediction of the two models, study which is more suitable for the financial early warning of small and medium-sized listed companies in China. Finally, the data of 15 sample enterprises and the remaining 27 normal enterprises from 2018 to 2020 are used to verify the model. After verification, the two models are suitable for financial early warning of small and medium-sized enterprises, and the prediction level of PLS Cox model is better.

Keywords: SME financial crisis · Financial indicators · Early warning model · Cox model

1 Introduction

In recent years, the problems of small and medium-sized enterprises in the operation process are increasing, mainly reflected in: China's economic growth gradually slows down to a stable, aging makes the advantages of labor resources weaken, and the cost of raw materials increases, leading to the development bottleneck of small and medium-sized enterprises [1, 2]. E-commerce encroaches on the original sales share of small and medium-sized enterprises in the Internet era; we expect that the technology quality and innovation ability of small and medium-sized enterprises can meet the needs of

B. J. Jansen et al. (Eds.): *International Conference on Cognitive based Information Processing and Applications (CIPA 2021)*, LNDECT 85, pp. 445–453, 2022.
https://doi.org/10.1007/978-981-16-5854-9_56

economic globalization recognized by the world [3, 4]. A systematic and perfect financial management system can promote the growth of enterprises at the beginning of the establishment of enterprises, and detect and prevent the possible crises of enterprises. Therefore, it is necessary to establish a financial crisis early warning model which is in line with the characteristics of small and medium-sized enterprises [5, 6].

In recent years, many scholars have put forward their own opinions on the early warning model of small and medium-sized enterprises. For example, when foreign scholars establish the enterprise financial risk model, they first propose that the multivariate model is further established by the single variable model, then the regression model and ANN model are introduced to lay the foundation for the application of financial early warning field. Ayvaz E and other scholars take the listed companies in manufacturing industry as the research object, select health and crisis enterprises to establish logistic model, then establish PCA logistic model and pls logistic model through variable screening, which proves the superiority of PLS in variable screening [3, 4, 7, 8].

This paper takes 12 listed companies of SMEs marked ST or * ST in 2015–2020 and 24 financial normal companies randomly selected by 1:2 ratio for financial abnormal in 2015–2020. 10 indicators which can distinguish crisis enterprises and normal enterprises through significance analysis are taken as independent variable data [9]. First, the model group was constructed by taking the data of 9 enterprises marked in 2016–2017 and 18 enterprises randomly selected from 24 enterprises as training samples. Then, according to the highly irrelevant nature of Cox model, PCA and pls are used to extract factors from data, and PCA Cox model and pls Cox model are built respectively by using PCA and pls instead of original index and dependent variable respectively model, because the new factors and the original indicators have a clear linear relationship, we can use the data of model validation group to carry out model prediction verification [10].

2 Overview of Financial Crisis of Small and Medium-Sized Enterprises

2.1 Financial Crisis Overview

2.1.1 Financial Risk

There are risks in every step of enterprise operation, which can be divided into financing risk, investment risk, operation risk and income distribution risk. For different stakeholders, the positioning of the enterprise is different, so the financial risk of the enterprise is complex and changeable. In different industries, the financial risks faced by enterprises are different, and the results of the risks are also different. We should seize the opportunity from the risks to obtain greater profits.

2.1.2 Definition of Financial Crisis

Foreign scholars believe that financial crisis can be summarized as business failure, enterprise technology failure, accounting failure and enterprise bankruptcy. Domestic scholars believe that the financial crisis is a process from business failure to inability to repay debts and then to bankruptcy.

2.2 Survival Analysis

In the research of financial crisis, survival analysis regards it as a dynamic process, which is more practical, accurate and practical.

2.2.1 Survival Time

The survival time is the time that the research object passes from the beginning of the study to the end event. This paper sets the research starting point as the model construction group at the end of 2015, and the model verification group is the end of 2018. The listed companies of small and medium-sized enterprises are marked by the CSRC for "abnormal financial situation" as the death time of the individual, that is, the starting point of the research determined in this paper is the starting point of the survival time, the time marked ST or *st is taken as the termination time and the year as the time measurement scale. The functions describing the distribution of life time are risk function h(t), probability density function f(t) and survival function s(t).

(1) The risk probability H (T) is the probability that an individual survives at t time, but dies at $t + \Delta t$, namely:

$$\mathrm{h(t)} = \lim_{\Delta t \to 0} \frac{p(t \le T \le t + \Delta t | T \ge t)}{\Delta t} \tag{1}$$

Then the cumulative risk function is the sum of the probabilities of risk occurrence between 0 and T, that is:

$$\mathrm{H(t)} = \int_0^t h(t)dt \tag{2}$$

(2) The density function is the probability of an event occurring between $(t, t + \Delta t)$:

$$\mathrm{f(t)} = \lim_{\Delta t \to 0} \frac{p(t \le T \le \mathrm{t} + \Delta t)}{\Delta t} \tag{3}$$

Suppose that the distribution function exists, then the distribution function is the probability that the individual survival time is less than or equal to t:

$$\mathrm{F(t)} = \mathrm{P(T \le t)} = \int_0^t f(t)dt \tag{4}$$

(3) The survival probability is the probability of individual survival time > t

$$\mathrm{S(t)} = \mathrm{P(T \ge t)} = 1 - \mathrm{F(t)} = \int_t^{+\infty} f(t)dt \tag{5}$$

2.2.2 Survival Data Classification

The data of survival time analyzed by survival analysis is called survival data, which can be divided into complete data and incomplete data according to the outcome of individual. Complete data refers to the data that can obtain all observation information, while incomplete data refers to the data that cannot obtain all observation information.

2.2.3 Methods of Survival Analysis

Survivability analysis techniques include nonparametric, parametric and semi parametric methods. There are two common nonparametric models: life scale analysis, multiplicative limit analysis and life scale analysis. The multiplicative limit rule is to take the complete data and the deleted data into account, and analyze the survival data to build a time event model. The common methods of parameter distribution model are exponential distribution model, log logistic distribution model, log normal distribution model and Weibull distribution. Semi parametric method is a kind of analysis method which has the properties of parametric method but is more flexible than non-parametric method, and it is easier to analyze and explain the results than non-parametric method.

2.3 Variable Screening Method

2.3.1 Principal Component Analysis

In this paper, we select eight financial indicators to analyze the data, that is, multiple variables. In order to obtain as much information as possible, we should select as many indicators as possible. On the other hand, the increase of indicators will increase the load of research issues. Because of the correlation between indicators, information overlap may affect the analysis process, so we consider using fewer indicators to replace most of the original indicators, but at the same time we can retain most of the information.

2.3.2 Partial Least Squares

The derivation process of the algorithm is to standardize the set of independent variables and the set of dependent variables, extract the components of the set of independent variables and the set of dependent variables respectively, and maximize the correlation between the two components, and then calculate the parameter vector in the regression model.

3 Financial Early Warning Research Design

3.1 Research Objects

Based on the data of a database, the index data of small and medium-sized listed enterprises are selected according to the stock market classification. Because this paper needs to forecast the financial situation of T-Year with T-3, T-2 and T-1 data, therefore, considering the distribution of samples and data availability comprehensively, taking 2015–2020 as the observation period, the listed SMEs marked ST or *ST year as the crisis boundary time point, 43 enterprises marked by the CSRC as ST or *ST in 2015–2020. The number of times of Tianjin prin and East China CNC, which is marked twice within three years, is subject to the first time of marking, but not within 3 years. Therefore, 12 enterprises marked as ST or *ST and 24 small and medium-sized board enterprises that have not had financial crisis during the random observation period are taken as samples.

3.2 Data Selection

This paper is based on the time requirements of financial crisis early warning model and the dynamic grasp of time by Cox model. Assuming that the year of ST or * ST is t, then the previous year is T-1 and the first two years is T-2. So on, this paper attempts to select the end of 2015 as the starting point of survival time, and the focus of the survival time is the enterprise falling into crisis, and the survival time of the enterprise is 1 respectively In, 2 and 3 years, the model was built with the marked enterprises in 2016, 2017 and 2018 as samples, and the marked enterprises in 2018, 2019 and 2020 were used to verify the model. The sample size that can be studied is shown in Table 1.

Table 1. Number of sample groups

	Model building group	Model validation group
Fall into crisis group	9	15
Normal enterprise	18	27
Total	27	43

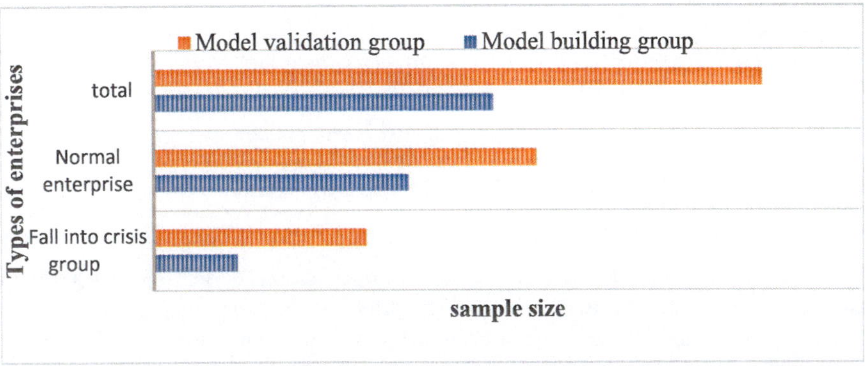

Fig. 1. Number of sample groups

As shown in Fig. 1, the data sample selected in this paper is that there are 9 crisis groups and 18 normal enterprises in the model construction group. In the model validation group, there are 15 crisis groups and 27 normal enterprises.

3.3 Research Variables

The research variables should be able to comprehensively and objectively reflect the financial situation of the enterprise, and in order to facilitate the research and analysis and the results accurately conform to the Cox model. The model requires variable, we need to remove the repeated or strongly related indicators to make these indexes distinguish financial crisis enterprises and financial normal companies obviously. At the same time, we should pay attention to the availability of data, and we should choose the data available from the public data of enterprises. To sum up, this paper selects 10 indicators from eight aspects, among which the corporate governance capability index is non-financial information, which can reflect the status of enterprises from the perspective of corporate governance framework.

3.4 Research Methods

This paper deals with the original data with Excel. Firstly, the indexes which can distinguish financial crisis enterprises and healthy enterprises are screened out by IBM SPSS statistics 23. Then, the extraction factors of SPSS are used as the extraction factors for Cox regression analysis, and PCA Cox financial crisis warning model is constructed. Then, the extraction components of Matlab function are used to participate in Cox regression analysis, and pls Cox is constructed In the last part of model validation, SPSS is used to do Kaplan Meier analysis to estimate the benchmark survival rate of enterprises, so as to further verify the prediction ability of the two early warning models.

3.5 Financial Index Screening

In this paper, six enterprises marked in 2016, 2017 and 2018 and 18 enterprises randomly selected as samples were selected as samples. The indicators data of the end of 2015 were tested by using IBM SPSS statistics 23 tool and Mann Whitney U test to test whether there are significant differences between health enterprises and crisis enterprises.

4 Model Construction and Empirical Research

4.1 Construction of Financial Crisis Early Warning Model

4.1.1 Construction of PCA Cox Model

Kmo and Bartley spherical test were used to extract factors by PCA. The maximum variance method was used to do rotation. The orthogonal rotation method was used to minimize the number of variables of each factor with maximum load. SPSS software was used to process. Because kmo $= 0.702 > 0.6$, it is suitable for principal component analysis. Under the 95% confidence level, the significance value of Bartley sphericity test is close to 0, that is to say, each index has correlation, which meets the condition of principal component analysis. Therefore, these data are suitable for principal component analysis, and the results of common factor variance extraction are shown in Table 2.

As can be seen from Fig. 2, the regeneration commonality of the extracted factors by the principal component method is almost greater than 60%, which indicates that the principal component extraction is meaningful and extracts most of the information.

Table 2. Results of common factor variance extraction

Index	Initial	Extract
X1	1.00	0.378
X2	1.00	0.837
X3	1.00	0.673
X4	1.00	0.952
X5	1.00	0.941
X6	1.00	0.925
X7	1.00	0.621
X8	1.00	0.523
X9	1.00	0.941
X10	1.00	0.942

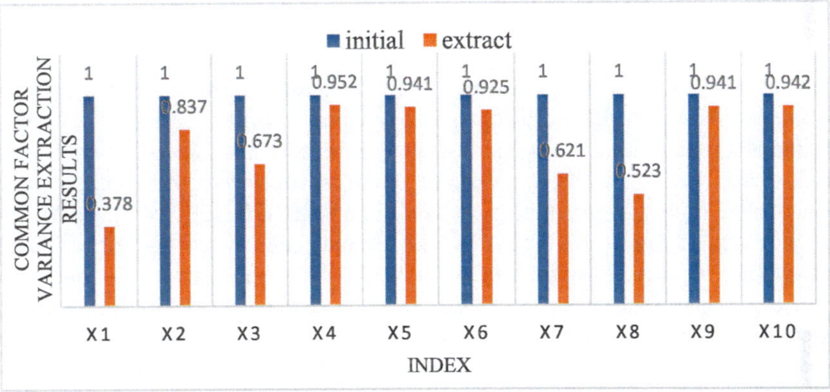

Fig. 2. Results of common factor variance extraction

4.1.2 Pls Cox Model Construction

In this paper, the original index is processed by the function of MATLAB. According to the method of cross validation, when the third pls factor is extracted, $Q_3{}^2 = 0.0083 < 0.0997$, the model meets the accuracy requirements, so this paper determines three pls alternative components. First, the PLS vector $p = (p_1, p_2, p_3)$ Compared with the original index $X = (X_1, X_2, \ldots, X_{10})$ $P = XB + C$. The specific expression of PLS Cox financial early warning model is as follows:

$$h(t) = h_0(t)\exp(0.528p_1 + 0.01p_2 - 0.352p_3 \tag{6}$$

4.2 Verification of Model Prediction Ability

4.2.1 Model Prediction Accuracy

In order to determine the critical value of the survival rate of an enterprise as its critical point, this paper mainly selects the percentage of the normal enterprise sample size in the total survival rate of the whole enterprise as the calculation method of its critical value. Because the normal enterprise size accounts for 18/(9 + 18) of the total survival rate of the whole enterprise in the construction process of the model, this paper selects 66.67% as the critical point of its survival rate. By substituting the original data to be analyzed and predicted and the benchmark survival rate of the corresponding time into the following formula, the average survival rate of each sample in year t can be directly calculated. If the average survival rate to be predicted is 66.67%, it is a normal enterprise, otherwise it is a crisis enterprise.

5 Conclusion

This paper selects 12 enterprises marked with ST or * st by CSRC from 2015 to 2020 and 24 enterprises with normal finance from 2012 to 2017 randomly selected by 1:2 ratio as samples, and takes 9 enterprises marked with ST or * ST from 2012 to 2014 and 18 normal enterprises as model construction group samples, and the other 15 Enterprises and 27 enterprises as model verification group samples. In order to describe the financial situation of an enterprise more comprehensively, this paper first selects 10 financial and non-financial indicators which can describe the risk level, development ability, market value, relative value and corporate governance of an enterprise. Firstly, Mann Whitney method was used to test the indicators, and six indicators were obtained to identify the two types of enterprises. Then PCA Cox and pls Cox models were established by combining 7 factors extracted from PCA and 3 factors extracted from pls with Cox model Financial crisis early warning model, and through the fitting and prediction of the two models, study which is more suitable for the financial early warning of small and medium-sized listed companies in China. Comprehensive comparison results show that PLS Cox model has higher accuracy for both st group and non ST group. PCA does not consider dependent variables in extracting factors, and partial least squares method has stronger variable interpretation ability. This paper makes good use of the feature that Cox model can consider censored data and can conduct dynamic research on samples, which further verifies the applicability of COX in financial early warning.

References

1. Early warning of financial crisis of colleges and Universities Based on the BSC perspective. J. Guangxi Univ. Financ. Econ. 030(002), 88–94 (2017)
2. Chang, A.Y., Cheng, Y.T.: Analysis model of the sustainability development of manufacturing small and medium- sized enterprises in Taiwan. J. Clean. Prod. **207**, 458–473 (2019)
3. Zhang, Y.: Financial system crisis early warning model analysis based on neural network with entropy method, pp. 560–563 (2018)
4. Coudert, V., Idier, J.: Reducing model risk in early warning systems for banking crises in the euro area. Int. Econ. **156**(DEC.), 98–116 (2018)

5. Zheng, B.: Problems and Countermeasures of financial management in small and medium sized enterprises. J. Hebei Univ. Technol. (Soc. Sci. Edn.) **035**(004), 46–47 (2018)
6. Zheng, C.: Analysis of innovation measures of enterprise economic management in the post financial crisis era. SME Manage. Technol. (034), 26–27 (2017)
7. Li, G.: An integrated model of rough set and radial basis function neural network for early warning of enterprise human resource crisis. Int. J. Fuzzy Syst. **21**(8), 2462–2471 (2019). https://doi.org/10.1007/s40815-019-00758-z
8. Ayvaz, E., Kaplan, K., Kuncan, M.: An integrated LSTM neural networks approach to sustainable balanced scorecard-based early warning system. IEEE Access **PP**(99), 1 (2020)
9. Bahraminejad, M., et al.: Proposing an early-warning system for optimal management of protected areas (Case study: Darmiyan protected area, Eastern Iran). J. Nat. Conserv. **46**, 79–88 (2018)
10. Catullo, E., Palestrini, A., Grilli, R., Gallegati, M.: Early warning indicators and macro-prudential policies: a credit network agent based model. J. Econ. Interac. Coord. **13**(1), 81–115 (2017). https://doi.org/10.1007/s11403-017-0199-y

Automatic AI Detection Technology of Foul Shot Scanning in Simulated Basketball Match Under the Background of AI

He Li(✉)

Physical Education Institute, Bohai University, Jinzhou 121000, Liaoning, China

Abstract. The current development speed of AI technology is following fast. It can track and analyze a single sports target in the foul shots of simulated basketball games, and use convenient automatic AI detection technology to achieve the best results. And based on this, it also analyzes the current situation of AI in sports competitions at home and abroad, and also finds whether there is research in AI simulation basketball through domestic and foreign websites. The relevant literature on the lens scanning technology of the regulations in the competition, and then conduct in-depth analysis and research on it. And this purpose is to help study the analysis of foul shots in modern basketball games and make an important pavement, and the ultimate goal is to provide theoretical support and guidance for the further development of foul shots in modern simulated basketball games. AI uses foul action detection technology to accurately capture and detect foul actions in basketball games, which is an important auxiliary technology for accurate judgment in fierce competitions. Through experimental research, we propose a foul shot scanning automatic detection technology in the field of AI. The technology uses the Baum-Welch algorithm to obtain the action contour of the basketball player, and uses the AdaBoost algorithm to extract the key frames of the basketball player's action image, and finally uses the LLE algorithm to extract feature information to determine whether the action is foul. Finally, through experimental analysis of the research results, we can see the use of AI technology can not only provide basketball players with convenient and effective strategies, but also develop different training plans according to each basketball player's physical fitness.

Keywords: AI basketball game · Foul mode · Intelligent automatic detection · Intelligent algorithm

1 Introduction

A large amount of multimedia data (including audio, video, pictures, language, etc.) can be accurately captured using modern intelligent technology. The precise capture of AI technology and the analysis and prediction of different foul times of athletes in a simulated basketball game make it more accommodating and convenient [1]. However,

B. J. Jansen et al. (Eds.): *International Conference on Cognitive based Information Processing and Applications (CIPA 2021)*, LNDECT 85, pp. 454–462, 2022.
https://doi.org/10.1007/978-981-16-5854-9_57

due to the different results observed from different angles, it is easy to be affected by subjective judgment and easy to make mistakes. Therefore, scanning automatic detection technology gradually replaced pure visual observation [2, 3]. This automatic detection technology can automatically analyze all the basketball players' sports in the game, select fouls and possible fouls, and then the referee decides [4].

In recent years, with the development of science and technology, the problem of referees' misjudgment due to their subjective consciousness has been solved [5]. The technology of automatic lens scanning detection has been successfully used in various competitions, and then it has become an important technology for detecting foul actions of athletes in major competitions [6]. In basketball games, common misjudgments and misjudgments often cause conflicts between players and referees. However, it is difficult to detect the foul behavior of basketball players in the fierce competition by using the foul action detection technology in the current basketball game. Therefore, how to reduce errors and misjudgments and accurately detect athletes' foul actions has become a major problem. The requirements of the basketball game action specifications are very high. In the game, whether an athlete's action fouls is mainly judged by the referee's visual observation [7].

Through the comparative analysis of the literature, we can know that the wide application of AI and its intelligent factors have become a very common phenomenon in the capture of foul shots in modern basketball games. The emergence of this technology has been adopted in the field of basketball games. Continuous experimental analysis to improve the ability to make it more perfect to enter the field of basketball foul shot capture detection [8]. And perform big data analysis on a single opponent player, which can improve the player's predictive ability and make corresponding responses [9, 10]. The automatic detection of foul action lens scanning is a challenging task, including operations such as motion contour extraction, motion feature extraction, and motion analysis. The following are the difficulties encountered in the research process and the current status of the research on automatic detection of foul shots [11, 12].

2 Method

2.1 Baum-Welch Algorithm

So it has the following formula:

$$\mathrm{P}(\mathrm{O}, \lambda) = \sum\nolimits_{I} P(O|I, \lambda)P(I, \lambda) \tag{1}$$

$$\mathrm{Q}(\lambda, \overline{\lambda}) = \sum\nolimits_{\mathrm{I}} \log \mathrm{P}(\mathrm{O}|\mathrm{I}, \lambda)\mathrm{P}(\mathrm{I}, \lambda) \tag{2}$$

$$\mathrm{Q}(\lambda, \overline{\lambda}) = \sum\nolimits_{\mathrm{I}} \log \mathrm{P}(\mathrm{O}|\mathrm{I}, \lambda)\mathrm{P}(\mathrm{I}, \lambda) + \sum_{\mathrm{I}} \langle \sum\nolimits_{\mathrm{t}=1}^{\mathrm{T}-1} \log \mathrm{a}_{\mathrm{I}_t \mathrm{I}_{t+1}} \rangle \mathrm{P}(\mathrm{O}|\mathrm{I}, \overline{\lambda})$$
$$+ \sum\nolimits_{\mathrm{I}} \langle \sum\nolimits_{\mathrm{t}=1}^{\mathrm{T}} \log \mathrm{b}_{\mathrm{I}_t}(\mathrm{O}_t) \rangle \mathrm{P}(\mathrm{O}|\mathrm{I}, \overline{\lambda}) \tag{3}$$

$$\mathrm{b}_{\mathrm{j}}(\mathrm{k}) = \frac{\sum_{\mathrm{t}=1}^{\mathrm{T}} (\mathrm{O}, \mathrm{i}_t = \mathrm{j}|\overline{\lambda})\mathrm{I}(\mathrm{O}_t = \mathrm{v}_k)}{\sum_{\mathrm{t}=1}^{\mathrm{T}} \mathrm{P}(\mathrm{O}, \mathrm{i}_t = \mathrm{j})|\overline{\lambda}} \tag{4}$$

2.2 Learning Algorithm of RBF Neural Network Parameters

Take the performance index function of the network as:

$$E_k = \frac{1}{2}(y_k - y_{m,k})^2 \tag{5}$$

The gradient processing algorithm is used to modify the hidden layer neuron center, baseline parameters and exit weights of the RBF neural network. The specific iterative algorithm is as follows:

$$\Delta c_{ji} = (y_k - y_{m,k})\omega_j h_j \frac{x_i - c_{ij}}{b_j^2} \tag{6}$$

$$\Delta b_j = (y_k - y_{m,k})\omega_j h_j - \frac{\|X - C_j\|^2}{2b_j^3} \tag{7}$$

$$\omega_j(k) = \omega_j(k-1) + \eta(y_k - y_{m,k})h_j + \alpha(\omega_j(k-1) - \omega_j(k-2)) \tag{8}$$

Through RBF neural network identification, the sensitivity of target input is:

$$\frac{\partial y_k}{\partial u_k} \approx \frac{\partial y_{m,k}}{\partial u_k} = \sum\nolimits_j \frac{\partial \omega_j h_j}{\partial u_k} = \sum\nolimits_j \omega_j h_j \frac{c_{1j} - x_1}{b_j^2} \tag{9}$$

2.3 Extraction of Action Features of Basketball Players

(1) Initialize the characteristic action image of the basketball player

$$D(i) = \frac{1}{Z} \tag{10}$$

(2) Among the T samples of basketball player actions, select some known data as the Thunder samples, and use the weak classifier to find suitable values for the training samples. The error rate of the training samples is:

$$g_i = P_D(h_i(x_i) \neq y_i) \tag{11}$$

The weight of the weak classifier is:

$$a_i = \frac{1}{2}\ln\left(\frac{1 - \varepsilon_i}{\varepsilon_i}\right) \tag{12}$$

The sample weight of the updated training set is:

$$D_{I+1} = \frac{D_i \exp(-a_1 y_i h_i(x_i))}{G_i} \tag{13}$$

2.4 AdaBoost Algorithm

First, to learn to get the basic processor:

$$T_m(x) : X \to \{-1, +1\} \tag{14}$$

$$e_m = P(T_m(xi) \neq yi) = \sum_{i=1}^{N} wmiexp\left(-a_m yiT_m(xi)\right) \tag{15}$$

Calculate the coefficient of Tm(x):

$$a_m = \frac{1}{2}\log\frac{1 - e_m}{e_m} \tag{16}$$

Update the probability distribution of the training data set:

$$D_{m+1} = (W_{m+1}, 1, \ldots, W_{m+1}, i, \ldots, W_{m+1}, N)$$

$$W_{m+1,i} = \frac{W_{mi}}{Z_m}exp(-a_m y_i T_m(x_i)), (i = 1, 2, \ldots, N) \tag{17}$$

After completing the above steps, construct a linear combination of classifiers:

$$f(x) = \sum_{m=1}^{M} a_m T_m(x) \tag{18}$$

3 Experiment

3.1 The Establishment of Two-Dimensional Animation Interactive Demonstration Model

In the process of using two-dimensional animation to interactively demonstrate the teaching process. After I_0^'s low-frequency sub-image I_0LL^', next, use the SIFT feature extraction method to extract the feature points of I_0LL^', which mainly includes the following steps:

(1) Extremum detection in scale space. As shown in formula (19):

$$L(x, y, \sigma) = G(x, y, \sigma) * I'_{0LL}(x, y) \tag{19}$$

It can describe the overall content of the two-dimensional animation interactive demonstration teaching process. The smaller the value, the better it can be. The detailed information of the three-dimensional animation interactive demonstration teaching process is described in detail. The expression of the shown in formula (11):

$$G(x, y, \sigma) = \frac{1}{2\pi\sigma^2}e^{-(x^2+y^2)/2\sigma^2} \tag{20}$$

SIFT feature point generation.

The operation defines the scale space of the image:

$$L(x, y, \sigma) = G(x, y, k\sigma) * I(x, y) \tag{21}$$

$$D(x, y, \sigma) = G(x, y, k\sigma) - G(x, y, \sigma) * I(x, y) = L(x, y, k\sigma) - L(x, y, \sigma) \tag{22}$$

Basic Principles of PCA.

After maximizing the data mapping, the variance formula can be obtained as:

$$\max_w \frac{1}{m-1} \sum\nolimits_{i=1}^{m} \left(w^T (x_i - \overline{x})^2 \right) \tag{23}$$

Definition W is a matrix composed of the column vectors of all feature map vectors. This matrix can better retain the information in the data. The matrix can obtain an optimized objective function through linear transformation, which is

$$\begin{aligned} &\min_w tr\left(W^T A W\right), \\ &s, t, W^T W = I \end{aligned} \tag{24}$$

$$A = \frac{1}{m-1} \sum\nolimits_{i=1}^{m} (x_i - \overline{x})(x_i - \overline{x})^T \tag{25}$$

4 Evaluation Results

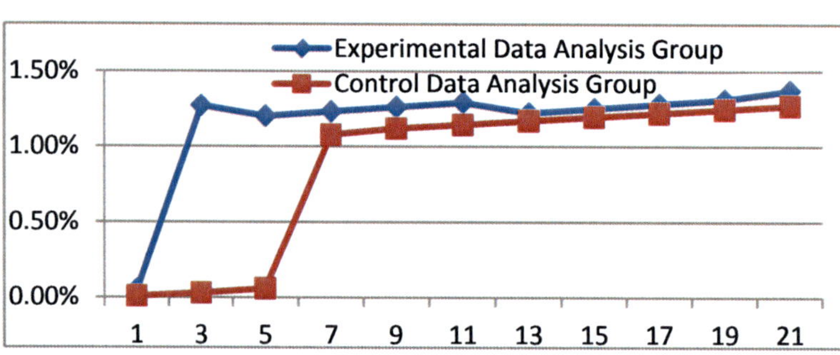

Fig. 1. The increase of the intelligent foul detection system in the two experimental groups

From Fig. 1 and the comprehensive data analysis with Table 1, it is understood that the control group A of the basketball special sports game after the traditional coaches do referee training, the increase in the check rate of 1 foul action is 3.07%, compared with the test score of the control group 4.12%. It can be said that the increase is obvious, which also proves that artificial intelligence technology can greatly promote the improvement of the training method and strategy of the foul detection system. In terms of basketball in the

Table 1. Experimental statistical data analysis based on feature algorithm

Experimental data			Experimental result			
Serial number	Ball	Foul	Ball	Misjudgment	Recall rate%	Accuracy%
1–50	74	32	72	7	95.8%	97.2%
51–100	65	29	69	5	96.9%	98.3%
101–150	32	25	21	3	86.3%	92.6%
151–200	21	12	23	2	93.5%	91.2%
201–250	79	43	3	3	96.2%	99.5%

current basketball game, basketball coaches need to be able to provide systematic sports for players, just like other coaches. First of all, the first is to conduct a comprehensive assessment of the athlete's physical fitness and stamina. And the second is to analyze the main foul actions of basketball players. Can have a substantive supervision training for basketball players, but also can carry out this training. And the last is to be able to make appropriate adjustments and optimization analysis to the system through the overall substantive situation of the basketball players.

This paper proposes an artificial intelligence-based method for extracting foul action features of basketball players. By acquiring and analyzing image features, the extraction of foul action features of basketball players is completed. The simulation results show that the method in this paper can effectively perform foul actions on basketball players.And the foul action recognition parameters are large, which has certain practical significance. Basketball competition is the world's largest sport, and it is a sports event with physical contact and fierce confrontation. With the continuous improvement of basketball players' level, both offensive and defensive The contradiction has become more and more intense. With the continuous improvement of the skill and tactics of the basketball game, the rhythm of the basketball game has become faster and faster, resulting in more intense fighting behavior in the basketball game. The physical contact between the players on both sides continues to increase, and the frequency of foul actions by basketball players is getting higher and higher.

Athletes have been injured frequently. In order to ensure the physical safety of basketball players and make them play the highest level, FIFA has continuously revised and improved the rules of basketball games. The traditional method of extracting foul action features of basketball players is to describe the foul action as a series of symbols and decompose each symbol.To identify foul action features. However, this method has the problem of low accuracy of foul action feature extraction in high-level basketball games. In order to solve the above problems, a research on the extraction method of foul action features of basketball players based on artificial intelligence is proposed. Artificial intelligence is artificial A branch of intelligent development that uses machines to replace human eyes for judgment and measurement. It mainly uses artificial intelligence products to convert targets into image signals, which are transmitted to special image processing systems, and the color, brightness, and distribution of image pixels are converted into digital signals. The calculation extracts the target feature information. The simulation results show that the proposed method can effectively improve the accuracy of the extraction of the foul action feature of the basketball player, and the time-consuming is shorter.

4.1 Development Direction of Basketball Player Training Strategy Based on AI

Next, we will select and analyze the data in Table 1 and we can see that when the data frame error detection is relatively large, this situation will basically not occur, but the response speed of the algorithm will be relatively slow. The test result is to use the successfully trained neural network to test and verify 10 test samples, and obtain relatively ideal results. The error value of the prediction result is basically between 0.02 and 0.04. Among the subjects, the risk of athletes' injury during exercise also requires the accuracy of early warning, and the accuracy of sports foul detection reached 99.5%.

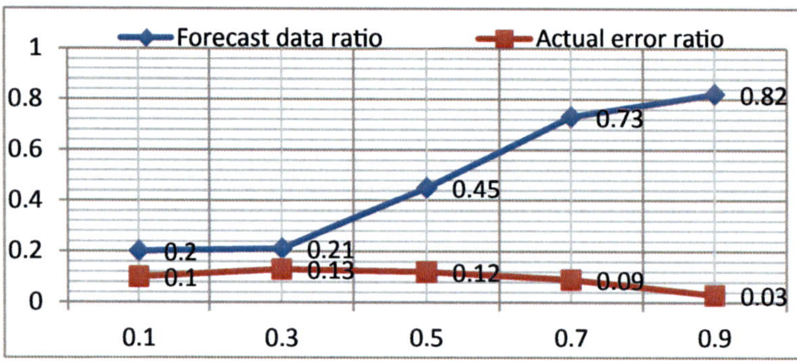

Fig. 2. Comparative analysis of comprehensive evaluation of target images and videos of difficult multiple tests

According to the comparative analysis of the comprehensive evaluation of target images and videos of the highly difficult multiple tests in Fig. 2, we can see that in our test results in the detection of non-target action images, all instances tend to develop a negative trend, and we can know the fouls of non-target athletes. Action images can be classified as positive data images according to certain instances. Calculate the movement time and position of the athletes based on the images of these biological data. Even if there are some obstructive movements during the movement, the technology researched in this paper can still detect the movement time and the position of their targets. And more importantly, it can detect whether the action of the renamed athlete is foul. Through actual experimental process, we can prove that the technology of this article can detect the foul action of basketball players in the game through AI intelligence.

Studies have confirmed that the aggressive behavior of basketball players is related to the level of male hormones in the body. Obviously, male players have higher levels of male hormones than female players, so aggressive fouls are often seen in men's basketball games. Due to the different temperament types of basketball players, there are also differences in aggressiveness among players of the same sex. Basketball offensive foul behavior often occurs in athletes who are impatient, have poor self-regulation ability, and have stubborn personality. Aggressive foul behavior in basketball games can be understood as the athlete's emergency or venting. Aggressive foul behavior often has an adverse effect on athletes or the group, and hinders physical and mental health. Athletes who are deliberately fouled or injured by their opponents in the game will often have thoughts of revenge, and will make more aggressive fouls than the opponents in the next game. By observing and analyzing the offensive fouls in basketball games, it can be concluded that the offensive fouls in basketball games can be roughly divided into two categories: one is basketball reactive offensive fouls, which are generated by the anger of the attacker., Has an aggressive foul that does not care about the consequences, harms or pains the opponent; the other is a basketball purposeful offensive foul, which is for the purpose of winning the game or receiving a reward, and strengthens the offensive behavior that is limited to certain conditions.

4.2 The Impact of the New Rules on Basketball Tactics

In order to make basketball develop in a formal and reasonable direction, the physical fitness requirements of athletes are more stringent, the speed of offensive and defensive conversion is emphasized, and the role of offensive and defensive balance in the game is emphasized. Basketball teaching and training should be combined with the newly introduced basketball rules, in training methods and methods. It is necessary to make timely adjustments, strengthen the training of individual and team skills and tactics, and strengthen the learning and practice of the new rules. The modification of the new rules optimizes the quality of the game. The evolution of basketball rules. With regard to the technical and tactical changes brought about by ball movement, dribbling skills, defensive skills, long-range shooting techniques, and restrictions on tall athletes, research on the evolution of basketball rules should be strengthened, and more attention should be paid to the ability of basketball rules in teaching and training. Improve the innovation of skills and tactics. The implementation of new basketball rules has significantly increased the number of free throws. Free throws have become an effective way to win. Therefore, the free throw percentage has also become an effective way for the team to win the final victory. The revision of basketball rules It greatly improves the viewing of the game, greatly improves the rhythm and speed of the offensive and defensive transition on the basketball court, and emphasizes movement in the offense, which increases the difficulty of defense, and puts higher requirements on the physical fitness of the athletes and requires coaches. Players effectively use the new rules to conduct basketball technical training and promote the rapid development of basketball.

5 Conclusion

This research analyzes the technical system of intelligent AI-assisted game foul shot scanning based on AI, evaluates the mode of foul action detection in functional basketball games, and builds a multi-target feedback training method, which is effective in assisting the coach's foul action. It is difficult to detect the foul action of a basketball player in a fierce game by using the foul action detection technology in the current basketball game. Therefore, this article proposes a technology for automatic detection of athletes' foul actions in basketball games. Experiments show that this technology can accurately detect the foul action of basketball players in the game.

References

1. Licato, J., Zhang, Z.: Correction to: evaluating representational systems in AI. AI Rev. **52**(4), 1 (2019)
2. Ström, P., Kartasalo, K., Olsson, H., et al.: AI for diagnosis and grading of prostate cancer in biopsies: a population-based, diagnostic study. Lancet Oncol. **21**(2), 222–232 (2020)
3. Prem, N.R.M.B.A., Jaret, M.K.M.S., Benedict, U.N.M.B.A., et al.: Regarding "editorial commentary: AI in sports medicine diagnosis needs to improve". Arthrosc.: J. Arthrosc. Relat. Surg. **37**(5), 1365–1367 (2021)
4. Watson, D., Womack, J., Papadakos, S.: Rise of the robots: is AI a friend or foe to nursing practice? Crit. Care Nurs. Q. **43**(3), 303–311 (2020)
5. Santhosh, P.K., Kaarthick, B.: An automated player detection and tracking in basketball game. Comput. Mater. Continua **58**(3), 625–639 (2019)
6. Yoon, Y., Hwang, H., Choi, Y., et al.: Analyzing basketball movements and pass relationships using realtime object tracking techniques based on deep learning. IEEE Access **PP**(99), 1 (2019)
7. Diaconu, O., Fotin, A., Cosereanu, C.: Beech wood (Fagus sylvatica L.) flooring structures designed to increase the basketball bounce height. Holz als Roh-und Werkstoff **77**(5), 853–860 (2019)
8. Alves, P.N., Foulon, C., Karolis, V., et al.: An improved neuroanatomical model of the default-mode network reconciles previous neuroimaging and neuropathological findings. Commun. Biol. **2**(1), 1–14 (2019)
9. Rachakonda, L., Mohanty, S.P., Kougianos, E.: iLog: an intelligent device for automatic food intake monitoring and stress detection in the IoMT. IEEE Trans. Consum. Electron. **PP**(99), 1 (2020)
10. Lin, S.F., Sun, Y.C., Liao, J.R.: An automatic detection algorithm of abandoned objects and abandoner for surveillance systems. J. Electr. Electron. Eng. **9**(1), 1–8 (2019)
11. Nagarajan, D., Lathamaheswari, M., Talea, M., et al.: Intelligent algorithm for trapezoidal interval valued neutrosophic network analysis. CAAI Trans. Intelli. Technol. **5**(2), 88–93 (2020)
12. Fetouh, T., Elsayed, A.M.: Optimal control and operation of fully automated distribution networks using improved tunicate swarm intelligent algorithm. IEEE Access **PP**(99), 1 (2020)

One-Cycle Control Strategy for Active Power Filter Based on Current Feedback

Yiyuan Shao(✉) and Xueqin Wu

School of Computing, Ezhou University, Ezhou, Hubei, China

Abstract. A new one-cycle control scheme for Active Power Filter (APF) is raised to improve its performance, which is enlightened by the discovery that source current should oscillate near zero point by applying traditional one-cycle control strategy for APF based on unipolar operation. It uses ip-iq based algorithm to isolate the load foundational current and treats the foundational component of load current as tracking target. So the circuit of one-cycle control for APF is composed of foundational component isolation of load current, integrator, comparator and trigger to implement the function of hormonic compensation of APF, by which the dependence on system parameters can be lowered. It establishes the mathematical model of one-cycle control for APF based on current feedback, analyses the working process of the system, and compares the raised strategy with traditional one-cycle control strategy based on unipolar operation by employing simulation experiment. Computer simulation results show that the raised strategy is with better stability and tracking capacity, and can acquire lower THD.

Keywords: One-cycle control · Single-phase APF · Unipolar operation · Foundational current isolation

1 Introduction

The active power filter (APF) is a new kind of power electronic device that can dynamically suppress harmonic waves and compensate reactive power. Nowadays the control strategies of an APF employs are mainly hysteresis comparison control, triangular wave control, and spatial vector modulation. Due to having to use the high-speed digital microprocessor and the A/D converter, these control strategies suffer from such disadvantages as beat, high cost and increased circuit complexity [1]. One-Cycle Control (OCC) was proposed in 1991, by K. M. Smedley and S. Cuk who were from California Institute of Technology [2]. The basic idea of OCC can be described as follows. The switching frequency of the converter is a fixed value. The duty ratio of the switch is controlled such that in each cycle the average value of a switched variable is equal to or proportional to the control reference signal thus eliminating errors in transient or steady state. For application of this method to APF, the analog circuit is used, featuring simple circuit structure, fast response and fixed switching frequency, to meet the control requirements of high accuracy, high speed and high immunity to interference.

B. J. Jansen et al. (Eds.): *International Conference on Cognitive based Information Processing and Applications (CIPA 2021)*, LNDECT 85, pp. 463–473, 2022.
https://doi.org/10.1007/978-981-16-5854-9_58

Application of OCC to the single-phase APF involves mainly two modulation methods: bi-polar and uni-polar. In most cases, the actual application is improvement and derivation of the two methods. The OCC of bi-polar modulation cannot only bring about the DC component offset of power supply current but also affect the local stability in the system. In the Reference [3], the control equation of the OCC of bi-polar modulation was derived. Taking into consideration the DC component offset in the output of bi-polar control, the ripple was introduced replacing the power supply current and used as the input to the comparator. Although the DC component was removed, the width of chopped wave peak increased. In the Reference [4], a selector was added so that the power supply voltage adopted the original control strategy at its positive half cycle and the symmetrical strategy at its negative half cycle. The benefit was the elimination of the DC component due to the control strategy. Nonetheless this method failed to improve wave peak chopping and produced much greater circuit complexity. In the Reference [5], a method so-called bi-loop control was proposed for improving bi-polar modulation. The current integral functionality was incorporated in the control circuit, to realize the closed-loop regulation of the DC component and eliminate the DC component offset. However, the incorporation of current integral functionality led incidentally to poor stability. In the Reference [6], based on the analysis of the stability of bi-polar modulation, another improvement strategy known as bi-directional complementary control was put forward such that in the positive half cycle of the voltage the original control strategy was complied with but in the negative half cycle of the same the current waveform was made symmetrical to that in the positive half cycle. As a result, the DC component disappeared and the global stability was accomplished. Different from the bi-polar modulation, the uni-polar modulation modulates the positive and negative half cycles separately, thus creating no DC component offset. Furthermore, for the uni-polar modulation only two switches were turned ON and OFF at the switching frequency, greatly reducing the switching loss. In the Reference [7], the uni-polar control equation was derived, the complete topological structure of circuit was proposed, and the method for determining the ranges of the circuit parameters was determined. In the Reference [8], the uni-polar modulation was inferior to the bi-polar modulation in terms of the OCC stability. In particular, the uni-polar modulation even experienced steep voltage drop on the DC side within the positive half cycle of power supply voltage. The uni-polar modulation had such shortcomings as the system instability and the excessive dependence on parameter variation.

To eliminate the DC component offset resulting from the bi-polar modulation, reduce the dependence on parameter variation in the case of uni-polar modulation as well as enhance the system stability, a unit-polar modulation strategy based on the fundamental wave current detection is devised with the addition of the fundamental wave current detection process. Taking the place of the power supply current, the fundamental wave current becomes the main component of the input to the comparator. The power supply current is hence capable of tracking the fundamental wave current of the load, with the system stability enhanced and the dependence on parameter variation reduced.

2 Methodology

2.1 APF OCC Model Based on UNI-Polar Modulation

Shown in Fig. 1 is the topological structure of the main circuit of a typical parallel-connection single-phase APF. In the APF, a nonlinear load is in parallel connection with a voltage-type full-bridge converter that consists of an IGBT, four backward diodes and a capacitor. The converter is equivalent to a controlled harmonic wave source. Implementing the control algorithm, the controller generates PWM waves, to control the ON and OFF of the IGBT and create a current in the same magnitude and opposite direction as the power supply harmonic wave current.

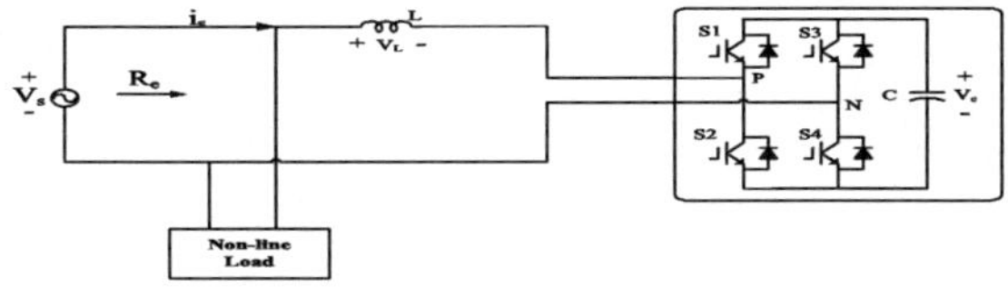

Fig. 1. Main circle of shunt APF

Generally the OCC is achieved through the Unified Constant-Frequency Integration (UCI). The OCC hardware includes mainly a controller, a comparator, an integrator with reset, and a clock. Among them, the controller is either an RS flip-flop or a D flip-flop. The purpose of the OCC is for the average value within a switching cycle of the output from the controlled switch to track the reference variable. Generally, the controlled switches under uni-polar modulation operate as follows within a power-frequency cycle T_S. In the positive half cycle of power supply voltage, S4 remains ON while S3 keeps OFF, and S1 and S2 turn ON alternatively but stay at different states. In the negative half cycle, the operation is symmetrical to that in the positive half cycle. S3 remains ON while S4 keeps OFF, and S1 and S2 turn ON alternatively but stay at different states. The switching converter has a total of 4 operation states as shown in Table 1, where D represents the duty ratio.

Table 1. Switching operation under unipolar operation

State	V_S	t	S1	S2	S3	S4
1	$V_S > 0$	$[0, DT_S]$	0	1	0	1
2	$V_S > 0$	$[DT_S, T_S]$	1	0	0	1
3	$V_S < 0$	$[0, DT_S]$	1	0	1	0
4	$V_S < 0$	$[DT_S, T_S]$	0	1	1	0

It can be seen from Table 1 that under the uni-polar modulation the polarity of the AC side voltage of the converter remains unchanged within the same half cycle. Within each power-frequency half cycle, only two switches operate at the switching frequency f_s, while the remaining two operate at power frequency. Compared to other modulation techniques, the uni-polar modulation incurs a much smaller switching loss.

Within the positive half cycle of power supply voltage ($V_s > 0$), the inductor voltage V_L relates to the DC side voltage V_c as follows:

$$V_L(ON) = V_s; \quad 0 < t < DT_s \tag{1}$$

And

$$V_L(OFF) = V_s - V_c; \quad DT_s < t < T_s \tag{2}$$

According to the volt-second balance principle, the average value of inductor voltage within a control cycle can be thought to be invariable if the switching frequency is far greater than power frequency.

$$V_s \cdot DT_s + (V_s - V_c) \cdot (1 - D)T_s = 0 \tag{3}$$

Likewise, within the negative half cycle of power supply voltage ($V_s < 0$) the following equation applies:

$$V_s \cdot DT_s + (V_s + V_c) \cdot (1 - D)T_s = 0 \tag{4}$$

Combining (3) and (4), the control objective equation is obtained as follows:

$$|V_s| = V_c \cdot (1 - D); \quad 0 < t < T_s \tag{5}$$

The control objective is that the APF is able to fully compensate the system for the harmonic wave current produced by the nonlinear load and the power supply current i_s becomes the sine wave having the same phase as the voltage. Thus the parallel connection of the nonlinear load with the APF can be equivalent to a pure resistor R_e as shown in Fig. 1.

$$V_s = R_e \cdot i_s \tag{6}$$

Let the reference variable

$$V_m = \frac{R_s}{R_e} \cdot V_c \tag{7}$$

where R_s is the internal resistance of the power supply current sensor, and V_m is the DC voltage feedback signal from the control loop. Substituting (7) into (5) yields

$$V_m(1 - D) = R_s \cdot |i_s| \tag{8}$$

Considering the switching frequency is typically above 10 kHz and the integration time DT_s is close to zero, (8) can be transformed approximately into:

$$V_m - \frac{1}{T_i}\int_0^{DT_s} V_m dt = R_s \cdot |i_s| \tag{9}$$

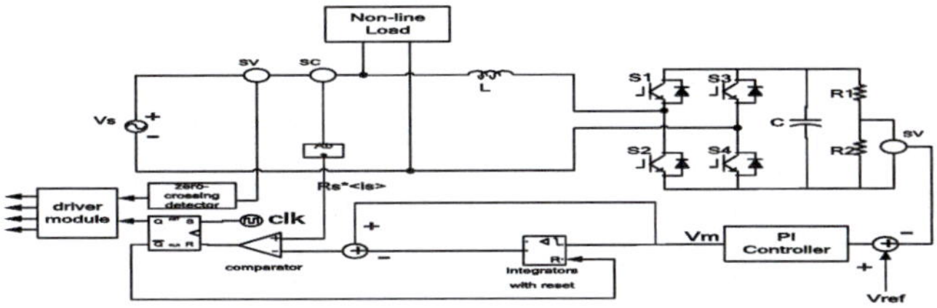

Fig. 2. Realization diagram for unipolar operation of one-cycle control

Adhering to the OCC principle, the circuit consisting of an integrator, a comparator and a flip flop is used to realize the compensation current control of the APF as shown in Fig. 2.

Vm is obtained by the PI controller from the sampled DC side voltage. Let the integral time constant Ti = Ts, the control objective (8) can be realized. Within each clock cycle ($0 \leq t \leq T_s$), the integrator starts calculating integral since the instant $t = 0$, and $R_s \cdot |i_s|$ is compared with the result of $V_m - \left(\frac{1}{T_s}\right) \cdot \int_0^t V_m d\tau$. If $R_s \cdot |i_s|$ is greater than the result of $V_m - \left(\frac{1}{T_s}\right) \cdot \int_0^t V_m d\tau$, with the integral proceeding, the comparator will flip, the RS flip-flop in conjunction with the logic module will change the states of the switches, and the integrator will reset to zero, waiting for the next cycle. Therefore within each control cycle the value on the right side of (9) can track that on the left side fulfilling the control objective.

The OCC under uni-polar modulation converts the negative half cycle of waveform into positive half cycle, thus dismissing the possibility of DC component offset. From (9), in the vicinity of $i_s = 0$, when i_s changes in the positive direction, a switch could operate in the same manner as in the positive direction. It is the cause of oscillation even out-of-control around the zero point.

2.2 Improved UNI-Polar Modulation

To prevent the occurrence of oscillation in the vicinity of zero point of power supply current, consideration is given to addition of a fundamental wave current detection module, treating as the tracked object the fundamental wave component of load current. Then (9) turns into (10):

$$V_m(1 - D) = R_s \cdot \left|i_{Lf}\right| + k\left|i_s - i_{Lf}\right| \tag{10}$$

Where i_{Lf} is the fundamental wave component of load current, k is produced by the PI controller and $k < R_s$. The control objective changes into making $\left|i_s - i_{Lf}\right|$ approach zero. Figure 3 shows the circuit realization of improved uni-polar modulation.

Under the original uni-polar modulation, in the vicinity of zero point, i.e. when D approaches 1, the power supply current jittering is liable to occur on account of the uncertainty of $R_s \cdot |i_s|$.

The fundamental wave current detection module operates with the *ip-iq* detection based on the instantaneous reactive power theory [9, 10] See Fig. 4.

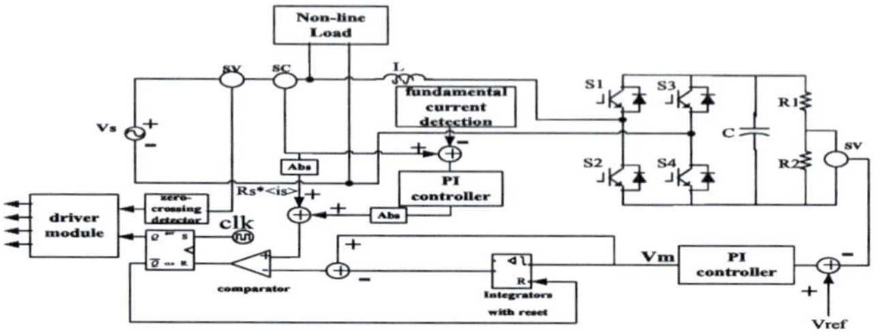

Fig. 3. Realization diagram for improved unipolar operation of one-cycle control

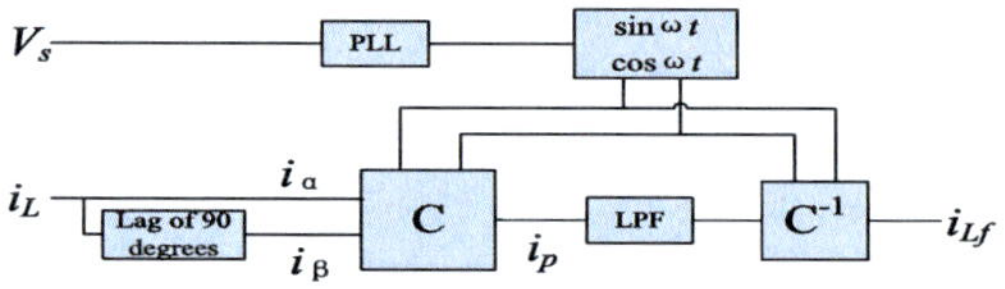

Fig. 4. ip-iq algorithm based harmonic isolation diagram

Lagging the load current $i_\alpha(i_L)$ by 90° obtains i_β. Apply Fourier resolution to i_α and i_β, then:

$$i_\alpha(t) = i_L(t) = \sum_{n=1}^{\infty} I_n \, sin\,(n\omega t + \varphi_n) \tag{11}$$

$$i_\beta(t) = -I_1 \, cos\,(\omega t + \varphi_1) + \sum_{n=2}^{\infty} I_n \, sin\left(n\omega t + \varphi_n - \frac{n\pi}{2}\right) \tag{12}$$

Where I_n is the magnitude of the nth-order harmonic wave current and φ_n is the initial phase of the nth-order harmonic wave current. The initial phase of fundamental wave is supposed to be zero, i.e. $\varphi_1 = 0$.

With the power supply voltage V_s as the input, the phase-locked loop produces the sine signal $sin\,\omega t$ and the cosine signal $cos\,\omega t$ that are in the same phase as V_s, constituting the matrix $\boldsymbol{C}$:

$$C = \begin{bmatrix} sin\,\omega t & -cos\,\omega t \\ -cos\,\omega t & -sin\,\omega t \end{bmatrix} \tag{13}$$

The C-transformation of $\boldsymbol{i_\alpha}$ and $\boldsymbol{i_\beta}$ separates the active component i_p from the reactive component i_q. Omit i_q because merely the detection is required of the fundamental wave active current component. Application of the instantaneous reactive power theory gives

$$i_p = i_\alpha \, sin\,\omega t - i_\beta \, cos\,\omega t = \overline{i_p} + \tilde{i_p} \tag{14}$$

Where $\overline{\boldsymbol{i_p}} = \boldsymbol{I_1}$, the fundamental wave active DC component, and $\tilde{\boldsymbol{i_p}}$ is the fundamental wave active AC component.

With i_p as the input, the low-pass filter generates $\overline{\boldsymbol{i_p}}$. The inversion transformation of $\overline{\boldsymbol{i_p}}$ obtains the load fundamental wave active current:

$$i_{Lf} = i_{\alpha f} = \overline{i_p} \, sin\,\omega t \tag{15}$$

3 Results and Discussion

The operating process and performance of the control scheme is simulated in the Matlab/Simulink environment. Referring to Fig. 1, the simulation conditions are as follows: the power supply voltage V_s is 220 V at the frequency of 50 Hz, the nonlinear load consists of an uncontrolled rectifier bridge and a resistor-capacitor circuit in parallel connection, and the load current i_L has a maximum of 22 A and an effective value of 12.2 A.

3.1 Analysis of the Operating Process

The derivation of the control equation of uni-polar modulation demonstrates that, to ensure the realization of control algorithm and the excellent current-tracking performance, the voltage across the DC capacitor of the converter must be increased to then held at an appropriate value. According to (7) and (10), the excessive fluctuation of the voltage across the capacitor on the DC side will certainly pose excessive fluctuation in the power supply current. The consequence is failure to realize compensation for harmonic wave, and the creation of even more harmonic waves. Thus it is of great importance to maintain the DC side voltage stability. While the APF is operating normally in the grid having harmonic waves, energy is exchanged between them as a result of the harmonic waves. Furthermore power loss exists in any power electronic device in operation. The DC side voltage fluctuates accordingly. Unlike a majority of other control algorithms of APF, the OCC requires no additional control algorithm to maintain the DC side voltage stability.

Immediately after the APF is connected to the grid, the DC side voltage must ascend from zero to an appropriate value and then remain stable. The OCC under uni-polar modulation can be divided into two stages as shown in Fig. 5.

The first stage serves for rectification and voltage rise. In this stage, the main circuit is equivalent to a single-phase full-bridge rectification circuit, and the DC side voltage can rise to the peak value of the grid side AC voltage. The second stage is for the voltage to rise to an appropriate value and then remain stable. At the beginning of this stage, V_c reaches the peak value of the grid side AC voltage, lower than the stable value. Generally V_{ref} should be greater than the maximum of stable V_c. V_m is produced by the PI controller from the difference between the sampled capacitor voltage and the reference voltage. Nonetheless V_m is primarily associated with *p* parameter. Considering only the *p* parameter establishes

$$\mathrm{V_m} = \mathrm{K_p}(\mathrm{V_{ref}} - \mathrm{K_c V_c}) \tag{16}$$

Where $\mathrm{K_p}$ is the *p* parameter of the PI controller, and $\mathrm{K_c}$ is the sampling coefficient of capacitor voltage.

According to (10), the duty ratio D before stabilization is smaller than that after stabilization and at the same power-frequency phase, and the converter stays at States 2 and 4 longer than at the states after stabilization. It is good for charging, and V_c can continue to rise following a gradually gentle slope and eventually remain stable.

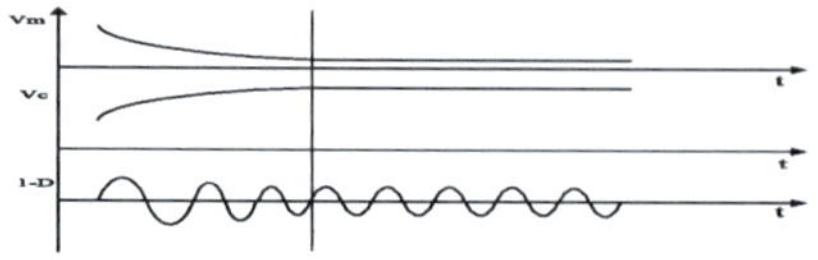

Fig. 5. The waveform graph of relevant variables

According to (7), V_m will diminish to a value near $\left(\frac{R_s}{R_e}\right) \cdot V_c$ and subsequently keep invariable.

When V_c becomes stable, APF can normally operate. The waveform graph of the control module under uni-polar modulation is shown in Fig. 6. In each switching cycle, the duty ratio D is determined mainly from the intersection of $R_s \cdot |i_{Lf}|$ and $V_m \cdot \left(1 - \frac{t}{\tau}\right)$, and adjusted in view of the difference between i_s and the fundamental wave component of load current. As a result, no oscillation occurs in the vicinity of the zero point of power supply current, and it is possible to track, within an ON-OFF cycle, the fundamental wave component of load current and finally eliminate the harmonic waves.

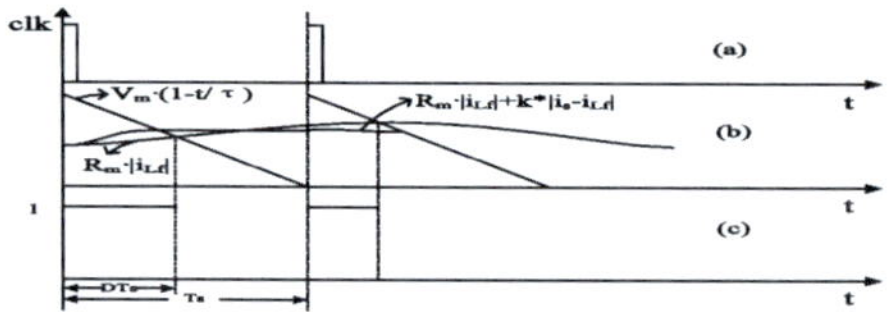

Fig. 6. The waveform graph of control-block

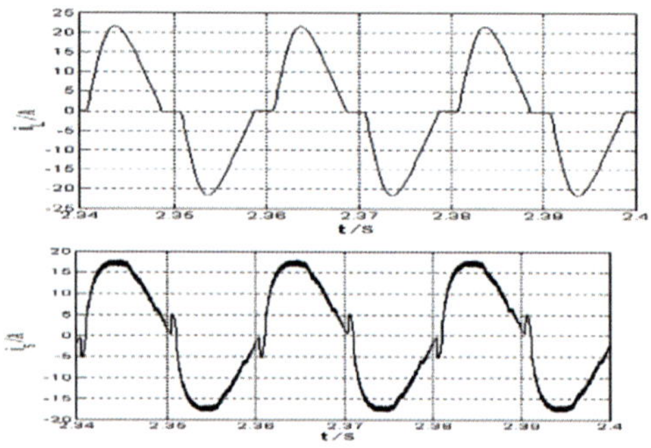

Fig. 7. The simulation waveforms under original scheme

3.2 Comparison Experiment

The total harmonic distortion (THD) is 58.7% for the load current. Under the original uni-polar modulation, the parameters of the main circuit are as follows: the AC side inductance $L = 8$ mH, the DC side capacitance $C = 6800$ μF, the clock frequency f_s

= 10 kHz, and the internal resistance of current sensor $R_s = 1\ \Omega$. The DC side voltage V_c is increased to 450 V. From (7), $V_m = 36.9$ V, and let $V_{ref} = 70$ V. From (16), K_p = 1.47. After compensation, the THD of current shrinks to 19.6%. Shown in Fig. 7 are the simulation results. Figure 7 reveals that the power supply current jittering at the zero point. In this sense the compensation is far from ideal.

Simulation for the improved uni-polar modulation is performed with the circuit parameters unchanged. The results are shown in Fig. 8. After compensation, the THD of current drops to 9.4%. For purpose of comparison, Fig. 9 and Fig. 10 depicts the harmonic spectrum of current, where n is the number of harmonic waves. Thanks to the improved control algorithm, the power supply current is able to track the fundamental wave component of load current, no power supply current jittering at the zero point occurs, and the APF compensation performance is improved, with the influence of parameters on the compensation performance mitigated.

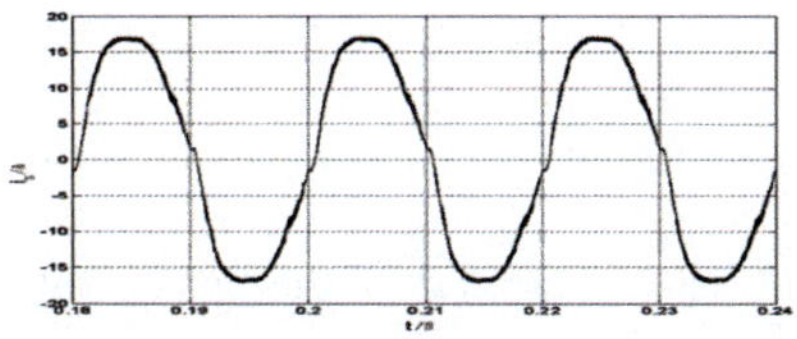

Fig. 8. The simulation waveforms under modified scheme

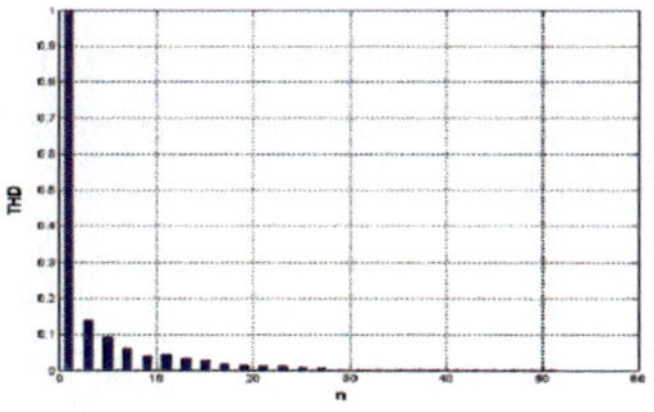

Fig. 9. The THD after compensation under original

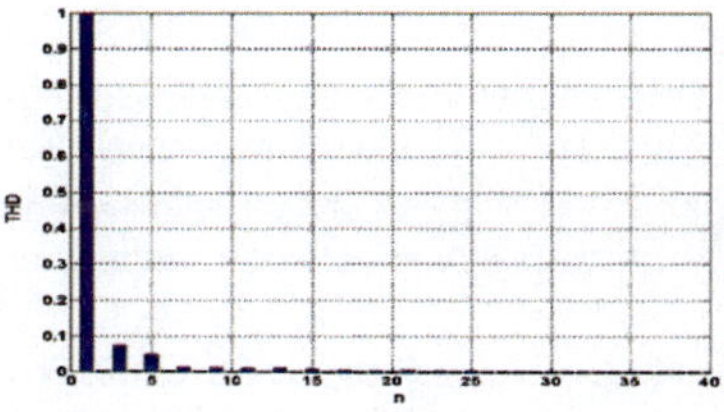

Fig. 10. The THD after compensation under modified scheme

The control algorithm is realized in the DSP. Since the OCC imposes strict requirements on real-time response, and taking into account the DSP resources, the means of interruption may be employed for purpose of realization. The software system flow chart is shown in Fig. 11.

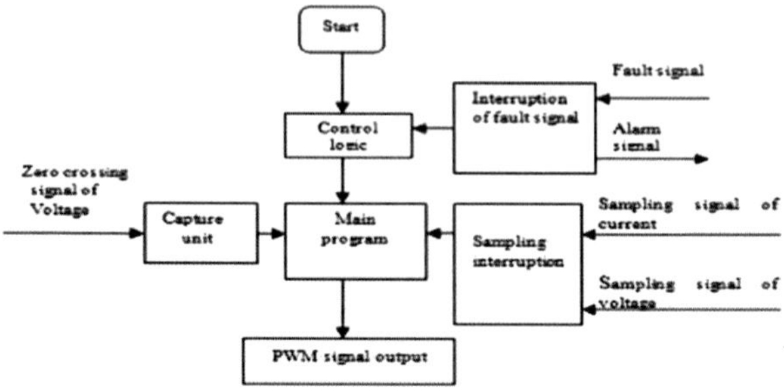

Fig. 11. Software system flowchart

4 Conclusions

Based on the analysis of the OCC principle and uni-polar modulation, improvement has been made to the uni-polar modulation single-phase APF. The power supply current is hence capable of tracking the fundamental wave current of the load, the influence of system parameters are mitigated, and no power supply current jittering at the zero point occurs. The control equations are derived through circuit analysis. The simulation model is established and simulation analysis is performed. The results demonstrate that the improved algorithm can provide excellent uni-polar modulation with decreased instability hence be substantially meaningful.

Acknowledgements. This work was supported by science and technology research project of educational commission of Hubei Province in China (B2020404).

References

1. Hu, Z., Hu, M., Gui, W., et al.: Novel self-tuning predictive control method of shunt active power filter. J. Electr. Mach. Control **13**(1), 99–103 (2010). (in Chinese)
2. Smedley, K.M., Cuk, S.: One-cycle control of switching converters. In: IEEE Operation Center Applied Power Electronics Conference, pp. 888–896 (1991)
3. Smedley, K.M., Zhou, L., et al.: Unified constant-frequency integration control of active power filters-steady-state and dynamics. IEEE Trans. Power Electron. **16**(3), 428–436 (2000)
4. Yu, S., Zhou, Z.: The modified strategy of active power filter based on one-cycle control. In: IEEE Pacific-Asia Workshop on Computational Intelligence and Industrial Application, pp. 170–174 (2008)
5. Qian, T., Lv, Z., Hu, J., et al.: Dual-loops scheme for unified constant-frequency integration control of active power filter. J. Proc. CSEE **23**(3), 34–37 (2003). (in Chinese)
6. Fei, W., Lv, Z., Qian, T.: A novel one-cycle control active power filter based on complementary control strategies. J. Autom. Electr. Power Syst. **27**(3), 50–54 (2003). (in Chinese)
7. Qiao, C., Smedley, K.M., Maddaleno, F.: A single-phase active power filter with one-cycle control under unipolar operation. IEEE Trans. Circuits Syst. Part 1: Regular Papers **51**(8), 1623–1630 (2004)
8. Xu, M., Zhou, L., Wang, W., et al.: A survey on one-cycle control strategy for single-phase active power filter. J. Power Syst. Technol. **30**(22), 81–86 (2010). (in Chinese)

9. Akagi, H., Nabae, A.: Instantaneous reactive power compensators comprising switching devices without energy storage components. IEEE Trans. Ind. Appl. **20**(3), 625–630 (1984)
10. Hu, Z., Meng, Y., Hu, M., Gui, W., et al.: An improved ip-iq detection approach of positive fundamental active and reactive current based on p-q transformation. J. Central South Univ. (Sci. Technol.) **41**(3), 10–14 (2010). (in Chinese)

The Qualification Verification of Athletes in Sports Competition by Using HDFS Distributed File Storage System with MySQL Relational Database

Ruoguo Li(✉)

Pu'er University, Pu'er 665000, Yunnan, China

Abstract. In sports competitions, it is an important basic work to verify the qualifications of participants. With the rapid growth of data, it is too cumbersome to manually verify the qualifications of participants. Based on this, we need to build a complete set of population resources data analysis service platform through the technology of big data storage, calculation, analysis and excavation. The purpose of this paper is to use big data analysis technology to build an intelligent personnel information management system to verify the qualification of participants in sports competitions, and realize the function of centralized and intuitive display of electronic files of participants. This paper uses HDFS distributed file storage system with MySQL relational database to store data, uses spark distributed file computing engine to analyze and process data, and then uses web front-end technology and echart data visualization technology to design a personnel information management system with strong expansibility, high efficiency and fast data processing speed, strong practicability and simple operation. This paper describes the design of the system from the server and the client, and finally tests the function and performance of the system to ensure that it meets the requirements of users. After testing, it is found that when the test times of the system is 600, the connection time is less than 1 s, and the system passes the connection speed performance test.

Keywords: HDFS distributed file storage system · MySQL relational database · Contestant qualification · MYSQL

1 Introduction

With the rapid development of computer, as one of the core technologies of information system in China, analytical database technology has become a tool and method for the research of auxiliary management data. It has realized various functions of analyzing, storing and managing data, acquiring and analyzing data [1, 2]. The application of network technology has realized the communication and resource sharing of terminal equipment such as computer. In recent years, the database technology and network technology of computer have been greatly developed. By using the combination of

B. J. Jansen et al. (Eds.): *International Conference on Cognitive based Information Processing and Applications (CIPA 2021)*, LNDECT 85, pp. 474–481, 2022.
https://doi.org/10.1007/978-981-16-5854-9_59

the two, an information management system can be established perfectly. In sports competitions, such a system is needed to review the qualifications of the participants [3, 4].

So far, there are many research results on personnel verification system [5, 6]. For example, some scholars have established a simple and high security passenger monitoring and screening system for the fugitive personnel under the premise of ensuring the perfect function for the complexity, insecurity and lack of pertinence of the finished software in the current market. From the perspective of staff assessment, combined with the existing RFID technology and database technology, the enterprise personnel information is counted in a data table, and an employee information management system is established to realize the identification function of employees [7, 8].

The main research work of this paper is to design the technology of high expansibility, high efficiency and fast data processing, and strong practicability by using HDFS distributed file storage system with MySQL relational database, using spark distributed file computing engine to analyze and process data, and then combining with web front-end technology and ECHART data visualization technology simple operation of personnel information management system [9, 10]. This paper focuses on the realization of the basic information module of the contestants, the realization of the information module of the participants' control and the realization of the interface of the control system. Finally, the system is tested for function and performance to ensure that it meets the user requirements [11, 12].

2 Contestant Qualification Verification System

2.1 System Requirement Analysis

Technical Requirements. This system is based on Java programming language and developed in eclipse integrated development environment. As a complete object-oriented language, Java can achieve module division more succinctly. In terms of database, MySQL is a powerful relational database, and its capacity for the database can fully support the data storage generated by the comprehensive screening system of athletes in sports competitions.

Operational Requirements. This system is a small sports event personnel management system with clear functions, clear division of rights and very little consumption. The basic computer can set up the client of the comprehensive screening system for railway passengers at large. The system has a friendly interface, which is completely windows visual operation interface and does not need professional operation training for users. Therefore, the operation requirements are satisfied.

Function Requirement Analysis. Data must be unified and shared in this wide range. Therefore, there must be a central database and server to process these data. For multiple stadiums, we need a special client to interact with the server data. Therefore, the functional requirements analysis is divided into two modules according to the division of modules.

2.2 Application of Random Forest in Data Verification

Random forest can calculate the importance of a single variable, and can be used to screen variables and reduce dimensions. The importance scoring algorithm of random forest is based on GINI index and permutation method. The formula is as follows:

$$\mathrm{VIM}_{jm}^{Gint} = GI_m - GI_r - GI_t \tag{1}$$

Denotes the importance score of variable *Dj* in node M. *GIt* and *GIr* represent the Gini index of left and right new nodes after node m is divided.

VIM_{jm}^{Gint} represents the variable the importance score of X_j in node m, GI_t and GI_r represents the Gini index of left and right new nodes after node m is divided.

$$VIM_j^{gini} = \frac{1}{n}\sum\nolimits_{i=1}^{n} VIM_{ij}^{gini} \tag{2}$$

Among them, VIM_j^{gini} represents the variable x in the whole forest_J, n is the number of trees in the random forest.

Another method to calculate the importance score is the replacement method, which is based on the error rate of data outside the bag. The formula is as follows:

$$\mathrm{VIM}_j^{OOB} = \frac{1}{n}\sum\nolimits_{i=1}^{n}(ER_j - ER_j') \tag{3}$$

3 The Implementation of the Management and Control Information Module of the Participants

3.1 Detailed Design and Implementation of Server

Design of Server Program Module. According to the functional requirements and functional division of the server in the overall design, the functions of the server can be encapsulated into the following categories: database operation class (responsible for the connection of the database, executing the request for the operation of the database); Operation center class (processing all kinds of user requests from the client); Port monitoring class (responsible for monitoring the communication port of the server in real time. After receiving the client connection request, the class object of operation center is established, and then the corresponding processing is carried out); Communication (communication with client, data encapsulation and disassembly); Log class (responsible for recording system operation information and user management operation information); Encryption and decryption (encryption and decryption functions of transmission data); File operation class (read and write physical files); Initialize static classes (reset and provide static resources); Program entry.

Design and Implementation of Server Operation Process. After the server program is started, the program will initialize the system, initialize the server by calling the init() method of static variable class static, read the configuration file, connect with the server, and then instantiate the service monitoring class to open the port monitoring of the server. When the system receives the connection request, the server will instantiate the operation center object and pass the socket object connected with it, so that the operation center can start a new thread to run, and the main thread will continue to run and accept the next connection request. When the new thread runs, it will continuously obtain the request information from the client. After decrypting and analyzing the information, it will obtain the type of the request content, and process the corresponding data according to the request type. After the processing, it will return the processing result to the client, and then accept the next request information until it sends the close request.

3.2 Client Detailed Design and Implementation

The client program uses MVC. In the client, the model is the server of the comprehensive screening system for the participants of sports events, which is used to process the user's data logic. In the client, the view is part of the user interface, which directly displays the information users need. In the client, it means to get the background data of each user operation and pass it to the model. According to the requirements, it is divided into the following modules: communication module (for data transmission), data encryption and decryption module (for data transmission and data encryption and decryption), user interface module, function control (client running core, which processes the request information of user interaction through the interface and the information transmitted by the server for user interface display) Static class (storing static resources of the system), program entrance and other modules.

Design and Implementation of Basic Client Functions. To display different pages for different users, we use the jtabbedpane control in Java Swing toolkit to encapsulate each function page into a basic JPanel. According to the user ID in the result returned by the server after the user sends the login request, we add the specified JPanel to the main jtabbedpane, So as to achieve the display function of different function pages. At the beginning, all clients are the most basic login interface, so here we use the CardLayout layout layout in the swing framework to put the login and registration system and jtabbedpane in a CardLayout layout frame, and login the system and jump as needed.

The realization of function control, that is, the control module in MVC structure, is mainly used to operate the user's request data. After packaging, it is sent to the model module, that is, the server of the comprehensive screening system for railway passengers at large. After the data processing of the server, it obtains the return data of the server, and then pushes the data to the view module for users to display. In this process, we need to encrypt and decrypt the information.

Design and Implementation of User Client. According to the determination and arrangement of functions, the exclusive functions of gymnasium staff are log view function, administrator account management function and unit management function.

Design and Implementation of Registration Function. The registration function is the basic function required by the client. When the user clicks the registration interface to jump, the control module will send a request to the server to obtain the station list and the station police station list. After obtaining these data, it will jump to the page for the input and selection of attributes. After the user confirms the input, the control module will send the data. It is processed by the server, and the registered user can take effect only after the administrator approves the registration.

The Function and Realization of Personal Management. Personal management is mainly divided into basic information browsing, information modification and password modification. When the user clicks the load button in the personal management page, the user interface sends a request to the control module.

3.3 Implementation of Management and Control System Interface

Interface Call. Taking the Java language commonly used in external platform as an example, this paper introduces how to call the external platform through the interface. The implementation process of the call is: create a call object in the service instance object, set the name and URL interface address of the calling service, execute the call through the invoke method, and get the result set.

Write Class Interface. Get the instance object of query service queryservice from servicelocator static class, parse the received write request (XML string) into the written fields and field values into the list through the defined insertutil class methods getopterfiledsbytagname and getoptervalues, and then write the list data into the map one by one, then call the insertobjectbymap method of the object to store the data in the database, and assemble the processing result set into an XML string resultset to return.

4 Test of the Management System of the Participants

4.1 System Hardware Configuration

At present, the amount of data access to the system is not large, so three servers are temporarily used to build a cluster. The specific server configuration and client configuration tables are shown in Table 1.

Due to the strong scalability of HDFS storage system, when the amount of data exceeds the bearing range of servers, it is only necessary to increase the number of servers, which will not affect the use of the system.

4.2 System Environment Construction

All servers of the system use Linux system, Linux operating system itself has high security and performance, and can support multi-user and multitask processing well. Create app folder under the root directory of system user, which can be used to store the installation directory of all software; Create a data folder to store all test data; Create lib file to store jar package needed by all environments; Create a software folder

Table 1. Server and client configuration table

Terminal category	Machine name	Equipment number	Configuration description
Server side	Nothing	Nothing	CPU 2.0 GHz, 10 core/20 thread memory DDR4 16g hard disk 1TB 2-port gigabit network card
Client	Nothing	Nothing	Any computer that can access the Internet and browse the web

to store the directory of the software installation package; Create the source folder to store our framework source system all servers configure JDK environment. After creating the folder, download the JDK jar package matching the hardware and software environment on Oracle official website. Extract the jar package in the software directory to the app folder directory. After decompression is completed, configure Java into the system environment variable and use the source command to make it effective. After the operation, use the Java version command to detect and verify whether the environment is successfully configured. If the configuration is successful, the Java version number and other information will appear after the command is used.

4.3 System Performance and Safety Test

System Performance Test. The purpose of the system connection speed test is to ensure that the connection time can meet the user's requirements. The test results are shown in Fig. 1.

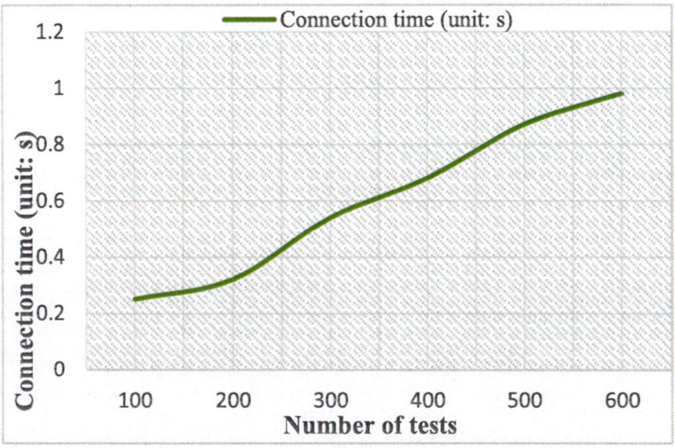

Fig. 1. System connection speed test

As can be seen from Fig. 1, although the average connection time of the system is gradually increasing with the increase of the number of tests, when the number of tests of the system is 600, the connection time does not exceed 1 s. It can be seen that the connection speed of the system fully meets the needs of users.

The performance test results of the unified SQL system are shown in Fig. 2.

According to Fig. 2, for traditional relational databases, when the data scale is relatively small, the average time consumption of the unified SQL system is not much different from that of the original database. Because the time consumption itself is relatively small, the unified SQL system will not cause significant sensory impact on users. When the data scale gradually increases, the time consumption of the unified SQL system is far less than that of the original database; therefore, its time-consuming can be ignored.

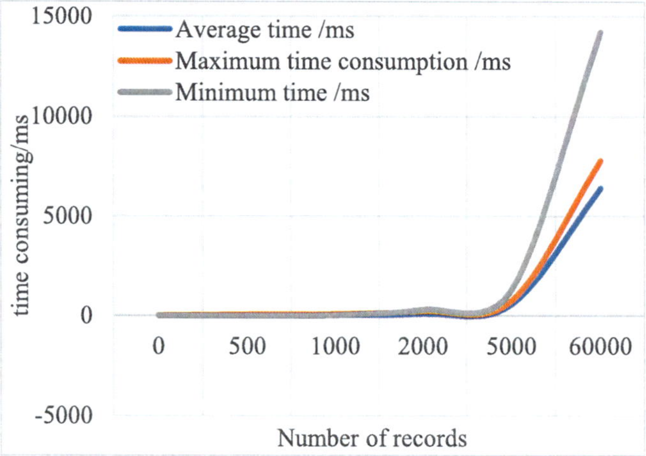

Fig. 2. Database performance test

Load Test of the System. Load test is to test the performance of the system at different load levels, so as to ensure the system can run normally at different load levels. The test results are shown in Table 2:

Table 2. Statistics of system load test

Number of loads	System operation	Error type
0–100	The system runs stably without any adverse conditions	Nothing
100–200	The system runs stably without any adverse conditions	Nothing
200–300	The system runs stably without any adverse conditions	Nothing
300–380	The system runs stably without any adverse conditions	Nothing
Over 380	The system runs stably without any adverse conditions	Nothing

5 Conclusion

In this paper, through the design of key personnel management and control system database, the establishment of data association layer, and the selection of appropriate and efficient database table building method, to solve the key personnel of different data sources, heterogeneous data how to effectively integrate the problem. By establishing a reasonable integral model according to the categories of key personnel, the basic data and performance of key personnel can be automatically integrated and summarized step by step to form the final comprehensive integral. At the same time, through the design and implementation of key personnel management and control system data interface, it provides application access support for the external platform. Through the research of this paper, the personnel information scattered in different types of business systems and functional modules is integrated, and the personnel information is fully displayed.

References

1. Aihara, M.: Practical study of the sports garage pick tour in Japan sports events to solve social problems. Sports Sci. (English Version) (2), 56–64 (2019)
2. Allen, W.: The importance of: cable verification, qualification and certification. Whats New Electron. **38**(5), 26–29 (2019)
3. Wagner, L., Mebsout, A., Tinelli, C., Cofer, D., Slind, K.: Qualification of a model checker for avionics software verification. In: Barrett, C., Davies, M., Kahsai, T. (eds.) NFM 2017. LNCS, vol. 10227, pp. 404–419. Springer, Cham (2017). https://doi.org/10.1007/978-3-319-57288-8_29
4. Machalek, N.S., Li, C., Barton, C., et al.: Universal qualification of near ultraviolet circular dichroism spectroscopy: method performance parameters and limits of detection for structural differences. J. Pharm. Sci. **109**(12), 3728–3733 (2020)
5. Br Own, W., Perivilli, S., Podolsky, D., et al.: The critical role of the USP performance verification test in dissolution testing and qualification of the paddle apparatus. Dissolut. Technol. **26**(1), 6–12 (2019)
6. Bartels, N., Allenspacher, P., Bauer, S., Rödiger, B., Sproll, F., Riede, W.: Design and qualification of a recessed satellite cornercube retroreflector for ground-based attitude verification via satellite laser ranging. CEAS Space J. **11**(4), 391–403 (2019). https://doi.org/10.1007/s12567-019-00255-x
7. Calvi, A., Bastia, P., Suárez, M.P., et al.: Mechanical verification logic and first test results for the Euclid spacecraft. Adv. Aircr. Spacecraft Sci. **7**(3), 251–269 (2020)
8. Pluta, P.L.: CQV #5: problems in aseptic technique qualification. J. Validation Technol. **25**(2), 1–2 (2019)
9. Kapoor, D., Vyas, R.B., Dadrwal, D.: An overview of analytical instrument qualification with reference of pharmaceutical industry. J. Drug Deliv. Ther. **8**(5), 99–103 (2018)
10. Sklyar, V., Kharchenko, V.: The equipment qualification testing framework: model driven development for design and model-based testing for verification. WSEAS Trans. Syst. Control **15**, 647–654 (2020)
11. Urban, B., Musa, T., Smith, R., et al.: Enhancing compliance through novel approaches to commissioning & qualification. Pharmazeutische Industrie **79**(9), 1304–1310 (2017)
12. Kukoyi, P.O., Osuizugbo, I.C., Yohanna, H.S., et al.: Pre-qualification of selecting construction project contractors using health and safety criteria. J. Eng. Project Prod. Manag. **11**(1), 30–36 (2021)

Development Trend of Smart Leisure Tourism Based on Big Data Analysis

Xia Xu(✉)

Guilin Tourism University, Guilin 541000, Guangxi, China
xx2006@gltu.edu.cn

Abstract. Today's information technology is highly integrated with the tourism industry, and the Internet, big data and artificial intelligence have quickly become the hot spots of industrial development. Therefore, this kind of "Internet + Tourism"-"Smart Leisure Tourism", which can improve the efficiency of the tourism industry and increase the overall benefits of the industry, came into being. This article aims to discuss the development trend of smart leisure tourism based on big data analysis. This article first analyzes the significance of the development of smart leisure tourism, and summarizes and analyzes the problems existing in the construction of smart leisure tourism through questionnaire surveys and big data. Combining the status quo of smart leisure tourism development, research the goals, ideas and realization paths of smart leisure tourism development, and propose optimization strategies for the development of smart leisure tourism and the promotion of the upgrading and development of the tourism industry in my country. This article takes H city as the research object to explore the development trend of smart leisure tourism in this city. The survey results show that from 2016 to 2020, all indicators of smart leisure tourism in City H have shown a substantial increase. In 2020, the operating income of travel agencies will reach more than 4.3 billion, indicating that the revenue of smart leisure tourism in City H has grown rapidly.

Keywords: Big data analysis · Smart tourism · Development trend · Tourism

1 Introduction

With the rapid development of our country's tourism industry, there is a trend of integration of tourism and information technology [1]. Smart leisure tourism is the product of the integration and development of information technology, tourism and service innovation. It is also the latest achievement in the application of "Internet+", big data and artificial intelligence in the tourism industry. It can solve the structural contradictions of the tourism industry and promote the further upgrading of the tourism industry. Avoid the "involution" of the tourism industry and promote its evolution and upgrading [2, 3]. The driving force of smart leisure tourism lies in the upgrading of the original industry triggered by the changes in tourist behavior patterns and demand. At the same time, the new generation of information technology also plays an important role in guiding and

B. J. Jansen et al. (Eds.): *International Conference on Cognitive based Information Processing and Applications (CIPA 2021)*, LNDECT 85, pp. 482–490, 2022.
https://doi.org/10.1007/978-981-16-5854-9_60

catalyzing [4, 5]. The connotation, development trend and service mode of smart leisure tourism have always been hotspots in the academic and business circles.

In the research on the development trend of smart leisure tourism based on big data analysis, many scholars at home and abroad have conducted different perspectives on it, and have achieved good results. For example, Chang SE takes the development status of "smart leisure tourism" as the background, Highlight the prominent role of government departments in the development of "smart leisure tourism" [6]; Pai C describes the business components of smart leisure tourism as dynamically interconnected stakeholders, digital core business processes, and organizational flexibility [7]; Lee H studied the problems faced by the development of smart leisure tourism under the background of big data, and constructed a smart leisure tourism application model based on big data platforms [8]; Pai CK proposed based on the construction of "smart city" How to develop a new strategy of "smart leisure tourism" [9].

This research aims to analyze the development trend of smart leisure tourism based on big data. Based on a full understanding of the related concepts of smart leisure tourism and tourism, this article analyzes the significance of the development of smart leisure tourism to the construction of smart cities, tourism management, tourism product upgrades, and tourism service quality. Then study the status quo of our country's smart leisure tourism development from two aspects: the current status of smart leisure tourism network platform construction and smart leisure tourism service capabilities. Finally, a case analysis method is used to analyze the development trend of smart leisure tourism in our country's H city through questionnaires and big data, and on this basis, the development of smart leisure tourism service module and model implementation strategies are proposed.

2 Development Trend of Smart Leisure Tourism

2.1 Development Significance

(1) Improve the construction of smart cities
As a sub-project of smart city, smart leisure tourism occupies a very important position [10]. Smart city construction needs the help of smart leisure tourism, and the development of smart leisure tourism must also be based on smart city construction. At the same time, the construction of smart leisure tourism effectively combines the construction of smart transportation and the construction of smart medical services in a smart city to make the construction of a smart city more perfect.

(2) Promote innovation in urban tourism management
By constructing smart leisure tourism, on the one hand, tourism management departments can more effectively and accurately collect, organize, transmit and analyze various types of tourism data, which is conducive to tourism management departments to make tourism-related decisions. On the one hand, Tourism management departments can conquer the development of tourism-related industries more dynamically to maintain the order of the tourism market in a more timely manner and ensure the safety of tourists.

(3) Promote the upgrading of tourism products
Analyze the travel preferences of tourists in different tourism markets, and tailor personalized travel services for different types of tourists. Smart leisure tourism can effectively use various resources for integration, creating new tourism products and integrated tourism consortia. At the same time, smart leisure tourism can give tourists more travel options, more market competition and promote the upgrading of tourism products.
(4) Improve the service quality of the tourism industry
The gradual improvement of the level of tourism information has improved work efficiency, increased the possibility of data analysis, made decisions more scientifically, and further improved the management level of the entire tourism industry. The construction of the intelligent leisure travel system can provide general information services for all travel issues, make the needs of tourists more clear, provide more targeted services, and effectively improve the service quality of the tourism industry [11].

2.2 Status Quo of Smart Leisure Tourism

(1) Status Quo of Smart Leisure Tourism Network Platform
At present, information display network platforms have been established in many places in my country. At the same time, many major tourist agencies in tourist destinations have carried out extensive marketing activities on third-party travel service platforms such as Ctrip and Qunar. Third-party travel platforms rely on extensive traffic advantages and strength. The ability of resource integration basically solves the tourist information service needs of tourists [12]. However, due to the inadequacy of self-built network platforms and third-party tourism platforms in certain regions in the mining of global tourism resources and related service supply, tourists have no way to obtain complete tourism service solutions, which affects the regional tourism brand building and industry development of.
(2) Tourism information integration capability
At present, tourism platforms built in more developed tourist regions serve as consumer portals to provide a series of services such as smart travel, smart maps, smart weather, smart food, and smart hotels. The platform itself has a certain ability to integrate tourism elements. According to the travel needs of consumers, it can provide standard element services including the six elements of "food, lodging, travel, shopping and entertainment", as well as accurate recommendation services and customized services. However, in some areas, due to the lack of software technology development capabilities and other reasons, the integration capabilities of tourism elements need to be improved.
(3) Smart leisure tourism service capabilities
Many regions in my country have a group of mid-to-high-end service organizations that provide leisure travel services, and they have formed certain industry norms and standards in terms of integration of tourism elements, and formed a complete set of hotels, transportation, accommodation, scenic spots, etc. in terms of tourism services.

2.3 Smart Leisure Tourism Development Strategy

(1) Improve the construction of smart leisure tourism platform
Smart leisure tourism relies on the service management of the scenic area to realize its own value, and it also adds convenience to the service management of the scenic area. However, at present, smart leisure tourism in scenic spots is still in the initial stage of construction, and its role in service management is not obvious enough. Therefore, the development of smart leisure tourism in scenic spots is not significant. Tourists' perception of smart leisure tourism is not deep enough. Tourists truly understand the "smartness" of smart leisure tourism. Scenic spots need to fully tap and play the value of smart leisure tourism, and upgrade smart leisure tourism to a strategic position, so that the service management of scenic spots can play a role in smart leisure tourism to better serve tourists.

(2) Strengthen the promotion of smart leisure tourism
The urban smart leisure tourism website should reposition its own functions and get rid of the shackles of simply providing tourists with tourism information, so as to comprehensively enhance the image of the scenic spot in the hearts of tourists. First of all, the scenic spot should upgrade the ticketing system so that tourists can use their mobile phones to automatically purchase tickets and use QR code scanning to enter the park automatically, freeing themselves from the shackles of paper tickets. If tourists need paper tickets, they can hold relevant proof of ticket purchase procedures. Second, turn on the comprehensive positioning system, visitors can check their location anytime and anywhere during the tour, so that they can plan for the formation. At the same time, the AR function can be incorporated into the map module to realize the mobile phone scanning to understand the relevant information of the nearby attractions. Then, the intelligent reminder function can be turned on. During the travel, tourists can check the related traffic, accommodation, catering and other services of the scenic spot at any time.
At the same time, they can also control the dense area of tourists in the scenic spot at any time, reminding tourists to go to places with less crowds first play, disperse the concentration of tourists. Finally, for tourists in the scenic spot, real-time group communication can be adopted, allowing visitors who are playing in the scenic spot to share their own experience and the local landscape at any time and anywhere, so that the tourists can arrange their own itinerary reasonably.

(3) Promote the construction of smart leisure tourism content
To a large extent, smart leisure tourism is still reflected through the improvement of service management level. Therefore, we must grasp the characteristics of smart leisure tourism serving tourists and build a service-oriented emerging smart leisure tourism platform. Scenic spots can use the real-time community communication platform of mobile terminals, so that tourists can learn about the tourism information of the scenic spots through the platform, and at the same time, can realize the evaluation of the service management quality of the scenic spots through the platform. At the same time, the scenic spot can publish the distribution of tourists and traffic conditions in the scenic spot on the mobile terminal in a timely manner through the real-time positioning system, and tourists can directly query the most convenient way of travel through the terminal.

2.4 Data Clustering Algorithm Suitable for Tourism Industry-Spectral Clustering Algorithm

The similarity calculation of spectral clustering algorithm is generally calculated by using Gaussian kernel function. The element of similarity matrix S is s_{ij}, and the degree matrix D (degreematrix) formula is obtained by transforming the processed matrix as shown in (1):

$$D = (d_i) = \sum_{j=1}^{n} w_{ij} \quad \mathrm{i, j} = 1, 2, \ldots, \mathrm{n} \tag{1}$$

The matrix formed by the difference between the degree matrix D and the weight connection matrix is the Laplacian matrix L. The non-normalized Laplacian matrix is shown in formula (2):

$$\mathrm{L = D - W} \tag{2}$$

The normalized Laplacian matrix has the following properties:

$$L_{sym} = D^{-12} L D^{-12} = I - L D^{-12} W D^{-12} \tag{3}$$

$$L_r w = D^{(-1)} L = I - D^{(-1)} W \tag{4}$$

Among them, L_{sym} is a symmetric matrix, and L_{rw} is a random walk matrix, usually asymmetric. The algorithm finally calculates the L matrix constructed according to the above formula, calculates the matrix to obtain k eigenvalues and obtains the eigenvectors. K-means algorithm is used to cluster the feature vectors to obtain k manifold structures according to the requested feature vectors.

3 Development Trend of Smart Leisure Tourism Based on Big Data Analysis

3.1 Research Methods

(1) Questionnaire survey method
Obtain the tourism development data of City H through questionnaire surveys and network data, and based on this data, use statistical analysis and econometric analysis to discuss the development status and trends of "smart leisure tourism".
(2) Comparative analysis method
Study the construction practice of smart leisure tourism in some regions, and explore the construction direction of smart leisure tourism suitable for the development of H city by comparing and analyzing the differences between the current situation of smart leisure tourism construction in H city and other cities.

4 Analysis of the Development Trend of Smart Leisure Tourism Based on Big Data Analysis

4.1 Scale of Smart Leisure Tourism Industry

The smart leisure tourism industry has a strong industry driving effect and has the foundation to become a pillar industry. According to the "Statistics of China's Smart Leisure Tourism Industry", the development of the entire smart leisure tourism industry in City H from 2016 to 2020 shows a good growth trend, and has a certain advantage in the number of receptions. Its convenient transportation network, abundant smart leisure tourism resources, the smart leisure tourism facilities of China and the government's vigorous promotion have promoted the development of the entire industry. The data of smart leisure tourism in City H is shown in Table 1: The number of travel agencies and star-rated hotels will increase from 2016 to 2020. The number of tourists received has become an increasing trend. Star-rated hotels, travel agencies, and the number of employees in the smart leisure tourism industry as key smart leisure tourism elements can show the development momentum and potential of smart leisure tourism, indicating that the smart leisure tourism industry in H City has developed well in recent years.

Table 1. Tourism Statistics of City H

Years	Total number of travel agencies (number)	Total number of receptions (million)	Star-rated hotel (number)
2016 year	182	94.87	103
2017 year	191	106.01	124
2018 year	198	113.28	131
2019 year	216	126.73	136
2020 year	218	138.43	138

Combining Fig. 1 shows that from 2016 to 2020, all indicators of smart leisure tourism have shown substantial growth. In 2020, the operating income of travel agencies will reach more than 4.3 billion, indicating that the revenue of smart leisure tourism has grown rapidly. At the same time, the development of smart leisure tourism has also made a great contribution to the economic development of City H, and the status of smart leisure tourism has gradually risen. It can be seen that the number of smart leisure tourism practitioners and the number of travel agencies in City H has shown a continuous increasing trend. In recent years, City H has also introduced a series of measures to encourage the development of smart leisure tourism. In the future, there will be more room for development in smart leisure tourism.

Fig. 1. The development trend of smart tourism industry in City H from 2015 to 2020

4.2 Transaction Scale of Online Smart Leisure Tourism Market

In the past two years, the scale of mobile phone online shopping users, the penetration rate of mobile phone online shopping, and the number of users of smart leisure travel products ordered by mobile phones have increased significantly, indicating that the online smart leisure travel market, especially the online smart leisure travel market on mobile clients, has a good development prospect. According to big data statistical analysis, the transaction scale of the online smart tourism market in 2020 is shown in Table 2: The number of online bookings for air tickets reached 169.4 million, and the number of bookings for hotels exceeded 70 million.

Table 2. Online market transaction scale

	Total (millions)	Proportion (%)
Air tickets	1694.8	55.1
Hotel	763.6	24.8
Vacation	620.5	20.1

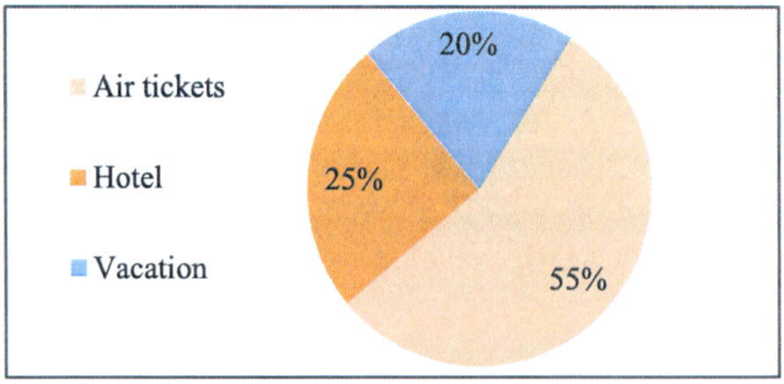

Fig. 2. Online market transaction scale

By observing Fig. 2, we can see that people's demand for tourism is great, the market share of online smart leisure tourism is gradually expanding, and the market scale of tourists who are able to conduct smart leisure tourism is also gradually expanding.

5 Conclusion

With the advent of the Internet+ era, information technologies such as big data, cloud computing, Internet of Things, artificial intelligence, and virtual reality are developing rapidly, and the tourism industry is rapidly upgrading. Tourists are also showing increasingly complex personalized needs and decentralized travel modes. The current service model of the tourism industry exposes certain structural contradictions. How tourism companies should develop and how to better meet the needs of tourists, provide them with a full-process, omni-directional, all-element perception and experience service, so as to realize the co-creation of service value between tourism companies and tourists, and become an urgent issue for the development of tourism companies problem. This paper systematically studies the development trend of smart leisure tourism based on big data analysis, and conducts applied research on the service model of smart leisure tourism in my country. It is of great significance to further enrich the theoretical system of smart leisure tourism service management. It is sufficient to improve the service level of smart tourism. Bringing into play the role of smart leisure tourism in driving and leading the upgrade of the entire tourism industry, and promoting the sustainable development of the tourism industry, provides theoretical method support and decision-making reference.

Acknowledgements. This paper is based on the following projects:

1. By Leisure Health Tourism Research Center, Key Research Base of Humanities and Social Sciences in Guangxi Higher Education University: Research on Supply Side Structural Reform of Guilin Community Pension Industry (No. YJZX2016YB03).

2. Research Fund of Guilin University of Tourism: Research on Construction of Internet-based Smart Hotel System (No. 2015ZD05).

3. First class course undergraduate courses of Guangxi Autonomous Region: Theory and Practice of Tourism Reception Industry (2020xjkc013).

References

1. Tan, W.K., Lu, K.J.: Smartphone use and travel companions' relationship. Asia Pac. J. Mark. Logist. **33**(4), 1088–1103 (2020)
2. Francesco, P., Antonio, B., Mara, G., et al.: Social innovation in smart tourism ecosystems: how technology and institutions shape sustainable value co-creation. Sustainability **10**(2), 140 (2018)
3. Koo, C., Cantoni, L.: Special issue on informatics/data analytics in smart tourism. Inf. Process. Manage. **55**(4), 1373–1375 (2019)
4. Maio, A.D., Violi, A., Santoro, F., et al.: Exploiting social data for tourism management: the SMARTCAL project. Qual. Quant. (8–9) (2020)
5. Lane, B.: Successful case studies in the Canary Islands' tourism industry. J. Sustain. Tour. **26**(4–6), 852–853 (2018)

6. Chang, S.E., Shen, W.C.: Exploring smartphone social networking services for mobile tourism. Int. J. Mobile Commun. **16**(1), 63–81 (2018)
7. Pai, C., Kang, S., Liu, Y., et al.: An examination of revisit intention based on perceived smart tourism technology experience. Sustainability **13**(2), 1007 (2021)
8. Lee, H., Hlee, S.: The intra- and inter-regional economic effects of smart tourism city Seoul: analysis using an input-output model. Sustainability **13**(7), 4031 (2021)
9. Pai, C.K., Liu, Y., Kang, S., et al.: The role of perceived smart tourism technology experience for tourist satisfaction, happiness and revisit intention. Sustainability **12**(16), 6592 (2020)
10. Koo, C., Cantoni, L.: Special issue on informatics/data analytics in smart tourism. Inf. Process. Manag. **57**(1), 102154 (2019)
11. Hassannia, R., Barenji, A.V., Zhi, L., et al.: Web-based recommendation system for smart tourism: multiagent technology. Sustainability **11**(2), 323 (2019)
12. Jeong, M., Shin, H.H.: Tourists' experiences with smart tourism technology at smart destinations and their behavior intentions. J. Travel Res. (1), 004728751988303 (2019)

A New Detection Method of Overhead Power Line Based on HED Algorithm

Haiping Liang[1], Yanzhe Yin[1](✉), Xinming Wang[2], Shaobo Li[2], Huayang Liang[2], and Shize Li[2]

[1] Department of Electrical Engineering, North China Electric Power University (Baoding), Baoding 071003, Hebei, China
2192213154@ncepu.edu.cn

[2] State Grid Hebei Electric Power Co., Ltd. Maintenance Branch, Shijiazhuang, Hebei, China

Abstract. Power line detection is an important content of power system inspection, which of important significance for UAV inspection, obstacle avoidance, and visual measurement. Traditional power line detection algorithms are mostly based on straight line detection, and the power lines are further extracted from the detected straight lines. Therefore, the disturbance of background information is large, and the detection effect for curved lines is not satisfied. Classical deep learning object detection algorithms are not suitable for the detection of elongated objects such as overhead power lines. Based on the Holistically-Nested edge detection (HED) algorithm, this paper uses the Tensorflow deep learning framework to build a HED network model and train it. Through pixel-level classification of the image, the pixel-level segmentation of the wires in the image is realized. Through experimental verification, the power line detection achieves a high accuracy rate.

Keywords: Power line detection · HED algorithm · Tensorflow · Pixel-level segmentation

1 Introduction

Overhead transmission line is the most important power transmission equipment. Because of their wide coverage and long transmission distances, the electric power bureau has to invest a lot of resources in line inspections every year. Complex natural environmental factors such as terrain and weather also increase the difficulty of power lines inspection and maintenance [1, 2]. With the development of artificial intelligence and UAV technology, using UAV for power lines inspection has become a new method. UAV combines image processing, computer vision, GPS, image encryption and other technologies [3], which effectively overcomes the shortcomings of manual line inspection and improves inspection efficiency. Computer vision is the core of UAV inspection. Its main function is to replace human vision to check whether there are defects and safety hazards in transmission lines, and to measure wire diameter and sag, etc. Therefore, overhead power line detection is an important task of UAV line inspection. First,

B. J. Jansen et al. (Eds.): *International Conference on Cognitive based Information Processing and Applications (CIPA 2021)*, LNDECT 85, pp. 491–499, 2022.
https://doi.org/10.1007/978-981-16-5854-9_61

it is of great significance for detecting the running condition of the wire and avoiding obstacles of the UAV. Modern computer vision is often combined with deep learning to realize automatic recognition and detection of objects. Based on the HED deep learning algorithm, this paper uses the overhead power line dataset to retrain and adjust the original network to realize the automatic identification of overhead power lines.

2 Present Condition

Early power line detection based on computer vision basically used the straight line detection algorithm to detect the straight lines in the image, and then used other identification methods to extract the power lines. For example, literature [4] detected the straight lines in the picture based on the Hough straight line detection algorithm and view them as power lines directly; the literature [5] uses motion estimation at the pixel level, combined with edge detection, followed by a windowed Hough transform to detect power lines; the literature [6] compares the differences of power line detection in different line detection methods. These methods essentially detects a straight line, which is dependent on the camera's shooting position and background information, and is greatly affected by the disturbance of background. As shown in Fig. 1, because the background of the picture contains clouds, the Hough line detection cannot get satisfied effect. Then some scholars began to study how to separate the power lines from the background. For example, literature [7] used manual selection of ROI area to remove the background, combined with binocular vision technology to achieve power line detection and 3D reconstruction; literature [8] used Otsu-2D of Simulate Anneal Arithmetic (SAA) to separate the power lines and background in the picture, and uses an improved Freeman chain code table to extract the power lines, and finally uses the fitting method to restore and extract the missing part of the lines.

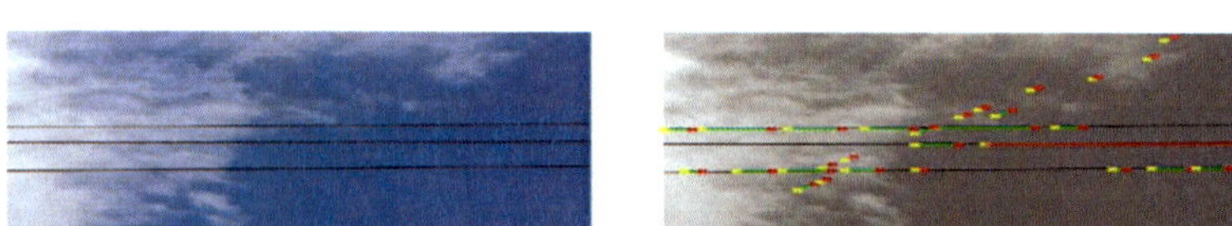

Fig. 1. Hough line detection

In recent years, the innovation of algorithms and the improvement of hardware computation have continuously promoted the development of deep learning, and many deep learning algorithms have begun to be applied to various fields in the power system. Overhead power line detection belongs to the field of object detection in computer vision. At present, many kinds of object detection algorithms based on deep learning have been proposed, which are mainly divided into two categories. One is a two-stage detection method, such as the R-CNN series, and another is the one-stage detection method, such as YOLO and SSD. Many studies have used these detection algorithms to achieve accurate detection of various electrical equipment, equipment defects, and fault locations in the power system [9–16]. There are also some studies on power line detection [17, 18], but many of them focus on the detection of wire defects, insulators, towers that is closely related to the power lines, and there are few studies on the detection of power lines directly. Unlike other electrical equipment, overhead power lines account for a small proportion of pixels in the entire picture, and due to the long camera shooting distance, the wires on the picture lack texture, color and other information, so there is not much feature information that can be used to extract. In addition, the object detection mostly uses rectangular boxes to mark the detected objects. For the wires in the picture, because the shapes are thin strips and the directions are different, it is difficult to mark with the rectangular boxes. In addition to object detection, there is another important task in computer vision, that is, semantic segmentation. Semantic segmentation needs to classify and display each pixel in the image. Therefore, each pixel must be labeled in an image and rectangular boxes are no longer needed. The detection level is more refined than the object detection task. Therefore, this paper uses a semantic segmentation algorithm—HED edge detection algorithm to build a neural network and trains with the marked overhead power line images. The recognition result only displays the pixels of the overhead power lines.

3 Holistically-Nested Edge Detection

HED [19] is a new image edge detection algorithm proposed in 2015. Different from traditional methods using edge detection operators, HED is a semantic segmentation algorithm essentially. HED uses the concept of multi-scale and multi-level feature learning and can give an image-to-image prediction. Because the model can automatically learn the rich features in the image, it is benefit to solve the problem of boundary blur in object edge detection.

3.1 Network Structure

HED takes VGG-16 [20] as the backbone network, and uses transfer learning to initialize the network weights. The schematic diagram of the HED network structure is shown in Fig. 2. As is shown in Fig. 2, on the basis of VGG-16, the fully connected layer and the last pooling layer of VGG-16 are deleted, and the remaining network layers are divided into 5 groups in the order of "convolution + ReLU layer"-"pooling layer". After an image is sent to the backbone network for feature extraction, the last layer of each group is extracted from the "convolution + ReLU layer" for 1 × 1 convolution, and then deconvolute it for upsampling to obtain 5 single-channel feature maps with the same size as the input image. So that there are 5 more side outputs on the basis of the original backbone network. Finally, the five feature maps are fused to one single-channel feature map that can be viewed as prediction result.

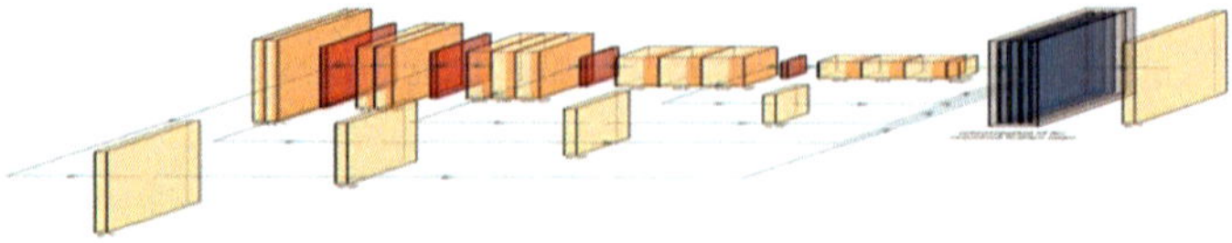

Fig. 2. HED neural network structure

4 Loss Function

Suppose the input training set during network training is $S = \{(X_n, Y_n), n = 1, \ldots, N\}$, where $X_n = \left\{x_j^{(n)}, j = 1, \ldots, |X_n|\right\}$, represents original input image, $Y_n = \left\{y_j^{(n)}, j = 1, \ldots, |Y_n|\right\}$, $y_j^{(n)} \in \{0, 1\}$, represents the binary image with marked edges corresponding to the original image. The original weight parameter of the VGG-16 network is W. There are five side outputs in the HED network, and each output also has a corresponding weight parameter, which is combined into a matrix $w = \left(w^{(1)}, \ldots, w^{(5)}\right)$. The overall objective function of the five side outputs is

$$\mathcal{L}_{\text{side}}(W, w) = \sum\nolimits_{m=1}^{5} \alpha_m \ell_{\text{side}}^{(m)}\left(W, w^{(m)}\right) \tag{1}$$

where α_m represents the weight of the loss function of each side output that can be adjusted according to the training log or set to 1/5. ℓ_{side} represents the cross entropy loss function of each side output, that is

$$\begin{aligned} \ell_{\text{side}}^{(m)}\left(W, w^{(m)}\right) = -\beta \sum\nolimits_{j \in Y_+} \log \Pr\left(y_j = 1 | X; W, w^{(m)}\right) \\ -(1-\beta) \sum\nolimits_{j \in Y_-} \log \Pr\left(y_j = 0 | X; W, w^{(m)}\right) \end{aligned} \tag{2}$$

where $\beta = |Y_-|/|Y|$, $1 - \beta = |Y_+|/|Y|$, $|Y_-|$ and $|Y_+|$ respectively represent the number of edge pixels and non-edge pixels in the labeled image. $\Pr\left(y_j = 1|X; W, w^{(m)}\right) = \sigma\left(a_j^{(m)}\right)$ is the predicted probability that the pixel j is a positive sample, using the sigmoid function, and the output range is [0, 1]. The five side output images will finally be fused into one image as the output image. This step is implemented by the fuse layer. Therefore, the weight parameter of the fuse layer is introduced. Add five side output with weight, which is equivalent to 1×1 convolution. The loss function output by the fuse layer is

$$\mathcal{L}_{\text{fuse}}(W, w, h) = \text{Dist}\left(Y, \hat{Y}_{\text{fuse}}\right) \tag{3}$$

where $\hat{Y}_{\text{fuse}} = \sigma\left(\sum_{m=1}^{5} h_m \hat{A}_{\text{side}}^{(m)}\right)$, $h = (h_1, \ldots, h_5)$ are the weight parameters of fuse layer. $\hat{A}_{\text{side}}^{(m)} = \left\{a_j^{(m)}, j = 1, \ldots, |Y|\right\}$ are predicted pixel values of each side output. The Dist() function reflects the difference between the final predicted image and the labeled image, which is still represented by the cross-entropy loss function. So the final objective function is

$$Loss = \mathcal{L}_{\text{side}}(W, w) + \mathcal{L}_{\text{fuse}}(W, w, h) \tag{4}$$

We only need to continuously change parameters according to the gradient descent method to reduce the *Loss* value until within the allowable range.

5 HED Network Training

This chapter created a HED network, and used power lines and their labeled images as training data to train the network that can individually identify power lines and achieve pixel-level segmentation. The specific settings of the experiment are as following.

5.1 Network Structure and Hyper-parameter Settings

The experiment is based on the Tensorflow deep learning framework, combined with the Keras deep learning library to build a HED neural network model. The HED network structure is shown in Fig. 2. The backbone network is VGG-16, so the weight parameters of the trained VGG-16 are used as the initial parameters of the backbone network. Training is performed on this basis to adjust the parameters. Before training, adjust the size of the imported picture to $480 \times 480 \times 3$. According to Sect. 3.1, the HED backbone network can be divided into 5 modules. The last layer of modules is the max pooling layer except the last module. The size of the pooling operator is 2×2, so the output image size is reduced into half through this layer but the depth increases due to the existence of the convolutional layer. Table 1 shows the output image sizes of the four pooling layers. The model produces 6 feature maps, including 5 side outputs and 1 fuse output. The loss function values of the 6 outputs will be calculated separately and then added as the overall loss function value. We use Adam as the optimizer of the model. Adam optimization is a stochastic gradient descent method based on the first and second order of random estimation, and the learning rate is set to 0.001. The data set is divided into two parts, the first 80% is used for training, and the remaining 20% is used for validation. The images are trained in batches, and the number of images for each batch is 10. This experiment includes 1500 images totally, so 1200 images are used for training, 300 images are used for validation, and a total of 120 batches are trained. The number of training epoch is set to 100.

The prediction effect of the model is evaluated by the pixel error, that is comparing the final predicted image with the labeled image, then calculate the ratio of the number of pixels that are predicted incorrectly in the entire image, the formula is

$$error = \frac{|Y_1|}{|Y|} \tag{5}$$

where $|Y_1|$ is the number of pixels that are predicted incorrectly, and $|Y|$ is the number of pixels of entire image.

Table 1. Pooling layer output image size

Pooling layer	Output image size
block1_pool	240 × 240 × 64
block2_pool	120 × 120 × 128
block3_pool	60 × 60 × 256
block4_pool	30 × 30 × 512

5.2 Dataset and Augmentation

The dataset used in this experiment consists of 71 images of power lines downloaded from the Internet and taken shooting from outdoor. Use Labelme image labeling software to label the power lines on the image. There are only two types of labels: power lines and background. Since the original image is large, and the input image size required for network training is only 480 × 480, the original image and its corresponding labeled image are randomly cropped into 500 × 500 pictures, the pictures without power lines are removed, and the dataset is expanded to 350 images. Next, augment images on the dataset, including: 1) Random rotation, executed at a probability of 0.8, with a maximum rotation angle of 25 degrees; 2) Random image flipping, including up-down flipping and left-right flipping; 3) Random image zooming, executed according to the probability of 0.8, and the area is 0.82 times of the original image. After image augmentation operations, the dataset was expanded to 1500 images.

5.3 Result

Figure 3 shows the training loss function value curves of the six outputs during the training process, and Fig. 4 shows the sum of the loss function values of the six outputs during the training process and the validation process. It can be seen from Figs. 3 and 4 that the training process reached converge, and the fusion layer got the smallest loss function value. The model achieved a better prediction effect after the fusion of five side output layers.

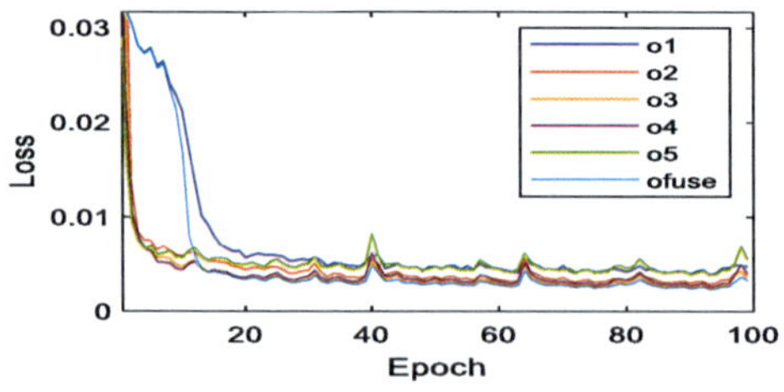

Fig. 3. Six layers output loss function curve

Figure 5 shows the predicted pixel error in the validation process after each training epoch is completed. It can be seen that the initial error is not very large, because the

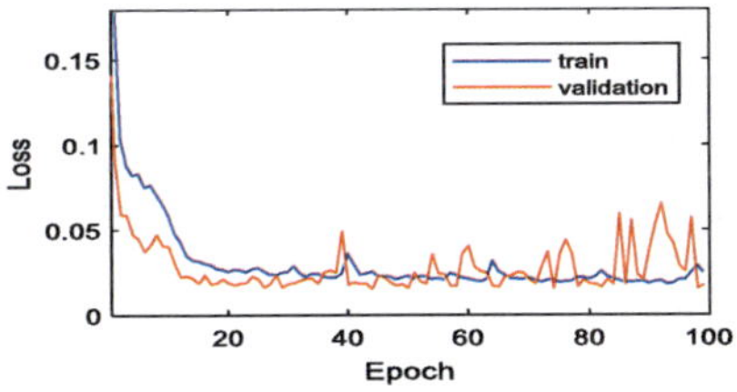

Fig. 4. The sum of the 6 layers output loss function values

initial training parameters of the network is set as the same as VGG-16 network, so it has a better ability to extract the edges of objects in the image. After training, the network prediction error is continuously reduced, and the final error is less than 3%.

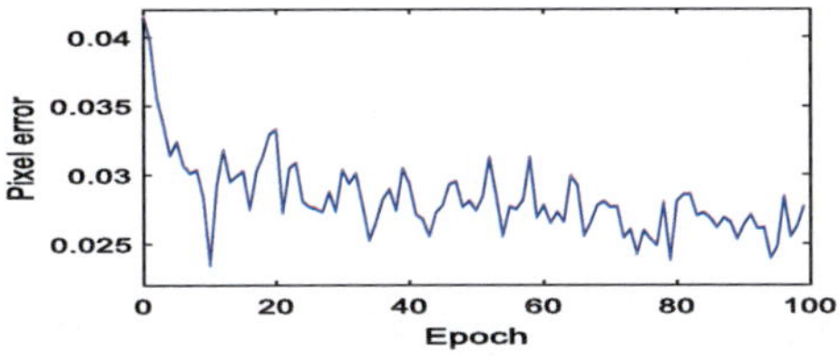

Fig. 5. Predicted pixel error

Fig. 6. Network prediction results

Figure 6 shows the prediction results of three power line pictures. From the images, it can be seen that the network has achieved good results in background filtering and power line extraction. For an image with an complex background, there are still errors in the predicted output picture. For example, the pixels that do not belong to the power line are divided into it. but in general, most power lines on the image can be extracted on the whole.

6 Conclusionss

This paper builds a multi-level convolutional neural network based on the HED algorithm. The initial parameters of the network are the VGG-16 network parameters. On this basis, input the images of overhead power lines and corresponding labeled images into the model to train and adjust it. The network can achieve pixel-level segmentation of the power line image and realize the extraction of power lines, with a prediction error of less than 3%. Compared with the algorithm based on straight line detection, the network in this paper can filter out the disturbance of the background in the image to the greatest extent, and also has a better extraction effect for the non-straight line part of the power lines, which improves the accuracy of extraction. It has high application value for the subsequent development of UAV inspection and visual measurement.

References

1. Xiangyang, P., Zhengjun, L., Xiaoming, M., Zhibin, L., Ke, W., Xiaowei, X.: A transmission line inspection system based on remote sensing: system and its key techonologies. Remote Sens. Inf. **30**(01), 51–57 (2015)
2. Guiwei, S., Liu Zhuang, F., Jing, T.J., Yi, C., Liwei, Z.: Research progress in unmanned aerial vehicle inspection technology on overhead transmission lines. High Voltage Eng. **46**(01), 14–22 (2020)
3. Kun, W., Wenkai, K., Nan, L., Hao, Y., Sheng, S.: Aerial encryption image hardware encryption method for inspection of overhead transmission lines. Inf. Secur. Commun. Priv. **02**, 75–82 (2021)
4. Golightly, I., Jones, D.: Visual control of an unmanned aerial vehicle for power line inspection. In: ICAR 2005 Proceedings of the 12th International Conference on Advanced Robotics 2005, pp. 288–295 (2005). https://doi.org/10.1109/ICAR.2005.1507426
5. Candamo, J., Kasturi, R., Goldgof, D., Sarkar, S.: Detection of thin lines using low-quality video from low-altitude aircraft in urban settings. IEEE Trans. Aerosp. Electron. Syst. **45**(3), 937–949 (2009). https://doi.org/10.1109/TAES.2009.5259175
6. Yetgin, Ö.E., Şentürk, Z., Gerek, Ö.N.: A comparison of line detection methods for power line avoidance in aircrafts. In: 2015 9th International Conference on Electrical and Electronics Engineering (ELECO), pp. 241–245 (2015). https://doi.org/10.1109/ELECO.2015.7394489
7. Xiong, C., Xiaobo, L., Shaohua, H., Ning, Z., Zhimin, H.: Research on transmission line identification and 3d reconstruction technology considering real-time channel monitoring. Electric Power Sci. Eng. **36**(03), 30–37 (2020)
8. Fengjie, S., Zhenxun, Y., Yuanyuan, L., Jieqing, F.: Methods of transmission line target recognition. J. Image Graph. **17**(03), 349–356 (2012)
9. Gong, X., Yao, Q., Wang, M., Lin, Y.: A deep learning approach for oriented electrical equipment detection in thermal images. IEEE Access **6**, 41590–41597 (2018). https://doi.org/10.1109/ACCESS.2018.2859048
10. Research on Infrared Image Faults Recognition for Electrical Equipments based on Dual Supervision Signals Convolution Neural Network. Tianjin University of Technology (2018)
11. Liu, N.: Research on Electric Power Equipment Image Classification based on Deep Learning. North China Electric Power University (Beijing) (2019)
12. Li, W., Xie, K., Liao, X., Li, X., Wang, H.: Intelligent diagnosis method of infrared image for transformer equipment based on improved faster RCNN. South. Power Syst. Technol. **13**(12), 79–84 (2019)

13. Hua, S.: Research on Electrical Equipment Detection Method Based on Deep Learning. Anhui University (2020)
14. Xue, Y., Sun, Q., Dang, W.: Infrared image defect recognition method of distribution network equipment based on Faster RCNN. Inform. Technol. **44**(07), 79–83+91 (2020)
15. Jiang, A., Yan, N., Shen, B., Gu, C., Zhu, H., Huang, H.: Research on infrared image recognition method of power equipment based on deep learning. IEEE Int. Conf. High Voltage Eng. Appl. (ICHVE) **2020**, 1–4 (2020). https://doi.org/10.1109/ICHVE49031.2020.9279631
16. Da, C., Wenhu, T., Zhewen, N.: Fault diagnosis method for power equipment infrared images based on deep learning. Guangdong Electr. Power **34**(01), 97–105 (2021)
17. Zhou, Z., Gao, J., Zhang, W., Wang, X., Zhang, J.: Object detection of transmission line visual images based on deep convolutional neural network. Chinese J. Liq. Cryst. Display **33**(04), 317–325 (2018)
18. Zheng, W., Li, C., Cui, X., Shang, P.: A multi-objective parallel detection algorithm for images of power transmission line corridors. In: 2020 IEEE 5th Information Technology and Mechatronics Engineering Conference (ITOEC), pp. 432–436 (2020). https://doi.org/10.1109/ITOEC49072.2020.9141566
19. Xie, S., Tu, Z.: Holistically-Nested Edge Detection. IEEE Int. Conf. Comput. Vis. (ICCV) **2015**, 1395–1403 (2015). https://doi.org/10.1109/ICCV.2015.164
20. Simonyan, K., Zisserman, A.: Very deep convolutional networks for large-scale image recognition. Comput. Sci. (2014)

Study on Algorithm and Parameters of Seismic Design Code for Medium and Long Span Bridges

Jiangping Li(✉), Ping Zhang, and Zhijun Meng

Xinjiang Vocational and Technical College of Communications, Urumqi 831401, Xinjiang, China

Abstract. China is an earthquake prone country, ensuring the safety of bridges under earthquake is of great significance to the protection of people's lives and property. With the rapid development of medium and long-span bridge construction, the research of bridge seismic analysis method has become a research hotspot in academic circles at home and abroad. It is generally believed that the spatial variation effect of seismic excitation should be considered in the seismic analysis of medium and long-span bridges. But the response spectrum method, which is widely used in current design codes, can not deal with this kind of problem. The time history analysis method is difficult to be popularized in engineering application because of its huge amount of calculation. The random vibration method has great advantages in theory in dealing with this kind of problem, but it has not been applied in practical engineering because of the complexity of its traditional calculation method. The virtual excitation method breaks through the bottleneck of calculation efficiency and makes it possible to introduce the random vibration method into the new code for seismic design of highway bridges.

Keywords: Response spectrum · Power spectrum · Random vibration · Virtual excitation · Seismic action combination

1 Introduction

Earthquake is one of the biggest natural disasters faced by life on the earth. The instant damage caused by every big earthquake brings serious disaster to human beings, and causes long-term psychological fear. In fact, the direct cause of death in the earthquake is mainly caused by the collapse of buildings caused by the earthquake. The resulting secondary disasters such as fire and traffic disruption are also very serious, and may even exceed those caused by direct damage [1].

At present, the response spectrum method is still the most basic method in the seismic design codes of bridges at home and abroad. It assumes that all bearings move according to the same law (that is, it assumes that all bearings are rigidly connected). This is OK for the structure with small span, but it is not suitable for the bridge with hundreds of meters or even larger span, In this case, the relative motion between supports due to ground deformation must be considered, that is, the multi-point non-uniform excitation must be

B. J. Jansen et al. (Eds.): *International Conference on Cognitive based Information Processing and Applications (CIPA 2021)*, LNDECT 85, pp. 500–505, 2022.
https://doi.org/10.1007/978-981-16-5854-9_62

considered; And the partial coherence effect between the support excitations caused by the complex mechanism of seismic source and the irregular reflection and refraction of seismic wave in the inhomogeneous soil medium; sometimes the local effect caused by the different properties of the site soil at different supports should be considered. These complex factors are difficult to be calculated by the conventional response spectrum method.

2 Review of Structural Seismic Analysis Methods

The random vibration method not only can realize the function of response spectrum method, but also can be applied to the seismic analysis of long-span structure under multi-point seismic excitation, so the function is more powerful. In the past, due to the complexity and low efficiency of the calculation method, it is difficult to be applied to engineering design. With the development of computer technology and computational mechanics, the distance between random vibration theory and engineering design is becoming shorter and shorter [2].

Ground motion is a non-stationary random process. Even if it is a linear system, the response to such an excitation is also a non-stationary random process. However, for non-stationary random vibration, although the virtual excitation method can solve the structural response analysis, there is still a long way to go from engineering practice. In our current revised code, the ground acceleration is assumed to be a stationary random process, and the basic calculation method is the virtual excitation method.

2.1 Response Spectrum Method

The motion equation of the single degree of freedom system excited by the uniform ground is as follows:

$$m\ddot{y} + c\dot{y} + ky = -m\ddot{x}_g(t) \tag{1}$$

Or expressed as:

$$\ddot{y} + 2\zeta\omega_0\dot{y} + \omega_0^2 y = -\ddot{x}_g(t) \tag{2}$$

According to the response spectrum method based on envelope control, the solution is as follows:

$$y = \alpha g/\omega_0^2 \tag{3}$$

Where α is the seismic influence coefficient given by the code and g is the acceleration of gravity.

2.2 Basic Principle of Virtual Excitation Method

In engineering applications, if the external excitation is assumed to be a stationary random process (usually also assumed to obey normal distribution), then the self power spectral density function of random excitation should be given. The main calculation of structural analysis is to calculate the power spectral density of important displacement, internal force and other responses. Then the corresponding spectral moments (especially the variance and the second moment) are calculated. According to these spectral moments, various statistics directly applied to engineering design can be calculated, such as the first exceedance probability or fatigue life leading to structural failure, and the indexes for evaluating vehicle ride comfort. Obviously, it is of great significance to improve the calculation method of structural response power spectral density to make it convenient, efficient and accurate for promoting the practicability of random vibration results. Virtual excitation method is a new accurate and efficient algorithm developed for this purpose. The following is the basic principle of single point incentive.

3 Multi Dimensional Seismic Analysis by Virtual Excitation Method and Research on Seismic Action Combination

Ground motion is the vibration of the soil layer near the surface caused by the seismic wave released from the source, which is the main basis for the seismic calculation and design of engineering structures. The propagation of seismic wave shows that the ground motion is multi-dimensional, both theoretically and practically. Therefore, it is not enough to consider only one-dimensional seismic action in structural seismic analysis.

The virtual excitation method solves the problem of low calculation efficiency when random vibration method is used for seismic analysis of long-span structures for a long time. However, there are still a lot of problems to be solved to apply the random vibration method to engineering design. The processing of multi-directional earthquake is one of the problems [3]. In the outline of the new code for seismic design of highway bridges, the provisions on the combination of seismic actions are mainly based on the provisions of the European code. We do not fully understand the error estimation and applicability of the provisions. Sometimes it's really helpless to copy the provisions of foreign norms. But now that we have the powerful weapon of virtual incentive method and the developed software, we should test and even amend these "import" regulations, so as to truly realize the "foreign for China" in line with China's national conditions. Therefore, this chapter makes a lot of calculation, analysis and comparison with some typical engineering examples, and basically clarifies the applicability of these provisions and the degree of error with the exact solution. Moreover, the applicability of these Provisions is tested when the traveling wave effect is considered. The research shows that the conclusions of these Provisions are not valid for the case of traveling wave effect. In engineering, the calculation should be carried out according to the earthquake in multiple directions, and then the most unfavorable response should be extracted for design.

3.1 Research Status of Time History Analysis Method for Multidimensional Seismic Analysis

Time history analysis is a kind of direct integration method. The commonly used methods are linear acceleration method, wiom day method and Newmark method. Compared with the response spectrum method, the time history analysis method has more extensive applicability, such as considering the nonlinearity of the structure, determining the order of plastic hinges and the location of structural weak links. Its disadvantage is that it lacks statistical significance and can only select certain seismic waves. The results obtained from different seismic waves may have large deviation. Therefore, it is generally used as a supplement to the response spectrum method and for nonlinear analysis of structures. The elasto-plastic torsional response of asymmetric spatial frame structures under bidirectional seismic waves is studied by using time history analysis method. In this paper, an inter story calculation model of bar system is established, and the elastic-plastic torsion of irregular steel frame structure under bidirectional earthquake is analyzed by using time history method. In this paper, the nonlinear response of Luoyang TV Tower under three-dimensional earthquake is analyzed by time history method, which is compared with that under one-dimensional earthquake. The elasto-plastic seismic performance of spatial reticulated shell structure under three-dimensional earthquake is also studied by time history method when the elasto-plastic seismic response of spatial reticulated shell structure is studied. Most of the time history analysis methods are real seismic waves, which can be used to test the rationality of other methods and provide a basis for further research.

3.2 Random Vibration Method for Multidimensional Seismic Analysis

It has been half a century since earthquakes were regarded as a random process. In recent decades, scholars at home and abroad have carried out extensive research in the field of random vibration, and achieved fruitful results in the random model of ground motion input and the random response analysis method of structures under earthquake action [4]. However, the calculation efficiency of random vibration method in multi-dimensional seismic analysis of large and complex structures is low, which seriously affects its application in engineering. Virtual excitation method is an efficient and accurate method for random vibration analysis, and it is easy to program. It can deal with all kinds of complex random problems and solve the limitations of traditional CQC algorithm in computational efficiency. It is suitable for application in the field of multi-dimensional random vibration. In this paper, the concept of virtual excitation method is used to derive the power spectrum of internal force and displacement response of the structure under the action of three-dimensional translational seismic component.

4 Conclusion

At present, the response spectrum method is still the most basic seismic analysis method in most bridge seismic design codes at home and abroad. The current research shows that the seismic spatial effect can not be ignored in the seismic analysis of medium and long span bridges, but the response spectrum method is limited because it can not analyze the problem of multi-point nonuniformity. At present, the time history method used to analyze this kind of problems can not be widely used because of its large amount of calculation. Random vibration method has great advantages in dealing with this kind of problems in theory. The appearance of virtual excitation method breaks the bottleneck of calculation efficiency of random vibration method. Therefore, one of the main tasks in the revision of the new code is to give some guiding suggestions for the seismic analysis of medium and long span bridges using random vibration method, which requires a lot of theoretical research and calculation analysis. For example, the determination of the input power spectrum of the random vibration method, the study of the influence of the spatial effect, the determination of the integral step size, the upper and lower bounds of the integral of the random vibration method and so on need to be solved. After repeated research, many of these problems have achieved better solutions, but there are still many problems to be solved. This paper mainly makes a supplementary research on two aspects.

New York's Brooklyn Bridge across the New York East River, connecting Brooklyn and Manhattan Island, officially put into use on May 24, 1883. With a total length of 1834 m, the bridge is 41 m above the water by tens of thousands of steel cables. It was the longest suspension bridge in the world at that time and the first steel bridge in the world. When it was completed, it was regarded as the eighth wonder after the seven ancient wonders of the world and was hailed as one of the seven epoch-making architectural wonders in the world in the era of industrial revolution. When the bridge celebrated its centennial, the United States issued a 20 cent stamp to commemorate it, showing the majestic and elegant demeanor of the bridge. Hart Crane, a modern American poet, wrote a long poem for it, which is called bridge.

The designer of the bridge is John Roblin, who came to the United States to start a business with the bridge technology he learned in Germany. In 1869, his plan to build the Brooklyn Bridge was approved, but he himself died in an accident during a riverside survey. His 32 year old son was appointed chief engineer of bridge construction immediately, but because of the use of pneumatic caisson method to build piers, he suffered from caisson disease and had to rely on his wife Emily to command the construction site. It took 14 years from 1869 to 1883, with a capital investment of 25 million US dollars. During the construction period of this bridge, besides John A. Rowling, 20 construction workers were killed (the sign board on the bridge tower was attached to mourn them). Finally, this monument in the history of bridge in the world was built. When completed, the pier was 87 m high, which was one of the tallest buildings in New York at that time. At that time, the three major landmarks of New York were the Empire State Building, the statue of liberty and the Brooklyn Bridge.

The existing westbound bridge is designed and rebuilt with open frame separation and reinforced pillars, and ventilation holes are opened to allow the wind to pass through. Opened to traffic on October 14, 1950, it is 5979 feet (1822 m) - 40 feet (12 m) longer than the original bridge. It and the parallel eastbound bridge together form the fifth longest suspension bridge in the United States. Because the resonance problem that caused the collapse of the front bridge has been solved by the new design, local residents have given the bridge a new nickname - "strong Gerty". When the westbound bridge was built, it was the third largest suspension bridge in the world. Like other modern suspension bridges, westbound bridges are made of sharp steel plates instead of the original flat edge steel plates. The designed daily traffic flow of the bridge is 60000 vehicles. After completion, it will be opened to traffic in both East and west directions until the eastbound bridge is officially opened to traffic on July 15, 2007.

References

1. Lichu, F.: Earthquake Resistance of Bridges. Tongji University Press, Shanghai (1997)
2. Xian, Y.: Seismology. Seismological Press, Beijing (1988)
3. Jiahao, L., Yahui, Z.: Virtual Excitation Method of Random Vibration. Science Press, Beijing (2004)
4. Zi, C., Suduo, X., Yigang, Z., et al.: Random response analysis of single-layer reticulated spherical shell under multi-dimensional earthquake action. Spat. Struct. **8**(2), 3–11 (2002)

Using Object-Oriented Programming Method to Realize the Control of PLC in the Raw Material Conveying System

Yan Qin(✉)

ShaZhou Professional Institute of Technology,
Zhangjiagang 215600, Jiangsu, People's Republic of China

Abstract. With the continuous development of society, the market competition is more and more fierce, so each production enterprise urgently needs to improve the production technology. In the enterprises that need to transport products, manual handling method has been used in the past, which leads to low efficiency and poor competitiveness. In view of the above problems, PLC technology is used to design the conveying system, which achieves good control effect in the process of raw material conveying. In this paper, the object-oriented programming method is used to realize the automatic control of PLC conveying system, which has the advantages of fast response, high accuracy and reliable performance.

Keywords: Object oriented programming method · PLC · Raw material transportation · Control system

1 Introduction

After decades of development, China's modern logistics has developed rapidly. With the advent of the Internet era and industrial automation, the dissemination of information, communication and control system automation has undergone tremendous changes. Information is the soul of logistics system. Internet technology promotes the information revolution. At present, commercialization of logistics, automation of logistics management, digitalization and coding of logistics information collection, electronization of logistics information processing, standardization and real-time of logistics information transmission, digitalization of logistics information storage have become the mainstream.

Using programmable logic controller (PLC), design low cost, high efficiency of raw material automatic conveying device. PLC as the main controller, combined with pneumatic device, sensing technology, position control technology, on-site control of automatic transportation of raw materials. The system has the characteristics of high degree of automation, stable operation, high precision and easy control. According to different objects, the system can be slightly modified to meet the requirements of different objects. Raw material transportation is the center of logistics. Since its birth,

B. J. Jansen et al. (Eds.): *International Conference on Cognitive based Information Processing and Applications (CIPA 2021)*, LNDECT 85, pp. 506–511, 2022.
https://doi.org/10.1007/978-981-16-5854-9_63

logistics science has shown great vitality and become one of the most active and influential new disciplines. Logistics science takes the dynamic flow process of goods as the main research object, and reveals the internal relationship between logistics activities (transportation, storage, packaging, loading and unloading, distribution, circulation and processing, logistics information, etc.) [1].

2 Object Oriented Programming

The basic starting point of object-oriented method is to simulate human's way of thinking as much as possible, and pursue to describe and solve problems in the real world naturally; It is based on the concepts of information concealment and abstract data type. In the traditional programming method, the separation of function abstraction and data abstraction leads to the inconsistency of software structure and problem structure, which makes it difficult to maintain and understand the software, Each object encapsulates a group of data and a group of procedures, so that this group of procedures can process this group of data, and when defining an object, it can specify the permission of the outside world to operate on it. This method is direct and natural, which enables designers to focus on the system level and pay less attention to the details. Object oriented programming method can be summarized as the following five basic concepts, which constitute the core of object-oriented programming language; Object, class, instance, method and message [2].

Object is the general name of the independent unit composed of data and the operations performed on it. Here, data is the instance variable and operation is the method. The problem and its elements can be abstracted by object. It encapsulates its data structure definition and function realization, realizes information hiding and data abstraction, and embodies modularization. A class is a description of a group of objects with the same data structure and operation. A class defines an object type and describes the properties of all objects belonging to the object type. A class is equivalent to the type in traditional programming languages. The difference is that a class describes the data structure and internal operation as a whole, which fully embodies the idea of abstract data type. To solve a problem with object-oriented method, it is necessary to determine which objects are involved in the problem. The commonness and individuality between objects are represented by the public and private members of a class, and the dependency of a class = inheritance. The inheritance of a class enables a subclass to inherit the characteristics and abilities of its parent class. Message transmission can occur between different objects (different instances) of the same class, or between instances (objects) of different classes. Only through message transmission can the corresponding actions in the class be activated, requiring an object to perform an operation defined in its class. An instance is an object described by a specific class. Because of the universality of the object-oriented method, the concept of class is a powerful tool in modern programming. The object-oriented design method can be divided into the following steps.

(1) Problem definition: ① object requirement analysis; ② object partition. It is required to divide the problem to be solved into object units and analyze the requirement of the whole object

(2) Object design and definition: ① object retrieval; ② object definition; Detailed design and definition.
(3) The connection of objects: ① message passing between objects, and i-System coordination. We should connect the various objects to form a harmonious whole.

3 Characteristics of PLC

3.1 High Reliability and Strong Anti-interference Ability

Some of the various types of PLC used in the automation system are centrally installed in the control room, and some are scattered in the single equipment on the production site. Although most of them are in the electromagnetic environment formed by the strong current circuit and strong current equipment, PLC is a control device specially designed for the industrial production environment, In the design and manufacturing process, multi-level anti-interference and component selection measures are adopted, so it has strong ability to adapt to poor industrial environment, stable operation and high reliability. Therefore, it can be directly used in industrial environment without taking any special measures [3].

3.2 Strong Versatility, Variable Control Program, Easy to Use

PLC has a wide range of hardware devices. It can form a control system that can meet various requirements. Users do not have to design and make their own hardware devices. After the hardware is determined, the user does not need to change the hardware of PLC when the production process is changed or the production equipment is updated. Only need to adapt the program to meet the requirements. Therefore, PLC is not only used in single machine control, but also widely used in factory automation.

3.3 Strong Function and Wide Adaptability

Modern PLC not only has the functions of logic operation, timing, counting, sequence control, but also has the functions of digital and analog input and output, power drive, communication, man-machine dialogue, self inspection, record display, etc. It can not only control a production machine, a production line, but also control a production process.

3.4 Programming is Simple and Easy to Master

At present, most PLCs still adopt "ladder diagram programming" in the form of relay control. It not only inherits the intuitive clearing of traditional control circuit, but also takes into account the drawing reading habits and programming level of most electrical technicians in factories and enterprises, so it is very easy to accept and master. The symbols and expressions of programming elements in ladder diagram language are quite close to the schematic diagram of relay control circuit. Through reading PLC user manual or short-term training, electrical technicians and technical function diagram, statement table and other programming languages [4].

4 Object Oriented Programming Method to Realize the Control of PLC in the Raw Material Conveying System

The main program steps of PLC control software are as follows:

4.1 High Speed Counter Initialization

In order to realize the conveying function more accurately, it is necessary to add a high-speed counter to count the pulse signal. After defining the relevant information of the high-speed counter, the problem to be considered is how to call the subroutine. First of all, we should consider that the high-speed counter must be effective at the beginning of program execution. Therefore, we use sm0.1 instruction to call subroutine at the beginning of program execution. Secondly, in the process of running the program, the high-speed counter needs to be called continuously and initialized. In order to make the program work more orderly, the high-speed counter needs to be called once after the conveyor belt runs for one process to reload the preset value and current value, so as to prevent the running confusion caused by the constant growth of the high-speed counter. The corresponding program ladder diagram is shown in Fig. 1.

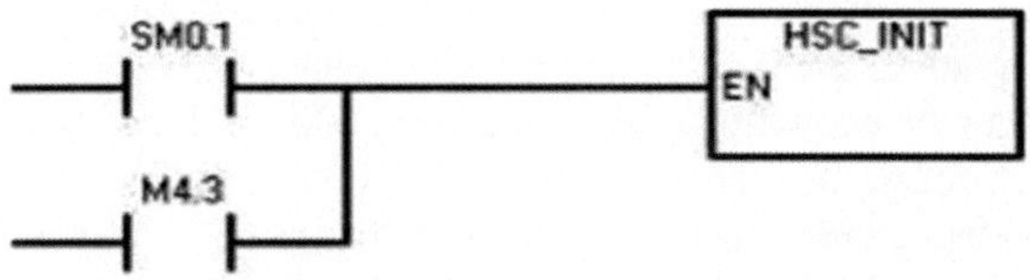

Fig. 1. Calling high speed counter ladder diagram

4.2 Operation and Stop of Conveyor Belt

In this system design, an important link to be considered is to control the operation and stop of the conveyor belt. First of all, the operation of the conveyor belt must be in the return state of the discharge cylinder, because when the material is ejected, it must be ensured that the material can enter the conveying link smoothly, so as to prevent falling from the conveyor belt during the conveying process and affecting the sorting effect. Secondly, if there is no material in the material tank, then in order to avoid waste of resources, the conveyor belt will stop automatically after the corresponding process. When there is an accident in the sorting system, in order to avoid greater losses, the conveyor belt should be stopped immediately to facilitate timely maintenance. The ladder diagram of the corresponding program is shown in Fig. 2.

4.3 Setting Gear and Process

In this design, we first use tape measure and vernier caliper to measure the radius of the shaft and the distance from the discharge point to the last cylinder. The measured radius

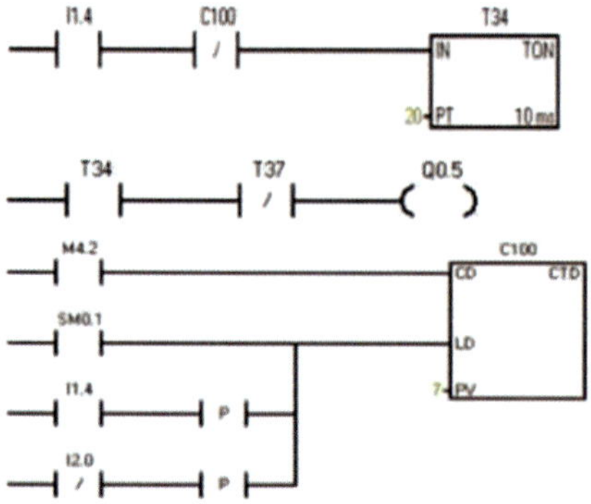

Fig. 2. Conveyor control ladder diagram

is about 22.85mm, the diameter is about 45.7mm, the perimeter is about 143.5mm, and the distance from the discharge port to the last cylinder is about 518mm. Considering that the material must pass through all cylinders and cannot slide out of the conveyor belt, the material must travel a moderate distance on the conveyor belt. At the same time, in order to achieve this function, it is necessary to measure the distance between two adjacent pulses detected by the encoder, and the distance between two adjacent pulses is about 1.4mm. According to the calculation, it takes about 370 pulses from the outlet to the last cylinder. As the resolution of the encoder is 100p/R, it needs to be divided into several gears to complete this function. According to the analysis, the material should pass through the last cylinder position without sliding out of the track. Therefore, 385 pulses are selected as the walking distance of the material. Within the range of 0 to 100, the first integer that can divide 385 is 55. Therefore, 55 pulses are selected as a gear, so that 7 such gears are needed to complete an operation process to realize the above functions, The corresponding program ladder diagram is shown in Fig. 3.

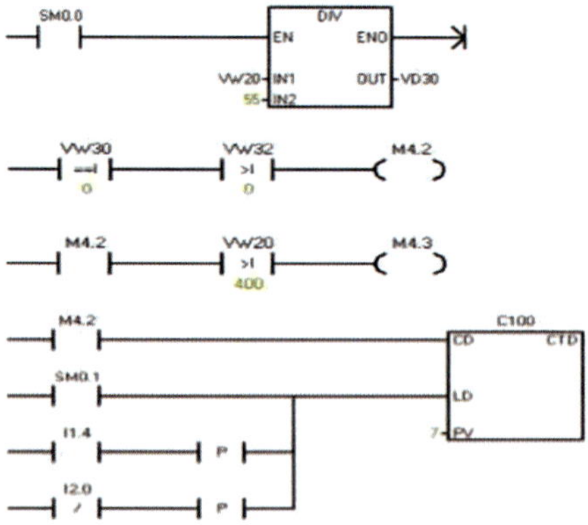

Fig. 3. Ladder diagram of gear design

5 Conclusion

The automatic conveying system of raw materials adopts the object-oriented programming method to realize the PLC control, which is monitored by the rainbow control configuration software, and the running picture is displayed by the upper computer. The raw materials are identified by sensors on the conveyor belt and classified. Then through the conveyor belt transportation, this recognition technology has the characteristics of fast response, accurate and reliable, simple operation, good control effect, stable operation and so on. It has three main features: it can transport raw materials continuously and in large quantities, and basically realizes unmanned operation. The use of personnel in the automatic conveying system is only limited to the receiving of goods, system control, system operation, management and maintenance, which just meets the requirements of enterprises to reduce the use of personnel, reduce the labor intensity of employees, and improve the use efficiency of personnel. Therefore, the market prospect is broad and the application scope is wide.

References

1. Yang, C.: Fundamentals and applications of PLC. Chongqing: Chongqing University Press 200
2. XueGuo, W.: Programmable Controller. Henan Mechanical and electronic publishing house, Henan (2001)
3. Goffrey, A.: Pascoe, computer science of object oriented programming, (6), 39–43 (1988)
4. Liao Chang Chu: PLC Programming and Application, p. 9. China Machine Press, Beijing (2002)

Analysis of Configurable Human Resource Management System Based on Cloud Computing

Wei Wu(✉)

Sichuan University Jinjiang College, Meishan 620860, Sichuan, China

Abstract. With the continuous expansion of the scale of enterprises and the increasingly fierce market competition, how to give full play to the core resources of enterprises, the efficiency of one person has become the key to improve the efficiency of enterprises. A human resource management system suitable for the development mode of enterprises can not only effectively avoid the defects of manual management mode, such as low efficiency, strong randomness and many human factors, but also have a good understanding of people's thinking and thinking In order to achieve the goal of the enterprise, we should properly guide, control and coordinate the psychology and behavior, give full play to people's subjective initiative, make the best use of people's talents, make things suitable for people and personnel. Based on the analysis of the function modules of human resource management system under cloud computing, this paper analyzes the establishment and operation of the database of human resource management system, hoping to provide reference for relevant people.

Keywords: Human resource management · Cloud computing · Cloud Security

1 Introduction

With the wide application of information technology, especially the Internet, information technology represented by management information system and e-commerce is playing an increasingly important role in enterprise management. At present, MRP, ERP, CRM and other management systems, CAD, cam and other professional technical systems, EDI, etc. EC is a representative of e-commerce system in enterprises, especially in developed countries, are widely used. Because of the dual role of business competition and technological progress, enterprises rely more and more on information technology, and become one of the indispensable conditions for their survival and development.

On the other hand, among many enterprise resources, the importance of human resources is undoubtedly not doubt. Even some scholars call human resources "human capital", and put it on the technology and capital. Especially, the human resources represented by knowledge workers will become the main force of economic development. Based on this background, human resource management will become an important part of enterprise management, and human resource management system (HRMS) has become

B. J. Jansen et al. (Eds.): *International Conference on Cognitive based Information Processing and Applications (CIPA 2021)*, LNDECT 85, pp. 512–517, 2022.
https://doi.org/10.1007/978-981-16-5854-9_64

a very important part of many enterprise management. As a computer tool of human resource management, human resource management system can manage almost all information related to one of the most important assets in an enterprise from recruitment, post and organization setting, training, skills, salary and welfare, performance to resignation management, The problems of information incompatibility, updating and sharing caused by the discrete storage of human resource related data are avoided effectively, which makes the management of human resources in enterprises go to standardization, scientificalness, digitization and networking [1].

2 Introduction to Human Resource Management System

Human resource management system is a complete solution of human resource management in enterprises realized by information technology. It is a new human resource management mode based on advanced software and high-speed and large capacity hardware. Through centralized information base, automatic processing information, self-service and service sharing, the system can reduce costs, improve efficiency The purpose of improving the staff service mode. It can ensure the synchronous development of human resources and the changing technological environment by connecting with the existing network technology of the enterprise.

2.1 Analysis of the Current Situation of Human Resources Management System

There are many software in human resource management system at home and abroad, and the competition is fierce. There are HRMS of UFIDA, e-HR of Kingdee, jinyikang human resource management information system of jinyikang company, Huizhi human resource management system of Shanghai radio and television software information Co., Ltd., Jiayang human resource management system of Shanghai Jiayang Information System Co., Ltd., and human resource management system of huilinke software (Shenzhen) Co., Ltd. Foreign companies have Oracle human resources management system, SAP human resources management system, PeopleSoft launched the global human resources management system, etc. Generally speaking, domestic manufacturers such as jinyikang have a high degree of standardization in China's personnel information management, while Kingdee's k/3 human resource system has the advantages in overall planning, integration with other application systems, humanized design, and based on.Net development. The human resource management system of foreign manufacturers is simple to operate, friendly in interface, clear business process and standardized in management. Because the management system of foreign companies is different from that of our country, many foreign human resource management software is not suitable for our country [2].

2.2 Characteristics of Modern Human Resource Management System

The modern human resource management system should be based on the inter / Intranet. Inter / Intranet makes human resource management system break through the closed state, information flow breaks through the limitation of department, and makes human resource management system an open and interactive technology platform. Modern human resource management system emphasizes the self-service of employees and managers. Employee self-service is a real-time and efficient data access mode. Employees can browse, establish and maintain their own data in human resource management through web technology or interactive voice response technology. To benefit all from human resources management. If the personal information of an employee changes, he or she can update his or her information and the approved procedure will take effect. Similarly, the daily administrative affairs such as training, vacation application and reimbursement can be handled similarly. This not only reduces the workload of human resource management personnel used for data collection, confirmation and update, but also ensures the quality of data and the speed of data update. And because the Internet is not limited by time and geographical location, even if the manager is far away from abroad, he can handle all kinds of applications of his employees in time, and will not affect his work because he is not in the company. And can effectively improve the internal communication channels of the company [3].

3 Function Modules of Human Resource Management System Under Cloud Computing

The human resource management system based on cloud computing technology includes eight parts.

First, training management. This module is to train enterprise employees, including on-the-job training, pre job training, academic education and so on, and establish targeted training files. The contents of the file include training category, name, training time, training method, training cost, training place, training personnel, training unit and assessment results.

Second, personnel information management. The module is to maintain and display the employee information of the whole enterprise, including query, input, modification and other functions. The specific information includes employee's gender, home address, e-mail, name, nationality, contact number, specialty, employee type, date of birth, native place and contact number.

Third, attendance management. The module includes overtime record and daily attendance management. Daily attendance includes addition and modification. Query and delete the attendance information of each employee, and the attendance management personnel can report the company's operation status according to the attendance of each department, which is conducive to improving the work enthusiasm of employees.

Fourth, file management. This module mainly manages the information of each person's position and resignation, including the functions of querying, inputting, modifying and deleting employee information. The main information includes file number, learning experience, employee basic information, training process, performance information and attendance information.

Fifth, performance evaluation management. This module mainly carries on the performance appraisal to the entire enterprise staff, implements according to the quarter, and establishes the performance appraisal form combining with the rewards given to the staff, including the name of the appraiser, each part of the members, the appraisal standard, the appraisal time, the appraisal result and the specific appraisal content.

Sixth, contract management. This module is to query, input and modify all contract information of employees. Relevant departments record employee contract information, including contract renewal time, contract type, employee name and probation salary. Contract period and termination time.

Seventh, salary and welfare management. This module mainly manages the employee's welfare and salary, including overtime salary, basic welfare, basic salary, payment time and reward salary, etc., and realizes the modification, query and deletion of information.

Eighth, system management. This module is mainly used to add, delete, modify and check employee information, including user management, system data and department management. The information is saved in the form of a form, and needs to be backed up regularly to provide the security of system data. The whole process of software system construction is shown in Fig. 1.

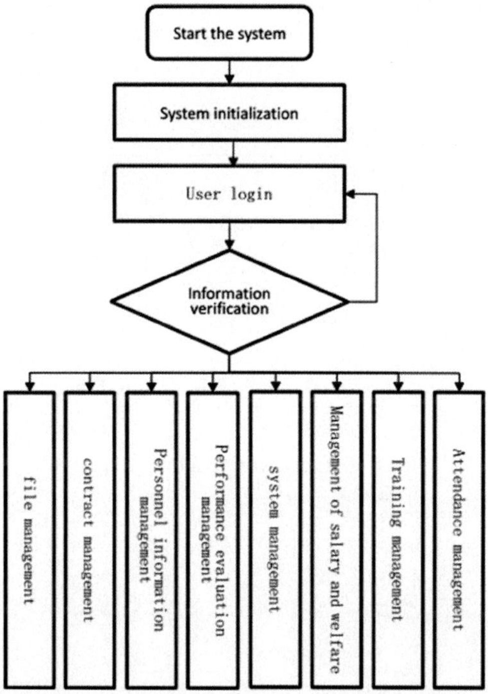

Fig. 1. Software system construction process

4 Database Establishment of Human Resource Management System

4.1 Database Analysis

General web applications need to save information in time, so as to synchronize with the request. At present, the most common use of database is to store the database. Multiple records are formed into a table, and the index is used to search the data. You can query according to the association or table, and process multiple records as a whole. Data storage system includes relational database, file system, object database and so on. Each kind of database has its own advantages and disadvantages. And select the appropriate database according to the application access program [4].

Gae data storage system is datastore. It is built on the basis of big table. It is similar to object database. It can query and has high reliability. The storage object of datastore is entity, which is similar to relational database table, but it is very different from previous relational database. The relationship between objects is established by using gae program, and data query is carried out according to index, so as to realize database function. The storage database is suitable for web system, as shown in Table 1.

Table 1. Comparison between relational data and data storage

Comparison content	Data storage	Relational database
SQL support	Only basic queries are supported	All support it
Affair	Execute on only one entity	Support
Main structure	Arrangement	Relationship
Expansibility	Pretty good	It needs a lot of modification and is difficult
Average execution speed	Less than 200 ms	Small confession 100 ms
Indexes	Some of them are created by themselves	Create manually

4.2 Data Protection

For the data in the human resource management system, generally the data stored in the cloud are more important. These data are protected by the cloud service provider. To ensure the security of data, the first thing is to use encryption technology, especially a large amount of data in cloud computing. There are many encryption methods, mainly including DES, AES, MD5, etc. DES is suitable for use here.

4.3 Access Control System

The system analyzed in this paper is a multi-user system, which can improve the security of the system. Users need to use user identity to access and other operations. At the same time, this paper uses servlet filter to deal with the request of each department, the user identity is different, the access folder is also different. Servlet filter is mainly responsible for the interface of java.servlet.filter program class. It can process requests in JSP and servlet modules. It performs corresponding operations according to these two pages and intercepts illegal requests. The servlet filter completes the permission verification through session. When accessing the web program, the system will create a session object according to the user's request and save the user's related information state until the user disappears from the session in the web point. To establish a session object is to obtain the user information through the servlet filter, so as to judge whether the operation is legal or not.

5 Conclusion

The emergence of cloud computing technology has greatly reduced the degree of segmentation in the past it field. It can directly face Internet users and integrate the segmentation fields, so as to provide targeted services for users. This paper mainly analyzes the human resource management system based on gae cloud platform, and introduces the method of data store and operation data of CAE data storage area based on big table, as well as the establishment of cloud platform environment, so as to realize data access and data protection.

References

1. Caiyun, Y., Shiyu, L.: Analysis and design of enterprise human resource management system. Comput. Appl. Res. **2**, 186–188 (2004)
2. Wang, F, Zhang, J., He, W.J., et al.: Development of human resource management system based on Intermet **203**(9), 95–97.
3. Lian, K.: Enterprise informatization and human resource management informatization Shandong electronic **200**(3), 1–3
4. Weiguo, M.: How to choose HR management software in the Internet age. China Hum. Res. Dev. **1**, 46–47 (2002)

Research on Forest Fire Recognition Based on Computer Vision Technology

Kai Yao, Jinglan Zhang(✉), and Jiuqing Liu

College of Mechanical and Electrical Engineering, Northeast Forestry University, Harbin 150040, China

Abstract. In order to solve the problems of high false alarm rate and poor real-time performance in forest fire monitoring system due to the large amount of sample data and multi dimension of forest fire behavior monitoring, a support vector machine algorithm based on computer vision was proposed for forest fire monitoring to improve the recognition accuracy and realize all-weather forest fire automatic monitoring and early warning. Firstly, the image is obtained for pre-processing, and whether there are pyrotechnic areas in the image is preliminarily identified; Then, the feature extraction of forest fire is carried out, the feature vector is generated by training samples, and the SVM algorithm based on radial basis kernel function and polynomial kernel function is used to identify the fire-works, which puts forward a new idea for the intelligent exploration of forest fire prevention equipment.

Keywords: Computer vision technology · Forest fire · Fire identification

1 Introduction

Forest fire is a serious natural disaster in nature, which does great harm to the ecological environment and poses a great threat to human property and personal safety. Forest fire monitoring and prevention is an effective means to reduce the incidence of disasters. The early monitoring methods are mainly manual patrol and staring tower. Although the effect is obvious, it costs manpower and material resources. With the development of science and technology, more and more new technologies are gradually applied in the field of forest fire monitoring, including satellite monitoring, infrared image recognition monitoring, fire detector monitoring and so on. However, due to the constraints of climate, technical conditions, monitoring costs and other factors, satellite monitoring can not be used as a real-time and reliable means of forest fire monitoring and early warning; The infrared image monitoring is mainly aimed at the open fire monitoring, and the smoke is earlier than the open fire in the early stage of forest fire, so the use of infrared image monitoring leads to the delay of the best time to put out the fire; Although smoke and temperature detectors have a good effect on the detection of buildings, cabins and other small space, the early forest fire signals have the characteristics of randomness and non structure. The forest environment is complex and changeable, and the interference of air

B. J. Jansen et al. (Eds.): *International Conference on Cognitive based Information Processing and Applications (CIPA 2021)*, LNDECT 85, pp. 518–523, 2022.
https://doi.org/10.1007/978-981-16-5854-9_65

flow, water droplets and electromagnetic field will also produce similar characteristics, which easily leads to false alarm and missing report of fire detectors [1].

In order to solve the above problems, reduce the monitoring cost, improve the monitoring accuracy, and meet the needs of real-time and efficient forest fire monitoring. In the event of fire, it can quickly identify and locate the fire point. According to the principle of "fight early, fight small, fight", this paper based on computer vision forest fire identification analysis and recognition research, can effectively reduce the forest fire risk level and bring significant social and economic benefits.

2 Computer Vision Technology

Computer vision is an interdisciplinary and interdisciplinary field, and its range includes artificial intelligence, signal processing, statistics, computational mathematics, information theory and cognitive science. Through the detection, recognition and analysis of digital images or videos, the computer has high-level understanding and decision-making ability for the scene information captured by the camera. From the perspective of technology level, computer vision can be divided into low-level vision and high-level vision. Low level vision includes image or video shooting, signal preprocessing, segmentation and so on; high level vision includes scene information detection, recognition, tracking and so on.

The core task of computer vision is to extract the features of the image captured in the actual scene, get the specific value or symbol information, and then make further detection or recognition according to these values or symbols, and express the results in the form of decision-making. Therefore, it can be considered that computer vision technology uses geometry, physics, statistics and learning theory to build models, and then decomposes the discrimination information from the image data. The image here is generalized, and its data sources include photos, videos, medical images, multi-modal images taken by UV-IR camera and so on. The sub fields of computer vision include scene reconstruction, event detection, video tracking, target recognition, 3D pose estimation, motion estimation and image restoration, etc. [2].

3 Image Preprocessing

3.1 Image Acquisition

The image is extracted from the video captured by the high-resolution industrial CCD (charge coupled device) camera. According to the requirements of the video decoder, the video image is captured in 0.5 s step, and the acquired image is saved in the memory of the industrial control computer. Because the industrial control computer has a good video capture interface, strong anti-interference ability, and the system is stable and reliable, it can meet the requirements of accurate and rapid recognition of fireworks images in the system.

The height and width information of the acquired image and the RCB (red green blue) image of the original image are extracted for image color pre recognition to generate a binary image, and the image edge is smoothed by the second difference operation of the binary image. Then, through the median filtering of the binary image, the fireworks region and the region boundary can be distinguished [3].

3.2 Image Color Pre Recognition

Scan the acquired image, obtain the RGB value of any point, binarize the image, and obtain the RGB space component maps. This paper gives two groups of RGB space component maps of smoke and flame with better recognition effect of test samples, as shown in Fig. 1, which are used for pre recognition of fireworks based on color detection.

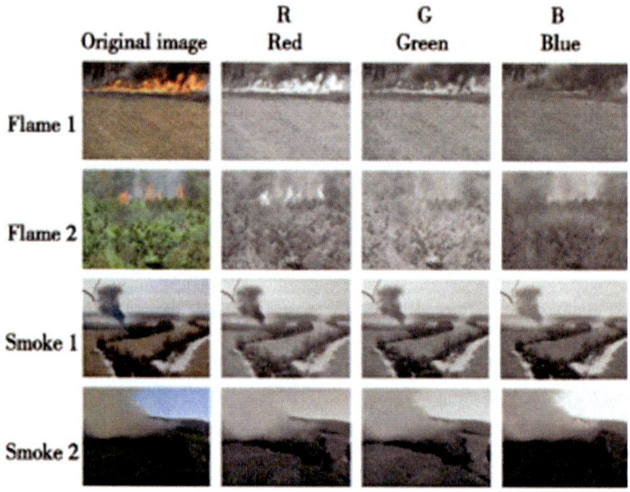

Fig. 1. Schematic diagram of RGB spatial components

3.3 Image Smoothing

The binary difference operation is performed on the image generated by color pre recognition, and the image is scanned from top to bottom and from left to right to determine whether any point *P (x, y)* in the image is a pyrotechnic area. If point *P (x, y)* is a non pyrotechnic area (mark bit 0), then the upper and lower six adjacent points are judged. If the adjacent points are all fireworks area (marked as 1), Mark the point as 1. Then, the second scan is carried out according to the same principle in the order of bottom up, right to left, so as to achieve image smoothing.

3.4 Median Filtering

After image smoothing, the image is scanned from top to bottom and left to right again to determine whether point *P (x, y)* is a pyrotechnic area (marked as 1). If the pyrotechnic area is used, the point marked as 1 is found on the left and right. If no point marked 1 is found in the 3 points, the area is not closed, the area is deleted and the tracking is finished. If it is judged to be area closed, mark the area boundary as 1 end trace. In order to improve the confidence of the processing results, the closed region is compared with the given threshold value, and the area below the threshold is deleted and the median filter is completed.

Although the false report rate of the area of fireworks is very low by color pretreatment in practical application, many objects (such as trees and grass) in forest area will show the color of flame and lead to high false alarm rate due to the reflection of sunlight. Therefore, it is necessary to further identify the other characteristics of fireworks.

4 Image Feature Extraction

The typical phenomenon in the process of forest fire is the continuous smoke and flame until it goes out. Smoke is accompanied by the whole process of fire, its movement is from bottom to bottom, spread with wind or irregular diffusion. The flame is produced later than smoke, and the range of purple to red changes in visible light range, while its spreading speed is slightly slower than that of smoke, which is usually spread in a radial way, but the overall situation is irregular. In addition, the complex natural environment of forest also has some influence on the visual perception of fire smoke and flame. Firstly, forest dense planting has a certain visual shielding effect on the early flame, secondly, the monitoring time line will have a certain impact on the identification of flame, and it is difficult to find the fire Source Flame earlier.

Aiming at the complexity of forest fire behavior, this paper mainly aims at the visual characteristics of forest fire smoke and flame, overcomes the influence of light and shadow changes, weather changes and various interference targets in complex background, extracts and recognizes the texture and dynamic characteristics of forest fire to improve the recognition accuracy and recognition efficiency of forest fire [4].

4.1 Texture Feature Extraction

Texture feature is also a global feature, which also describes the surface properties of the scene corresponding to the image or image area. Unlike color features, texture features are not based on pixel points, and they need to be calculated in the regions containing multiple pixel points. In computer vision, this kind of regional feature has great superiority, and it can not be recognized successfully due to local deviation. As a statistical feature, texture features are always rotation invariable and have strong resistance to noise.

4.2 Dynamic Feature Extraction

Based on the non directionality and randomness of the flame and smoke, the movement of the highlight object that interferes with the recognition of fireworks is usually directional and regular. According to the characteristics of fireworks, dynamic features can be extracted to distinguish most of the highlight interference sources.

The former frame of fireworks brightness image is defined as $I_1(x,y)$, and the latter frame of fireworks brightness image is defined as $I_2(x,y)$. $d_{max}(x,y)$ represents the relative movement of I_2 to I_1. By calculating each frame of the image, we can get the moving center point sequence $(X d_1, Y d_1)$, $(X d_2, Y d_2)$,... Because of the randomness and non directionality of the fireworks, we can recognize the image fireworks area through the moving center point sequence morphology.

5 Image Feature Recognition

Because of the particularity of the natural environment, forest fire is difficult to be captured in time, so using the above features and according to the forest environment, climate and vegetation type to identify forest fireworks, using industrial CCD camera to collect video data, through the characteristic value of smoke and flame constitute a feature vector for feature extraction. After judging the pyrotechnic area preliminarily identified by the system, if it is a real pyrotechnic area, it is divided into positive samples, and the rest of the misjudged pyrotechnic areas are taken as negative samples, and the feature vectors and sample types of these samples are stored in the sample training database.

Support vector machine (SVM) algorithm is used to solve the classification problem, which is similar to neural network in function form. Its output is a linear combination of the sum function of input samples and support vector corresponding to some intermediate nodes, as shown in Fig. 2.

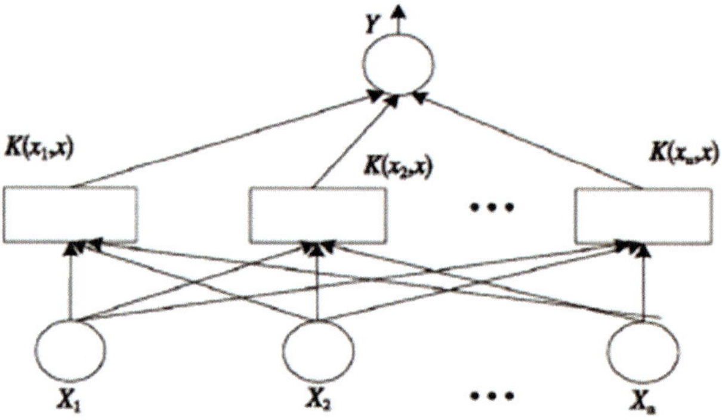

Fig. 2. Support vector machine schematic

In Fig. 2, the input vector is $X = (X_1, X_2, \ldots, X_N)$, and the output based on the nonlinear variation of n support vector machine vectors $x_1, x_2, \ldots, x_N$ is as follows:

$$Y = \mathrm{sgn}\left[\sum_{i=1}^{n} \alpha_i y_i K(x_i, x) + b\right] \tag{1}$$

Where $K(x_i, x)$ is the kernel function; α_i, y_i is the weight.

The radial basis kernel function is used to identify the smoke

$$K(x_i, x_j) = \exp\left\{\frac{|x_i - x_j|^2}{\sigma^2}\right\} \tag{2}$$

The polynomial kernel function is used to identify the flame

$$K(x_i, x_j) = [(x_i . x_j) + 1]^q \tag{3}$$

In order to reduce the response time of the system, before the implementation of classification and recognition, the data are saved in the database, and the support vector is normalized (range [−1,1]). The support vector machine specially designed for forest fire behavior is used to process two groups of support vectors (smoke and flame) according to the actual environment.

6 Conclusion

This paper presents an algorithm for forest fire identification based on computer vision and comprehensive feature recognition. Firstly, the image is preprocessed, and the special characteristics of forest fire behavior and forest environment are taken into account. Then, SVM algorithm, which takes radial basis function and polynomial function as kernel function, is used to classify and identify forest fire behavior. It can effectively warn forest fire risk and has a good application prospect in forest fire protection field.

References

1. Xueying, D., Chang, L., Jian, S., et al.: Study on forest fire loss assessment technology. For. Eng. **31**(2), 42–45 (2015)
2. Zhang, Z.: Overview of image target recognition methods based on machine vision [J]. Science and technology and innovation (14) (2016)
3. Xiaowen, G.: Research and Implementation of Forest Fire Detection Technology based on Forehead Sequence. Donghua University, Shanghai (2017)
4. Liu, K., Shu, L., Yu, Z., et al.: Application and discussion of forest fire prevention video monitoring technology. For. Mach. Woodworking Equip. **46**(2), 10–12 (2018)

Application of Decision Tree Algorithm in Mental Health Evaluation

Guo Ping(✉)

People's Public Security University of China, Xicheng District, Beijing 100038, China

Abstract. In recent years, a variety of social problems caused by college students' mental diseases occur frequently. The introduction of data mining technology into the diagnosis of College Students' mental health has incomparable advantages over other technologies. It can mine the hidden laws in things. This paper expounds the principle of C4.5 algorithm of decision tree, constructs decision tree by data preprocessing, and predicts mental health by extracting rules. The experimental results show that the algorithm can classify students' mental health correctly. The mining results can guide mental health educators to make correct counseling plan, which is helpful for decision-making.

Keywords: Decision tree · Mental health · C4.5 algorithm

1 Introduction

College Students' mental health is not a simple mental disorder. Mental health can improve their ability and make them easier to achieve their goals in life. The mental health of college students should be the focus of attention, not only those with mental disorders. Mental health is not only a small and isolated problem, it affects the whole society and is a great challenge of global development. No group has immunity to mental health diseases, but the risk of mental health diseases is greater among the poor, the unemployed, the people with low educational background, and today's college students. Mental, physical and social health are closely intertwined, which is an important part of all individual life. In recent years, the use of data mining technology to explore college students' mental health and psychological crisis intervention mode has become a hot spot of domestic and foreign scholars. Data mining technology is used to mine the relationship between the factors that lead to students' psychological problems from the survey data, and the main factors that affect students' psychological problems, so as to provide the basis for the school to carry out the planning and decision-making of mental health education, and further explore the new methods of early prevention and intervention of students' psychological obstacles, so as to make the school's psychological counseling work more targeted, Improve the level and efficiency of psychological counseling. Foreign research on data mining technology has achieved fruitful results, and there are more successful application cases in the retail industry, financial insurance and medical service industry. Many software companies have developed data mining software products, which have

B. J. Jansen et al. (Eds.): *International Conference on Cognitive based Information Processing and Applications (CIPA 2021)*, LNDECT 85, pp. 524–529, 2022.
https://doi.org/10.1007/978-981-16-5854-9_66

been applied in North America, Europe and other countries. For example, quest and ileligent miner developed by IBM, knowledge seeker based on rules and decision tree developed by angoss software, advanced software, etc. Domestic research on data mining started late, and the application of data mining technology to assist business activities is still in its infancy. There are few successful cases, and the research on the application of data mining technology to college students' mental health is rarely involved [1].

2 The Concept of Decision Tree

From the perspective of personnel training, relying on MOOC MOOC to build micro specialty has obvious advantages. First, we can make full use of the existing MOOCS resources, gradually shift the focus of online course construction from "construction" to "use", and show the gold content of the course in the process of use. At the same time, promote the deep integration between universities and enterprises, and promote the combination of industry, University and research. As shown in Fig. 1. The second is to promote the sustainable development of MOOC through the use of users. Online open courses need continuous investment of funds and personnel [2]. Yuan Songhe and others believe that in addition to the incentive of education policy, China's open education can only rely on mature business operation mode to support and maintain its sustainable development.

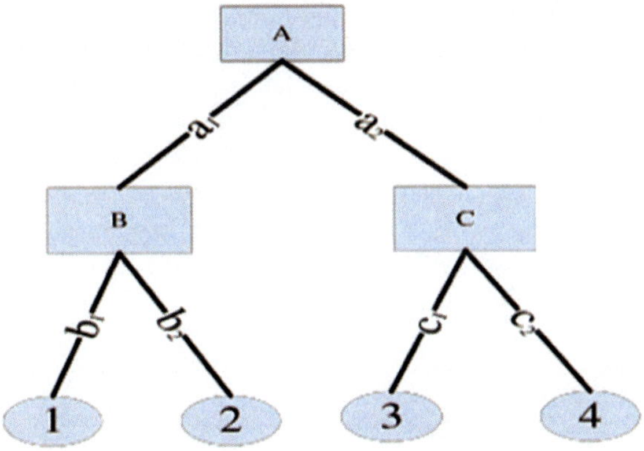

Fig. 1. Simple strategy tree

As shown in Fig. 1, a simple decision tree 3, in which nodes a, B and C represent the subset of the sample data set respectively, corresponding to the set of attributes in the data set, each branch represents the value of an attribute in the current subset of the sample set, a1, a2, b1, b2, c1 and c2 represent the value of split attributes, and leaf nodes 1, 2, 3 and c2 represent the value of split attributes Four represent the types of decision attributes in each sample set.

3 Common Algorithms of Decision Tree

3.1 ID3 Algorithm

The basic strategy of ID3 algorithm is as follows:

(1) The tree starts from a single node to represent the training sample. It has no in edges but zero or more out edges;
(2) If the samples are all in the same class, the node becomes a leaf node and is labeled as the class, that is, an internal node, which has exactly one human edge and two or more outgoing edges;
(3) If the value in the test attribute is known, the correct branch is established and the samples are divided according to it;
(4) use the same process, top-down recursion, until all the samples of a given node belong to the same class, and there are no residual attributes that can be used to divide the samples and branches. When none of the three conditions of samples is true, the recursion will stop. During the construction, the part of the online courses of mental health service micro specialty played a positive role. During the anti epidemic period, nearly 20000 people took courses in the MOOC platform of Chinese University, such as introduction to mental health education, psychological consulting theory and technology, communication and popularization of mental health, and practice of mental health education, etc., to disseminate and popularize mental health knowledge and improve people's mental health literacy and psychological protection skills. At the same time, the advantages of these online teaching are fully exerted, which make up for the training needs of the existing mental health service talents, and combined with the "demand of psychological anti epidemic", the professional development ability of learners has been rapidly improved. In this epidemic resistance, the advantages of the micro specialty of mental health service have been initially played. In addition, the 10 micro specialty MOOCS, as the online resources of some professional courses, implemented the hybrid teaching mode of MOOCS and SPOC, which promoted the students' deep learning and strongly supported the construction and development of applied psychology.

3.2 C4.5 Algorithm

The most popular decision tree algorithm is C4.5 of J. R. Quinlan, which is Quinlan's improved version of ID3. C4.5 chooses the attributes to construct the decision tree according to the information gain rate of the attributes, which overcomes the problem of too many branches of the decision tree in ID3 algorithm to a certain extent. Information gain rate is equal to the ratio of information gain to split information.

$$GainRatio(A) = Gain(A)/SplitInfo(A)$$

Where *splitinfo (A)* is the "split information" used to normalize the information gain, that is, the entropy of the training set S with respect to each value of attribute A.

$$\text{SplitInfo(A)} = -\sum_{i=1}^{m} \frac{|S_j|}{|S|} \times \log_2 \frac{|S_j|}{|S|} \tag{1}$$

3.3 C5.0 algorithm

C5.0 is extended on the basis of C4.5. It belongs to one of the classical decision tree models, and it is also an algorithm in clementinell2.0 decision tree model. The decision tree generated by it has many branches, and the target variable is classified variable. C5.0 algorithm can not only generate decision tree but also generate inference rule set. Generally speaking, the C5.0 model is divided into samples depending on the attributes that can bring maximum information gain. C5.0 algorithm can process data incomplete, mark unsuitable attributes, keep the data intact, and also handle a variety of data types, such as date, times, timestamps, etc., the data processing speed is faster and memory usage performance is greatly improved. Because of the promotion method, the decision tree generated is smaller and has higher classification accuracy [3]. The advantages of the system are as follows: it is very robust in the face of data omission and many input fields; It is not usually necessary to estimate the number of training; C5.0 model is easier to understand than one or some other models, and the rules introduced by the model have a very intuitive explanation; Provide powerful technology to improve classification accuracy. But C5.0 algorithm is difficult to predict the continuity field.

4 Application of Decision Tree Algorithm in College Students' Psychological Data

First, the training sample set is selected. The data of this paper is from the mental health scale of 2108 college students after entering the school. The symptom Self-evaluation Scale SCL-90, the most famous mental health test scale in the world, is used to test the data. The students covered nearly 20 major in science, engineering, literature, body and art, among which 856 boys and 1252 girls were girls.

4.1 Data selection and Preprocessing

According to the results of Chinese norm, if the total score of SCL90 is less than 160, it is divided into "Health (a)"; More than 160 points and less than 200 points are divided into "further inspection (b)"; If the score is more than 200, it is equal to 250 and it is classified as "obvious (c)"; Greater than 250 is classified as "more severe" and the four criteria are classified as attributes. The score table of each factor in the training sample set of decision tree (as shown in Table 1).

Table 1. Score table of decision tree training sample set

Number	Somatization	Obsessive compulsive symptoms	Interpersonal sensitivity	Depressed	Paranoia
1	1.56	1.15	1.70	2.00	1.14
2	1.56	1.38	1.60	1.67	1.57
3	1.57	1.46	1.20	1.33	1.00

The most important step of data mining is data preprocessing. In order to prepare the effective data set for modeling, this step is the most time-consuming process in mining. Data preprocessing is divided into sub tasks of data selection, data cleaning, data synthesis and data format. Because of the large number of data selected, there are clutter, repeatability and incomplete, so the data must be preprocessed to ensure the quality of data. When selecting the attributes that affect psychological symptoms, the data table is analyzed and the data is complete and special. For the attribute with continuous value in the table, if there is a missing value, fill in the average value of the attribute value; If the value of the attribute is discrete, the value of the attribute with the highest frequency is found and the missing value is filled with this value [4].

4.2 The Development of Railway Construction Industry Needs High-Quality Psychological Service

For example, to guide railway construction personnel to overcome their panic in the construction area, to pay attention to the impact of the psychological health of the drivers and passengers on passenger transport and its service quality, and to implement talent assessment for employees to promote personnel post matching. Southwest Jiaotong University has distinctive characteristics of rail transit, prominent highlights of mental health education, and good construction foundation. Micro specialty connects with the development of rail transit industry, focuses on the development of mental health education and service industry, and cultivates mental health service talents with sound personality, virtue and talent, and all-round development in accordance with the requirements of four-dimensional innovative talents cultivation system of value shaping, personality cultivation, ability cultivation and knowledge exploration. Through a series of MOC learning of mental health service, learners can not only master the basic theory and skills of mental health service, but also meet the requirements of education and teaching, psychological education activities, consultation service and prevention intervention ^ At the same time, it can also be competent in the planning, construction and management of psychological services in the field of rail transit.

5 Conclusion

Using decision tree method to classify the mental health status of college students, the classification accuracy is high. The method has good application value in predicting the mental health status of college students. The classification rules can be used as the basis for the school to carry out mental health counseling. Data mining is feasible in the research of College students' mental problems. However, how to prune the decision tree generated by C4.5 algorithm remains to be further studied, and the information gain rate function also has defects. If the information entropy is very small, the information gain rate will be unstable.

References

1. Li, D., Guo, M., Duo, Q., Yu, B.: Research and improvement of decision tree pruning algorithm. Comput. Eng. **31**(8), 19–21 (2005)

2. Zhang, M.: Handbook of psychiatric assessment scale. Changsha: Hunan Sci. Technol. Press, 17–19 (1998)
3. Li, Q., Di, W., Lou, X.: Two algorithms for improving the performance of decision tree. Comput. Eng. Des. **29**(15), 3989–3991 (2008)
4. Xiyu, J., Qiuming, H., Wei, L.: Examples of Data Mining Technology. China Machine Press, Beijing (2009)

Research on the Innovation and Development of Data Mining Algorithm

Na Wei(✉)

Zhengzhou Normal University, Academy of Fine Arts Henan, Zhengzhou 450044, China

Abstract. The era of big data poses a severe challenge to education. In order to keep up with the development of the times, future education must go with the times. Schools in the 21st century should teach students to master and use 21st century skills to understand and solve the challenges of the real world. This paper studies the innovation and development of art teacher education under the data mining algorithm. The Art Literacy with vision, modeling, space, creativity and aesthetic feeling as the core corresponds to the 21st century skills. Through school art education, every teenager can obtain the essential core literacy for survival and communication in a highly complex world - Art Literacy, Through school art education, we can help young people master 21st century skills to base themselves on life and make positive contributions to their community and society.

Keywords: Data mining algorithm · Art education · Innovation-driven development

1 Introduction

With the progress of society and the development of science and technology, people's lives are increasingly networked and digitized. Information management system has replaced the traditional way of text recording, followed by the linear growth of data recording the trajectory of human daily behavior. In education, especially in school education, students' behavior data has increasingly become the most significant indicator of teaching improvement. As a technology of mining potential information from massive user learning behavior data, educational data mining (EDM) has been widely used in scientific research, business, finance and other fields. With the deepening of information construction in Colleges and universities, universities have a higher demand in the use of information gathered and accumulated student behavior data. At the same time, education informatization itself has become a necessary measure and means of education management in Colleges and universities [1].

China's art education curriculum standard has clearly proposed to integrate new media technology into classroom teaching. In the compulsory education art curriculum standard (2011 Edition), it is proposed that "according to students' learning needs, carry out computer and network art teaching, encourage them to actively retrieve art information, use digital cameras and computers to create art works and interact." In

B. J. Jansen et al. (Eds.): *International Conference on Cognitive based Information Processing and Applications (CIPA 2021)*, LNDECT 85, pp. 530–535, 2022.
https://doi.org/10.1007/978-981-16-5854-9_67

the new media information age, the rise of the Internet has brought us huge information resources, especially the wide application of various social media tools nowadays, which has greatly expanded the channels of art communication. People can accept and learn art knowledge and information from all over the world through more and more ways of communication. New media technologies such as computer technology, video, audio and other forms of new media technology are beginning to enter the classroom. Students can constantly update their knowledge in all fields related to art through these media. Under such circumstances, art teachers can update these new knowledge and ideas through certain teaching means, so that they can filter, store and extract the essence of students in their minds. To eliminate the dross, the positive aesthetic elements suitable for students' physical and mental health development are transmitted to students through new media technology, so that they can learn how to appreciate and inherit excellent art learning elements, so as to conform to the knowledge and technology and development of the times.

2 Overview of Data Mining Algorithm

Data mining appeared in the late 1980s, and has been developing rapidly in the field of artificial intelligence in the past decade. It has become a hot spot in all walks of life in the era of big data. The so-called data mining technology refers to the non trivial process of revealing hidden, noisy, random, previously unknown and potentially valuable information from large databases [2]. Through data screening and data preprocessing, it can help decision-makers to analyze historical data and current data, analyze original data with high automation, and carry out inductive reasoning. In the research content of this paper, the use of data mining technology to explore the innovation and development of art teacher education.

In the case of linear separability. Given training sample set: $Z = (x_1, y_1), (x_2, y_2), \ldots, (x_n, y_n), y_i \in \{+1, -1\}$。 The decision classification function can be described by the linear discriminant equation $w^T.x + b = 0$ in n-dimensional space. As shown in Fig. 1, suppose there are three hyperplanes a, B and C, and the cross sign and circle represent positive and negative samples respectively. SVM tries to find an optimal hyperplane to separate positive and negative classes, while minimizing the classification error and maximizing the margin. In this case, hyperplane B should be selected, because hyperplane B satisfies the conditions of maximum classification interval and minimum empirical risk.

In most cases, training data is linearly inseparable, such as face image, text document data and so on. A mapping function from low dimensional feature space to high dimensional feature space can be defined by adding relaxation variable and penalty factor to the condition.

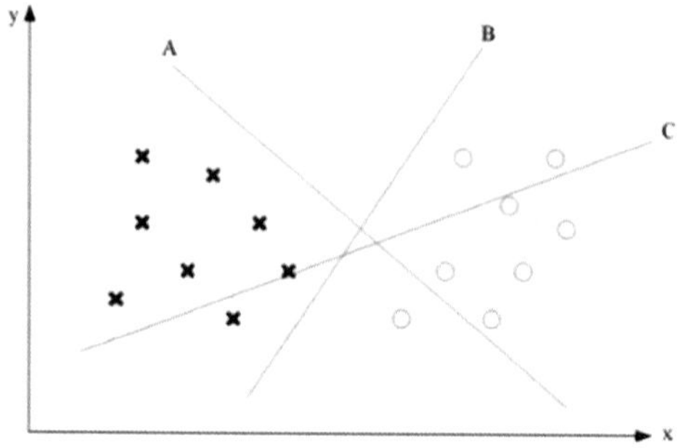

Fig. 1. Hyperplane of vector machine

3 Unique Contribution of Art Education in School

The progress of science and technology and the development of media have changed the form of information communication, and also changed the way people accept information. Especially since the era of big data, ubiquitous images have changed the way people live, learn and perceive the world. Alvin Toffler, an American futurist, pointed out in the third wave as early as 1980 that human society is breeding three kinds of illiteracy, namely, the literacy of the text culture, the illiteracy of computer culture and the blindness of visual culture. This is because in today's world, people use a variety of communication types to transmit information, such as oral or written language, images, formulas, symbols, sounds, gestures, charts, cultural relics, etc., language is not the only communication system available, many types of visual images and symbols have specific significance. In traditional sense, reading and writing ability is limited to literacy. Nowadays, we need to read or write different types of text with different types of reading and writing abilities. Among them, art literacy is also called "visual literacy" or "visual literacy".

Nowadays, in the visual culture background of big data age, all young people have a new need to use image skills. New ability needs not only change learning environment, but also change the workplace. No matter now or in the future, people who are engaged in any work should rely on the artistic literacy, which has become the basic cultural literacy of all the people. People who have not received art education in schools, can not recognize and understand the information transmitted by images and who can not express and communicate with images will be difficult to carry out high-quality learning, work and life, and make outstanding contributions to their own communities and society.

We believe that the artistic literacy with vision, modeling, space, creativity and aesthetic feeling as the core corresponds to the skills of the 21st century. Through the school art education, every young person can obtain the essential core literacy of survival and communication in a highly complex world - Art Literacy.

(1) Art course promotes the development of students' learning and innovation skills
Art is unique, and provides different thinking modes from science, mathematics, philosophy and literature. It is an important part of all the national cultures on the earth. Art tells us the meaning of life and promotes the development of social culture,

economy and science and technology with creativity and knowledge and skills. Art is extremely rich, and the school art curriculum is also rich and colorful, covering painting, sculpture, design, technology, calligraphy, seal cutting, new media art, etc., and also includes various tools, technologies and processes.
Brain research has confirmed that background and multi sensory instructions are easier to learn and remember, and it is the background and multi sensory instruction that school art courses provide. Art is an important language. When we interpret, transmit and create image information, we can easily activate the creative, imagination, thinking, emotion and ideas that can not be replaced by the examination, and promote the smooth thinking mode [3].

(2) The art course expands students' digital literacy skills Internet is an amazing source of excellent image information. It will bring art world from all over the world into the classroom, which is beneficial to improve students' global awareness. Since 1990s, the rapid development of new media art is a creative and challenging art category which appears at the intersection of art and science.

The new media art course integrating art literacy and media literacy is an effective way to cultivate students' skills of using, understanding and creating digital literacy. The new media art units in the school art course include the interpretation and creation of photography, camera, computer painting, computer design, visual notes, animation, small films, etc.

In the era of big data, high quality school art curriculum and teaching more and more combine media technology with art knowledge and skills. It is worth mentioning that the primary and secondary school students in the era of big data have few technical bottlenecks in the use of digital cameras, digital cameras, computer software, network, etc. they are passionate and fast to enter into the creative activities, and can last for a long time. The arts quality, media literacy and comprehensive ability can be significantly improved. The teaching of digital world art in the future is endless. As art educators, students should learn how to construct the digital art world and make them realize the possibility of digital art world being put into the real world (such as using 3D printing technology).

4 The Innovation and Development of Art Teachers' Education Under Data Mining

This paper puts forward the research on the innovation and development of art teachers education under the data mining. Specifically, SPC uses two-stage classifier, which consists of neural attention encoder decoder structure and SVM classifier. The former consists of RNN based encoder, attention mechanism, sequence feature generator and decoder. The main idea of this task is to construct the implicit representation of behavior sequence, and then classify and predict based on the characteristics of behavior sequence. RNN is a neural network with short-term memory ability. Compared with other neural networks (such as convolutional neural network (CNN), RNN can process different length sequences. In this network structure, neurons can not only accept information from other neurons, but also accept their own information. As shown in Fig. 2:

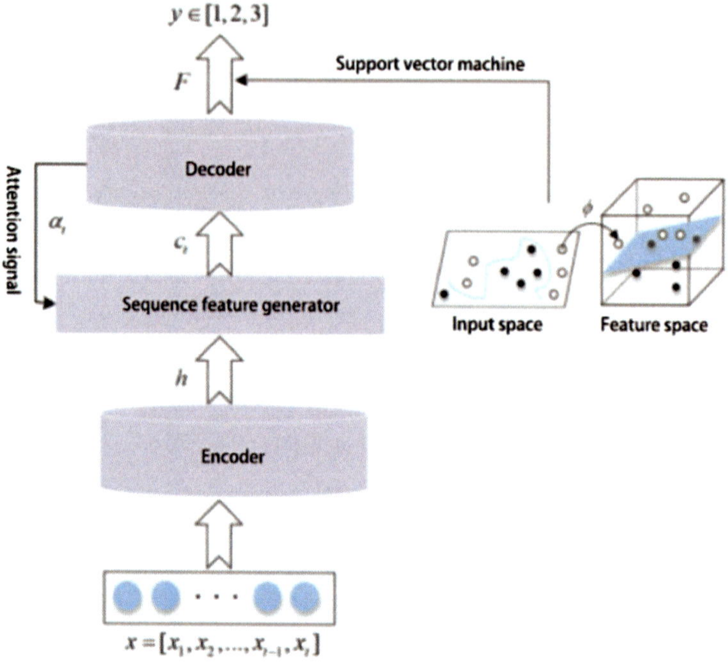

Fig. 2. Data flow framework

In the past decade, with the advent of the era of big data and the trend of cultural diversification, the visual culture of countries around the world, especially Europe, America, Australia and other countries, has undergone a significant transformation in history. This transformation is mainly manifested as follows: first, the focus of creativity has shifted to the thinking of creative objects and the concern of conceptual art; Second, photography, film and print media have become independent art forms; Third, digital technology has been combined with traditional art forms, and the field of electronic creativity has been developed; Fourthly, it advocates the aesthetic ideal of visual culture based on the nation, and has a new interest in and recognition of indigenous traditional arts and crafts; Fifthly, the context of art has also changed. Through the global movement, refugees and immigrants have changed the cultural integration of society [4].

In the era of big data, we need to provide learners with the necessary knowledge, skills, values and attitudes to build a more equitable, peaceful, tolerant, inclusive, safe and sustainable world, which are essential elements for contemporary and future generations to cope with future global challenges. The starting point of art education is the same as that of general education, which recognizes the relevance of education in understanding and solving global social, political, cultural, economic and environmental problems. It also means recognizing the role of education in transcending knowledge and cognitive development, building learners' values, social and emotional skills and attitudes, and promoting international cooperation and social change.

5 Conclusion

Under the background of big data, the application of data mining algorithm in art courses is the inevitable requirement of social development and education. Education is to cultivate talents at all levels in various fields. If such requirements are not met, education will lose its significance. When data mining algorithm is applied in art courses, students can be exposed to more technology and culture, Learn more abundant and widely used knowledge, cultivate students' innovative spirit, and truly achieve the combination of innovation and practice. In this era of keeping pace with the times, science and technology are constantly developing forward. Combining learning with the times can not only enable students to learn more knowledge and skills closely related to future life, but also promote the innovation, progress and development of the whole era, and keep pace with the times.

References

1. Qing, Z., Chao, M., Dan, Y.: Review on research progress of educational data mining. Acta Softw. Sinica **26**(11), 3026–3042 (2015)
2. Art Curriculum Standard for compulsory education (2011). Beijing Normal University Publishing Group (2011)
3. Chuxi, Q.: Art Teaching Theory and Method. Higher education press (2013)
4. Liu, N.: Research on decision support method of university teaching management based on data analysis. Zhejiang University (2019)

Application of Cognitive Environment and Data in Urban Sewage Biological Phosphorus and Nitrogen Removal System

Bingbing Jin(✉)

School of Environmental and Municipal Engineering, Qingdao University of Technology, Qingdao 266525, China

Abstract. As an important water conservancy project, reservoir plays a positive role in flood interception, water supply guarantee and water resources allocation, and also provides a significant help for social and economic development and ecological civilization construction. Reservoir water storage is one of the important indicators for the healthy operation of the reservoir, which is closely related to the regional climate, humidity and precipitation. It is necessary to study the reservoir storage capacity for the rational protection and utilization of water resources and the formulation of relevant policies. This paper collects and reviews a large number of relevant literature, discusses how to apply data mining technology and data analysis of reservoir storage problem.

Keywords: Data mining algorithm · Data preprocessing · Decision tree · Reservoir storage capacity

1 Introduction

In recent years, due to climate change and the increase of extreme weather, flood disaster is still the biggest natural disaster that brings loss to human society. According to the latest data from the United Nations Office for disaster reduction, the flood in 2018 alone caused property damage to 35.4 million people around the world, resulting in 2859 deaths [1]. On the other side of the flood, some areas are suffering from severe drought and lack of fresh water resources, which not only increases the pressure on domestic water for residents, but also increases the water consumption for Industry nd agriculture. To this end, the governments of various countries have taken a lot of countermeasures, and the reservoir, as one of the active water conservancy projects, effectively alleviates the above situation. The reservoir has the function of intercepting flood, ensuring water supply and allocating water resources, which can not be ignored for social and economic development and ecological civilization construction. The reservoir is one of the important indicators of water storage and operation, which is closely related to the reservoir's water volume and humidity. Therefore, it is necessary to study the water storage capacity of the reservoir for the rational protection and utilization of water resources and the formulation of relevant policies.

B. J. Jansen et al. (Eds.): *International Conference on Cognitive based Information Processing and Applications (CIPA 2021)*, LNDECT 85, pp. 536–540, 2022.
https://doi.org/10.1007/978-981-16-5854-9_68

2 Data Mining Technology

2.1 Association Rules Analysis

Association rule is a kind of relationship or mutual influence between the values of two or more data items in the data set. For example, the change of the value of one or more data items will cause the change of the value of another or more data items. Association mainly includes: simple relation Association, causality Association and time series Association. Association rule analysis is to find the association relationship hidden in the object data set, so as to mine the knowledge. The task of. Association rules can also find rules in the occurrence time or sequence of events, that is, the occurrence of one or some events will cause another or some events to occur in sequence.

2.2 Neural Network

Neural network is a widely used and very important method in data mining. This method is based on the theory of neurobiology, which imitates the working mechanism and structure of human brain neural network [2]. Neural network presents many characteristics of human brain, which is composed of many interconnected neurons. It can carry out nonlinear operation and complex logic operation, and has some functions of human brain. Because the neural network is a distributed matrix structure, the neural unit has high parallelism and distribution, and can parallel calculate and store the data, which makes the neural network has strong self-adaptive, self-organizing and self-learning ability. The intelligence of neural network is mainly reflected in nonlinear mapping function, classification and recognition, associative memory, optimization calculation and knowledge processing. Based on Hebb learning rules and M model, neural network consists of feedforward neural network. This method mainly uses learning extraction method and decomposition extraction method to extract rules from neural network. The most outstanding advantage of neural network is that it can still achieve high precision prediction in the case of complex problems. However, this method also has some defects, such as: easy to be affected by excessive training; training time is long; it is not easy to be understood.

3 Classification in Data Mining

3.1 Purpose of Classification

Firstly, the data in the training sample set is analyzed. According to the characteristics of the training samples with known class labels, the description (model) of known categories is constructed. Then, the unknown data are classified by the constructed class description (model), and the class label of the unknown data is predicted. Data classification can be summarized into two steps: building a model and using a model for classification, as shown in Fig. 1.

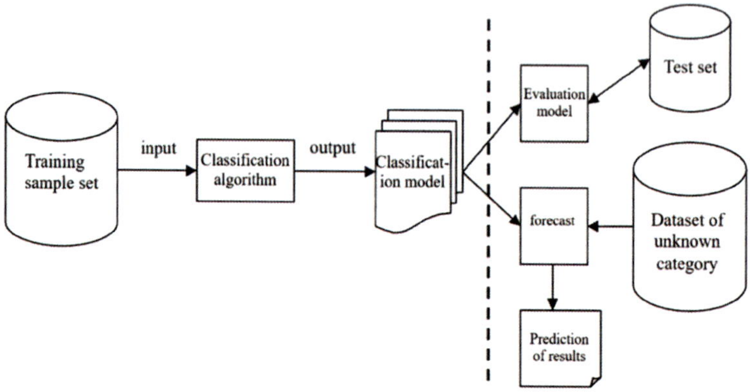

Fig. 1. Steps of classification

3.2 Classification Algorithm Based on Decision Tree

Decision tree classification algorithm is a case-based inductive learning algorithm. The algorithm can generate decision tree model based on given data set and extract rules from it directly [3]. attributes of the data set by using the attribute measurement, selects the root node, and the data set is divided into several sub datasets. Then recursively use the same method to divide the sub dataset, and finally reach the leaf node, each leaf node represents a category. Classification algorithm based on Bayes (Fig. 2).

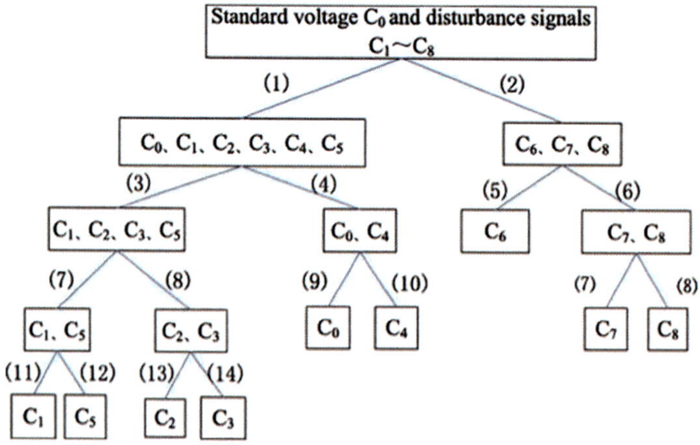

Fig. 2. Classification algorithm based on decision tree

Using Bayesian algorithm for classification is to construct a statistical classifier based on Bayesian theorem, namely Bayesian classifier. The posterior probability of each pattern is obtained from the samples, and it is used to predict the probability

Classification distribution of each category in the sample set. Then, the probability c_i is:

$$P_r(C = c_i|A_1 = a_1 \wedge \cdots \wedge A_k = a_k) \tag{1}$$

The to which class is the highest, it belongs to which class. Bayes theorem is as follows:

$$P_r(C = c_i \backslash A_1 = a_1 \wedge \cdots \wedge A_k = a_k) = \frac{P_r(A_1 = a_1 \wedge \cdots A_k = a_k \backslash C = c_i)}{P_r(A_1 = a_1 \wedge \cdots \wedge A_k = a_k)} P_r(C = c_i) \tag{2}$$

$P_r(C = c_i)$ is the prior probability of each class, which can be obtained from the training sample set.

However, in terms of time and space, how the agricultural phenology and climate conditions change in the reservoir area and surrounding areas before and after the completion of the Three Gorges dam has been paid close attention by scholars at home and abroad, but has not yet been studied. At the same time, the rice phenology is affected by the change of local climate conditions. In view of the background of the change of climate conditions in response to the reservoir, the research on rice phenology in the reservoir area will play a guiding role in agricultural production.

4 Estimation Model of Reservoir Storage Capacity

The reservoir water storage model is a mathematical model established by relevant researchers in order to quickly and accurately estimate the reservoir storage capacity. The reservoir can be generally divided into lake type reservoir and river type reservoir (Fig. 3).

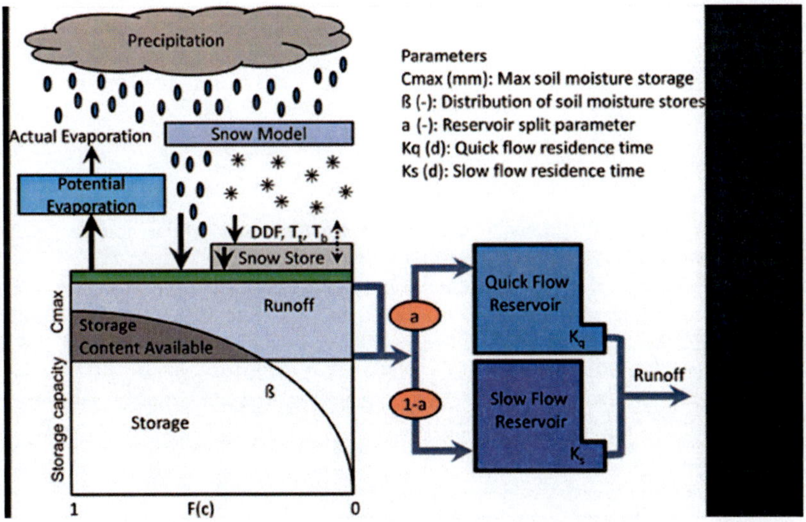

Fig. 3. Estimation model of reservoir storage capacity

The characteristics of lake type reservoir are that the water area is wide, the water flow is gentle, and the water surface of each part is basically kept at the same height; the characteristics of river type reservoir are that the reservoir body is narrow and long, the water flow area is small, and the water flow is fast, and there is an obvious water level difference between the inflow area and the outlet area, As a result, the whole reservoir will frequently submerge the surrounding fluctuating areas [4].

According to the above two characteristics, the lake type reservoir can be regarded as the static storage reservoir, and the river type reservoir can be regarded as the reservoir with the sum of static and dynamic storage capacity. Among them, the estimation of water storage capacity of river type reservoir is more complex, and the early method is difficult to accurately estimate the dynamic storage capacity of river type reservoir, and the key of research is whether the dynamic storage capacity of river type reservoir can be estimated more accurately. On the basis of literature and data, this paper studies the mainstream and classic reservoir storage estimation models, including hydrological formula method, data assimilation method and accumulation method. There are five models: method, mean value method and integration method. In practical application, the above models can be used either alone or in combination.

The application of data assimilation method in reservoir water storage estimation is to combine water level, water area and water storage with relevant mathematical models to achieve reservoir water storage estimation. The common models include polynomial model, exponential model and power function model.

5 Conclusion

In view of the current situation and shortcomings of reservoir water storage estimation, this paper proposes to use data mining algorithm to extract reservoir parameters. After verifying the accuracy of the extracted water area, the quantitative function relationship of water area difference of data mining algorithm is constructed by using data assimilation technology, Finally, a reservoir water storage estimation system based on data mining algorithm is designed and implemented.

References

1. Huizhong, W., Anqun, P.: Research status and development trend of data mining. Ind. Min. Autom. **2**, 33–36 (2011)
2. Zhang, L., G., Zhang, Li, F.: Study on remote sensing analysis method of Laolongkou reservoir capacity based on landsa8 and DEM. Sci. Technol. Innovation Appl. **252**(32), 126–127 (2018)
3. Yonghai, T., Zhihui, Z., Songyang, Y.: Design of flood regulation system of qinghe reservoir based on web. J. Water Resour. Water Eng. **21**(1), 168–170 (2010)
4. Qin, Z., Ling, H.: Design of automatic detection system based on window message mechanism. J. Guangxi Acad. Sci. **26**(4), 497–499 (2010)

Test-Bed, Prototype Implementation and Applications

Application of Machine Learning in Finance

Dipeng Song(✉)

School of Economics, Shanghai University, Shanghai, China

Abstract. The rapid development of information technology and big data has provided a fertile ground for the application of machine learning in the financial field. This article introduces machine learning, a powerful tool for processing big data, and analyzes its application in different aspects in the financial field. Finally, some suggestions have made on how to better use machine learning in the financial field.

Keywords: Machine learning · Investment · Risk management

1 Introduction

The development of information technology and the availability of big data have aroused widespread interest among researchers, especially in the financial field. Face of changing market, relying solely on traditional data collection is often difficult to make accurate interpretation of complex real world, the available data of a large property so that people understand the financial markets closer to the real world. Although data sources have been greatly expanded, it is still difficult to play the value of these data without suitable data processing technology. For example, the processing of hypertext data such as voice, picture, video, etc., often requires artificial intelligence to complete.

The combination of artificial intelligence and machine learning technology with the financial field provides financial decision makers with new methods of modeling and forecasting, and has corresponding applications in investment, asset pricing, and risk management. This article from the three aspects were applied in the financial sector of machine learning in specific analysis, and compared with traditional methods, pointed out the advantages of machine learning. Before the analysis, this article first introduces the classification and advantages of machine learning technology and its algorithms.

2 Machine Learning

2.1 What's the Machine Learning?

Machine learning is a technology that focuses on digital processing in the field of artificial intelligence, and is often used for data modeling and prediction. As the availability of big data has increased, machine learning has increasingly exerted its powerful data processing capabilities. Another key technology related to artificial intelligence is natural

B. J. Jansen et al. (Eds.): *International Conference on Cognitive based Information Processing and Applications (CIPA 2021)*, LNDECT 85, pp. 543–549, 2022.
https://doi.org/10.1007/978-981-16-5854-9_69

language processing. It takes the understanding and processing of text data as the core to mine information that is difficult to capture by traditional data. Machine learning can be applied to natural language processing to help learning and prediction A large amount of integrated text data.

2.2 Classifications and Advantages

Machine learning can be divided into supervised learning and unsupervised learning. These two learning methods have their own advantages for different data processing. According to different data processing purposes, suitable learning methods can be selected for processing, which can almost meet all data processing needs.

There supervised learning is already training samples for training in order to obtain an optimal model, and then use the model to predict the unknown data and classification. In supervised learning, the classification information of the data is known before processing, and the data in the training sample contains both feature and label information, and the corresponding output is obtained based on this information. After continuous training, when the model prediction accuracy reaches a certain level When the time, it is applied to the forecast of the new data. According to the different characteristics of input data and output data, supervised learning can be divided into regression problems (input variables and output variables are continuous variables), classification problems (output variables are finite discrete variables), and labeling problems (input variables and output variables both are variable sequences). Common supervised learning algorithms include linear regression algorithms, BP neural network algorithms, decision trees, support vector machines (SVM), and KNN.

In unsupervised learning, the labeled information of training samples is unknown. This type of learning mainly reveals the internal laws and properties of the data through the learning of samples labeled with unknown information, and provides a support basis for further data analysis. The most widely used unsupervised learning should be the "clustering" problem, which is to gather similar things together by calculating the distance between the sample and the group. When we have very little information about a data set, we can implement clustering through unsupervised learning, so as to extract the inherent information and laws contained in the data set. Common unsupervised learning algorithms include: density estimation, anomaly detection, hierarchical clustering, EM algorithm, K-Means algorithm, and DBSCAN algorithm.

Especially in the field of financial research market anomalies, since the relevant economic theory is not perfect, traditional methods often rely on personal experience and subjective judgments, and machine learning, especially unsupervised learning to overcome this shortcoming, greatly reducing the subjective Judgment, on the contrary, relatively independent and objective judgments are given through algorithms. Machine learning and its different algorithms can also meet different applications in the financial field, such as investment. Asset pricing and risk management, etc. The potential for ML in financial decision making was first explored from a research perspective in Hawley et al. (1990) [3] with a focus on neural networks as an aide to financial decision-making.

More details about some algorithms of machine learning are in the Table 1.

Table 1. Details about some algorithms of machine learning

Techniques	Merits	Limitations	Typical applications
SVM	– Accuracy in general	– Speed of learning	– Classifification
	– Speed of classifification	– Tolerance to missing values	– Regression
	– Tolerance to irrelevant attributes	– Tolerance to noise	– Time series forecasting
	– Tolerance to redundant attributes	– Dealing with danger of overfifitting	
	– Tolerance to highly interdependent attributes	– Attempts for incremental learning	
	– Model linear and non-linear problems	– Explanation ability	
	– Dealing with danger of overfifitting	– Model parameter handling	
Decision Trees	– Speed of learning	– Accuracy in general	– Classifification
	– Speed of classifification	– Tolerance to redundant attributes	– Regression
	– Tolerance to missing values	– Tolerance to highly interdependent attributes	– Credit risk models
	– Tolerance to highly interdependent attributes	– Tolerance to noise	– Market segmentation
	– Explanation ability	– Attempts for incremental learning	– Time series forecasting
	– Model parameter handling	– Upper bound on error	
	– Dealing with danger of overfifitting		
	– Dealing with binary and continuous attributes		
	– Handles attributes with different costs		
Random Forest	– No overfifitting	– Black box	– Classifification
	– Combination of several tree predictors	– Explanation ability	– Regression

(*continued*)

Table 1. (*continued*)

Techniques	Merits	Limitations	Typical applications
	– Upper bound on error		– Time series forecasting
	– Robust to noise		
	– Fastest tree-based technique		
	– Not very sensitive to node split variable		
Adaboost	– No overfifitting	– Increase variance of responses	– Classifification
	– Feature selection capability	– Performance depends on the quality and quantity of data and weak learner	– Ranking
	– Minimize bias of the error term	– Tolerance to noise	
	– The tree version (alternating decision trees) facilitates interpretation of results		
	– Upper bound on error		
ANN	– Speed of classifification	– Tolerance to missing values	– Time series forecasting
	– Dealing with binary/continuous attributes	– Tolerance to irrelevant attributes	
	– Attempts for incremental learning	– Tolerance to redundant attributes	
		– Tolerance to noise	
		– Dealing with danger of overfifitting	
		– Explanation ability	
		– Model parameter handling	
		– Black box	
Deep Learning	– Same as ANN	– Black box	– Vision and voice processing
		– Heavy data requirement	– Time series forecasting

Source: H. Ghoddusi, G. Creamer and N. Rafizadeh / Energy Economics 81 (2019) 709–727 [1]

3 Machine Learning in Finance

3.1 Investment

Machine learning is the most widely used in investment analysis, which is mainly reflected in the following aspects: 1) investor sentiment analysis; 2) technical analysis; 3) investor behavior analysis; 4) investor decision support analysis.

In terms of investor sentiment analysis, the most popular machine learning method is Support Vector Machine (SVM), which can classify data well. Support vector machines (SVM) can usually be used to classify words related to public sentiment in financial reports and news reviews on social media to explore the relationship between them and stock prices. For example, Smailović et al. (2014) [2] showed that public sentiment based on twitter information sources can be used to predict the trend of stock prices. The text is classified by support vector machine, and the sentiment is divided into negative, neutral and positive according to the twitter content, and then these tweets with different sentiments are connected with the stock market of the same period to study investment How the mood of the investor affects the stock price. In addition, there is research to obtain unlabeled text from stock message boards, and also mark its emotional color, thereby revealing the potential relationship between it and stock prices.

In terms of technical analysis, it is mainly based on quantitative analysis to reduce the error of human judgment and achieve profitability. Through the combination of radial basis function (RBF) neural network and improved error minimization technology, the classic K-line analysis in technical analysis is realized, and the real and false bull markets are further distinguished. And the application of cyclic neural network has significantly improved the forecasting ability of financial time series, which has great advantages in forecasting.

In terms of investor behavior analysis, it is mainly based on behavioral finance theory, using machine learning algorithms to simulate investors' decision-making. For example, based on prospect theory, emotions are incorporated into predictive models through human-computer interaction, and these emotional factors as well as cognitive and behavioral deviations are considered when simulating decision-making, so as to improve the model and be closer to real investors.

Finally, in terms of investor decision support, it is mainly based on the calculation and iteration capabilities of machine learning algorithms, which can help investors make decisions in a short time. According to their own risk tolerance and income needs, investors can obtain suitable trading strategies through algorithms, and do not need to invest a lot of time, which greatly reduces the time cost.

3.2 Asset Modeling

In addition to investment, machine learning has a wide range of applications, and it also has important applications in asset modeling. For example, investment portfolio, volatility modeling, financial forecasting technology, etc., these are based on the powerful forecasting capabilities of machine learning. Shen and Wang (2017) [4] solved a classic problem of portfolio selection based on markovitz, that is,mean-variance portfolio estimation which requires long time data to estimate input parameters by neural networks. Ma et al. (2014) [5] used genetic algorithms, and Chela Boiner et al. (2013) [6] applied both genetic algorithms and artificial neural networks to hedge option portfolios.

In terms of portfolio optimization, it takes a long time for the classic Markowitz portfolio to calculate the mean-variance portfolio parameter estimation in practice, which brings great inconvenience. The integrated learning algorithm in machine learning greatly simplifies the calculation process and provides more possibilities for the construction and optimization of investment portfolios.

In terms of volatility modeling, traditional methods rely on standard technologies such as the ARCH model, and machine learning provides an alternative method for this, and the ARCH model is improved. It is mainly reflected in the following two points. One is to use high-frequency and ultra-high-frequency data to replace daily returns in the ARCH model to predict stock price volatility. High-frequency and ultra-high-frequency data contain far more information than daily data; Consider other signals and join the traditional ARCH model to predict. Machine learning predicts the volatility of stock prices by training algorithms such as random forests and neural networks, making full use of high-frequency data and ultra-high-frequency data, as well as other signals that exist within the data but are difficult to capture.

In financial forecasting, a continuous time series is often required, which limits the forecasting effect when there is disordered financial data. Recurrent neural networks in machine learning can solve this problem. It allows some form of memory in the neural network. Therefore, even if the entire time series is chaotic and disorderly, there are temporal connections. In addition, cyclic neural networks can also model chaotic time series through multi-layer perceptrons by establishing nonlinear models.

3.3 Risk Management

Machine learning has great application prospects in risk management, including risk assessment, risk prediction, operational risk management and credit risk modeling. Thanks to the powerful clustering method of machine learning, risk modeling is one of the earliest focus of attention in the application of machine learning in the financial field.For example, Marmier et al. (2014) [7] using a decision tree model in new product development based on project risk. Son et al. (2016) [8] show that nonparametric ML model with deep learning is superior to the traditional benchmark model in price prediction accuracy and practical hedging measures.

The outbreak of the global financial crisis shows the fragility of the financial system and the existence of systemic risks, and also shows the failure of traditional risk assessment and risk prediction methods. Machine learning algorithms, such as decision trees, have huge advantages in understanding the inherent complexity of the financial system and managing this complexity. When using decision trees, the integration process can be introduced to consider a variety of different risks at the same time. The neural network and game theory models simulate the risks brought about by different choices, and are suitable for risk assessment and management of long-term high investment projects. In addition, machine learning also improves the accuracy and practicality of risk assessment and prediction.

Operational risk management refers to the possibility and loss of operating activities that fail to meet expectations due to the complexity and change of the external environment and the limited cognition and adaptability of the subject to the environment during the operation of an enterprise. The complexity of the environment makes operational risks inevitable, but with the help of artificial intelligence such as machine learning, operational risks can be better identified, evaluated, and managed. This is because machine learning can learn from known risk cases to build appropriate models.

Traditional credit risk modeling mainly focuses on linear regression, logistic regression and unit probability regression. These modeling methods have more or less the risk of misclassification. Machine learning is not an isolated application in credit risk modeling, but is combined with classic modeling methods, using the excellent characteristics of machine learning, and applying neural networks and other algorithms to improve the accuracy of traditional models. Scholars have tested and verified that when the traditional credit modeling method is combined with the neural network-based credit risk model, it has been significantly improved in predicting default.

4 Conclusion

As mentioned above, machine learning has a lot of advantages, such as excellent prediction accuracy, the ability to process large amounts of unstructured data, the ability to solve complex relationships, insensitivity to data quality, no need for excessive data preprocessing, and flexibility in application For classification, regression, and sorting problems. With the availability of big data and the gradual decline of traditional analysis in the financial field, machine learning is bound to provide new solutions for the financial field. However, we still have to prevent falling into the trap of convenience brought about by technology. Although machine learning has many advantages, it is still insufficient for explaining the laws of things, mainly because the algorithm itself is a black box. The application of machine learning in the financial field far exceeds that described in this article. We expect that this technology can better serve financial research and provide more possibilities for exploring the financial field.

References

1. Ghoddusi, H., Creamer, G.G., Rafizadeh, N.: Machine learning in energy economics and finance: a review. Energy Econ. **81**, 709–727 (2019)
2. Smailović, J., Grčar, M., Lavrač, N., Žnidaršič, M.: Stream-based active learning for sentiment analysis in the financial domain. Inform. Sci. **285**, 181–203 (2014)
3. Hawley, D.D., Johnson, J.D., Raina, D.: Artificial neural systems: anew tool for financial decision-making. Financ. Anal. J. **46**(6), 63–72 (1990)
4. Shen, W., Wang, J.: Portfolio selection via subset resampling. In: Proceedings of the Thirty-First AAAI Conference on Artificial Intelligence, pp. 1517–1523 (2017)
5. Ma, Y., Gong, X., Tian, G.: A mean-semi-variance portfolio optimization model with full transaction costs. In: 2014 International Conference on Computational Intelligence and Communication Networks (CICN), pp. 623–627. IEEE (2014)
6. Chellaboina, V., Bhatia, A., Bhat, S.P.: Explicit formulas for optimal hedging stratergies for European contingent claims. In: 2013 IEEE Conference on Computational Intelligence for Financial Engineering and Economics (CIFEr), pp. 122–127. IEEE (2013)
7. Marmier, F., Ioana, F.D., Didier, G.: Strategic decision-making in NPD projects according to risk: application to satellites design projects. Comput. Ind. **65**(8), 1107–1114 (2014)
8. Son, Y., Byun, H., Lee, J.: Nonparametric machine learning models for predicting the credit default swaps: an empirical study. Expert Syst. Appl. **58**, 210–220 (2016)

Virtual Reality and System Simulation Technology Under the Background of Network Informationization

Yongming Pan(✉)

Department of Digital Media and Design, Neusoft Institute Guangdong, Foshan, Guangdong, China
panyongming@nuit.edu.cn

Abstract. Virtual reality technology is one of the most anticipated and promising new technologies in the 21st century. Simulation system is one of the main applications of virtual reality technology. It can provide users with a good virtual environment. Even if they don't go to the scene, they can get the environment information of the scene, and they can operate freely according to their needs, and carry out good scenes. At the same time, with the rapid development of science and technology and the further needs of society, simulation experiment technology has become increasingly mature, and it is moving from theoretical research to practical application. The purpose of this article is to study the exploration of virtual reality and system simulation technology under the background of network informationization. This article introduces and discusses virtual reality technology to a certain extent, analyzes the current types of virtual reality technology and the actual situation in terms of the degree of simulation. The current deficiencies and practical applications of virtual reality technology are studied. It conducts experimental research and analysis on other virtual reality simulation systems, and combines multiple virtual reality software and hardware technologies to study the constructed simulation system, and conducts research and analysis on the data through real-time collection of relevant data in the research process. Experimental research results show that: simulation technology is the most widely used in simulation-assisted production, up to 27.19%, followed by simulation-assisted research, with 26.2%, and finally, the least-applied simulation-aided design, with only 21.24%. All in all, it is necessary to increase efforts to realize the maximum value of simulation technology in practical applications, so that it can develop faster and better.

Keywords: Information age · Virtual reality · System simulation · Exploration

1 Introduction

Virtual reality technology is one of the most anticipated and promising new technologies in the 21st century [1, 2]. Simulation system is one of the main applications of virtual reality technology. It can provide users with a good virtual environment. Even if they

B. J. Jansen et al. (Eds.): *International Conference on Cognitive based Information Processing and Applications (CIPA 2021)*, LNDECT 85, pp. 550–557, 2022.
https://doi.org/10.1007/978-981-16-5854-9_70

don't go to the scene, they can get the environment information of the scene, and they can operate freely according to their needs, and carry out good scenes [3, 4]. At the same time, with the rapid development of science and technology and the further needs of society, simulation experiment technology has become increasingly mature, and it is moving from theoretical research to practical application [5, 6].

In the exploration and research of virtual reality and system simulation technology under the background of network informationization, many scholars at home and abroad have conducted research on it and achieved good results. Li A shows that the current simulation technology at home and abroad can be divided into centralized Simulation technology and distributed simulation technology [7]. The simulation of large-scale complex systems at home and abroad mainly uses advanced distributed simulation technology [8, 9]. Yap H J put forward new ideas, new methods and new technologies, which have a wide range of applications, and their field of action is not limited to supporting object-oriented analysis and design [10]. It also supports the entire process of software development starting from requirements analysis, and has received extensive support, becoming the de facto industry standard for visual modeling languages [11, 12].

The purpose of this article is to study the exploration of virtual reality and system simulation technology under the background of network informationization. This article introduces and discusses virtual reality technology to a certain extent, analyzes the current types of virtual reality technology and the actual situation in terms of the degree of simulation. The current deficiencies and practical applications of virtual reality technology are studied. It conducts experimental research and analysis on other virtual reality simulation systems, and combines multiple virtual reality software and hardware technologies to study the constructed simulation system, and conducts research and analysis on the data through real-time collection of relevant data in the research process.

2 Research on the Exploration of Virtual Reality and System Simulation Technology Under the Background of Network Informationization

2.1 Types of Virtual Reality Technology

(1) Desktop virtual reality system
Users use their own computers to carry out simulation experiments, use computer screens and use various devices to achieve interaction. Its cost is relatively low, so the desktop virtual reality system is the foundation and core of a large-scale virtual reality system, which has a great role in promoting the actual teaching, research and development and application of technical workers.

(2) Immersive virtual reality system
Refers to the user with some special equipment (such as helmet-type equipment, etc.) that can provide the user with a full range of immersion and interaction environment, and the user can achieve the effect of complete immersion. In the process of participation, the user can use some equipment to completely seal off his sense of hearing, vision, touch, etc., so as to provide the user with a brand-new, interactive virtual space.

(3) Augmented virtual reality system
It is the direction of the future development of virtual reality technology. It integrates the virtual objects and information it builds into the real environment that users live and can perceive. The effect it wants to achieve is not only a virtual and simulated real world, but also it also enhances the user's experience of the real world.
(4) Distributed virtual reality system
It refers to the joint development and design of virtual reality technology application services under the network environment, making full use of resources and technologies distributed in various places. This is a branch of the development of virtual reality technology, because it connects immersive virtual reality systems distributed in different places by means of the network, so as to achieve a certain effect together.

2.2 Virtual Reality Technology Simulation Degree

(1) Simulated virtual reality system
Based on the principles of real physics, a computer is used to simulate the real world environment, which is a virtual environment simulation that conforms to objective laws. The concept of multi-thinking is recreated and visualized, with certain material conditions assisted by some methods and means, so as to realize all that is seen and simulated.
(2) Transcendent virtual reality system
It is also based on real physical laws to simulate the real-world environment, but the simulated objects are inaccessible to us in daily life or cannot be perceived by the human five senses. But the system is also based on objective reality, and its main purpose is to create scenarios that transcend reality and build an environment that transcends reality.
(3) Fantasy virtual reality system
There is no need to obey the objective and true laws of physics, and use computers, graphics, images, audio and other equipment to transform the imagined environment into audible, seeable, and perceptual multimedia works. Therefore, the fantasy virtual reality system can give users a broader imagination space, and can further promote the cultivation of users' thinking ability and creativity.

2.3 Current Deficiencies

(1) It is difficult to develop simulation experiments and not easy to spread
The knowledge in this area is not popular enough, and there is very little communication between various schools at present, resulting in the development of experiments that cannot communicate with each other, the content is fixed in packaging, and it is difficult to modify. This kind of consideration for short-term needs, without considering long-term benefits, sharing with peers, and not providing convenience for others to supplement and modify content, leads to a lack of openness and makes it difficult to disseminate simulation experiments.

(2) Blindly promote virtual reality technology in the teaching process
Virtual reality technology has a considerable auxiliary effect on part of the learning content. This is unquestionable, but this does not mean that virtual reality technology must be promoted in the entire physics subject or in all teaching fields. It takes a lot of manpower and material resources to develop a set of courseware based on virtual reality technology. Therefore, it is necessary to select appropriate content for development. There is no need to design and develop teaching content that rely on virtual reality technology to not play an obvious role.
(3) The experimental effects of some experiments need to be strengthened
From the perspective of virtual reality technology, it mainly refers to the lack of credibility of the entire environment created, which does not conform to the understanding and experience of beginners, and lacks physical reality and behavioral reality. For learners who just want to learn this part of the content, everything is novel and everything is unknown. At this time, what they need is objective, vivid, and real content. Many current experiments all use abstract geometric figures. Replacing real experimental instruments makes beginners confused, which undoubtedly increases the difficulty of learning.
(4) Lack of timely help and guidance
Current developers often know very little about teaching and ignore the most fundamental topic that is developed for teaching. In the development process, there is a lack of links for students to answer questions. Students will inevitably encounter doubts when using the courseware to learn, and they cannot get timely guidance from the teacher, which often makes them frustrated, and will affect their enthusiasm for learning in the long run.

2.4 Population and Sample

Suppose a finite population contains N population units, and the k-th population unit is identified by k, where the finite population can be expressed as:

$$U = \{1, \cdots, k, \cdots N\} \tag{1}$$

Let y denote a research variable, which is the value of the research variable of the k-th overall unit in the population. Usually before the investigation, the value is unknown. And what we need to estimate is the total value of the research variable y

$$t = \Sigma_u y_k \tag{2}$$

Or study the overall mean of variable y

$$\overline{y}_u = t/N = \Sigma_u y_k/N \tag{3}$$

3 Experimental Research on the Exploration of Virtual Reality and System Simulation Technology Under the Background of Network Informationization

3.1 Experimental Subjects and Methods

In this experiment, the virtual reality simulation system and simulation technology are the main research objects, and the functional modules of the system and the application range of the technology are studied. Conduct research through actual testing and investigation, and perform statistical analysis on the data in the process.

3.2 Data Collection

Investigation task decomposition makes a detailed breakdown of the total amount of investigation activities, assigns investigation tasks to investigators or investigation teams, and assigns tasks equally by default. Specific task assignments, such as how many pieces of data each investigator or survey team needs to collect, the time range for data collection, and the system to be tested. The survey tasks can also be manually assigned. If the system automatically divides them into an inappropriate way, you can modify the assigned tasks.

4 Experimental Research Analysis on the Exploration of Virtual Reality and System Simulation Technology Under the Background of Network Informationization

4.1 Module Analysis of Virtual Reality Simulation System

This experiment focuses on the virtual reality simulation system. By taking two virtual reality simulation systems as examples, the system is analyzed in five aspects: detection module, feedback module, control module, modeling module and sensor. The experimental results are shown in Table 1:

Table 1. Module analysis of virtual reality simulation system

	Virtual reality simulation system A	Virtual reality simulation system B
Detection module	41.37	56.63
Feedback module	51.94	48.06
Control module	46.19	53.81
Modeling module	55.13	44.87
Sensor	53.94	46.06

As shown in Fig. 1, the most functional is the detection module of System B, with 56.63%, followed by the modeling module of System A, accounting for 55.13%, but the overall gap is not very large, and the average is maintained at Around 45%.

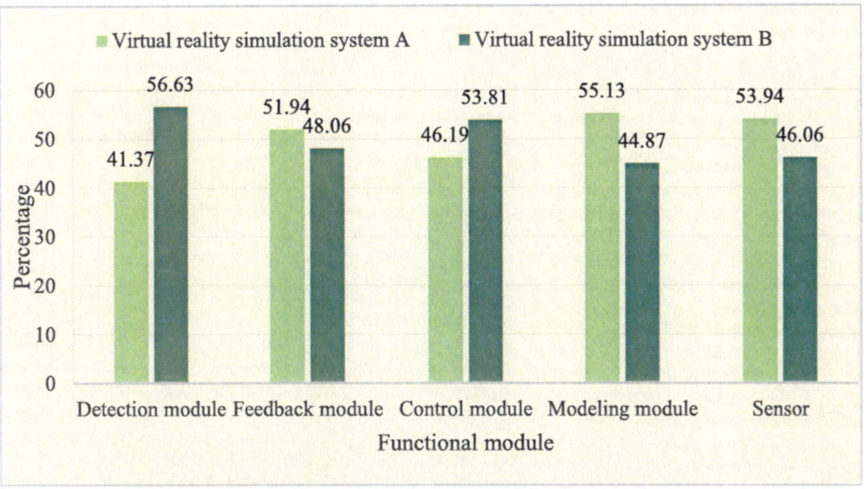

Fig. 1. Module analysis of virtual reality simulation system

4.2 Application Analysis of Simulation Technology

This experiment analyzes the simulation technology in four aspects: simulation-assisted training, simulation-assisted design, simulation-assisted production and simulation-assisted research. The experimental research results are shown in Table 2:

Table 2. Analysis of the application scope of simulation technology

	Simulation Technology
Simulation-assisted training	25.37
Simulation aided design	21.24
Simulation-assisted production	27.19
Simulation aided research	26.20

As shown in Fig. 2, simulation technology is the most widely used in simulation-assisted production, up to 27.19%, followed by simulation-assisted research, with 26.2%, and finally, the least-applied simulation-aided design, with only 21.24%. All in all, it is necessary to increase efforts to realize the maximum value of simulation technology in practical applications, so that it can develop faster and better.

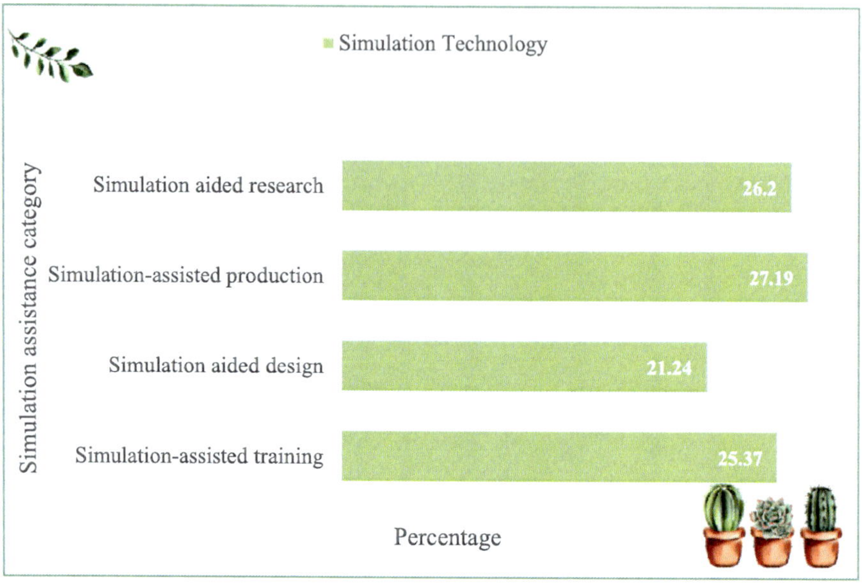

Fig. 2. Analysis of the application scope of simulation technology

5 Conclusion

At present, virtual reality technology has been widely used in various fields and has attracted people's attention. However, in the field of simulation, progress is relatively slow. Therefore, whether it is now or in the near future, the application of virtual reality technology in simulation systems will become a hot spot. This article introduces and discusses virtual reality technology to a certain extent, analyzes the current types of virtual reality technology and the actual situation in terms of the degree of simulation. The current deficiencies and practical applications of virtual reality technology are studied. Virtual reality technology can support almost all human activities, but the software and hardware environment, and the shortage of funds limit the practical engineering application of virtual reality technology to a certain extent. Virtual reality technology mainly uses computers to virtualize objects in reality. Virtual reality technology is one of the most anticipated and promising new technologies in the 21st century. Simulation system is one of the main applications of virtual reality technology. It can provide users with a good virtual environment. Even if they don't go to the scene, they can get the environment information of the scene, and they can operate freely according to their needs, and carry out good scenes.

References

1. Chen, W., Liu, X., Qiao, L., et al.: Construction of virtual reality-interactive classroom based on deep learning algorithm. Wirel. Commun. Mob. Comput. **2020**(8), 1–9 (2020)

2. Wang, V., Liu, L., Xu, J.: Study on the design of the railway vehicle running control system based on the virtual reality technology. Revista de la Facultad de Ingenieria **32**(13), 799–808 (2017)
3. Lei, Z., Shimizu, S., Ota, N., et al.: Construction of urban design support system using cloud computing type virtual reality and case study. Int. Rev. Spat. Plann. Sustain. Dev. **5**(1), 15–28 (2017)
4. Stylopoulos, N., Cotin, S., Dawson, S., et al.: CELTS: a clinically-based computer enhanced laparoscopic training system. Stud. Health Technol. Inform. **94**(5), 336–342 (2016)
5. Dong, L., Wang, Z.: Plant layout and simulation roaming system based on virtual reality technology. MATEC Web Conf. **214**(11), 04001 (2018)
6. Dai, H., Yang, Z., Xue, Z., et al.: Research on key technologies of substation 3D interactive scene simulation based on virtual reality technology. J. Phys. Conf. Ser. **1345**(6), 062036 (5pp) (2019)
7. Li, A.: A study on tennis competition enhancement technology on the basis of virtual reality technology. Int. J. Multimedia Ubiquit. Eng. **11**(12), 57–70 (2016)
8. Tsao, H.H., Lei, X., Rau, P.: Design, development, and evaluation of a virtual aging simulation system. Educ. Gerontol. **46**(4), 1–14 (2020)
9. Ma, J., Miao, Z., Tian, J., et al.: Design of evacuation simulation system for subway station personnel in fire environment based on SPH. IOP Conf. Ser. Earth Environ. Sci. **218**(1), 012090 (2019)
10. Yap, H.J., Tan, C.H., Phoon, S.Y., et al.: Process planning and scheduling for loop layout robotic workcell using virtual reality technology. Adv. Mech. Eng. **11**(9), 168781401987832 (2019)
11. Wang, C., Zhang, X., Liu, L.: The framework of simulation teaching system for sports dance based on virtual reality technology. Rev. de la Fac. de Ing. **32**(15), 530–536 (2017)
12. Phommaly, K., Yu, L.: Research of Virtual reality technology in home design. J. Phys. Conf. Ser. **1624**(6), 062017 (9pp) (2020)

The Study of the Innovation of Commercial Operation Pattern of Chinese Design Industry Under the Influence of 3DP

Xinjue Zhang and Qinghua Lin(✉)

School of the Art, Northeast Petroleum University, Daqing 163318, Heilongjiang, China
hhn@nepu.edu.cn

Abstract. There are some problems with the Chinese traditional commercial operation pattern of the design industry, such as high cost, high risk, difficulty in meeting individualized needs. Recently, with the development of Internet, the 3D Printing technology has already entered into the design industry and proposed some solutions to solve the problems with which today's design industry's traditional commercial operation are facing. Believing that the situation of mutual development of the mass production and the custom-tailored production will come true in the future of Chinese design industry.

Keywords: Internet · 3D printing · Design industry · Commercial operation pattern

1 The Technology of 3D Printing

1.1 The Outline of 3D Printing

3D Printing was created at the end of last century in America. This is a creation technology which based on the digital file model then use the bonding materials, such as powder metals, plastics, under the control of computer, 3D Printing could create a three-dimensional object layer by layer. 3D Printing also be called additive manufacturing or rapid manufacturing. We can get some information from its name. Firstly, 3D Printing is different from the traditional subtractive manufacturing which means modelling by removing any unnecessary parts from a bulk material, it could use some materials to build products directly layer by layer. 3D Printing is an additive manufacturing method, so it with no need for mould and will not waste a great deal of materials. In the field of metalworking, the waste caused by the 3D Printing account for merely 60% of the conventional way. And the 95%–98% waste caused by the additive manufacturing could be sent for recycling during the other 3D Printing process, this is a great help to reducing the cost of production. Secondly, the time spend to producing a product by 3D Printing could be shorter than the traditional way. 3D Printing could make a product in an incredibly short period of time; shorten production cycle, save the cost in time. In addition, 3D Printing combines high precision and personalized customization with whole

B. J. Jansen et al. (Eds.): *International Conference on Cognitive based Information Processing and Applications (CIPA 2021)*, LNDECT 85, pp. 558–564, 2022.
https://doi.org/10.1007/978-981-16-5854-9_71

shaped technology and on line volume production. If the traditional technology want to build something, the company have to follow the cumbersome process, such as select the factory, send the layout design, make the prototype, modify the sample. Now the 3D Printing could achieve manufacture products according to the requirements of the customer, the customer just need to upload the data of its ideal, then the computer could generate the image of the sample, the customer could see every details before building a material object and modify it until the customer feel satisfaction with the product, then we can print if by 3D Printer directly and quickly [1].

Meanwhile, the developments of the 3D Printing are facing a plenty of troubles. First of all, there are some limitations in the field of the materials which be used to do 3D Printing. At this stage, only plastics, metals, resins could be used to 3D Printing. On account of the limitation in the field of materials, lot of industries could not adopt this new technology – 3D Printing, so we have to try our best to develop the materials which are suitable for the 3D Printing. Then, the price of the high-level 3D Printers still is too expensive for most companies. The technology of 3D Printing cannot be spread to almost any industry in China, due to the limitation of the cost of high-level 3D Printers. Whats' more, although the 3D Printing technology could achieve high precision in theory, this technology have high demand for the operation, not only in the part of preliminary data processing, but also in the stage of medium-term process and later maintenance. So China have to cultivate more and more talents in the filed of 3D technology. Last but not the least, the operation of the online devices which used to achieve the cheap, high-volume production is too difficult for the ordinary being. The barriers to entry this industry this industry are higher than some plain industries. Therefore we need professionals [2].

1.2 The Development and Application of 3DP

The technology of 3D Printing technology relies on the Internet technology's development and prosperity, it also began to boom. Today, Europe and America are the most advanced regions in the field of 3D Printing technology. American companies 3D Systems and Stratasys owned the most advanced 3D Printing technologies among all of the same type companies and they occupied the highest market shares in the present world [3]. In the fourth quarter of 2019, 3D Systems held a news conference to announce that they already invented nearly ten kinds of new materials which are fit for 3D Printing. And in these new materials, there have some materials could adapt to heat much more than others and some materials owned biocompatibility, these new inventions provide more possibilities for the industries which need high temperature resistant products and healthcare industry. In 2020, Stratasys announced F370 3D Printer, this new type 3D Printer improved the level of automation, therefore it saved the cost of production mostly [4].

In recent years, 3D Printing technology not only penetrated into a wide range of fields in the world, but also joint develop with lots of fields in China and already deliver results. In the field of medical, in 2016, Wang Huaigui professor from Tsinghua Changguang Medical College, adopt the 3D Printing technology to "clone" pathological entity's lumbar vertebra model to help patient, then remove the tumor accurately. This is the first time use 3D Printing technology to do this kind of operation in the world. In the course of this operation, 3D Printing technology help to shorten nearly half of the time of patient's operative and anesthesia, shorten about one-third of time of patient's wound healing, and saved probably a quarter of the fees for operation. This prove that 3D Printing technology could real help to benefit patients not only in the expense also in the lessen the pain of patients [5]. In the field of education, in some Chinese elementary schools already started several practical activities about 3D Printing such as "Luban Lock", "my campus", via these activities students begin to realize what is 3D Printing technology and what effects 3D Printing technology will bring to us. These 3D Printing activities let us know that on the one hand, 3D Printing technology could help students to improve their creative abilities and actual operation abilities, on the other hand, 3D Printing technology could help to make up for the shortages which the traditional educational model have, could help to develop education better [6]. In the field of mould, Chinese Haomai company in Shangdong Province already used metal 3D Printing technology in the mould manufacturing of tyre to product tyre and try their best to achieve volume production. In the 2018 Pyeongchang Winter Olympics, Hunan Farsoon High-Technology company combined metal 3D Printing with mould to customize athletic sports equipment [7]. This approach achieve shorten development cycles and lower manufacture costs, is better than the traditional way. The laboratories in HUST (Huazhong University of Science and Technology) and SCUT (South China University of Technology) have produced some metal parts which have nearly 100% density, already be used to aviation and automotive field. In the field of automobile, Wuhan Sapu Technology company has took advantage of the 3D Printer which are made by Hunan Farsoon High-Technology company to manufacture the air conditioning equipment [8]. By this way, we cut the cost of production and improve the speed of production. In the field of design, 3D Printing technology already be used in the design industry widely, from the industrial design to today's fashion design, the use of 3D Printing technology still play an essential role in the develop of the design industry [9]. In the field of aviation, Hunan Farsoon High-Technology company issued a new kind of 3D Printer named FS121M in March 2016, this kind of 3D Printer make a big contribution to the parts manufacture in the field of aviation.

2 The Traditional Commercial Operation Model in Design Industry

2.1 Profile

The traditional commercial operation model is an essential part in any firm. Every industries' traditional commercial operation model are made up of product design, product production and product market. In design industry, traditional commercial operation model has the following characteristics.

The design is "from up to down". In the traditional design model, caused be limited by the cost and technological means, designers and their teams hard to investigate into the target audience members, therefore, in the traditional design industry, at the beginning of design, the questions are raised by the designers and their teams, such as what is audiences' demand? Audiences have what kinds of troubles? How can make audience pay attention to this product? As for how to answer these questions, on the one hand, designers and their teams will give the answers according to their experience or some existing product, on the other hand, company will held a small-scale study and then according to this small-scale study to speculate the most likely answer. So far, the traditional design industry formed its own "from up to down" model.

The productive process are "product-centric". The designers cannot make more innovations on products' form due to the limitation of traditional handicraft and cost of economics. After sending the design drawing to the production plant, during the course of manufacturing, every designs have to compromise to the old approach and old materials, in order to ensure the quantity of sale, between the practicability and innovation, companies often choose the former one.

Sales models most are "shopkeeper model". The most primitive and most fundamental sales model is the "shopkeeper model", this model is choose a place which you think have the most potential customers for your products to set up a brick-and-mortar store to sell your products. The "shopkeeper model" would be divided into three main patterns: dealer, direct business and franchisees. The dealer means the shopkeeper reach an agreement with the enterprise, the relationship between them is the buyer-seller relationship, the changes in the prices of the products will only be carried by the shopkeeper, will be irrelevant to the enterprise. Therefore this pattern will cause lower risk to the enterprises. The direct business means the enterprise choose the place which is suitable to set up a shop by itself and run this shop by itself. This pattern could be more convenient to manage the shop but the construction cost will be more and the risk will be higher. The franchisees model means the enterprise will choose the shopkeepers directly, and then this shopkeeper will choose location by itself and this model require every shopkeepers have to according to the rules to operating. So this model are benefit for the enterprises to create a brand image but the risk and the cost both are high for enterprises, meanwhile this model will impose more restrictions on shopkeepers.

2.2 The Limitation of Traditional Commercial Operation Model

The misunderstanding to the demand of products. Due to saving the cost of economics, Chinese traditional commercial operation model form a "from up to down" pattern, this kind of pattern is more easier to make inaccurate judge about what is customers need, even will give an opposite forecast, as a result, the enterprises usually cannot reach the desired result.

Inefficiency in the products' innovation. In order to ensure the sales volume, when the enterprises cannot guarantee the innovation on the product's form could be accepted by the public, they will claim that the product have to pay more attention to its practicability. In the traditional model, a product's sample need go through long time and many links. During this progress, although the designer's design drawings have a certain degress of innovation and creation, these creation maybe could not be totally understood by the plant, so the product maybe different from what designer want. Therefore in traditional design industry we usually need plant to create the first version, the second version, the third version even the fourth version. Due to the lengthy process, the companies' cost will be added, so the companies in order to avoid high cost, they will limit the creation of products.

Against the protection of copyright. During the process of send the design drawings to the plant, the product proofing and send back to the plant to modify, all exist the possibility to let out the data of product. So the traditional commercial operation model is not good at the protection to the copyright.

The products reach the customers need too much time. The traditional commercial operation model give rise to that is spend too much time and go through so many links when a company delivers their products to their customers. Under this circumstance, the products will become more easier to be hysteretic nature, then will lead to the products' sales condition cannot reach the expectation. In the Disney's production chain, the Walt Disney Company usually need to design the cartoon peripheral products about the movie which will be released in the next six months. This means that they have to prepare the products before nearly half a year. In this way, the cartoon peripheral products about this movie could be reached audiences during the film showing, this also be called "golden time". During this period, the products could be sold at the top price and reach even exceed the expected sales. However, this method do not have referential meaning to the middle and small-sized enterprises, because this method require that the companies have to have the ability to estimate the sales of the products and own a certain extent economic strength, otherwise, misjudgment will be easier to cause the unmarketable products, even cause the company near to bankruptcy.

Difficult to satisfy the personalized demand. In order to ensure the profit of the companies, the traditional commercial operation model must pay more attention to the first half of the demand curve, be more likely to choose to product the products which are in larger demand. Therefore this will lead to some minority needs be neglected and could not be satisfied.

3 The Innovation of Commercial Operation Model Under the Influence of 3D Printing

It's generally accepted that business model exists ten elements, such as value proposition, target consumer groups, distribution channels, cost structure, customer relations, core competencies, fission mode. The customer value, enterprises resources and ability, profit model consist the three dimensional model of all kinds of business model. Internet technology and 3D Printing technology both give rise to big influence to the three aspects, and today, we believe that the combination of Internet technology and 3D Printing technology could bring new blood into the older model.

Accurate prediction. In the Internet age, through the technologies of "the Internet of things" and "big data", we already achieve that get the first hand information via a wide range of market survey at much lower cost. Then according to these information, the company could predict more accurate about what kind of products the audiences will prefer and what kind of products the audiences will pay more attention to. By these new technologies, we could turn the traditional "from up to down "model into a new method which be supported by accurately predict. Compared with the older one, the new one could reduce the companies' risk mostly.

Help to spark ideas. 3D Printing technology owned so many advantages, such as low-cost, high-precision. Due to these characteristics, the companies could allow to create some innovative products, and before some products which owing to the traditional craft and ineffective communication could not be allowed to product, now they could be come true. When the companies would not due to the pressure from economic cost to restrain the designers' ideas, more and more unrestrained creativity will come true, and will give a new way of looking at tradition.

Protect the security of data information. If you want to product a product, you just need to upload the data information about your design into your own computer then operate it by yourself. When your computer generate a image that you think is perfect, you could manufacture the product by a 3D Printer directly. The whole process could be proceed in a little space, therefore this could highly protect the security of the new product and reduce the cost of copyright protection.

More time-sensitive. After combining the Internet technology with the 3D Printing technology, we could achieve that transform an uncomplicated product's drawing into a material object in fifteen minutes, and could control the error of the material object within 0.3 mm [10]. In addition, using the 3D Printing technology could save the time for sending the design drawing to the plant and modify, so the products could be faster to reach the customers. This could resolve the trouble that traditional operation model consume too much time.

It could be more convenient in market and payment. With Internet technology joining, the development of electronic commerce become more and more prosperous. This also exerts a powerful influence to the design industry. Many small and medium sized companies initiate a new pattern of receiving orders via the web. Customers could select the company which is the most satisfied from all the country even all over the world, and every company could make deals with anyone in this world. As a result, this could promote the design industry prosperity. Today's online payment platforms all are becoming more and more convenient and becoming safer. For example, Alipay already built a protecting net between the consumers and the merchants, only when the customer click the "confirm receipt" button, then Alipay will gives the money to the merchant.

Personalized customization. 3D Printing technology is different from the traditional craft, compared with the traditional one, the new one's cost is lower, and the forming speed is faster, so 3D Printing technology could make the material object according to the customers' needs. For example, using the 3D Printing technology, the manufacturers of shoes could customized the shoes according to every customer's data, the" 098 XYZ "shoes made by the Earl Stewart are the products of 3D Printing [11]. Using the 3D Printing technology to customize the products could on the basics of traditional form to catch the profit from the latter part of curve.

Decreasing operating costs. Compared to the traditional commercial operation model, the new commercial operation model do not need so many links, buyers and sellers could communicate through the Internet directly, therefore, the cost of economic could be reduced, and for buyers, this means they could buy more things use the same amount, for sellers, this means they could earn more money at the same price.

When the 3D Printing technology and Internet technology bring some good influences to the design industry, they also will cause some problems. It's hard to control the credibility of enterprises and customers. The seller may use some illegal ways to improve their good reputation, the buyer may exploit the advantage of seven days of no reason to return. The 3D Printing products mostly are customized, once be malicious return, the companies could not sell it again, so the companies will take a big risk. The customers could not see or touch the material objects, they just see the products online, so customers maybe more easier to be cheated by the unscrupulous businessmen.

4 Conclusion

The development of Internet technology and 3D Printing technology makes some big differences on Chinese design industry, especially for the traditional commercial operation model. It's important to note that the 3D Printing technology used in the design industry is a supplement not an alternative for the traditional commercial operation model. The 3D Printing technology and Internet technology make up for the deficiencies caused by the traditional one, letting the design industry could become more prosperous. Believing that the situation of alone big will be declined, the situation of mutual development of the mass production and the custom-tailored production will come true in the future of Chinese design industry.

Acknowledgements. Project source: The Influence of Internet Plus 3D Printing Technology on the Traditional Design and Its Business Model Guiding innovation fund of Northeast Petroleum University Project no.: 2018YDW-01.

References

1. Hao, H.: Analysis of the impact and challenge of 3D Printing technology on traditional art design industry in the field of computer electronics. In: Journal of Physics: Conference Series (2020)
2. Guan, Y., Wang, F.: The study of 3D Printing technology's status quo and future. Sci. Technol. Innov. (18), 78–79 (2020)
3. Zhang, F.: The study of 3D Printing's business model and current situation. News Transm. 53–55 (2021)
4. An, G.: The current situation and prospect of the 3D Printing technology. Shandong Ind. Technol. (04), 48 (2019)
5. Yu, B., Jin, L.: The study of 3D Printing's application in the field of Spinal neurosurgery. Chinese J. Invasive Neurosurg. 30–31 (2021)
6. Lei, S.: The Empirical Study of 3D Printing Promote the Development of Pupils' Innovation Ability. Central China Normal University (2020)
7. Jiang, M., Chen, J.: The 3D Printing help to make Olympic Winter Games' sports goods. Appl. Laser 120–125 (2017)
8. Lan, B.: Analyzing the application foreground of automobile industry. 79–80 (2017)
9. Xu, Y.: 3D Printing's effect on the development of industrial design. Sci. Technol. Innov. Appl. 79–80 (2020)
10. Li, P.: The Study of Modeling Approach Based on the 3D Scanning and 3D Printing. Hebei University of Engineering (2016)
11. Wang, X.: 3D-printed insoles change the way you move. Fintech Era **23** (2015)

Spread Model of Vespa Mandarinia through Machine Learning

Qiming Zhuang(✉)

Department of Mathematical Science, Soochow University, Suzhou, Jiangsu, China

Abstract. Vespa Mandarinia, the largest species of hornet discovered in the world, is viewed as a kind of invasive organism which would threaten people's life and health and have a devastating blow to the survival of European honeybees by preying on them as well as destroy their nests in USA. Therefore, an alarm was sounded to the local people and government upon its emergence. In this paper, we mainly discussed the prediction of the spread of the pest and miss-classification possibility as well as coming up with a strategy for identification.

Keywords: Vespa Mandarinia · Stochastic time-step difference equation · Convolutional neural network

1 Introduction

Being one of the largest species observed in our planet, Vespa Mandarinia with their queens able to grow in excess of 2 in. with a wingspan of 3 in. which appear and nest in spring hibernate in the soil. This kind of insect has been witnesses in Canada and the nest of which was destroyed immediately based on the perception that bumblebees are an invasive pest that preys on European honeybees in September 2019. Under this circumstances, a helpline and website established by the Washington government to facilitate the provision of reports by Hornet witnesses and victims. However, considering that people may mistake other bee species for bumblebees, this report has certain errors.

We mainly need to solve how to interpret the known data given by public reports consisting of image and text and provide the analysis of whether the spread of pest colony in a certain period of time can be predicted, and give the prediction accuracy.

2 Assumptions and Justifications

Assume that the spread of the hornet when it first invades is uncontrolled.

Assume that the queen bee has left the nest before the primary nest is destroyed.

Assume that each primary hive will produce a certain number of progeny that obey the Poisson distribution.

Assume that the climate conditions required by all Asian Hornets are similar.

Assume that as the latitude changes the number of successful nests decreases linearly.

B. J. Jansen et al. (Eds.): *International Conference on Cognitive based Information Processing and Applications (CIPA 2021)*, LNDECT 85, pp. 565–572, 2022.
https://doi.org/10.1007/978-981-16-5854-9_72

3 Stochastic Time-Step Difference Equation Model

We established a stochastic time step difference equation model to describe the spread of the pest. This is a simulation model. Without loss of generality, following the life cycle of a bumblebee, we have determined a time step. The question tells us that the life cycle of the bee colony can be expressed as follows: the queen bee normally starts to build a nest (primary nest) in the spring, and as the colony expands, the primary nest will be abandoned; Then the queen bee will build a secondary hive that can carry more Asian hornets in summer; since all worker bees except the queen bee are annual, as new queen bees appear in autumn, they will eventually die before overwintering. It can be seen that the life cycle of this bee colony presents a rigid seasonal behavior and can be described by a discrete time model.

3.1 Data Processing

The official data set has a total of 4,441 observations, and after analysis and screening, we got only 14 Positive IDs, and most of the remaining data are Negative ID and Unverified ID. In other words, only through data preprocessing can we find possible correct sighting reports. Considering that there are a lot of Negative ID data given, we use the TD-IDF process to extract 50 main features, as shown in the Fig. 1 below. Therefore, we removed 1123 observations from the data set. In this way, we made a simple prediction of the propagation of the colony without the image data.

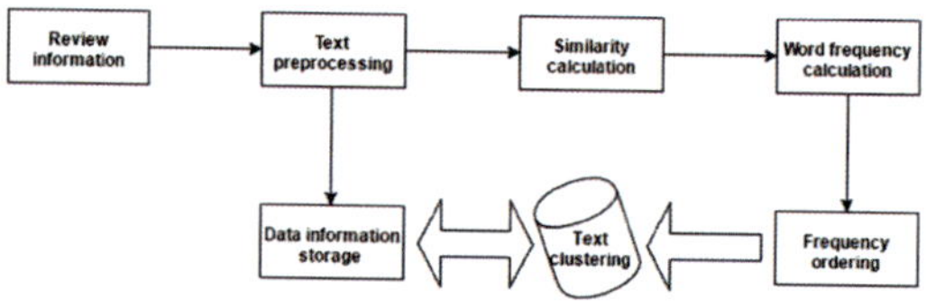

Fig. 1. TD-IDF process

Fig. 2. Word cloud

Figure 2 above vividly shows the correlation between the 50 main keywords and the accuracy of the witness report. Among them, the larger the fonts, the higher the relevance of the keyword.

3.2 The Number of Successful Queens Expressed by Possion Distribution

First of all, according to the above data processing, we assume that each primary hive will produce a certain number of progeny that obey the Poisson distribution. These offspring will build new sub-hive on this basis and continue to multiply. We believe that the mean of this Poisson distribution will gradually decrease under the influence of certain factors [1].

These factors include but are not limited to local factors, that is, local competition C, local environmental suitability E [2–6], and the latitude where the hive is located. Among them, the latitude of the hive can well describe the impact of temperature changes on the success of the hive. Based on this, we have established a simple random discrete time model, and this model can calculate the number of honeycombs well. We believe that the equation for the intensity of competition for honeycomb i (C_i) is as follows. Among them, Δ_{ij} is the distance between hive i and hive j, and the constant σ_c is a parameter measuring the local interaction distance [7–10].

$$C_i = \sum\nolimits_{j=1}^{n} \exp(-\Delta_{ij}^2/2\sigma_c^2) \tag{1}$$

The above result is a valid measure of the possibility of interaction between two nests. Assuming that this interaction is a competition of wasps on the same food source, the standard deviation is σ_c. Let $\sigma_c = \sigma_f = 2$ km be twice the average foraging distance of wasps. This indicates that the foraging circles of the wasps from the two hives must overlap. Therefore, C_i can reflect the intersection of foragers from nest i and foragers from other nests.

Considering that the environmental adaptability of a given area is determined by the Gaussian average of the habitat type, we determined the following equation to describe the environmental adaptability E_i of an area. Among them, the following cumulative summation is the summation of all habitats within the predation range.

$$E_i = \xi \sum\nolimits_{j=1}^{n} \exp(-\Delta_{ij}^2/2\sigma_f^2) \tag{2}$$

Table 1. Terrain value of different habitat type

Habitat type	Terrain value
Artificial surfaces	1 (by default)
Agriculture areas	0.87
Forest and semi-natural areas	0.14
Wetlands	0.02
Other areas	0

The terrain-specific values for each location list the appropriate weightings for each terrain-type table. In order to simplify the model, we assume that the climate and geographical conditions required by all Asian Hornets are similar. The report we provided through the title described the distribution of Asian Hornets, and thus provided local environmental suitability indicators for the region (Table 1).

Considering that there is little data about the impact of climate or latitude, we assume that as the latitude changes, the number of successful nests decreases linearly. This decreasing situation can be represented by the parameter f. Then, we comprehensively consider the estimated reproductive potential, regional carrying capacity, the influence of latitude changes and density-dependent effects and other factors, and determine how many successful queens (Q_i) the nest i will be produced based on the above factors.

Among them, we use the expression of Poisson distribution to describe the number of queen bees. In the equation, γ represents the strength of the density-dependent effects, and r is the expected reproductive ratio.

$$Q_i = \text{Possion}(\frac{r \times E_i \times f(\text{lattitude}_i)}{1 + \gamma C_i}) \tag{3}$$

3.3 Description of the Space Spread of the Asian Giant Hornet

According to the analysis and calculation above, we can find the number of successful queens. Then, considering the dispersion ability of the queen and topographic factors, we placed the hive in a certain area according to these parameters. We use Q_i to represent the number of queen bees in area i, and Q_j to represent the number of queen bees in area j. The ratio of the two is denoted as P_{ij}. In particular, the probability of a successful queen bee migrating from i to j next year and building a nest is proportional to P_{ij} expressed by Eq. (4) below.

$$P_{ij} = \exp(-\frac{\Delta_{ij}}{Q_j}) / \sum_j \exp(-\frac{\Delta_{ij}}{Q_j}) \tag{4}$$

It is worth mentioning that we adopt this form of expression to fully take into account the influence of C, E and distance while satisfying the prerequisite of simplicity.

3.4 Results of Our Model

First, we found two active nests of the Asian Giant Hornet and test the establishment and spread of nests. Without loss of generality, we make the following assumption: (1) The spread of the hornet when it first invades is uncontrolled, (2) The queen bee has left the nest before the primary nest is destroyed.

It can be seen from Fig. 3 that there were two gathering points in 2010, mainly in the northern part of Washington State. By 2015, there were three clusters, and this number was slightly higher than the real data. From the distribution point of view, two new areas, east and coastal, appeared on the basis of the original location of the hive. By 2020, the scale of Asian hornet invasion has increased exponentially, and the area where Asian hornet appears has greatly increased, and the overall trend is to spread to the west and north. Also, the predicted nest density is high in some areas.

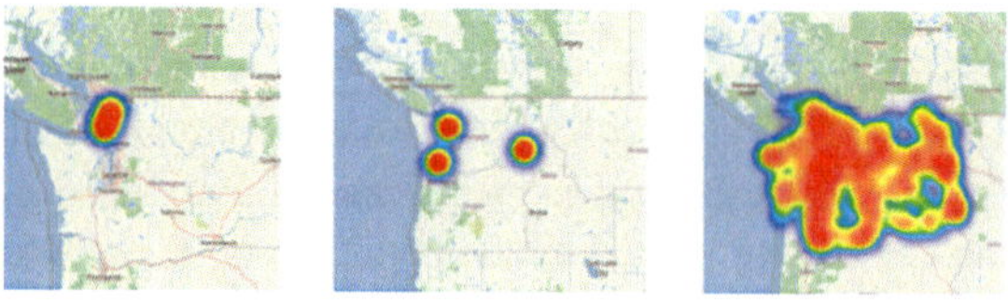

Fig. 3. Distribution in 2010, 2015, 2020

These simulations well demonstrate the reproduction mechanism of the pest. As we expected, the early dynamic propagation of it mainly depends on the spread of nests and the expected number of successful founders of each nest, and has certain randomness. The long-term dynamics of the bee species' spread are determined by latitude changes and the suitability of the local habitat. After that, we will further focus on the influence of latitude on the distribution. However, we must pay attention to the uncertainty and random variability of parameter variables in the process of analysis and calculation, which may have a significant impact on the final size of the pest, as is shown in Fig. 4.

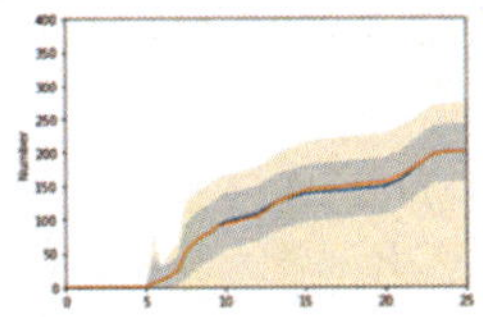

Fig. 4. Impact of parameters on colony size

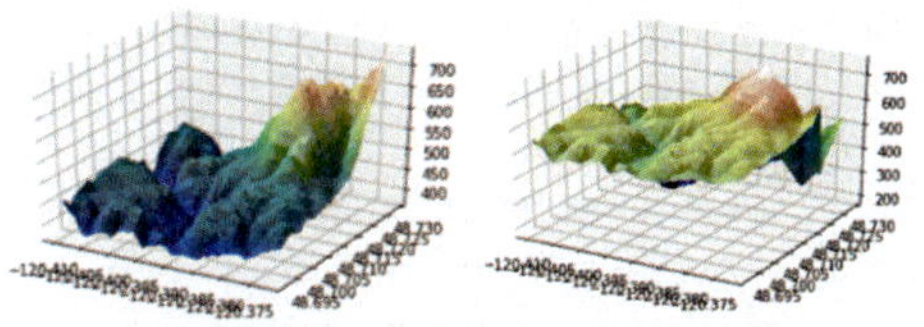

Fig. 5. 3D map of correlation

Then, we made a pair wise comparative analysis of the correlation between C, E, and latitude factors and the results. We can see that the distribution of honeycombs is mainly manifested in the trend of spreading from various centers to surrounding areas, which is mainly caused by factor C. By comparing the correlation between the distribution and E, we found that they are highly correlated. At the same time, we use Python to visualize the data, and get Fig. 5 (left) (Correlation between longitude & distribution) and Fig. 5 (right) (Correlation between latitude & distribution).

We can clearly see that the influence of latitude is significant. We know that the WSDA took a specimen of the pest in Washington in December. This is the first time this invasive bee species has been discovered in the United States. This shows that Washington is probably the starting point for its invasion, and it is mainly caused by human activities and other factors. In the subsequent invasion process, the nest distribution center gradually shifted to low-latitude areas, indicating that latitude has a significant impact on the reproduction of wasps. According to the literature we searched, the suitable temperature for the Asian hornet is about 102 °F, which supports the fact that the colony migrates to low latitudes.

Therefore, in the short term, the prediction of the model will fluctuate to a certain extent due to the probability, but the model results are good in the long term.

3.5 Prediction Accuracy of the Model

With the model we build above, we can successfully make the prediction of the spread of Asian Giant Hornet in a specific period of time. And to test the accuracy of our prediction results, it is a necessity that we calculate the MSE of latitude and longitude.

In Fig. 6, we found that through simulation, the number and location of nests did not deviate much from the real situation at the beginning of the simulation. But as the simulation progresses, this deviation will gradually increase. The initial bias is mainly

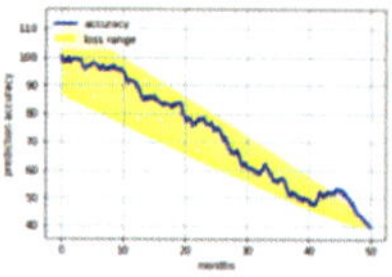

Fig. 6. Mean square error of latitude and longitude

caused by the randomness of the simulation, while the later bias is mainly caused by the delay of the report and the lack of initial intrusion reports. In the reference literature, the probability that the nest is not found is also proved to be very high, reaching 96%. This also provides reason and motivation for our future traceability research to improve the model results.

At the same time, based on calculations, we found that the accuracy of the model reached at about 46.54746%, which shows that the model is able to explain and predict the spread of the Asian Giant Hornet.

4 Image Identification Based on Convolutional Neural Network

To judge the correctness of the images, we use CNN to recognize them.In Convolutional Layer we extract image features. In order to identify the image information, we use the classic method of extracting the information of the three elements of RGB from the image. The convolution kernel slides on the image and performs a two-dimensional convolution operation. Each convolution kernel can be extracted from the output image. Features are called feature maps. In this way we can successfully extract the information of each color of the image. Figure 7 below simply shows a schematic diagram of this process.

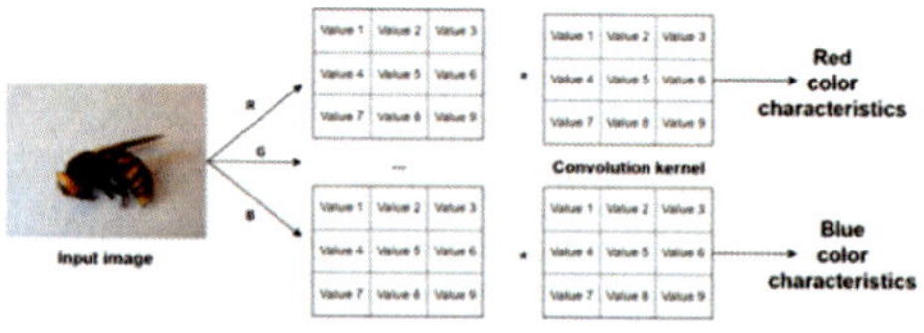

Fig. 7. Schematic diagram of process in convolutional layer

Let x_j^l be the jth feature map of the lth layer in the network, and it can be expressed by the following formula.

$$x_j^l = f(\sum\nolimits_{i \in M_J} x_j^{l-1} \times W_{ij}^l + b_k^l) \quad (5)$$

In equation, $l-1$ represents the previous layer of the l layer, M_J is the set of feature maps x_j^{l-1} in the $l-1$ layer, W represents the convolution kernel x_j^{l-1} is the weight matrix of the feature map x_j^l to the feature map x_j^l, b_k^l is the offset of the output feature map, and f is the activation function. Here we use the *ReLU function* by default in the j activation layer.

4.1 Results of the CNN Model

According to the above analysis, we use CNN to establish a model for recognizing the correctness of images and got an image that matches the basic characteristics of the Asian Hornet. In this part, we use the loss function to quantify the difference between the prediction and the real situation, and get the difference map (Fig. 8). Additionally, each epoch contains 3000 parameter updates.

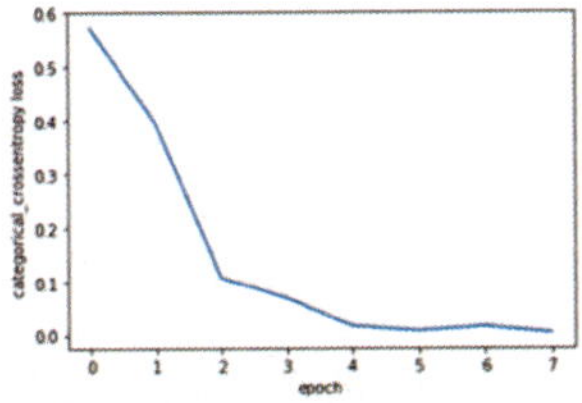

Fig. 8. Difference map

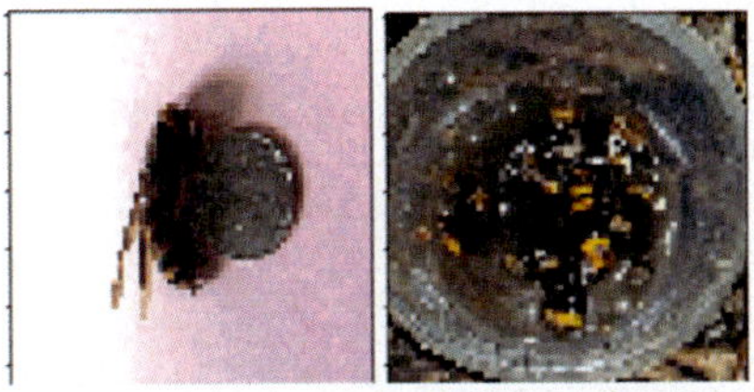

Fig. 9. Extracted features of the pest

It can be seen from the *Categorical cross-entropy loss-Epoch* line chart obtained from the operation that after one iteration, the loss reduction rate is still fast, and it is higher than the loss reduction rate generated in the first iteration; after the second iteration, The downward trend of loss has slowed down, and after the third and fourth iterations, the rate of loss decline has stabilized, and the loss value has remained at a very low level. The final accuracy rate we obtain can be as high as 99.48%.

In order to better show the image recognition effect of our model, we also extracted the Asian Hornet features that were considered the most significant after training shown in Fig. 9 (Left). It is worth mentioning that the image we output is compressed (64 * 64), so the resolution will be low and the image will be blurred.

It can be seen from picture above that after feature extraction, color and size of them are the two main factors to determine whether the picture meets the requirements, with its body length is approximately greater than the diameter of a coin, which is between 1 and 2 in. This matches the basic characteristics of the Asian Hornet. We compared the information here with the teleprompter result obtained in the data processing stage of the first model, and found that the wrong description often contains wrong features such as 'big' and 'black'. This also illustrates the consistency of the results.

References

1. Keeling, M.J., Franklin, D.N., Datta, S., et al.: Predicting the spread of the Asian hornet (Vespa velutina) following its incursion into Great Britain. Sci. Rep. **7**, 6240 (2017). https://doi.org/10.1038/s41598-017-06212-0
2. Alaniz, A.J., Carvajal, M.A., Vergara, P.M.: Giants are coming? Predicting the potential spread and impacts of the giant Asian hornet (Vespa Mandarinia, Hymenoptera: Vespidae) in the united states. Pest Manag. Sci. **77**(1), 104–112 (2020)
3. Zhu, G., Illan, J.G., Looney, C., Crowder, D.W.: Assessing the ecological niche and invasion potential of the Asian giant hornet. Proc. Natl. Acad. Sci. **117**(40), 24646–24648 (2020). https://doi.org/10.1073/pnas.2011441117

4. Nuñez-Penichet, C., Osorio-Olvera, L., Gonzalez, V.H., et al.: Geographic potential of the world's largest hornet, Vespa mandarinia Smith (Hymenoptera: Vespidae), worldwide and particularly in North America. PeerJ **9**, e10690 (2021). https://doi.org/10.7717/peerj.10690
5. Mohammed, B.M., Mohammed, Y.: Eigen-CAM: visual explanations for deep convolutional neural networks. **2**(1) (2021)
6. Krizhevsky, A., Sutskever, I., Hinton, G.: ImageNet classification with deep convolutional neural networks. Commun. ACM [Serial Online] **60**, 84–90 (2017)
7. Smirnov, E.A., Timoshenko, D.M., Andrianov, S.N.: Comparison of regularization methods for ImageNet classification with deep convolutional neural networks. AASRI Procedia **6**, 89–94 (2014)
8. How Integrated analytic hierarchy process and its applications – a literature review. Eur. J. Oper. Res. **186**, 211–228 (2008)
9. Vidal, L., Marle, F., Bocquet, J.: Using a Delphi process and the Analytic Hierarchy Process (AHP) to evaluate the complexity of projects. Expert Syst. Appl. **38**, 5388–5405 (2011)
10. Perrard, A., et al.: Geographic variation of melanisation patterns in a hornet species: genetic differences, climatic pressures or aposematic constraints? PloS One **9**, e94162 (2014)

Application of ZigBee Wireless Sensor Network Technology in the Cultivation of Innovative Talents

Mingxiang Cao(✉) and Guoxiong Xiang

School of Humanities, Guangdong Peizheng University, Guangzhou 510830, Guangdong, China

Abstract. In the era of knowledge economy, a country's international competitiveness will be marked by its creativity, and a country's overall creativity will be determined by the quantity, quality and structure of its innovative talents. Innovative talents have become a symbol of the country's comprehensive national strength and a core element of national competitiveness. The cultivation of innovative talents (CoIT) is a systematic project. In recent years, the use of ZigBee wireless sensor network technology (WSNT) to cultivate innovative talents has been rapidly developed. ZigBee is an emerging short-distance, low-speed, low-cost, and low-power wireless network technology. It has strong self-organization, self-healing capabilities and high security, and is very suitable for wireless sensor networks. The purpose of this article is to study the application of ZigBee WSNT in the CoIT. By studying relevant domestic and foreign documents, this paper constructs the ZigBee WSNT in innovative talent training system from a multidisciplinary perspective, analyzes the CoIT in universities, and puts forward corresponding training strategies and suggestions. The results of the study showed that 7.6% of the people said that there were "a lot of", 70.1% of the people said that "Yes, but not much", 12.7% of the people said that "Basically not", and 9.6% of people said they were not very clear. On the whole, our country's higher education has begun to recognize the importance of innovative talent training, but it still needs to increase its efforts.

Keywords: ZigBee · Wireless sensing · Network technology · Innovative talent training

1 Introduction

Today in the 21st century, knowledge is power, and innovation determines overall national strength. To enhance the overall national strength and core competitiveness, the key lies in cultivating outstanding talents, and ultimately in the development of education [1, 2]. However, talent is currently the most lacking resource in the world. China is now the largest developing country in the world [3, 4]. Due to the influence of many factors such as history, although China is a populous country, it still lacks human resources and talents [5, 6]. While the population quality is relatively low overall, and talents are

B. J. Jansen et al. (Eds.): *International Conference on Cognitive based Information Processing and Applications (CIPA 2021)*, LNDECT 85, pp. 573–583, 2022.
https://doi.org/10.1007/978-981-16-5854-9_73

very scarce compared to developed countries, my country's effective utilization of existing talents is not high. Therefore, the use of ZigBee WSNT is of great significance to the CoIT.

In the application research of ZigBee WSNT in the CoIT, many scholars have conducted research on it and achieved good results. Internationally representative and influential wireless sensor network practical and R&D projects include remote battlefield sensor system, network center warfare and smart sensor network, smart dust, Intel & reg; Mote, Smart-Its project, Sens IT, Sea Web, Behaviour Monitoring Project and Royal Network, etc. [7]. In particular, the recently successfully tested low-cost US military "wolf pack" ground wireless sensor network marks the latest breakthrough in the field of electronic warfare [8]. In terms of talent training, American colleges and universities have adopted a model that emphasizes both research-type and expert-type doctoral training, and has established extensive cooperative relations with all parties in the society. American universities have trained many scientists for the United States. Among the Nobel Prize winners in natural sciences, the number of American scientists ranks first in the world [4]. There are many well-known universities in the United States. Harvard University not only pays attention to the teaching of knowledge, but also pays more attention to the cultivation of students' innovative and practical abilities, and at the same time enhances their personality charm. These all provide important conditions for cultivating innovative talents [9].

Through reading literature and consulting related materials, this paper compares the current scientific and technological innovation talent training models between some domestic institutions of higher learning and some developed countries, and finds out the differences, so as to put forward suggestions and countermeasures for cultivating innovative talents in our country. Through questionnaire surveys, the current situation of innovative talent training is investigated and summarized, and conclusions are drawn based on data statistics, which provide more powerful data support for the discussion of this article, making it more convincing and credible [10, 11].

2 Application of ZigBee WSNT in the Cultivation of Innovative Talents

2.1 Characteristics of Scientific and Technological Innovation Talents

Innovative talents are the all-round development of people. They have innovative basic systems, innovative power systems, innovative intelligent systems and innovative realization systems. That is, to have a broad knowledge system, outstanding personality, innovative thinking mode and key ability to practice. The specific characteristics are as follows:

(1) Have a broad knowledge system
In the era of knowledge economy, having rich and comprehensive scientific knowledge is an essential element for a person to become a talent. Abundant and comprehensive knowledge not only requires the study of theoretical foundation and professional knowledge, but also requires knowledge summary and accumulation

on this basis. Innovation is a pioneering activity on the basis of the predecessors. It is based on the theoretical knowledge and practical experience of the predecessors to put forward innovative viewpoints and obtain new results that are conducive to the development of human society.

(2) Have outstanding personality
Outstanding personality character refers to the unique character characteristics of a person when he grows up. Through the study of the biographies of some Nobel Prize winners and academicians of the two academies, it can be seen that these technological innovation talents basically have tenacious willpower, a strong sense of social responsibility, and high-intensity resistance to pressure. These are all personalities. From some relevant literature in psychology, we can see that a person's personality directly affects his behavior and thus his creativity. Therefore, we should pay attention to the cultivation of individual character.

(3) Have an innovative way of thinking
Thinking is a unique talent of mankind, which is produced and developed by mankind in the social practice of labor activities and language communication. Innovative thinking is a systematic, complex, high-level thinking of human beings, and it is the core influencing factor in the formation of innovative ability. Without innovative thinking, it is impossible for people to form an awareness of innovation, nor can they have the ability to innovate. In teaching, the cultivation of scientific and technological innovation talents should pay special attention to the cultivation of thinking, and cultivate students' efficient use of thinking, so that students will eventually have good creative thinking ability.

(4) Possess the key ability to practice
The key ability of practice refers to the ability to use the existing knowledge system, combined with innovative methods, to propose new inventions, new schemes and new results on the basis of previous research. It refers to the ability to raise questions and the ability to analyze and solve problems. The key to scientific and technological innovation is the realization of innovative activities. It is not enough to have extensive knowledge and innovative thinking. It also requires keen observation, strong reasoning ability, and the ability to transform innovative results. Therefore, we must pay attention to the cultivation of the key ability of students' innovative practice.

For scientific and technological innovation talents, it is not enough to have a broad knowledge system. They also need to have excellent personality, innovative ways of thinking and key abilities in practice. These four points range from innovative basic systems, power systems, intelligent systems and achieve a systematic perspective and comprehensively define scientific and technological innovation talents.

2.2 Use ZigBee Wireless Sensor Network Technology to Create an Innovative Talent Training System

(1) ZigBee WSNT innovation talent training environmental system

1) Educational environment
The educational environment in a broad sense is a multi-dimensional and multi-level system. The system is centered on education, and the natural environment, humanistic environment, and social environment all have varying degrees of impact on education. This is what Mengmu's three relocation describes. The educational environment mentioned here mainly refers to the internal educational environment of colleges and universities, including the physical environment, cultural environment, and psychological environment of college campuses.
2) Educational concept
Innovative educational concepts are the ideological guide for colleges and universities to train innovative talents in science and technology. Innovation is the core competitiveness between the country and the country, and its specific implementation is implemented in the training of innovative talents in various universities, especially those with the quality of scientific and technological innovation talents.
3) Education system
System belongs to the category of production relations, and it is a procedure that all members of a certain group must abide by. In a broad sense, the education system takes the country as a unit. Different countries have different education systems. It is an organic combination of various organizations and institutions related to education and their rules and regulations. It is an aspect of the social system in the social structure. The training of scientific and technological innovation talents in colleges and universities depends to a certain extent on the education system.

(2) ZigBee WSNT innovation talent training structure system

1) Teachers
The faculty is the basis for the training of scientific and technological innovation talents in colleges and universities. Technological innovation is not only for students, but also a requirement for college teachers. Innovative education is the foundation of the cultivation of scientific and technological innovation talents, and teachers play an important role in innovation education. The teaching staff directly affects the quality of the cultivation of scientific and technological innovation talents. The more prominent the creativity of teachers, the stronger the creativity of students, and the ability of teachers to innovate the level of education directly affects the teaching effect, thus affecting the CoIT.
2) Curriculum
Generally speaking, the definition of curriculum design refers to the setting of all courses and the arrangement of class hours in the school teaching process.

People's understanding of "curriculum setting" still stays in the arrangement of subject courses, and this understanding is one-sided. On the one hand, considering that the curriculum of colleges and universities reflects the focus of teaching work to a certain extent, on the other hand, because the cultivation of innovative scientific and technological talents is one of the important teaching goals of colleges and universities.

3) Teaching methods
The teaching method is a set of flexible teaching system formed according to the teacher's teaching ideas and subject characteristics, including guiding ideology, teaching methods, basic theories and implementation methods. The teaching method is to complete the teaching task with high efficiency and high quality, and the training method and the best path to achieve the educational goal.
4) Subject structure
The scientific, standardized, and rational discipline structure is a common requirement for the cultivation of scientific and technological innovation talents, in the process of building innovative scientific and technological talents.

(3) ZigBee WSNT innovation talent training management system

1) Teacher management
Teachers are the main element of human resources in colleges and universities. The key to cultivating scientific and technological innovation talents lies in improving the quality of teaching, that is, the overall quality of the teaching staff and the management level of managers. Therefore, colleges and universities must promptly discover their problems in teacher management and innovate teachers. Management concept and implementation of innovative management.
2) Student Management
College students are a group of young people with active thinking, open personality and open-mindedness. Colleges and universities should pay attention to the management of students, and their management should involve all aspects and flexibility, actively explore, and provide a brand-new management model for the cultivation of talents.
3) Performance management
Performance management belongs to the category of human resource management. It is a comprehensive system activity. It is more common in the management activities of enterprises and organizations. The goal is to improve individual capabilities while achieving corporate goals. The methods include communication and counseling. The purpose of university performance management is to improve its own competitiveness.
4) Coordination management
The objects of coordinated management include teachers, students, leaders, departments and other links. The ability to coordinate management is the ability to resolve contradictions.

(4) ZigBee WSNT innovation talent training input system

1) Scientific research investment
   Scientific research investment is a material prerequisite for cultivating talents in universities, and the intensity of scientific research investment is closely related to the effect of cultivating scientific and technological innovation talents.
2) Teaching investment
   Teaching work is the backbone of all tasks in any university, and scientific teaching input can promote the potential of teachers' teaching work. Colleges and universities should balance the investment in scientific research and teaching, and rationally and flexibly arrange the proportion of investment in scientific research and teaching.

(5) ZigBee WSNT innovation talent training and incentive system

1) Target incentives
   The object of college goal motivation is teachers. Colleges and universities take the cultivation of scientific and technological innovation talents as their goal, and fully mobilize teachers' innovative teaching enthusiasm by scientifically satisfying teachers' reasonable needs.
2) Material incentives
   The objects of material incentives in colleges and universities include students and teachers. The various scholarships, competition bonuses, project funds, and patent bonuses provided by colleges and universities to students are material incentives; for teachers, the improvement of housing conditions, wages, and scientific research equipment are all material rewards.
3) Emotional motivation
   The objects of emotional incentives also include students and teachers. It is complementary to material incentives. It belongs to the category of respect and self-realization among the five major needs. Colleges and universities create a good and comfortable learning and research environment for students and teachers, care for teachers and students, and postgraduate tutors trust students, give them challenging scientific research tasks.
4) Evaluation rewards
   The object of college appraisal incentives is teachers, including positive incentives and reverse incentives. Colleges and universities should be rewarded for innovative evaluations, and point out the staged results and existing staged problems that teachers have achieved in the teaching process through multiple dimensions of teacher evaluations, so as to encourage teachers to find their own shortcomings from the fundamental point of view, and strive to improve their innovative teaching Level.

(6) ZigBee WSNT innovation talent training and evaluation system

1) Teaching evaluation
Teaching evaluation in a broad sense includes the evaluation of teachers' work and the evaluation of students' learning. Here, it mainly refers to the evaluation of teachers' work. The subject of teaching evaluation should include groups that have a direct or indirect relationship with the teacher's teaching; evaluation methods should also be diversified, especially in the process of innovative teaching, avoiding past unilateral and formal evaluation forms and evaluations. The phenomenon of the subject's passive evaluation.
2) Evaluation of learning outcomes
The evaluation of learning achievement is the most basic evaluation of the school's learning situation of the students. However, although it is basic, it cannot be equated with singularity. It should be a diversified evaluation method that covers basic professional knowledge, comprehensive literacy, and innovation capabilities.
3) Evaluation of innovation achievements
The object of evaluation of innovation achievements here is college students. When evaluating innovation achievements, colleges and universities should follow scientific principles, classify the results, set up evaluation grades, implement corresponding incentive measures based on the evaluation results, encourage students to actively participate in innovation activities, and guide students have produced more innovative achievements, which has promoted the formation of a strong atmosphere of technological innovation on the university campus.

3 Application Experiment of ZigBee WSNT in the Cultivation of Innovative Talents

3.1 Questionnaire Distribution and Collection

The total number of sample cases planned to be completed in this questionnaire survey is 400, a total of 396 questionnaires were returned, and 2 questionnaires were invalid such as unanswered questions. Therefore, the actual number of completed survey cases was 394, and the response rate of the survey was 98.5%. Both the recovery rate and the effective recovery rate are relatively high, which meets the requirements of the questionnaire survey.

3.2 Statistics

This article uses SPSS 22.0 software to count and analyze the results of the questionnaire, and conduct a t test. The t-test formula used in this article is as follows:

$$t = \frac{\overline{X} - \mu}{\frac{\sigma X}{\sqrt{n}}} \tag{1}$$

$$t = \frac{\overline{X_1} - \overline{X_2}}{\sqrt{\frac{(n_1-1)S_1^2+(n_2-1)S_2^2}{n_1+n_2-2}(\frac{1}{n_1} + \frac{1}{n_2})}} \quad (2)$$

$$t = \frac{\overline{d} - \mu_0}{s_d/\sqrt{n}} \quad (3)$$

4 Application Experiment Analysis of ZigBee WSNT in the Cultivation of Innovative Talents

4.1 Development of Innovative Education and Related Activities in Colleges and Universities

The results of whether colleges and universities have carried out activities related to innovation education are shown in Table 1.

Table 1. Whether the university has carried out the results of activities related to innovation education

	Number of people	Percentage (%)	Effective percentage (%)	Cumulative percentage (%)
A lot	30	7.6	7.6	7.6
Yes, but not much	276	70.1	70.1	77.7
Basically no	50	12.7	12.7	90.4
do not know much	38	9.6	9.6	100.0

From Table 1 and Fig. 1, it can be seen that 7.6% of the people said that there are "a lot", 70.1% of the people said that "Yes, but not much", 12.7% of the people said that "Basically not", and 9.6% of people said that not sure. On the whole, our country's higher education has begun to recognize the importance of innovative talent training, but it still needs to increase its efforts.

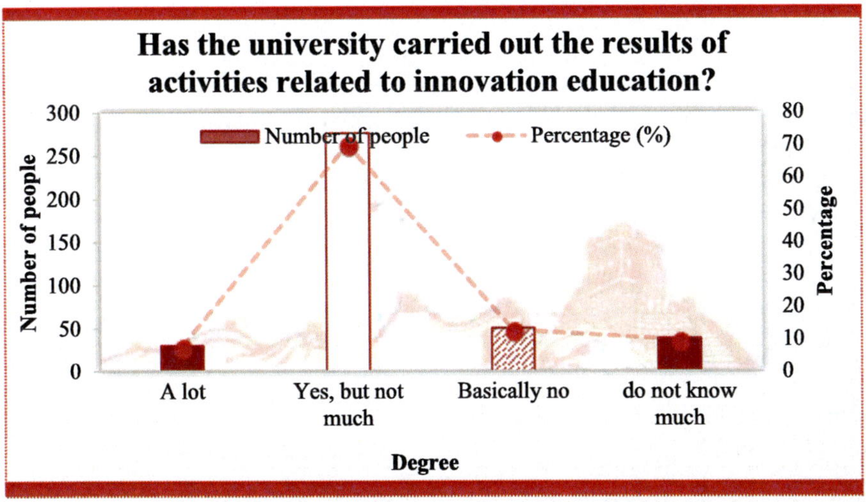

Fig. 1. Has the university carried out the results of activities related to innovation education

4.2 Situation of College Students' Awareness of Innovation

In terms of the subject consciousness of students, the question set is "Do you usually pay attention to the cultivation of your own innovative thinking and innovative consciousness?" The specific results are shown in Fig. 2.

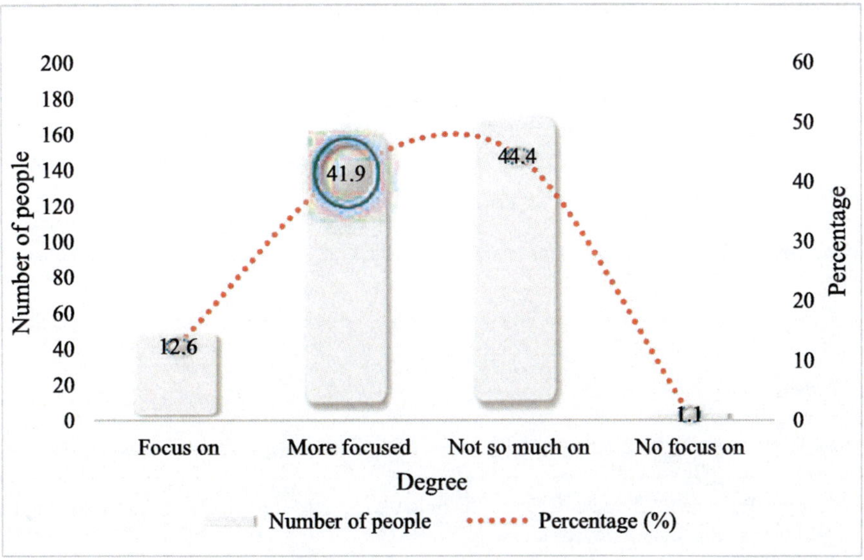

Fig. 2. Do you usually pay attention to the cultivation of your own innovative thinking and awareness

Regarding the subject consciousness of students, the question set is "Do you usually pay attention to the cultivation of your own innovative thinking and innovative consciousness?" 12.6% of the people "pay attention", 41.9% of the people "pay more attention", and 44.4% of the people "do not pay attention" Too much attention", 1.1% of people choose "not to pay attention".

My country began to put forward the slogan for the CoIT in the 1980s, and the domestic call for the CoIT has always been high. At present, many colleges and universities have also carried out useful exploration and practice in the CoIT, but not many colleges and universities have truly realized the CoIT.

5 Conclusions

In the 21st century today, the international competition for comprehensive national strength is becoming more and more fierce. The essence of competition is the competition of talents. There is a saying that says: the prosperity of the country depends on talents, the cultivation of outstanding talents depends on education, and the sustainable development of education depends on reforms. In 1999, in order to meet the society's demand for talents, colleges and universities began to expand their enrollment and vigorously set up higher vocational education. However, due to the constraints of the system and mechanism, the positioning of higher vocational education is not clear enough, and the target positioning of talent training is not clear enough. Scholars once called it the "compressed biscuit" of undergraduate education, and pointed out the current shortcomings of higher vocational education sharply. Colleges and universities follow the example of undergraduate colleges in terms of professional construction and curriculum setting. Therefore, there is still a considerable gap in meeting society's demand for talents. At present, there are quite a lot of achievements in researching innovative talents and applied talents training model, but there are few researches on training models of applied innovative talents. In order to further optimize the college's talent training model, this article uses ZigBee WSNT to make suggestions for the college to build an application-oriented innovative talent training model from several important elements of the talent training model.

Acknowledgement. Education Department of Guangdong Province 2017 Young Innovative Talents Program: 2017WQNCX151.

References

1. Meka, S., Fonseca, B.: Improving route selections in ZigBee wireless sensor networks. Sensors **20**(1), 164 (2019)
2. Kaliyar, P., Lal, C., Choudhary, C.M., Sharma, L., et al.: Multi-constraint Zigbee routing to prolong lifetime of mobile wireless sensor networks. Int. J. Ad Hoc Ubiquit. Comput. **31**(4), 244 (2019)
3. Bai, N., Gong, S., et al.: Improving throughput of communication link in IEEE 802.15.4 based energy-harvesting wireless sensor network. Chin. J. Electron. **28**(04), 179–187 (2019)

4. Truong, V.T., Lone, S.A.: System performance of wireless sensor network using LoRa-Zigbee hybrid communication. CMC **68**(2), 1615–1635 (2021)
5. Nourildean, S.W.: ZigBee-based wireless sensor network topologies using one and multiple coordinators. Periodicals Eng. Nat. Sci. (PEN) **8**(3), 1625–1640 (2020)
6. Hasan, A.K., Kadhim, F.A.: Solving isolated nodes problem in ZigBee pro for wireless sensor networks. Cihan University-Erbil Sci. J. **3**(2), 31–36 (2019)
7. Zhu, W., Qu, Q., et al.: Study on problems and countermeasures of cultivation of marine science and technology talents in China. Meteorol. Environ. Res. **10**(06), 61–63+67 (2019)
8. Wang, H.: Precision agriculture monitoring model based on wireless sensor network. Revista de la Facultad de Agronomia **37**(1), 526–537 (2020)
9. Bianchi, V., Ciampolini, P., De Munari, I.: RSSI-based indoor localization and identification for ZigBee wireless sensor networks in smart homes. IEEE Trans. Instrum. Meas. **68**(2), 566–575 (2019)
10. Zeng, Z., Zeng, F., Han, X., et al.: Real-time monitoring of environmental parameters in a commercial gestating sow house using a ZigBee-based wireless sensor network. Appl. Sci. **11**(3), 972 (2021)
11. Deshmukh, P., Ladgaonakr, B.P., Adat, D.M., et al.: Monitoring and control of gas leakages of industrial sector using pic 18f4550, ZigBee and wireless sensor actuator network. I-manager's J. Electron. Eng. **8**(3), 5–13 (2020)

New Media Technology System and Development Path of Broadcasting and Hosting Style in the New Media Era

Yini Chu(✉) and Yang Sun

College of Journalism and Communications, Bohai University, Jinzhou, Liaoning, China

Abstract. In the new media era, the diversity and multi-level structure presented by the media have formed the multi-level main diversification of public opinion and circulation channels. The technical system of new media consists of the application system and the supporting system. Application systems include digital magazines, digital newspapers, digital broadcasting, social software, mobile TV and touch media, etc.; supporting systems include digital technology, Internet technology, and communication technology. Broadcasting and hosting in the new media era presents new characteristics such as authenticity, diversity, interactivity, uniqueness, timeliness and aesthetics. Based on the current situation, characteristics and skills of broadcasting and hosting in the new media era, this paper proposes innovative development paths for the broadcasting and hosting style to adapt to the development needs of the new media era.

Keywords: New media technology system · New media era · Broadcasting and hosting style · Development path

1 Introduction

New media is a new form of media that emerges under the support of information technology. Compared with the traditional media of newspapers, outdoor, radio and television, it specifically includes digital magazines, digital newspapers, digital TV and touch media. The most notable features of new media are interactivity and immediacy, mass and sharing, multimedia and hypertext, personalization and socialization [1]. New media can provide users with personalized or customized information; in new media era, users are more active in choosing information content and form; new media can meet the needs of users for instant interaction, and users are also producers of information, which eliminates the boundary of information between the communicator and the receiver; the new media caters to the fragmented needs of people's attention and time in the new era. New media have a fast spreading speed, a wide range of audiences, diverse channels, accurate promotion information, and convenient information operations. As a result, they are showing a relatively rapid development trend. The emergence of new media has a greater impact on traditional radio and television media.

B. J. Jansen et al. (Eds.): *International Conference on Cognitive based Information Processing and Applications (CIPA 2021)*, LNDECT 85, pp. 584–591, 2022.
https://doi.org/10.1007/978-981-16-5854-9_74

Broadcasting and hosting is an important part of radio and television programs. Only through broadcasting hosting can the program effectively convey relevant information to the audience and achieve the desired effect. With the improvement of new media technology and the enhancement of information service capabilities, and the continuous emergence of video apps and programs, it is difficult for broadcasting and hosting to gain higher influence with platform advantages, and even show a downward trend. Under the new media environment, more oral and online language broadcast method allows the audience to feel more intimacy and realism, and there are fewer limitations in the communication process [2]. The traditional serious and rigid way of broadcasting and hosting lacks entertainment and flexibility, and has not adapted to the needs of the new media era. Therefore, broadcasting and hosting must achieve style innovation, meet the aesthetic and information dissemination needs of audiences in new media era, and promote the innovative development of news dissemination.

2 Technical System of New Media

In the era of new media, the diversity and multi-level structure presented by the media have formed the multi-level main diversification of public opinion and circulation channels. The technical system of new media consists of application system and support system, as shown in Fig. 1 [3, 4].

3 Characteristics of Broadcasting and Hosting in New Media Era

Broadcasting and hosting in new media era presents new characteristics. Compared with the traditional media era, the new features are mainly reflected in the following aspects:

- Authenticity. Broadcasting and hosting need to disseminate the most authentic and fresh information to the audience. The authenticity of language is the basic characteristic and the foundation of language artistry. In the era of new media, audiences have more channels for obtaining information, and untrue information may soon be seen through, affecting the reputation of the host or the media. Broadcasting and hosting is required to have a high level of professionalism, with third-party perspective [5], comprehensively report the content of the event, and not distort the facts, nor delete or supplement the true information.
- Diversity. The basic work content of traditional broadcasting and hosting is to control the time and sequence according to the requirements of the director or the program group, and connect all links in a reasonable manner. In new media era, the form of broadcasting and hosting is becoming more and more diversified. It is possible to fully grasp the trend of the program and formulate special reporting forms according to the audience's interests and hobbies. Higher requirements are put forward for the adaptability of the broadcasting and hosting, which needs to be re-created according to the on-site situation to enhance the freshness and attractiveness of the program [6].
- Interactivity. Early broadcast interactive forms were more limited. Usually, listeners used letter delivery to pass information to the program group. After screening, the staff edited the corresponding content into the program to achieve simple interaction.

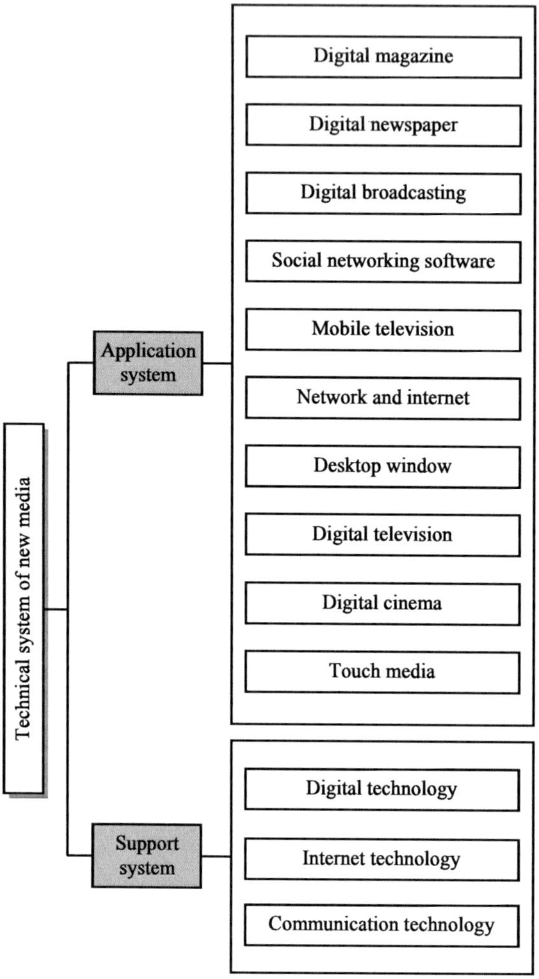

Fig. 1. Technical system of new media

Later, the interactive method represented by hotline communication achieved good interactive effects. In new media era, modern interactive methods such as messages, forums, Weibo and WeChat have broadened the channels for audiences to participate in interaction and stimulated their enthusiasm for interaction.

- Uniqueness. The traditional broadcast host has little room to play, and the age level is relatively high. In the long-term work, an inherent working mode is formed, and it is difficult to form a personalized host style. In the era of new media, if broadcasters want to gain advantages, they must show their uniqueness and present the most prominent advantages to the audience. Under the premise of ensuring the truthfulness and reliability of information, the form of expression is diversified, changing the dignified and rigid image of traditional broadcasting and hosting, highlighting their own characteristics, optimizing program effects, and attracting audiences.

- Timeliness. Timeliness is an important feature of news. Without timeliness, news loses its appeal and even its value. In the face of the rapid development of new media, traditional paper media has been severely impacted. In new media era, the latest situation of the event can be disseminated in a timely manner through live broadcast, telephone, and the Internet, and valuable information can be delivered to the audience as soon as possible. For some emergencies, we can combine the advantages of new media to produce short and concise videos to disseminate information.
- Aesthetics. The aesthetics of broadcasting and hosting is an important feature that attracts audiences, meets their aesthetic needs, and adapts to the development of new media. The audience's perception, emotion, imagination, understanding and many other psychological functions in the audiovisual process are mobilized to obtain information and cognition, enjoy pleasure and beauty, and gain experience and insight in psychological expectations and aesthetics. The audience's changing aesthetic needs for broadcasting and hosting have made the artistic beauty of language organization and the natural beauty of body shape more important.

4 Skills of Broadcasting and Hosting in New Media Era

Broadcasting and hosting in new media era have put forward higher requirements and need to master certain skills, as shown in Fig. 2:

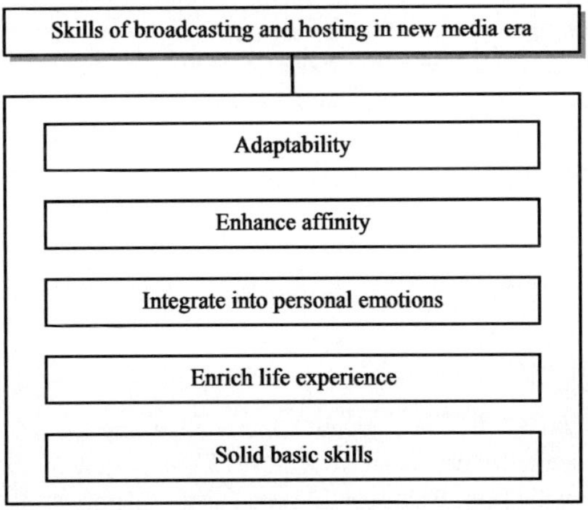

Fig. 2. Skills of broadcasting and hosting in new media era

4.1 Adaptability

The ability to adapt to changes is the basic requirement of broadcasting and hosting, and they need to calmly deal with changes. Whether it is an entertainment program or a news program, the broadcasting and hosting must have the ability to respond to various emergencies, play a unique transitional role, and make the link between the program more flexible. In new media era, live broadcasting has become the main form, and the unexpected factors buried in the live broadcasting scene will also increase. The difficulty of managing and controlling on-site guests will increase, and the temporary changes of plans will increase. Therefore, the broadcasting and hosting is required to adapt and be able to grasp various situations. Resolve various temporary problems in the live broadcast process in time [7].

4.2 Enhance Affinity

Affinity is a metaphor for bringing people close or willing to contact. Affinity stems from people's recognition and respect for people. It does not express the physical distance between people, but the accessibility of the soul. It is a kind of mutual benefit conversion based on equal waiting for people. The affinity of broadcasting and hosting can be divided into two aspects, one is the affinity of external performance, and the other is the affinity of internal emotions. In new media era, only with a strong affinity can the broadcasting and hosting close the distance with the audience, develop good communication with the audience, encourage the audience to become interested in the program, and achieve the purpose of fully disseminating information.

4.3 Integrate into Personal Emotions

Emotion is a positive or negative psychological response to external stimuli, including love, anger, sadness, fear, admiration and disgust. Emotion is the basis of broadcasting, and the ability to grasp emotion is the basis of success or failure of broadcasting. In order to increase the appeal, broadcasting and hosting needs to deeply understand the manuscript, refine the emotion of the manuscript, and recreate it artistically, integrate the true feelings, and play a correct guiding role. In the era of new media, the audience's awareness of participation has increased. Broadcasters need to take into account the emotional needs of the audience, integrate personal emotions in the process of conveying information, resonate with the audience, and strengthen the appeal of language [8].

4.4 Enrich Life Experience

Life experience is the positive and negative knowledge or experience that individuals experience from personally experienced events. Life experience takes time to temper. Abundant life experience is the wealth of the broadcaster, an important part of knowledge, the solvent for the broadcaster to dissolve theoretical knowledge, and an important factor to enhance the sense of reality and credibility. Lack of knowledge accumulation or insufficient life experience will result in empty words in the host process, lack of appeal, and unable to meet the audience's information acquisition needs. Broadcasting and hosting in new media era requires richer experience and knowledge reserves to increase the content and appeal of the program.

4.5 Solid Basic Skills

Only with solid basic skills can the broadcasting and hosting clearly display important information. Specifically include: correct and standardized language expression, solid language logic thinking, standard mandarin, clear oral expression, quick response ability and on-site control ability [9]. Broadcasting and hosting in new media era face a more complex working environment. To complete their work in a tense working atmosphere, they need solid comprehensive skills, including writing, logical analysis, editing, planning, and live hosting. Only with solid basic skills can successfully complete the program hosting and win the recognition and favor of the audience.

5 Innovative Development Paths of Broadcasting and Hosting Style in New Media Era

New media has brought new development opportunities to broadcasting and hosting. At the same time, it is also facing huge challenges. It is necessary to adapt to the information dissemination forms and characteristics of new media, grasp the development opportunities in new media era, and realize the innovative development of broadcasting and hosting styles, be invincible in the media competition [10]. The specific innovation development paths includes the following aspects:

5.1 Strengthen the Use and Integration of Language and Emotion

Emotional expression is the soul of hosting a program. Whether there is emotion in the process of hosting a program directly affects the broadcast effect. Emotional expression is the basic skill of broadcasting and hosting. If you lack emotional input and broadcast the program stiffly, you cannot express the true inner emotions and cannot enhance the appeal [11]. Incorporating real emotions into language expression, enriching the artistic charm of the broadcasting and hosting, and gaining the audience's psychological and emotional identification. When integrating emotional experience and personal opinions, it is necessary to reasonably grasp the scale of emotional integration, cooperate with humorous language, and reasonably control the fluctuations of tone and intonation to provide good emotional and aesthetic experience for the audience. Only by continuously strengthening the integration of language and emotion can the audience experience more pleasant and comfortable emotions and ensure the social benefits and influence of the broadcasting and hosting.

5.2 Focus on the Standardization of Broadcasting and Hosting Language

Broadcasting and hosting need to discover the unique laws of language, and use voice to express the emotion of the language content, vividly convey it to the audience, and arouse resonance. Language standardization is the basic requirement of broadcasting and hosting, and language artistry is the embodiment of the high level of broadcasting and hosting. New media has enriched the expressive power of language, but also challenged norms, exposed many problems, and undermined the rules of language and communication. Affected by this, broadcasting and hosting often have language irregularities, resulting in a bad cultural atmosphere, affecting the quality of broadcasting and hosting and hindering the promotion of mandarin. Therefore, language standardization is the professional standard that broadcasters must fulfill. It is necessary to actively participate in professional business training, compare with the pronunciation of professionals, accumulate life and daily work experience, persistently train pronunciation standard, and form the standardized broadcasting and hosting language [12].

5.3 Improve the Professional Skills of Broadcasting and Hosting

In new media era, broadcasting and hosting must not only be competent in basic tasks, but also have the ability to organize, plan, and compile, to make programs innovative, and to resonate with the audience through personalized hosting styles and win their love. This requires broadcasting and hosting with extensive knowledge and keen insight, through deep thinking and insights, to form unique views and insights; with good language literacy, vivid language expression, clear organization, strong logic, and superb control ability; with organizational ability and adaptability, program hosting is personalized and flexible, and the hosting method can be adjusted at any time according to needs; full of personality and innovative spirit, fully display the personality charm, and give the audience a refreshing feeling. This requires broadcasting and hosting to read extensively, think diligently, strengthen basic language skills training, open up ideas, be creative, and continuously improve their professional skills in social practice.

5.4 Shape the Personality Charm of the Broadcasting and Hosting

Personality charm is the attractive power of an individual in terms of personality, temperament, ability and moral quality. The innovation of the broadcasting and hosting style needs to enhance the personal charm of the broadcasting and hosting, especially the personality charm with highly personalized characteristics. As the controller of the entire show, broadcasting and hosting exerts his personality charm, improves his ability to respond to emergencies, and interacts with guests through language skills, creating the harmonious atmosphere on this basis. The formation of unique personality charm depends on the following four aspects: first, good character cultivation, the basis for the formation of personality charm, includes a positive attitude towards life and a noble state of mind. Second, firm belief, correct values, correct style and life habits. Third, peaceful, kind and natural personal image, distinctive personality, unique hosting style, and fresh language expression [13]. Fourth, through the spread of pictures and sounds, the connotation of the program is combined with the emotion of the broadcasting and hosting, so that the audience can enjoy the beauty in the process of watching and listening.

5.5 Enhance the Information Literacy of Broadcasting and Hosting

Information literacy refers to the reflective exploration of information, the understanding of information generation and evaluation, and the use of information to create new knowledge, and a series of comprehensive abilities to reasonably participate in the learning community, including information awareness, information knowledge, information skills, and information ethics. In new media era, broadcasting skills are still the core, but the lack of information processing capabilities has become a common problem. As the main body of information literacy, broadcasting and hosting, according to the needs of public information dissemination behavior, "determine information problems independently, judge and select related information sources to obtain relevant information, and therefore have strong pertinence and direction" [14]. Information has become the main content of broadcasting and hosting work. To efficiently select, understand and process information, we must possess excellent information literacy. To improve the information literacy of broadcasting and hosting needs to start from two aspects: first, humanities, literature, art, history, and philosophy are the basic foundations, and sociology, psychology, communication, and law are the foundations of information interpretation; second, information technology, master a variety of tools, proficiently use information technology to obtain information, effectively analyze, select, summarize, classify and store information, conduct comprehensive applications and abstract expressions, and grasp the development direction and characteristics of new media in real time.

References

1. Xi, H.: Innovation is the only way of broadcasting and hosting style in the new media environment. Media Forum **3**(6), 46–47 (2020)
2. Xu, B.R.: Innovative exploration of broadcasting and hosting in the new media environment. Satell. TV IP Multimed. **17**(3), 89–90 (2020)
3. Baidu Knows. What are the main contents of the new media? https://zhidao.baidu.com/question/556020469247065772.html. Accessed 24 Mar 2021
4. Baidu Wenku. Experience of introduction to new media. https://wenku.baidu.com/view/2e00857e52ea551810a6878a.html. Accessed 24 Mar 2021
5. Chen, Z.: Research on the skills and characteristics of broadcasting and hosting in radio and television stations from the perspective of new media. J. Commun. **36**(24), 101–102 (2020)
6. Jiang, M.M.: The art of broadcasting and hosting in the new media era. Radio TV J. **12**(5), 233–234 (2017)
7. He, C.: Probe into the strain capacity of announcer for multi-role host under new media environment. J. Beijing Inst. Graph. Commun. **26**(4), 32–34 (2018)
8. Lu, Y.: Analysis on the strategies of speech communication and emotional expression of broadcasting hosts in the new media era. Sci. Technol. China's Mass Media **30**(8), 102–104 (2019)
9. Ou, H.L.: The importance of broadcasting and hosting basic skills in radio and TV programs. J. Media Stud. **27**(1), 78 (2017)
10. Xin, Z.Y.: Innovation path of broadcasting and hosting in new media era. Res. Transm. Competence **4**(19), 112–113 (2020)
11. Li, Y.P.: Thoughts on the emotional expression of broadcast hosts. J. News Res. **7**(23), 138 (2016)
12. Zhang, W.H.: The way to innovate the host language of radio and television in the new media era. West China Broadcast. TV **41**(16), 184–186 (2020)
13. Ma, C.: Discussion on the basic quality and personality charm of broadcasting host. J. News Res. **6**(2), 22 (2015)
14. Su, Z.X.: The cultivation of information literacy in the professional education of broadcasting and hosting. Charming China **19**, 30 (2017)

Three-Dimensional Design of IMP Propeller Rotor Blade Based on Lift Method

Qiao Li(✉)

Department of Industrial Automation, Guangdong Polytechnic College, Zhaoqing, Guangdong, China

Abstract. Integrated motor pump-jet propeller, as a new type of underwater propulsion device for underwater vehicles, among them, the rotor blade is one of the core components of the IMP propeller, which is related to the stability and safety of the IMP propeller. Therefore, this article mainly focuses on the design technology of the rotor blade. The lift method is used to establish the rotor blades, and the rotor blade profile and blade three-dimensional model with the best annular distribution are obtained.

Keywords: Integrated motor pump-jet propeller · Lift method · Rotor blade design · Three-dimensional design

1 Introduction

IMP (Integrated Motor Pump-jet Propeller) is a new type of underwater power propulsion equipment that integrates the motor body, stator (guide vane), rotor (impeller) and duct. The application efficiency of this equipment is relatively better, and the radiation noise is smaller, and it has more ideal anti-cavitation and critical speed performance [1–4]. The rotor blade is one of the core components of the IMP propeller, which is related to the stability and safety of the IMP propeller. It is extremely important to obtain the rotor blade structure with the best performance. Therefore, this article mainly focuses on the design technology of the rotor blade. The lift method is used to establish the rotor blade, and the rotor blade profile and blade three-dimensional model with the best annular distribution are obtained.

For rotor blades, the design methods mainly include lift method, arc method and singular point distribution method [5–7]. Among the three methods, the lift method is already a relatively mature and complete design method. The lift method was also the first consideration when designing rotor blades. It has the widest range of use. The calculation formula and calculation cycle in the lift method design process are relatively simple and fast. Many rotor blades designs have been successfully completed with this method before. Based on this, this paper uses the lift method to design the rotor blades.

B. J. Jansen et al. (Eds.): *International Conference on Cognitive based Information Processing and Applications (CIPA 2021)*, LNDECT 85, pp. 592–598, 2022.
https://doi.org/10.1007/978-981-16-5854-9_75

2 The Design of Rotor Blade

2.1 The Selection Principle and Determination Method of the Structural Parameters of Rotor Blades

(1) Rotor diameter

The diameter of the rotor is mainly determined by the speed of the axial surface. In order to facilitate the selection of the airfoil of the blade, it is necessary to calculate the correct blade placement angle and the appropriate axial surface speed. The shaft surface speed before the rotor inlet is determined by formula (1) [8, 9], and the most suitable blade placement angle can be obtained.

$$v_m = (0.06 - 0.08)\sqrt[3]{n^2 Q} \quad (1)$$

From the formula, n is the speed of the rotor and Q is the flow rate of the pump. For the coefficient range in the formula, the low specific speed coefficient selects a small value, and the high specific speed coefficient selects a large value.

According to the continuity condition of the fluid, the axial velocity in the rotor area can be obtained by the formula:

$$v_m = \frac{4Q}{\pi D^2 (1 - (\frac{d_h}{D})^2)} \quad (2)$$

From the formula D—Rotor diameter, m
d_h—Hub diameter, m
$\frac{d_h}{D}$—Hub ratio.
Then by formula (1) and formula (2):

$$D = (4.0 - 4.6)\sqrt{\frac{1}{1 - (\frac{d_h}{D})^2}}\sqrt[3]{\frac{Q}{n}} \quad (3)$$

(2) Hub ratio, Number of the rotor blades Z

The specific rotation number of the axial flow pump affects the selection of the hub ratio d_h/D. The specific rotation number is obtained when the total efficiency of the axial flow pump is the maximum under the design of the axial flow pump and the corresponding design unit working parameters and optimal geometric parameters, generally expressed. It is generally selected according to the statistical relationship curve of wheel hub ratio and specific speed. Because the specific rotation speed is limited by hydraulic friction loss and the structural arrangement of the blades, the hub ratio of the impeller with a lower rotation number is generally larger, and vice versa.

The number of rotor blades can be selected according to the number of specific rotations, and the statistics are shown in the table below (Table 1).

Table 1. The relationship between the number of blades and the specific speed

n_s	<500	500–800	>800
Z	6	4–5	3–4

2.2 Designing of Rotor Blades by Lift Method

Basic parameters for designing rotor blades by lift method: (1) The flow rate Q; (2) The specific revolution n_s; (3) The hub ratio d_h/D; (4) The number of the rotor blades Z; (5) The impeller diameter D and hub diameter d_h; (6) The cascade density l/t and l; (7) The relative thickness of airfoil $y_{\max}/l$ and $y_{\max}$; (8) The airfoil thickness distribution; (9) The airfoil placement angle β_e.

Before designing, given the working conditions of the propeller, the pump flow rate $Q = 0.012\,\mathrm{m}^3/\mathrm{s}$, the speed $n = 750r/\min$, and the pump head $H = 0.4$ m, the specific speed $n_s = 600$ is calculated according to the following formula [8, 9].

$$n_s = 3.65n\sqrt{Q}/H^{3/4} \tag{4}$$

(1) Calculating the design power p_d of the rotor and selecting the power p_e of the motor

The input power of the thruster in the design working condition p:

$$p = Q\rho gH/1000\eta(kw) \tag{5}$$

In the formula: η—The total efficiency of the propeller.

Considering the actual work of the pump, it will be affected by various factors, so taking a design actual working condition coefficient to obtain the design power of the propeller. $p_d = 1.2p$, the power of the motor $p_e \geq p_d$.

(2) Hub ratio d_h/D, Number of the rotor blades Z

Hub ratio d_h/D is related to the specific speed of the rotor, and its value can be selected according to the following table, here $d_h/D = 0.55$, and according to the size of the specific speed, the number of blades is $Z = 5$ (Table 2).

Table 2. The relationship between the hub ratio and the specific speed

n_s	500	600	700	800	900	1000	1100
d_h/D	0.5–0.63	0.46–0.59	0.44–0.56	0.4–0.53	0.37–0.5	0.35–0.48	0.33–0.46

(3) The impeller diameter D and hub diameter d_h

According to the formula (1) the speed of the shaft surface v_m can be obtained, and then the diameter of the impeller D can be obtained, and then the hub diameter can be obtained according to the hub ratio d_h.

(4) The cascade density l/t and l

Using the lift coefficient curve $(l/t)_D = f(K_H)$ to determine the rim of the wheel l/t. $K_H = H/n^2D^2$, Taking 1.2–1.3 times that of the hub's l/t at the rim's l/t, and the other radii's l/t change linearly. The empirical relationship curve with the head coefficient and the cascade density can be found in the vane pump designing manual [10].

(5) The relative thickness of airfoil $y_{\max}/l$ and $y_{\max}$

The NACA4409 airfoil was selected in this paper, the maximum thickness of the blade at the hub can be estimated as follows:

$$y_{\max} = (0.012 - 0.015)KD\sqrt{1.5H} \tag{6}$$

In the formula, $K = \sqrt{[\sigma_1]/[\sigma_2]}$ is the material factor, σ_1 is the allowable stress for stainless steel materials, σ_2 is the allowable stress of the material used for the pump impeller.

The airfoil thickness of the other blade cascades between the two blade cascades at the outer edge of the impeller and the hub is between the thickness of the airfoil at both ends, and the thickness and the radius are linear. The values of the airfoil thickness at the radius of 38.5 40, 45, 50, 55, 60, 65, 70 mm are shown in Table 3.

Table 3. The maximum airfoil thickness at each section $y_{\max}$

Project	Unit	The aspect of calculating							
Radius	mm	38.5	40	45	50	55	60	65	70
$y_{\max}$	mm	5.9	5.7	5	4.3	3.6	2.9	2.2	1.5

(6) The airfoil thickness distribution coordinates of each section of the blade

According to the maximum thickness of the airfoil at each cutting surface mentioned above, the distribution coordinates of the airfoil thickness of each cutting surface of the blade can be obtained, as shown in Table 4. In the table, h and b are the coordinates of two points at the two ends of the airfoil at the L% of each hub on each section in the parameter coordinates x, oy, respectively.

Table 4. The cascade airfoil thickness distribution coordinates at each tangent plane

r (mm)	38. 5		40		45		50		55		60		65		70	
%L	h	b	h	b	h	b	h	b	h	b	h	b	h	b	h	b
0	0	0	0	0	0	0	0	0	0	0	0	0	0	0	0	0
1. 25	−1. 19	0. 69	−1. 15	0. 66	−1	0. 58	−0. 86	0. 5	−0. 72	0. 42	−0. 58	0. 34	−0. 44	0. 26	−0. 3	0. 17
2. 5	−1. 71	0. 9	−1. 65	0. 87	−1. 45	0. 76	−1. 25	0. 65	−1. 04	0. 55	−0. 84	0. 44	−0. 64	0. 33	−0. 43	0. 23
5	−2. 45	1. 08	−2. 37	1. 04	−2. 08	0. 92	−1. 78	0. 79	−1. 49	0. 66	−1. 2	0. 53	−0. 91	0. 4	−0. 62	0. 27
7. 5	−3. 04	1. 14	−2. 94	1. 1	−2. 57	0. 97	−2. 21	0. 83	−1. 85	0. 7	−1. 49	0. 56	−1. 13	0. 42	−0. 77	0. 29
10	−3. 52	1. 13	−3. 4	1. 09	−2. 98	0. 96	−2. 56	0. 83	−2. 15	0. 69	−1. 73	0. 56	−1. 31	0. 42	−0. 89	0. 29
15	−4. 33	1. 01	−4. 19	0. 98	−3. 67	0. 86	−3. 16	0. 74	−2. 65	0. 62	−2. 13	0. 5	−1. 62	0. 38	−1. 1	0. 26
20	−4. 8	0. 85	−4. 64	0. 82	−4. 07	0. 72	−3. 5	0. 62	−2. 93	0. 52	−2. 36	0. 42	−1. 79	0. 32	−1. 22	0. 22
25	−5. 17	0. 67	−5	0. 65	−4. 38	0. 57	−3. 77	0. 49	−3. 16	0. 41	2. 54	0. 33	−1. 93	0. 25	−1. 32	0. 17
30	−5. 4	0. 5	−5. 22	0. 48	−4. 58	0. 42	−3. 94	0. 36	−3. 3	0. 3	−2. 66	0. 24	−2. 01	0. 19	−1. 37	0. 13
40	−5. 47	0. 23	−5. 28	0. 22	−4. 63	0. 19	−3. 99	0. 17	−3. 34	0. 14	−2. 69	0. 11	−2. 04	0. 09	−1. 39	0. 06
50	−5. 15	0. 05	−4. 98	0. 04	−4. 37	0. 04	−3. 76	0. 03	−3. 14	0. 03	−2. 53	0. 02	−1. 92	0. 02	−1. 31	0. 01
60	−4. 58	−0. 09	−4. 43	−0. 09	−3. 88	−0. 1	−3. 34	−0. 1	−2. 8	−0. 1	−2. 25	−0. 1	−1. 71	−0	−1. 17	−0. 02
70	−3. 77	−0. 17	−3. 64	−0. 16	−3. 2	−0. 1	−2. 75	−0. 1	−2. 3	−0. 1	−1. 85	−0. 1	−1. 41	−0. 1	−0. 96	−0. 04
80	−2. 76	−0. 17	−2. 66	−0. 16	−2. 34	−0. 1	−2. 01	−0. 1	−1. 68	−0. 1	−1. 36	−0. 1	−1. 03	−0. 1	−0. 7	−0. 04
90	−1. 53	−0. 09	−1. 47	−0. 09	−1. 29	−0. 1	−1. 11	−0. 1	−0. 93	−0. 1	−0. 75	−0. 1	−0. 57	−0	−0. 39	−0. 02
95	−0. 83	−0. 02	−0. 8	−0. 02	−0. 7	−0	−0. 6	−0	−0. 5	−0	−0. 41	−0	−0. 31	−0	−0. 21	0
100	0	0	0	0	0	0	0	0	0	0	0	0	0	0	0	0

3 3D Design to Generate the Rotor Blades

Because the propeller rotor blade is a typical part with a special profile, this paper is based on the 3D modeling function of Solid works, and generates the rotor blade through the relationship between the airfoil profile value points and the tangent contour projection points.

3.1 Coordinate Conversion of Blade Shape Value and Solid Works

Since there are 8 radius tangent planes of rotor blades, here is a list of the model value point coordinate conversion table on each tangent plane of the hub with a radius of 40 mm. As shown in Table 5, the parameters in the second row of the table are the parameter coordinate system x, oy, and Calculate the coordinates of the two points at the two ends of the airfoil at the two ends of the hub L% of each tangent plane in the coordinate system xoy and their projection point coordinates. The conversion table of the type point coordinates of the other tangent planes is not listed. Due to the complex calculations, excel is used to process a large amount of data.

3.2 The Specific Generation of the Leaves

The single airfoil cut surface of the 38.5 mm cut surface at the hub and the single airfoil cut surface at each cut surface are shown in Figs. 1 and 2. Finally, the curved surface of a single rotor blade can be obtained, as shown in Fig. 3.

Table 5. Coordinate conversion of various value points at 40 mm of the cut surface

r (mm)	40										
%L	x'	y' (a)	y' (a')	x (a)	y (a)	x (a')	y (a')	x (a)$_t$	y (a)$_t$	x (a')$_t$	y (a')$_t$
0	−14. 88	0	0	−12. 6327	−7. 86316	−12. 6327	−7. 86316	−12. 4237	−7. 86316	−12. 4237	−7. 86316
1. 25	−14. 26	−1. 15	0. 66	−11. 4986	−8. 51185	−12. 4551	−6. 97521	−11. 3409	−8. 51185	−12. 2548	−6. 97521
2. 5	−13. 64	−1. 65	0. 87	−10. 7081	−8. 6087	−12. 0397	−6. 46929	−10. 5806	−8. 6087	−11. 8587	−6. 46929
5	−12. 4	−2. 37	1. 04	−9. 27485	−8. 5647	−11. 0768	−5. 6697	−9. 19196	−8. 5647	−10. 9358	−5. 6697
7. 5	−11. 16	−2. 94	1. 1	−7. 92092	−8. 39335	−10. 0558	−4. 9635	−7. 86925	−8. 39335	−9. 95022	−4. 9635
10	−9. 92	−3. 4	1. 09	−6. 62511	−8. 12861	−8. 9978	−4. 31673	−6. 59486	−8. 12861	−8. 92211	−4. 31673
15	−7. 44	−4. 19	0. 98	−4. 10219	−7. 48877	−6. 83422	−3. 09959	−4. 09501	−7. 48877	−6. 80102	−3. 09959
20	−4. 96	−4. 64	0. 82	−1. 75895	−6. 56028	−4. 64422	−1. 9249	−1. 75838	−6. 56028	−4. 63379	−1. 9249
25	−2. 48	−5	0. 65	0. 536742	−5. 55539	−2. 44893	−0. 7587	0. 536726	−5. 55539	−2. 44741	−0. 7587
30	0	−5. 22	0. 48	2. 758448	−4. 43163	−0. 25365	0. 407506	2. 756262	−4. 43163	−0. 25365	0. 407506
40	4. 96	−5. 28	0. 22	7. 001054	−1. 86152	4. 094643	2. 807828	6. 965363	−1. 86152	4. 087496	2. 807828
50	9. 92	−4. 98	0. 04	11. 05342	1. 014229	8. 400662	5. 276067	10. 91328	1. 014229	8. 339043	5. 276067
60	14. 88	−4. 43	−0. 09	14. 97368	4. 102218	12. 68026	7. 786755	14. 62641	4. 102218	12. 46894	7. 786755
70	19. 84	−3. 64	−0. 16	18. 76711	7. 39396	16. 92815	10. 34838	18. 08613	7. 39396	16. 42735	10. 34838
80	24. 8	−2. 66	−0. 16	22. 46014	10. 84701	21. 13905	12. 96944	21. 29838	10. 84701	20. 16872	12. 96944
90	29. 76	−1. 47	−0. 09	26. 0422	14. 47834	25. 31296	15. 64992	24. 24104	14. 47834	23. 65697	15. 64992
95	32. 24	−0. 8	−0. 02	27. 7936	16. 35767	27. 38142	17. 01987	25. 6105	16. 35767	25. 29253	17. 01987
100	34. 72	0	0	29. 4763	18. 34738	29. 4763	18. 34738	26. 88004	18. 34738	26. 88004	18. 34738

Fig. 1. Single-wing section of the 38.5 mm section

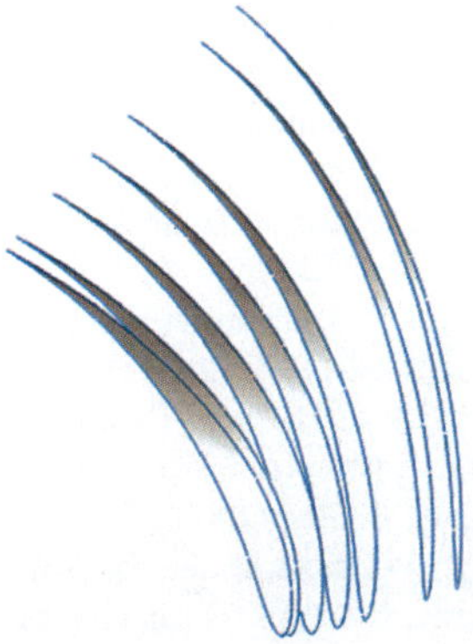

Fig. 2. The single airfoil section at each section

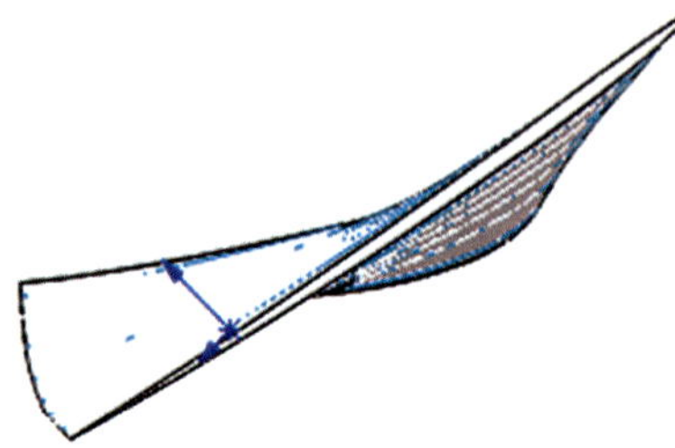

Fig. 3. The surface of the blade

4 Conclusions

This article mainly focuses on the design technology of rotor blades. The lift method is used to establish the solution method of the rotor blade. According to the design steps and key points of the lift method, the 3D modeling software Solid works is used to obtain the best annular distribution of the rotor blade profile and the 3D model of the blade.

Acknowledgements. This work was supported by Youth Innovative Talents Project (Natural Science) of Ordinary Universities in Guangdong Province (2020KQNCX112) and Innovation and Strong School Project of Guangdong Polytechnic College(2019GKJZK004).

References

1. Li, X., Wang, Z.: Introduction to integrated motor/propulsor (IMP). Mar. Electr. Electron. Eng. **23**(2), 23–25 (2003). (in Chinese)
2. Yang, L.: Acoustic stealth technology of submarine. Mod. Ships (11), 23–25 (2009). (in Chinese)
3. Cheng, B., et al.: Analytical design of the integrated motor used in a hubless rim-driven propulsor. IET Electr. Power Appl. **13**(9), 1255–1262 (2019). (in Chinese)
4. Yuan, J., et al.: Dynamic analysis of cavitation tip vortex of pump-jet propeller based on DES. Appl. Sci. **10**(17) (2020). (in Chinese)
5. Guan, X.: Modern Pump Technical Manual. Astronautic Press, Beijing (1995). (in Chinese)
6. Wang, Y.: Research on the Operating Characteristics of Axial Flow Pumps. China Agricultural University, Beijing (2005). (in Chinese)
7. Wan, Q.: Principle Design of Axial Flow Water Jet Thruster. Northwestern Polytechnical University, Xian (1995). (in Chinese)
8. Chasen: Principle and Hydraulic Design of Vane Pump. Machinery Industry Press, Beijing (1988). (in Chinese)
9. Ding, C.: Centrifugal Pump and Axial Flow Pump. Machinery Industry Press, Beijing (1981). (in Chinese)
10. Shenyang Machinery Research Institute et al.: Vane Pump Design Manual, pp. 365–385. Machinery Industry Press, Beijing (1983). (in Chinese)

Construction of Practical Courses of Financial Sharing Center Under the Background of Cloud Computing

Yapeng Wang and Ge Guo(✉)

School of Management, Wuhan Donghu University, Wuhan, Hubei, China

Abstract. In the information age, cloud computing and big data technology have been widely used in all walks of life. Data information is the core content, and cloud computing is a special tool for computing. The combination of cloud computing and financial accounting produces the financial sharing center model, which is a new product of the combination and application of the two in the whole financial accounting field. Since 2016, universities have cooperated with many accounting software companies to launch financial sharing service courses, and the training mode of financial talents is in urgent need of reform and transformation. Based on the background of cloud computing, this paper designs the framework of practice course of Financial Sharing Center, proposes measures to improve the construction of practical courses of financial sharing center, and provides some references and suggestions for promoting the construction of practical courses of financial sharing center in colleges and universities.

Keywords: Cloud computing · Financial sharing center · Practical courses · Measures

1 Introduction

In 2015, the government work report made the decision and deployment of the "Internet+" action. In response to the call of the country and the needs of The Times, the finance and accounting industry deeply integrates and interacts with the Internet technology. The combination of cloud computing, big data and finance promotes the integration of business and finance, and forms a financial sharing model. Financial sharing mode can integrate data and materials for easy use, reduce enterprise costs and reduce financial risks. Many enterprises have used financial sharing service center to achieve financial standardization and process, and internal management is more standardized, thus realizing enterprise value increment. Therefore, the talents needed by the financial sharing center are comprehensive, and they are no longer accounting personnel with simple accounting and repetitive work. The talents needed by the financial sharing center are required to have management, big data processing and foreign language skills.

As the base of talent training, colleges and universities have provided a large number of professional talents for the society, and are the link between the need of social and

B. J. Jansen et al. (Eds.): *International Conference on Cognitive based Information Processing and Applications (CIPA 2021)*, LNDECT 85, pp. 599–605, 2022.
https://doi.org/10.1007/978-981-16-5854-9_76

economic development and talent training. For this reason, colleges and universities should actively promote the transformation of talent training mode, realize the efficient docking between graduates and enterprises, and cultivate high-quality applied talents with the practical ability of cloud financial management integrated with new technologies such as shared services and smart finance.

2 Research Status at Home and Abroad

Pantelie O, Pajica and Nikolica (2016) studied the advantages and disadvantages of cloud computing and decision-making model, and believed that enterprises could promote their development by using cloud computing [1].

Quinn (2016) defined cloud accounting as a new accounting processing system formed by the integration of cloud computing technology and traditional accounting [2].

Schneiderjans (2016) explores the impact of cloud accounting on economic and environmental performance based on transaction costs and economics [3].

Jun Guo (2017) proposed that under the background of "Internet+", the "Internet" should be introduced into the classroom to promote the improvement of teaching level with information technology [4].

Danzhou Wang (2017) analyzed that the application of Internet technology can promote the reform of teaching mode, realize the innovative case teaching mode, and further improve the teaching quality [5].

Hongyan Ding (2015) believes that there are problems between education under the background of "Internet+" and the existing talent training program. The training of accounting personnel should not be limited to the study of theoretical knowledge, but should pay attention to the improvement of practical ability, and run through the whole teaching process [6].

Lin Qi (2017) proposed that the emergence of financial robots would promote colleges and universities to cultivate versatile accounting talents. In addition to mastering the most basic financial accounting skills, they should also have financial management and enterprise management abilities to help companies operate. In this way, they will not be eliminated by the society, conform to the trend of The Times and meet the needs of social development [7].

In the era of big data, cloud computing technology has received great attention, all walks of life are actively exploring its application, it also promotes the accounting industry to enter a new stage. For cloud finance, most foreign countries focus on cloud computing and improve the various functions of cloud services. There are few theoretical studies on the combination of cloud technology and education, especially on the teaching design of financial sharing center and cloud finance [8].

3 Teaching Framework Design of Practical Course of Financial Sharing Center

Before designing the teaching framework of the practice course of financial sharing center, we should first meet the requirements of the construction of the practice course of financial sharing center. First, realize platform sharing. Based on the financial sharing course platform, it simulates the sharing of intelligent finance related platforms, data, personnel and organizations by enterprises. Second, achieve efficient integration. On the basis of unified policies, rules, standards and other contents, complete the integration of business, finance and management in the simulation enterprise, and update the fiscal laws and regulations and policies at any time; Third, achieve a high degree of coordination. With the financial shared course platform, the division of labor is clear and the business and financial parts are highly coordinated. Fourth, realize intelligent management. The technology of cloud computing and big data is used to realize the design and application of intelligent scenes. After the above conditions are met, the framework of the practice course of financial sharing center is shown in Fig. 1.

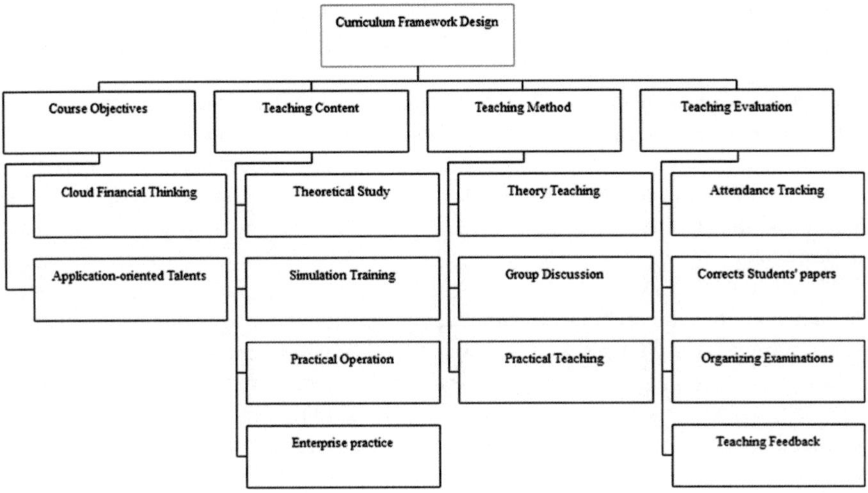

Fig. 1. Curriculum framework design

3.1 Determine the Teaching Objectives

The teaching objective is the embodiment of the best learning effect of the course. First of all, to grasp and understand the concept of sharing financial center, the future development direction and path of implementation. It mainly includes the organizational form of the enterprise, job division and corresponding responsibilities, business and financial management and other aspects of the content. Second, be proficient in the specific procedures of the enterprise financial sharing center, such as service acceptance, task assignment and transmission, operation and management, business process optimization and maintenance, etc. Third, use the financial sharing center platform to find out the problems existing in the process of enterprise operation, and provide solutions to optimize the collaborative relationship between financial sharing and business, finance and management. Fourth, evaluate the feasibility and operability of the financial sharing center, and write the construction plan of the financial sharing center. Fifth, students should be equipped with humanistic spirit, scientific literacy and integrity quality, cultivate students to become applied talents with the practical ability of cloud financial management integrating new forms of business and new technologies, such as shared services and smart finance.

3.2 Develop Teaching Content

Learning financial sharing theory. The study of financial sharing theory mainly includes related concepts, characteristics, feasibility, framework and implementation path of Financial Sharing Center construction. At the same time, we can introduce relevant cases for specific analysis, determine the address selection, business scope, main business process of the simulation enterprise, build the financial sharing center platform, understand the whole process of the Financial Sharing Center, and make preparations for the later practical operation.

Financial sharing virtual simulation operation. After understanding the relevant content of the financial sharing center simulation enterprise, let the students learn the specific operation process in the virtual environment by grouping, allocating posts and dividing them into different roles, so as to cultivate the students' handling ability of business, finance and management and improve the students' application ability of the financial sharing center.

Financial shareware operation. To operate the content of Financial Sharing Center, we should choose the appropriate financial sharing software. Now, the financial sharing software used in universities is mainly UFIDA, Kingdee and wave. There are many design contents of financial sharing software, which are mainly reflected in four modules: first, cloud reimbursement module. The cloud computing technology is used to connect the reimbursement platform with the commercial platform, so that the reimbursement business can be directly transferred to the reimbursement platform of the financial sharing center, and the integrated process of automatic reimbursement, online payment, voucher generation and report form generation is realized. Second, the cloud accounting module. Integration of business and financial processing, promote efficient integration of business and financial, automatic generation of receipts for procurement, payable and sales, receivable and other businesses, transfer to the financial sharing center, complete accounting and generate vouchers. Third, share operating modules. Achieve sharing of images, expenses, receivables, payables, operating reports, audits and evaluations. Fourth, budget control module. Through the financial sharing platform, it can realize the control before, during and after the event. If the budget is exceeded, it will be automatically warned and controlled.

Financial shared center practice. After students have skillfully used financial sharing software, they should cooperate with enterprises that have financial sharing platform if permitted, and let students visit and practice in enterprises to feel the application of financial sharing center in actual work, so as to better achieve the docking between university talents and enterprises.

3.3 Choosing Teaching Methods

In the practical courses of financial shared center, we generally adopt the case teaching method. The financial sharing center case of a simulated enterprise is introduced, and the relevant materials and data of the enterprise are integrated into the teaching content, and boring theoretical knowledge points are interspersed in the teaching process to enrich the course content. At the same time, through the method of role-playing, students can find their own positions and operate them, so as to master the process of business, finance, sharing and management. These can help students find the weak links in their study of this course through the operation of cases, and find solutions, which is conducive to the improvement of students' vocational ability, improve students' independent thinking ability and cultivate students' teamwork ability [9].

3.4 Establish an Assessment and Evaluation System

Evaluation is the summary of the teaching process and teaching effect and feedback, in the teaching process, teachers can by tracking attendance, check students' homework, refer to students' classroom feedback, correcting students test paper method is used to determine the teaching process [10]. At the same time, students can evaluate each other, better participate in the whole process of course assessment, and change the single assessment method, to achieve multi-dimensional assessment form.

4 Measures to Perfect the Construction of Practical Courses of Financial Sharing Center

4.1 Diversified Teaching Methods

Financial Shared Center is a practical teaching course. If only teaching method is adopted, students will not have a good grasp effect and not be skilled enough in practical operation. Teachers should take advantage of the Internet platform to achieve the integration of various teaching methods. For example, teaching methods like SPOC, flipped classroom or PBL. Through SPOC, we can obtain high-quality course videos, ppts and practice questions through screening. At the same time, teachers can answer students' questions online and discuss students' common problems offline. Through flipped classroom, students can use the time after class for data collection and sorting, display in the classroom, and teachers can comment on the content explained by students. In general, the application of various teaching methods plays a positive role in promoting students' self-completed learning and teamwork.

4.2 Achieving School-Enterprise Cooperation and Joint Training

At present, most private colleges and universities are in the process of transforming into application-oriented universities. At present, most private colleges and universities are in the process of transformation to application-oriented universities. They can cooperate with UFIDA and other enterprises and institutions to adapt to the digital learning mode under the new form of "intelligence + education" and jointly formulate a characteristic talent training plan integrating industry and finance, theory and practice. Digital analysis courses such as financial sharing center training are integrated into the course. Integrate teaching resources and strive to build digital intelligence business training center.

University-enterprise cooperation not only can improve the practical ability of the teacher, the teacher can also bring in enterprises, further improve students' vocational skills, under the guidance of the instructor, students can use the enterprise financial sharing center software, or into enterprise practice, operating the real financial sharing center, improve the students' practical ability.

4.3 Optimize the Construction of Teaching Staff

Introduced and hired a industry background of scholars and experts, financial sharing center technical staff teaching practice courses, practical courses of training of the school teachers at the same time, can help the teacher credentials the way through to the enterprise financial sharing center field, optimize the structure of teachers, and increase the proportion of "double qualified" teachers, and invite them to participate in the project of talent fostering and the textbook compilation, Better to help finance and accounting graduates with a good connection with enterprises.

5 Conclusion

With the continuous development of cloud computing, higher requirements are put forward for financial and accounting professionals, making standardized accounting accounting personnel transform into management accounting personnel. At the same time, financial personnel should not only master accounting knowledge, but also strengthen the processing and analysis ability of big data, and have a strong foreign language level. Therefore, colleges and universities should optimize the talent training program, establish the teaching system of practical courses of financial sharing center, and improve students' practical ability to cope with the new requirements of the society on accounting personnel under the background of cloud computing.

Acknowledgements. This work was supported by letter from the Higher Education Department of the Ministry of Education on the publication of the list of the first batch of industry-university cooperation and education projects in 2019.

References

1. Pantelic, O., Pajic, A., Nikolic, A.: Analysis of available cloud computing models to support cloud adoption decision process in an enterprise. In: International Conference on Computers Communications and Control, vol. 8, pp. 135–139 (2016)
2. Cleary, P., Quinn, M.: Intellectual capital and business performance: an exploratory study of the impact of cloud-based accounting and finance infrastructure. J. Intellect. Capital **17**(2), 255–278 (2016). https://doi.org/10.1108/JIC-06-2015-0058

3. Schniederjans, D., Hales, D.: Cloud computing and its impact on economic and environmental performance: a transaction cost economics perspective. Decis. Supp. Syst. **86**, 73–82 (2016)
4. Guo, J., Wu, W., Zhang, X.: Exploration on the composition and cultivation path of MPAcc students' core literacy under the background of "Internet +". Ind. Technol. Forum **7**, 122–123 (2017). (in Chinese)
5. Wang, D., Liu, A., Tang, S.: Research on innovation of full time MPAcc case teaching driven by Internet. Commun. Finan. Account. **22**, 61–63 (2017). (in Chinese)
6. Ding, H., Liu, G.: Reform of MPACC training mode under the guidance of accounting practice. Finan. Account. Monthly **27**, 124–126 (2015). (in Chinese)
7. Qi, L.: The influence of the emergence of financial robot on the reform of efficient accounting personnel training mode. Manager's J. **24**, 179–180 (2017). (in Chinese)
8. Cheng, P., Tao, S.: Teaching design of "internet + accounting" MPAcc financial sharing course based on ADDIE. Finan. Account. Monthly **22**, 24–29 (2018). (in Chinese)
9. Gu, Z.: Thoughts on offering financial sharing service training course for finance and economics majors. Lab. Res. Explor. **38**(7), 265–267 (2019). (in Chinese)
10. Wang, C.: Research on the construction of financial sharing course in Colleges and Universities under the background of financial transformation. Bus. Account. **1**, 124–126 (2020). (in Chinese)

Application of Machine Learning in Financial Asset Price Prediction and Allocation

Yunfei Cao(✉)

Sichuan University, Chengdu, Sichuan, China

Abstract. In today's era, the wave of artificial intelligence has swept across, injecting unparalleled vitality into various industries. The financial field is no exception. Artificial intelligence has provided this industry with new and colorful technologies and brought new opportunities for its further development. This article introduces several basic machine learning algorithms from the perspective of underlying machine learning algorithms, reveals the underlying algorithms used by more popular technologies, and explains the application of machine learning in financial asset price prediction and allocation.

Keywords: Machine learning · Financial assets · Price prediction · Asset allocation

1 Introduction

Compared with machines, brain capacity and computing power have severe limitations on human behavior and thinking. The machine-learning algorithm is used to design a method for the computer, so that the computer can automatically make decisions on similar problems after learning by itself, and it has a speed far exceeding human. Therefore, in the financial field, some tasks with huge data volume and needle finding in a haystack are suitable for using machine learning to complete.

There are two categories of machine learning algorithms: supervised learning and unsupervised learning, as shown in Fig. 1. Supervised learning is particularly widely used in the financial field. The general steps are as follows. In the learning and training stage, the machine learns a class of problems to train a model. At this stage, a test set with answers is first needed, and samples with correct decisions have been made for the computer to learn. In the testing phase, the test set is used to test the model that has been trained, and to further modify the model to get better results.

2 Machine Learning Algorithm

2.1 Decision Tree Algorithm and Random Forest

The decision tree algorithm is a classification algorithm based on the 'tree' data structure [1]. A 'tree' is composed of nodes and branches, where nodes are used to store information and data, and branches are used to represent the relationship between information and data.

B. J. Jansen et al. (Eds.): *International Conference on Cognitive based Information Processing and Applications (CIPA 2021)*, LNDECT 85, pp. 606–612, 2022.
https://doi.org/10.1007/978-981-16-5854-9_77

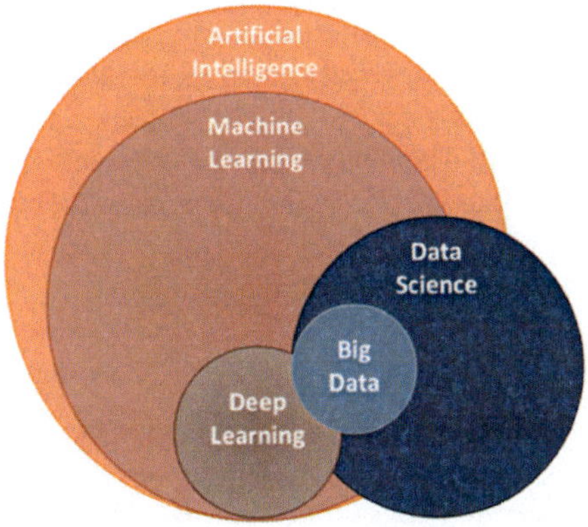

Fig. 1. Types of mechanical learning

In this algorithm, how to construct a decision tree scientifically and reasonably is the key to this algorithm. Considering the information gain when selecting the features of non-leaf nodes and the feature division method. From the top to the bottom of the decision tree, selecting the feature with the largest information gain and the feature division method as non-leaf nodes, so that the diversity of features can be utilized as much as possible [2].

2.2 Neural Networks

In the financial industry, neural networks are essential factors. In banks, handwritten signature recognition and face recognition are already key technologies for identifying identities. The application of them is based on a special neural network structure, called convolutional neural network. The human eye's vision of each pixel of the picture is not independent, and has a lot to do with the surrounding pixels [3]. The convolutional neural network contains a convolutional layer. When the matrix data of the image is forwarded through the convolutional layer in the convolutional neural network, the pixel matrix is divided into blocks. After that, convolutional operation is performed. The result of the operation continues to the last output is passed before. In this way, the convolutional neural network combines each pixel in the picture with the surrounding pixels, and after calculation, they jointly form the characteristics that conform to the observation of the human eye, which greatly improves the recognition and processing capabilities of the image [3].

2.3 Support Vector Machines

Support vector machines have powerful classification functions and are widely used in the financial industry. As data breaches become more common, fraud has become a major problem for banks and credit card companies. Due to the huge amount of data, it is impossible to manually identify fraudulent behaviors. At this time, it is necessary to ask for machine learning algorithms [4]. Fraud detection is actually a classification problem that divides transactions into non-fraud and fraud. Therefore, vector machine algorithm becomes a perfect tool for fraud detection. Its realization process is also very simple. First, it is needful to train a detection model based on a known sample set where both fraud and non-fraud exist. Then the data of the transaction to be detected need to be extracted and input into the detection model based on the support vector machine [4]. Computers can detect a large number of transactions at an extremely fast speed and complete tasks that cannot be achieved by human resources.

3 Application of Machine Learning in the Field of Financial Configuration

The following are the main areas of current machine learning applications. Remember that many application models use multiple artificial intelligence methods—not just machine learning.

3.1 Improve the Corporate Structure

At this stage, the speed of economic development is very fast. If you want to reduce corporate financial assets and financial asset risks, you need to improve the internal structure of the enterprise to meet the needs of social development and reduce financial risks. According to the requirements of modern development, enterprises should standardize the corporate governance structure and allocate certain internal responsibilities to ensure that every employee can have tasks to do [5]. At the same time, it is necessary to establish a scientific and reasonable supervision system. The senior leaders of the company need to supervise the employees in the company at any time, so that they can understand the company's development status. The main important thing is to ensure communication between various departments. Only smooth communication can ensure unimpeded information exchange, and enterprises can develop steadily in a competitive market and reduce financial risks [5].

3.2 Establish a Financial Risk Management System Within the Enterprise

In the process of enterprise development, it has always been necessary to face risks. Combining this situation, it is necessary to establish a certain financial risk management system within the enterprise, so that financial risks can be controlled to a certain extent, and at the same time, risks can be reduced [5]. To the minimum and reduce the loss to the enterprise. Enterprises should make formulations based on relevant standards issued by the state. Financial asset management enterprises should regard risk management as the center of development, and can comprehensively conduct financial risk management on enterprises. In the management, you can't just squander time and perfunctory management, you must manage the enterprise in an all-round way [6]. This includes accounting, finance, and various departments within the enterprise, and at the same time formulating financial risk management. The most important thing is to let the employees in the enterprise have the awareness of risk crisis, and they can have a certain sense of prevention when facing financial risks, and at the same time solve them according to the corresponding risks [6].

3.3 Enterprises Must Control Some External Risks

Enterprises must also determine the source of funds when financing, and ensure that the sources of funds are in compliance with legal procedures [7]. According to this situation, the government should be qualified, but relatively small-scale enterprises to enter the loan work, this approach can ensure the rationality of the source of funds, while preventing the occurrence of vicious incidents. The most important thing is to ensure the credit of the borrower. Before proceeding with the borrowing operation, an in-depth investigation should be conducted on the borrower to ensure that the borrower is creditable [7]. This will ensure that the funds can be returned in full on a regular basis and avoid unnecessary production.

3.4 Loan/Insurance Underwriting

Underwriting is the most suitable job for machine learning in the financial industry. In fact, many people in the financial industry are worried that machines will replace most of the current underwriting jobs. Especially in large institutions. Is the default rate among specific groups of people in the past 15 years rise)? These results have an important impact on the company-but currently those large companies that have the resources and strength to hire data scientists and have many transaction data can get this information [8].

3.5 Customer Service

Chatbots and interactive interfaces are the areas that venture capital focuses on, and they are also one of the areas where companies have a relatively large amount of capital budget investment (in the 2016 artificial intelligence executive survey, this area was regarded as the most potential direction for short-term artificial intelligence consumer applications) [8].

4 The Application of Finance in Price Prediction

4.1 Three Key Points for Constructing Asset Allocation Portfolio

Now we determine to use Markowitz's mean variance-efficient frontier theory as the basis of asset allocation portfolio. In addition, derive three key points of the construction model: (1) how to estimate the risk, (2) how to estimate the return, and (3) how to correctly classify user attributes.

(1) Estimated risk
(2) Estimated income

For the second part of the income estimation part, the benefits are: lower input data form restrictions; linear/non-linear learning; self-evolution, correction, etc. Machine learning can make supervised learning of model factors [9]. The selection of factors can include fundamental, technical, and bargaining data, and make corresponding estimates

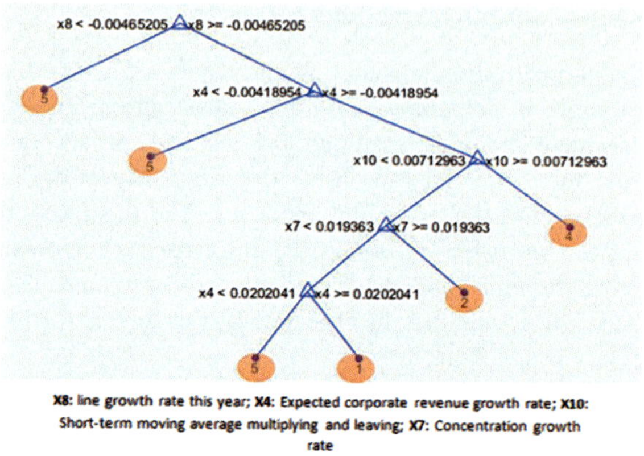

Fig. 2. U.S. stock market using decision tree factor analysis case

of market returns. Figure 2 shows a case in which the U.S. stock market uses the decision tree factor model to analyze returns.

There are many ways to estimate earnings. We put the asset returns calculated by several different return estimation models as input parameters into Markowitz's effective frontier model or Black-Litterment model to calculate, and analyze its impact on the investment portfolio, that is, the return on investment. Which performed [9].

Below we use three methods to estimate earnings, one is the support vector machine regression model (SVM), the other is the linear regression model, the analysis factors of these two models include interest rate, price-earnings ratio, price-to-book ratio, dividend ratio, enterprise earnings, transaction value, implied volatility, MACD, KD, etc.[10] In addition, the third is the most commonly used by ordinary people, directly using the market trend in the past few years to predict the future market. Please see Fig. 3:

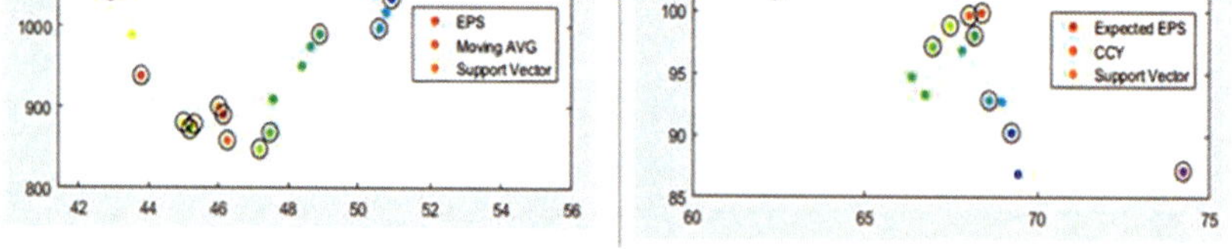

Fig. 3. MACD Regression factor analysis

4.2 Data Fitting and Prediction

A relatively direct application of artificial intelligence methods in the financial field, machine learning and deep learning can process higher-dimensional data, have better representation capabilities, and generally have stronger predictive capabilities [10]. There are related studies in both low frequency and high frequency. The more interesting direction may be to develop more targeted models and structures in combination with the characteristics of financial data.

4.3 Derivatives Pricing

The use of deep learning for the pricing of derivatives is considered to be an early research. It is mainly used to solve some difficulties in numerical algorithms, such as high-dimensional data. Derivative directions include the use of deep learning, reinforcement learning for hedging, etc. [11]. There are teachers in our department who are working on this direction recently.

4.4 Credit Review, Anti-fraud, Insurance Underwriting, etc.

The powerful predictive ability of artificial intelligence technology is naturally suitable for credit review tasks and can help banks in the front-end risk control. However, because this type of business involves the evaluation of people, it will involve a series of social issues, like fairness, racial discrimination, etc.; the fairness of neural networks is a hot trend recently [11]. In addition, this type of task often requires that the results given by the model are interpretable, and neural networks are generally difficult to interpret, causing certain application difficulties.

5 Conclusion

In summary, the speed of social and economic development is very fast, and machine learning and AI have been well developed. However, in the process of development, capital is the most critical and important. In the process of capital management and allocation, how to maximize the use of capital while reducing risks is worth discussing. The application of machine learning for asset prediction and allocation combination has been very rich. For financial institutions and asset management companies, strengthening the cultivation of financial technology R&D talents may be an important task in the next few years.

References

1. Yan, W.M., Li, D.G., Wu, W.M.: Application of machine learning in the financial field. Comput. Educ. **8**(08), 87–89 (2014)
2. Chen, H.T.: Research on the application of machine learning in home finance. Fintech Era **11**(08), 11–13 (2015)
3. Wu, Y.X.: Application of machine learning algorithms in financial market risk prediction. Inf. Syst. Eng. **13**(02), 114–116 (2019)

4. Li, Y.N.: Application of neural network model in anti-fraud in banking internet finance. Era Finan. Technol. **11**(08), 24–28 (2018)
5. Ju, Z.: Application of machine learning algorithms in financial market risk prediction. Finan. Econ. **9**(32), 22–24 (2017)
6. Zhao, H.T.: A review of research on the application of machine learning in financial asset price prediction and allocation. J. Manag. **9**(21), 216–219 (2018)
7. Zhang, K.M.: The application of machine learning in the financial field, the era of wisdom and wealth, **12**, 120–123 (2019)
8. Li, Z.: Application of machine learning algorithms in financial market risk prediction. Financ. Econ. **13**(03), 400–402 (2019)
9. Lu, S.X., Hu, G.F.: Discussion on application experience and suggestions of machine learning technology. Finan. Econ. **8**(11), 102–105 (2018)
10. Wang, G.Q.: Application of machine learning algorithms in financial market risk prediction. Zhifu Times **10**(18), 272–275 (2017)
11. Wang, Y.J.: Research on bank personal customer credit scoring model. Mod. Bus. Ind. **36**(19), 64–65 (2015)

Application of BIM Technology in the Informatization of Construction Management

Dun Miao(✉)

Civil Engineering, Tongji Zhejiang College, Jiaxing, Zhejiang, China

Abstract. The whole life cycle concept proposed by BIM technology is to realize the closed-loop management of engineering projects from planning, design, construction, to operation and maintenance through information technology and Internet technology. BIM technology not only changes the underlying logic of the design, but more importantly makes it possible to realize the informatization of construction management by relying on BIM technology to improve the quality, cost, and efficiency of engineering projects. Building a BIM collaborative management platform on the basis of the BIM model can make the project construction management work visible and three-dimensional. Meanwhile, it can also optimize the construction plan, formulate management procedures and responsibilities, and improve the efficiency of implementation of various measures in construction management. Through the Internet technology, the traditional one-to-one information communication method can be transformed into a one-to-many information sharing and collaborative mode, breaking the constraints of time and space. BIM technology can greatly simplify the "three controls" workload of schedule control, quality control, and safety control, and improve the degree and efficiency of management information.

Keywords: BIM · Construction · Informatization · Collaboration

1 Introduction

BIM technology is the core technology to promote the informatization upgrade of the engineering industry. It realizes the unification of the associated visualization of the geometric information of engineering projects and the associated storage of the non-geometric information, and realizes the mapping, control and management of the entire life cycle of engineering projects in the information dimension [1]. BIM technology not only changes the underlying logic of the design, but more importantly, it is based on the informatization of vertical industries and the integration of horizontal data and services to realize the informatization of construction management under the Internet mode, and improve the quality, cost, and efficiency of engineering projects [2].

The characteristic of modern engineering construction is "large size and heavy workload", which not only requires high quality and rapid construction, but also requires

B. J. Jansen et al. (Eds.): *International Conference on Cognitive based Information Processing and Applications (CIPA 2021)*, LNDECT 85, pp. 613–621, 2022.
https://doi.org/10.1007/978-981-16-5854-9_78

economic benefits and environmental protection. It greatly increases the workload and difficulty of project construction management [3]. The traditional manual management model has problems such as poor timeliness, low work efficiency, information communication barriers, and unclear management responsibilities. It has been difficult to meet the needs of modern engineering construction management. Therefore, it is an urgent need and an inevitable trend for the engineering industry to use BIM technology to promote construction management informatization. This paper takes a part of bid lots of a road project in Zhejiang Province as a case to initially explore the application and effect of BIM technology in construction management informatization.

2 Project Overview and the Goal of Construction Management Informatization

This project is a road project located in Jiaxing, Zhejiang Province. The total length of the road is 3.6 km. The construction route of the road project is long, involving a wide range of areas, and the construction management and coordination are difficult. The bid lot is located in a dense residential area with a complex flow of people. The total length is 476.5 m, and the red line of the road is 18 m wide. It adopts urban branch road standards, and the design speed is 30 km/h. Along the line includes a 5 + 6 + 5 m reinforced concrete hollow slab simply supported girder bridge. This bid lot of the project includes road engineering, bridge engineering, drainage engineering, lighting engineering, etc., as shown in Fig. 1.

Fig. 1. The roaming map and panoramic QR code of the bid lot of the project

Using BIM technology for construction management, on the one hand, it can make the whole construction process more standardized and assign responsibilities to specific posts and personnel. On the other hand, it can guarantee construction quality and construction safety, and achieve project-level cost reduction Increase efficiency:

(1) In the early stage of the project construction, according to the surrounding environment and traffic requirements of the project, carrying out temporary construction model establishment, temporary facility layout simulation, special construction plan

simulation, etc., and then reasonably preparing construction organization design based on the simulation effect, improving the quality of construction group, in order to create conditions for orderly and effective development. (2) In the middle and late stages of project construction, using the high integration and multi-dimensional characteristics of BIM technology, through the development of BIM three-base (web, mobile app, PC) collaborative management platform, the BIM and the database are linked to achieve the visualization of the model and construction. It can help locate, search, calculate, and share, so as to better exert its multi-dimensional application effect [4].

3 BIM Basic Model Construction and Its Application at the Initial Stage of the Construction

3.1 Road BIM Full Model Construction

In this paper, Revit was used to create a full model of the first section of the road according to the design goals and the accuracy requirements of the construction model. The Revit software in the road and bridge professional family library construction is not complete, so the project built a bridge family library, a sign family library, a lighting family library, etc., on this basis, a bridge model, a pavement model, a steel bar model and a drainage model were constructed. As shown in Fig. 2.

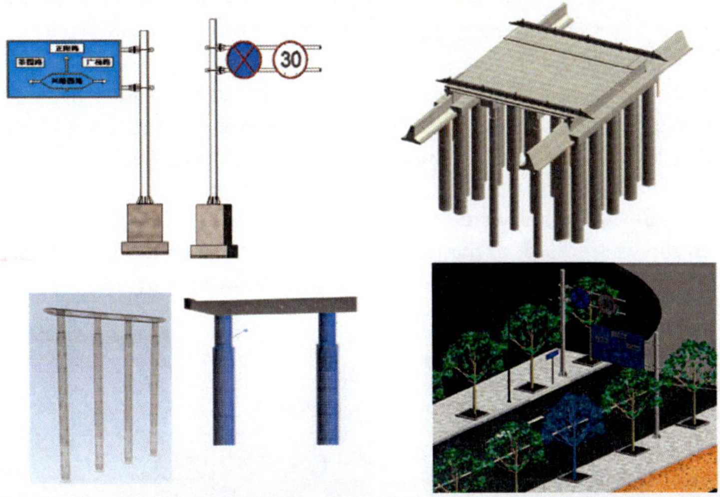

(a) sign family (b)bridge model (c)rebar model (d)road full model (partial)

Fig. 2. Partial display of BIM model

3.2 BIM Application at the Initial Stage of the Construction

Using BIM's three-dimensional visibility and system dynamic management, based on the actual situation of the site, plan the site construction plane, simulate the construction progress, design feedback and technical disclosure can be done.

3.2.1 Plan the Site Construction Plane

The on-site construction site planning can visually show whether the traffic organization planning, the use of large-scale equipment, the material storage yard and the processing site, and the use of temporary construction facilities in each major stage are reasonable. Through the combination of BIM model and GIS geographic environment, engineers can effectively plan and manage the construction site under the visualized status in three dimensions and ensure that the construction is carried out in a reasonable and orderly manner [5].

3.2.2 Simulation of Construction Progress and Green Construction Plan

Through the implementation of 4D construction simulation [6], engineers can formulate a reasonable construction plan for the project, accurately grasp the construction schedule, optimize the use of construction resources, optimize the construction process, predict and avoid risks, and ensure construction safety [7].

The traditional construction progress and plan are in text mode, which brings difficulties to the investigation and comparison of front-line workers. This project can transmit and share the progress and plan to the constructors in a three-dimensional visualization through the mobile phone APP, so that the constructors can have the most intuitive understanding about the overall situation, including scheduling and plans.

3.2.3 Design Feedback

The phenomenon of "concurrently designing, constructing, constructing, and designing" is common in the engineering construction process. Through BIM modeling, the construction drawings can be reviewed more accurately and quickly, which saves the time of project personnel reviewing drawings and communicating with designers. For example, a total of 11 collision points were detected during the design and construction of this project, and timely feedback to the responsible person through the BIM collaboration platform, as shown in Fig. 3, increased the drawing review time by more than 50%.

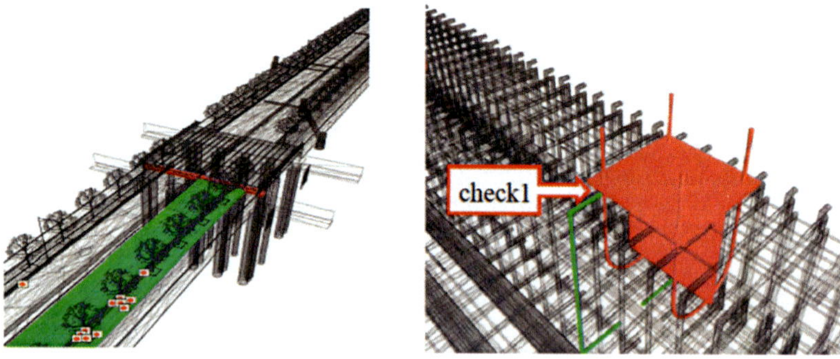

Fig. 3. Design collision point feedback

3.2.4 Technical Disclosure

The project used Revit, Navisworks4D simulation and 3DMax to make animations for construction plan simulation and comprehensive visualization. In practical applications, Internet technology can be used to visualize the BIM technology on the web and mobile terminals (as shown in Fig. 4). This has changed the past based on scripted confession, which increased the interest of the confessed and increased the participation of the confession. After the confession, the on-site technicians and construction workers were assessed, and it was found that the quality of the confession had been significantly improved, which effectively guaranteed the quality of the on-site construction.

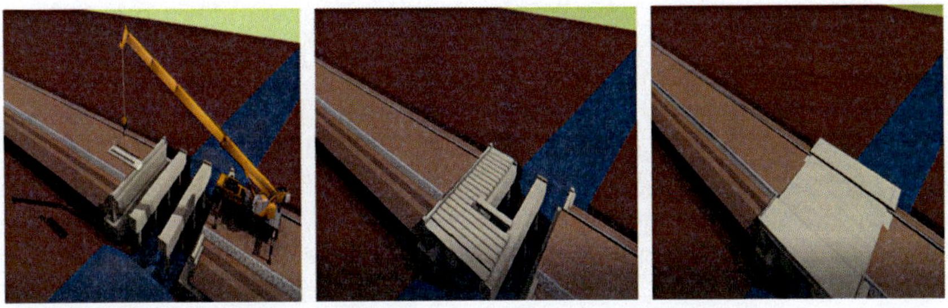

Fig. 4. Animation of technical clarification using BIM technology

4 BIM Collaborative Platform Construction and Mid- and Late-Stage Application

There is a big difference between engineering construction and industrialized assembly lines. Due to different projects, large process changes, and design changes, a lot of on-site information and data will be generated during the construction process, which makes communication and coordination costs relatively high, and construction management productivity cannot be effective improve. Based on the BIM lightweight model and construction management collaboration platforms developed on the Internet and mobile APP [8, 9], the traditional one-to-one information communication method has been transformed into a one-to-many information sharing collaboration mode. Fixed time-space communication has become a mode that is not restricted by time and space. This project has developed six sub-modules on the collaborative platform: graph viewing, quality management, safety management, regulatory library management, data analysis and cloud storage. By using the Internet of Things technology, it has established a project management information system that combines virtual and real, which takes visualization model as a carrier for the purpose of construction process control [10].

4.1 The Sub-module of Graphical Model View

The lightweight BIM model realized 2D + 3D integration. Through this module, the BIM model and construction drawings can be integrated and viewed, and some key viewpoints can be annotated to assist construction site management and improve management and control efficiency.

4.2 The Sub-module of Quality Management

Developing quality problem traceability, quality inspection, rectification and tracking management mechanisms, and forming an online technical plan for review, cloud technical file management, and a new information management model that combines online and offline process quality acceptance to ensure the implementation of problems, efficient resolution, complete storage and acceptance specification.

4.3 The Sub-module of Safety Management

The traditional safety inspection process is still in the paper approval stage, the process is not standardized. It is also difficult to trace, and difficult to manage. Through this module, design safety management and inspection procedures can be designed so that site safety and quality personnel can promptly initiate rectification notifications on the mobile phone when they discover problems. The corresponding subcontractors receive APP notifications to initiate rectification, which greatly improves the efficiency of on-site safety management. As shown in Fig. 5. The labors and management personnel of this project have formed the habit of using mobile phones to inspect and take pictures during the construction of the project.

4.4 The Sub-module of the Law and Regulations Library

The enterprise security database is configured by citing the group enterprise-level inspection regulations database to form the core competitiveness of the enterprise. The final construction system of this project has a built-in quality and safety inspection standard library of 775, with over 3000 quality and safety inspection specifications. The law database can be searched during quality management and safety management rectification initiation, and it can be automatically quoted at the front end to improve work efficiency and quality.

4.5 The Sub-module of Data Analysis

Based on the construction process quality and safety data, carrying out statistical analysis of the quality and safety issues of the project, and expanding the application of the data. On the one hand, it can formulate corresponding measures for the parts with frequent problems, and on the other hand, it can provide materials for the formulation of later project construction plans. As shown in Fig. 6, a two-week data statistic was carried out. The number of inspection items was 22 (blue line in the figure), 18 unqualified items (red line in the figure), and 4 qualified items (blue line in the figure). According to statistical analysis, the most unqualified items are safety belt construction, which is a key concern of construction management.

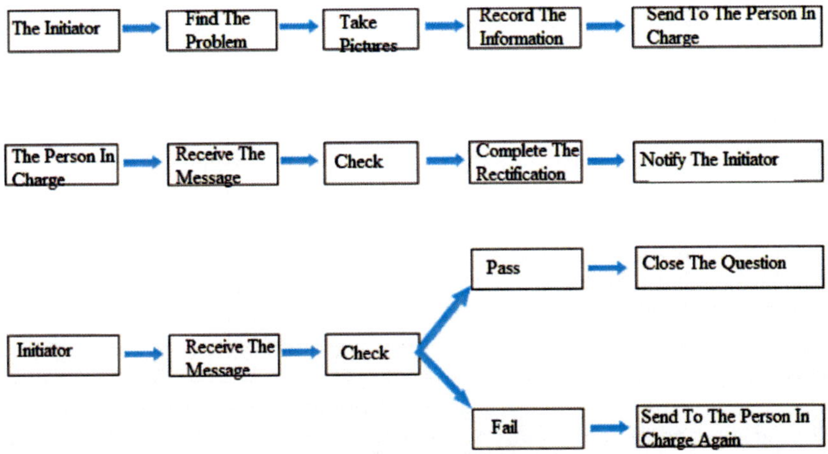

(a) inspection process

(b)mobile phone on-site inspection

Fig. 5. Safety management and inspection pictures

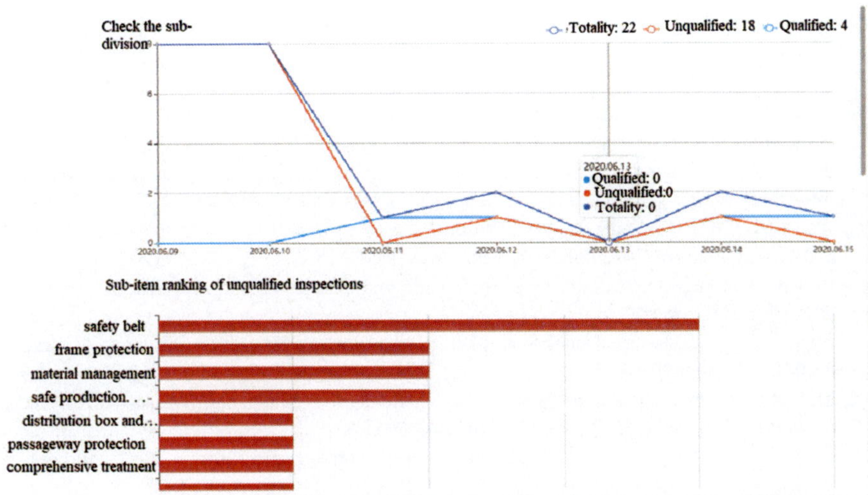

Fig. 6. Screenshot of the data analysis page

5 Conclusion

It is important to integrate BIM technology into project production organization and management, make BIM technology a soft landing, continue to exert its effectiveness, realize closed-loop control of design, construction, and management, improve project quality and efficiency, and reduce costs and energy consumption.

First, BIM technology and the collaboration platform enable the BIM model to have multiple dimensions in addition to the three-dimensional space, such as time, engineering, cost, quality, and safety, integrate model information. It can turn complex management into management based on visualized three-dimensional models, make management have public media to rely on, simplify management, organize information, and clarify the process, which can improve the efficiency of project operation and management.

Second, for BIM technology, what you see is what you get. It has overturned the traditional construction method that based on two-dimensional construction drawings. Managers can directly view the three-dimensional image of the structure through the mobile terminal of the mobile phone, obtain its three-dimensional size information, view the construction plan and technical disclosure information online, realize paperless office on-site, save costs and improve the sharing of information, and enable project management Increased participation in personnel projects.

The Last but not the least, the collaborative management platform makes safety management, schedule management, and quality management transparent and traceable. It uses models as the carrier to form a complete database. At the same time, the authenticity of construction information is improved, and the actual traceability is stronger, which has changed the traditional paper-based information storage phenomenon such as "error, chaos, and fake".

References

1. Liu, C., Chai, G., Zhan, K., et al.: The application of BIM collaborative platform in construction management of water transportation projects. Port Eng. Technol. **57**(01), 67–70 (2020)
2. Liang, H., Guo, S., Tian, Z., et.al.: The implementation and application of BIM technology in the engineering construction process. Constr. Technol. **47**(S4), 61+982–985 (2018)
3. Wei, Z.: Research on the application of BIM technology in safety production management. Intell. City **28**(02), 67–70 (2021)
4. Wang, M., Gong, X., Wang, L., et al.: Research on construction knowledge base management and sharing platform based on BIM technology. Intell. City **28**(01), 38–40 (2021)
5. Luo, Y.: The large scene coordinates conversion and 3D visualization application of BIM + GIS. China Highw. **09**(24), 117–119 (2021)
6. Zhang, C.: 4D-BIM based quality and safety management for Qugang highway. J. Graph. **40**(04), 778–782 (2019)
7. Li, S., Lin, Q.: Construction progress management method of tall buildings based on P6+BIM technology. Constr. Technol. **49**(24), 34–37 (2020)
8. Fang, Y., Song, S., Shang, H., et al.: Research on digital project management and control platform of construction works. Mod. Tunnel. Technol. **58**(01), 92–98 (2021)

9. Kou, Y., Liu, K.: Research on refined construction management of prefabricated buildings based on BIM technology. J. Eng. Manag. **34**(06), 125–130 (2020)
10. Geng, L., Deng, D.: BIM application: collaboration makes high efficiency. Inf. China Constr. **25**(18), 38–41 (2020)

Guangxi University Scientific Research Management Based on Neural Network AHP and BP

Manying Huang(✉)

GuangXi Vocational Normal University, Nanning, Guangxi, China

Abstract. At the shortcomings of low convergence efficiency and weak generalization performance in current scientific research management, a scientific research management model based on neural network AHP-BP is proposed. Construct the performance evaluation index system, collect the corresponding performance evaluation data, use neural network AHP-BP to simulate the human brain to train the performance data, and use the gradient method to determine the parameters of the neural network AHP-BP to establish a performance evaluation model. Finally, its performance is tested through simulation experiments. Actually, this paper studies the management of colleges and universities in Guangxi based on neural network of AHP and BP.

Keywords: AHP · BP · Neural network · Guangxi universities · Scientific research

1 Introduction

Current knowledge has high application value in human capital and technology, and it is the key for universities to enhance their own competitiveness [1]. Knowledge management is the implementation of university scientific research management personnel to control internal and external knowledge of universities and the use of effective management research models to implement accurate real-time evaluation of the innovation performance of university knowledge management can find problems in innovation and improve efficiency and quality [1].

The traditional innovation management scientific research model has the defects of low convergence efficiency and weak generalization performance. Therefore, an innovation management scientific research model based on the neural network AHP-BP is proposed to improve its capabilities [2]. The experimental results show that the knowledge management innovation management scientific research error of the evaluation model of this article is small, and it has high practical application value [2].

B. J. Jansen et al. (Eds.): *International Conference on Cognitive based Information Processing and Applications (CIPA 2021)*, LNDECT 85, pp. 622–627, 2022.
https://doi.org/10.1007/978-981-16-5854-9_79

2 Management and Scientific Research Neural Network AHP-BP Model

2.1 Neural Network AHP-BP Management Principle

There are many interference factors in management evaluation, and there are more complicated mapping relationships, which affect the evaluation effect of knowledge management [3]. The neural network AHP-BP is used for evaluation and management, and the neural network simulate the operation of the human brain to implement parallel operation and dynamic transformation of information. First, clarify the input and output of the neural network AHP-BP, which are used to describe the information of the information management of the university to be evaluated and the final evaluation goal; then use many known samples to train the neural-network AHP-BP and adopt adaptive learning. Obtain weights and thresholds, and accurately describe the structure of the neural network AHP-BP; the trained neural network AHP-BP has human brain memory performance and recognition performance [3], and can simulate the relationship between evaluation indicators and management. In addition, use the test sample to implement the test until an accurate result is obtained. The finally obtained neural network AHP-BP is an effective evaluation tool. Input an unknown sample to obtain the sample evaluation result.

The evaluation indicators of knowledge management are input in the input layer of the neural network AHP-BP and transmitted to the output layer through the hidden layer. The output result is compared and analyzed with the expected output result. If the accuracy requirement is not met, the error back propagation process is entered. In this process, the network weight threshold is adjusted according to the output data to ensure that the network error is continuously reduced and meet the demand, and the accuracy of the neural network's mapping of the input vector is improved [4].

2.2 Management Research of Neural Network AHP-BP

The search in the neural network AHP-BP process algorithm is based on the gradient method, using the parallel processing characteristics of the neural network; there are two processes of input forward transmission and error backward transmission. The input data is operated by the hidden layer and the neural unit of the output layer to obtain the evaluation result [5], which is the output result of the network. If the difference between the output result and the expected result is within the allowable error interval, the error signal is propagated back and the weight threshold of the neuron is adjusted and controlled, and the output of the neural network AHP-BP is consistent with the expected value through the iterative process, so that the average square error of the network is the expected minimum value, and the neural network AHP-BP learning process is terminated [5].

The realization of the neural-network AHP-BP process algorithm is as follows:

The weights are adjusted by batch processing methods. After the input of all samples is completed, the total error is calculated, and the weights are adjusted based on the total error. This control method can ensure operation in the direction of error reduction [6], and has high processing efficiency. The management and scientific research process of BP neural-network is shown in Fig. 1.

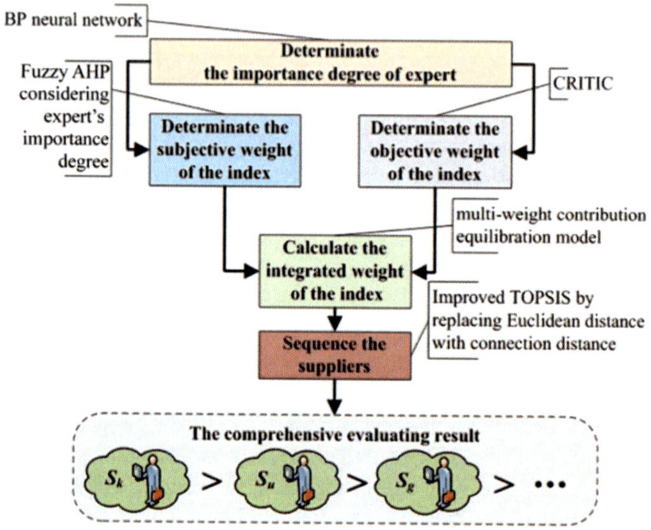

Fig. 1. Management and scientific research process of BP neural-network

3 AHP and BP Neural-Network are Used in the Management of Colleges Guangxi

3.1 Neural Network Model of BP Management Research

Set the output layer and the output layer nodes of BP neural-network according to the actual situation, and management scientific research has determined 8 evaluation indicators, then there are 8 nodes in the input layer of BP neural-network, and the output layer includes a neuron, which is used to describe the results of management scientific research. Algebraic value, its value range is [0, 1]. Obtain the optimal hidden layer nodes by trial and error. First, use a smaller number of nodes to train the network. If the obtained error is higher than the expected target [7], the nodes is continuously increased based on the same sample set, and trial and error are repeated. Until the network error is the lowest, the final number of hidden layer nodes is obtained, as shown in Fig. 2. By comparing and analyzing the number of training times and errors, it can be seen that when the number of hidden layer nodes is 10, the network has the highest convergence efficiency and stability, so this article chooses the number of hidden layer nodes to be 10. Set the error performance index of BP neural-network convergence to MSE, and the accuracy of the network is 0.0001. The maximum training step size is 10 000, and the training set training can be terminated when the network training reaches the maximum step size [7]. Clarify the momentum factor of 0.96 to avoid local minimum problems in network training.

Based on the specification requirements of the BP neural-network, the S-tangent function tensing and the logarithmic function logsig are regarded as the transfer functions of the layer and the output layer neurons respectively [7]. If the BP neural-network structure and training algorithm have been clarified, the network training samples have

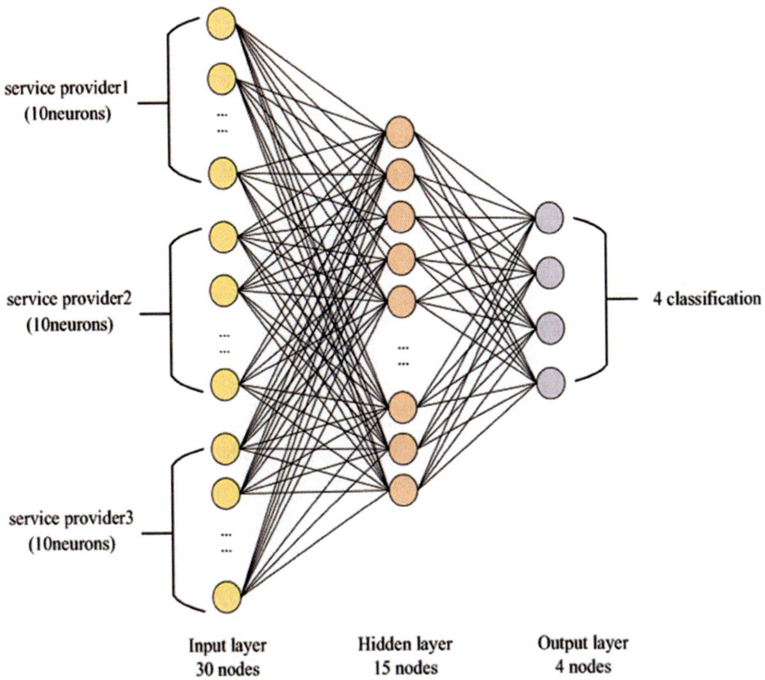

Fig. 2. Neural network model of BP management scientific research

a decisive effect on the network mapping and generalization performance [8]. The BP neural-network samples are divided into training samples and test samples. The test samples perform detection on the BP neural-network of the training function [8], and analyze whether the detection results meet the expected goals. Usually, each of the two samples Accounted for 80% and 20% of the total sample number. The knowledge management scientific research indicators of many universities are used as the input data of the BP neural-network, and the learning samples are used to train the BP neural-network to adjust the network weights to ensure that the network output error is minimized and the optimal management scientific research results are obtained [8].

3.2 Determine the Weight of the Training Quality Demand Index Based on the Analytic Hierarchy

The analytic hierarchy process refers to a complex multi-scheme decision-making problem as a system, decomposing it into multiple goals and criteria, and performing qualitative and quantitative analysis to calculate the system method for multi-objective [9]. The characteristic of this method is not to directly make decisions based on the overall goal, but to split the overall goal into multiple levels of indicators for specific implementation. By solving the eigenvectors of the judgment matrix, the priority of each level to the elements of the previous level is obtained. Finally, re-weighting method is used to recursively merge the final weight [9].

3.3 University Training Process Model and Demand Hierarchy Model

The AHP management talent training process model is shown in Fig. 3. The core layer is the stakeholder's demand for the quality of entrepreneurial talents, and is the ultimate goal. The functional layer is mainly training methods and approaches, which mainly include four aspects: innovation education, system guarantee, personal trait education, and practice platform provision [10]. The evaluation layer is to evaluate the quality demand system of entrepreneurial talent training. The training model starts from the needs of stakeholders, and creates four training methods and approaches in the functional layer based on their needs. Finally, it evaluates the quality demand system for entrepreneurial talent training in order to make timely adjustments to the current training methods and models. It also proposes improvement and remedial measures for the training model that has appeared problems, and finally completes the transition from "what is needed" to "how to do it" [10]. The cultivation and promotion of entrepreneurial talents' ability is largely achieved by the entrepreneurial talent training process, so it is very important to understand the training process model of entrepreneurial talents.

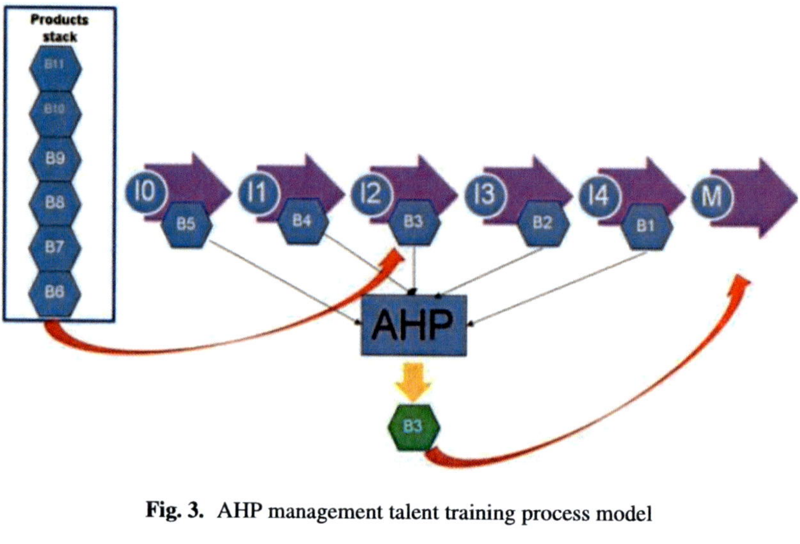

Fig. 3. AHP management talent training process model

4 Conclusion

This paper proposes a management scientific research model based on the neural network AHP-BP, which is composed of two processes: the forward propagation of information and the back propagation of errors. The search process of the neural network AHP-BP algorithm is realized by the gradient method. This management mode solves the shortcomings of low convergence efficiency and weak generalization performance in the traditional Guangxi university scientific research management scientific research model, and has high regression performance and generalization ability.

Acknowledgement. Project: 2020 Guangxi Philosophy and Social Science Planning Research Project: Research on the Evaluation of Scientific Research Performance of Guangxi Higher Education Institutions Based on AHP and BP Neural Network (Project Number: 20FGL026).

References

1. Liu, X.F., Li, Ch.F.: Scientific research of logistics university service management based on matter element analysis. Logist. Technol. **34**(21), 94–97 (2015). (in Chinese)
2. Wei, Q., Cai, Y.L.: An improved algorithm for middle guidance based on neural network. J. Xi'an Jiaotong Univ. **50**(07), 125–130 (2016). (in Chinese)
3. Wang, X.: Research on lean supply chain collaborative management based on bp neural network. Mod. Bus. Ind. **36**(05), 63–64 (2015). (in Chinese)
4. Shan, D.H., Cai, Q., Lu, S.X.: Research on harmonic current detection method based on BP neural network. Electron. Design Eng. **24**(10), 42–44 (2016). (in Chinese)
5. Zhang, Z.Sh., Li, X.T.: Research on evaluation of entrepreneurship policy management based on DEA model. J. Tianjin University (Soc. Sci. Ed.) **18**(05), 385–389 (2016). (in Chinese)
6. Zhao, Q.Ch.: Research on evaluation of agricultural supermarket docking supply chain management based on BP neural network. Value Eng. **35**(22), 88–90 (2016). (in Chinese)
7. Zhang, T., Li, Zh.F.: The construction of the evaluation model of learning management under the flipped classroom. Mod. Educ. Technol. **26**(04), 74–78 (2016). (in Chinese)
8. Li, P., Xu, A.L.: Image recognition technology of intelligent manufacturing system based on BP neural network. Mod. Electron. Technol. **39**(18), 107–109 (2016). (in Chinese)
9. Wang, B., Guo, D.Q.: Research on comprehensive evaluation of knowledge transfer management in universities based on BP neural network. Inf. Sci. **34**(01), 141–145 (2016). (in Chinese)
10. Wang, Y., Wang, Ch.T., Sun, X.Z.: Jiangsu provincial economic evaluation based on AHP-BP neural network. Sci. Technol. Manag. Res. **36**(09), 68–72 (2016). (in Chinese)

Design of IIR Digital Filter Based on FPGA

Chuang Wang(✉), He Xi, Tienian Chen, and Jiawei Liu

CRRC Dalian R&D Co., Ltd., Dalian, Liaoning, China

Abstract. A method of implementing high-speed IIR digital filter in FPGA is proposed, and the stability problem of IIR digital filter coefficients after rounding is theoretically analyzed; the filter is designed using FDATool, and the simulation is programmed in Matlab; the method of experimental simulation is used. Determine the quantization word length of the IIR filter coefficients to ensure the optimization of IIR filter performance and hardware resources, so that the IIR filter can be applied to high-speed occasions. The calculation characteristics of the calculation components in the FPGA are studied, and the Verilog hardware description language is used to implement iterative calculations and Symbol number multiplication; Finally, the IIR digital filter is programmed to be realized by simulation and implemented on FPGA. It is verified by experiments that the IIR digital filter designed by this method converges and can be applied to systems that require high real-time performance.

Keywords: FPGA · IIR · Digital filter · Filter design

1 Introduction

Digital filters have been widely used in many fields such as communications, automatic control, radar, military, aerospace, medical treatment, and household appliances. Among them, the IIR digital filter and the FIR digital filter are the two most commonly used by people at present. Digital filters are usually implemented by computer software, dedicated digital filters, DSP devices or programmable logic devices (such as FPGA). Because the FPGA implementation of digital filters has the advantages of strong real-time performance, high flexibility, fast processing speed, and low cost of small batch production, it has been widely used [1]. This article takes the Butterworth digital bandhas filter as an example, and introduces its design and implementation methods in more detail. The sampling frequency of a given Butterworth digital bandhas filter is 500 Hz, and the upper and lower sideband cut-off frequencies are 150 Hz and 30 Hz, respectively [1].

2 The Basic Design Principle of Digital Filter IIR

The so-called digital filter is an important function for processing digital signals. It filters, detects and estimates the parameters of the signal, that is, eliminates the noise in the digital signal, so that the valuable signal can be retained. The application of digital filters

B. J. Jansen et al. (Eds.): *International Conference on Cognitive based Information Processing and Applications (CIPA 2021)*, LNDECT 85, pp. 628–634, 2022.
https://doi.org/10.1007/978-981-16-5854-9_80

is more widely. The advantages embodied in the application of digital filter and analog filter are more obvious, with high precision and stability, small size of the equipment and flexible use, and the special filtering function of analog filter can be realized without matching impedance [2]. The digital filter is actually a discrete system. From the actual network structure or the corresponding classification of the unit pulse, it can be divided into two types: unlimited impulse response (IIR) and finite impulse response (FIR). The basic principles of IIR digital filter design are as follows:

From the point of view of the filtering process, both the input and output are digital signals. When passing through the filter, a certain calculation is used to compare and analyze the frequency contained in the input signal, so as to filter out those "noise" parts. IIR filtering the design principle of the filter is based on an analog filter, and then it is converted into a digital filter by equivalent [2]. Among them, the impulse response invariant method is a relatively common design method, and the process is:

$$\mathrm{H(s)} \rightarrow \mathrm{h(t)} \rightarrow \mathrm{h(n)} \rightarrow \mathrm{H(z)}$$

Considering that the traditional filter has a large amount of calculation and the filter characteristics are not easy to adjust, in order to solve this problem, the powerful digital processing and calculation ability of MATLAB has solved the problem of large calculation amount in design, not only simplifies the calculation amount, but also can Adjust the characteristic parameters of the Gini filter according to the design requirements, making it more flexible [3]. The specific steps are as follows: Determine the basic indicators of the filter according to the actual requirements of the project, such as the boundary frequency; the minimum attenuation and maximum attenuation of the stop band; convert the technical indicators of the digital filter into the analog filter indicators, and the pulse response invariance method is commonly used. Designing an analog filter is to select the model of the analog filter according to the index, such as Butterworth or Chebyshev filter, etc.; select the appropriate transformation algorithm to convert the analog filter into a digital filter; finally use software or hardware technology to make it run [3].

3 FPGA-Based IIR Digital Filter Design

3.1 Filter System Model

The bilinear transformation method uses frequency transformation on the basis of the low-pass filter to convert the low-pass filter into the required filter [4]. According to this method, we get the following filter system model:

$$H(z) = \frac{0.1514(1 - 3z^{-2} + 3z^{-4} - z^{-6})}{1 - 1.8195z^{-1} + 1.3322z^{-2} - 0.8195z^{-3} + 0.6167z^{-4} - 0.2152z^{-5} + 0.0105z^{-6}}$$

3.2 MATLAB Design of Filter

Since this article uses Butterworth filter, the calling format of the two functions buttord and butter that need to be called in the toolbox is:

$$[\mathrm{N, wc}] = \mathrm{buttord\ (wp, ws, Rp, As)}$$
$$[\mathrm{b, a}] = \mathrm{butter(N, wc)}$$

Where N is the filter order; wp and ws are the passband cutoff frequency vector and stopband cutoff frequency vector respectively, the unit is π, generally the analog frequency index is normalized to half of the sampling frequency; Rp and As are respectively pass with maximum attenuation and stopband minimum attenuation, the unit is dB; wc is the 3 dB edge frequency vector; b and a are the coefficient vectors in equation [4].

After obtaining the coefficients 6 and a, call the function freqz(b, a, k, Fs) to calculate the complex frequency response vector H at point k according to the following formula:

$$H(e)^{jw} = \frac{B(e)^{jw}}{A(e)^{jw}} = \frac{b(1) + b(2)e^{-jw} + \cdots + b(m+1)e^{-jmw}}{a(1) + a(2)e^{-jw} + \cdots + a(n+1)e^{-jnw}}$$

Then the amplitude-frequency and phase-frequency characteristic curves of point K can be drawn to check whether the calculated coefficients meet the required filter indicators.

3.3 Filter Structure

The IIR filter of the same system function can be realized with different structures, such as direct I type, direct II type (canonical type), cascade type and parallel type. Different structures have different characteristics [5]. In order to facilitate hardware implementation and easily determine the number of data bits, this text chooses the direct I-type structure.

The direct I-type structure of the system is shown in Fig. 1.

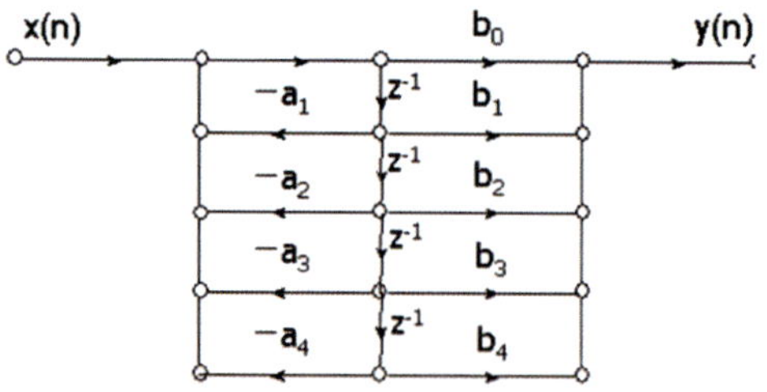

Fig. 1. The direct I-type structure of filter

3.4 FPGA Main Modules

(1) D flip-flop module.
Use D flip-flop to achieve the required delay function in Fig. 2. Whenever the rising edge of the clock arrives, the input data is latched [5]. D flip-flops can be customized and implemented using the LPM macro module in the Quartus II software.
(2) Multiply and accumulate module.
From the filter system model described by Eq. (1), the corresponding difference equation can be obtained as follows:

$$\begin{aligned} y(n) = 0.1514x(n) - 0.4542x(n-2) + 0.4542x(n-4) \\ - 0.1514x(n-6) + 1.8195y(n-1) - 1.3321y(n-2) \\ + 0.8195y(n-3) - 0.6167y(n-4) + 0.2156y(n-5) - 0.0105y(n-6) \end{aligned}$$

The multiply and accumulate module undertakes the calculation work of multiply and accumulate in above formula.

3.5 System Block Diagram and Experimental Waveform Diagram

The block diagram of the system structure using digital filters to filter analog signals is shown in Fig. 2.

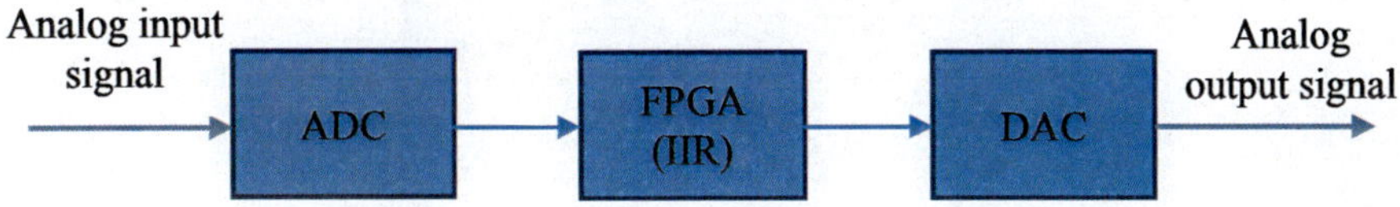

Fig. 2. The hardware framework of the filter system

The analog input signal is converted into a digital signal by an A/D converter, filtered by an FPGA digital filter, and then converted into an analog output signal by a D/A converter to complete the filtering process of the analog signal [6]. What the FPGA chip uses is Cyclone II series EP2C5T144C8 FPGA. A/D and D/A converter adopts Kangxin high-speed ADC/DAC module.

Input a sine signal with a peak-to-peak value of 5V, with periods of 10 Hz, 100 Hz, and 200 Hz. The output signal waveforms obtained after processing by the filter system are shown in Figs. 3 and 4, and the output signal amplitudes are 400 mV and 2.32 V respectively.

4 Precautions for Writing VHDL Language Code

The data in the process of multiplication and addition is signed two's complement. Usually, the header of the VHDL source file created under the Xilinx ISE integrated development environment will have "use IEEE.STD_LOG-IC_UNSIGNED_ALL;", such as changing it to "use IEEE.STD_LOGIC_SIGNED.ALL;" that should include a signed

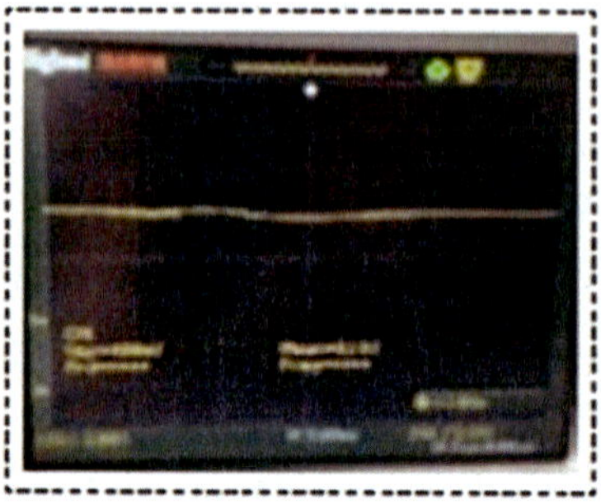

Fig. 3. Filter output signal waveform(Input 10 Hz sine signal)

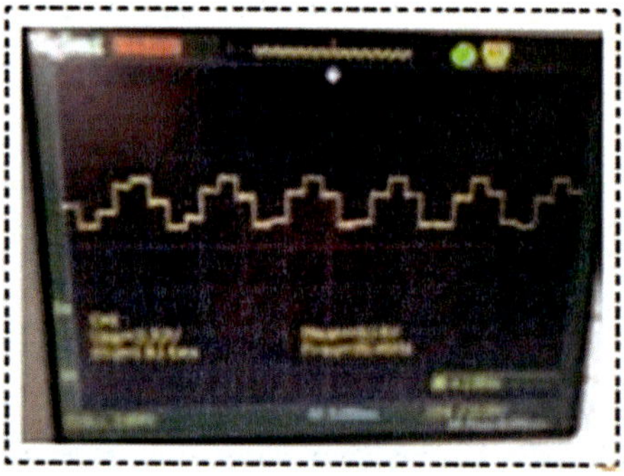

Fig. 4. Filter output signal waveform(Input 100 Hz sine signal)

number arithmetic package, so as to ensure that all std_log-ic_vector type data in the code are operated according to the rules of signed two's complement [7].

Since the FPGA cannot represent floating-point numbers, it can only use limited precision methods to implement data operations, that is, use the integer part of the data (including the input and output of equation and coefficients) (the decimal part is truncated) for approximate operations, and The std_log-ic_vector data type is required to represent the twos complement form of the integer part of the data, but this will cause truncation errors. In order to reduce the truncation error, the data should be enlarged by an appropriate multiple (usually 2L times, L is a positive integer), so that the fractional part can be ignored [8]. The larger the expansion, the smaller the truncation error and the more accurate the data obtained. However, the std_logic_vector type data length used to represent the integer part of the data will be larger, which will occupy more FPGA internal resources. Therefore, appropriate The choice of data expansion is the key. In addition, the length of the conversion of various data into std_logic_vector data should be at least sufficient to represent the two's complement code (including the sign bit). If the FPGA internal resources are sufficient, the std_logic_vector type data length can be increased to reduce the truncation error and improve the calculation accuracy [8].

Usually the coefficient b obtained by MATLAB simulation is far less than 1, so the positive integer L should be selected appropriately. In the calculation, the coefficients b and a (except the first coefficient of a) can be multiplied by 2L at the same time, and then rounded to get B = round(b*2L) and A = round(a*2L).

The x(n) input at the current time may sometimes be too small. In order to reduce the truncation error, an appropriate integer M should be selected to multiply x(n) by 2M, that is, to add the std_logic_vec-tor variable that represents the current time input M '0's. In this way, the output y (n) obtained at the current moment is the data expanded by 2L+M times, and the real output at the current moment should be divided by 2L+M. The VHDL language does not support division operations, so the method of truncating the end (L+M) digits should be used to approximate the division operation [9]. This approach is equivalent to truncating the decimal part after dividing the original output y by 2L+M.

When observing with an oscilloscope, the output waveform of the filter may have many large sharp burrs, which seriously affects the performance of the filter. A glitch is a phenomenon in which a circuit output has an instantaneous error due to the competition of a combinational circuit. The usual way to eliminate the glitch is to add a latch to the specific circuit. This article adopts another optimization method, that is, assign the output signal to an intermediate signal through the symbol "<=" in the source code, and then use the intermediate signal as the output, which is equivalent to delaying the signal and then outputting it. This method does not need to know the specific circuit structure, and also need to write other code modules, so the optimization is simpler and faster, and the optimization effect is very good [10].

5 Conclusion

Taking the IIR filter as an example, this paper describes the entire process of IIR digital filter from MATLAB design to FPGA implementation, discusses some key problems encountered in the design, and makes different levels of simulation on MATLAB. The technical indicators of the filter design are finally verified on the hardware platform.

References

1. Chen, J.N., Hu, J.T.: Application of FIR digital filter in digital profiling data processing. Precis. Manuf. Autom. **32**(03), 257–261 (2018)
2. Ma, J., Dong, J.M.: Implementation of IIR digital filter on fixed-point DSP. Acoust. Electron. Eng. **12**(34), 150–153 (2019)
3. Yang, Y., Li, C., Wang, K.: Design method of FIR digital filter. J. Guilin Aerosp. Ind. Coll. **9**(06), 102–106 (2019)
4. Guo, H.L., Wang, Z.T.: Design and implementation of FIR digital filter based on Matlab/Simulink. Electr. Appl. **9**(11), 43–46 (2019)
5. You, W.J.: IIR digital filter design. Guangxi Light Ind. **9**(15), 5–7 (2019)
6. Zhu, X., Xu, W.B.: Analysis and improvement of multiplication module in FIR digital filter. Microcomput. Inf. **19**(04), 69–72 (2019)
7. Gu, W.Y.: Optimization and verification of FIR digital filters. Electron. Measur. Technol. **15**(05), 104–106 (2017)

8. Wei, R.: Simulation and realization of FIR digital filter. Sci. Technol. Plaza **11**(06), 23–25 (2019)
9. Ru, X., Li, G., Tang, M.: Optimal design of FIR digital filter. Caizhi **7**(12), 29–31 (2014)
10. Dong, Y.P.: Multi-objective optimization design of IIR digital filter. Televis. Technol. **31**(01), 11–13 (2018)

Application of Computer Aided Language Learning in College Russian Teaching

Congmin Wang(✉)

Mudanjiang Normal University, Mudanjiang 157011, Heilongjiang, China

Abstract. With the development of computer technology. Computer Assisted Language Instruction (CALL) system is becoming more and more mature. More and more CALL systems have been applied in teaching and examination to assist teachers in teaching and improve students' learning efficiency. This paper explores a more efficient pronunciation error detection algorithm under the framework of deep neural network (DNN). In view of the particularity of oral learning, this paper firstly improves the original acoustic model, and then studies the pronunciation error detection algorithm from three different perspectives, namely, posterior probability calculation, hypothesis testing model and supervised learning classification, expands and improves the traditional method, and proposes a new method under the framework of DNN. The deep neural network model was built to model the eigenvalues to get the final score.

Keywords: Computer aided language · Russian teaching · Neural network · Speech recognition

1 Introduction

With the rapid development of computer technology, its application has penetrated into various fields including the education industry, which has had a huge impact on the traditional work, study and life style, and the traditional teaching method has also undergone a huge change. Education informationization is the requirement of The Times, and the integration of information technology and subjects is the inevitable trend. As an important part of the integration of information technology and classroom teaching, the computer-aided instruction system for subject courses has changed the traditional teaching mode and met the requirements of the development of current education informationization [1]. "Internet+ Education" has transformed the traditional education model, and education has adapted to the requirements of The Times. In recent years, computer aided instruction system has become a hot topic in the field of computer and education. In particular, it has gradually replaced the teacher in correcting the language test in the high grade section, which has become a great revolution in the education circle. This kind of system is called the Computer Assisted Learning Call system [2]. At present, there have been a lot of computer-aided assessment systems in the actual teaching, examination process

B. J. Jansen et al. (Eds.): *International Conference on Cognitive based Information Processing and Applications (CIPA 2021)*, LNDECT 85, pp. 635–642, 2022.
https://doi.org/10.1007/978-981-16-5854-9_81

large-scale application, such as Russian composition correcting system, computer program correcting system, etc. This system is more accurate and, most importantly, can save human resources and improve efficiency. In recent years, with the development of speech recognition technology and the maturity of essay correcting system, it is expected to break through the automatic correcting technology of open spoken English questions from the technical level and reach a practical level [3].

The CALL system dates back to the 1960s, when it was first built into mainframe computers as a course project for college students. The PLATO project of the University of Illinois is one of the representative examples of the early CALL system [4]. In the late 1970s, the rise of the microcomputer greatly promoted the development of CALL [5]. After the 1990s, it is called the comprehensive stage. Thanks to the popularity of the Internet and the development of multimedia technology, images, audio and video are also integrated into the CALL system. Call creates a virtual social environment for language learners, thus making language learning more vivid and interesting. Another significant advance in Call's development during this period was the application of artificial intelligence technology. Especially, the application of speech synthesis, speech recognition and other technologies makes the CALL system tend to be intelligent, and the interaction becomes more friendly and convenient [6].

In this paper, the work has an important significance to the development of the CALL system, improve the credibility and accuracy of the automatic grading open spoken, in Russian can also play a large role in oral English teaching, students can exercise of Russian spoken language, more comprehensive evaluation of students' oral English level, teacher's daily teaching, ease the pressure on teacher corrects the spoken language.

2 Application of Computer-Aided Language Learning

2.1 Application of the Deep Neural Network in Speech Recognition

In recent years, deep learning has been widely applied in the field of speech recognition. Various deep models, such as deep neural network, recursive neural network and recursive neural network based on field short-term memory unit, have replaced the traditional Gaussian mixture model and greatly improved the accuracy of speech recognition. In this paper, deep neural network is used to model the acoustic model of speech recognition, and the obtained acoustic model is expressed as DNN-HMM [7, 8].

The acoustic model training based on HMM is mainly to learn the feature spatial distribution of each senone. In a traditional GMM-HMM system, the observation probability space of SENONE is described by a GMM containing more than a dozen Gaussian components. In DNN-HMM, a DNN model is used to directly learn the posterior probabilities of each Senone, so it is a discriminant learning process. The DNN model takes acoustic features as input and learns the posterior probabilities of each Senone through multiple hidden layers and their nonlinear transformations [9, 10]. Its essence is a multi - category classification problem. Within DNN ssnone - HMM system s characteristics of input x the likelihood function value of p (x I s) calculated by type (1).

$$p = (X|S) = \frac{p(S|X) \cdot p(X)}{p(S)} \propto p(S|X)/p(S) \quad (1)$$

Where P (s) is the prior probability of Senone s, which is statistically obtained from the training data set.

Compared with the traditional GMM. HMM acoustic model, the DNN-HMM model has the following advantages:

(1) Support a wider observation range. When classifying the Senone of the current frame, DNN not only accepts the features of the current frame, but also cascades the observations of several adjacent frames as input to obtain a wider observation range [11]. In this way, more accurate classification results can be obtained. At the same time, some phoneme or syllable information can be learned, such as fundamental frequency curve, tone, stress and so on.
(2) Highly nonlinear and strong feature learning ability. Compared with GMM, which is a shallow model, DNN has multiple hidden layers, each layer includes thousands of nodes, and there is an activation function between each layer to nonlinear transform input features, so it has more powerful feature learning and expression ability.
(3) Support any form of input characteristics. DNN makes no underlying assumptions about the input features, and its input can be continuous, discrete, Boolean, or any combination of the three. There is also no requirement for the prior distribution of features. In the GMM-HMM system, in order to reduce the complexity of the model, the input feature dimensions are usually required to be relatively independent, so the input feature also needs to be de-coupled in advance [12]. DNN, on the other hand, does not have this requirement and can instead learn from more primitive features higher-level features that are more discriminative than those of traditional manual design.
(4) Discriminating the nature of learning. Different from GMM, DNN, as a discriminant model, can directly learn the posterior probabilities of samples for each Senone, so as to obtain higher classification accuracy at the frame level and higher discriminability of each Senone question.

2.2 Design of Voice Intelligent Scoring Mode

(1) Voice noise reduction design

This paper will adopt a different method from the traditional noise reduction: keep the signal processing part of the traditional noise reduction algorithm unchanged, do not use the neural network, but let the neural network learn all the parts that need to repeatedly debug the parameters after the signal processing. In this way, we can prepare our noise data set to train RNN according to usage scenarios, without repeatedly debugging the parameters of the signal processing part.

In the traditional noise reduction methods, the most difficult part is the noise spectrum analysis part, and this paper will use RNN to replace this part, the other parts are unchanged. Recurring neural networks (RNN) can play a very important role in speech denoising because they can model time series rather than just considering input and output frames. This is especially important for noise suppression because we need time series to get an estimate of the noise. For a long time, the capabilities of RNNs have been greatly limited because they cannot hold information for a long time and because the gradient descent process involved when propagating back through time is very inefficient (gradient vanishing problem). Both of these problems are addressed by new gated units such as Long Term Memory (LSTM), Gated Circulation Unit (GRU) and many of its variants. Compared with LSTM, GRU lacks a gate function, so the number of parameters is also less than that of LSTM. Therefore, the training speed of GRU is faster than that of LSTM on the whole. Therefore, this paper will adopt RNN network in the form of GRU.

Our noise reduction module design includes not only noise reduction, but also end-to-end detection (VAD), which is a necessary step in speech noise reduction. The so-called end-to-end detection is to detect the beginning and end time of speech and to cut off the mute time, which is of great help to the subsequent noise reduction training.

First of all, we use time window to sample the input audio signal with noise. We use 20 ms long time window with 50% overlap to sample the audio signal, which conforms to Princen-Bradly criterion. At the same time, we carry out pitch filtering on the original audio to extract the envelope of the speech. We perform a fast Fourier transform on the sampled audio to transform the audio data from time domain to frequency domain. Then feature extraction is carried out, and the most commonly used Mel cepstrum coefficient (MFCC) is used for feature extraction, and the eigenvalue is a 34-dimensional vector. In addition, we also extracted 6 frequency band tonal gain (sound intensity), pitch period, which helps us to track the noisy part of speech. These feature vectors are input into the RNN network, and the output is the frequency band gain interpolation of the noise in the speech, and then it is multiplied with the envelope of the original speech. Finally, the inverse Fourier transform and window overposition operation are carried out, and the output is the speech after noise reduction.

(2) Voice based scoring design

First, the speech file is cut and processed to produce a number of speech fragments, which may be one word or several words, ready for later grading using speech recognition patterns. After that, we carried out speech recognition on these speech fragments respectively. After identifying the text, we found the corresponding words in the standard pronunciation template library, calculated the similarity between them and the standard word pronunciation template, and got the pronunciation score. Finally, the pronunciation score of all the speech fragments was averaged. To make our score faster, we only rate the pronunciation of keywords such as nouns and verbs, so we use a stop list that removes meaningless words and does not rate the pronunciation of those words.

For the fluency rating, we used classic speed of speech (ROS) as the fluency rating criterion. The calculation formula is as follows: where Nwords is the total number of words spoken by the examinee, t is the total time of the audio file, and tvad is the mute time of the voice, namely the time of endpoint detection.

$$ROS = \frac{N_{words}}{t - t_{vad}} \tag{2}$$

3 Simulation Experiment of Sound Quality Evaluation Model in Russian

3.1 Data Sources

The data set used in this paper comes from the test data collected by the language training system of a certain university in this city in recent four years. These data are collected from the oral Russian test of undergraduates. These data come from different test papers. In this paper, 500 data from the same test question are selected as the experimental data set of this paper, from 500 examinees. In the following experiments, some of them required manual transcription of speech, so this paper manually transliterated 100 of the 500 pieces of data. These 100 pieces of data will be used in the experiments that required manual transcription.

The dataset contains three parts: audio files of students' spoken language recorded on the test site; Expert rating data; Students evaluate each other's grades.

3.2 Data Handling

Data preprocessing includes the following items: audio format conversion, removal of dirty data, and expert marking.

(1) Audio Format Conversion: We use FFMPEG to format raw audio. FFMPEG is an open source library for recording, converting and streaming digital audio and video formats. The original audio format in the data set is MP3, 16 kHz, and we convert it to PCM, 16 kHz format, which is for the convenience of our voice noise reduction and speech recognition later, because PCM format is a binary format, which is convenient for computer recognition and processing.
(2) Removal of dirty data and empty data: in our original data set, there are some dirty data. Generally speaking, the voice quality is very low and even manual grading cannot complete the data. We remove these data to eliminate the impact of these dirty data on our subsequent processing. In addition, there are some empty data in the data set, which also need to be deleted in advance. All these works rely on manual screening.
(3) Expert marking: In the original data set, the experts score is kept in a XLS file, file has both oral topic grading and other topics in the training system of rating data, and no corresponding to voice alone, so we need to artificial expert oral annotation on audio, particular way is to save every examinee's score alone in a TXT file, put together with the audio data of the examinee.

4 Simulation Experiment Results

4.1 Results of Different Acoustic Models on Test Datasets

As shown in Tables 1 and Fig. 1, after embedding F0 information, the frequency of FER on the test dataset decreased from 45.8% to 38.7%. The third column shows the recognition error rate at the phoneme level. Here, instead of using a complex phonetic

Table 1. Identification results of different acoustic models on test datasets

	Frame Error Rate (FER)	Phone Error Rate (PER)
GMM-HMM(MFCC)		40.21%
DNN-HMM(MFCC)	45.8%	27.15%
DNN-HMM(MFCC + F0)	38.7%	23.04%

model, we use a free phoneme cycle without grammatical constraints, in which each phoneme is followed by any phoneme with equal probability. As can be seen from the table, PER decreased from 40.21% to 27.15% after DNN differentiated training. By embedding F0 information into DNN model, PER was further reduced to 23.04%.

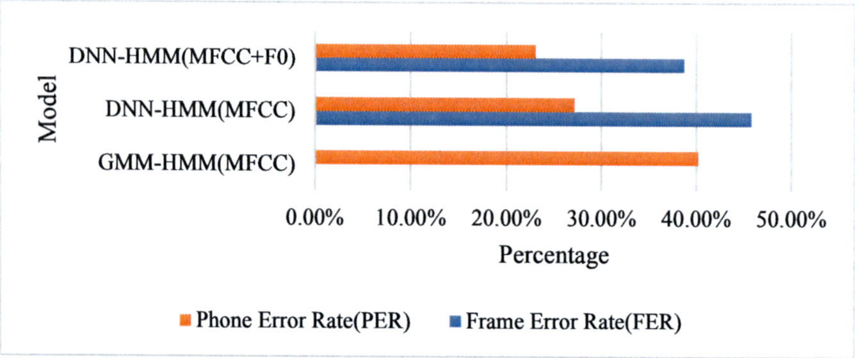

Fig. 1. Identification results of different acoustic models on test datasets

4.2 Diagnostic Results of Mispronunciation

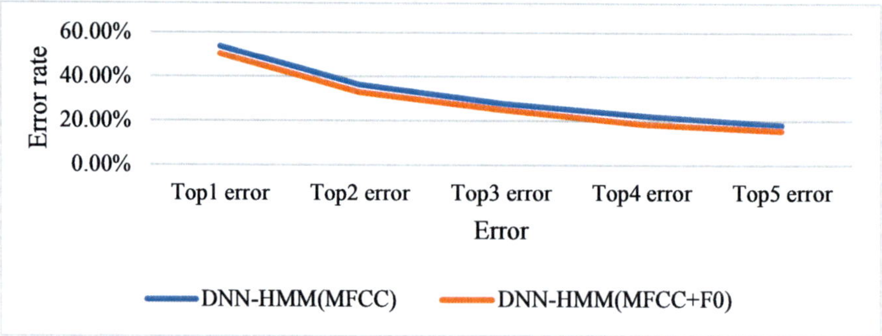

Fig. 2. Diagnostic results of mispronunciation

As shown in Fig. 2, the horizontal axis is N value, and the vertical axis is the corresponding TOPN error rate. As can be seen from the figure, DNN-HMM(MFCC + F0) is superior to DNN-HMM(MFCC) at any N value, and can reduce the error rate of TOPN by about 3%. The specific values of the error rates from TOPL to TOP5 of the two systems are given in the table. By using DNN-HMM(MFCC + F0) algorithm, the error rate of TOPL is 50.4%, and the error rate of TOP3 is less than 25%.

5 Conclusions

This paper investigates the current status of automatic correcting systems at home and abroad, and makes an in-depth investigation of the current research status of automatic correcting systems for spoken foreign languages according to the research content. The technical characteristics and limitations of these systems are discussed. Combined with the application background of this paper, the research background, significance and research content of this paper are summarized. This paper studies the manual scoring standard of open spoken questions, and makes full use of the manual scoring model in the design, and strives to approach the effect of manual scoring, and enhances the credibility of the model. In this paper, deep learning technology is studied and applied to the automatic correcting model of spoken language. It has achieved good results in the experiment and improved the accuracy of the open automatic correcting model of spoken language.

References

1. Syaad, P., Hidayat, W.N.: The effectiveness of learning management system (LMS) on computer assisted learning course for informatics engineering education students. J. Comput. Theor. Nanosci. **24**(4), 2642–2645 (2018)
2. Leathard, H.L., et al.: Comparison of the cost-effectiveness of a computer-assisted learning program with a tutored demonstration to teach intestinal motility to medical students. ALT-J **3**(1), 118–125 (2016)
3. Zhuhadar, L., Carson, B., Daday, J., Thrasher, E., Nasraoui, O.: Computer-assisted learning based on universal design, multimodal presentation and textual linkage. J. Knowl. Econ. **7**(2), 373–387 (2016). https://doi.org/10.1007/s13132-016-0371-y
4. Rum, S., Ismail, M.A.: Metacognitive support accelerates computer assisted learning for novice programmers. Educ. Technol. Soc. **20**(3), 170–181 (2017)
5. Costello, E., Ruckert, E., Lyons, L., et al.: To treat or not to treat: the use of computer assisted learning to enhance clinical decision making and self-efficacy of student physical therapists in the acute care setting. J. Phys. Ther. Educ. **31**(3), 27–36 (2017)
6. Sengupta, P., Sharma, A., Das, N.: Is there any benefit of integrating computer-assisted learning with conventional teaching format in pharmacology to demonstrate the effects of different drugs on mean arterial blood pressure in an anesthetized dog?: A comparative study. J. Nat. Sci. Biol. Med. **8**(2), 181–185 (2017)
7. Zhao, T., Zhao, Y., Chen, X.: Ensemble acoustic modeling for CD-DNN-HMM using random forests of phonetic decision trees. J. Signal Process. Syst. **82**(2), 187–196 (2015). https://doi.org/10.1007/s11265-015-1001-9

8. Fahad, M.S., Deepak, A., Pradhan, G., Yadav, J.: DNN-HMM-based speaker-adaptive emotion recognition using MFCC and epoch-based features. Circuits Syst. Signal Process. **40**(1), 466–489 (2020). https://doi.org/10.1007/s00034-020-01486-8
9. Ochiai, T., Matsuda, S., Watanabe, H., et al.: Speaker adaptive training localizing speaker modules in DNN for hybrid DNN-HMM speech recognizers. IEICE Trans. Inf. Syst. **E99.D**(10), 2431–2443 (2016)
10. Tan, X., Xie, Y., Ma, H., et al.: Recognizing the content types of network traffic based on a hybrid DNN-HMM model. J. Netw. Comput. Appl. **142**(Sep), 51–62 (2019)
11. Rath, S.P.: Scalable algorithms for unsupervised clustering of acoustic data for speech recognition. Comput. Speech Lang. **46**(Nov), 233–248 (2017)
12. Joy, N.M., Umesh, S.: Improving acoustic models in TORGO dysarthric speech database. IEEE Trans. Neural Syst. Rehabil. Eng. **26**(99), 637–645 (2018)

Application of Semantics Analysis in Text Classification of Computer Technology

Ziqi Wang(✉)

College of Foreign Languages, Jilin University, Changchun, Jilin, China
wangzq0318@mails.jlu.edu.cn

Abstract. Text classification is a basic problem in the field of natural language processing. It is a very active research direction in the field of machine learning, and has many important practical applications. This paper mainly studies the application of semantic analysis in text classification of computer technology. In view of the fact that most current methods ignore the important correlation between features, this paper adopts the LSA method, which fully considers the semantic relationship between features and reduces the dimension of feature space, so as to construct a new semantic space. Experimental results show that the proposed method can effectively reduce the dimension of feature space and improve the performance of text classification.

Keywords: Semantic analysis · Computer technology · Text classification · Feature extraction

1 Introduction

With the progress of society and the development of science and technology, social networks have gradually become People's Daily life circle, followed by the explosive growth of all kinds of data information, and the Internet has gradually stepped into the "era of big data". The rise of social media makes more and more students love to express their emotions, opinions and comments on platforms such as post bars, microblogs and forums, which makes text information more and more difficult to deal with. However, the potential value hidden by these massive text information cannot be ignored. Is the so-called "evaluation" when the group, "theory" to group points, once the negative emotions or false content will reflect the students from the side of the potential psychological problems. The large number of texts and various types of texts make it difficult to supervise the relevant platforms. Text classification has also become one of the key projects in the field of artificial intelligence to continuously explore and study [1]. Text classification is helpful to organize and manage information, which is of great significance to the discovery and prediction of abnormal behavior of students. With the rapid development of machine learning technology, a combination of artificial feature engineering and shallow classifier is gradually formed in text classification methods, but text representation still has problems of high dimension and sparse, which leads to weak

B. J. Jansen et al. (Eds.): *International Conference on Cognitive based Information Processing and Applications (CIPA 2021)*, LNDECT 85, pp. 643–650, 2022.
https://doi.org/10.1007/978-981-16-5854-9_82

text feature expression ability [2]. In recent years, thanks to the great progress made by deep learning technology in the fields of machine vision and speech recognition, some scholars have begun to integrate the idea of deep learning into the field of NLP, and it has become the current mainstream research direction [3].

With the rapid development of text classification technology, more and more fields have been involved in and applied to text classification technology, such as medical application, social media analysis, financial analysis and other fields. Because it involves a wide range of fields, the methods of text classification emerge in an endless stream. At the beginning of machine learning becoming the mainstream technology, many scholars have used this technology to classify texts and achieved good results. Naive Bayes (NB) is the most widely used text classification task based on machine learning in the early stage [4]. Traditional text classification technologies are mostly inclined to traditional statistical language processing technologies. In recent years, with the rapid development of machine learning and deep learning, more and more deep learning-based methods have achieved good results in text classification [5]. Convolutional Neural Network (CNN) is one of the most widely used deep learning structures. It has the characteristic of sharing weight, which makes the training parameters relatively small, the network structure efficient, the dimension-reduction speed fast, and the problem adaptability strong [6]. CNN first excelled in image processing problems, and researchers have successfully applied it to many other areas.

This paper will study the general text representation model on the basis of deep feature extraction, word meaning and text sequence information, learn good text representation and apply it to text classification task, and improve the effect of text classification.

2 Text Classification Based on Semantic Analysis

2.1 Text Preprocessing

The Chinese text preprocessing part usually includes the steps of word segmentation and word removal. The words in the Chinese text are continuous, so the text in the sentence needs to be divided into discrete word streams. In addition, in the face of some special types of corpora, special processing may be required according to the requirements of the experiment [7]. However, in the experiment of this paper, since such factors as title and time are not needed to be considered, only steps such as word segmentation, stem extraction and stopping words are needed.

(1) participles

At present, the commonly used methods in the research are generally based on dictionaries or statistics. Word segmentation method based on dictionary, usually on the basis of pre-built dictionary, will need to deal with the text data and carried out in accordance with the order entry words in dictionary one by one, if it is found that in the current need of existence and the dictionary word in the sentence segmentation matching string, the string will be extracted from the sentence, until the end of the sentence. The principle of the statistics-based word segmentation method is to judge whether a fixed word item can be formed according to the adjacent position relationship between words and the frequency of their common occurrence [8].

(2) word stem extraction
Stem extraction is mainly aimed at English language text. Because some words actually have the same root but have different forms due to different subjects or tenses, the process of combining these words together is called stem extraction. Chinese class text does not need this step. Such as: while the words "listen," "listen," "listen," and "listen" are brightly different, "listen" are brightly different because of the tenses, the words "listen" are different words because the subject is different, but they represent the same meaning – "listen". Therefore, it is not necessary for these words to appear in the feature set at the same time. Instead, choose the word "listen" as the feature item. At present, in the research field of stem extraction, the classic rule-based stem extraction methods mainly include Porter algorithm, Lancaster algorithm, Lovin algorithm and Dawson algorithm [9]. This paper adopts the commonly used Porter word stem extraction algorithm.

(3) Go to the park
After word segmentation, the text data should be processed to stop words, the main purpose of which is to remove those useless words for text classification, so as to reduce the number of feature words in feature set and the error rate of classification. Stopping words mainly refer to various special symbolic characters and function words in languages, etc., which usually appear in the text set at a high frequency and have the same probability of appearing in each document. However, they have almost no effect on text classification, that is to say, their ability to distinguish categories is poor [10]. It usually includes some adverbs, prepositions, modal words and other function words with no real meaning. In English, such words as "the", "a", "in", "about", "of", "after" and "Oh"; In Chinese, such as "di", "we", "not only", "ah" and other words, these words belong to stop words, they should be filtered out in the pre-processing stage. In this step to stop words, stop words are usually removed by using a stop word list, which can be existing on the Internet or can be created by oneself.

2.2 Selection Method of Text Characteristics Based on Potential Semantics Analysis

Semantic analysis technology refers to the comprehensive use of various methods to establish a computational model, so that it can learn and understand natural language like human beings [11, 12]. Semantic analysis is generally divided into word semantic analysis, sentence semantic analysis and text semantic analysis.

For most of the current supervised feature selection algorithms, some pay more attention to the document frequency, but ignore the word frequency of the feature, and some are just the opposite. Other methods only pay attention to the influence of the distribution of features among different categories of the data set on its importance, while ignoring the influence of the distribution of features within a certain class. Similarly, some methods are just the opposite. Aiming at the shortcomings of existing methods, this paper combines document frequency, word frequency and inverse category frequency, which are three important factors, to propose a new feature selection method based on semantic analysis, and comprehensively considers the influence of the distribution of features within and between classes on the ability to distinguish categories.

Latent Semantic Analysis (LSA), also known as Latent Semantic Index (LSI). Initially applied in the field of information retrieval, LSA has been widely used in text data mining, such as text classification, text clustering, information filtering and dictionary construction, along with the continuous research and development of LSA.

Assuming there are n documents and m features in the text data set, it can be constructed into an m × n word-document matrix A = [AIJ], and the element AIJ in the matrix represents the weight of the ith feature word in the JTH document. In this paper, TF-IDF algorithm is used to calculate the weight, and the matrix will be sparse. Let R be the rank of the matrix. The matrix can be converted into two orthogonal matrices and a diagonal matrix by singular value decomposition technique, as shown below:

$$A = U\sum V^T \tag{1}$$

Wherein, U is the orthogonal matrix of M × R, and also the left singular value vector of matrix A, and U is the orthogonal eigenvector of ATA. The sigma is A diagonal matrix of R × R, which is also the singular value matrix of matrix A, and the elements in the sigma are arranged in descending order. In order to reduce the dimension, the first K largest singular values in the diagonal matrix sigma are selected, the rest elements are set to zero, and the first K columns of U and V are retained. In this way, the k-rank approximate matrix of A can be obtained, as shown in Eq. (2):

$$A_K = U_K \sum\nolimits_K V_K^T \tag{2}$$

In this way, the "noise" problem in the original word-document matrix can be well reduced, and the rank of the matrix is reduced from r to K, and the underlying semantic structure in the original matrix is described in the k-dimensional eigenspace. When the LSA method is applied in the feature selection stage of text classification, the document can be mapped into the semantic space, and the original document can be mapped using Eq. (3).

$$d_{lsa} = d^T U_K \sum\nolimits_K^{-1} \tag{3}$$

3 Simulation Experiment

3.1 Experimental Environment

This paper mainly focuses on text classification algorithms based on semantic analysis, and carries out relevant experiments and comparative experimental analysis. Model coding is mainly completed by TensorFlow, the mainstream deep learning framework, and Python development language. TensorFlow deep learning framework is the most popular deep network model in the field of natural language processing. It has the advantage of distributed computing of heterogeneous devices and can realize the automatic running of the model in a distributed system built by multiple CPUs/GPUs. The experimental environment of this paper is as follows:

Operating system: WindowsL0 64 bit
Hardware and software environment:
CPU: Intel (R) Core (TM) i5-6500
RAM: 8.00 GB
Development language: Python
Integrated development environment: PyCharm.

3.2 Experimental Data of Text Classification

The data in this part mainly includes sentence classification data, text classification data and user dictionary data. The sentence classification data includes two English corpus and one Chinese corpus: TREC question type classification, Standard Sentiment Treebank (SST-1), and hotel review Sentiment polarity corpus.

4 Simulation Experiment Results

4.1 Text Classification Results of Different Algorithms

Table 1. Experimental comparison of text classification with different algorithms

	MP	MR	MF1
SVM	0.854	0.846	0.851
CNN	0.877	0.869	0.873
CNN+LSA	0.913	0.911	0.908

As shown in Fig. 1 and Table 1, three kinds of experimental results are presented, in which Mp represents the macro accuracy of multi-classification problems, MR represents the macro recall rate of multi-classification problems, and MFL represents the macro F1 value of multi-classification problems. Compared with traditional machine learning, CNN neural network was higher than SVM in the three experimental results, the accuracy rate of CNN neural network reached 0.877, the recall rate was 0.869, and the F1 value was 0.873. Compared with other feature extraction methods, the CNN neural network using latent semantic analysis for feature extraction is beneficial to improve its classification hours. Compared with CNN, the CNN + LSA method has a certain improvement in the three experimental results.

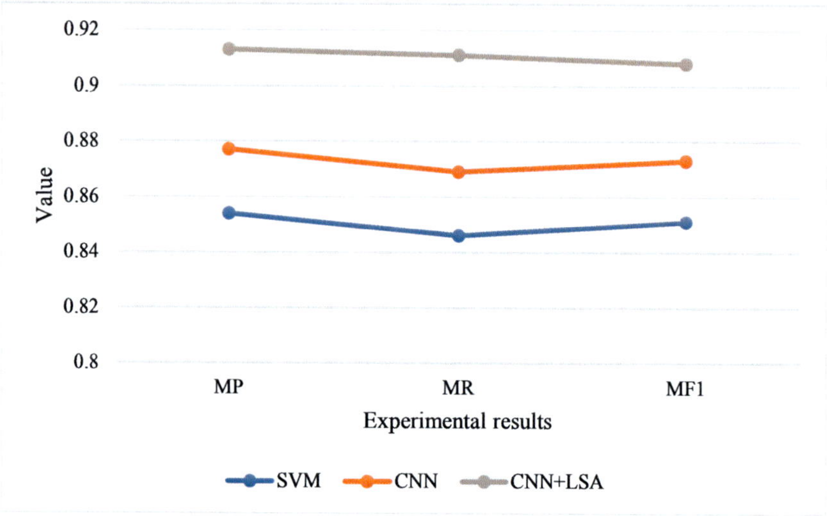

Fig. 1. Experimental comparison of text classification with different algorithms

4.2 Accuracy Rate of Different Learning Modes

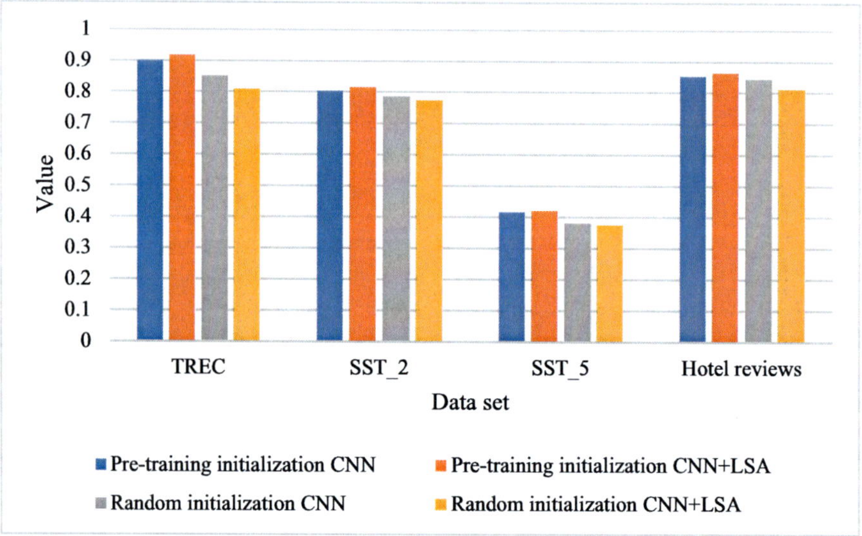

Fig. 2. The accuracy of three algorithms on different data sets

As shown in Fig. 2, the effect of the model is better than that of the randomly initialized word vector when the pre-trained word vector is used. Words in the process of the training vector studied its environment of distributed information, word coding on the essence of the general language features information (such as: semantic), said the study to understand a text, the text is very good, when learning a language with people to learn the word meaning, usage, and then according to the syntax and grammar rules such as the process of construct sentences and articles are very similar. At the same time, it should be noted that the improvement of the model by pre-training word vectors is limited, and the effect of the model based on randomly initializing word vectors is not too bad. In general, the experiment shows that the pre-training of word vectors (learning semantic information) can help the classification model to better classify.

5 Conclusions

Natural language processing technology is a very practical branch of artificial intelligence. It realizes the interaction between human and computer through the input, recognition and understanding of text content. However, with the continuous improvement of people's production and living standards, a large number of new text types continue to emerge, making the data types become more and more diversified. As an important part of text classification, feature selection has attracted more and more attention. In this paper, LSA method is adopted in feature extraction. Considering the semantic relationship between features and reducing the dimension of feature space, a new semantic space based on LSA method is constructed on the basis of the feature space formed in the first stage. The feature selection method proposed in this paper is superior to other methods in improving the accuracy of text classification in most cases.

References

1. Jiang, L., Li, C., Wang, S., et al.: Deep feature weighting for naive Bayes and its application to text classification. Eng. Appl. Artif. Intell. **52**(Jun.), 26–39 (2016)
2. Zablith, F., Osman, I.H.: ReviewModus: text classification and sentiment prediction of unstructured reviews using a hybrid combination of machine learning and evaluation models. Appl. Math. Modell. **71**(JUL.), 569–583 (2019)
3. Yan, K., Li, Z., Zhang, C.: A New multi-instance multi-label learning approach for image and text classification. Multimed. Tools Appl. **75**(13), 7875–7890 (2015). https://doi.org/10.1007/s11042-015-2702-6
4. Goudjil, M., Koudil, M., et al.: A novel active learning method using SVM for text classification. Int. J. Autom. Comput. **15**(03), 44–52 (2018)
5. Pavlinek, M., Podgorelec, V.: Text classification method based on self-training and LDA topic models. Expert Syst. Appl. **80**(SEP.), 83–93 (2017)
6. Uysal, A.K.: An improved global feature selection scheme for text classification. Expert Syst. Appl. **43**(JAN.), 82–92 (2016)
7. Hu, R., Namee, B.M., Delany, S.J.: Active learning for text classification. Expert Syst. Appl. **45**(MAR.), 438–449 (2016)
8. Schmidt, S., Schnitzer, S., Rensing, C.: Text classification based filters for a domain-specific search engine. Comput. Ind. **78**(C), 70–79 (2016)
9. Liu, C.L., Hsaio, W.H., Lee, C.H., et al.: Semi-supervised text classification with universum learning. IEEE Trans. Cybern. **46**(2), 462–473 (2017)
10. Mironczuk, M.M., Protasiewicz, J.: A recent overview of the state-of-the-art elements of text classification. Expert Syst. Appl. **106**(sep.), 36–54 (2018)

11. Günther, F., Dudschig, C., et al.: Latent semantic analysis cosines as a cognitive similarity measure: evidence from priming studies. Quart. J. Exp. Psychol. **69**(4), 626–653 (2016)
12. Boulares, M., Jemni, M.: Learning sign language machine translation based on elastic net regularization and latent semantic analysis. Artif. Intell. Rev. **46**(2), 145–166 (2016)

Droop Control Strategy of Microgrid Anti-island Based on Virtual Impedance

Huaizhong Chen(✉), Riliang Xu, and Jun Meng

Zhejiang Industry Polytechnic College, Shaoxing, Zhejiang, China

Abstract. In the process of switching from grid-connected to isolated island mode, photovoltaic microgrid is intermittent and uncertain, which is often improved by traditional droop control. Under this control mode, the adjustment and distribution ability of micro-power supply load power is insufficient, which leads to the system power distribution not being optimized, and the grid frequency fluctuation deviation is large. This paper analyzes the basic principle of increased droop control efficiency method by using virtual impedance, designing the optimized droop control method. By introducing virtual impedance, the control strategy realizes automatic power decoupling control. The optimized droop control method can effectively enhance the power control sharing effect of microgrid, and improve the power quality. Simulation result proves not only the effectiveness but also the practicability of the proposed control method.

Keywords: Microgrid · Virtual impedance · Droop control · Anti-island

1 Introduction

Photovoltaic microgrid has attracted wide attention from the international community for its advantages of cleanness and low energy consumption. However, due to the intermittence and uncertainty of photovoltaic microgrid, the safe and reliable operation of microgrid system is greatly affected in the transition from grid connected mode to island mode. Because photovoltaic power generation is very sensitive to the change of environmental conditions, it raises higher requirements in accuracy and rapidity. Thus, it is very necessary to research this control and other parameters of photovoltaic microgrid switching from grid-connected to isolated island. In microgrid, the inverter adopts droop control, which makes it convenient for microgrid to switch to isolated island mode, and keeps the output voltage of inverter unchanged. Sag control of photovoltaic microgrid is mainly to realize decoupling control of each coupling parameter in the system, so as to realize intelligent average distribution of reactive and active load, and automatically keep the output voltage and frequency of the system stable. However, the power distribution influence of traditional droop control strategy is influenced by line impedance and other factors. The intelligent adjustment and distribution ability of reactive and active load is insufficient, which leads to the failure of intelligent sharing of system power distribution and large deviation of grid frequency fluctuation when switching between isolated island and grid-connected operation modes. By introducing this improved strategy, the sharing effect of reactive and active load of the system can be further enhanced [1, 2].

B. J. Jansen et al. (Eds.): *International Conference on Cognitive based Information Processing and Applications (CIPA 2021)*, LNDECT 85, pp. 651–656, 2022.
https://doi.org/10.1007/978-981-16-5854-9_83

2 Establishment of Photovoltaic Microgrid Model

Two groups of PV (Photovoltaic arrays, PV) with equal capacity are selected as distributed generation power sources. The topology of photovoltaic microgrid is shown in Fig. 1 [3].

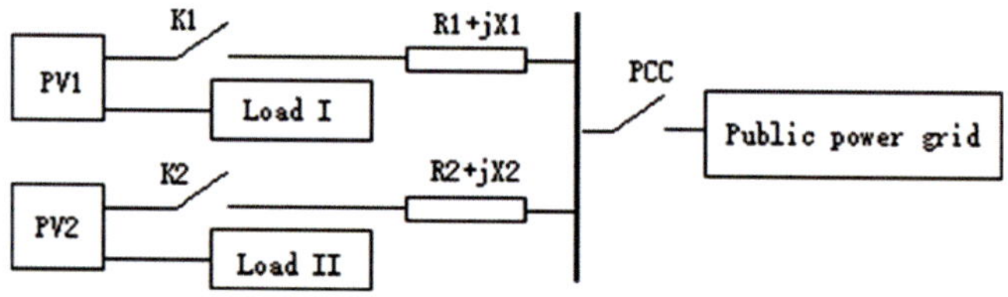

Fig. 1. Topology diagram of photovoltaic microgrid

In the Fig. 1, the photovoltaic microgrid is merged into the distribution network at PCC(Point of Common Coupling, PCC) through the static transfer switch. Load I and load II are connected to PV1 and PV2 as local loads respectively. Microgrid inverter generally adopts reactive current regulation control based on active current. When microgrid operates in island mode, PV is disconnected from local load network, and the power of photovoltaic microgrid is determined by load. When PV rated load is greater than load power, droop control strategy is generally adopted to reduce the impact of switching between isolated island and grid-connected operation modes of photovoltaic microgrid.

3 Droop Control Principle

Sag control simulates the power and phase angle characteristics of traditional generators, so that the microgrid can obtain stable voltage, power and frequency. The expression of sag control is [4]:

$$P_i = \frac{U_i U}{X_i}\theta_i \tag{1}$$

$$Q_i = \frac{U_i(U_i - U)}{X_i} \tag{2}$$

U represents the virtual value of the voltage at the point of common coupling; U_i represents the virtual value of the output phase voltage of the ith distributed micro power supply; θ_i is the phase angle difference between the output electric tension of the ith micro power and the electric tension at the point of common coupling; X_i is the reactance of the transmission line from the ith micro power supply to the load; Q_i is the reactive power output; P_i is the active power. Equation 1 and 2 shows that the phase angle difference θ_i determines the transmission of active power P_i of micro-power, P_i is proportional to θ_i, and P_i is transmitted from the leading point to the lagging point. The amplitude difference between the output electric tension of supply and the electric tension in common coupling point determines the transmission of reactive power Q_i of distributed micro-power, and Q_i is transmitted from the point with high voltage to the point with low voltage.The relationship between frequency f and phase angle θ of microgrid system is:

$$f = \frac{\omega}{2\pi} = \frac{1}{2\pi} \bullet \frac{d\theta}{dt} \tag{3}$$

It can be seen from formulas 1–3, By adjusting the active power of microgrid, the phase angle and frequency of output voltage can be changed. By adjusting the reactive power of microgrid, the amplitude of output voltage can be changed.

In droop control, micro-power is transmitted to the load in a given way of reactive power and active power. The expression of droop control is [5]:

$$f = f_n + m(P_n - P_s) \tag{4}$$

$$U = U_0 - nQ_s \tag{5}$$

In which: P_n is rated active power generated by micro power supply; f_n is rated frequency; U_0 is the voltage value of the microgrid when the output reactive power of the micro power supply is zero.Ps and Qs are the average active and reactive power output from the microgrid to the load. m is Ps-f sag coefficient, and n is Qs-U sag coefficient. The expression of m and n is:

$$m = \frac{f_n - f_{\min}}{P_{\max} - P_n} \tag{6}$$

$$n = \frac{U_0 - U_{\max}}{Q_{\max}} \tag{7}$$

In which: f_{min} is the minimum value of system frequency decline according to Ps-f droop characteristic when the average active power output by microgrid is maximum; P_{max} is the maximum average active power output by microgrid; Umin is the minimum value that the voltage value of microgrid drops according to Qs-U droop characteristic when the maximum average reactive power output by microgrid is maximum; Q_{max} represents the maximum average reactive power output by microgrid [6, 7].

When the active power generated by micro-power supply is too large, the active power produced through micro-power supply will be reduced with the increase of system frequency.

When the reactive power generated by micro-power supply is too large, increasing the voltage amplitude output by micro-power supply will reduce the reactive power generated by micro-power supply. or vice versa, Dallas to the auditorium Through this control, not only the reactive but also the active power are distributed reasonably.

4 Improved Droop Control Strategy

The power distribution effect of traditional droop control strategy is affected by line impedance and other factors. The intelligent adjustment and distribution ability of reactive and active load is insufficient. In case of isolated island mode and grid connected operation mode are switched, the system power allocation cannot realize intelligent average allocation, and the fluctuation deviation of power grid frequency is large.

The virtual impedance control strategy is introduced to solve the above problems in this paper. Virtual impedance in virtual impedance control does not actually exist in the circuit, but is equivalent in the circuit.

In the inverter circuit, an impedance is added virtually.This changes the output impedance of the inverter, but because the impedance is virtual, there will be no voltage drop in the circuit and no increase in energy loss in the circuit. After the virtual impedance is introduced, the impedance of the whole circuit changes.The principle of its working structure is shown in Fig. 2 [8].

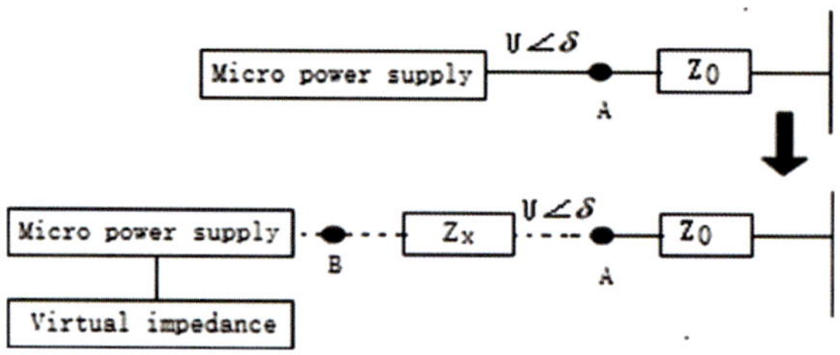

Fig. 2. The principle structure diagram with virtual impedance

The original impedance value of the line is Z_0, by pulling in virtual impedance Z_x into A-B circuit, the total impedance in the circuit becomes $Z_x + Z_0$. After this improvement, the impedance in the circuit is inductive, not only the reactive but also the active power both can be distributed and optimized reasonably.The reference voltage value of inverter of microgrid is the voltage value obtained by droop control characteristic minus the voltage value drop on virtual impedance. The virtual impedance loop is connected in parallel to conventional droop control system. The difference between the voltage signal measured in the droop control system and the voltage signal in the impedance loop is the reference voltage signal input to the double-loop control, which can make the impedance in circuit appear inductive and meet the requirements [9].

5 System Simulation

To further verify the effectiveness of the proposed improved droop control strategy, a simulation model of photovoltaic microgrid operation is designed. The model of photovoltaic microgrid operation includes G1 and G2 microgrids, which are simulated by conventional droop control and improved droop control. In the simulation example, the two micro-power supplies G1 and G2 are equally replaced by DC sources.Load I and load II power load values are as follows [10]:
Load I: active power P1 = 220 kw; reactive power, Q1 = 80 kvar;
Load II: active power P2 = 100 kw; reactive power Q2 = 40 kvar;

The simulation process of photovoltaic microgrid is as follows: firstly, two inverters are set to run stably with load I, with active and reactive power of 220 kw and 80 kvar, and load II is put into operation in 0.25 s.

The active and reactive power are 320 kw and 120 kvar after stable operation.When the load II is cut off in 0.6 s, the active and reactive power of the system are reduced to 220 kw and 80 kvar. Based on the conventional droop control, the reactive power output by two micro power supplies G1 and G2 is shown in Fig. 3.

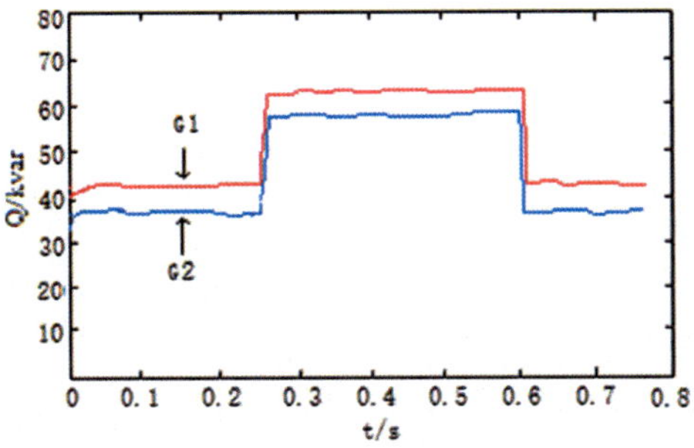

Fig. 3. Reactive power curve of power supply under traditional droop control

Figure 3 shows that with the conventional control strategy, when the load variation increases or decreases, the output reactive power of PV microgrid cannot be evenly distributed. There is a big difference between the reactive power output values of the two micro-power sources G1 and G2 after the Load II is switched in at 0.25 s and switched off at 0.6 s.

After introducing virtual impedance, the reactive power output by two micro-power supplies G1 and G2 is shown in Fig. 4.

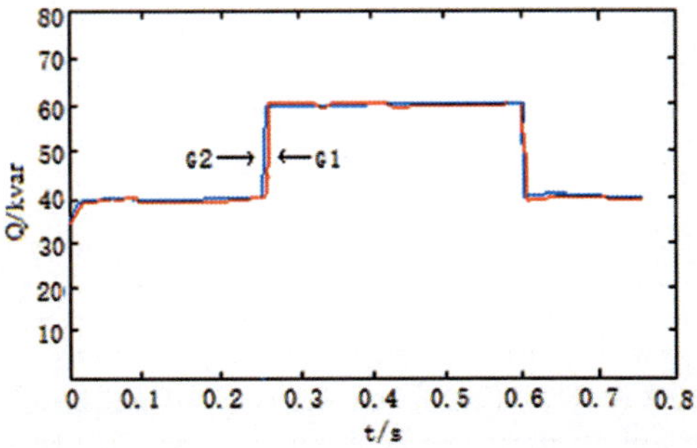

Fig. 4. Reactive power curve of power supply by introducing virtual impedance

Figure 4 shows that after the virtual impedance is introduced, when the load variation increases or decreases, the output reactive power of the system can be evenly and accurately distributed according to. The reactive power of PV microgrid outputs of the two micro-power sources G1 and G2 are close to the same after switching in Load II at 0.25 s and switching out Load II at 0.6 s, and the average distribution effect of reactive power is obvious [1, 11].

Therefore, when the load fluctuation increases or decreases, the system can realize the average distribution of reactive and active power.

6 Conclusion

Throughout the switching between isolated island and grid-connected mode, photovoltaic microgrid adopts traditional droop control, and the intelligent adjustment and distribution ability of reactive and active power of load is insufficient. This deficiency will lead to the system power allocation is not optimized.This paper analyzes the basic principle of improved droop control method by introducing virtual impedance. This control strategy can reconstruct the line impedance and realize automatic power sharing control. By introducing and optimizing the virtual harmonic impedance control strategy, not only the reactive but also the active power sharing effect of microgrid can be make further enhanced.The power quality can be further improved. It lays a good foundation for the follow-up study of photovoltaic microgrid operation optimization control.

Acknowledgements. This research was supported by Zhejiang Provincial Basic Public Welfare Research Project of China under Grant No. LGG21F030001.

References

1. Li, X., Zhao, S.: Reactive power distribution strategy and droop control method based on virtual impedance in microgrid. IAENG Int. J. Appl. Math. **50**(3), 678–689 (2020)
2. Guerrero, J.M., Uasquez, J.C.: Control strategy for flexible microgrid based on parallel-connected UPS inverters with wireless load-sharing control. IEEE Trans. Ind. Electron. **52**(4), 1126–1135 (2014)
3. Meng, F., Meng, J.: Research on control strategy of PV micro-grid connected based on improved droop method. Renew. Energy Resour. **34**(5), 660–665 (2016)
4. Wang, H.: Control strategy of microgrid based on droop control. J. Xuchang Univ. **37**(12), 77–81 (2018)
5. Nie, X., Gao, J.: Reactive power distribution strategy and droop control method based on virtual impedance in microgrid. Res. Explor. Lab. **39**(8), 107–111 (2020)
6. Zhang, L., Zheng, H.: An adaptive droop control strategy for islanded microgrid based on improved particle swarm optimization. IEEE Access **8**, 3579–3593 (2020)
7. Liu, Z., Su, C., Høidalen, H.K.: A multiagent system-based protection and control scheme for distribution system with distributed-generation integration. IEEE Trans. Power Deliv. **32**(1), 536–545 (2017)
8. Zishun, P., Jun, W.: Reactive power distribution strategy and droop control method based on virtual impedance in microgrid. IEEE Trans. Sustain. Energy **9**(3), 1157–1168 (2018)
9. Rahul, S., Sathans, S.: Virtual impedance based phase locked loop for control of parallel inverters connected to islanded microgrid. Comput. Electr. Eng. **73**, 58–70 (2019)
10. Xue, C.: Study on improved droop control and harmonic suppression method of microgrid. Master's Thesis of Shenyang University of Chemical Technology, pp. 22–32 (2019)
11. Wenyang, D., Yi, D.N.: A virtual-impedance droop control for accurate active power control and reactive power sharing using capacitive-coupling inverters. IEEE Trans. Ind. Appl. **56**(6), 6722–6733 (2020)

Application of Oblique Photogrammetry in 3D Modeling

Weihua Qiao(✉)

Shandong University of Science and Technology, Qingdao, Shandong, China

Abstract. In the measurement process, the application of high-tech has successfully ensured the traditional aerial photography technology. With the advancement of modern science and technology, new requirements have been put forward for the measurement technology in engineering. Among them, oblique photogrammetry is a new technology in development, and it plays a particularly important role in urbanization. In the construction of urbanization, the application of oblique photogrammetry technology can provide powerful guidance for the three-dimensional model in the city and provide an important basis for subsequent project construction. Based on this, the article analyzes and studies the fast three-dimensional modeling method of oblique photography.

Keywords: Tilt · Photogrammetry · Three-dimensional · Modeling

1 Introduction

Oblique photogrammetry technology, as a new and widely watched technical method, is no longer limited to the traditional or thophoto shooting method that can only be taken from a vertical angle. It uses multiple sensors placed on the same flight platform, and then from the vertical at the same time. Front view, rear view, left view, right view and other different angles to collect images, so as to reflect the actual situation of the ground objects more realistically, and obtain more accurate data information of ground buildings, which is convenient for the insiders to analyze and sort the data [1]. Thus greatly supplementing the shortcomings of traditional photographic image applications, so it has a wide range of application prospects in urban three-dimensional modeling. However, the oblique photogrammetry technology in the domestic urban 3D modeling is still in the experimental stage, and further research and further promotion are needed. Accordingly, this article will focus on the related technology of oblique photogrammetry and carry out related research work [1].

2 Related Technology of Tilt Photogrammetry

The three-dimensional measurement technology based on oblique photographic data mainly studies the basic principles of oblique photogrammetry, the characteristics of oblique photogrammetry, and the related technologies of oblique photogrammetry.

B. J. Jansen et al. (Eds.): *International Conference on Cognitive based Information Processing and Applications (CIPA 2021)*, LNDECT 85, pp. 657–663, 2022.
https://doi.org/10.1007/978-981-16-5854-9_84

2.1 Basic Principle of Tilt Photogrammetry

By placing multiple sensors on the same flight platform, in addition to using the vertical angle of traditional photogrammetry, traditional aerial photogrammetry is processed by advanced technology to produce 4D products. In order to obtain the texture information of the side of the building on the ground, it also adds Front view, rear view, left view, and right view capture images from four other different angles, and four oblique photography images for front view, rear view, left view and right view [2]. The tilt angle of the captured images is required to be between 15° and 45° between. In a period of time, in order to select the clearest one to make the detailed texture, the insider needs to take several sets of overlapping photos in succession, and the same building can be found in these photos, so that the insider can correct the tilt at the end the images are analyzed and sorted [2], as shown in Fig. 1.

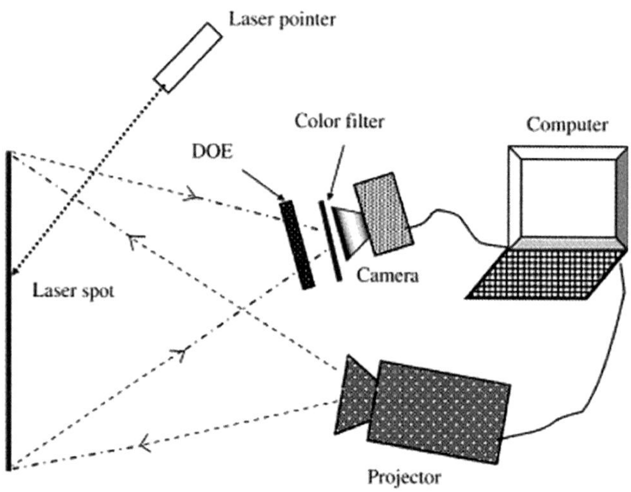

Fig. 1. Basic principle of tilt photogrammetry

2.2 Features of Oblique Photogrammetry

Feature 1: The surrounding conditions of the features are truly reflected. The three-dimensional data obtained by oblique photogrammetry can truly reflect the appearance, position, height and other attributes of the features; enhance the sense of reality brought by the three-dimensional data, and make up for the shortcomings of traditional artificial models with low degree of simulation. Feature 2: Single image measurement can be achieved through tilted images [3]. By using the corresponding application software, the measurement including length, height, angle, slope, etc. can be directly realized, which makes the application results can be used in public security, emergency response, surveying and mapping, tourism, environmental protection and other industries. It can be seen that the application of oblique photography technology in various industries has been widely expanded. Feature 3: Can collect the texture of the side of the building [3]. Aiming at various applications of urban 3D modeling, using the huge scale of oblique photography in space and the advanced method of batch extraction can greatly increase the raw material utilization rate of urban 3D modeling, reduce pollution and reduce costs.

2.3 Related Technology of Tilt Photogrammetry

Technique 1: Multi-view image joint adjustment. The joint adjustment of multi-view images needs to fully consider the azimuth elements of the mutual occlusion relationship between the images and the deformation of the space scene. In order to obtain a relatively better matching result of the same name points, it is necessary to adopt coarse to precise image processing at all levels [4]. Method, using free web beam method and automatic matching adjustment of points with the same name.

Technique 2: Multi-primitive, multi-view image matching. In recent years, with the rise of computer vision, multi-element, multi-view image matching has gradually developed. One of the basic elements of tilt photogrammetry is image matching. Multi-view images have the characteristics of high resolution and large shooting range. From this, by searching for features on the multi-view images, the two-dimensional data of ground buildings can be determined. The two-dimensional features of different perspectives can transform the three-dimensional features of the image. For example: the determination of the wall surface can be given a certain weight first, and the wall surface is classified according to the type [4]. To obtain the side structure of the building, then the various walls of the ground building are scanned and divided into planes, and extracted Figure out the height and contour of the building's roof.

Technique 3: Correction of photographic images.

First, according to the solution results of the azimuth elements of each image, an appropriate image-matching target is selected for dense matching and feature matching. In order to improve the calculation efficiency, parallel algorithms can also be introduced [5]. Then the different matching targets are merged and filtered to form a unified DSM. Finally, according to the spatial distribution and shape characteristics of the continuous terrain and discrete ground building objects displayed by the photographic images, the information is re-collected through methods such as contour drawing, one-sided fitting, and roof reconstruction, and then the image is established based on the results of dense matching and joint adjustment [5]. The corresponding relationship of the same name between the square and the object square is then established to correct the global optimization sampling, as shown in Fig. 2.

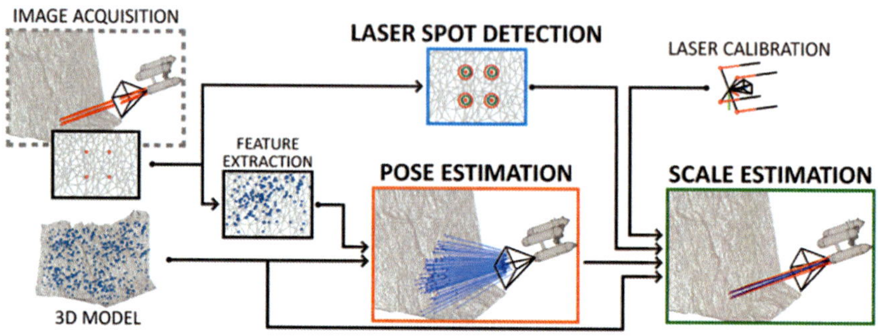

Fig. 2. Correction model of azimuth elements of each image

3 The Application of Oblique Photogrammetry in Basic Information Collection

3.1 Digital Line Drawing

In digital mapping, the most common type is the digital line drawing, which more comprehensively describes the visual effects and surface phenomena, as well as the colors it has. In order to express the geographic vector data set of terrain and ground buildings, digital line drawing is usually in the form of specific graphic symbols or points, lines, and surfaces [6]. The coordinate point is used to represent the point element in the vector data, the coordinate group is used to represent the line element; the area element is represented as a series of coordinate groups with the first and last points and the corresponding attribute values, through the outer frame and coordinates of ground building Information can enable insiders to quickly generate white molds [6].

3.2 Division of Urban Measurement Areas

First collect data and analysis data, and then divide similar or identical models into a survey area. Finally, the four side tilt images of the ground objects are named and texture numbered. This facilitates the sharing of similar textures and is more conducive to the sharing of similar textures [7].

3.3 Tilted Image

The tilted image of the ground feature can not only truly reflect the situation of the ground feature, but also through the use of advanced positioning technology, embedded accurate geographic information, richer image information, more advanced user experience, as shown in Fig. 3, greatly expanded the remote sensing image [8]. The application field of remote sensing image and make the industry application of remote sensing image more in-depth.

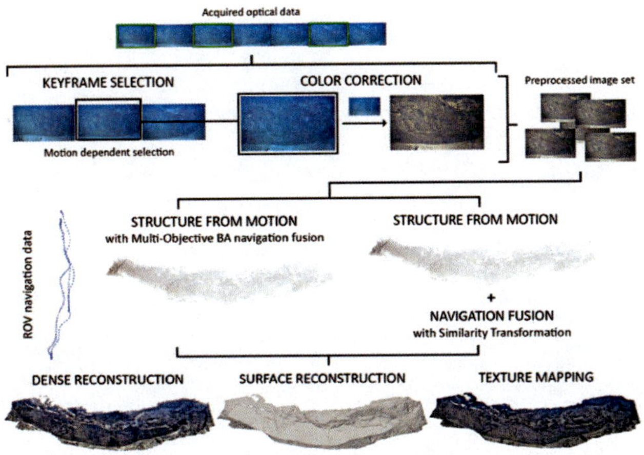

Fig. 3. Tilted image of ground features

3.4 Feature Attribute Table

The feature attribute table can record the specific name of the feature in detail, and corresponds to the model name, coordinates, texture number, heading overlap, side overlap, speed, altitude and other information [7].

3.5 Combination Table

The combined table records the information of the model name, coordinates, number, texture, etc., which is convenient to find out which operator made a certain model and which inspector inspected [8]. It is the basis for the sub-map of the model, and it is also convenient for the insider to check the surveying and mapping items or Project management.

4 Method for Establishing Three-Dimensional Model of Oblique Photography

4.1 Detailed Analysis of Building Model

The architectural model can intuitively express the designer's design intent, and make up for the flatness and inactivity of the drawing in the performance. It is not only a part of the designer's design process, but also a form of design. It is widely used in urban construction, real estate development, commercial housing sales, design bidding and investment cooperation [9]. After the white model has preliminary information such as the frame and elevation of the building, the insider analyzes the specific details of the tilted image, and builds the building's roof, bay window, dormer window, viewing platform, balcony, door bucket and other detailed information [9], shown as Fig. 4.

Fig. 4. Analysis of building model

4.2 Model Texture Making

Use oblique images and image processing software to texture the fine white model of the ground building. Since the texture can be changed multiple times, only representative building textures can be made [10]. The insider finds a certain detailed texture of the building on the oblique image, and accurately produces the texture of the building by means of vanishing point and transformation.

4.3 Model Merger

Insiders are usually used to adjust the model to the direction of true south and true north, and move the model to the origin of the coordinate for production. After the production is completed, the model is placed back to the original position, and the normal line is aligned and the axis point is rotated by moving [10]. Wait for operations, put the model back into place, and merge the models made by multiple people into one model.

5 Conclusion

Oblique photography is a new aerial photography technology that has emerged in recent years. It is a revolutionary breakthrough to traditional orthophoto photography. Because it can collect the side texture information of surface objects from low and medium altitudes, it greatly improves the efficiency of rapid and automatic modeling of large-area continuous images, so that automatic modeling can replace the costly and inefficient manual modeling quickly. This article describes the basic principles, characteristics, key technologies of oblique photogrammetry and the application of oblique photogrammetry in urban 3D modeling.

References

1. Zhang, Y.Q.: Application of aerial photogrammetry technology. China's New Technol. New Prod. **11**(21), 144–148 (2018)

2. Zhao, Zh.N.: Analysis of 3D modeling technology of UAV tilt photogrammetry. Technol. Innov. Prod. **11**(07), 64–68 (2019)
3. Li, F.L., Yang, Y., Li, S.G.: Application of oblique photogrammetry technology in urban 3D modeling and 3D data update. China Stand. **10**(04), 88–91 (2018)
4. Duan, Z.G., Zhang, H.Sh.: Low-cost 3D modeling with single-lens tilt photography. Urban Surv. **7**(04), 129–131 (2019)
5. Guo, Y.H.: Factors affecting the accuracy of 3D modeling of UAV tilt photography. Archit. Eng. Technol. Design **16**(15), 29–30 (2019)
6. Chen, L.: Analysis of 3D modeling technology of UAV tilt photogrammetry. Archit. Eng. Technol. Design **35**(08), 68–70 (2019)
7. Yu, Zh.R.: Research on UAV single-lens tilt photogrammetry and 3D modeling technology methods. Software, **4**(12), 51–55 (2019)
8. Yang, Ch.M.: UAV tilt photogrammetry technology for smart city 3D modeling. China Real Estate **13**(08), 106–109 (2019)
9. Lu, Y.H., Hao, B.M.: Three-dimensional modeling of drone tilt photogrammetry data based on context capture. Jiangxi Build. Mater. **13**(10), 63–65 (2019)
10. Li, F., Liu, Y.Y., Li, T.X.: Three-dimensional modeling and accuracy analysis based on drone tilt photogrammetry technology. Jiangxi Surv. Mapp. **11**(03), 40–42 (2019)

Influence and Strategy of 5G Technology on E-commerce Marketing and Operation Mode

Yongliang Li[1](✉) and Yanli Wang[2]

[1] College of Electronic and Commerce, Luoyang Normal University, Luoyang, Henan, China
[2] School of Software, Henan University of Science and Technology, Luoyang, Henan, China

Abstract. This paper analyzes the high-quality communication network based on 5G, uses the support of big data, artificial intelligence and virtual reality technology, uses the Internet of Things technology to optimize the supply chain, better integrates online and offline data, and realizes the e-commerce marketing mode of online and offline resource sharing. The countermeasures and suggestions of e-commerce in 5G era are discussed. The application of VR technology based on 5G communication is bound to become the key of the emerging e-commerce platform. The paper presents influence and strategy of 5G technology on E-commerce marketing and operation mode.

Keywords: 5G · Electronic commerce · New retail sales · Live broadcast marketing · E-commerce operation

1 Introduction

With the breakthrough of 5G technology, it will break the shackles of 4G network technology on artificial intelligence technology, Internet of Things technology, virtual reality big data and other technologies, and the development of technology is an important factor to promote the huge changes of the retail model.

5G mobile communication technology is the current hot research topic. China's e-commerce is still making steady progress after the continuous high-speed appreciation [1]. Studying the impact of 5g on e-commerce live broadcasting can let us have a systematic and comprehensive understanding of 5g and e-commerce live broadcasting, and estimate the opportunities and challenges, so as to make full preparations for the opportunities and challenges we will face.

5G Smartphone will become the first focus of the public products, mobile phone network download speed will be calculated by GB/S, projection, 3 D imaging technology application performance will be improved, the major entertainment function software APP will also appear, disappear or restructuring, it will mean that for new media marketing is a rare opportunity and challenge, but also have to face the tide of The Times [2]. Therefore, new media marketing will have a great possibility to carry the 5G this fast train, towards a more new, multi, three-dimensional, sensory, real marketing way.

B. J. Jansen et al. (Eds.): *International Conference on Cognitive based Information Processing and Applications (CIPA 2021)*, LNDECT 85, pp. 664–670, 2022.
https://doi.org/10.1007/978-981-16-5854-9_85

Electricity as an important part of the global economy, needs to take use of the once-in-a-lifetime opportunity, actively participate in it, using 5g high-tech high momentum, ride the wind, develop and promote new operation mode suitable for new economic development, penetrate the market, stimulate its potential, make the electricity industry through 5g, make its model more innovative and efficient, convenient and perfect.

With the development of mobile wireless network, people are more and more keen to watch short video and live in fragmented time, so derived electricity live marketing, especially the past two years, appeared a large number of web celebrity live with goods, live with goods also become more and more popular, there are many entertainment stars also participate in live marketing camp, web celebrity and entertainment stars in screen interaction with the audience let consumers can quickly into the shopping scene, can shorten the distance between buyers and merchants, to promote consumption.

With the continuous improvement of 5G mobile wireless network technology and enhancing people's experience of live broadcasting, the popularity of e-commerce live broadcast marketing will also become higher and higher.

New media marketing, from his name is not difficult to understand, refers to the use of new media platform for enterprise marketing way. Therefore, to understand what new media marketing is, the first thing is to understand the concept of new media. First, new media is a relative concept. It refers to forms of content other than traditional media, where all content media we often reach except radio, radio, television, newspapers, magazines are considered new media. Second, in a broader sense, new media uses two digital and network application technologies, and through the Internet, broadband LAN, wireless communication network and satellite channels, in the Internet, mobile phones and digital TV terminals to provide information and entertainment services. The paper present.

2 Analysis of 5G Mobile Communication Technology and E-commerce Live Broadcast Marketing

The Internet of Things has extremely high requirements for the speed and accuracy of information exchange between devices [3]. Although the existing network technologies such as Wi-Fi and 2G, 3G, 4G can meet the basic needs of the Internet of Things applications, with the increase of data volume, it needs to transmit data with the help of faster and more reliable networks.

The best 4 will delay under the G network, which makes our call quality, unable to achieve good call effect, and under the 5G network can well avoid this phenomenon, the empty mouth delay down to 1 ms, people's voice and video calls will become smoother.

The era of 5G will inject fresh blood into the world. In addition to the more mature characteristics such as high transmission speed and low network delay, as well as providing the technical foundation for artificial intelligence and big data, leading the Internet of Everything is the beginning of the new era. 2019 is the year of 5G worldwide. In mid-2019, the Ministry of Industry and Information Technology issued 5G commercial licenses to China Mobile, China Unicom and China Telecom, which officially opened the development of 5G commercial licenses in China.

There are six marketing methods for the development of new media marketing. One is WIKI marketing, that is, customers after a certain platform on the Internet, add their own products and services to promote, and then through the corresponding links, pictures or text to achieve the effect of publicity and promotion, which is mainly a keyword.

E-commerce can better develop and remain invincible. The arrival of 5g is an opportunity and a challenge for the e-commerce industry [4]. I believe that the future e-commerce, under the active promotion of 5g, as is shown by Eq. (1), where c will be more efficient and more efficient mode on a larger scale, and the operation mode will be more diversified and more technological.

$$c(\alpha) = \frac{\rho'\theta'' - \rho''\theta'}{((\rho')^2 + (\theta')^2)^{\frac{3}{2}}} \tag{1}$$

Using the Internet of Things technology, we can realize the high data management of store products and product inventory, understand the information of goods and real-time sales information, realize the control of single product flow, and predict the commodity demand of different categories, reasonably arrange the procurement frequency of goods, and realize the efficient management of store display and grouping, and improve the operation efficiency of the overall supply chain of the retail industry.

The large promotion of 5G needs to establish a lot of small base stations to meet the needs of network use, which is the characteristic that 4G lacks, because the wavelength of millimeter waves is short, so the transmission distance is not so far, so it needs to build multiple small base stations to connect with each other, but the establishment of small base stations will not consume too much energy power.

The arrival of G technology will be a new technological revolution is expected to change the whole network structure, carry out a new revolution, the development of information and communication will also usher in an era of rapid development. The 5G network will provide us with download speeds up to 10 Gibbits/s (Gbps), regional 2 ms delay, as is shown by Eq. (2), where x is 1 million ultra high density connections per km^2 and p is mobility over 500 km per hour [5].

$$\|x\|_p := \left(\sum_{k=1}^{\infty} |x_k|^p \right)^{1/p} \tag{2}$$

E-commerce, the electronic nature of such business activities, is e-commerce in a broad sense. The business activities of an enterprise is the conceptual model diagram of the enterprise business activities and e-commerce involved in the enterprise management. Concept model of enterprise business activities and e-commerce. As can be seen from the picture, the composition of the enterprise is people and things, and the main activities of the enterprise are procurement, production and sales.

Big data is those digital information with certain common characteristics that has the characteristics of large data volume, various data types, low value density and low data fast aging. Using data services to get through online and offline services can realize the deep integration of "online and offline logistics". Finally, from suppliers to retailers are the core to serve consumers.

E-commerce has expanded to many areas such as life services, healthcare, entertainment, and expanding business in social affairs and finance. E-commerce has also played an important role in promoting the adjustment of sales structure, and has provided a new platform for technological innovation and creation.

5G networking is expected to enter commercial deployment in late 2020. As a leading enterprise with a large domestic market share, China Mobile has taken Shanghai, Guangzhou and other cities as the 5G development route. China Unicom and China Telecom followed, also testing 5G networks in several cities [6]. However, it is not enough to rely on operators, requiring domestic communication industry giants as the mainstay, such as industry leaders such as Huawei and ZTE, who have more research and development strength.

No matter how the business markets, its purpose is to get more customers and to find ways to improve customer loyalty. Now WeChat fire we said the WeChatB 1 marketing, Weibo fire when we say Weibo marketing, blog fire when we also said blog marketing, so it is not difficult to see that only from the perspective of marketing tools to discuss marketing itself, this is not the significance of this attempt.

The new generation of e-commerce shopping model based on VR scenarios will shift consumers from passive consumption to active consumption. Customers can communicate closely with merchants through VR virtual scenarios, and customize personalized services according to their own needs. In addition, sellers can modify the items they sell in the VR scenarios for many times, so as to meet the personalized needs of consumers more quickly.

3 Countermeasures and Suggestions for E-commerce Operation Mode in the Era of 5G

5G is through the upgrading of technology and the expansion of the transmission channel, the problem of 4g era is completely solved, so as to achieve a lower delay. Low delay means faster data transmission and greater time shortening and precision, a feature that makes it possible for many future AI scenarios, such as driverless driving and industrial automation.

Logistics distribution is a crucial link in the sales link. It can improve the efficiency of logistics distribution and improve the shopping satisfaction of consumers. With the help of big data, we can realize intelligent logistics distribution, improve the informatization and automation level of logistics distribution link, realize real-time monitoring of logistics distribution link, improve logistics efficiency, optimize logistics distribution process, and meet consumer diversified and high-quality distribution needs in an all-round way. Effectively reduce logistics costs.

At present, under the fierce competitive e-commerce market, a large number of web celebrity anchors and more and more stars have flocked to enter the live broadcast platform [7]. Especially in the past two years, live streaming e-commerce in full swing, a large number of Li Jiaqi, Weiya and other web celebrity live streaming traffic big coffee quickly fire up, with the popularity of live streaming marketing, the increase of traffic, not only B anchors in the broadcast room.

Promoting the development of industrial informatization is the most important content. The new characteristics of 5G with large bandwidth, ultra-low delay and ultra many links can fully support the transformation of traditional manufacturing industry and service industry. As the first step of the service industry, the logistics industry will surely lead the popularization of the application of 5G and enter the era of intelligent logistics with automation and network into typical characteristics.

It has always restricted the development of the e-commerce industry. 4G network does not effectively deal with these security risks. In the 5G era, while documents and regulations issued by relevant national departments regulate network transactions, e-commerce enterprises began to rely on their technical means to escort e-commerce transactions. From the perspective of the construction objectives and the technology core, the 5G technology coordinates the security system with the working system.

$$W_j = \overline{span\{\psi_{j,i} : i \in Z\}},\ j \in Z \tag{3}$$

In the process of new retail development, as is shown by Eq. (3), where W is the most important thing and Z is how to realize the transformation from a traditional retail to a new retail model. Promoting the sharing of data resources, strengthening the communication and cooperation between enterprises, and the establishment of shared data resource databases among industries are conducive to the better capture of consumer needs [8]. At the same time, strengthen the application of big data, artificial intelligence, the Internet of Things, virtual reality technology and the development of new products.

Platform live to alleviate the bottleneck in traffic, in the process of live, users traditional shopping visual senses is impacted, online real-time interaction into live process has injected great vitality, technical language recognition input function and even wheat function is convenient interactive communication, bring users good shopping experience, in the next 5G era, we look forward to have more technology and platform innovation to promote the communication between businesses and customers, facilitate customers' online shopping.

When the logistics industry started, we cannot get effective and rapid solutions and feedback on the problems arising in the logistics process, or the real-time location of the goods. With the advent of big data in the 5G era, the service quality was greatly improved [9]. From the traditional logistics focuses on the point-to-point or line-to-line service, to the modern logistics began to build a global service network, the service area radiation is more extensive.

New media marketing has a strong interaction, which is that the main and the most advantageous feature is compared with the traditional marketing methods. Enterprises use the way of new media marketing to convey corporate culture and their own product information to the majority of consumers in the form of pictures, videos and other forms. Compared with traditional marketing, the product information released by new media marketing will be more intuitive.

4 Influence and Strategy of 5G Technology on E-commerce Marketing and Operation Mode

In the context of the 5G era, mobile e-commerce will bring more innovation. The comprehensive nature and diversity of 5G technology provides unlimited possibilities to optimize e-commerce shopping mode, and provides reliable technical support for various software of e-commerce platforms, thus ensuring the vigorous development of VR e-commerce shopping mode.

5g technology based on big data intelligent computing center, will gradually replace artificial, can in just a few minutes to the platform various information accurate screening, further standardize the electricity market, if electricity business enterprise lack of attention to products and crisis consciousness, a little careless, will be found, and by the relevant punishment [10].

This paper uses the changes brought by 5G network to the market, constantly develops and upgrades to build equipment and products that meet the needs of consumers' shopping scene, makes full use of virtual reality technology to develop interactive consumption scenarios, enhance the friendliness of the interactive interface, improve the shopping experience, better meet the needs of the market and consumers, and constantly reduce the operating cost of enterprises.

Network security technology needs to be continuously developed and improved, so as to provide an effective guarantee for the network security of 5G live broadcast technology. 5G live broadcast technology network should not only pay attention to its own functionality, as is shown by Eq. (4), where f is flexibility and x is scalability, but also pay close attention to the security of the development of 5G live broadcast technology network.

$$f_i(x) = \frac{1}{\pi^M \det\{R\}} \exp(x_i^H R^{-1} x_i) \tag{4}$$

Through the mode of new media marketing, the unique enterprise culture is completely presented to the majority of consumers. Therefore, the new media marketing will also open up a new road for the brand marketing of enterprises. The advent of the 5G era will lead the new media to a new road, and the new media marketing will also usher in a new marketing model in the 5G era.

In the era of 5G, the application of VR technology is bound to become the key to the victory of the emerging e-commerce platform, and also one of the major breakthroughs for various e-commerce related industries to seek benign development. Although, the application of VR technology in the e-commerce platform is not mature, but the 5G era is bound to achieve the new pattern of the Internet era, and the e-commerce industry will also be re-interpreted and defined.

Taobao also use its huge users into social electricity, even electricity live with goods, in 5g era, technology progress will taobao field potential, thus cause geometric multiplication, with the technology, and have a very large user traffic, 5g will bring greater income and business advantages for taobao. Taobao also takes advantage of VR and AR technology in the 5g field to cooperate with Microsoft to launch the MR mixed reality shopping model, subverts the traditional e-commerce shopping mode, actively integrates with technology, and brings immersive shopping experience to users.

5 Conclusion

With the advent of the era of 5G big data, connect artificial intelligence, collect consumer purchase behavior big data, establish consumer purchase behavior database, and analyze consumer purchase behavior with the help of the data in the database, so as to meet the purchase needs of consumers more accurately. The paper studies the impact of the implementation of 5G network technology on the development of new retail, and makes bold speculation and inference on the possible direction of new retail in the 5G era. The realization of 5G technology is of great significance to the development of new retail. In the 5G era, big data, Internet of Things and virtual reality technology will break through the shackles brought by the previous 4G network, break through the technical problems, and develop more mature.

References

1. Huang, L.: Development prospect of E-commerce in 5G era. Shanxi Agric. Jing (23), 22–23 (2019)
2. Ren, F.: 5G Promote the intelligent development of logistics industry. Logist. Technol. Appl. **24**(07), 66–69 (2019)
3. Zhou, Y.: Analysis on intelligent development of logistics industry in 5G era. New Bus. Week (021), 250–251 (2019)
4. Wang, L.: See cross-border E-commerce from 5G. Bus. Cult. (07), 48–51 (2020)
5. Zhang, Y.: Opportunities and challenges of China's 5G development in the new era. Mod. Manage. Sci. **38**(8), 1–2 (2019)
6. Shi, L.: Characteristics, trend and countermeasures of retail reform in China under the background of E-commerce. Bus. Econ. Res. (02), 39–42 (2020)
7. Tian, C.: Research on electronic commerce development based on 5G model. Natl. Circul. Econ. (31), 16–17 (2019)
8. Chen, F.: 5G helps new upgrade of e-commerce. China Foreign Invest. (23), 82 (2019)
9. Ge, H., Zheng, Y.: Application of 5G network in logistics industry. Logist. Eng. Manage. **42**(1), 73–74 (2020)
10. Wang, Z.: 5G network opportunities to the delivery industry. Logist. Technol. Appl. (12), 151–154 (2019)

Principle and Application of Peak Force Tapping Mode Atomic Force Microscope

Shoujin Wang(✉) and Chunxue Hao

School of Shenyang Jianzhu University, Shenyang, Liaoning, China
wangshoujin@sjzu.edu.cn

Abstract. Unlike conventional optical and electron microscopies, atomic force microscopy (AFM) characterizes the properties of samples at high resolution by controlling and detecting the interaction between the tip and the sample. The basic working principle of AFM is briefly described. Quantitative nano-scale morphology, mechanical properties and corresponding test methods of mechanical parameters of AFM with peak force tapping mode (PFTM) are described. Finally, the practical applications of peak force tapping mode atomic force microscopy (PAM) in analysis and measurement are reviewed, showing the advantages and broad application prospects of its PFTM.

Keywords: Atomic force microscope · Peak force tapping · Nano-characterization · Force curve

1 Introduction

In recent years, the ever-booming nanotechnology has been gradually perfected with the continuous exploration of the micro-world. Scanning Probe Microscope (SPM), an important technical tool in the field of micro-nano science, mainly includes Scanning Tunnel Microscope (STM) and AFM. This new type of microscopy has a magnification of 300 million times and a minimum resolution of 0.01 nm. In 1982, Binnig and Rohrer from IBM's Zurich Laboratory co-developed the Scanning Tunnel Microscope and won the Nobel Prize in Physics in 1986. STM is the first instrument that can truly reflect material external information at the atomic scale. It enables people to observe for the first time the state of a single atom and the surface of a substance and its related physical and chemical properties in real time. However, the working principle of STM determines that it can only study the surface of conductive samples, not the external insulators. To make up for this shortage of STM, in 1986 Binnig of IBM co-invented the Atomic Force Microscope [1] with Quate and Gerber of Stanford University. Because of its low requirement for working environment, and its ability to operate in vacuum, atmosphere or even liquid, it can detect not only conductor, semiconductor surface but also insulator surface, so it has rapidly developed into an important tool for the study of nanoscience.

The earliest AFM was mainly worked on observing the external morphology of samples, which can obtain very accurate morphology information (accurate to nanometers). However, in the measurement of nano-micro-scale forces, because the complex

B. J. Jansen et al. (Eds.): *International Conference on Cognitive based Information Processing and Applications (CIPA 2021)*, LNDECT 85, pp. 671–679, 2022.
https://doi.org/10.1007/978-981-16-5854-9_86

force and deformation of the microprobe, as well as the law of motion are not clear, in dynamic scanning, the NM can only be observed qualitatively but not provide quantitative information. In this case, the contact mode is used as the working mode. During the scanning process, the tip always contacts the sample, which produces a large lateral force [2, 3], resulting in damage to the tip and the sample. Until the advent of Young Knock Mode Atomic Force Microscope in 1991 [4–6], the problem of injuring the tip and sample by too much lateral force of the contact mode was solved. During tap mode scanning, the cantilever beam vibrates at a resonance frequency, and the tip contacts the sample intermittently. When the surface of the sample fluctuates, the amplitude of the cantilever beam changes. The morphology and properties of the sample can be characterized by using the amplitude of the cantilever beam as the feedback quantity. In addition, the phase imaging principle of tapping mode can also give a comprehensive characterization of the mechanical properties of the sample [7–9]. With the development of devices and technological advances, until 2009 Veeco (now Bruker) proposed the Peak Force - QNMTM nano-scale mechanical properties quantification module for PFTM. The method of controlling and detecting the interaction between them, can not only characterize the external morphology of the sample with high resolution, but also quantitatively analyze the external properties corresponding to the interaction force. Since its launch in 2009, PAM Because of its good force control and the ability to effectively eliminate some environmental disturbances during imaging, high resolution imaging of samples in liquid environment has been applied in materials, chemistry, biology and other fields by many universities and scientific research institutions in China, and a lot of achievements have been made [10, 11].

The paper is classified into six parts, each of which is arranged as follows: the first part discusses the research background and significance of the application of AFM, introduces the main work and innovation points, outlines the structure of the paper; the second part summarizes the relevant research progress at home and abroad; the third part describes the main components and working principle of AFM, including the basic working principle of AFM and the interaction between tip-sample of AFM model; Part IV, introduces the measurement methods of nano-scale morphology, mechanical properties and corresponding mechanical parameters of PAM; Part V, introduces the application status of PAM; Part IV summarizes and prospects the full text.

2 Related Work

PAM has many advantages, but there is still room for improvement. Slower scanning speed is an important issue. In 2017, M.E. Dokukin and I. Sokolov of Tufts University invented the ring mode atomic force microscope [12] based on the peak force tap mode. During the PFTM, the adhesive force disappears instantaneously when the tip leaves the sample, which causes the cantilever beam resonance (ringing signal).In previous peak tap modes, the ring signal was generally considered useless and tried to avoid it. In ring mode technology, it is precisely by analyzing the ring signal to obtain more properties of the sample, such as more accurate adhesion and viscoelastic properties, and at the same time faster than traditional PFTM.

In 2015, Adrian P. Nievergelt et al. designed a high-speed and large-scan AFM [13].The high frequency stacked piezoelectric actuator is assembled onto a piezoelectric tube with a high scanning range to move the sample over to achieve high bandwidth and large scanning range. However, too fast scan speed can still cause problems with imaging. During the scanning process, the fast turning of the piezoelectric scanning head causes the piezoelectric scanning tube to resonate, which affects the imaging. In addition, when characterizing the elastic modulus of a sample, a cantilever beam similar to the elastic modulus of the sample is required, which limits the application of PFTM to samples with wide variation of the elastic modulus. This is because when the elastic coefficient of the cantilever is much higher than the elastic modulus, it will be difficult for the cantilever to produce enough deflection to accurately detect the interaction force between the tip and the sample. However, if the elastic coefficient of the cantilever beam is much smaller than the elastic modulus, it will be difficult to generate enough elastic force to generate sufficient indentation for detection.

In order to solve this problem, Xie Hui et al. improved it and proposed a magnetic excitation PAM [14]. The mechanical characterization of the large elastic modulus range can be achieved by using a single cantilever beam using a magnetically excited peak force modulated atomic force microscope. On the back of the cantilever, a ferromagnetic bead was pasted above the tip of the pin, and a solenoid was assembled under the sample table to generate an AC magnetic field in the Z direction, which causes the ferromagnetic bead to drive the cantilever to sinusoidal vibration under the magnetic moment. Before starting the scan, when the cantilever beam vibrates under the action of magnetic moment, the position detector records the free vibration signal as the background signal. During the scanning process, the force curve at each pixel point can be obtained by subtracting the deflection signal detected by the position detector from the background signal using the background subtraction algorithm. The nano-mechanical properties of the sample are derived by analyzing the force curve at each pixel point, and a soft cantilever beam is used to image when scanning multiphase samples with large modulus changes. When scanning high elastic modulus areas, the magnetic force of the magnetic beads is increased by adjusting the magnetic field strength, resulting in a sufficiently large effective indentation on the driving pin tip. The magnitude of the magnetic force can be calculated, so a soft cantilever beam can also be used to image samples with large elastic modulus.

In 2017, Xie Hui et al. proposed a high resolution multi-parameter Kelvin Probe Force Microscope [15] based on the Kelvin Probe Force Microscope with PFTM. High frequency piezoelectric ceramics are added to traditional PAM. Under normal operation of PFTM AFM, a second-order resonance frequency vibration is applied to the cantilever beam, which makes the cantilever beam vibrate at the same time with sinusoidal motion of low frequency and large value. When the tip of the pin approaches, contacts and departs from the surface of the sample from a distance. Because of the electrostatic force, the phase of the high frequency vibration of the cantilever changes. The phase difference of the high frequency vibration obtained by the lock-in amplifier can reflect the properties of contact potential difference of the sample crystal. At the same time, the nanomechanical properties of the sample can be obtained by analyzing the force curve in each period.

Since it was developed in 2009, the PFTM has been applied in many fields such as materials, chemistry, biology and many other domestic universities and scientific research institutions because of its good force control, effective removal of some environmental factors in the imaging process. Even with the rapid force array mode AFM developed by Bruke in 2017, the imaging speed is still not as fast as that of the peak force tap mode AFM, which has become one of the essential tools in the field of nanoscience.

3 Principal Composition and Working Principle of Atomic Force Microscopy

3.1 Fundamental Working Principle of AFM

Figure 1 illustrates the basic working principle of nano-characterization using an atomic force microscope, a force-sensitive cantilever with a tiny probe at its tip. When a probe scans a sample, laser beam is hit on the cantilever beam and reflected onto the four-quadrant position detector to form a spot. The laser path constitutes a light lever to amplify the deflection of the cantilever beam. The four-quadrant position detector can detect the deflection size of the spot and then determine the deflection size of the cantilever beam by converting the light signal into an electrical signal. The interaction force signal between the tip and the sample is obtained.

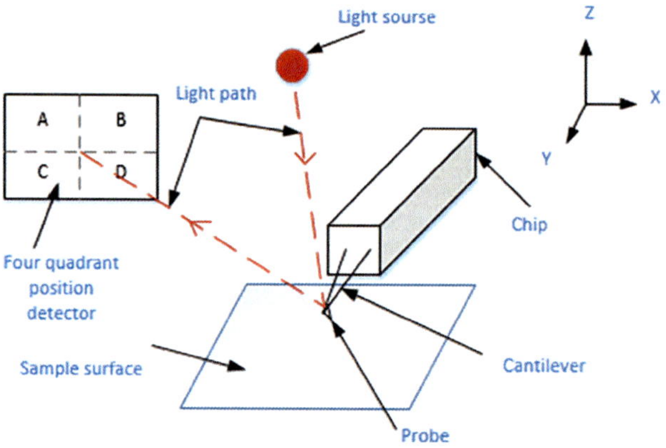

Fig. 1. Schematic diagram of basic working principle of AFM

3.2 Model of Interaction Force Between Tip and Sample of AFM

In order to obtain physical parameters from the interaction between the probe and the sample, we must consider theoretical models and formulas to describe these physical tests. There are two basic models to describe their interaction. The first model is based on the spherical structure pressing into a solid surface. The Hertzian probe model is constructed by combining the pressing depth, spherical radius, pinpoint force, modulus, and ratio of cross-section to axial strain, as shown in Fig. 2.

δ represents the depth of the indentation, R represents the radius of the sphere, F represents the force acting on the pinpoint, and the mathematical expression of the force acting on the pinpoint is:

$$F = (4/3)(\frac{E}{1 - V^2})\sqrt{R}\delta^{3/2} \tag{1}$$

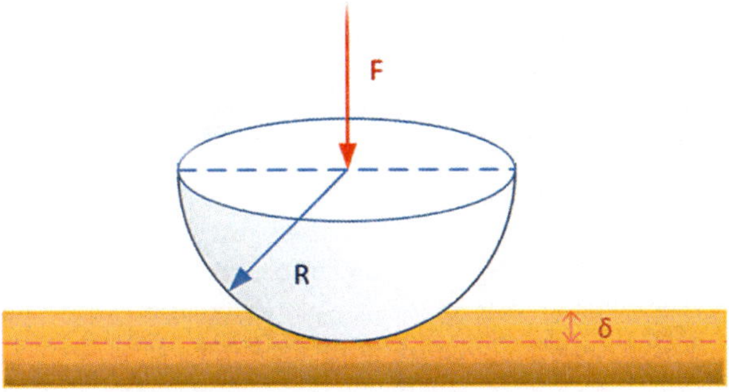

Fig. 2. Hertzian probe model diagram

The second model is based on the conical structure pressing into the solid surface, and combines the indentation depth, conical half-angle, pinpoint force, modulus and Poisson's ratio to construct the Sneddon probe model, as shown in Fig. 3.

δ represents the depth of indentation, α represents the half-angle of the cone, F represents the force acting on the tip, and the mathematical expression of the force acting on the tip F is:

$$F = (2/\pi)(\frac{E}{1 - V^2})\tan(\alpha)\delta^2 \tag{2}$$

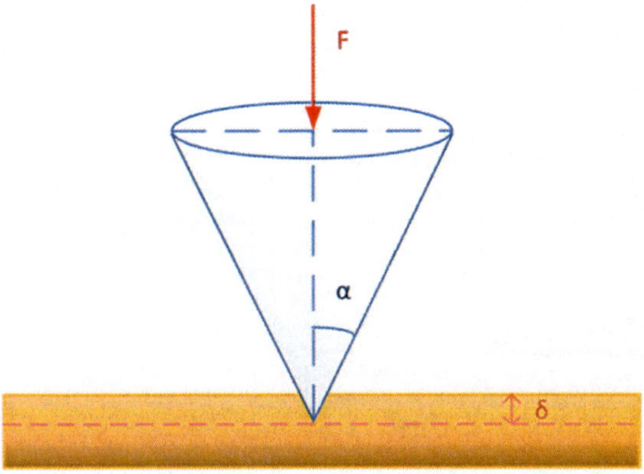

Fig. 3. Sneddon probe model diagram

4 Quantitative Nano-microscale Morphology, Mechanical Properties and Testing Methods of Corresponding Mechanical Parameters by PFT-Atomic Force Microscopy

Peak Force Tap Mode Atomic Force Microscope drives the scanning head with a sinusoidal signal, making the scanning head move up and down sinusoidally at each pixel point at a vibration frequency far below the resonance frequency (0.5–8 kHz). The tip approaches and touches the sample from a distance. The interaction between the tip and the sample then reaches a normal peak at the half-cycle, and the tip then moves away from the case. As the tip moves away from the sample, the adhesive force acts as an impediment until the tip is about to leave the sample. When the tip leaves the sample, the cantilever beam will attenuate the vibration due to the disappearance of the adhesive force, thereby generating a ring signal. At each pixel point, the process is recorded as a force curve. The external morphology of the sample can be obtained by using the peak normal force at half cycle as feedback. It can be seen that the PFM controls the normal peak force as feedback, so the their interaction can be precisely controlled. Meanwhile, since the background subtraction algorithm of the PAM can effectively eliminate the influence of parasitic deflection signals such as resistance on the imaging, the PFM has a smaller error than other modes.

The sinusoidal drive probe pushes into the sample to produce a force-time curve to monitor the peak force at each scan point, as shown in Fig. 4(a). The sample morphology can be obtained by processing and analyzing the force-time curves at each scanning point. The data in the force-time curve image of a single scan point can be divided into two parts, proximity and contraction, so that the force curve with Z distance can be obtained, as shown in Fig. 4(b). The Z distance is a summary of the bending amount of the probe and the depth of sample injection. The Z-direction data can be converted into a seperation to reflect the distance between the probe and the maximum pressure point, as shown in Fig. 4(c). The deformation, peak force, adhesion force, elastic modulus and dissipated energy are extracted from these curves and reflected in the corresponding image channels. The transverse axis of the force curve reflects the deformation, the maximum bending value reflects the peak force, and the shrinkage curve reflects the adhesion force. The Young's modulus is obtained by fitting the contact part of the shrinkage curve with the DMT model, and the area between the approximation and the shrinkage curve reflects the dissipation energy.

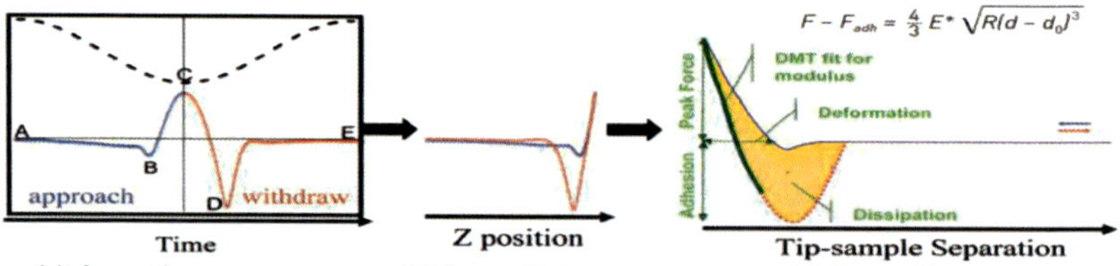

(a) force-time curve (b) force-distance curve (c) force-distance curve

Fig. 4. Curve diagram of interaction force and distance between tip and sample

5 Application

Over the past 20 years, PAM has been widely used in the field of nano-biological materials. PAM can effectively reveal the external structure and mechanical properties of nano-biological materials, and can be manipulated and processed as nano-processing tools. PAM can be worked on observing the external morphology of nano-biomaterials, analyzing their feature height and external roughness, and pay attention to their dynamic process in situ. Phase maps also provide some external features that are sometimes not available from height maps. The PAM curve can be used to value the adhesion force between the tip and the nano-biological material and the interaction force inside and outside the molecule. In addition, PAM has also been explored for the accurate, controllable and reproducible processing of nano-biological materials. In summary, as a powerful nanotechnology tool, PAM has become an ideal external analysis and external processing tool in the related study field of nano-biological materials.

PAM provides a more precise and effective means for scientific research to advance to nano-scale. Its appearance has made it difficult to achieve previous research techniques to solve, not only to obtain high-resolution images, but also to observe the interaction between samples, and to analyze the mechanical properties of the surface of materials, even to process nano-biological materials on the molecular scale, which has been widely used in the field of nano-materials, especially nano-biological materials. From Table 1, it can be seen that the fastest mechanical measurement mode of force measurement by force microscopy is PFTM, which has obvious advantages over force volume mode. The advantages of PFM in the study of nano-biological materials are as follows: (1) it has a very high resolution, with a transverse resolution of 0.1–1 nm and a longitudinal resolution of 0.01–0.2 nm; (2) it has a relaxed environment and can work in vacuum, atmosphere, ambient temperature and even solution, and PFT-atomic microscope. Sample preparation is simple without conductivity requirement; (3) Viscoelastic and other mechanical properties of the sample can be obtained accurately at nano-scale using PAM, and the detection process is basically non-destructive; (4) PAM is used. Nanomanipulation and processing of materials surface enables controlled manipulation of samples; (5) real-time, in-situ imaging observation; (6) easy combination with other external optical and spectral instruments, such as optical microscopy Fourier transform infrared spectroscopy, probe-enhanced Raman spectroscopy, and fluorescence correlation spectroscopy, which complement each other in research to better play a role in each other. Advantages. With the continuous development of over 20 years, PFTM microscope has gradually become a multi-functional tool, bringing great convenience to scientific research.

Table 1. Comparison of atomic force m force microscope measurement techniques

	Single force curves	SCFS	Force volume	Peak force
Ramp rate	0.1–10 Hz	0.1–10 Hz	0.1–10 Hz	0.25–2 kHz
Benefits	Highly accurate discrete force measurements can be precisely targeted	The probe can be modified with cells for the measurement of interactions between cell and cell, or between cell and substrate	Mechanically map the sample by collecting a series of force curves at an array of points on the sample automatically	High resolution morphology imaging and high accurate force measurement can be done simultaneously for the same area in minutes
Disadvantages	No	It is technically difficult to attach a cell to the AFM cantilever. Thermal drift during the measurement is hard to control	The lateral resolution is low; it is time consuming to complete a measurement; the tip is easily contaminated	No

6 Conclusion

In a word, PAM has a more mature physical measurement mechanism, can concisely characterize the external morphology of the sample, and can quantitatively analyze various external properties corresponding to the forces. In addition to its high resolution, it has some advantages that traditional microscopy does not have. It can be predicted that, as the continuously growing of computer systems, it can use better cantilever and pinpoint, introduce more accurate laser systems, piezoelectric drivers and feedback systems. And PAM will be greatly improved in scanning speed and resolution used as a basic and common means. It will be popularly used in chemistry, materials, and so on. Biology and other fields have wide application prospects.

References

1. Binnig, G., Quate, C.F., Gerber, C.: Atomic force microscope. Phys. Rev. Lett. **56**, 930–933 (1986)
2. Bloo, M.L., Haitjema, H., Pril, W.O.: Deformation and wear of pyramidal, silicon-nitride AFM tips scanning micrometre-size features in contact mode. Measurement **25**, 203–211 (1999)
3. Kimura, K., Kobayashi, K., Yamada, H.: Orientation control of molecular chains in polymers using AFM. Jpn. J. Appl. Phys. **43**, 4575–4579 (2004)
4. Zhong, Q., Inniss, D., Kjoller, K.: Fractured polymer/silica fiber surface studied by tapping mode atomic force microscopy. Surf. Sci. Lett. **290**, L688–L692 (1993)

5. Magonov, S.N., Elings, V., Whangbo, M.H.: Phase imaging and stiffness in tapping-mode atomic force microscopy. Surf. Sci. **375**, L385–L391 (1997)
6. Putman, C.A., van der Werf, K.O., de Grooth, B.G., van Hulst, N.F., Greve, J.: Viscoelasticity of living cells allows high resolution imaging by tapping mode atomic force microscopy. Biophys. J. **67**, 1749–1753 (1994)
7. Bar, G., Thomann, Y., Brandsch, R.: Factors affecting the height and phase images in tapping mode atomic force microscopy, study of phase-separated polymer blends of poly(ethene- co -styrene) and poly(2,6-dimethyl-1,4-phenylene oxide). Langmuir **13**, 3807–3812 (1997)
8. Mclean, R.S., Sauer, B.B.: Tapping-mode AFM studies using phase detection for resolution of nanophases in segmented polyurethanes and other block copolymers. Macromolecules **30**, 8314–8317 (1997)
9. Schmitz, M., Schreiner, G.: Friedbacher, Phase imaging as an extension to tapping mode AFM for the identification of material properties on humidity-sensitive surfaces. Appl. Surf. Sci. **115**, 190–198 (1997)
10. Hu, S., Mininni, L., Hu, Y.: High-speed atomic force microscopy and peak force tapping control. In: Proceedings of Spie the International Society for Optical Engineering, vol. 8324, pp. 56–64 (2012)
11. Shi, J., Hu, Y., Hu, S.: Method and apparatus of using peak force tapping mode to measure physical properties of a sample. US9291640B2 (2012)
12. Dokukin, M.E., Sokolov, I.: Nanoscale compositional mapping of cells, tissues, and polymers with ringing mode of atomic force microscopy. Sci. Rep. **7**, 11828 (2017)
13. Nievergelt, A.P., Erickson, B.W., Hosseini, N.: Studying biological membranes with extended range high-speed atomic force microscopy. Sci. Rep. **5**, 11987 (2015)
14. Meng, X., Zhang, H., Song, J.: Broad modulus range nanomechanical mapping by magnetic-drive soft probes. Nat. Commun. **8**, 1944 (2017)
15. Xie, H., Zhang, H., Hussain, D.: Multiparametric kelvin probe force microscopy for the simultaneous mapping of surface potential and nanomechanical properties. Langmuir Acs J. Surf. Colloids **33**, 2725–2733 (2017)

Innovation of Supply Chain Finance Model Based on Blockchain Technology

Jianfei Ke(✉) and Zhengyi Tang

Fujian University of Technology, Fuzhou, China
kejianfei@fjut.edu.cn

Abstract. Taking the whole industry as the main body and providing financial services to the upstream and downstream of the whole supply chain (SC) is the main feature of SC finance. However, there are some problems in the process, such as poor information communication for financing enterprises and backward management mode, which pose certain risks to banks and other financial institutions. In recent years, blockchain technology has been developed rapidly, especially in financial services. However, there is little research on it in China. Therefore, this paper puts forward the innovation mode of blockchain technology in SC finance. This paper makes an in-depth research and Analysis on the blockchain technology and SC finance, and believes that the main problems in the current SC finance can be better solved with the assistance of blockchain technology, which has a positive role in improving the overall service capacity of China's SC finance. In this paper, the characteristics of the application of financial blockchain technology are analyzed in detail. In order to further study, this paper investigates the current situation of the application and development of blockchain technology in China. Through sorting out the public data, it can be seen that by 2019, the number of supporting policies for blockchain in China has reached 135, and the number of relevant blockchain entrepreneurship activities has also reached 176, which shows that blockchain technology has a good prospect for the use of SC finance. This paper studies the innovation mode of blockchain technology in SC finance, and analyzes that blockchain technology is an indispensable important tool in the development of SC finance and plays a key role in the development of SC finance.

Keywords: Blockchain technology · Supply chain finance · Mode innovation · Financial services

1 Introduction

Since Japan put forward SC logistics management in 1980s, its technical development and history mainly focus on three stages: paying attention to logistics management, pursuing value-added, and replacing single line with network [1–3]. Due to the fine division of labor and diversification of operation in modern SC management (SCM), the extension of SC and the positioning of enterprise participation nodes are not accurate.

B. J. Jansen et al. (Eds.): *International Conference on Cognitive based Information Processing and Applications (CIPA 2021)*, LNDECT 85, pp. 680–686, 2022.
https://doi.org/10.1007/978-981-16-5854-9_87

If the traditional management methods and technical means are adopted, it is difficult to coordinate the interests of all parties, which becomes the difficulty and focus of SCM [4, 5].

Taking blockchain technology as a technical tool to maintain the development of SC can fundamentally solve the difficulties of SCM. The technical basis of blockchain is distributed accounting. Each entity can participate in the bookkeeping process and form a unique bookkeeping. With the gradual improvement of the SC process, participants can ensure the authenticity and reliability of transactions by querying account books, and transaction information can be smoothly transmitted along the SC [6–8]. Moreover, the service system based on blockchain technology can effectively restrain the financing behavior of enterprises in the supply chain, integrate relevant transaction data, and form online payment information, providing formal and reliable evidence for transaction behavior. The financing services of banks and other financial institutions can cover multi-level suppliers, greatly reducing the investment costs of banks and financing costs of enterprises [9, 10].

This paper makes an in-depth study on the practical application of the current national blockchain technology, realizes that the blockchain technology attaches great importance to China's SC finance, understands the blockchain technology is highly concerned, and has issued a large number of relevant supporting policies, and the number of blockchain financial service entrepreneurship has reached 176, which is in a period of rapid development. However, the existing SC financial cooperation mode of blockchain still has some typical problems, such as backward mode and poor management mode, which cannot give full play to the maximum effect of blockchain technology in SC finance. Therefore, this paper puts forward the research on the innovation mode of blockchain technology in SC finance, hoping to further improve the application and innovation ability of blockchain technology in the field of SC finance. In view of the existing problems, this paper gives specific optimization and improvement scheme, especially the practical impact of blockchain technology on SC finance, and the future development direction. The analysis shows that SC finance plays an important role in training and supporting the development of China's industry, but the actual effect is not ideal due to the defects of its mechanism and technology. The blockchain technology proposed in this paper, to a certain extent, can make up for the main problems in this field. At the same time, it improves the overall service ability of SC finance and avoids most financial risks.

2 SCFinance and Blockchain Technology

2.1 Concept of Supply Chain Finance

SC financing refers to the comprehensive financial products and services provided by banks or financial institutions. SC finance is a kind of traditional financial mode, which is different from the traditional financial way. It no longer pays attention to the enterprise assets, production scale and profitability. Due to the lack of internal conditions, small scale and weak operation ability, it is difficult for SMEs in the upstream and downstream of the SC to obtain loans. At this time, we need to pay more attention to the operation of the whole supply chain. According to the actual financial transaction operation status, the overall market competitiveness of core SC enterprises and the credit management qualification of core financial enterprises, we provide professional and high-quality corporate financial services with the majority of core financial enterprise customers as the service center. From the perspective of the whole industry chain, financial service providers integrate logistics, information flow and capital flow, and use rich capital operation mode of financial products.

2.2 Block Technology and Its Characteristics

Blockchain is a unique digital encryption method, which uses distributed accounting to store, verify and exchange data. There are three aspects of technology implementation: first, blockchain technology realizes the storage of information and the information can be traded; secondly, each block records all the data of the previous block; finally, it is composed of a node, and each node stores a copy of the same block to ensure the security and traceability of the node.

(1) Dispersion characteristics. The feature of decentralization is that the data is stored in the way of point-to-point access, avoiding relying on the third-party management organization to achieve data storage, verification and transmission.
(2) Trust characteristics. From the current development status of blockchain application technology and its application, the rapid development of bitcoin, digital currency and crypto virtual currency reflects a strong trust. The biggest feature is that all transaction tasks do not need credit guarantee from the third party.
(3) The characteristics of traceability. A real-time, non-tampered data storage platform based on blockchain technology is implemented. The data storage logic in the blockchain is based on the block chain storage.

3 Main Problems in the Development of Supply Chain Finance Based on Blockchain Technology

3.1 Market Risk Uncertainty Increases

The financial service industry of SC enterprises has a high industry complexity, which often involves the capital flow of enterprises and many problems related to the economic income and even the survival of enterprises through direct factors. The strategic cooperation between enterprises and some financial institutions and suppliers of upstream and downstream products naturally needs to be cautious. Even many new uncertainties may directly lead to huge trust problems.

3.2 The Amount of Information is Large and Difficult to Process

Capital flow record is an important indicator to measure the SC finance, because many enterprises exist in the supply chain, but not all have business relations with core enterprises, and these enterprises cannot get corresponding financial support. To some extent, the competition between enterprises is the competition between information acquisition and information processing. With the help of information technology, the analysis, processing and transformation of information will become the key to an enterprise's invincible position. But at this stage, although financial institutions have mastered a lot of detailed information, the phenomenon of Information Island still exists, and is particularly prominent.

4 Discussion

4.1 Survey on the Development of Blockchain Application

In the development and application of blockchain technology, China and Western countries are basically synchronized. Through the strict supervision of the central government, the constantly updated blockchain technology has received policy support measures and project subsidies from many local governments, and the central government's encouraging attitude towards blockchain has also been revealed.

In China, especially in coastal areas, block chain industrial development parks have been established and policy support has been strengthened. According to the survey statistics in Table 1, Fig. 1 and Fig. 2, China has issued 135 (62) block chain support policies by 2019. Beijing, Shanghai, Guangzhou and other places have released the policy guidance information of blockchain, and adopted the ways of increasing project subsidies and financial investment to promote industrial development. There are many ways of policy support, including the establishment of industrial parks, technological incentives and so on. Among the provinces carrying out blockchain entrepreneurship activities in China, Beijing ranked first with 176, followed by Shanghai, Guangdong and Zhejiang, with 99, 73 and 41 respectively, while the number of other provinces was relatively small.

Table 1. Number of supporting policies for blockchain in provinces (autonomous regions and municipalities directly under the central government) (unit: item).

	Guangdong	Zhejiang	Beijing	Shandong	Guangxi	Chongqing	Hebei	Jiangsu	Xinjiang
Number	19	13	9	6	5	4	3	2	1

Data source: chain tower think tank

4.2 Necessity of the Combination of SC Finance and Blockchain Technology

SC finance is that financial institutions, according to different SC nodes, introduce new risk control variables, such as core enterprises, provide comprehensive financial services such as closing accounts, and construct SC structure. The fundamentality of SC finance is to rely on risk control variables to help enterprises revitalize their current assets. The whole business process is usually opaque and it is difficult to ensure the normal operation. Blockchain is an underlying technology of bitcoin, which is essentially a decentralized database. The core feature of blockchain technology is decentralization, trust and traceability. The application of blockchain technology is based on its science and technology

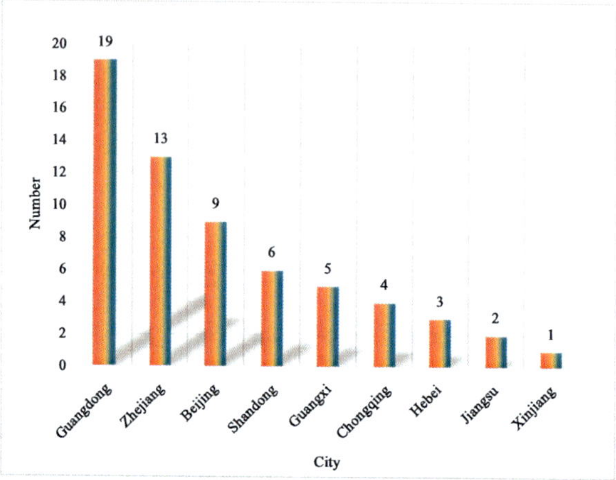

Fig. 1. Number of supporting policies for blockchain in provinces (autonomous regions and municipalities directly under the central government) (unit: item). Data source: chain tower think tank

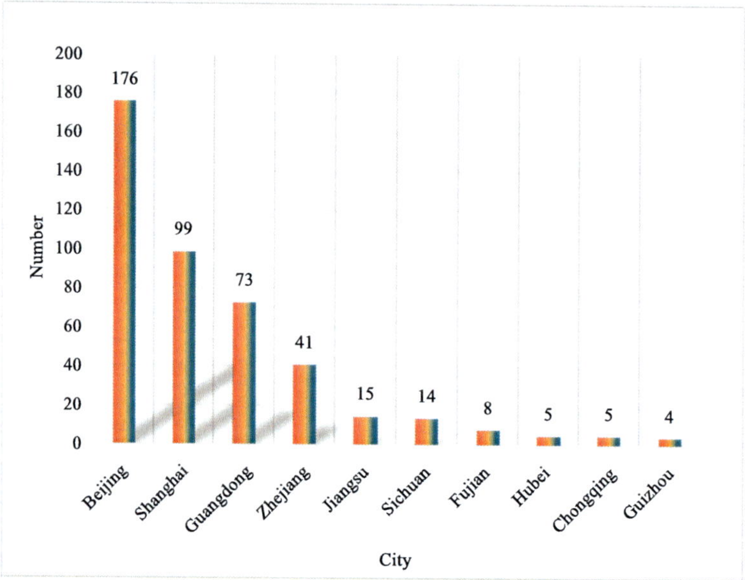

Fig. 2. Top 10 provinces with blockchain entrepreneurial activity in China (unit: piece). Data source: chain tower think tank

itself. The transmission of information and the traceability of information in the whole process are the innovative development of SC financial business. The application of blockchain technology effectively solves the problem of information asymmetry in different SC links such as warehousing and supervision. It not only provides a convenient financing platform for China's warehousing enterprises, but also provides an effective financing platform for China's enterprises. The successful establishment of credit digital financial asset management system of blockchain financial SC can further promote the reconstruction of credit management system in traditional commodity trading and enterprise financing, and promote the rapid development of transaction industrialization and standardization.

4.3 Innovative Research on the Application of Blockchain Technology in Supply Chain Finance

(1) SC financing platform based on blockchain technology. This paper introduces blockchain technology to SC finance, establishes capital platform, and records all information of blockchain in all levels of transaction process. Through the smart contract, the SC enterprises and financial institutions automatically clear the funds after the transaction. On the final service platform, financiers can apply for financial assets online, and the blockchain system will automatically approve credit and provide services.

(2) Blockchain application technology has solved many complex financial problems of SC enterprises. After the blockchain transaction technology is successfully injected into the financial service trading platform of China's SC enterprises, all open enterprise financial transaction data information will be automatically recorded on the enterprise blockchain, and the transaction information will also be automatically disclosed and shared. However, only those copyright advocates with detailed information are allowed to view any details of stock trading here. The capital market flow mechanism can effectively monitor the whole financing process, thus effectively reducing the potential risk of bank loans. Small loans have been provided to more and more small and medium-sized enterprise customers in China, which has alleviated the problem of urgent financing for small and medium-sized enterprises.

4.4 Path of Financial Information Transmission in Supply Chain from the Perspective of Blockchain Technology

The operation of SC finance needs to generate corresponding information records, because the activity information of SC finance is the two main bodies to establish point-to-point, and focuses on the information of core enterprises, upstream and downstream enterprises. The information obtained by banks and financial institutions is a process disclosed by limited core enterprises. In the process of SC finance, all subjects participate in value creation. Each topic, especially the core enterprises or commercial banks, needs to clearly define the "responsibility, rights and interests", and disclose the responsibilities and information of each subject, so that the bottom SMEs can also obtain information. Visual information recording and information transmission process is an indispensable convenience for SC finance brought by blockchain technology. With the continuous development of computer network technology, the efficiency of financial information transmission in SC has been greatly improved.

5 Conclusions

This paper studies the application of blockchain technology in SC finance in China. The analysis shows that even though SC finance is promoting the positive development of the industry, there are still some technical problems in the actual operation process, such as poor information communication and weak supervision.In recent years, blockchain technology has been widely used to solve the main problems in SC finance, which plays an important role in improving the relevant policies of SC finance and sustainable development in the field of SC finance. At present, SC finance is still in the initial stage of development, and it is still not perfect. After being integrated with blockchain technology, it can further improve the overall service capacity of SC finance and maximize the interests of all parties.

Acknowledgments. Research on the path and strategic choice of China's industrial upgrading base on the perspective of trade imbalance and trade friction (GY-S18101).

References

1. Lemieux, V.L.: Trusting records: is blockchain technology the answer? Rec. Manag. J. **26**(2), 110–139 (2016)
2. Zhang, Y., Wen, J.: The IoT electric business model: using blockchain technology for the internet of things. Peer-to-Peer Netw. Appl. **10**(4), 983–994 (2017)
3. Miraz, M.H., Ali, M.: Applications of blockchain technology beyond cryptocurrency. Ann. Emerg. Technol. Comput. **2**(1), 1–6 (2018)
4. Wang, J., Peng, W.U., Wang, X., Shou, W.: The outlook of blockchain technology for construction engineering management. Front. Eng. Manage. **4**(01), 71–79 (2017)
5. Eyal, I.: Blockchain technology: transforming libertarian cryptocurrency dreams to finance and banking realities. Computer **50**(9), 38–49 (2017)
6. Gelsomino, L.M., Mangiaracina, R., Perego, A., Tumino, A.: Supply chain finance: a literature review. Int. J. Phys. Distrib. Logist. Manag. **46**(4), 348–366 (2016)
7. Su, Y., Lu, N.: Simulation of game model for supply chain finance credit risk based on multi-agent. Open J. Soc. Sci. **03**(1), 31–36 (2015)
8. Hofmann, E., Zumsteg, S.: Win-win and no-win situations in supply chain finance: the case of accounts receivable programs. Supply Chain Forum **16**(3), 30–50 (2015)
9. Brunet, P.Y., Babich, V., Aouam, T.: Supply chain finance: overview and future directions. Found. Trends Technol. Inform. Oper. Manage. **10**(3–4), 237–252 (2017)
10. Lekkakos, S.D., Serrano, A.: Supply chain finance for small and medium sized enterprises: the case of reverse factoring. Int. J. Phys. Distrib. Logs Manage. **46**(4), 367–392 (2016)

Application of Blockchain Technology in Practical International Technology Trade

Dandan Ren(✉)

College of Arts and Information Engineering, Dalian Polytechinic University, Dalian, Liaoning, China

Abstract. Since the reform and opening up, China has maintained rapid economic growth for more than 30 years and has achieved world-renowned achievements. Economic development is inseparable from the promotion of science and technology. Since the new century, competition between countries has developed into a contest between science and technology. As a result, the single tangible goods transaction between international trade in the economic field gradually changes to a diversified direction. The link between international trade and technological change is complementary and mutually reinforcing. Innovative technology is an imperative trend, which can improve the competitiveness of products and make the country dominate in international trade. Therefore, promoting technological progress and enhancing its role in trade has become a top priority. International technology trade is an important way to open up the international market, improve trade structure, and enhance international competitiveness. Blockchain technology will change the production relations and productivity of human society. On the other hand, the principle of blockchain technology solves the problems of trust and agreement, in order to enable unfamiliar nodes on the blockchain to carry out economic activities smoothly, reduce credit fees and transaction costs, and improve production efficiency and production efficiency. In addition, the urgently needed solution in the application of blockchain is to design a corresponding reward mechanism. The value transmitted by the blockchain has the function of benefit distribution belonging to the category of production relations. With the deepening of blockchain applications, the wave of "blockchain+" gradually began. The advantage of blockchain technology is to reduce the risk of settlement and improve settlement efficiency. Finally, in terms of technical level, legal supervision level and industrial standards, provide target proposals for the country to adopt blockchain technology in trade settlement.

Keywords: Blockchain technology · International technology trade · Technological characteristics · Technological innovation

1 Introduction

The research on technology trade theory first appeared in the 1960s, and after years of development, it has gradually spread to the world economic system [1]. And the theory

B. J. Jansen et al. (Eds.): *International Conference on Cognitive based Information Processing and Applications (CIPA 2021)*, LNDECT 85, pp. 687–694, 2022.
https://doi.org/10.1007/978-981-16-5854-9_88

of technology diffusion shows that if technological innovation is to truly affect the real economy, it should expand its own application range, radiate as wide a region as possible, and obtain substantial technological progress through continuous technological imitation and advancement [2]. Although technology trade has made considerable progress abroad, the research on my country's technology trade is still in its infancy, so there are still many areas to be improved and developed [3]. How big is the impact of technology trade on China's trade? These issues are waiting for us to continue to explore. In real life, especially in international economic exchanges, most of the transfers related to products and production technologies are the transfer of personally owned technologies through paid methods. Private technology has the important characteristics of private property, which is private and exclusive. The property right is protected by law. Without the permission of the right holder, no one can enjoy and use the right, otherwise it will be regarded as an infringement of the intellectual property rights of the right holder. And to bear legal responsibility, which is the legal basis for international technology trade [4].

Blockchain technology is the biggest core technology of the fifth wave of revolution after steam engine, electricity, information and network technology [5]. The application of blockchain technology extends to digital finance, Internet of Things, intelligent manufacturing, supply network management, digital asset trading and other fields. In order to promote the innovation of blockchain technology and theory, it is necessary to combine actual and new development practices and actively promote it. Blockchain is connected with the integrated development of economic society, which makes blockchain technology play a greater role in building network power, developing digital economy, and promoting economic and social development [6]. In the context of the rapid development of economic globalization, international trade is showing a trend of expansion. With the development of big data technology, the application fields of big data technology are becoming more and more extensive, bringing new directions to the development of international economy and trade. Exploring the new low-cost and low-risk international trade payment model has become a new topic. With the proposal of blockchain, many industries have begun to explore this emerging technology, hoping to apply it to the industry to achieve technological breakthroughs [7].

In the past two years, as the topic of blockchain continues to heat up, under the background of economic globalization, some researchers have begun to explore opportunities for blockchain in international technology trade. Researchers have now linked blockchain to international trade and explored the tremendous changes that blockchain has brought to trade finance, customs procedures, and logistics tracking [8]. In the final analysis, the competition between international trade is the relative advantage of each trade subject. Data is an important information transmitter and a source of profit and value for an enterprise. Taking into account the economic background of this era, we must fully pay attention to technological property rights, integrate corresponding protection policies, and combine actual experience and opportunities in the economic globalization environment to adjust and integrate new ideas [9]. In addition, with the development of computer network technology, the degree of social information is getting higher and higher. The network contains a lot of information and data resources. These data and information include personal information protection, etc., and may be attacked by viruses such as hackers or Trojan horses, leading to the disclosure of personal privacy and confidential information. Block chain technology has the characteristics of programmable, traceable and non-tamperable. Based on block chain technology, the security of network data transmission can be improved, and the personal privacy of customers can be protected safely [10].

2 Method

2.1 Give Full Play to the Role of the Market

In the process of realizing international technology trade, the market is a factor that cannot be ignored. The essence of international technology trade is to find a balance between technology and the market to obtain greater benefits in the market, which requires us to be market-oriented. With international technology trade as the center, improve the core competitiveness of enterprises and realize rapid development of enterprises. According to the status quo of international technology trade and intellectual property control strategies, the WTO's market maintenance and market construction should be improved, actively mobilized to maintain the competitiveness of enterprises in the international market, and adjust and optimize the trade model mechanism of enterprises. Second, the government must strengthen market supervision and management, do a good job in building market integrity, and provide strong conditions for the development of international technology trade. Finally, the government can provide economic support to small and medium-sized trading enterprises by relaxing loan conditions and other policies to promote the development of my country's international economy and trade.

2.2 Enterprises Must Establish Their Own Core Competitiveness

With the use of blockchain technology, both parties of trade can trade and circulate commodities on the Internet, saving transaction costs and time, and reducing trade risks. Therefore, companies must cultivate blockchain technology talents, optimize their working environment, improve corporate salaries and benefits, and retain big data talents. Secondly, encourage talents to innovate, and attract outstanding external talents through holding some innovative and entrepreneurial activities. The introduction of technology and talents is not temporary, but a relatively long-term plan for the future. Therefore, we should not only focus on the present, we should look to the future and make a long-term plan to avoid repetitive resource consumption. Cultivating and introducing a large number of technical and technical talents will bring huge benefits to the development and development of the enterprise. Use a large number of talents to analyze and discuss plans that are friendly to the development of the enterprise, accelerate the development of the enterprise, and enhance the core competitiveness of the enterprise.

2.3 Promote the Implementation of Blockchain Technology

The application mode of blockchain technology has three types: public chain, private chain and client chain. There are seven types of blockchain technology: open agreement, decentralization, trustlessness, anonymity, non-tampering, traceability, and programmability. Although the development of international blockchain technology has undergone an obvious acceleration process, there are no substantial industrial applications in the blockchain technology industry. The country should actively promote in-depth research on the level of blockchain technology, provide good blockchain technology application research for international technology trade, and will conduct research and investigations through a variety of methods. Blockchain technology needs a reliable network to drive. It needs a reliable identity, reliable ledger, reliable calculation and reliable storage technology to realize the application of blockchain technology in international technology trade. Therefore, asset equity can realize fast and safe circulation and transactions on the blockchain, and quickly confirm its price in the market. Through the application of blockchain technology, it not only promotes the development of international technology trade, but also increases the quantity and quality of international technology trade transactions.

3 Experiment

3.1 Subject

There are many ways to measure technological innovation, and they are often measured by technological innovation performance. Gao Jian defines "Technical Innovation Performance" (TIP) as "the efficiency of the enterprise's technological innovation process, the output results and its contribution to the business success of the enterprise, including the performance of technological innovation output and the performance of technological innovation process". From the perspective of investment, it includes the investment of R&D expenses and the input of R&D personnel; from the perspective of output, it includes new products, new processes, and new technologies. The measurement of new blockchain technology in the actual international technology trade is nothing more than the measurement of the number of patents granted. Using patents as a measure of technological innovation performance can avoid the difficult-to-transformation problems caused by using inputs as technological innovations, so as to better measure the effect of blockchain technology on international technology trade.

3.2 Experimental Method

This experiment uses a combination of literature review methods and empirical analysis methods to study the impact of blockchain technology on the application of international technology trade. The specific methods used in this experiment are as follows:

(1) Literature review method:
The literature review method refers to a method of summarizing and sorting out the literature related to experimental research by collecting and sorting out. Through the literature review method, we can understand the status quo of the current research problems and the current research development, so as to pave the way for further experimental research.

(2) Chart analysis method
When studying the development status of different technological spillover channels and technological innovations, this experiment uses a chart analysis method to show the changes in international technology trade in recent years and the growth and changes of related patents in the form of charts. So as to be able to intuitively understand the development of international technology trade.

(3) Statistical analysis
In this experiment, statistical analysis was used in the descriptive statistical analysis of each variable. By analyzing the maximum, minimum, median and standard deviation, it revealed some characteristics of the variables themselves. And understand the general situation of different variables.

(4) Metrological analysis method
Quantitative analysis is an analysis method that uses measurement software to perform additional processing on data. This experiment uses Eviews8.0 to test the normality of the reported data and regression analysis, so as to further obtain experimental empirical results to test the purpose of this experiment.

4 Results

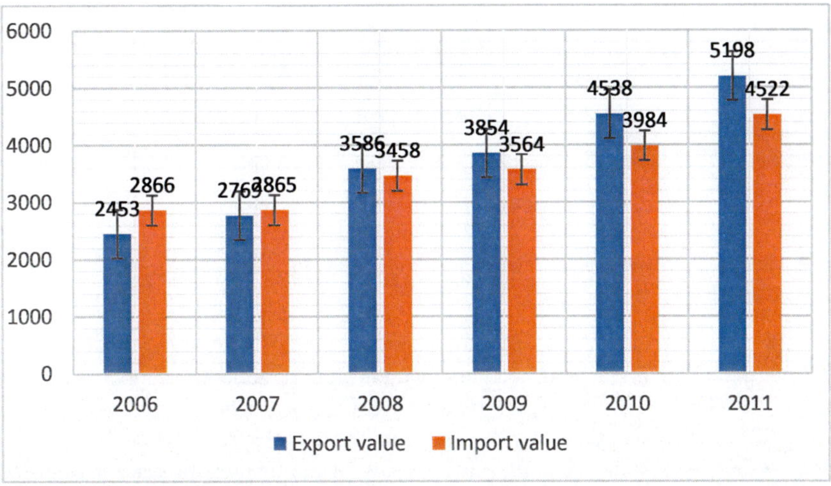

Fig. 1. Import and export of high-tech products (unit: 100 million US dollars)

In recent years, the actual situation of using blockchain technology to conduct technology trade can be seen from the import and export situation of the international technology trade industry as shown in the figure above. It can be seen from the graph that the international technology industry trade has grown rapidly: From 2006 to 2011, the import and export trade volume of high-tech industries maintained growth at an average annual growth rate of 30%. High-tech industry exports accounted for 35.8% of merchandise trade from 23.4% in 2006, and imports increased from 31.2% in 2006 to 35.4%.

From 2006 to 2011, the proportion of technology industry trade in exports of manufactured goods increased from 24.9% to 38.5%, and imports increased from 37.5% to 49.7%. Compared with overall commodity trade, the growth rate of high-tech industry trade is obviously faster. In addition, the proportion of imports and exports of high-tech products in the trade of commodities and industrial finished products has continued to increase. In 2011, this proportion reached 30.5%, an increase of 11.5% over the same period last year. The rapid development of the high-tech industry in 2006 is evident from this (Figs. 1 and 2).

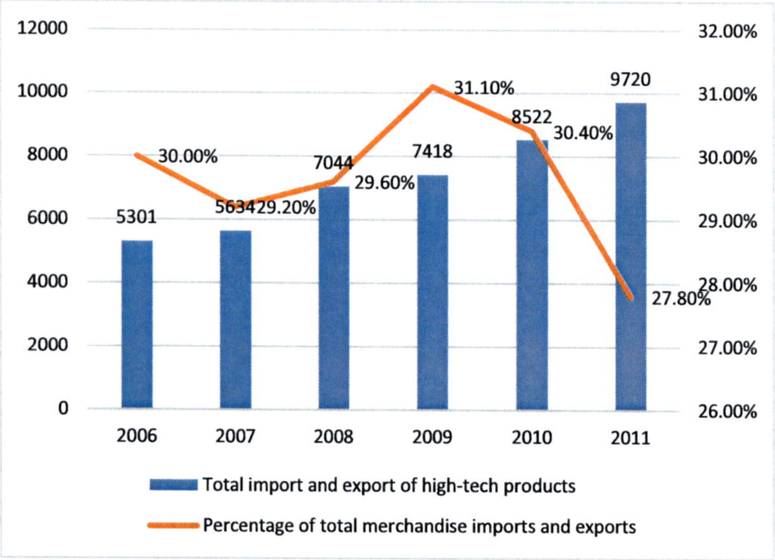

Fig. 2. Total imports and exports of high-tech products and their proportion in total imports and exports of goods

After the actual international technology trade after the adoption of blockchain technology, in the high-tech industry, the import and export trade conditions of different technical fields may be very different. As shown in Table 1, electronic technology exports were US$86.584 billion, accounting for 15.78% of the total exports in the high-tech sector, while exports in the fields of biotechnology, material technology, and aerospace technology developed relatively slowly, accounting for 0.08% of the total exports of high-tech industries. 0.86% and 0.84%. The high-tech industry imported a total of 463.225 billion U.S. dollars, of which the imports of electronic technology and optoelectronics technology ranked the top two, respectively 213.975 billion U.S. dollars and 54.203 billion U.S. dollars, accounting for 46.19% and 11.7% of the total imports respectively. The area with the fastest year-on-year increase in imports is life science and technology. Before and after the use of blockchain technology, the volume of international technology trade has been greatly improved, and the import and export volume has been greatly improved compared with before using blockchain technology.

Table 1. High-tech import and export distribution by technical field (before and after the implementation of blockchain technology) Unit: USD million

	Export value	Total (%)	Increase over the previous year	Import value	Total share (%)	Increase over the previous year	Difference
Aerospace technology	4599	0.84	31.62%	19015	4.1	13.87%	−14416
Biotechnology	414	0.08	16.45%	449	0.1	6.29%	−35
Electronic technology	86584	15.78	11.77%	213975	46.19	9.05%	−127391
Life science technology	17843	3.25	28.73%	15796	3.41	35.21%	2047
Material technology	4716	0.86	6.60%	5933	1.28	2.35%	−1217
Optical technology	32114	5085	12.19%	54203	11.7	3.63%	−22089
Other technologies	677	0.12	60.48%	1287	0.28	44.40%	−610

5 Conclusion

Blockchain technology is a new theory and technology emerging with the application of international technology trade. The current blockchain technology research is at the initial stage. Although all aspects need to be improved, good results have been obtained in the process of applying international technology trade. In other words, since economic integration is an indispensable step in social development, competing platforms and sanctions supported by technical means are just a bargaining chip for the platform to implement alternative actions, not a weapon. The sanctions themselves will also be affected. However, as a new thing, blockchain also has institutional and technical deficiencies. Therefore, in the current development, we should continue to explore how to solve the regulatory and technical application problems of blockchain in international technology trade. The study found that blockchain technology has become a new driving force for the development of international technology trade, promoted the diversification of international technology trade development methods, changed the pattern of world technology trade, and brought new trade risks. Put forward specific countermeasures for the development of international technology trade.

References

1. Radziwill, N.: Blockchain revolution: how the technology behind bitcoin is changing money, business, and the world. Qual. Manage. J. **25**(1), 64–65 (2018)

2. Miraz, M.H., Ali, M.: Applications of Blockchain Technology beyond Cryptocurrency. Ann. Emerg. Technol. Comput. **2**(1), 1–6 (2018)
3. Möser, M., Soska, K., Heilman, E., et al.: An empirical analysis of traceability in the monero blockchain. In: Proceedings on Privacy Enhancing Technologies, vol. 2018, no. 3, pp. 143–163 (2018)
4. Falkowski, K.: Competitiveness of the baltic states in international high-technology goods trade. Comparat. Econ. Res. **21**(1), 25–43 (2018)
5. Pastpipatkul, P., Saeor, S.: Analysis of relationship between digital technology and international trade of Thailand. Int. J. Intell. Technol. Appl. Stats **11**(3), 155–166 (2018)
6. Jinji, N., Zhang, X., Haruna, S.: Do deeper regional trade agreements enhance international technology spillovers? World Econ. **42**(8), 2326–2363 (2019)
7. Wang, W., Pan, F., Li, M., et al.: Extraction of characteristic technical parameters in cut strip processing of different production lines contributing to sensory quality variation. Tobacco Sci. Technol. **51**(8), 92–99 (2018)
8. Kisilowski, J., Kowalik-Adamczyk, E.: Test of quantity characteristic technical condition of railway turnouts. AUTOBUSY – Tech. Eksploatacja Syst. Transp. **19**(12), 438–444 (2018)
9. Thiamwong, L., Sole, M.L., Ng, B.P., et al.: Assessing fall risk appraisal through combined physiological and perceived fall risk measures using innovative technology. J. Gerontol. Nurs. **46**(4), 41–47 (2020)
10. Solomon, A., Polievoda, Y.: Innovative technology of production of fermented milk drinks for nutrition elderly people. Eng. Energy Transp. AIC **1**(108), 65–74 (2020)

The Influence of Calculating and Simulating Talent Filtering Index Model on the Activation of Talents

Chen Zhang(✉), Xiaoyan He, and Qi Yang

Haojing College of Shaanxi University of Science and Technology, Xi'an, Shaanxi, China

Abstract. The continuous development and wide application of big data technology has led to profound changes in our thinking, behavior and communication methods, and has promoted changes in our lifestyle. Big data has the characteristics of large-scale, fast, differentiated and value-oriented data collection, processing and analysis, which has profoundly affected all aspects of our lives and made our lives easier. Especially in the era of globalization, the overall big data governance must also be carried out under the intervention of multiple global forces. In this way, based on the effective supervision and guidance of the development of big data, it can better serve the promotion of human development and the satisfaction and realization of human values. In short, analyze the reasons, nature, obvious contradictions, problems and consequences of this change, understand its changing trends, and propose necessary countermeasures and suggestions to deepen the social impact of big data. Exploring specific news about the importance, value and development of changes in human lifestyles, and the mechanisms and mechanisms of changes in the technological environment, has promoted thinking about the relationship between technological innovation and social development. In the era of big data, people's lifestyles and relationships are coordinated and developed to help. This article systematically analyzes the activation of talent vitality through analytic hierarchy process and fuzzy comprehensive evaluation method, analyzes the impact of talent vitality on us and enhances talent vitality. And on this basis, we designed the evaluation criteria for the talent vitality model, and constructed the comprehensive index model of talent development. Based on theory and practice, the calculation simulation talent filtering index was put forward. Through the method of this article, the vitality of talents has increased by 7%.

Keywords: Big data · Talent filter index model · Analytic hierarchy process · Fuzzy comprehensive evaluation method

1 Introduction

1.1 Background and Significance

The arrival of the era of big data is closely related to the rapid development of science and technology [1]. In the era of big data, data no longer simply refers to "data", and its scope

B. J. Jansen et al. (Eds.): *International Conference on Cognitive based Information Processing and Applications (CIPA 2021)*, LNDECT 85, pp. 695–702, 2022.
https://doi.org/10.1007/978-981-16-5854-9_89

includes various formats such as text, images, audio and video. Comparing data in the era of big data with traditional data, we find that traditional data is structured, while data in the era of big data includes structured and semi-structured data as well as unstructured data. Historically speaking, data is structured data [2]. When people analyze data, they can refer to experience and practice to discover and explore data laws. For example, based on the linear correlation with the report, you can predict the income and expenses of the business or unit in the next month. The direction of consumer demand can also be calculated through the search trajectory of network users [3]. The advent of the era of big data enables people to access information faster, and has also discovered new ideas that are changing people's current economic, political and cultural aspects. At the same time, the era of big data will also affect people's lives in the era of data. The country produces shocks and revolutions [4].

This article systematically examines and analyzes the overall influence and influence of the era of big data on the evolution of people's lifestyles. It not only helps to distinguish the essence of the big data era and its inherent contradictions and conflicts, but also helps the data era to be an example of concrete exploration of importance, value and development as well as mechanisms and mechanisms [5, 6]. The evolutionary mechanism of the new technological environment has changed people's lifestyles to promote people's thinking about the inner connection between scientific and technological innovation and social development. Enrich and deepen research in related fields. At the same time, the research in this article can comprehensively and systematically explore the multi-dimensional challenges brought by the era of big data to human social life, communication, behavior and way of thinking, and understand a series of contradictions, essence and challenges [7]. The reasons for the paradox not only help deepen the crisis and determine the meaning, value and role of the big data era itself, but also provide specific practical reports and reference for exploring and solving specific issues such as privacy and privacy. Social interaction in the data age [8].

1.2 Related Work

Because of the importance of big data research, more and more scholars and research teams have begun to study big data and have made very little progress. Alatan once conducted research on enhancing the vitality of talents and proposed respecting the development of talents' personality [9]. But this topic is too broad, so this article proposes based on the influence of the big data era on the activation of talent vitality. Use big data to study the vitality of talents [10].

1.3 Main Content

In order to solve the problem of talent vitality, this article uses the literature research method to study the past methods of activating talent vitality, and summarize and summarize the past methods of activating talent vitality. Sum up the methods for activating the vitality of talents. Through the method of this article, the vitality of talents has increased by 7%.

2 On the Application of Computational Simulation Talent Filter Index Model to Activate Talent Vitality

2.1 Research Methods on the Impact of Talent Incentives in the Era of Big Data

(1) AHP
The process of analyzing the hierarchical structure is mainly based on the relationship between the evaluation topic decomposition indicators at a fairly smooth level. Each level of the indicator has roughly the same state and function, and then the indicators of the same level are compared in pairs to determine multiple episodes., Use the characteristic root corresponding to the characteristic value as the weight of the corresponding level indicator, and finally synthesize the total weight. The advantage is the combination of quality and quantity, and high reliability; the disadvantage is that when multiple indicators are at the same level, the crisis matrix is prone to inconsistencies. It is suitable for horizontal and vertical comparison of multiple samples and complex decision-making samples, such as cost-benefit decision-making.

(2) Fuzzy comprehensive evaluation method
The fuzzy comprehensive evaluation method must first determine the evaluation target's index and evaluation set, determine the weight of each index and its average score, obtain a fuzzy evaluation table, and then perform fuzzy evaluation. And use the total weight of the indicator to calculate, and finally standardize the Total Vague evaluation results. Its advantage is the combination of qualitative and quantitative, which avoids the unification of results; its disadvantage is that the operation is complicated and it cannot solve the problem of information overlap caused by the correlation between indicators. It is suitable for horizontal and vertical comparison of multiple samples, such as determining the best consumers and analyzing the total amount.

(3) Principal component analysis and factor analysis
The fundamental component analysis method must transform a given set of multi-dimensional relative variables into a set of smaller-dimensional side variables through linear transformation, and arrange them in order according to the change contribution rate. The main feature is derived from the requirement of combining feature value and cumulative change contribution rate. The method of factor analysis is to group variable data according to the correlation size of each index, so that the correlation between variables in the same group is higher, rather than the group correlation is lower. Factor analysis achieves the purpose of reducing the dimension by evaluating the number of initial variables in the category. The advantages of both are strong objectivity, effectively solve the problem of information overlap between different indicators, simplify the structure of the indicator system, and have a wide range of applications; the disadvantage is that the calculation process is cumbersome and requires high indicators. The quantity and quality of samples. The actual concept represented by the key ingredients to be extracted will change, and its meaning needs to be refined and summarized. It is suitable for horizontal and vertical comparison of multiple samples, such as the classification of evaluation items.

(4) Cluster analysis method
The cluster analysis method must standardize the variance of the data set to be tested through a series of tests, and then compare its linear coordinates, and then perform hierarchical grouping according to the overall linear correlation of the data set. Its advantage is that it solves the evaluation objects with high correlation between the indicators; the deficiency of land is that it has a high demand for sample size and requires a lot of statistical data. It is suitable for complex large-sample decision-making schemes, such as full portfolio decision-making.

2.2 Construction of Talent Vigor Index Model and Dependence Degree Model

$$\alpha_t = \sum_{k=1}^{s} W_k \sum_{j=1}^{m} W_j \sum_{i=1}^{n} W_i \frac{P_{i,t} - P_{i,t-1}}{P_{i,t-1}} \tag{1}$$

The comprehensive impact of various influencing factors on the development of talents is shown in (1), where

α_t—Comprehensive index of talent development:

W_k, W_j, W_i—factors, index categories, and index weights to the upper level:

P—index value;

N—the number of index classes in a certain index class;

m—the number of index categories for a certain factor;

s—the number of factors affecting the development of talents.

A comprehensive index of the influence of various factors on the development of human resources.

Get the value of the two-by-two comparison between the factors, and the weight of each factor at this level can be obtained through calculations. The specific calculation method is as follows:

First calculate the sum of the importance of each factor to other factors, namely

$$R_i = \sum_{j=1}^{n} R_{ij} \tag{2}$$

Then calculate its proportion in the overall, that is, get the weight of the factor at this level, expressed by the formula:

$$H_i = \frac{R_i}{\sum_{i=1}^{m} R_i} \tag{3}$$

3 Research Experiments on the Impact of Activating Talents in the Era of Big Data

In the process of comprehensive assessment of talents, the construction of talent evaluation index system is the foundation. The quality of the selected evaluation index will affect the accuracy, rationality and scientificity of the analysis object. The urban talent vitality evaluation index system can closely link the evaluation with the person in charge of the research object, and it can also link the evaluation method with the research object, and establish and implement a scientific and practical system. Evaluation to obtain scientific accuracy. To establish important conditions for evaluation, the following principles must be followed in the process of establishing a comprehensive evaluation index system for urban talent vitality:

3.1 Scientific Principles

In the urban talent vitality evaluation system, the selection of evaluation indicators should be able to scientifically and comprehensively reflect the concept of urban talent vitality, and should follow scientific principles, that is, to guide ideology with scientific theories and through theoretical analysis, to better integrate with reality. The introduction of the talent vitality evaluation index system must meet the following three points: First, the selection of evaluation indicators must have scientific and logical basic concepts, definitions and strict logical structures; second, it is necessary to understand and master the evaluation goals. Under certain conditions. Its practical significance: Third, it is necessary to use the correct scientific theoretical basis and objective reality to ensure the rationality and scientificity of the selection of evaluation indicators.

3.2 Principle of Integrity

In order to collect rating indicators, it is inevitable to find the complexity of the rating indicator system. Therefore, it is necessary to ensure that the urban talent vitality index system can systematically and systematically reflect the true state of urban talent vitality, and to ensure that different factors and evaluation indicators are relevant. It can fundamentally reflect the basic concept of urban talent vitality.

3.3 The Principle of Expediency

The expedient principle mainly reflects the practicality, expediency and functionality of the selected evaluation indicators. In other words, the concept of urban talent vitality index must be clear and usable. Each grade index can be represented by a numerical value, which is feasible and can be used for actual statistics and calculations. The beginning of the hierarchy when establishing the urban talent vitality evaluation index system, there must be a certain level and target level evaluation index.

3.4 The Principle of Simplicity

There are many evaluation indicators for the vitality of urban talents, and the selection of concise, clear and representative evaluation indicators should be considered when collecting. In other words, the selection of urban talent vitality indicators must reflect the concept of urban talent vitality, avoid duplication of evaluation indicators, and maintain the relative independence of evaluation indicators.

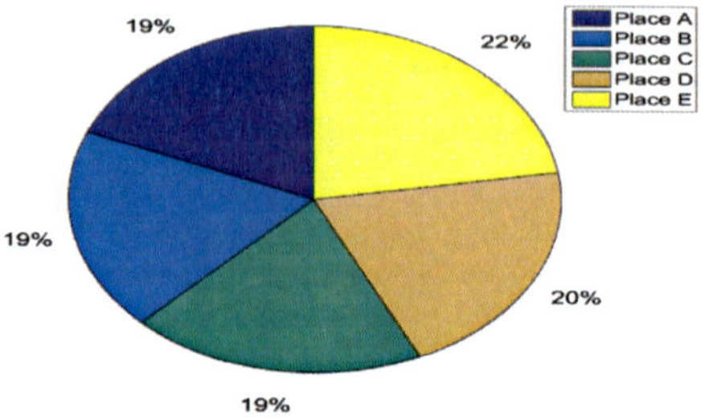

Fig. 1. Weight analysis chart of the five major factors of urban talent vitality

4 Research and Analysis on the Impact of Activating Talents in the Era of Big Data

4.1 Talent Vitality Structure

Through the analysis of Fig. 1, it is found that among the five major factors affecting the vitality of urban talents, talent growth and talent mobility account for only 10% and 10%, respectively. Compared with the other three factors, they have no impact on urban talent vitality. Big. However, it does not affect our research on talent growth and talent mobility. The result of weighting is affected by subjective factors. Therefore, we need to research and analyze the evaluation indicators of each factor. Only in this way can we find ways to improve the vitality of talents in each city and ensure the scientific and comprehensive analysis results.

4.2 Comprehensive Evaluation and Analysis of Urban Talent Vitality

Table 1. Talent vitality index value and ranking in 2019

City	Vitality index	Rank
Shanghai	0.754	1
Shenzhen	0.557	2
Beijing	0.477	3
Guangzhou	0.101	4

From this table, we can find that Shanghai, as a key development city in my country, has a talent vitality index of 0.754 far ahead of Shenzhen, which ranks second; Shenzhen, Beijing, and Guangzhou rank second, third, and fourth respectively. From the side, it can explain the development speed of the city. The talent vitality index can help cities develop further. For each city, the demand for talent is huge. Only when the city attracts enough talents can the city develop better. This is also the reason why first-tier cities such as Shanghai are developing rapidly (Table 1).

5 Conclusions

Today, big data has become one of the main internal driving forces for the development of technology and the times. Whether it is technological creation, industrial innovation, social management, and government decision-making, everything is closely related to big data and requires public participation and integration. In this process, big data has become more and more obvious in all aspects of daily life and changes in people's lifestyles. It does not change the specific manifestations of people's work, leisure, consumption and communication methods. It promotes differentiation and convenience in life, and also brings a negative impact on people's spiritual life. Reasonable consumption, etc. In fact, the revolutionary change caused by big data is not how many advanced technologies it creates, but the impact and changes it brings to people. We not only regard big data as simple data and data exchange and transmission, but we must also closely connect big data with people who use big data, return to the individual, give play to people's initiative and move on. Enhance the positive impact of big data on lifestyle, and actively create a healthy and reasonable lifestyle based on its own development.

Acknowledgments. Project Number: No. 20JK0078.

Project name: Research on the driving factors of talent vitality and talent retention mechanism in Xi Xian New Area under the "Five New" strategy.

Project type: Scientific research project of the Education Department of Shaanxi Province.

References

1. Deng, X., Zhong, W., Wenjun, Z., et al.: Research on the innovation of Ideological and political education mode in universities in big data Era. In: 2017 3rd International Conference on Economics, Social Science, Arts, Education and Management Engineering (ESSAEME 2017), vol. 9, no. 2, pp. 56–84 (2017)
2. Yuan-Xiang, D., Fan, Y., Jing-Yi, M.: Research on the talent training mode of management science in colleges of finance and economics in the era of big data—guided by the national standards for teaching quality. J. High. Educ. Financ. **5**(1), 452–468 (2019)
3. Tang, S., Zhang, W., Xu, Z., et al.: Research on talents cultivation of applied smart tourism specialty in the global tourism era. Jiangsu Sci. Technol. Inform. 036(017), 75–77 (2019)
4. Hui, X.: Challenges and countermeasures of management accounting in the era of big data. World Sci. Res. J. **5**(10), 115–121 (2019)
5. Cheng, Y.: College English teaching reform based on the cultivation of applied talents in the new era. Theor. Res. Pract. Innov. Entrep. **002**(009), 68–69 (2019)
6. Chang-Xi, B.I., Bing-Hui, L.U., Marxism, S.O., et al.: A research on the practical effect of the ideological and political education of colleges in the era of big data. J. Xi'an Univ. Archit. Technol. (Soc. Sci. Edn.) **3**(3), 69–92 (2018)
7. Xie, M., Meng, F., Feng, W., et al.: Education reform of theoretical mechanics for civil engineering in big data era. J. Phys: Conf. Ser. **1624**(3), 032–037 (2020)
8. Chen, J., Hu, X.: Research on the cultivation of accounting informatization talents in colleges and universities in the internet+ era. Educ. Teach. Forum (005), 254–256 (2017)

9. Alatan, Zhang, B.: Research on legal protection system of personal information in big data era. In: IOP Conference Series: Materials Science and Engineering, vol. 806, no. 1, pp. 012–032 (2020)
10. Qing, S.: Research on the competency model of the introduction of high-end medical talents in public hospitals in the context of big data. China Health Stand. Manage. **6**(1), 152-155 (2019)

Human Resources Development and Management System Based on "Five New" Strategy

Qi Yang(✉)

Haojing College of Shaanxi University of Science and Technology, Xi'an, Shaanxi, China

Abstract. Human resources are the first to be determined in the production factors of various enterprises and units. Nowadays, human resource management is getting more and more attention. Human capital is business profit and the driving force of development and competitiveness. Therefore, effective human resource management has become the focus of the normal development of enterprises. The purpose of this article is to study the development and management of current human resources from the perspective of the "five new" strategy, determine the development and management of new human resource concepts, and encourage new trends in innovation and sustainability. According to social and local needs, training human resources, focusing on new human concepts. This article starts from the perspective of the "five new" strategies for human resource management. These strategies are increasingly important on the basis of relevant realistic assumptions and theories, and aim to provide new ideas for the development and management of human resources. The "Five New" strategy is based on the effective development and breakthrough of talent training in applied universities as specific research topics, and has created a framework to research development and human resource management systems from all aspects of education, technology, science, talent, and local economy.. The experimental results show that based on the new perspective of the "five new", more effective means can be used to develop and manage human resources. In the future, there are reasons to believe that all enterprises and departments in various industries will be under the "five new". Human resources management and talent development must be integrated, especially application-oriented universities that provide the source of human resources are the top priority. Therefore, the current applied university must reorganize and reposition itself, looking for new ways of human resource development and management.

Keywords: "Five new" strategy · Human resources · Development and management · Applied university

1 Introduction

The 21st century is an era of fast-changing information. With the rapid development of computer technology and the popularity of smart phones, more and more computers

B. J. Jansen et al. (Eds.): *International Conference on Cognitive based Information Processing and Applications (CIPA 2021)*, LNDECT 85, pp. 703–710, 2022.
https://doi.org/10.1007/978-981-16-5854-9_90

and mobile phones are used for human resource management [1]. The current era of big data can improve the convenience of human resource management, provide a more reliable guarantee for business development, can effectively manage the company's human resources, and obtain the latest and most accurate information about the company's human resources [2]. Human resources is to make full use of the huge potential of every talent, provide a reliable talent reserve for enterprises, become the main energy for enterprise development, and create efficient value [3].

China's human resource management system is still lagging behind developed countries. The biggest challenge facing China's internal human resource management system is: First of all, most of them are management information systems and services developed for specific business needs (such as human resource management). The purpose is specific automation needs. The idea of the project is mainly created according to the needs of specific small and medium-sized companies, and usually cannot adapt to the structure of cross-system, diversification and cross-organization [4]. Secondly, at present, the internal human resource management software has expanded its module functions to the entire human resource management or human capital management business field, but the human resource management of overseas resources and remote areas still lacks integrity, advanced technology and system integration [5]. The current advantages are mainly reflected in strong economic strength, training of technical personnel, huge investment in R&D and marketing, equipment manufacturers and database companies with specific advantages, and partners including consulting companies have formed a strong Combination of strength [6]. The "five new" strategy put forward by the 13th Shaanxi Gathering is to cultivate new kinetic energy, reach new heights, stimulate new vitality, jointly establish a new life and highlight a new image [7]. The five new strategies are Shaanxi's main measures to fully embody the important speech of Secretary-General Xi Jinping, especially the spirit of Shaanxi's important speech, and to fully implement the new development concept [8].

The proposal for the implementation of the "five new" strategy in the comprehensive strategic plan of Shaanxi Province is mainly based on a puzzle: Why can't we combine the benefits of science and education and turn them into the benefits of local economic and social development? How can local applied universities help the goal of local economic development? The research on these issues has very special practical significance in today's lack of human resources [9]. Applied universities play a very important role in promoting regional economic and social development. Regarding the "five new" strategy, applied universities should adapt to regional economic and social development strategies and regional development. Provide talents with corresponding needs and get local support to provide theoretical support for promoting regional economic and social development. For this reason, this paper has conducted necessary research on human resource development and management system on the basis of the "five new" strategy [10].

2 Method

2.1 Human Resource System Design Based on the "Five New" Strategy

As the premise of this research, the "five new" strategy occupies a primary role in the human resource development and management system of the "five new" strategy. It is more important to the requirements of regional and applied universities. As an extremely important part of local economic and social development, it has Certain analogies, miniatures and immutability.

Application-oriented universities are an important part of the local area, and their role in promoting local economic development, as the human resource point of the design, can be expressed as follows:

$$X = (w_1, w_2, \ldots, \mathrm{w_k}) \tag{1}$$

$$w_k = \{x^{(k,1)}, x^{(k,1)}, \ldots, x^{(k,N)}\} \tag{2}$$

$$X^{(k,i)} = (x_1^{k,i}, x_2^{(k,i)}, \ldots, x_d^{(k,i)})^T \tag{3}$$

2.2 Define the Inter-class Divergence Matrix

In the above formula, $N = \sum_{i=1}^{K} N_k$ Is the total number of samples, $m_{\mathrm{k}} = \frac{1}{N_k} \sum_{x=\partial_k} x$ Represents the mean value of the k-th sample, $m = \frac{1}{N} \sum_{K=1}^{K} \sum_{x=\partial_k} x$ Is the overall mean. Define the inter-class divergence matrix as:

$$S_b = \sum_{k=1}^{k} \frac{N_k}{N}(m_k - m)(m_k - m)^T = \Phi_b \Phi_b^T \tag{4}$$

First, define the inter-class divergence matrix when the criterion reaches the maximum, and then after orthogonal transformation, after calculation, the product of the overall defined inter-class divergence matrix also reaches the maximum value. And it is necessary to require each sample to be separated as far as possible, so as to calculate the variance of the sample and keep the mean square error of the original sample small, so as to get a proof $X^{(k,i)} = (x_1^{(k,i)}, x_2^{(k,i)}, \ldots, x_d^{(k,i)})^T$ Then the eigenvectors corresponding to the first d largest eigenvalues of the matrix obtain the degree of dispersion. In the process of data processing, the minimum value of the error function brought by the analysis of the sample data of computational application-oriented universities should be used, and the weights and thresholds need to be adjusted iteratively to improve the calculation accuracy.

3 Methods and Experimental Research Design

3.1 Do a Good Job in the Development of Refined System Management of Human Resources

The human resources part of the enterprise needs to analyze each position according to its own enterprise needs, establish and build the enterprise's own human resources structure framework organization, clearly distinguish the work purpose and nature of each job, and select jobs that are competent for different jobs. Personnel, make full use of their work talents to maximize human resources. Establishing and improving the refined human resources system is a prerequisite. Organize the current corporate human resource management system, improve the execution of the human resource system management system, and change the original corporate human resource system management model, according to the 'refinement' Management concept for the design of the human resource system, to implement the human resource system management system. The second is to establish a reasonable and complete human resource management evaluation standard. One is the good effect of human resource allocation, and the other is the result of employees' work. Make use of scientific methods and technologies to formulate a standardized and reasonable human resource evaluation mechanism suitable for themselves, and form a good corporate culture and atmosphere. Application-oriented universities must integrate into the new trend of the era, which is guided by innovative vitality and based on professional skills, and fully implements the education of "the common development of teaching and economy, the mutual support of majors and needs, and the collaboration between universities and localities" for each student. At the same time, the rational arrangement of professional settings, curriculum settings, teaching methods and other aspects of education can finally be implemented in the training goals of innovative talents, so as to achieve the goals of the "five new" strategic goals, promote local economic development, and produce new and effective human resource Management.

3.2 Establish New Development Ideas Focusing on Improving the Fit Between Talent Training and Social Needs

Applied universities provide the basic guarantee of talents to realize the "five new" strategic goals. From this perspective, applied universities are not enough to meet the actual needs of local governments, society and industries. Regarding talent management, it should be taken into account that the demand of the demand side should be adjusted in a timely manner to connect the talents with supply and demand in a simple way. The satisfaction of talent demand should also actively respond to the supply side. On the one hand, applied universities need to promote independent demand-side innovative talents, and local enterprises, industries and governments will also be driven by applied universities to drive talent management to achieve a win-win situation. After being developed locally, applied universities can improve school conditions, achieve development goals and improve talent training. It can be seen that cultivating applied innovative talents with applied skills and innovative qualities to promote and contribute to the development of regions and industries is not only an inevitable requirement for local economic and social development, but also for the survival of applied universities themselves.

And develop a proactive approach. Therefore, applied universities should put forward new development ideas to cultivate talents and improve social needs, and play a leading and core role in the human resource management system “local government-applied universities-industry and commerce”.

3.3 Experimental Research Objects

First of all, the investigation and research method is adopted, and detailed investigation and research, research data, research rules, master first-hand information, and refine and summarize the specific case must be carried out. This paper selects 200 students from a certain applied university, and conducts practical investigations and studies in terms of student enrollment, teaching professional training process, whereabouts and employment after graduation.

Table 1. Questionnaire survey report

Questionnaire issuance and recovery	Student cadre representative	Student representative	Total
Issue	40	160	200
Recycle	38	155	193
Effective	37	150	187
Efficient	97.4%	96.7%	96.9%

In the survey with the research report of the experiment, a total of 200 students were rationally selected and given an experimental research questionnaire, and a long-term survey of the 200 students (see Table 1) was conducted to evaluate each student’s admission, major and Employability. For this, it is necessary to firstly study the student’s movement from admission to a few years after graduation, and at the same time study the direct assimilation of data information in order to apply research and research methods, as well as to concretize and summarize specific cases. Secondly, it analyzes the use of case analysis method, for which a large amount of specific student trends and psychological information are needed to analyze and study the status quo of teaching methods of applied universities through case studies. According to the investigation and analysis in this article, according to the analysis of Chinese applied universities the education model, teaching method, teaching effect, employment direction and other aspects have also been evaluated.

4 Results

4.1 Based on the “Five New” Strategy Applied University Personnel Training Plan and the Urgency of Implementation

The data shown in Fig. 1, from the data shown in the figure, we can understand that the application-oriented university students are not very satisfied with the traditional training

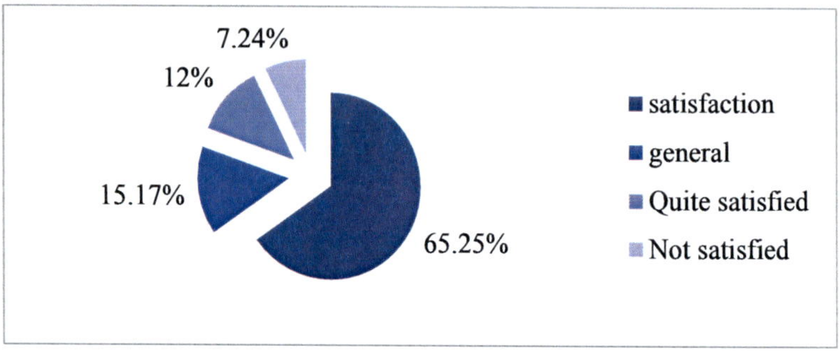

Fig. 1. Investigating the satisfaction of students in an applied university with traditional training methods

methods. The "satisfied" students only account for 65.25%, "comparatively satisfied" The combined ratio of "" and "unsatisfactory" is 19.24%. This is not a reasonable data. It shows that there are many problems in talent training in applied universities. Every company or place has efficient human resource management that can develop and create high economic value. As one of the several economic powers in the world, my country has always invested huge financial and material resources in human resource management to cultivate various kinds of related talents. From the perspective of the "five new" strategy, this poses a series of problems for the application-oriented universities to train outstanding talents, such as what knowledge level, views and ideas about society should be possessed by the talents cultivated by the applied university economy, and or it is the question of what qualities and abilities should be cultivated by applied universities. This has caused great difficulties for my country's applied universities to cultivate outstanding talents. It is a very difficult task to meet the needs of the times to cultivate relevant outstanding talents, and a long way to go.

4.2 Problems Faced by My Country's Applied Universities in Cultivating Outstanding Talents

It can be seen from Fig. 2 that there are some problems in the self-evaluation of college students in applied universities in my country. Development problems caused by their own professional abilities account for 41.3%. This is due to the impact of students' own problems, but due to political neglect and social The development restriction caused by demand accounts for 47.1%, which is enough to show that the current society has problems with the development and management of applied human resources. my country is a country that attaches great importance to the development of education. Educational issues have always been a development issue that the country attaches importance to. In many aspects, the country will implement reforms in education and strive to improve the overall education level of the country. As the advanced education stage of our country, applied university is an important link in cultivating talents that are beneficial to the development of various places in our country. As far as the majors of applied university in our country are concerned, it is an important birthplace for cultivating relevant talents for our country. However, for now, the economic management majors of ordinary

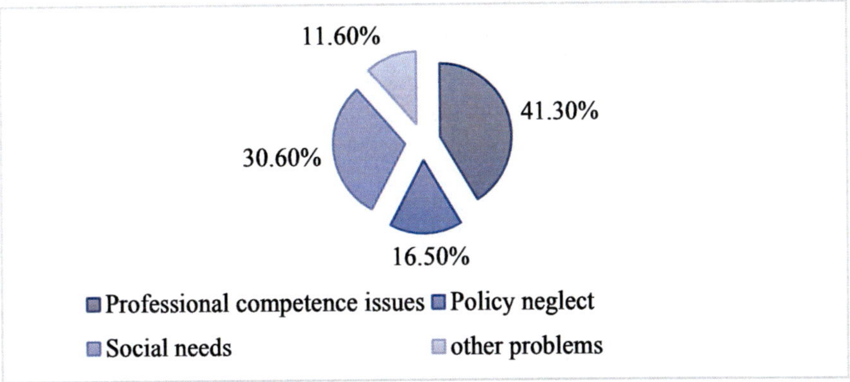

Fig. 2. Survey of the proportion of students in a certain applied university to the current biggest problems in their own development

applied universities in our country are still facing many problems in the cultivation of talents. For example, the concepts of talent cultivation and human resource management lag behind social practice, and the teaching mode of talent cultivation lags behind the development of practice.

5 Conclusion

On the basis of summing up multiple experiences and scientific research and judging the five new strategic tasks in the new era, applied universities need to maintain integrity and innovation in their own school positioning and service orientation, make clear and clear plans for cultivating talents, and be brave to surpass history and serve the country. Local supply of outstanding talents. From the perspective of the "five new" strategy, we should provide new ideas for the development of applied universities, make them an indispensable part of human resources, and provide new ideas for the development and management of human resources, and then develop into a human resources development and management system new model.

Acknowledgments. The scientific research projects of education department of Shaanxi: (research on the driving factors of talent vitality and talent retention mechanism in Xi Xian New Area under the "Five New" strategy) project number: 20JK0078.

References

1. Djedovic, A., Karabegovic, A., Avdagic, Z., et al.: Innovative approach in modeling business processes with a focus on improving the allocation of human resources. Math. Probl. Eng. **24**(2), 1–14 (2018)
2. Reis, L.F., da Costa Alves, F.J.: Brazilian sugarcane agro-industry human resources' management: strategies to increase work intensity. Gesto Produo **27**(2) (2020)

3. Parsehyan, B.G.: İnsan Kaynaklar Ynetiminde Dijital Dnüüm: K 4.0 [Digital transformation in human resources management: HR 4.0]. Turk. Stud. – Inform. Technol. Appl. Sci. **15**(2), 211–224 (2020)
4. Iarnien, R., Suprikien, R., Čiutienė, R., et al.: Managing human resources in nursing: the relationship of nurses' working time and patients' independence level. J. Bus. Econ. Manage. **20**(1), 192–207 (2019)
5. Sari, L.G.E., Yuesti, A., Sudja, I.N., et al.: Analysis of budget planning, competence of human resources and implementation of the Viii Lldikti region budget with organizational commitment as a moderation. Int. J. Contemp. Res. Rev. **10**(2), 21273–21288 (2019)
6. Bene, J., Hadzsiev, K., Melegh, B.: Role of carnitine and its derivatives in the development and management of type 2 diabetes. Nutr. Diabetes **8**(1), 8 (2018)
7. Enomoto, M., Kimura, F., Yamaguchi, T.: A crossountry study on the relationship between financial development and earnings management. J. Int. Financ. Manage. Acc. **29**(2), 166–194 (2017)
8. Chen, Y., Hong, W., Fang, Y.: Role of biological rhythm dysfunction in the development and management of bipolar disorders: a review. Rmd Open **33**(1) (2020)
9. Silva, W.R.D., et al.: Tripartite Influence Scale (TIS) applied to university students: validation study and application. Cadernos Saude Publica **35**(3), e00179318 (2019)
10. Alfrehat, B.M.: the efficiency of the psychological counseling field training program at Yarmouk University from seen by counselor trainees. Int. J. Res. Educ. **43**(1), 1 (2019)

Application of Image Feature Classification Method Based on Deep Learning

Jinhua Yang(✉)

College of Science and Technology, Dianchi College of Yunnan University, Kunming 650228, Yunnan, China

Abstract. This article mainly takes feature learning and classification as the main theme of the subject, introduces the method of single-level feature learning and the method of multi-level feature learning and classification. This paper mainly proposes the characteristics of learning discriminative layer by layer using a technical method based on manifold subspace learning, and combines this technical feature with convolutional networks to form a discriminative local registration network (DLANet). In this paper, DLANet feature is selected as a low-level feature and applied in the traditional machine learning framework. In the scene filter classification experiment based on the DLANet feature, it is found that the best scene filtering result is 85.14% when the number of scene filters is 9. The DLANet feature not only goes beyond the SIFT feature, but also goes beyond the PCANet and LDANet features. DLANet feature is an effective and robust feature on RGBD data.

Keywords: DL · Image features · Feature classification · Classification methods

1 Introduction

The information content contained in the image database is rich and colorful, and it has been widely used in all walks of life with its simple and intuitive functions and characteristics [1, 2]. With the popularity and application of machine learning, data mining, big data and modern artificial intelligence in modern computer applications, how to use high-performance computers and these new technologies to perform a large number of visual image feature recognition and classification operations has become an urgent need for exploration and research one of the topics [3, 4]. DL techniques and methods provide new opportunities for image feature recognition and classification. DL technology occupies a high position and role in the field of image classification technology [5, 6]. Especially in the era of big data, the close integration of image classification and DL technology has important practical significance and guiding value. Based on the full analysis of the theory of DL and its development status, this paper has thoroughly studied two typical DL network models, deep belief networks and convolutional NNs [7, 8].

B. J. Jansen et al. (Eds.): *International Conference on Cognitive based Information Processing and Applications (CIPA 2021)*, LNDECT 85, pp. 711–717, 2022.
https://doi.org/10.1007/978-981-16-5854-9_91

2 Research on Image Feature Classification Method Based on DL

2.1 Feature Learning and Classification Methods

Single-Layer Feature Learning Method. (1) Restricted Boltzmann machine model.The restricted Boltzmann machine model is a connection relationship. From the point of view of NN, this is a single-layer NN connection, that is to say, it can only be connected between input and output and cannot be connected between input and input, and between output and output [9]. From the perspective of the probability graphic model, this is an undirected probability graphic model [10]. (2) Automatic coding machine.The goal of the autoencoder is to reconstruct the original data through a feature transformation. In the framework of the automatic encoding machine, a parameterized feature extraction function is first explicitly defined. This function is denoted by f_θ, which is called an encoder, and it can calculate a feature vector from the input very directly and efficiently. (3) Sparse coding.On a set of unlabeled data sets, sparse coding hopes to obtain a set of over-complete basis vectors, so that all data can be expressed by a linear combination of this set of basis vectors, and hope that this expression uses fewer basis vectors It can be achieved. (4) Subspace learning.The dimensionality reduction algorithm transforms the metric of a high-dimensional space into a low-dimensional subspace through spectral analysis. It can be used to discover the essential structure of data in the original high-dimensional space and solve the problem of dimensionality catastrophe, so it can also be called subspace learning. Dimensionality reduction algorithms can generally be divided into two categories: traditional linear dimensionality reduction algorithms and learning-based manifold algorithms.

Multi-Level Feature Learning and Classification Method. (1) Deep Belief Network.The deep belief network mainly uses the Boltzmann computer with restricted technical conditions to predict and train the NN system layer by layer without supervision, and achieves the goal of analyzing and learning the basic characteristics of different levels of knowledge and language expression. The biggest advantage of this training method is that they can reasonably and fully utilize a large amount of unlabeled data for networked training. Unsupervised pre-training can make the entire deep network be in a better place when it is initialized. It is easier to get a better solution than the initialization obtained by the restricted Boltzmann machine pre-training through random initialization, without getting stuck in the local area. (2) Cascading automatic coding machine. The stacked autoencoder uses an autoencoder to replace the restricted Boltzmann machine to achieve unsupervised pre-training layer by layer. Every time you train an autoencoder, you need to use the parameters of the data itself as the training target to learn the transformation parameters. When the training is over, the connections and layers of the decoding part will be discarded, so the middle coding layer can be used as the input of the next level of automatic encoding machine. Such a loop can construct a deep structure by stacking the automatic program coding machine. (3) Deep Convolutional Network. The convolution operation is defined based on the two-dimensional structure of the image, which defines that each underlying feature of the local receptive field is only related to a subset of the input. The topological local restrictions in the convolutional layer will make the weight matrix very sparse, so the two layers connected by the convolution operation have only local connections. Calculating such matrix multiplication is more convenient and efficient than calculating a dense matrix multiplication. In addition, a smaller number of free parameters will make statistical calculations more beneficial.

2.2 Overview of DL

DL is considered a branch of modern machine learning. It is based on a multi-level representational learning model, and each level has corresponding features, factors or concepts in specific physics. The high-level concepts are based on low-level concepts, and the low-level concept itself helps define multiple high-level concepts. Deep NN can build a multi-level and complex original data feature display model. Unlike other traditional model design methods, deep NNs do not have to carry out a complicated artificial feature design process, which greatly simplifies feature design.

2.3 Convolutional NN

A convolutional NN is a multi-layered sensory layer, the buffer layer is composed of multiple 2D planes, and each level is composed of multiple neurons. The grid takes the original image as direct input, avoiding image preprocessing and data reconstruction. Detailed information can be extracted with a partial view. You can reduce the number of neurons by distributing weights, reducing network complexity, and reducing training time. It can reduce feature dimensionality by pooling downsampling, delete redundant information, and keep features unchanged in the case of translation, rotation, and scaling.

2.4 DLANet

DLANet uses the PCANet architecture. Here, PCANet learns the torsion filter bank by analyzing the main components. However, flaws in the principal component analysis affect network performance. To avoid these shortcomings and improve performance, this paper creates a DLANet network with identifiable local registrations.

First-level discriminative local registration (DLA).The output feature map of this layer is:

$$X_{i,d_1}^1 = X_i * K_{d_1}^1, \quad d_1 = 1, 2, \cdots, D_1 \tag{1}$$

Where $*$ represents a two-dimensional convolution with edge filling. This operation will keep the size of the image unchanged. In order to build a deeper network, this output feature map is input to the second DLA layer.

Second layer DLA. The output feature map of this layer is:

$$X_{i,d_1,,d_2}^2 = X_{i,d_1}^1 * K_{d_2}^2, \quad d_2 = 1, 2, \cdots, D_2 \tag{2}$$

Finally, there will be D_1D_2 output feature maps in the second layer. These feature maps can be used as input to the third DLA layer and can be used to build a deeper network.

Feature layer. Unlike ordinary convolutional NNs that directly pull the output feature map into the feature vector, a special block-by-block histogram feature is connected to the last DLA layer to convert the output feature map into a feature.

3 Application Experiment of Image Feature Classification Method Based on DL

3.1 Scene Classification Experiment Based on Dlanet Features

Experimental Data. In a scene classification experiment based on identifiable locally registered network function, this paper verifies the DLANeT function using three sets of scene data. These three datasets are the SIRI-WHU dataset, RSSCN7 dataset, and NWPU dataset. The SIRI-WHU dataset contains 12 categories of scene images, each category has 200 images, and each image has a pixel size of 200 * 200 and a spatial resolution of 2 m. The RSSCN7 dataset contains seven typical scene classes, each containing 400 images, and each image has a 400 * 400 pixel size. The NWPU dataset contains a total of 31,500 images of 256 * 256 pixels, covering 45 scene categories. There are 700 images in each category.

Data Preprocessing. Microsoft Kinect collects the depth images of all the images in the RSSCN7 data set, and fills in the blank areas in the bottom depth images, and uses a bilateral filter to smooth the noise areas to compensate for the vacancy of the infrared imaging jitter and light effects in Kinect. In addition, all images are converted into grayscale images, and all features are extracted on this grayscale image for comparison.

Experimental Design. In the experiment, the extracted DLANet features are used as basic features and put into the local constrained linear coding-spatial pyramid matching (LLC-SPM) recognition framework. The method proposed in this paper is compared with some widely used manual functions (such as SIFT, HOG, GIST) and some learnable features (such as PCANet and LDANeT). On the RSSCN7 data set, we spliced the features of the color image with the features of the depth image to obtain the final expression of the RGBD image pair. The last linear support vector machine is trained as a classifier, which is more suitable for the features generated by LLC-SPM. In addition, on the SIRI-WHU and NWPU datasets, this paper uses an enhanced Fisher vector, a Gaussian mixture model with 256 centers and two-layer average aggregation.

Experimental Method. When using the RSSCN7 dataset as the target of the experiment, repeat the experiment 10 times, similar to the general standard processing procedure. In each experiment, 100 images of each type are randomly selected as training samples, and the remaining images are used as test samples. Training sample sets are used to learn about features such as PCANet, LDANet, and DLANet. This experiment gives the mean and standard deviation of the recognition rates. When using the SIRI-WHU dataset as the experimental subject, the experiment is repeated 10 times, and each trial randomly selects 150 images of each type and divides them into a training set and a test set. The SIRI-WHU dataset contains 12 categories of scenes suitable for testing scene classification tasks. Therefore, in this experiment, we will not only compare the DLANet function suggested in this article to other functions, but also compare the best results with other methods such as VQ_SPM, ScSPM, Macro Function, FV-LRF.

When using the NWPU dataset as the target of the experiment, the experiment is repeated 10 times, and each experiment randomly selects 260 images of each type as training samples, and the remaining images are used as test samples.

4 Application Experiment Analysis of Image Feature Classification Method Based on DL

4.1 Scene Classification Based on DLANeT Features——Experimental Analysis of RSSCN7 Data Set

When the RSSCN7 dataset is used as the experimental object, the average accuracy and standard deviation of SIFT, HOG, GIST, PCANet, LDANeT and DLANet features on color images, depth images and color plus depth images are analyzed. The analysis results are shown in Table 1.

Table 1. Comparison of the accuracy of different features on the RSSCN7 data set

	Color	Depth of Field	Color + Depth of Field
SIFT	78.5 ± 1.6	68.23 ± 1.52	79.23 ± 1.6
HOG	74.01 ± 1.69	61.68 ± 1.79	75.94 ± 1.59
GIST	62.98 ± 1.30	49.26 ± 2.59	70.03 ± 1.61
PCANet	79.31 ± 1.49	73.01 ± 1.69	81.24 ± 1.47
LDANeT	77.03 ± 1.59	71.01 ± 1.74	80.01 ± 1.67
DLANet	81.45 ± 1.41	71.76 ± 1.54	82.71 ± 1.28

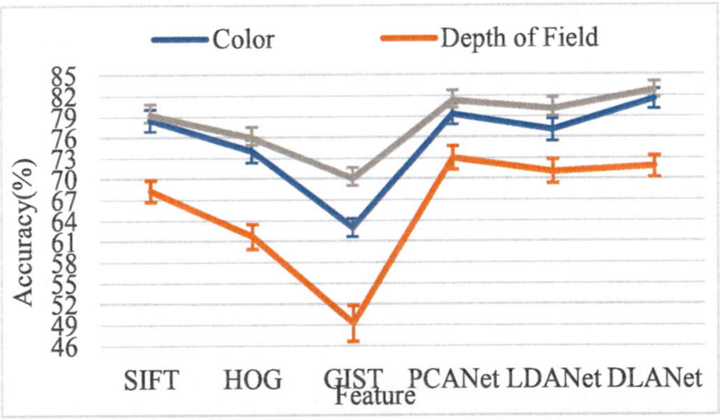

Fig. 1. Comparison of the accuracy of different features on the RSSCN7 data set

As shown in Fig. 1, among the three manual functions SIFT, HOG and GIST, the SIFT function has the best results. GIST is a lot worse than the others, with an average resolution of just 49.43%. For color images except for LDANet, the resolution of the other learnable features is slightly better than the three manual features, but for deep images, LDANet is 2.7% ahead of SIFT. The standard deviation of the DLANet recognition rate is too low. For all the features, the recognition rate can be greatly improved by incorporating color image features and image depth features, but there is no recognition rate particularly good when only using the image depth features. The above results show that the depth information can improve the accuracy of scene classification tasks. The DLANet feature proposed in this white paper not only exceeds the SIFT feature, but also that of PCANet and LDANet. DLANet is a powerful and powerful RGBD data feature.

4.2 Scene Classification Based on DLANet Features-Experimental Analysis of SIRI-WHU Data Set

When the SIRI-WHU data set is used as the experimental object, the accuracy of SIFT features is used as the benchmark accuracy, and the influence of the number of filters on the accuracy of DLANet, PCANet and LDANeT features is analyzed. The analysis results are shown in Table 2.

Table 2. The Accuracy of the SIRI-WHU data set varies with the number of filters

Number of Filters	SIFT	LDANet	DLANet	PCANet
5	82.25 ± 0.67	83.25 ± 0.76	83.76 ± 0.41	80.46 ± 0.37
6	82.25 ± 0.67	83.51 ± 0.34	84.26 ± 0.48	81.51 ± 0.40
7	82.25 ± 0.67	83.76 ± 0.51	84.31 ± 0.61	81.48 ± 0.64
8	82.25 ± 0.67	83.69 ± 0.62	84.43 ± 0.51	81.74 ± 0.31
9	82.25 ± 0.67	83.63 ± 0.47	85.14 ± 0.18	81.80 ± 0.15
10	82.25 ± 0.67	84.71 ± 0.58	85.12 ± 0.43	81.96 ± 0.31
11	82.25 ± 0.67	83.59 ± 0.67	84.97 ± 0.18	82.41 ± 0.16
12	82.25 ± 0.67	84.21 ± 0.46	84.89 ± 0.27	82.64 ± 0.49

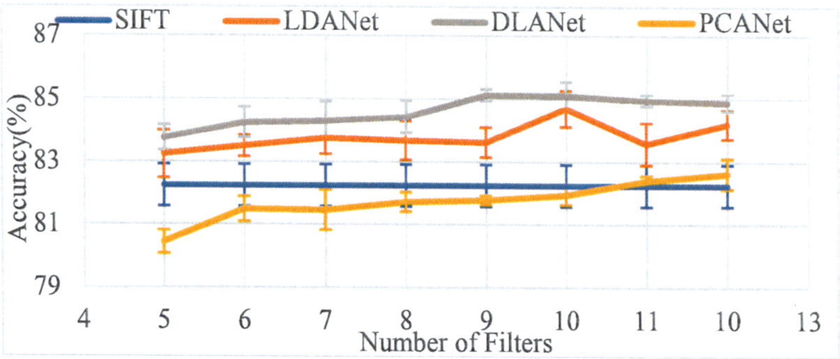

Fig. 2. The Accuracy of the SIRI-WHU data set varies with the number of filters

As shown in Fig. 2, when the number of filters varies from 5 to 12, the three learnable features have achieved their best results. The DLANet feature achieves the best result of 85.14% when the number of filters is 9, while the LDANet feature achieves the best result of 84.71% when the number of filters is 10. When the number of filters is less than 11, the recognition rate of PCANet features is lower than that of SIFT, but as the number of filters in the first layer of DLA increases, the recognition rate gradually increases. But no matter how the number of filters changes, the recognition rate of DLANet feature and LDANet feature is higher than that of SIFT feature. The above result identification information is very useful when learning features.

5 Conclusions

This article focuses on the feature learning problem in image classification along the route of extending from single-layer feature learning to multi-level feature learning and applying deep feature learning methods to practical problems. It is found that the learnable DLANet feature is better than other manual features. It is also superior to similar PCANet features and LDANet.

Acknowledgements. This work was supported by Scientific research project of Dianchi College of Yunnan University-"Intelligent campus system based on face recognition (Project's number: 2020XYB03)".

References

1. Pan, B., Shi, Z., Xia, X.: R-VCANet: a new deep-learning-based hyperspectral image classification method. IEEE J. Selected Topics Appl. Earth Obs. Remote Sens. **PP**(99), 1–12 (2017)
2. Gong, M., Yang, H., Zhang, P.: Feature learning and change feature classification based on DL for ternary change detection in SAR images. ISPRS J. Photogramm. Remote Sens. **129**(jul.), 212–225 (2017)
3. Yang, B.Q., Guan, X.P., Zhu, J.W., et al.: SVMs multi-class loss feedback based discriminative dictionary learning for image classification. Pattern Recogn. **112**(12), 107690 (2020)
4. He, C., Zhang, Q., Qu, T., et al.: Remote sensing and texture image classification network based on DL integrated with binary coding and sinkhorn distance. Remote Sens. **11**(23), 2870 (2019)
5. Prabhanjan, S, Dinesh, R.: DL approach for Devanagari script recognition. Int. J. Image and Gr. **17**(03), 1750016 (2017)
6. Demirci, M.Y., Beli, N., Gümüü, A.: Efficient deep feature extraction and classification for identifying defective photovoltaic module cells in electroluminescence images. Expert Syst. Appl. **175**(2), 114810 (2021)
7. Wen, T., Zhang, Z., Qiu, M., et al.: A two-dimensional matrix image based feature extraction method for classification of sEMG: a comparative analysis based on SVM, KNN and RBF-NN. J. Xray Sci. Technol. **25**(2), 287 (2017)
8. Jia Sen, W., Kuilin, Z.J., et al.: Superpixel-level Gabor feature fusion method for hyperspectral image classification. J. Nanjing Univ. Inf. Technol. **010**(001), 72–80 (2018)
9. Zhang, L., Jiao, L., Ma, W., et al.: PolSAR image classification based on multi-scale stacked sparse autoencoder. Neurocomputing **351**(JUL.25), 167–179 (2019)
10. Ren, Z., Hou, B., Wen Z., et al.: Patch-sorted deep feature learning for high resolution SAR image classification. IEEE J. Selected Topics Appl. Earth Obs. Remote Sens. **PP**(9), 1–14 (2018)

The Construction of the Virtual Simulation Experiment Platform Under the Background of Education Informatization

Zhihui Lin(✉)

Xi'an Siyuan University, Xian 710038, Shaanxi, China

Abstract. Under the guidance of the strategy of national informatization development, the application of modern information technology in the area of education and teaching is increasingly deepening. This has effectively promoted profound changes in educational concepts, educational models, learning and teaching methods. Virtual simulation teaching conforms to the development tendency and realistic needs of education informatization. It will definitely have a far-reaching and positive impact on the deepening of experimental teaching reform and the promotion of education quality. This paper studies the construction of the virtual simulation experiment platform. It explores the meaning, ideas, and implementation of platform construction. It aims to promote the quality of experimental teaching and practical education in universities and colleges under the background of information technology. The virtual simulation experiment platform is a completely fresh type of experimental teaching system based on network technology, multimedia technology and computer simulation technology. It uses a computer to simulate a real experimental environment and operates experimental equipment through an information network, which furnishes a brand-new teaching mode for experimental lecturing.

Keywords: Educational information · Virtual simulation · Experimental platform

1 Preface

The Ministry of Education clearly stated in the "Ten-Year Development of Education Informatization (2011–2020)": "Take education informatization as a strategic focus and priority area of national informatization in a comprehensive deployment and speed up implementation. It will take about ten years to initially establish an educational information system with Chinese characteristics. Make China's education informatization approach the international advanced level as a whole, and promote the scientific development of education."

The "Development Plan Outline and National Medium and Long-term Educational Reform (2010–2020)" puts forward: "By 2020, an education information system covering all types of schools at all levels in rural and urban areas will be mainly completed.

B. J. Jansen et al. (Eds.): *International Conference on Cognitive based Information Processing and Applications (CIPA 2021)*, LNDECT 85, pp. 718–724, 2022.
https://doi.org/10.1007/978-981-16-5854-9_92

The modernization of educational content, teaching methods and other methods will be promoted. Advanced technologies and high-quality resources will be fully utilized, management models and operating mechanisms will be innovated, and subsistent resources will be counted. Build efficient, practical and advanced digital education institution. Improve the construction of digital campus". At the same time, it is required to "enhance the application and development of high-quality educational resources, enhance the construction of a resource system for online teaching, combine international high-quality digital teaching resources, develop online learning courses, and build virtual laboratories and digital libraries."

In 2013, the Higher Education Department of the Ministry of Education issued the "Notice on Carrying out the Construction of a National Virtual Simulation Experimental Teaching Center". This marks the official start of the construction of the virtual simulation experiment teaching center. In 2017, the "Notice on the Construction of Demonstrative Virtual Simulation Experiment Teaching Projects from 2017 to 2020" issued by the General Office of the Ministry of Education stated that: (The Ministry of Education) based on the current layout and development of professional undergraduate disciplines, combined with the construction of professional laboratories in colleges and universities and the development needs of professional experimental teaching informatization, has planned the following goals as a whole:. By 2020, about 1,000 demonstrative virtual simulation experiment teaching projects will be recognized.

At present, China's virtual simulation experiment teaching work has transitioned from "central construction" to "project construction", and from "partial application" to "open sharing". The transformation of virtual simulation experiment teaching construction methods and construction goals reflects the urgent requirements of Chinese universities for the development of experimental teaching itself. At the same time, it is of great meaning to improve the construction and application of high-quality experimental teaching resources in universities and colleges [1–3].

2 Significance of Virtual Simulation Experiment Platform Construction

Virtual Reality: Simulation technology is the technology that only needs one system to imitate another real system. Virtual simulation is actually a computer system that can create and experience a virtual world (Virtual World). Virtual simulation experiment teaching is a real experimental teaching scenario constructed by using virtual simulation technology. The model of real system simulation through computer human-computer interaction technology allows students to complete experimental courses and independent learning. At the same time, it is also a vital content of higher education information and experimental teaching construction, a product of the information technology and the deep integration of disciplines, and the development trend of higher education in the 21st century [5]. It furnishes a brand-new teaching mode for experimental teaching in the information age. Not only can it reproduce the whole process of the real experiment, but also overcome the various constraints and drawbacks of traditional experiments. Effectively solve many problems existing in the current experimental teaching. As shown in

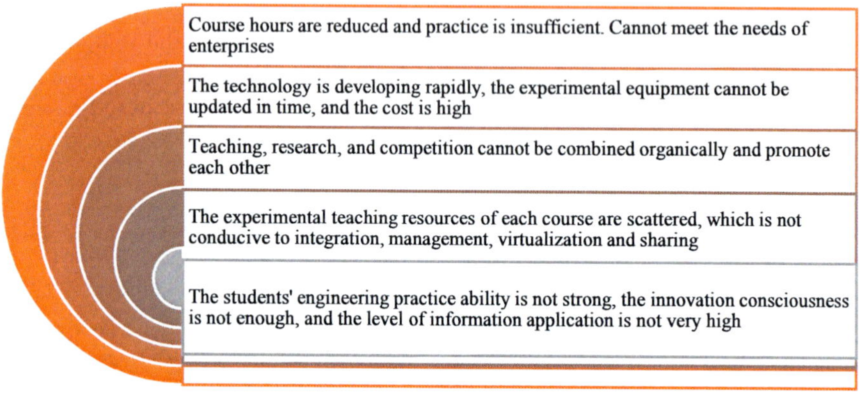

Fig. 1. Current status of experimental teaching

Fig. 1. To achieve the goal of enhancing educational resources and effectively making the quality of teaching better.

Combining with national policies and the reality of universities, it is essential to create an online virtual simulation experiment platform. It can be publicly released and integrated with the educational administration system, providing an open virtual simulation experiment teaching management platform and an experiment resource sharing platform. At the same time, it provides the online virtual experiment teaching auxiliary functions, provides services for the virtual experimental teaching environment and develops practical teaching applications. The platform aims to promote a new, personalized, intelligent, and ubiquitous experimental teaching mode that combines online and offline colleges and universities, form a virtual simulation experiment teaching platform for higher education informatization with reasonable professional layout, excellent teaching effect, open sharing and effective and support the overall improvement of the teaching quality of higher education [4].

Using this platform to carry out experimental teaching can significantly reduce the expenditure of experiments, expand the scope of benefits, and solve problems such as the sharing of high-quality experimental resources. The construction of the virtual simulation teaching platform aims to provide a comprehensive, shared and exemplary platform for professional construction, teaching team construction, and course construction. Through the combination of hardware and software construction and the integration of a variety of advanced technologies, a single advantage is turned into a system advantage, providing comprehensive teaching resources and resource sharing. At the same time, it reduces the purchase of hardware resources and reduces costs [7].

3 Ideas for the Construction of the Virtual Simulation Experiment Platform

The platform integrates advanced virtual simulation software, virtual simulation course resources, multimedia and network communication technologies. The application of virtual simulation technology to construct real experimental teaching scenarios, and the simulation of real experimental models through computer human-computer interaction technology can allow students to achieve task learning and autonomous learning in experimental courses [6].

The platform usually relies on the school's campus network and can be deployed in the same or different computer room with the school's portal website, the educational administration system and the third-party software system. Students, teachers and teaching assistants can access the experiment sharing platform online to complete corresponding teaching tasks and teaching assistants.

The architecture of the virtual simulation experiment platform is shown in Fig. 2.

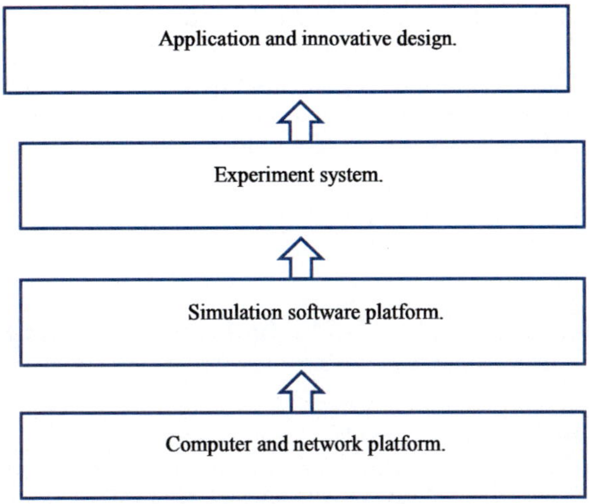

Fig. 2. Experimental platform architecture

(1) Computer and network platform
The platform can be applied to a local area network environment or a campus network environment. The former is aimed at sharing within a local scope. For example, laboratories or second-level teaching units. This mode does not require high network and server requirements. Only need to designate a PC as the LAN server; the latter is aimed at sharing in a wide area. All teachers and students of the school can use the platform. At this time, the server must use a professional server and be equipped with a network server operating system. It meets the needs of different types of clients to access at any time [8].

(2) The Simulation software platform
The core of the simulation software platform is mainly simulation software. Experimental teaching simulation software can be developed by teachers in the school or provided by a third party. Usually implemented by programming languages such as Java. The platform should provide an interface for the experimental teaching design module in the design, which is used to integrate the existing simulation software into the open virtual simulation experimental teaching management and the sharing platform.

(3) The experimental system
The system can be designed and developed according to actual teaching needs. Through the combination of the simulation design software and the experimental system, abstract theoretical teaching can be concretized and visualized. It improves students' interest in learning, but also exercises students' independent learning and application innovation capabilities. At the same time, it can effectively improve the efficiency and effect of the experiment, so that students can deeply understand the complete process from theory to practice.
(4) Application and innovative design
This part can be carried out according to the needs of the experiment. It can be divided into three levels: Namely: The basic experimental layer, to cultivate students' operational verification, analytical thinking ability; the comprehensive experimental layer, cultivate students' comprehensive design, solve and expand abilities; The innovative experimental layer, cultivating students' innovative thinking and engineering practice abilities. Systematized hierarchical teaching, focusing on quality and sustainability of teaching resources. Can attract students to participate in scientific research projects, corporate projects, and various design competitions. Realize a unified, open and shared comprehensive online service platform.

4 Application Examples of the Virtual Simulation Experiment Platform

Take the course design "monolithic water level control system" as an example. The traditional development progress is to purchase the device after the schematic design is completed, and then solder and debug. All software debugging and system debugging are based on the foundation of the hardware board. Before the hardware is completed, the corresponding software development and debugging cannot be carried out. And any error in any link will cause the hardware to be redesigned. The design efficiency is low, and manpower, material resources, financial resources, time, etc. are wasted.

Using the virtual simulation experiment platform, all hardware design debugging, software development, and system verification can be carried out on the simulation platform. Modifying hardware design errors is as simple as modifying software. The virtual simulation diagram of the water level control system is clearly exhibited in Fig. 3.

The students will complete the actual production, after the designed system is verified on the virtual simulation experiment platform, In this way, there will be basically no major problems. Figure 4 shows the actual system made by students. It achieves the functions and indicators required by curriculum design.

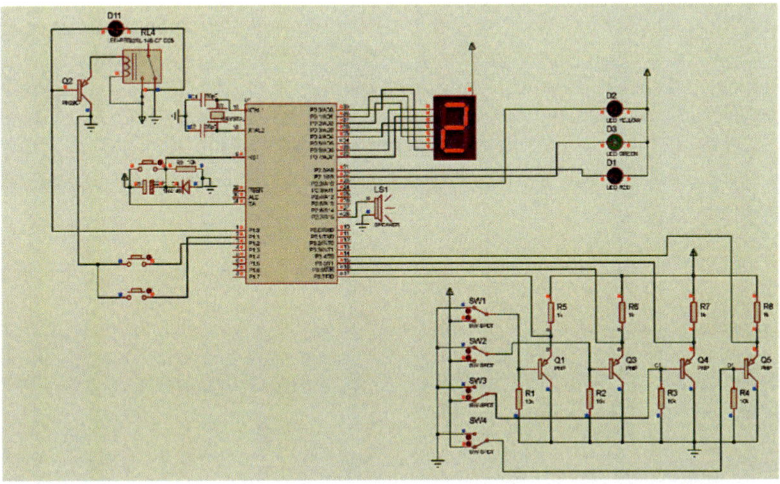

Fig. 3. The virtual simulation diagram of the water level control system

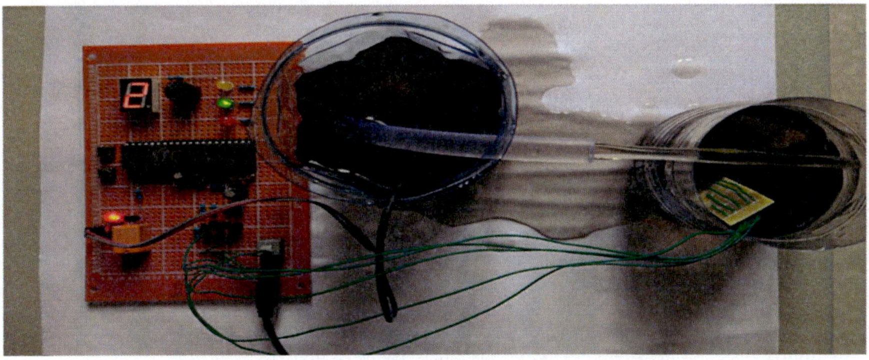

Fig. 4. Physical map of the water level control system

5 Conclusion

The virtual simulation practice teaching platform constructed by research is an advanced virtual simulation laboratory with network, information, virtualization, openness and economy. It can realize software sharing, instrument sharing and resource sharing. Make learning and experiment no longer limited by space and time. Students can do research and experiment anytime and anywhere. At the same time, it can also lay a solid foundation for participating in subject competitions, participating in corporate engineering projects, applying for patents, and writing research papers [9, 10].

In addition, because of the uneven regional economic development in China, the distribution of educational resources is uneven. It is extremely difficult for some universities and colleges to establish a good practice environment and high-quality teaching resources in a short time. Therefore, relying on the popularization of Internet applications, use emerging technologies such as the big data, Internet of Things, artificial intelligence, and cloud computing to promote the construction and application of virtual simulation platforms. Carrying out virtual simulation experiments in combination with reality and virtuality will greatly expand the benefits of practical courses and become a very rapid, effective and worthy of promotion of the radiation form of educational resources.

Fund Projects. 1. The 13th five-year plan of Educational Science in Shaanxi Province in 2020 (SGH20Y1434)

2. Research project of education and teaching reform in Xi'an Siyuan University (19SYGP002)

3. National and Provincial College Students' innovation and entrepreneurship training project of Xi'an Siyuan University in 2020 (S202013121004S, S202013121046)

References

1. Jin, X.: Summary of the construction of the national virtual simulation experiment teaching center in 2013 and recommendations for the application in 2014. Lab. Res. Explor. **33**(8), 1–5 (2014)
2. Hongqi, X.: New era teaching characteristics of the national virtual simulation experiment teaching project. Exp. Technol. Manag. **36**(9), 1–4 (2019)
3. Chun, S.: The distribution characteristics of the national virtual simulation experiment teaching center and the project proposal. Exp. Sci. Technol. **17**(2), 146–151 (2019)
4. Wei, Z., Jialu, Z.: Construction and practice of virtual simulation experiment teaching resource support platform. China Educ. Inf. **13**, 57–60 (2020)
5. Sen, W., Ping, L.: Analysis of the national virtual simulation experimental teaching center in 2014. Lab. Res. Explor. **35**(4), 82–86 (2016)
6. Wu, P., Yang, J., Jiang, Q.: Construction and practice of circuit experiment teaching combining virtual and real. J. Electr. Electron. Educ. **43**(02), 148–152 (2021)
7. Huiying, C., Mingxuan, C.: Research on the construction and development of digital education resources in the intelligent era. Mod. Distance Educ. **183**(3), 74–81 (2019)
8. Feng, S., Lanpeng, G.: Construction of virtual simulation platform for electronic communication majors in agricultural colleges and universities. Educ. Teach. Forum **43**, 252–254 (2018)
9. Wu, P., Yang, J., Jiang, Q.: Circuit experiment teaching construction and practice combining virtual and real. J. Electr. Electron. Educ. **43**(2), 148–152 (2021)
10. Liu, H., Shen, J., Wang, G.: Construction of virtual simulation experiment teaching resource platform from the perspective of engineering education. Exp. Technol. Manag. **36**(12), 19–22 (2019)

Application of G1-EWM-TOPSIS Method Based on MATLAB in the Evaluation of Social Contribution Benefits of Investment in a Certain Provincial Power Grid

Yan Lu[1,2](✉), Ke Lv[1,2], Jingzhong Yuan[1,2], Hongjian Li[1,2], Luofei Wu[3], and Sha Peng[3]

[1] Beijing Jingyan Electric Power Engineering Design Co., Ltd., Beijing, China
[2] State Grid Jibei Electric Economic Research Institute, Beijing, China
[3] School of Economics and Management, North China Electric Power University, Beijing, China

Abstract. With the acceleration of New Electric Power System Reform, the power grid is in the period of historical opportunity development, in the period of the most effective release of various advantages and potential, the comprehensive evaluation of the social contribution benefits of the provincial power grid investment has become an important research topic. In this paper, by constructing the social benefit evaluation index system of investment in the provincial power grid, establishing the G1-EWM-TOPSIS evaluation model, and using MATLAB to realize the high-speed calculation of the evaluation model, it provides powerful technical help for the social benefit evaluation of investment in the provincial power grid.

Keywords: MATLAB · G1-EWM-TOPSIS · Power grid benifits

1 Introduction

China's power grid is in the important construction period of "world first-class power grid" and "international middle and high-end power grid". The demand for power grid development continues to grow, and the demand for investment is still strong [1]. At present, the evaluation of power grid enterprise investment mostly focus on economic contribution benefits, ignoring the social contribution benefits generated by the public welfare nature of the power grid enterprise itself [2]. It is urgent to combine the characteristics of the power grid enterprise investment in the province region and conduct a specific assessment of its investment social contribution benefits [3].

B. J. Jansen et al. (Eds.): *International Conference on Cognitive based Information Processing and Applications (CIPA 2021)*, LNDECT 85, pp. 725–734, 2022.
https://doi.org/10.1007/978-981-16-5854-9_93

2 Evaluation Index of Social Benefit of Investment in the Provincial Power Grid

Combined with the characteristics of the provincial power grid, this paper summarizes the specific indicators reflecting the social contribution benefits of investment from the social economic benefits, social environmental benefits and social stability benefits, and quantifies the relationship between the social benefits and investment with the help of the input-output relationship [4].

(1) Social economic benefits
The driving influence of power grid investment on national economy refers to the influence of power grid investment activities on economic growth, which is driven by multiplier effect. According to the basic relation formula (2-1) of input and output, the relation formula (2-2) of power grid investment in this province on various industries can be obtained [5].

$$\boldsymbol{A}_{n+1}\boldsymbol{X}_{n+1} + \boldsymbol{Y}_{n+1} = \boldsymbol{X}_{n+1} \tag{2-1}$$

$$\Delta \boldsymbol{X} = (\boldsymbol{I} - \boldsymbol{A}_{n+1})^{-1} \Delta \boldsymbol{Y} \tag{2-2}$$

Where: A_{n+1} is the $n+1$ order direct consumption coefficient matrix added to the consumption sector. $\boldsymbol{Y}_{n+1}$ is the column vector of the final product. $\boldsymbol{X}_{n+1}$ is the $n+1$ order total product column vector added to the consumer sector. $\Delta \boldsymbol{X}$ is the output column vector of each industry. $\Delta \boldsymbol{Y}$ is the exogenous investment vector of the provincial power grid.

1) Additional gross production output for investment
The influence of grid investment on national economy in northern Hebei Province not only has direct impact on GDP but also indirect impact of other industries, which is mainly reflected in the positive impact of power grid investment on economic growth and industrial structure adjustment.

$$\Delta \boldsymbol{GDP} = \boldsymbol{A}_{nj} \cdot \left[(\boldsymbol{I} - \boldsymbol{A}_{n+1})^{-1}\right]_{n \times n} \Delta \boldsymbol{Y} \tag{2-3}$$

2) Additional labor remuneration for investment
In addition to providing more jobs to the society, the investment and construction of the power grid can also improve the average wage level of the employed people. The additional labor remuneration for investment driven by investment is the comprehensive result of the increase in the number of labor employed and the increase of per capita labor remuneration level generated by investment.

$$\Delta \boldsymbol{V} = \boldsymbol{A}_{tj} \cdot \left[(\boldsymbol{I} - \boldsymbol{A}_{n+1})^{-1}\right]_{n \times n} \Delta \boldsymbol{Y} \tag{2-4}$$

3) Additional taxes for investment

Additional investment first leads to economic growth and increases national income by increasing demand and improving production capacity. The increase of GDP is the source and driving force of tax growth.

$$\Delta T = A_{tj} \cdot \left[(I - A_{n+1})^{-1}\right]_{n\times n} \Delta Y \tag{2-5}$$

(2) Social environmental benefits

The provincial power grid has long valued and actively responded to the call for energy conservation and emission reduction, vigorously promoted the coal-to-electricity projects and the installation and grid connection of wind power, photovoltaic and other new energy. Coal consumption and carbon dioxide emissions have been reduced while the final energy consumption structure has been optimized [6].

1) Reduction of standard coal consumption for investment

The sources of reduction of standard coal consumption are mainly divided into two parts. The first part the standard coal consumption saved due to the integration of new energy installations to the grid instead of traditional thermal power generation, which can be calculated by multiplying the new energy generation capacity of the annual investment. The second part is the standard coal consumption which can be calculated by multiplying the replacement amount of electricity added in the annual coal conversion project by the conversion coefficient of standard coal.

$$\Delta R = \Delta R_1 + \Delta R_2 = \Delta Q_1 \times K_s + \Delta Q_2 \times K_s \tag{2-6}$$

2) Reduction of carbon dioxide emissions for investment

The sources of carbon dioxide emission reduction are mainly divided into two parts: one is the carbon dioxide emission reduction calculated by multiplying the new energy generation amount invested each year by the carbon dioxide emission coefficient; the other is the carbon dioxide emission reduction calculated by multiplying the new alternative electricity amount invested each year by the carbon dioxide emission coefficient.

$$\Delta C = \Delta C_1 + \Delta C_2 = \Delta Q_1 \times K_c + \Delta Q_2 \times K_c \tag{2-7}$$

(3) Social stability benefits

1) Additional electricity consumption for investment

As an important link between the preceding and the following in the power industry, the investment of the provincial power grid will also drive the increase of demand for related products, thus driving the increase of output of power and other related industries.

$$B(b_{ij}) = (I - A)^{-1} - I \tag{2-8}$$

$$\Delta X = (I - A)^{-1} \Delta Y \tag{2-9}$$

(2) Additional electricity consumption for investment
Additional electricity consumption for investment is selected as the evaluation index of social stability benefits, and the relationship between investment and regional electricity consumption increment is expressed by using the formula (2-10), which can reflect the contribution of enterprises to social stability through investment.

$$\Delta EC = \frac{\Delta GDP \cdot EC}{GDP} \tag{2-10}$$

3 G1-EWM-TOPSIS Model

3.1 Evaluation Index Weighting

(1) G1 method to determine the subjective weight
1) Establishment of the order relationship
Set the index set for comprehensive evaluation as $\{x_1, x_2, \ldots, x_m\}$, invite experts to make subjective evaluation of the indexes in the evaluation index set $\{x_1, x_2, \ldots, x_m\}$, and determine the order relationship of the evaluation indexes: $x_1 > x_2 > \cdots > x_m$.
2) Calculation of index importance
r_j is used to represent the relative importance between x_j and x_{j-1}. The calculation formula is as follows:

$$r_j = \tau_{j-1}/\tau_j \; (j = m, m-1, \ldots, 2) \tag{3-1}$$

3) Assignment of index importance
Calculate the importance between adjacent evaluation indexes according to the above order relation, and assign r_j value rationally.
4) Subjective weight of indicators
The weight τ_j of indicators is determined by rational importance r_j as follows:

$$\tau_j = \left(1 + \sum_{k=2}^{m} \prod_{i=k}^{m} r_i\right)^{-1} \tag{3-2}$$

(2) EWM method to determine the objective weight
In the decision matrix $D = \left[x_{ij}\right]_{m \times n}$, x_{ij} represents the index value of the i th programme under the j th index attribute. Indicates the probability of the i th programme under the j th index attribute [7].

$$p_{ij} = x_{ij} / \sum\nolimits_{i-1}^{n} x_{ij}; \quad 0 \le p_{ij} \le 1; \quad \sum_{i-1}^{n} p_{ij} = 1 \tag{3-3}$$

$$e_j = -k \sum_{i=1}^{n} p_{ij} \ln p_{ij}, \; k = 1/\ln n \tag{3-4}$$

Where: k is a constant, $e_i \in [0, 1]$.

The objective weight coefficient can be normalized by the following formula:

$$v_j = g_j \Big/ \sum_{i=1}^{m} g_j \tag{3-5}$$

Where: $g_j = 1 - e_j$, which indicates the coefficient of variance of each programme.

(3) G1-EWM method to determine combination weight

$$\omega_i = \upsilon_i \tau_i \Big/ \sum_{j=1}^{m} \upsilon_i \tau_i \tag{3-6}$$

3.2 TOPSIS Model Construction Based on G1-EWM Method

(1) Construct a weighted standardized matrix
According to the final weight determined by the G1-EWM method, weighting the standardized initial evaluation matrix, the weighted standardized matrix can be obtained:

$$R = \left(r_{ij}\right)_{m \times n} = \begin{bmatrix} w_1 x_{11}^* & w_2 x_{12}^* & \cdots & w_n x_{1n}^* \\ w_1 x_{21}^* & w_2 x_{22}^* & \cdots & w_n x_{2n}^* \\ \vdots & \vdots & \cdots & \vdots \\ w_1 x_{m1}^* & w_2 x_{m2}^* & \cdots & w_n x_{mn}^* \end{bmatrix} \tag{3-7}$$

Where: x_{ij}^* is the dimensionless data of the i th sample of the index x_j, and w_j is the corresponding weight of the j th index [8].

(3) Determine the positive and negative ideal solution
On the premise that the inverse index completes the positive processing, the positive and negative ideal solution are obtained as follows:

$$r_j^+ = \max_i r_{ij},\ r_j^- = \min_i r_{ij} \tag{3-8}$$

(4) Calculate the evaluation distance
The evaluation distance between the i th group of samples and the positive ideal solution is:

$$d_i^+ = \sqrt{\sum_{j=1}^{n} (r_{ij} - r_j^+)^2} \tag{3-9}$$

The evaluation distance between the i th group of samples and the negative ideal solution is:

$$d_i^- = \sqrt{\sum_{j=1}^{n} (r_{ij} - r_j^-)^2} \tag{3-10}$$

(5) Calculate the closeness of each sample

$$C_i = \frac{d_i^-}{d_i^+ + d_i^-}, \quad (C_i \in [0, 1]) \tag{3-11}$$

Therefore, the degree of closeness reflects the excellent social contribution benefits of the provincial power grid investment. The greater the closeness, the better the social contribution benefits of corporate investment [9].

4 Realization of Social Contribution Benefit Evaluation of Investment in the Provincial Power Grid Under G1-EWM-TOPSIS Model Using MATLAB

4.1 G1-EWM Determines the Weight of Social Contribution Benefits Evaluation Indicators

According to formulas (3-1) to (3-6), the combined weights of social contribution benefits evaluation indicators are obtained. And the final weight calculation result is shown in Table 1.

Table 1. The combination weight determined by G1-EWM

The evaluation index	Subjective weight	Objective weight	Combination weight
Additional gross production output for investment (100 million yuan)	0.1544	0.1239	0.1338
Additional labor remuneration for investment (100 million yuan)	0.1287	0.1293	0.1163
Additional taxes for investment (100 million yuan)	0.1287	0.1250	0.1125
Reduction of standard coal consumption for investment (Ten thousand tons)	0.1604	0.2540	0.2849
Reduction of carbon dioxide emissions for investment (Ten thousand tons)	0.1337	0.2540	0.2374
Reduction of carbon dioxide emissions for investment (100 million yuan)	0.1337	0.0677	0.0633
Additional electricity consumption for investment (100 million kwh)	0.1604	0.0462	0.0518

4.2 Comprehensive Evaluation of TOPSIS Method

According to the TOPSIS model based on combination weighting established in Sect. 3.2, MATLAB software is used to realize the comprehensive evaluation of the social contribution benefits of the provincial power grid investment in 2014–2018, which has the advantages of convenience and efficiency [10, 11].

(1) Calculation of evaluation index data

Collect the investment-related data of the provincial power grid from 2014 to 2018, and calculate the social benefit evaluation index data of the provincial power grid investment according to formulas (2-1) to (2-10) as shown in the following table (Table 2):

Table 2. Evaluation index of social contribution benefits of investment in the provincial power grid

The evaluation index	2014	2015	2016	2017	2018
Additional gross production output for investment (100 million yuan)	276.61	256.57	236.07	209.94	393.06
Additional labor remuneration for investment (100 million yuan)	137.36	130.75	120.31	106.99	200.31
Additional taxes for investment (100 million yuan)	34.12	31.83	29.28	26.04	48.76
Reduction of standard coal consumption for investment (Ten thousand tons)	136.59	161.36	238.58	767.63	1398.90
Reduction of carbon dioxide emissions for investment (Ten thousand tons)	337.09	398.20	588.78	1894.36	3452.24
Reduction of carbon dioxide emissions for investment (100 million yuan)	162.54	184.05	208.82	237.39	270.39
Additional electricity consumption for investment (100 million kwh)	32.68	29.81	27.07	23.68	35.89

(2) MATLAB calculation code

```
function [ output_args ] = TOPSIS(A,W,M)

A=input('Preliminary evaluation matrix=')
W=input('weight=')
M=input('Column where the positive indicator is located=')
N=input('column where the negative indicator is located=')
  [ma,na]=size(A);
  for i=1:na
          B(:,i)=A(:,i)*W(i);
end
V1=zeros(1,na);
V2=zeros(1,na);
BMAX=max(B);
BMIN=min(B);
for i=1:na
      if i<=size(M,2)
      V1(M(i))=BMAX(M(i));
      V2(M(i))=BMIN(M(i));
      end
      if i<=size(N,2)
      V1(N(i))=BMIN(N(i));
      V2(N(i))=BMAX(N(i));
      end
end

for i=1:ma
     C1=B(i,:)-V1;
     S1(i)=norm(C1);

     C2=B(i,:)-V2;
     S2(i)=norm(C2);
     T(i)=S2(i)/(S1(i)+S2(i));
end
output_args=T
```

(3) Evaluation results
After inputting the above code into MATLAB to run, the fast calculation of the G1-EWM-TOPSIS model is realized, and the sharing time is 5.28 s, and the comprehensive evaluation result of the social contribution benefits of the provincial power grid investment is shown in Table 3.

The closeness curve generally shows an upward trend. The social contribution benefits of investment in the provincial power grid in the first three years were relatively

Table 3. The output results of G1-EWM-TOPSIS model

Years	d_i^+	d_i^-	Closeness
2014	0.4005	0.0829	0.1714
2015	0.3999	0.0612	0.1327
2016	0.3890	0.0523	0.1186
2017	0.2856	0.1905	0.4002
2018	0	0.4340	1.0000

stable, and there was a significant improvement in 2018. The final ranking result of relative goodness is: 2018 > 2017 > 2014 > 2015 > 2016.

Among them, affected by the reduction of investment scale, the social and economic contribution of the provincial power grid company showed a declining trend from 2014 to 2017. However, due to the favorable external economic environment and the increase in the provincial power grid company's investment utilization rate, the social and economic contribution of the provincial power grid company's investment. The multiplier effect continues to expand, and the contribution of unit investment to the society and economy continues to increase. In addition, the implementation of policy projects such as wind power and photovoltaic projects and coal-to-electricity projects carried out by the provincial power grid company has long-term comprehensive social contribution benefits. It not only promotes the increase in the output value of related industries, but also plays an important role in energy saving and emission reduction. The social environmental benefits and social stability benefits are jointly optimized. In 2018, the provincial power grid company expanded its investment scale, and its social contribution benefits in terms of economy, environment and stability were greatly improved.

5 Conclusion

Based on the characteristics of the provincial power grid, this paper constructs an evaluation index system for the social contribution benefits of investment in the provincial power grid enterprise, establishes the G1-EWM-TOPSIS model, and uses MATLAB software to realize the rapid calculation of the model, which can provide powerful help for the provincial power grid decision-making.

References

1. Liangjie, T., Qing, X., Chunfang, L., et al.: Comprehensive benefit evaluation of distribution network planning based on the life cycle cost theory and the cloud matter element theory. Power Syst. Prot. Control **45**(19), 12–17 (2017)
2. Tian, S., Haozhong, C., Chang, H., et al.: Analysis and evaluation of social benefit from UHV power grid. Electr. Power Autom. Equip. **35**(2), 145–153 (2015)
3. Chao, X., Qian, G., Jian, Z., Gaofeng, S.: Research on the full cycle evaluation index system of social benefit of fixed assets investment in power grid enterprises. Eng. Econ. **26**(05), 44–48 (2016)

4. Jianjun, Z.: Comprehensive evaluation and analysis of social benefits of power grid construction projects. Shangqing **50**, 146 (2012)
5. Xiandong, T., Zhaoguang, H.: Study on the position and role of electric power industry in china's economy. Electr. Power Technol. Econ. **11**, 6–12 (2010)
6. Che, G., Dongxiao, N., Ming, M., et al.: Accommodating capability analysis and comprehensive assessment method of large-scale new energy areas interconnected. Electr. Power **50**(7), 56–63 (2017)
7. Tingcheng, H., Linhao, Y., Yongjun, Z.: Comprehensive evaluation on electricity consumption reliability of power distribution network based on AHP-entropy weight method. Guangdong Electr. Power **1**, 44–50 (2018)
8. Meng, C., Shuangzhi, Z., Lijie, W.: A comprehensive evaluation method for urban photovoltaic power generation project based on AHP-TOPSIS. Electr. Power **49**, 175 (2016)
9. Hongbo, G., Suping, M.: Application of MATLAB and TOPSIS in enterprise logistics service tender evaluation. Logist. Technol. **23**, 277–278 (2012)
10. Qingyou, Y., Xiaoya, W., Chao, J., et al.: External environment impact factors identification of asset management for power grid corporation. Water Resour. Power **29**(04), 182–184 (2011)
11. Zeng, S., Lu, H.: Milan. Comprehensive evaluation and international comparison of sustainable development of electric power enterprises – based on critical weighting and TOPSIS method. Enterp. Econ. **465**(05), 112–117 (2019)

Application of VR Technology in Interior Design

Yu Gui and Shuaiye Shi(✉)

School of Horticulture and Landscape Architecture, Tianjin Agricultural University, Tianjin, China

Abstract. With the rapid development of economy and technology in our country, people's requirements for interior design continue to increase, and traditional interior design methods have been difficult to meet the diverse needs of the market. Through the application of VR technology in interior design, an indoor VR technology experience hall is formed, which not only provides users with a more realistic visual experience, strengthens the real-time communication between users and designers, but also increases the convenience of design and injects new technology into traditional interior design. This article adopts the case analysis method and collects information extensively. The indoor VR technology experience hall uses the immersive, interactive, and imaginative advantages of VR technology to construct a three-dimensional simulation space, create a scene of tactile feedback technology, synthesize stereo, and interactively experience the stereo display technology, realizing the actual transformation of the virtual environment. Using VR technology are able to make up for the shortcomings and deficiencies of traditional interior design, help users get a better design and decoration experience, meet the diverse needs of users and the market, and provide beneficial scientific guidance for the development of interior design.

Keywords: Interior design · VR technology · Virtual environment · Interactivity

1 Research Background

With the development of economy and the advancement of technology, a good living environment is becoming more and more important. Traditional interior design methods can no longer satisfy the development of the design industry, and the integration of new technologies is urgently needed. Through the combination of VR technology and interior design, computer simulation of virtual environment and three-dimensional simulation of physical objects bring users a multi-sensory experience [1]. Under the current situation of the development of intelligent and informatized big data, the combination of VR technology and interior design also shows the humanization of VR technology, data optimization, and the high integration of VR and AI, which has become an emerging industry trend and a new form that conforms to future development trends.

B. J. Jansen et al. (Eds.): *International Conference on Cognitive based Information Processing and Applications (CIPA 2021)*, LNDECT 85, pp. 735–743, 2022.
https://doi.org/10.1007/978-981-16-5854-9_94

Humanization, under the new situation, continue to meet the user's pursuit of humanized decoration forms and home experience, constantly improve hardware equipment, and give consideration to subjective needs and aesthetic experience, and provide a more humane living space [2]. Data optimization, the combination of VR technology and interior design is not only a great opportunity for researchers to select information based on large databases, organize data, make up for gaps, and optimize data, but it is also an excellent time to incorporate more scientific and technological elements. VR and AI are highly integrated, and artificial intelligence technology is showing a diversified development trend. The integration of VR technology and AI improves the intelligence of VR interactive devices and enhances the interactive experience of vision and touch and so on. In the emerging industry trends, VR technology must be closely applied in practice to maintain new vitality in the wave of innovation, and VR technology can be popularized and applied by the public, making VR design a new trend.

2 VR Technology Conception

VR technology, namely virtual reality technology, relies on the Internet to create space scenes that people need, and users can obtain real space experience through the use of VR equipment and technical means. In the development of the interior design industry, in order to meet people's increasingly diverse needs, traditional drawing designs cannot show three-dimensional spatial details and restore the true emotional experience. As the application of VR technology in interior design becomes more mature, it not only satisfies users to provide more realistic visual experience, strengthens real-time communication between users and designers, increases convenience, and improves the work efficiency of interior design staff, which is of great significance to the development of the interior design industry [3].

3 Comprehensive Application of VR Technology in Interior Design Experience Museum

The VR interior design experience museum in this design is located in the Xidan Center business district of Xicheng District, Beijing, which has high technology level, convenient transportation, large flow of people and rich social culture. Many large and high quality residential areas are developed in the surrounding area, and the future development prospects are good. As a VR technology indoor experience positioning product, this project combines regional planning to carry out a more targeted, more humanized, more delicate design and analysis to meet the needs of users for culture and commercial services (See Fig. 1 and Fig. 2).

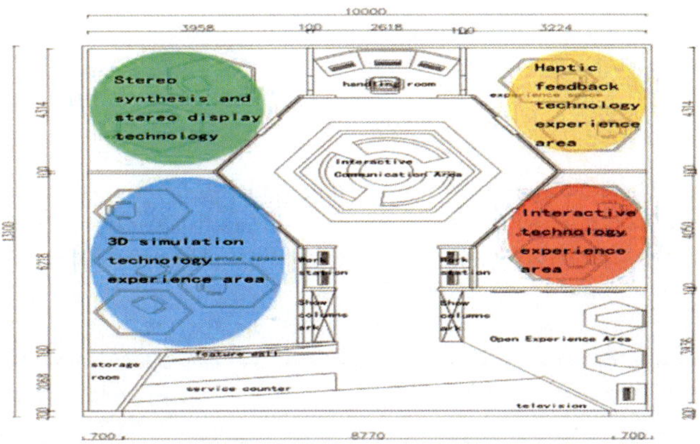

Fig. 1. Technical division diagram of VR Design Experience Pavilion

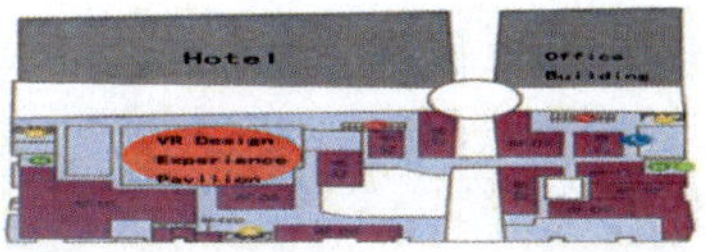

Fig. 2. Direction map of VR Design Experience Pavilion

4 Technical Application and Design Effect of VR Design Experience Hall

VR technology is an innovative extension on the basis of computer technology and information technology, which is closely related to people's life, study and work. Using VR technology can bring users a variety of spatial experience, optimize design schemes, promote the integration of interior design industry and VR technology, and develop toward a higher level. The VR design experience museum divides the area space from the three-dimensional simulation technology of VR technology, stereo sound synthesis and three-dimensional display technology, haptic feedback technology, interactive technology and other aspects, and makes reasonable application and promotion of VR technology, so as to bring people all-round and diversified space experience and feel the unique charm of VR design.

4.1 3D Simulation Technology

The architectural space in the real world is constructed by many material materials, and each space occupies its own three-dimensional environment, presenting a real and three-dimensional scene feeling [4]. The interior space environment and the entity are opposite to each other and influence each other. The entity can divide the space and the space is constructed by the entity. In the VR interior design experience hall, 3D simulation technology is one of the main application technologies of VR technology, and reasonable application and actual experience of virtual modeling technology and scene rendering technology are carried out in the 3D simulation space.

4.1.1 Virtual Modeling Technology

In the space of 3D simulation technology, virtual modeling technology is used to import the graphic design scheme, and the generated three-dimensional model can be put into the display area of the virtual model through VR equipment. At this time, users can also participate in it, communicate with the designer, and modify the scheme three-dimensional. The museum also provides users with pre-decoration experience, pre-decoration models, models of different design schemes for comparison and selection, virtual models of a variety of design styles are provided to users, designers can more intuitively modify according to the needs of users, to meet the needs of users, make users more satisfied. (See Fig. 3).

Fig. 3. Three-dimensional model diagram

In addition, virtual modeling is carried out for the space scene of the VR design experience hall, so that users can feel the effect of interior design scheme more intuitively by wearing VR equipment. Consistent with the design principle of the real world, the well-constructed model space is reasonably matched with the entity to obtain a complete virtual space scene. The use of virtual modeling technology to model the space scene of VR design experience museum belongs to interactive modeling. Users can change the perspective and sequence of viewpoints of the scene according to their needs, bringing users a sense of immersion and real sensory experience. (See Fig. 4).

Fig. 4. 3D model effect drawing

Virtual modeling technology is based on 3D software, which optimizes the model space, simulates the real environment, conducts modeling analysis based on the personal needs of users, and integrates and combs the user design information to build a VR design database to meet the personalized needs of more users.

4.1.2 Scene Rendering Technology

In the previous interior design and decoration, 3DMAX was used to make the model. Material selection and lighting rendering were performed through Vary. Finally, the camera Angle and distance were adjusted to render the renderings. The rendering model of traditional renderings is simple, visual monotonous, and all items are planar. In the VR design experience hall, scene rendering technology is adopted to solve the contradiction between the fidelity of the rendering scene and the rendering speed, and the interior design model scene is further rendered to improve the reality of the scene and items.

In the exhibition space, 3D rendering is done on the basis of virtual modeling, scenes of various design styles are rendered and colored, and lighting and object materials in the scene are processed in the form of images [5]. In the scene rendering, the user can select and edit the scene, model, material and lighting, etc., for one-button rendering, strengthening the three-dimensional and real effect of the space scene, highlighting the authenticity of the scene. When observing the model scene, experiencers can obtain a more realistic visual experience, which is different from the traditional renderings, and carry out multi-directional observation and rich visual perception experience. Therefore, it is an indispensable display way to use scene rendering technology to render space in the interior design experience museum. (See Fig. 5).

Fig. 5. Scene renderings

4.2 Stereo Synthesis and Stereo Display Technology

In the VR design experience hall, stereo sound synthesis and stereoscopic display technology are applied to simulate the auditory details and characteristics in daily life, so that users can feel the real sound effect when experiencing the VR design space.

4.2.1 Three-Dimensional Display Technology

Display area for thought in stereo space, on the basis of experience pavilion real space environment, the design of VR in the layout and details projected on a fixed space, use of VR simulation software to build three-dimensional entity model in driving and environment integration, build multiple dimensional space and virtual scene, including voice, environment, etc., showing a dynamic effect, What people feel by touch is the real scene [6]. At the same time, the indoor space of the experience museum will be extended and expanded in three dimensions, so as to enhance the three-dimensional and realistic spatial structure design of the museum.

For the existing space scenes in VR design experience museum, users can input panoramic photos into virtual equipment with the help of VR technology, and can automatically create three-dimensional models to restore the content details of the real space and scene. In addition, two-dimensional pictures are generated into three-dimensional space through virtual reality technology, so that users can create spatial roaming routes and roaming perspectives by themselves to obtain the best interactive experience, so as to promote the flow update of the design model of the experience museum.

Three-dimensional display technology can not only show the design scheme more intuitive and real, save the user time to choose their favorite design scheme, but also the model in the display space are real existing decoration materials and household items. The staff of the exhibition area will conduct on-site explanation to help users quickly build a three-dimensional space and generate a two-dimensional code for users to browse through electronic devices.

The VR design experience museum not only accurately shows the virtual space, but also constructs the things that do not exist in the real world and users' imagination in the experience museum. In the Experience Museum, users can open up a new way of thinking, better use VR technology, and get the best design scheme. The completion and experience of VR design experience museum also promotes the interior design industry to burst out new vitality and vitality.

4.2.2 The Technology of Stereo Sound Conduction

In real life, if the source of the sound and the distance from the human ear are different, the intensity and time of the sound transmission to the human ear will change [7]. Using VR technology in stereo synthesis of conduction gives a three-dimensional space scene rendering on hearing and vision, do what you see is what you get, and the users can feel the strong graphics and sound experience by VR technology in the scene. In order to get the best experience in the VR Design Experience Pavilion, users can also use VR devices to further enhance their body perception and enhance the reception and perception of stereo and stereo display objects, such as VR glasses, VR tactile sensors, VR headsets. For example, when resting on the outdoor balcony in the virtual space, users can hear the chirping of birds and see them gathering in the birch trees. They also could hear the rustling of the leaves as the wind blew through the branches, and see the branches swaying and shaking in the direction of the wind. Stereo acoustic transmission technology can build a real virtual display scene, for the user to create an immersive effect, personal experience of visual and auditory immersion.

4.3 Haptic Feedback Technique

Haptic feedback technology is also one of the technologies applied in VR design experience museum. By using this technology, users can contact and personally feel the virtual space design environment, and bring real haptic experience to themselves.

4.3.1 Motion-Sensing Interactive Technology

Motion-sensing interaction technology is an intelligent computer interaction technology. It captures the actions and images of experiencers through the sensing technology of VR camera, and uses infrared ray for induction positioning and motion capture [8].

In the human-computer interaction area, users can interact some simple body movements and gestures, such as 3D modeling and roaming animation, with the contents on the screen on the computer equipment, such as waving hands, turning forward, turning back and other simple actions. Users can also use speech recognition to communicate intelligently with virtual scenes, In the interactive system, voice recognition, microphone input and other functions are added, so that the audience can get rid of the boring operation mode and get different interactive feelings (See Fig. 6).

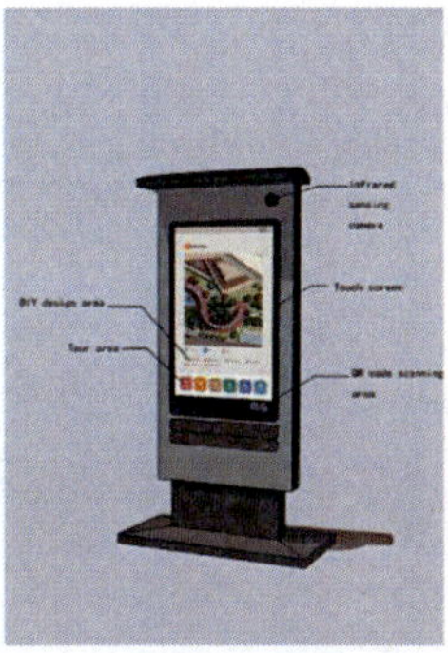

Fig. 6. Interactive interface

In addition to the technology of VR design and interactive technology, the interactive system also makes literature, leisure sports, etc. into combination, rendering themselves art, entertainment, technology and interactivity [9]. Experiencing the interactive technology in the VR design experience museum, when users read books and move objects in a simulated way, the objects will deform and move with the movements of their limbs. When a user touches an object in the virtual space, it will have a real sense of touching and feel the weight and shape of the object. It can also meet the user's personalized needs by changing the perspective, color, weight and structure of the object. By giving full play to the interactive characteristics of interactive technology in the experience hall, experiencers are attracted to participate in the interaction, which brings rich experience and expands the promotion of interactive technology.

4.4 Interaction Technology

Interaction technology refers to that in a virtual space environment, users can get a good sense of experience by interacting with the virtual space using VR equipment. In the past computer interaction process, mainly through the computer components to complete. But today's VR technology is different, requiring electronic sensing devices, digital trackers, force feedback systems and other tools to complete interactive tasks, to give users a more realistic body perception.

4.4.1 Scene Roaming Animation

In interior design, the use of real picture is very common. However, with the improvement of the level of science and technology and users' increasingly high decoration requirements, the real picture only brings the user a single sense of visual reality, and cannot increase the enjoyment of the experience, so the real picture slowly cannot meet the user's design experience needs.

As users become younger, more novel ways of presentation and intuitive picture stimulation have been added to the interior design experience hall, bringing real sensory experience. The display mode of scene roaming animation is applied in the interactive area, and the roaming animation and virtual reality technology are combined for application in the museum. The production cost of scene roaming animation can be reduced through different Settings of the model, such as material and lighting [10]. At the same time, it can also shorten the production time, improve the work efficiency, modify the scene for several times, enhance the interaction of the scene roaming animation, and bring more rich visual experience to users. In addition, users can conduct multi-angle analysis from the whole to the part in the museum, observe the scene image, and feel more real space experience.

4.4.2 Virtual Scene Interaction

In the interaction area, different design Spaces are created to create different experience scenes and perfect space models. By wearing VR motion-sensing devices, users can not only immersively integrate into the virtual scene for interaction and experience space design, but also get a good sense of experience. In VR scene interaction, the user's action route is the main body, which is transformed into the scene browsing mode. For example, when users pick up a book on the shelf in the virtual space, they can not only feel the weight of the book, but also map real actions such as flipping books into the virtual space through the VR device, as if they are really in the designed interior space. Through the tactile perception experience, the design scheme is optimized, and users are provided with real, varied and vivid experience effect and rich emotional experience.

5 Conclusion

To sum up, VR technology can make up for the defects and deficiencies of traditional interior design and help users get a better design and decoration experience. VR technology is applied in the interior design experience hall, which provides the best design service platform for designers to carry out spatial planning and design, meet the diversified needs of users, and bring users more intuitive process information experience.In the museum, VR technology is used to construct three-dimensional space, tactile feedback technology is used, stereo synthesis and three-dimensional display technology, interactive technology is used to create real space scenes, so as to enhance the user's experience of interior space design scheme, and obtain immersive sensory experience. Through VR interior design experience museum, the advantages and characteristics of VR technology can be reasonably and effectively applied, improve the design quality and design efficiency, provide a variety of browsing paths, satisfy people's all-round space experience, and provide power for the development of interior design industry.

References

1. Chen, X.: The new trend of home interior development under the background of big data - Smart Home. J. Hebei Univ. Eng. (Soc. Sci. Ed.) **36**(02), 29–31 (2019)
2. Kang, Q., Wang, X., Lin, Y.: Application of virtual reality technology in architectural interior design. Fujian Forestry Vocational Technical College, Nanping (2020)
3. Sun, Q., Zhi, Y.: Discussion on application of VR technology in architectural interior design. Art Panorama **12**, 148–149 (2018)
4. Bai, M., Gong, P., Dong, H.: Research on application of VR technology in interior decoration design and construction. Hous. Real Estate **27**, 94 (2019)
5. Yang, W.: Research on the generality of digital media art and virtual reality technology development mode. Appreciation **20**(4), 13 (2019)
6. Li, Y., Ji, K.: Discussion on the influence of the application of virtual reality technology on the future fields. Occupation **19**, 124–125 (2019)
7. Xue, Y.: Discussion on application of VR technology in architectural interior design. Build. Technol. Dev. **45**(7), 84–85 (2018)
8. Liu, J., Guo, S., Cai, H.: Research on application of VR technology in decoration design. Sci. Technol. Innov. Herald **16**(06), 48–49 (2019)
9. Man, W., Zhang, Q.: Application of virtual reality technology in interior space design. Furniture Inter. Des. **10**, 106–108 (2018)
10. Hu, X.: Research on Application for Virtual Reality Technology in Interior Decoration Design of Buildings. Guangxi University, Nanning (2019)

Classification and Grading Storage Method Based on Terminal Scanning

Weixiong Wang(✉), Gui Wang, Xiaodong He, Zhen Guo, Ziqin Lin, and Zhitao Wang

Airport Management College, Guangzhou Civil Aviation College, Guangzhou 510403, Guangdong, China

Abstract. Applications require storage systems to provide more and more storage space, at the same time; the performance of storage systems is also more and more demanding. In order to solve the contradiction between the cost and performance of storage system, hierarchical storage system composed of solid state disk and mechanical hard disk is widely used, which aims to obtain the performance of close solid state disk at the cost of approaching mechanical hard disk. Aiming at the problem of space utilization efficiency of solid-state disk, an adaptive hierarchical storage system is proposed ATS, ATS which can dynamically adjust the migration threshold according to the current access data scale of the load and the utilization of the storage device. The load can make full use of the storage space of solid-state disk without frequent data exchange at different levels. The experimental results show that compared with the static threshold mode and the MOD (Migrate on demand, access is migration) mode, ATS can obtain a better average response time regardless of whether the current load is in a case of a very large amount of data or a very small amount of data. ATS can reduce the average response time by up to 53% compared with the static threshold mode, and ATS can reduce the average response time by up to 50% compared with the MOD mode.

Keywords: Storage system · Hierarchical storage · Solid state disk · Threshold analysis

1 Introduction

As human beings enter the information age, the Internet, multimedia flourish, hundreds of millions of sensors, mobile phones, computer terminals continue to produce data. And humans are envisioning all the objects on earth being equipped with sensors, which will lead to explosive growth in data. According to international data and information companies, the amount of global data roughly doubles every two years. It is expected that by 2020, the world will have 35 ZB of data, and more than 85% of the data will exist in unstructured or semi-structured form [1–4].

At present, the most used storage system is mechanical hard disk HDD, mechanical hard disk. Because of its inherent hardware characteristics, magnetic disk is used to store data. Each access to data requires the head to rotate to the location of the data. Then the

B. J. Jansen et al. (Eds.): *International Conference on Cognitive based Information Processing and Applications (CIPA 2021)*, LNDECT 85, pp. 744–751, 2022.
https://doi.org/10.1007/978-981-16-5854-9_95

mechanical hard disk can be read and write operation. Now the manufacturing process of disk is increasing, the capacity of single disk and the speed of disk are rising, but the speed of disk development is far lower than that of computing power [5–7]. At present, the computing power of CPU is still developing according to Moore's law. In contrast, the development of mechanical hard disk is very slow, which leads to the gap between computer storage ability and computing ability. Storage performance has become the bottleneck of overall performance. The upper application produces more and more data, which requires the storage system to provide more storage space, and the application's access performance requirements to the data are also getting higher. Increasing the capacity of the mechanical hard disk alone can no longer meet the requirements of the upper application. For the storage performance of the storage system can meet the needs of the upper application, new storage media, such as phase change memory, solid-state disk SSD, have attracted wide attention. Solid state disk is widely used in storage system to improve the performance of storage system because of its advantages of fast reading and writing speed, non-volatile, anti-seismic and low power consumption. But because the solid-state disk is more expensive than the mechanical hard disk, it is too expensive to replace all the mechanical hard disks in the storage system with the solid-state disk. At the same time, the solid-state disk also has the disadvantage of limited erasure times. Therefore, at present, solid-state disk and mechanical hard disk are often used in storage system, and a small part of solid-state disk and most of mechanical hard disk are used to form a hierarchical storage system. This hierarchical storage system can performance close to solid-state disk at a cost close to mechanical hard disk. Solid state disk can not only form hierarchical storage system with mechanical hard disk, but also act as cache between DRAM and mechanical hard disk. Such cache scheme has FlashCache, LARC and so on [8–10].

A hierarchical storage system ATS.designed and implemented in this paper the hierarchical storage system faces a storage system, a solid state disk with a certain space and a mechanical hard disk. The existing load runs on the storage system. The data of the load can be distributed stably on solid state disk and mechanical hard disk to improve the overall performance of the system. ATS first analyze the trace of the load for a period of time with the available solid disk and select a threshold to avoid moving too much low value data into the solid disk. However, the access mode of the load may change, which is mainly reflected in the change of the amount of hot data. In order to deal with this situation, the hierarchical storage system uses a dynamic threshold strategy to adjust the migration threshold by combining the space of the solid disk.

2 Description Relevant Concepts and Definitions

2.1 Terminal Scanning

Terminal is just an abstract concept in software used to distinguish communication channels. Similar to the way IP addresses are used to identify computers on a network, terminals are used to identify specific applications used on a single computer. Your Web browser, for example, will connect to TCP port 80 in the HTTPURLs by default. If you specify a secure HTTPS protocol, the browser will try port 443 by default.

Terminal scanning is the behavior of remotely testing multiple terminals to determine what state they are in. Generally our most interesting state is open, which means that the application is listening and accepting connections on the port. While many terminal scanners traditionally centralize all ports to open or closed states, the tools we are going to use are Nmap finer, dividing ports into six states. These states are not inherent properties of the port itself, but describe how Nmap view them. Nmap scans from a unified target network, for example, may display port 135/tcp as a open, while scans using the same option from Internet may display the port as a filtered.

2.2 Classified and Classified Storage Schemes Using MOD Modalities

CEPH is a distributed file system. CEPH RADOS, as a storage cluster, can configure a solid-state disk cache to form hierarchical storage together with a mechanical hard disk. The cold data is placed on the mechanical hard disk and the hot data is placed on the solid state disk. If the data is not hit on the solid state disk, the data in the mechanical hard disk is first migrated to the solid state disk, and then the data is returned from the solid state disk. CEPH document does not mention the elimination strategy of solid state disk heat data, so the CEPH hot data elimination strategy is not discussed here.

HyCache is a FUSE based file system, which is mainly used in the data server of distributed file system. For the heterogeneous storage devices existing in the data server, HyCache can reasonably use the underlying heterogeneous storage devices. Separate cold and hot data. The cold data of the HyCachehierarchical storage model is stored on the mechanical hard disk, the hot data is stored on the solid state disk, the cold data is migrated up by MOD mode, and the hot data is migrated down HyCache The default LRU algorithm is adopted.

MOD advantage is that the use of this method can maximize the use of the space of the solid disk, there will be no idle solid disk, when the load access mode is very random, Or solid disk has a relatively large space, this way is better. The disadvantage of MOD mode is that when the space of solid disk is not very large, the data migrate frequently between solid disk and mechanical hard disk, consume a lot of IOPS and bandwidth, and even affect the performance of foreground load. At the same time, solid-state disk is a limited erasable device.

2.3 Classified Hierarchical Storage Schemes Using Static Thresholds

Different from the MOD mode, the static threshold hierarchical storage scheme chooses a more selective way to determine the cold and hot data and avoid moving a large amount of unnecessary data into the solid state disk.

EDT proposed a block-level hierarchical storage scheme for use in SAN environments. EDT mainly focuses on the implementation of a cost-effective hierarchical storage system. The main scenario is that when the storage medium used by the current system user is SAS disk, it needs to improve the performance of the system. EDT will recommend how many solid disk, SAS disk and SATA disk should be purchased by the user to achieve the cost-effective goal.

EDT whole system is divided into two parts, EDT-CA and EDT-DTM. EDT-CA analyze the load for a period of time trace, then recommend a device purchase scheme, using which the target performance can be obtained. EDT-DTM runs with the actual load, its main task is to place the data at the appropriate level when the load is actually running, and to reduce energy consumption as much as possible.

On the whole, the size of solid disk with fixed capacity D, N load, and the allocation of a solid disk space i the load need to be:

$$\max \sum G(x_i) \tag{1}$$

THTS recording the upgrade costs and benefits for each file, calculate for each file:

$$\text{priority} = \frac{access\_bytes \times access\_num \times avg\_\text{interval}}{access\_bytes} \tag{2}$$

2.4 Classified Hierarchical Storage Schemes Using Dynamic Thresholds

SANBoost is a block-level hierarchical storage scheme, which is used in SAN environment, using solid state disk as the cache of mechanical hard disk to realize hierarchical storage. SANBoost the migration block size is 256 KB, the SPC-1 is statistically analyzed. SANBoost statistics trace the number of visits per block in the life cycle, it is found that a large number of blocks only visit a few times. However, the commonly used MOD migrate all the accessed blocks to the solid disk, and there will undoubtedly be a lot of blocks after being migrated to the solid disk, but after a few visits, they will be migrated back to the mechanical hard disk, resulting in performance loss.

SANBoost propose a new standard:

$$G(threshold) = Hits/MigratedChunks \tag{3}$$

Until the SANBoost works, all blocks are stored on the mechanical hard disk, threshold means that a block is accessed on the mechanical hard disk threshold, and the block is migrated to the solid state disk. The Hits is the number of hits on the solid disk, MigratedChunks the number of blocks migrated to the solid disk. SANBoost goal is to find threshold to make the G (threshold) maximum. SANBoost designed two ways. The first is the static threshold mode, the trace is counted and analyzed in advance, and the threshold is obtained so that the G (threshold) is maximum, and then the trace. Is run the drawback of this approach is that.

(1) All trace of the load need to be statistically analyzed first to obtain the most reasonable threshold to actually run the load.
(2) A block accessed on a mechanical hard disk will always be placed on a mechanical hard disk if the number of accesses does not reach threshold. This results in a large number of solid disk access hits. Therefore SANBoost a dynamic threshold scheme is proposed. When the load is actually running, an initial threshold is set threshold, with the load running, if the G (threshold) becomes larger, it is considered that more blocks should be migrated to the solid disk threshold minus one.

3 Design of Classification and Hierarchical Storage Method Based on Terminal Scanning

3.1 Load Analysis

To realize a file-level hierarchical storage system, we first need to analyze the file-level load. Roselli et al. of the university of california, berkeley collected a number of different scenarios in 1996 trace, these trace were collected in the system call layer, recording all system calls. InstructionTrace is collected in one of the scenarios, because trace files record all system calls, it is necessary to filter all trace files before use, filter out all metadata operations, and only retain data operations. InstructionTrace recorded more than three months of system call information, this paper selects the InstructionTrace trace from November 1,1996 to November 30,1996 for analysis and testing.

3.2 Overall Design

(1) Initial threshold selector
ITC determines the initial migration threshold for the system to run. Before the load actually runs, the load is collected for a period of time trace, and the available storage device capacity of the data server is used as the input of the analyzer. The migration threshold is used to distinguish hot and cold data from solid disk and cold data from mechanical hard disk.

(2) Adaptive hierarchical processors
When the ITC calculates the initial threshold, the actual running load begins. At this time, with the actual load running ATP,ATP the access information of each file will be counted in real time. The priority of the file is calculated in some way and then compared with the threshold. At the same time, ATP will dynamically adjust the migration threshold according to the daily data size of the load, as well as the change of the available space of the solid disk. Avoid the occurrence of solid disk space utilization and migration of too many low reuse files resulting in a large number of files between solid disk and mechanical hard disk migration.

4 Analysis of Classification and Hierarchical Storage Method Based on Terminal Scanning

4.1 Selection of Initial Thresholds and Test Results

And this article uses a 30-day Instruction Trace to test, Analysing the Instruction Trace of the first 15 days, After 15 InstructionTrace for testing. Considering the size of the Instruction Trace data, Data server memory limited to 512 MB, Instruction Trace and 2 solid disks 15 days before ITC input, there is no limit to the size of the mechanical hard disk. After ITC treatment, ITC initial threshold for output is 10. Instruction Trace 15 days ago, Total size of files accessed more than 10 times a day, Data for day 7 totalled MB,1760 trace 15 days before Instruction Trace, The initial threshold is 10, Every day as a cycle, According to formula 3–4, Get a time window windows of 30 min.

The last 15 days of the Instruction Trace are tested. Figure 1 shows the total size of the file data with 15 days of access exceeding the initial threshold. It can be seen from the be seen from the diagram that the file size of more than 10 times is 2570 MB, and the capacity of the solid disk can no longer fully accommodate the data. Day 26 and 28 represent two cases of large and minimal load data, respectively. The trace of day 26 and 28 will be tested in this paper.

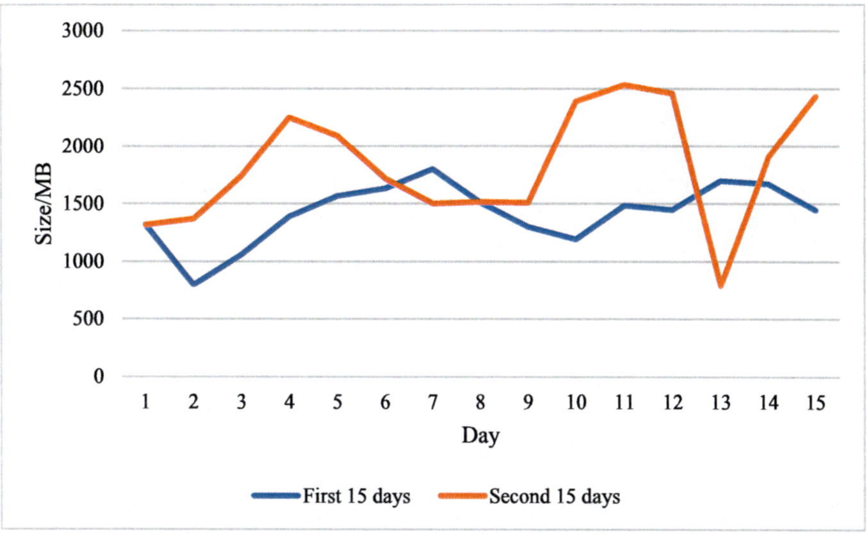

Fig. 1. Total Size of Files Accessed More than 10 Times in 30 Days

4.2 Testing When the Load Volume is Extremely Large

Figure 2 is the amount of data migrated. It can be clearly seen from the diagram that MOD migrated the most data, the solid disk can not hold so much data, resulting in the worst average response time of the MOD, ST the static initial threshold is used, the data is the least.

4.3 Testing with Extremely Small Load Volumes

Like DAY26, for clear comparison MOD, ST and ATS, this paper will be pairwise comparison. Table 1 shows the average response time and upgrade threshold change of ATS and ST. The average response time of the ATS is slightly lower than that of the ST in comparison with that of the ATS at the beginning stage, because the threshold will be reduced periodically, and the ATS threshold will be lower and lower as time goes by, accommodating more and more data. The average response time of the ATS is shorter and shorter; ATS and MOD have very close average response times at the beginning of the cycle, but as MOD migrate more and more data from solid-state disks,

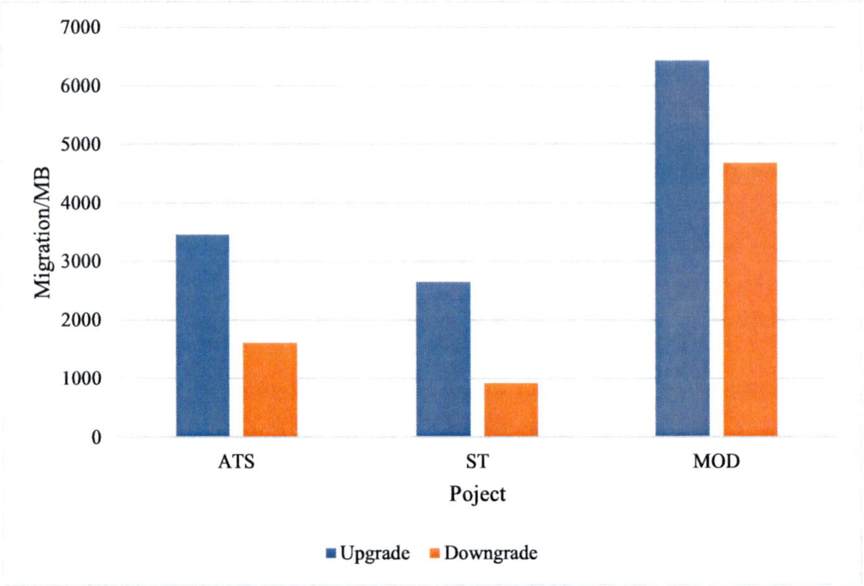

Fig. 2. Data Volume Migrated by Day 26

Table 1. Average Response Time of Day 28 ATS, MOD and ST(μ S)

Time/ms	30	90	150	210	270	330	390	450	510	570	630
ATS	912	721	941	732	1123	1351	623	403	904	412	398
ST	987	748	1210	871	512	1243	1156	1123	1510	1194	816
MOD	1320	884	498	986	1321	874	1220	456	615	487	654
Time/ms	690	750	810	870	930	990	1050	1110	1170	1230	1290
ATS	997	946	801	606	784	603	663	412	287	549	632
ST	832	1301	409	587	809	934	448	384	387	598	503
MOD	745	721	703	983	1325	841	2354	881	846	792	813

because of MOD implementation problems, when solid-state disk space is not enough, Need to synchronize the data migration to the mechanical hard disk to migrate, so the performance of the late MOD is getting worse and worse.

5 Conclusion

With the rapid development of information technology, the explosive growth of data produced by various electronic devices will double every two years, which brings severe challenges to the data storage system. When the speed of CPU increases according to Moore's law, the speed of mechanical hard disk of traditional storage device increases slowly, and the speed difference between mechanical hard disk and CPU becomes more and more, so the storage system becomes the bottleneck of the overall performance of the system. However, compared with mechanical hard disk, the cost of solid disk is higher, which makes solid disk can not completely replace mechanical hard disk. At the same time, the number of erasure of solid disk is limited, making the life of solid disk shorter. Because the cost of solid state disk is too high, the storage system composed of mechanical hard disk can no longer meet the performance requirements of the upper application, so the hierarchical storage scheme is proposed. The hierarchical storage scheme consists of a variety of storage devices with different performance and cost. This paper designs a hierarchical storage system based on the analysis of Instruction Trace ATS, which can achieve better overall performance than ST and MOD, ATS.

Acknowledgements. The project will be funded by the Special Fund for Science and Technology Innovation Strategy of Guangdong Province in 2020.

References

1. Han, S., Jiang, J., Huang, Y., et al.: Hierarchical TiO2–SnO2–graphene aerogels for enhanced lithium storage. Phys. Chem. Chem. Phys. Pccp **17**(3), 1580–1584 (2015)
2. Jia, Y., Wan, H., Chen, L., et al.: Hierarchical nanosheet-based MoS2/graphenenanobelts with high electrochemical energy storage performance. J. Power Sources **354**(JUN.30), 1–9 (2017)
3. Wang, P., Zhang, H., Wang, H., et al.: Energy storage: hybrid manufacturing of 3D hierarchical porous carbons for electrochemical storage (Adv. Mater. Technol. 6/2020). Adv. Mater. Technol. **5**(6), 13–15 (2020)
4. Xu, G., Shang, C., Fan, S., et al.: A hierarchical energy scheduling framework of microgrids with hybrid energy storage systems. IEEE Access **6**, 2472–2483 (2017)
5. Liu, H., Hu, H., Wang, J., et al.: Hierarchical ternary MoO2/MoS2/Heteroatom-doped carbon hybrid materials for high-performance lithium-ion storage. ChemElectroChem **3**(6), 922–932 (2016)
6. Dong, C., Jia, H., Xu, Q., et al.: Time-delay stability analysis for hybrid energy storage system with hierarchical control in DC microgrids. IEEE Trans. Smart Grid **9**(6), 6633–6645 (2018)
7. Chen, J., Huang, Y., Zhao, F., et al.: A hierarchical α-MoC1x hybrid nanostructure for lithium-ion storage. J. Mater. Chem. A **5**(17), 8125–8132 (2017)
8. Fang, S., Gou, B., Wang, Y., et al.: Optimal hierarchical management of shipboard multi-battery energy storage system using a data-driven degradation model. IEEE Trans. Transp. Electrification **5**(4), 1306–1318 (2020)
9. Wang, B., Jin, J., Rui, K., et al.: Scalable synthesis of hierarchical porous Ge/rGO microspheres with an ultra-long cycling life for lithium storage. J. Power Sources **396**(aug.31), 124–133 (2018)
10. Liu, Y., Yan, X., Xu, B., et al.: Tunable pseudocapacitive contribution in nanosheet-constructed titania hierarchical tubes to achieve superior lithium-storage properties by phase control. J. Mater. Chem. A **6**(47), 24298–24310 (2018)

Measurement of Low Carbon Effect of Industrial Structure Adjustment – Based on NSGA - II Genetic Algorithm

Wei Hu(✉)

University of the Chinese Academy of Social Sciences, Beijing 102488, China

Abstract. Nowadays, China is facing the deep contradiction between high-speed development and low-carbon development. Industrial structure adjustment is generally considered as an effective way of low-carbon development. In order to measure the low-carbon effect caused by the change of industrial structure, this paper constructs the input-output multi-objective optimization model, and uses the non dominated sorting algorithm with elitist strategy (NSGA-II) to obtain the Pareto optimal solution set; Focus on the factors of industrial structure. On the premise of ensuring stable growth and employment, industrial structure adjustment as a means of reducing carbon emissions is still effective in a certain range; We should promote high-tech industries and emerging industries with low energy consumption, low emission and high added value, and develop low-carbon tertiary industry including modern service industry; The key lies in the innovative development and application of energy-saving technology, the coordination of technological progress and structural adjustment.

Keywords: Industrial structure · Low carbon effect · Multi objective optimization · NSGA-II genetic algorithm

1 Introduction

With the trend of economic integration and industrial transfer strengthening, China's industrial economy has begun to accelerate development, and energy demand has shown a jump growth. In recent years, China has been considered to be the second largest carbon emission in the world in recent years, second only to the United States. It is predicted that China's CO emissions will reach 1.543-2174 million tons in 2020. Therefore, facing the severe situation of carbon emission, increasingly serious energy threat and environmental deterioration, low carbon development has become a necessary and urgent task in the development process of China and the whole world.

Economic growth, industrial structure, energy consumption, scientific and technological progress, resource utilization and other factors will affect carbon emissions. To some extent, carbon reduction can be promoted through technological reform, industrial structure upgrading, energy intensity reduction, and increasing utilization rate of clean energy and renewable energy. However, the influence degree, influence mode and

B. J. Jansen et al. (Eds.): *International Conference on Cognitive based Information Processing and Applications (CIPA 2021)*, LNDECT 85, pp. 752–758, 2022.
https://doi.org/10.1007/978-981-16-5854-9_96

mechanism of various factors are different, and the difference is more obvious under different time and space background. If the research stays on the overall analysis of multiple factors, it is difficult to go deep and specific, so it is necessary to select all the people to study deeply and finely; Moreover, at present, the differences between energy dependence and carbon emission in different industries in China can not be ignored [1]. The impact of industrial structure (i.e. input or output structure) on carbon emissions may be obvious, and will be restricted by some potential mechanism, and different from other factors. Therefore, this paper does not take the comprehensive analysis of various factors as the research goal, nor adopts the traditional "factor decomposition" thinking mode, but focuses on the "industrial structure" factors. Through the establishment of input-output multi-objective optimization model, the low carbon effect of industrial structure optimization is determined by genetic algorithm.

2 Model Construction and Parameter Determination

In order to scientifically measure the low-carbon effect of industrial structure optimization, this paper combines input-output technology with linear programming model, and constructs an industrial structure optimization model with input-output equation as the main constraint condition and a set of social and economic development goals as the objective function according to the basic mode of low-carbon economy.

2.1 Basic Assumptions

Assuming that the energy consumption and carbon emission of the living sector in the national economy are exogenous variables, and the carbon dioxide produced by the production department through the terminal consumption of petrochemical fuel accounts for the majority of the total emission (other emission sources can be ignored), suppose that the production department includes n types of industries, and the total input or output X of a certain year is known, and the specific input or output of various industries should be optimally allocated according to the target, That is, the optimization of industrial structure.

2.2 Establishment of Objective Function

According to the development goal of low-carbon economy, the adjustment of industrial structure is mainly to realize the sustainable development of economic, ecological and social benefits. Therefore, designing the objective function from three aspects of carbon emissions, national income and employment, as the core of the whole model, is the basic basis for evaluating the advantages and disadvantages of the scheme [2].

(1) Carbon emission targets. In the low-carbon economy mode, reducing carbon emissions is one of the main objectives of industrial structure optimization, which is expressed by minimizing the total carbon emissions of various industries. That is $\text{Min } Z_1 \mid = cX$.
Where, c is the row vector of carbon emission intensity of n industries, and X is the total output column vector of N industries $[X_1, X_2, \ldots, X_\text{n}]^t$.

(2) National income target. It is undeniable that while controlling carbon emissions, industrial structure adjustment should also be in line with the development direction of economic growth. Therefore, GDP growth as an important measure of economic development constitutes an objective function of the optimization model. That is, Max $Z_2 = i^T Y$.
Where, *i* is the *n*-order unit column vector, and *Y* is the column vector $[Y_1, Y_2,\ldots, Y_\text{n}]^T$ composed of the final usage of N industries.
(3) Employment target. Since the end of 1990s, the problem of employment in China has become more and more prominent; Under the concept of building a harmonious society, high employment rate should be maintained to ensure energy conservation and emission reduction and economic development. That is Max $Z_3 = \beta$X. Among them, β It is the row vector of labor coefficient of *n* industries.

2.3 The Definition of Constraints

There is an internal relationship among economic variables, which affects the real economic operation mechanism. To a certain extent, it determines the model structure and constrains the selection boundary of feasibility scheme. The main constraints include:

(1) Input output balance constraint. The input-output relationship among some variables reveals that there is a complex relationship between production departments, and the balance and coordination between them can ensure the healthy development of national economy. Therefore, the basic constraints are: $X - AX \geqslant Y$. Among them, a is the matrix of input-output direct consumption coefficient.
(2) Production investment constraints. Considering that the total investment volume is generally planned when new inputs are made every year, the following conditions must be met according to the basic assumption condition known total investor or output: $i^T X = \bar{X}$.
(3) Energy saving and carbon reduction target constraints. In order to achieve the goal of energy conservation and carbon reduction, the following conditions must be met: $cX \leq \bar{C}$, $\frac{eX+\bar{L}}{i^T Y} \leq \bar{E}$.
C is the row vector of carbon emission intensity in n industries, C is the upper limit of carbon emission target value, e is the line vector of energy intensity in class n industries, e refers to the upper limit of energy consumption index per unit GDP, and l is the total amount of energy consumption of the given life.
(4) Labor constraints. As the main factor of production, labor investment has a great influence on output. For China, the rich labor supply will not have too much restraint on economic growth. If structural shortage is not considered, we can think that there is no upper limit limit on labor demand, but there should be a lower limit to ensure that the employment situation will not deteriorate. Namely $w \leqslant \beta X$.
Where *w* is the lower limit of labor force [3].
(5) Non negative constraint. $X \geqslant 0$, $Y \geqslant 0$.

2.4 Algorithm Improvement and Analysis

In NSGA-II, the first generation of subpopulations is generated by selecting the league matches of the initial population and then performing the "Yijun" operation. It is a selection algorithm based on elite group reservation. Evolutionary algorithm is a randomized search method.

The generation of initial population can be regarded as a process of random search. It can be imagined that appropriately expanding the size of the initial population (search range) will certainly improve the quality of evolutionary computation, thus playing a positive role in the initial stage of evolutionary computation. Of course, we can not expand the size of the initial population without limit, otherwise it will be the same as pure random search. In this paper, let the population size be n and the initial population size be 2n. N individuals are selected by selection operator to participate in the future evolutionary computation. In the scale of the example in this paper, for the test function three, the calculation result is better than that of the original population size in the same algebra, as shown in Fig. 1.

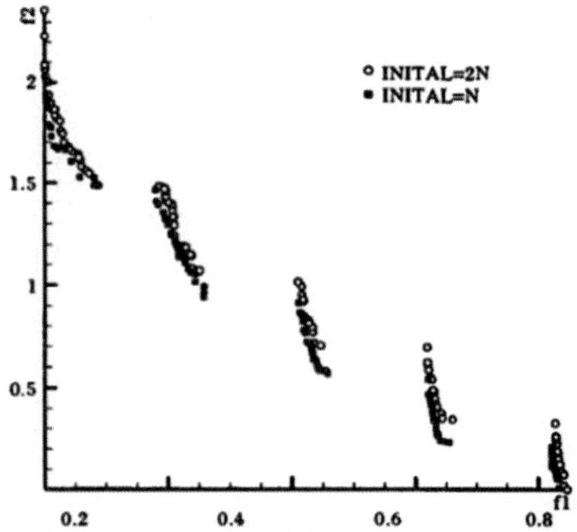

Fig. 1. Comparison of calculation results of the 300th generation

3 Model Solving Method Selection

Facing the era background of heavy pressure of growth, employment and pollution control, how to coordinate economic growth, full employment and carbon emission control as much as possible is the main task of structural adjustment and optimization, so the three objective functions need to be optimized at the same time. For the solution of the model, the most prominent problem lies in the contradiction and incommensurability between multiple objectives. It is impossible to achieve the optimal solution for all objectives. Only one set of optimal solutions can be obtained, namely the so-called Pareto optimal solution set or non inferior solution set [4].

In recent years, genetic algorithm has been widely used in various scientific fields. It has shown strong ability in solving some complex optimization problems. It has the characteristics of simple implementation, strong parallelism and wide adaptability. Multi objective genetic algorithm can be divided into two kinds: non Pareto based algorithm and Pareto based algorithm. Among the methods based on Pareto, the non dominated sorting genetic algorithm with elitist strategy (NSGA-II) is improved from the first generation of non dominated sorting genetic algorithm (NSGA), which has been proved to be the most effective by many scholars. NSGA algorithm was proposed by Srinivas and debl201 in 1994, and was first published in IEEE's evolutionary computation magazine. Based on this, NSGA-II algorithm, as an improved version of NSGA algorithm, was proposed in 2000. It has good convergence and distribution, fast convergence speed and good robustness, so it is used as a solution to the above optimization model.

The algorithm of gamultiobj function in MATLAB is a variant of NSGA-II, which provides a good way to solve the multi-objective optimization problem on the platform of MATLAB. This paper uses this analysis tool to solve the optimization model. The model has 3 objective functions, 12 endogenous variables and 8 sub modules; The control parameters of genetic algorithm are set as follows: Population Size 90, maximum evolution algebra 200, crossover probability 0.8, mutation probability 0.1.

4 Calculation Results and Analysis

In this paper, the multi-objective optimization based on NSGA-II algorithm does not need to change the multi-objective into a single objective, and it does not need to set the priority of each objective in advance. Instead, a Pareto optimal solution set or a non inferior solution set is obtained. There is no difference between the optimal solutions. According to the actual needs, a satisfactory solution that meets the preference can be selected as the basis of the design scheme. As shown in Fig. 2, the optimal Pareto solution set is obtained after multiple iterations. The population converges to a small area, and the individuals are evenly distributed in the solution set. Each point corresponds to a Pareto optimal solution, and the boundary individuals are well preserved. It can be seen that the optimization is relatively successful. If we start from the importance of the sub goals, we can choose three schemes, namely, low-carbon biased, growth biased and employment biased, according to the three different situations of "focusing on controlling carbon emissions", "focusing on promoting economic growth" and "focusing on expanding employment".

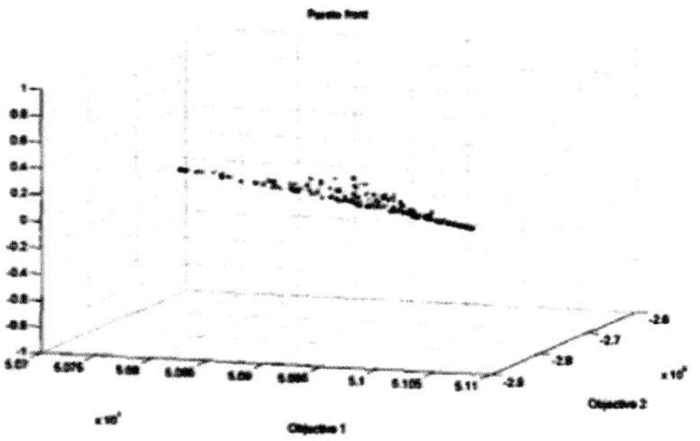

Fig. 2. Pareto front

5 Conclusion

In order to achieve the goal of low-carbon growth, industrial structure adjustment is still one of the effective ways. We should appropriately reduce the proportion of industry, support the development of agriculture, forestry, animal husbandry and fishery, maintain the low-level growth of transportation, warehousing, postal industry and construction industry, and vigorously promote the expansion of wholesale, retail, accommodation and catering industry, Objectively, it has become an important topic of economic development; There are many factors affecting the industrial or industrial structure adjustment, and the process is quite complex. We can adjust the internal structure and product structure of regional industries, such as reducing the proportion of high energy consuming industries such as steel, nonferrous metals, cement, eliminating backward production capacity and technology, and speeding up the construction of modern industrial system. We should also promote high-tech industries and emerging industries with low energy consumption, low emission and high added value, and actively develop low-carbon tertiary industries with low energy consumption, less pollution and more employment opportunities, such as modern service industry. The key lies in the innovative development and application of energy-saving technologies, and the transformation of traditional industries and production of traditional products with high-tech and environmental protection technologies. In a word, technological progress and structural adjustment are the two leading factors to balance energy conservation and emission reduction. They must develop in a coordinated way in order to promote the sustainable development of low-carbon economy more effectively.

References

1. Wang, Z., Zhu, Y.: Research on carbon emission situation and emission reduction countermeasures in China. J. Chin. Acad. Sci. (2), 109–115 (2008)
2. Wen, J.: Factor decomposition model and empirical analysis of carbon emissions in Jiangsu Province. East China Econ. Manag. (2), 29–32 (2010)

3. Liu, Z., Chen, C.: Research on low carbon economy and industrial structure adjustment. Foreign Soc. Sci. (3), 21–27 (2008)
4. Wang, W., Huang, X.: Factor decomposition model and empirical analysis of regional carbon emission intensity change: a case study of Jiangsu Province. Ecol. Econ. (12), 34–37 (2008)

Numerical Simulation and Multi-objective Genetic Algorithm Are Used to Optimize the Flow Field in the Slot

Jinnan Liu(✉) and Yihui Wang

Jiangyin Vocational and Technical College, Changzhou 214400, Jiangsu, China

Abstract. The change of velocity and temperature in melt blown flow field directly affects the fiber drawing. The structure of narrow die head is optimized by the combination of numerical simulation and multi-objective genetic algorithm. The teaching basis of optimization is based on the results of numerical simulation. The optimized structure parameters of narrow die head include: air slot neon degree, air gap angle, air gap angle In the 50th generation, the genetic algorithm of air return distance and jet depth stops operation to obtain the optimal die head structure parameters.

Keywords: Melt blown · Numerical simulation · Slot die head · Multi objective genetic algorithm

1 Introduction

In meltblown technology, molten polymer is extruded from the die head λ In a high-speed airflow field formed by two hot air streams, the polymer is rapidly stretched into ultra-fine fibers by hot air stream, and the formed fibers are directly integrated into the non-woven web. Non woven materials have great commercial applications, such as material passing, adsorption material, insulation material, medical material, sanitary material, oil absorption material, oil absorption material In recent years, numerical simulation has been widely used in the research of melt jet flow field, κ Rutka and others have simulated the flow field of the slot die, but the initial velocity used in the simulation is far lower than that of the commercial melt blowing; Moore et al. 8 considered the compressibility of air in the simulation, but the simulation was carried out under isothermal conditions, and the temperature attenuation was not considered; Krutka et al. 9 extended the simulation to non isothermal conditions, and simulated and analyzed the continuity field and temperature field of schwa ring die. 10,11 different from traditional melt spinning, the melt blown air not only provides tensile force, but also delays the solidification of polymer jet [1]. When the polymer is ejected from the nozzle, the air continuity causes the change of fiber momentum, which promotes the polymer to become thinner rapidly, At high temperature, the polymer maintains low viscosity, which prolongs the drawing time of the fiber. Therefore, in melt blown spinning, the flow field and temperature field affect

B. J. Jansen et al. (Eds.): *International Conference on Cognitive based Information Processing and Applications (CIPA 2021)*, LNDECT 85, pp. 759–765, 2022.
https://doi.org/10.1007/978-981-16-5854-9_97

the fiber forming process. In this paper, the structure parameters of the slot die are optimized systematically by combining the teaching value simulation and multi-objective genetic algorithm with the goal of speed and temperature.

2 Numerical Simulation and Experimental Verification

2.1 Numerical Simulation

The software used in the numerical simulation is fluent 63. The whole flow field calculation area is divided into two parts. The calculation area under the nozzle is 100 mm × Because of the symmetry of the calculation area, only 14 boundary conditions of the whole flow field need to be calculated. The cross grid area is set as the pressure inlet, the absolute pressure is 0.131 MPa, and the static temperature is 40315 K. under this pressure condition, the air is compressible, and the longitudinal grid represents the pressure outlet, The horizontal grid at normal temperature and pressure represents the symmetry plane, and other boundaries are the default walls. The turbulence model is the standard κ-ε Fluctuation model, fluctuation parameter C ε 1 and C ε 2 are 1.24 and 2.05 respectively η. The pulsation intensity of pressure inlet is 10% and the hydraulic diameter is 065 mm, while the pulsation degree of pressure outlet is 10% and the hydraulic diameter is 10 mm.

2.2 Configuration of Test Equipment

The high pressure gas supplied by the air compressor is heated by the meltblown testing machine and ejected from the two slots of the die head. At the same time, under the slot die head, the flow field is measured by the hot wire probe installed on the three-dimensional support. The hot gas flow has pressure and temperature loss in the pipeline, so the test set pressure and temperature are higher than the simulation set value, 2 MPa and 423.15 MPa respectively κ. The width of the air slot is 065 mm and the angle of the air slot is 30°, The air groove spacing is 1.28 mm, and the air groove height is 5.0 mm.

The hot wire anemometer of dandy company is used to measure the flow field. The diameter of the hot wire probe is 5 μm. The width of the probe bracket is 1.2 mm, and the small hot wire probe is convenient for measuring the flow field in the narrow space of the die, and the origin of the system coordinate is the center of the die face; The y-axis is parallel to the direction of air groove; The x-axis is perpendicular to the direction of air groove; The measurement area of the flow field of the slot die is: 6.0 mm $\leq$ x $\leq$ 6.0 mm; -: 30.0 mmsy $\leq$ 30.0 mm; 17.5 mm $\leq$ Z $\leq$ 100.0 mm. X = 0 in x-axis direction, ±3.0, ± There are 7 points at 4.5 and 6.0 mm; In the y-axis direction, due to the rapid change of the flow field near the central axis, the area of - 8.0 mmsy $\leq$ 8.0 mm is measured intensively, one point is measured every 1.0 mm, and one point is measured every 50 mm at other positions, a total of 27 points are measured; In the z-axis direction, 34 points were measured every 25 mm, and the distribution of the whole measurement point.

2.3 Comparison of Experimental Results and Simulation Results

In order to obtain the velocity and temperature decay curve at the center line (x = 0 mm) through experiment and simulation, that the gas velocity decays very fast near the die head, and then tends to slow down, which indicates that the velocity decay mainly occurs near the die head, and the attenuation is not obvious when it is far away from the die head. Compared with the velocity decay, the temperature decay is generally gentle, and it is still obvious when it is far away from the die head, The distribution of velocity and temperature along the y-axis at different Z positions shows a unimodal shape [2]. The position of Z is 40.0–100.0 mm, and the curve shape changes from sharp to blunt, indicating that the velocity and temperature change sharply near the central axis. It can be seen that the tensile action on the fiber mainly occurs near the central axis, and in the region of Weijin die head δ It can be seen that the attenuation curves of velocity and temperature obtained by the experiment are basically consistent with the simulation results, the experimental value of velocity field is slightly larger than the simulation value, and the temperature attenuation trend obtained by the simulation is smoother than the experimental result. In addition, from the experimental and simulation results, it can be seen that the attenuation curves of velocity and temperature in the gas groove area are different κ In order to save the calculation time, two-dimensional numerical simulation is used to optimize the die structure.

3 Parameter Optimization of Slot Die

3.1 Optimization Process of Genetic Algorithm

Genetic algorithm (GA) is an efficient global search algorithm based on natural selection and genetic theory, which combines the survival of the fittest rule in the process of biological evolution with the random information exchange system of chromosome in the population. In this paper, we use Vector Evaluated Genetic Algorithms (VEGA) 12: firstly, we use the Vector Evaluated Genetic Algorithms (VEGA) of multi-objective genetic algorithm, Vega generates an initial population (parent) randomly; Secondly, the parent population is equally divided into two subgroups: velocity and temperature. The subgroups are selected independently and then merged into one subgroup; The subpopulations were crossed and mutated to get the subpopulations; Finally, the next generation population is obtained by reinserting the progeny population. The optimal solution of speed and temperature can be obtained by continuously carrying out the "split select merge" operation. The calculation software of this paper uses matlab7.0, and the program file comes from the interval of die parameters selection of tool box of genetic algorithm developed by University of Sheffield in UK: the width of air slot e is 0.1–2.0 mm, and the angle of air slot is 0.1–2.0 mm α For 10° ~ eighty°, The optimization process is completed in a binary space. Each individual is represented by a binary string, and its length represents the accuracy. The accuracy of setting the width, angle, distance and depth of the air slot is 0.001, 0.01, 0.0001 and 0.001 respectively, The binary length is 51 bit, and the velocity V and temperature T are obtained by numerical simulation. In this paper, the optimization goal is to make the velocity and temperature values at 100 mm below the center line maximum, while the optimization goal of genetic algorithm is to make the target value of parameters minimum, so it is necessary to transform the target value.

$$\begin{cases} v = -v(e, \alpha, f, s) \\ T = -T(e, \alpha, f, s) \end{cases} \quad (1)$$

Negative value is not suitable to choose the low convergence rate linear scaling function (2) to calculate individual fitness:

$$F(x_i) = 2 - M + 2(M - 1)\frac{x_i - 1}{N - 1} \quad (2)$$

Where m is the upper limit of each generation and N is the population size; X is the individual phenotypic value; F(x) is the fitness value. Selection, crossover and mutation are the three basic operations of genetic algorithm. The corresponding generation gap rate, crossover rate and mutation rate are all limited [3]. The generation gap rate in this paper is 0.9, that is, the number of children is 90% of the number of parents; The crossover rate was 0.7, that is, 70% of the individuals crossed after the combination; The variation rate is 0.005, which means that 0.5% of the individuals in the combined subpopulation mutate. In addition, in order to maintain the stability of the population number of each generation, the re insertion function is used to retain the optimal individuals of the previous generation, and the optimization process is carried out circularly until the final conditions are met.

3.2 Optimization Results and Discussion

Genetic algorithm is a random search algorithm, so it is difficult to find a formal and clear convergence criterion. The common method is to use preset algebra and test the performance of the optimal individual according to the problem, In this paper, the minimum genetic algebra is set and the coefficient of variation (CV) of the target value is introduced to determine whether Vega is terminated or not. The termination condition is that the CV of speed and temperature should be in a downward trend at the same time when the cycle operation reaches the 50th generation. Otherwise, the operation will continue until the CV reaches the termination condition, There are 20 individuals in each generation. The optimization process includes the simulation of the flow field of 1000 slot die heads. The simulation results are all in dawning high performance computer group 32 × In this paper, the goal programming method 16 is used to deal with the goal value uniformly to obtain the optimal individual. The goal programming method first obtains the optimal value (x) of each objective function, and then according to the general requirements of the multi-objective function optimization design, the optimal value (x) of each objective function is obtained, Then the objective function f (x) is unified.

4 Coherent Anti Stokes Raman Scattering

A new detection method, namely surface enhanced coherent anti Stokes Raman scattering (SERS), is developed. The main purpose of this method is to design a new plasma enhanced substrate, which can detect the local electromagnetic field (EMF) of the surface plasmon resonance (SPR) between the nanoparticles on the substrate, By controlling the shape and structure of nanoparticles in the substrate, the near-field response of particles can be further enhanced. Nowadays, the commonly used reinforcing substrate materials are gold, silver and other metals. In recent years, there are also attempts to use special molecular structure instead of metal as the substrate materials.

A new detection method, namely surface enhanced coherent anti Stokes Raman scattering (SERS), is developed [4]. The main purpose of this method is to design a new plasma enhanced substrate, which can detect the local electromagnetic field (EMF) of the surface plasmon resonance (SPR) between the nanoparticles on the substrate, EM) to enhance the detection signal, the near-field response of particles can be further enhanced by controlling the shape and structure of nanoparticles in the substrate. Nowadays, the commonly used reinforcement materials are mainly gold and silver, and in recent years, special molecular structures have been used to replace metal as base material.

5 Research Method

5.1 Structural Design and Simulation Method

Figure 1 is a schematic diagram of the structure and parameters of the substrate. The radius of the middle disk is 72, the radius of the peripheral disk is 72, the distance d between the central disk and the peripheral disk is the same, all disks have the same thickness H1, the width of the conductive junction between the peripheral disks is l, and the angle between the conductive junction and the Y axis is 1 θ, The thickness of the conductive junction is H2, and the coordinate of the center of the conductive junction relative to the origin of the central disk is o0. The material used for the whole substrate structure is gold. The detected material uses tryptophan I, which is one of the three important amino acids of protein and plays an important role in biology and medicine. The Raman shift uses the Raman peak 26 at 1557 CM1.

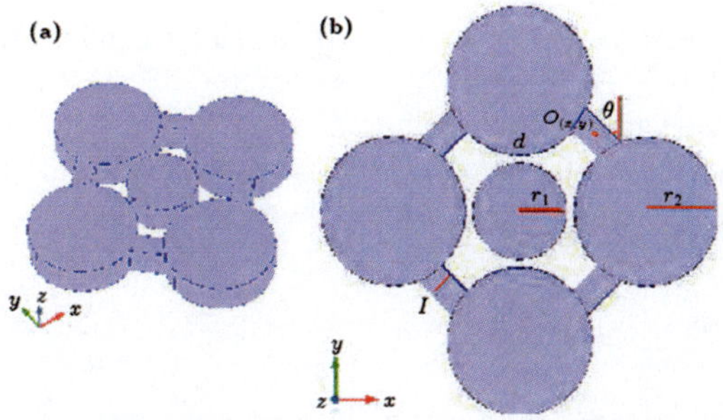

Fig. 1. Structural diagram of the base (a) and its corresponding parameter diagram (b)

The finite element method (FEM) is used in the numerical simulation, and the wave optics module of COMSOL multiphysics 55 is used as the calculation software, The PML with the same thickness as the wavelength is used to avoid false boundary reflection. The background refractive index of the simulation area is set to 1. The scattering cross section (q) in the simulation results of the model data provided by the software is used for the data of gold material Is the ratio of scattering coefficient to equivalent scattering area (SR) = π XRM, where the equivalent radius re is defined by the equivalent volume, v = 4 π × All the scattering cross sections are calculated by the finite difference time domain method (FDTD). In order to keep the stability of the simulation results, the size of the simulation area and the grid remain unchanged in all the calculation processes.

5.2 CTP and Fano Resonant Plasmon

The Fano resonance in oligomers is usually caused by the destructive interference between the super radiation (bright) and sub radiation (dark) resonance modes.2829 the wavelength position of Fano resonance can be adjusted by adjusting the distance between nanoparticles and the size of nanoparticles, CTP can adjust the resonance wavelength of nanostructures by changing the conductivity of the conductive junction between nanoparticles. The combination of CTP and the resonance mode of nanostructures can produce Fano resonance, which can make the resonance range of the substrate highly tunable, At the same time, CTP resonance can also broaden the resonance range of the original structure to the infrared wavelength. Changing the geometric parameters of the disk structure will change the position of its own hybrid resonance wavelength, so the geometric parameters of the disk remain unchanged in all simulations. The conductivity of the conductive junction can effectively control the resonance wavelength of CTP, It can be seen that the enhanced conductivity of the junction will cause the resonance peak of CTP to shift to higher energy, and its shielding effect will also cause the resonance position of sbdp to shift with the short wavelength position, so the first shoulder position will shift slightly blue. It can be seen that the change of junction tilt angle has little effect on the overall resonance, which is consistent with the characteristics of CTP and hybrid resonance, When the central coordinate (OA) increases from (70, 70) to (110, 110), there will be a blue shift at 80 nm and a red shift after 80 nm. This is because the size of the gold disk leads to the optimal conductivity position of 85 nm. When the conductivity increases from 70 to 80 nm, there is a blue shift. When the conductivity decreases from 80 nm, there is a red shift, The position of the second shoulder resonance caused by CTP will shift to the low energy position, but it has no effect on sbdp. Therefore, it can be seen that the position of the first shoulder in the, and the position of the second shoulder shifts to the long wavelength position.

6 Conclusion

In this paper, numerical simulation and multi-objective genetic algorithm are used to optimize the structural parameters of the melt blowing die. The objective is to obtain the die with the optimal structural parameters, so that the velocity and temperature attenuation of the melt jet flow field is the lowest. The optimized data are based on the results of numerical simulation. Therefore, the numerical simulation and experimental measurement of the flow field are carried out first, On this basis, the die structure, including the width of the air groove, the angle of the air groove, the distance between the air groove and the depth of the nozzle, is optimized. The following conclusions are obtained: (1) the optimal parameters are 1.988 mm for the width of the air groove and 12.28 mm for the angle of the air groove angle, The results show that the smaller the air groove angle and the larger the air groove width are, the more conducive to slow down the attenuation of speed and temperature; (3) the combination of numerical simulation and genetic algorithm provides an effective and low cost method for parameter optimization design.

Acknowledgements. Multi-objective Optimization Design of Melting Model Head and Rapid Commissioning of Melting Equipment of Jiangsu Province (No.: 20KJB530008).

"Optimization Design and Performance Research of Jiangyin Vocational and Technical College" (No.: XJ2020LG001).

References

1. Lei, Y.: Matlab Genetic Algorithm Toolbox and its Application, p. 102. Xi'an University of Electronic Science and Technology Press, Xi'an (2005)
2. Yu, C., Wang, J., Wang, X.: Optimal design of engineering parameters, pp. 213–215. Donghua University Press, Shanghai (2003)
3. Yang, S., Qian, J., Wang, M., Zhu, C.: Numerical simulation of blade icing based on XFlow. J. Guizhou Univ. (Nat. Sci. Ed.) **38**(02), 30–36 (2021)
4. Wang, M.: Multi objective optimization of hot stamping process of automobile B-pillar reinforcing plate based on genetic algorithm. Forging Technol. **46**(05), 46–52 (2021)

Application of Artificial Immune Algorithm in Evolutionary Creation

Shuyan Wu(✉)

Jiangxi Teachers College, Yingtan 335000, Jiangxi, China

Abstract. With the rapid development of modern information technology, the corresponding computer music is also produced. How to apply artificial intelligence technology in the field of music has become a new research field, which is a new challenge for music sociology. This paper first introduces the artificial immune algorithm, and then explains the application of artificial immune algorithm in the evolution of music creation, analyzes the trend and characteristics of this new way, hoping that through this kind of research, technology can better serve music.

Keywords: Artificial immune algorithm · Evolution music · Musical composition

1 Introduction

Artificial intelligence is not a new term. As early as the 1950s, the computer field caused the first upsurge of science and technology, and experienced many low ebbs. However, with alphago of Google's deepmind team beating the champion of human go, the artificial intelligence technology in a computer discipline quickly triggered the next wave of science and technology, and created new opportunities in various industries. Since the 1950s, artificial intelligence has developed rapidly and applied in engineering practice. The theory of artificial immune system (AIS) originates from the field of biology and medicine. At present, the definition of AIS is mainly as follows: "AIS is a strategy of data processing, classification, representation and reasoning, which is based on a specious biological paradigm, namely human immune system" [1].

With the development of artificial intelligence technology, it is gradually applied to music creation, production analysis and education. Among them, the use of AI composition or auxiliary composition has become one of the most concerned application directions, and many questions also arise. For example, can AI composing technology replace human composers to a certain extent? How will AI composition technology affect the thinking of traditional composition? These will become interesting and serious topics. It is not a new topic to compose music automatically with AI technology. The related research has been started many years ago, but it has been limited by technology.

B. J. Jansen et al. (Eds.): *International Conference on Cognitive based Information Processing and Applications (CIPA 2021)*, LNDECT 85, pp. 766–771, 2022.
https://doi.org/10.1007/978-981-16-5854-9_98

2 Artificial Immune Algorithm

The antigen and antibody are respectively corresponding to the objective function and feasible solution of the optimization problem in the artificial immune algorithm. The affinity between antibody and antigen is regarded as the matching degree between feasible solution and objective function: the affinity between antibody is used to ensure the diversity of feasible solution, the expected survival rate of antibody is calculated to promote the inheritance and variation of better antibody, and the memory cell unit is used to save the preferred feasible solution to inhibit the generation of similar feasible solution and accelerate the search for the global optimal solution, When the similar problem appears again, it can produce the better solution or even the optimal solution quickly [2].

2.1 Basic Immune Algorithm

The basic immune algorithm is based on the basic mechanism of biological immune system and imitates the human immune system. Inspired by somatic cell theory and network theory, the basic immune algorithm realizes the functions of antigen recognition, cell differentiation, memory and self-regulation similar to biological immune system. The basic immune algorithm is shown in Fig. 1. If the immune algorithm is compared with the general search method to solve the optimization problem, then the antigen, antibody, affinity between antigen and antibody respectively correspond to the objective function of the optimization problem, the optimization solution, and the matching degree between the solution and the objective function.

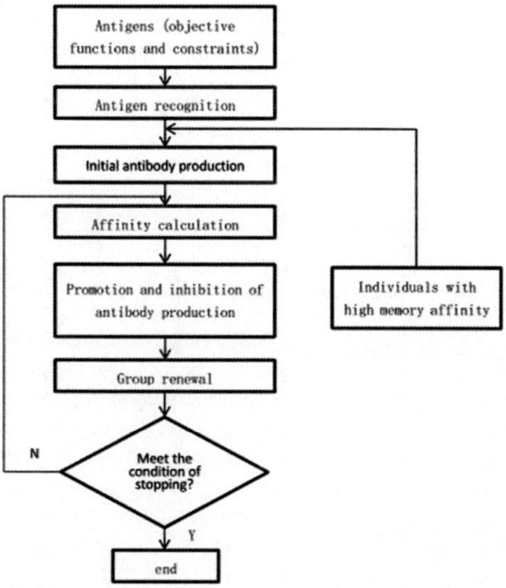

Fig. 1. Flow chart of basic immune algorithm

2.2 Negative Selection Algorithm

Negative selection algorithm is based on the specificity of biological immune system, and uses the "negative selection" process of thymus T cells generation in biological immune system for reference. Forrest studies a negative selection algorithm for detecting the change of variable data, which is used to solve the problems in the field of computer security. This algorithm makes the immune system play a role through the system's successful monitoring of abnormal changes, and the key to the success of monitoring is that the system can distinguish its own and non self information: randomly generate detectors, delete those detectors that have detected itself, so that those detectors that have detected non self can be retained. The flow of negative selection algorithm is shown in Fig. 2.

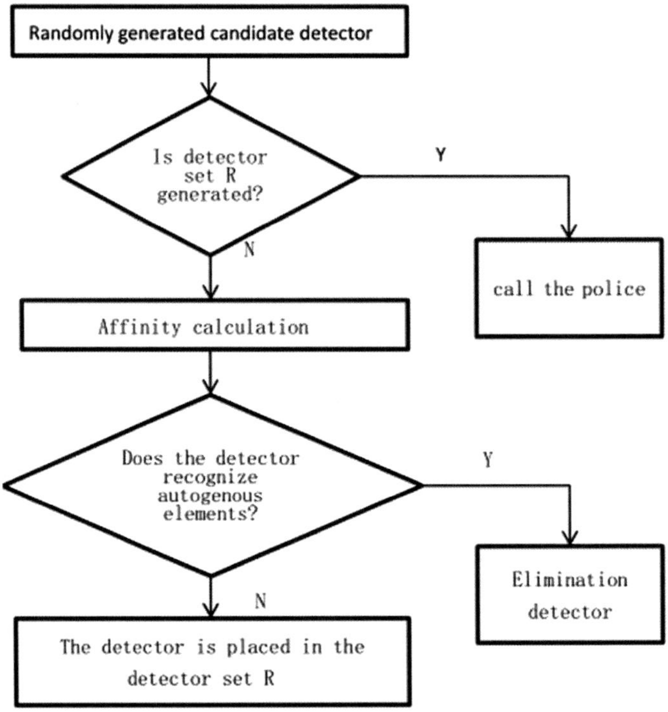

Fig. 2. Flow chart of negative selection algorithm

2.3 Clonal Selection Algorithm

In 1959, Burnet, a famous immunologist, put forward the clonal selection theory and explained the phenomenon of immune tolerance with the theory of clonal clearance. Burnet's clonal selection theory points out that there are about 10 * ~ 107 cell clones with immune activity in vivo, and each clone cell has a specific function to react with its corresponding antigenic determinants. However, the immature t.b-reactive cell lines were eliminated by antigen contact, which resulted in immune tolerance. It is known that a large number of immature autoreactive T cells in the thymus are cleared due to programmed death after contacting with the corresponding autoantigen, which is the most effective mechanism to maintain self tolerance [3].

In the experiment of using clonal selection algorithm to find the maximum of function $f(x, y) = xsen(4\pi x) - ysen(4\pi y + \pi) + 1$ the algorithm not only searches the global maximum of the function, but also obtains the local maximum. Genetic algorithm can only search the global maximum when solving the same problem. This shows the effectiveness of clonal selection algorithm in solving multimodal problems.

3 Traditional Algorithmic Composition

3.1 Translation Modes

This kind of mode is mainly to transform some other media information into music related information, which can be generated by certain rules or randomly. For example, the color and shade of a picture are mapped to the color and shade of a voice, but the results and forms of this direct and simple correspondence are often far fetched.

3.2 Mathematical Models

In this way, music is mainly produced by mathematical model. In this model, many elements of music are composed by non deterministic methods. This kind of random way has been applied in the creative process of many traditional musicians, and the random here is completely based on the mathematical way. In the creation process, the composer can participate in the parameter configuration related to the algorithm. The mathematical model is mainly realized by Markov chain, Gaussian distribution and fractal theory. The problem of this method is that the algorithm is completely based on the setting of rules. For the complex work of music creation, the formula implementation can only complete one of the processes, and the creation result completely realized by mathematical model is difficult to be perfect.

3.3 Grammar

In this way, we think that there is grammar in music, just like grammar in language. In this way, the concept of linguistics is used for reference, and music is treated as a special kind of voice. The research direction is how to extract algorithms from music in this way. At present, there are some research results, and we can also learn from other disciplines. The problems of this way are as follows: first, grammar is hierarchical, or the rules of language are relatively fixed, but the expression of music is impromptu or fuzzy; Second, when there are many grammar analysis and rules, especially in the case of fuzziness, the computational complexity is relatively high.

4 Application of Artificial Immune Algorithm in Music Evolution Creation

When the artificial immune algorithm is applied to music creation, the signal information or symbol information is usually used to determine the type of music signal to be processed. But in the process of deep learning, the use of symbolic information is relatively more common. The first kind of signal information is audio signal, which can be waveform file or audio spectrum information processed by Fourier transform. Symbol class information mainly uses MIDI information. MIDI is a mature and widely used format, which mainly uses note on information and note off information in note information, and uses the value between 0 and 127 to represent the main music information. Another information used for symbol information is the tick value in MIDI [4].

In immune algorithm, there is another way to exchange information with computer, which is to express music directly with text information. Of course, there are many rules. In addition, there are chords, rhythms or scores for deep learning. On the current level of research, first of all, the vast majority of studies do not consider the music expression of automatic composition, that is to say, the music produced is basically mechanical, or a combination of multiple audio sampling. Secondly, due to the relatively complete development of voice sampling and sound synthesis in commercial applications, there is no relevant AI research to focus on these aspects. Thirdly, because the structure of game music depends entirely on the game scene, most of the studies have not involved this type.

In fact, the related products with artificial immune algorithm as the core have begun to serve music. For example, the artificial intelligence composing system deepbach is used to create polyphonic music, especially chants. In order to ensure the final effect, the system mainly works around the four part chorus, and focuses on the creation of Bach's four part chorus works. The system uses a flexible and efficient artificial immune algorithm. In addition to generating by machine learning, users can also add note generation in the process, which is called adding unary constraint. The model is controlled by rhythm or related information. This way of allowing users to intervene in the process is often ignored in the music probability model.

5 Conclusion

From the above application examples, we can see that the combination of rules and related artificial immune algorithm model to achieve music creation is one of the relatively reasonable schemes. In the process of artificial immune algorithm participating in music creation, not only the melody and lyrics, but also the audience's user data will become its goal. It will fully participate in and change the music production process, from harmony, music arrangement to final sound synthesis, assist human work and improve efficiency.

There is no definite conclusion about the future of artificial immune algorithm in music, but we prefer to believe that it can solve the problems of traditional music industry, such as reducing the threshold of music education, improving the efficiency of teaching, solving the problem of music copyright and so on. It is believed that with the rapid increase of investment and attention of related research at home and abroad, more theories and applications will emerge.

Acknowledgements. Provincial subject of teaching reform in Jiangxi Province in 2020 JXJG—20—47—4 Research on College "Music+" Aesthetic Education Teaching.

References

1. Fernandez, J.D., Vico, F.: AI methods in algorithmic composition: a comprehensive survey. J. Arif. Intell. Res. **48**(1), 513–582 (2013)
2. Luo, X.: Artificial immune genetic learning algorithm and its engineering application. Doctoral Dissertation of Zhejiang University, Hangzhou (2002)
3. Xiao, R., Wang, L.: Artificial immune system: principle, model, analysis and prospect. Acta Computa Sinica (12) (2002)
4. Wang, C.: Humanized production of film and TV drama music and computer music. J. Hotan Teach. Coll. **25**(5) (2005)

Hidden Danger and Control of Construction Safety of Internet of Things Based on Cognition

Shihai Wang(✉)

Dongying Vocational Institute, Shandong 257091, China

Abstract. The development of scientific research, industrial production and residents' life all need to use various types of buildings, so the development of the construction industry at this stage is very fast. In the construction management, the importance of safety management is self-evident, it is not only the need of the sustainable development of the construction industry, but also the guarantee of social stability and economic development. Therefore, this paper analyzes the construction safety risks and control countermeasures based on association rules. With a sound and reasonable safety management system to effectively regulate all aspects of construction, reduce the incidence of safety accidents to the minimum, so as to realize the sustainable development of the construction industry and make contributions to the construction of a harmonious society.

Keywords: Association rules · Construction project management · Hidden danger · Countermeasures

1 Introduction

With the rapid development of China's economy and society, the construction industry has shown a trend of rapid development in recent years. At the same time, the construction industry has always been a high incidence of safety accidents, and the safety problems of the construction industry have attracted more and more attention of the society. There are all kinds of management work in the construction production process, among which safety management is one of the important work. The occurrence of production safety accidents will not only cause huge property loss to the state and the collective, but also affect the life safety and vital interests of employees, the social reputation and economic benefits of enterprises, and also cause adverse effects on society. To strengthen the safety management in the production process of enterprises and effectively guarantee the life safety of construction workers in the construction process is the safety production policy put forward by the party and the state. In the specific safety production process, all departments must fully cooperate, and all staff should actively participate. Safety management should run through the whole process of construction. Only by using scientific management means, management methods and relevant laws, regulations and policies, can safety production in the process of construction be finally realized [1].

B. J. Jansen et al. (Eds.): *International Conference on Cognitive based Information Processing and Applications (CIPA 2021)*, LNDECT 85, pp. 772–776, 2022.
https://doi.org/10.1007/978-981-16-5854-9_99

2 Association Rules

2.1 Basic Concepts of Association Rules

Association rules represent the rules of a certain association relationship between a group of objects in a database. For example, association rules can be expressed as "95% of the customers who bought items a and B buy items C and D again". From these rules, we can find out the customer's purchase behavior pattern, which can be applied to commodity shelf design, production arrangement, targeted marketing and so on [2].

Definition 1 (association rule) assumes that *X*,*Y* and y are nonempty itemsets. When $X \subseteq 1, Y \subseteq 1$, and $X \cap T = \phi$, the implication $X \Rightarrow$ Y is called association rule. Where x is called the rule antecedent and *Y* is called the rule antecedent. In order to make the rule $X \Rightarrow$ Y hold in transaction set *D*, the occurrence of itemsets *X* and *Y* in transaction set *D* should meet the threshold of minimum support s and minimum confidence *c* at the same time.

Definition 2 (support) is the percentage of both transaction X and transaction y in transaction database d in rule $X \Rightarrow$ Y, that is, support ($X \Rightarrow$ Y) $= P$ ($X \cup Y$). Where P ($X \cup Y$) is the probability that *X* and *Y* appear simultaneously in the data set *D*. The minimum support threshold s refers to the fact that transaction item $X \cup Y$ appears at least s% in transaction database *D*, which is recorded as minsuppo.

Definition 3 (confidence) in rule $X \Rightarrow$ Y, the percentage of transaction *Y* in transaction database *D* that contains *X* is also included in transaction *Y*, i.e.

$$confidence(X \Rightarrow Y) = P(Y|X) = P(X \cup Y)/P(X) \quad (1)$$

Where P ($Y|X$) is the conditional probability, which refers to the probability of *Y* transaction when *X* transaction appears in transaction database *D*. The minimum confidence threshold *c* means that at least *c*% of the transactions containing *X* in transaction database *D* contain transaction *Y*, which is recorded as minconf.

2.2 Smart Construction Site Platform for Workers' Credit Management and Wage Management

It has talent assessment function module, which can assess employees from multiple dimensions such as attendance, operation skills and work efficiency. According to the evaluation standard, the management staff can weight and score the staff, and generate the evaluation results into the personnel file information database. At the same time, a blacklist function is set in the system to include dishonest employees or employees who seriously violate regulations into the blacklist system, share it with authorized enterprises, and reduce the employment risk of enterprises. Another important part of the intelligent construction site platform is the wage management module, which is directly related to the interests of employees and the stability of enterprise employment. The wage management module can automatically check the account with the attendance record and the reward and punishment record, and timely send out the alarm for the situation that the wage is inconsistent with the attendance, the actual wage is inconsistent with the payable wage, and the account fund is insufficient. Through the wage management module, we can timely alert and remind the management personnel to solve the wage issue, so as to avoid the employment and salary disputes.

In Fig. 1, D is the data set, algorithm-1 is the search algorithm of frequent itemsets, algorithm-2 is the generation algorithm of association rules, and R is the excavated association rule set. Users interact with algorithm-1 and algorithm-2 by specifying minsupport minconfidence, and interpret and evaluate the mining results by interacting with R [3].

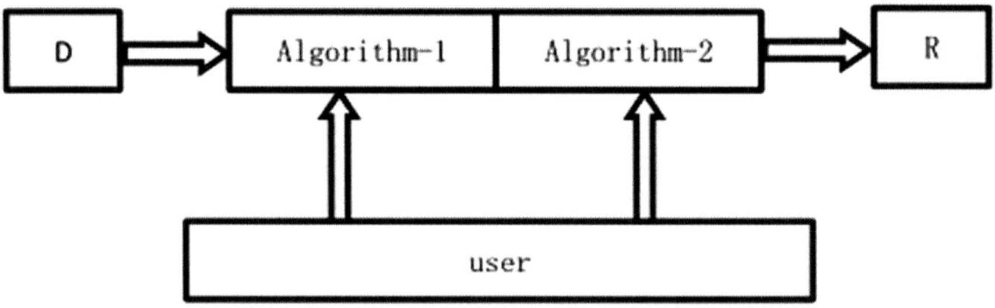

Fig. 1. Basic model of association rule mining

3 Hidden Danger in Construction

3.1 The Safety Protection Measures of Construction Engineering Are Not Perfect

Safety is an important foundation for stable construction of construction engineering. When some construction units are working, they often don't pay enough attention to the safety problems. They often don't set up a protective fence on the construction site, and they don't have a simple screen. When some construction units are working at night, in order to save costs, they don't equip the safety indicator lights. When they are working at height, they don't set up a protective net. In this case, construction workers, pedestrians or vehicles passing on the ground may fall and hit, causing casualties.

3.2 Excessive Labor of Construction Personnel

Generally speaking, the best time for construction is summer, which leads to the construction workers have to work in the hot season. In this case, the construction unit does not provide the necessary safety protection measures for the construction workers. What's more, the safety helmet is not distributed to the construction workers, and the relevant management personnel still pay attention to the progress of the project, Regardless of the actual conditions of the construction site and the actual labor ability of workers, careful construction in this environment will inevitably lead to safety accidents.

3.3 The Safety Awareness of Management Personnel is Weak, and the Safety Awareness of Construction Personnel is Relatively Low

In construction enterprises, the safety awareness of some management personnel is still very weak, in the construction site is often only concerned about the surface work, management construction personnel work, can not find the problems in the construction process in time, the relevant safety system has become empty talk, which makes the construction project left security risks. The safety awareness of construction personnel is also relatively low. The main reason is that most of the construction workers are migrant workers, their own knowledge and skills are often unable to meet the safety needs of construction projects, which requires training in related technology and safety [4].

4 Hidden Danger Control of Construction Safety Based on Association Rules

The characteristics of construction engineering determine the long-term and complexity of control and prevention of potential safety hazards. If this work is not carried out smoothly, it may have a great impact on the whole construction project. In this case, construction enterprises should focus on the prevention and control of potential safety hazards to avoid the occurrence of various safety accidents. The following describes the prevention and control principles of potential safety hazards based on association rules.

4.1 Prevention and Control of Security Risks by Association Rules

① In order to completely eliminate the potential safety hazard or reduce the risk to the greatest extent, the construction engineering system with perfect safety should be established on this basis, and all kinds of potential safety hazard information should be mined. At the same time, it is also necessary to protect individuals. ② The work of controlling potential safety hazards should run through the whole construction project, and be combined with relevant preventive means to establish a perfect system of preventing and controlling potential safety hazards. ③ Key safety information investigation should be carried out for potential safety hazards, and corresponding countermeasures should be formulated to minimize the negative impact of safety accidents.

4.2 Prevention and Control of Potential Safety Accidents

Accident prevention is mainly divided into two points. One is to analyze the information and data of possible safety accidents before the accident, and prevent them in advance. The other is to take a series of measures when the accident occurs. Based on association rules, the impact of the accident can be minimized, and the possible casualties and property losses can be reduced. In general, the main work sequence of accident prevention is as follows: To investigate the potential safety hazards: to control the factors of potential safety hazards; to avoid the chain reaction of safety accidents; to have perfect safety measures; to reduce the negative impact of accidents.

4.3 Safety Technology and Management Measures

When introducing construction raw materials, construction engineering units must ensure that the raw materials meet the corresponding safety standards. To avoid the occurrence of construction safety accidents caused by unqualified materials. In addition, the construction materials should be inspected regularly, combined with the relevant safety management system, to ensure the stable supply of raw materials to the greatest extent.

In view of the safety accidents that may occur in the construction of construction projects, the relevant units and departments have taken certain measures. Before the accident, the relevant enterprises and departments need to actively combine the relevant safety and measures, so as to avoid the accident as far as possible, and reduce the loss in time. Mining information data based on association rules, security protection in accordance with a certain order, the security risks are isolated; The safety weak link should be protected; Personal safety protection: life saving measures.

5 Conclusion

In short, at this stage, China is in a period of rapid economic development, engineering construction is increasing, and engineering safety management is one of the most important problems in engineering construction. In recent years, engineering safety accidents have occurred from time to time, which has brought serious social impact and casualties. It is closely related to the quality of the project, the safety of the construction personnel and the stability of the society. In the project construction management, the word "safety" must be strictly integrated into the whole project construction process, and effective prevention and control measures should be formulated and taken. At the same time, the training and education of the construction personnel should be paid attention to, so as to promote their attention to and implementation of the concept of safe construction, Further ensure the safety and stability of engineering construction, and realize the healthy development of construction enterprises on the premise that the interests of the state and the public are not infringed.

References

1. Song, R.: Discussion on construction safety hazards and countermeasures in construction management. Real Estate Guide **11**, 282 (2013)
2. Xu, J., Yu, G. Research and application of association rule mining algorithm. Microcomput. Inf. (12) (2009)
3. Sun, N.: Research on association rules mining. Comput. Learn. (1) (2009)
4. Li, Q.: Discussion on the potential safety hazards and improvement measures in architectural structure design. Theoret. Res. Urban Constr. (Electronic Version) (2014)

Application of Precise Mining Optimization Algorithm in Enterprise Economic Management

Maonian Lin(✉)

School of Economics and Management, Hainan College of Vocation and Technique, Hainan, Haikou, China

Abstract. In order to adapt to the development of today's society, how to better improve the economic management level of enterprises has become our current topic. Through research and innovation, we found that the cost optimization algorithm can be well applied to the application of enterprise economic management. It opens up a new application field for enterprise economic management. At the same time, according to the actual situation of the optimization and innovation, by the user's praise.

Keywords: Cost optimization algorithm · Enterprise economy · Business management

1 Introduction

In the current environment of China's socialist market economy, how an enterprise can survive better in the fierce competition and gradually expand itself is a major challenge for many enterprises. In addition to optimizing management and technological innovation, opening up a new market and a new field is also indispensable in enterprise economic management.

2 Introduction of Optimization Algorithm

2.1 Differential Evolution Algorithm

Differential evolution (DE) is a heuristic random search algorithm based on population difference, which is proposed by R storn and K, price to solve chebysher V polynomial. DE algorithm also belongs to intelligent optimization algorithm, similar to previous heuristic algorithms, such as ABC, PSO, etc., all belong to heuristic optimization algorithm. DE algorithm is mainly used to solve real optimization problems. The algorithm is a group based adaptive global optimization algorithm, which belongs to the evolutionary algorithm. Because of its simple structure, easy implementation, fast convergence and strong robustness, it is widely used in data mining, pattern recognition, digital filter design, artificial neural network, electromagnetics and other fields. In essence, differential evolution algorithm is greedy. If the new individuals generated after each mutation

B. J. Jansen et al. (Eds.): *International Conference on Cognitive based Information Processing and Applications (CIPA 2021)*, LNDECT 85, pp. 777–783, 2022.
https://doi.org/10.1007/978-981-16-5854-9_100

and crossover are not as good as the original individuals, they will not be updated for a long time [1].

There are many mutation mechanisms in the algorithm, and the most commonly used methods are as follows: M represents the current original population, n represents the mutation population, J represents the j-th individual in the population, and s represents the s-th generation population. The variation vector is calculated as follows.

$$N_j(s) = M_{r1}(s) + K(M_{r2}(s) - M_{r3}(s)) \tag{1}$$

The K mutation operator determines the difference step size of individual population. The smaller K will affect the differences among individuals and make the algorithm result fall into local optimum; the larger K will enhance the global search ability of the algorithm, which is conducive to the search of the optimal solution, but will affect the convergence speed of the algorithm [2].

2.2 Artificial Bee Colony

Artificial bee colony (ABC) algorithm is a global optimization algorithm based on colony, which was proposed by karaboga in 2005. The most intuitive background comes from the behavior of bees collecting honey. Bees act according to their different division of labor, and realize the sharing and communication of swarm information, so as to find the best honey source [3]. This algorithm can solve the problem of function solving very well. Because of its powerful performance, it is favored by the majority of researchers and has been widely studied and applied.

In the artificial bee colony algorithm, each hired bee first remembers a honey source location, and then learns from randomly selected neighbors to update the honey source location.

$$V_i^j = X_i^j + rand(-1, 1) \times (X_i^j - X_n^j) \tag{2}$$

In particular, the traditional ABC algorithm only updates a single dimension, and dimension J selects one of all dimensions. The artificial bee colony algorithm adopts greedy search strategy. If and only if the updated honey source location is better than the original honey source location, the updated honey source location will be accepted and the honey source location will be updated.

The artificial bee colony algorithm has strong global search ability and slow convergence speed compared with other algorithms. When it is close to the global optimal solution, the algorithm is easy to fall into the local optimal solution, the later search speed is slow and the accuracy is low. The algorithm is suitable for solving multivariable function optimization problems. In addition, the algorithm has less control parameters, strong robustness, flexibility and easy implementation. It is better than other algorithms in solving multimodal problems and is more suitable for solving those traditional complex optimization problems.

The population distribution of the algorithm in 300, 500, 800 iterations of sphere function (f1) is listed as shown in Fig. 1.

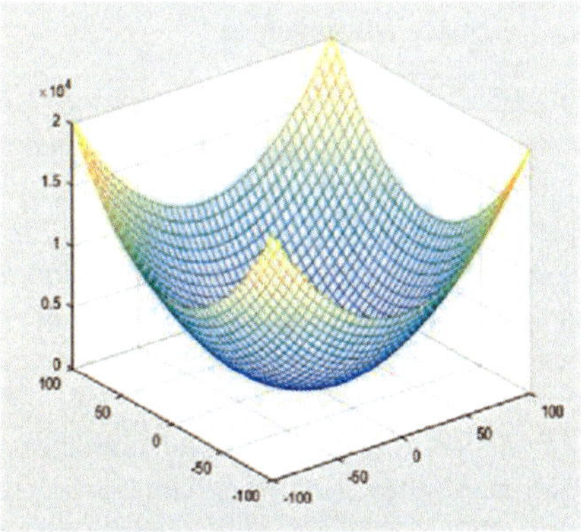

f1 Sphere function

Fig. 1. Typical standard test function diagram

3 Analysis of Countermeasures and Suggestions for Enterprise Economic Management

3.1 Uphold the Basic Concept of Innovation of the Times

As the saying goes, "innovation is the soul of a nation". Only through continuous innovation can each enterprise obtain a large amount of excess value in daily operation, which can promote social development and economic progress [4]. Nowadays, innovation capability, to a large extent, refers to the ability to constantly adapt to the needs of users and the basic ability of the development of the times. In the process of future development, enterprises need to follow the trend of the development of the new era, firmly grasp the pulse of the development of the times, gradually change the concept of development with a global perspective and strategic thinking, and gradually shift the core of development to optimize the enterprise structure and optimize the enterprise structure Scientific management and control, brand building and talent team, at the same time to gradually subvert the past, so as to obtain satisfactory development results. Specifically, enterprises need to ensure that in the process of production and operation, they constantly cultivate a stable overall view and forward-looking vision, make full use of the current convenient Internet, and closely link cultural facilities, production lines, factory suppliers, users, and products. This not only puts forward higher challenges to the decision-makers of enterprises, but also puts forward more requirements to the employees of various enterprises. In the follow-up work, it also requires enterprises to carefully analyze and judge their own decisions, activities and results before and after major decisions, detect the existing defects through reasonable examination, and then gradually correct them, Only in this way can we ensure the sustainable and healthy development of enterprises.

3.2 Strengthen Human Resource Management

Enterprises have accumulated a lot of resources in the process of long-term operation, which usually comes from the gradual growth of knowledge economy. Therefore, enterprises need to fully realize the key role of knowledge economy, and then apply it scientifically and efficiently. Moreover, they can gradually store a lot of knowledge power, enrich their own cultural connotation, and improve production technology, In order to improve the development of human resources imperceptibly. Nowadays, in the environment of China's knowledge economy, the competition among enterprises is becoming increasingly fierce. The focus of the competition is not only reflected in products, but also in the competition of talent team. In view of this situation, if enterprises want to get more long-term development, they need to enhance their comprehensive strength, and they need to systematically recognize the key significance of human resources in personnel training. In the process of human resource management, the relevant person in charge should firmly adhere to the basic principle of people-oriented, flexibly adopt the basic way of emotional management, and create a harmonious enterprise operation environment, so that the working ability and achievements of the employees can be recognized and encouraged by the decision-making level of the enterprise, and fully respect the individual wishes of the employees, We should mobilize their work enthusiasm and vitality to the greatest extent, stimulate the innovative ideas of all employees, constantly tap their innovative potential, and finally realize the innovation and reform of enterprise economic management. In addition, relevant enterprise leaders also need to enhance the strength of personnel training, organize employees to receive training and education on time, improve their working ability, broaden their horizons, let them fully grasp the overall situation of the current market economy development, clarify the differentiation demand standard proposed by professional talents, and timely let employees of various enterprises adjust their working methods, Innovate the concept of work [5].

3.3 Building a Sustainable Salary And Welfare Incentive System

Compensation and welfare incentive system has two significant characteristics, one is transient, the other is cyclical. At a certain stage, employees' enthusiasm for work is rising because of the salary and incentive system. However, society is progressing and the times are changing. If the salary and incentive system is not adjusted, employees will have slack psychology and lose their enthusiasm after a certain period of time. It is not only the ordinary employees of the enterprise, but also the managers, It is not conducive to the development of enterprises, the sustainable development of enterprises can not be guaranteed. Therefore, the establishment of a long-term salary and welfare incentive system in line with employees will mobilize the enthusiasm of all levels of employees, from grass-roots employees to managers, and promote the sustainable development of enterprises.

3.4 Overall Development of Internal Management Innovation Mode

First of all, the relevant person in charge should do a good job in information communication within the enterprise, so as to form a joint force of forward development.

Specifically, enterprises should flexibly use information systems, such as CRM and OA, to achieve the goal of internal information sharing. Only in this way can information be transmitted in time, and it also helps to speed up the approval of internal processes. With the gradual improvement of the level of technological development, mobile online approval can also be completed by means of mobile client. Therefore, the timeliness of information sharing can be improved, and the sense of post responsibility of staff can also be strengthened. It is worth noting that by setting up relevant systems and clarifying the responsibilities of each post, a number of business and economic resources inside and outside the enterprise can be more reliably transformed into various information data, and transmitted to each responsible person at a more efficient speed. In addition, during the production and operation of the enterprise, the relevant staff also need to integrate various data information in time, analyze the crisis behind the data information as a whole, maximize the profit and avoid risks, and make the development of the enterprise tend to be stable. Secondly, we should constantly improve the enterprise industrial chain.

Considering that enterprises are often in the established economic environment, from the perspective of enterprises, we need to consider the vertical and horizontal direction, that is, we should pay full attention to the development of the industrial chain. And the connection between the upstream and downstream enterprises refers to the upstream and downstream of the industrial chain. Under the background of sharing economy, it is not feasible for enterprises to develop independently. At this time, we should actively promote the coordination and cooperation between upstream and downstream enterprises to achieve mutual benefit and win-win situation among enterprises.

Generally speaking, within the enterprise, it will be subdivided into several industrial chains and value chains. And each process is extremely important, the relevant staff should pay enough attention to it, otherwise it will lead to unnecessary consumption of resources and loss of project business. For example, if we want to build an efficient budget system, we need to start from the sales department and integrate the capital budgeting of enterprise financial department. Therefore, in the process of carrying out the internal economic management of enterprises, it is necessary to abandon the small team management and control mode, and build a interlocking industrial value chain.

4 Innovating the Organizational Structure of Economic Management

Under the influence of the traditional enterprise model, the current economic management of many enterprises is not reasonable, lack of a perfect structural mechanism. In view of this situation, the relevant staff can start from the following aspects: first, break the traditional multi-layer vertical management and control mode, reduce the management level, and compress the company's core departments, so as to take an elite road. In this way, it will also help the enterprise to obtain accurate information in the operation process and realize the efficient transmission of information, To ensure the rationality of business decisions; secondly, to enhance the flexibility of the economic management structure, and constantly update the inherent mode of the past organizational structure, which is conducive to enterprises more quickly adapt to the new market environment; As shown in Fig. 2. finally, to promote the informatization process of enterprise economic management, optimize the division of labor and cooperation within the organization, and ultimately improve the actual efficiency of work.

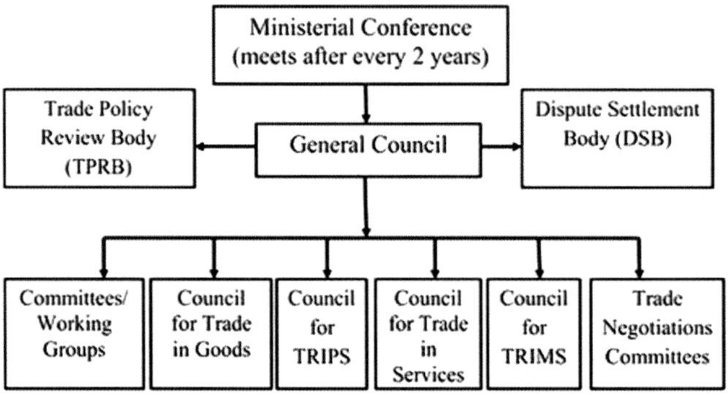

Fig. 2. Organization structure of economic management.

5 Conclusion

With the increasingly complex market environment in our country, at this time, in order to comply with the trend of market development, enterprises should constantly innovate their own economic management mode, take into account the actual development of enterprises, master the law of development, attach importance to the cultivation of talent team, and strengthen internal management, so as to help enterprises realize the maximization of economic and social benefits. This paper first introduces several common optimization algorithms. Based on the research of these theories, aiming at the slow convergence speed of the artificial bee colony algorithm and the low accuracy of the later algorithm, this paper mainly improves these two aspects of the algorithm, and applies the improved algorithm to practical problems, such as applying the mnabc algorithm proposed in this paper to the power economic dispatch problem, and applying the eabc algorithm proposed in this paper to the truss structure design problem, To verify the effectiveness of the proposed algorithm. Artificial bee colony algorithm has the characteristics of slow convergence speed and low precision in the later stage, so it is not easy to fall into local optimum when dealing with complex optimization problems. Inspired by particle swarm optimization algorithm, it provides an intelligent learning strategy for the employment bee colony stage and observation bee stage of artificial bee colony algorithm. In addition, according to the characteristics of the Scout bees, a novel learning strategy is proposed for the Scout bees.

References

1. Na, L.: Research on effective innovation measures of enterprise economic management under the new situation. Sci. Res. **34**(12), 176 (2015)

2. Chunli, Z.: Research on effective innovation and measures of enterprise economic management under the new situation. Chinese Market **21**(18), 108–110 (2018)
3. Na, L.: Research on effective innovation measures of enterprise economic management under the new situation. Sci. Res. **22**(01), 257–258 (2015)
4. Zhao, D., Wu, S.Y., Zhang, T.: Innovation strategy of enterprise economic management under the new situation. Bus. Exhib. Econ. **34**(12), 56–57 (2020)
5. Zhou, C.: Core exploration of enterprise economic management and management promotion activities. Fortune Today **23**(04), 45–46 (2021)

Application of AI Aided Cognitive Computing Method in CAI System

Tuo Niu(✉)

Sports Department of Shaanxi Institute of International Trade and Commerce, Xi'an 712046, Shaanxi, China

Abstract. This paper mainly studies the application of intelligent computer-aided teaching system in college physical education. Through the analysis of the necessity and current situation of computer-aided teaching system and efficient physical education, this paper puts forward the application strategy of computer-aided teaching system in college physical education, in order to effectively apply computer-aided teaching system in college physical education, To improve the quality and efficiency of College Physical Education and promote the healthy development of modern college physical education.

Keywords: Physical education teaching · Computer aided instruction system · Intelligence

1 Introduction

With the rapid development of modern science and technology, especially multimedia technology, the modern educational technology with computer as the core plays an increasingly important role in the teaching reform of our country. The birth of computer aided instruction system (CAI) in the 1950s marks the beginning of the era of computer teaching [1]. Cai is a new interdisciplinary subject, which integrates computer science, pedagogy, psychology, information theory, cybernetics, system theory and other disciplines and technologies. It integrates text, image, sound, animation, image and other media to organize teaching, and embodies a new educational idea and teaching method. It breaks through the linear limitation of traditional media and presents information knowledge vividly to learners, which has and will continue to have a profound impact on the reform of traditional education and the realization of teaching modernization. As an important part of education, physical education teaching reform is imperative. This paper will discuss the application and development of CAI in physical education.

2 The Application of CAI in Physical Education

In the traditional physical education teaching, the technology class is mainly "demonstration + explanation + ten exercises", while the theory class purely depends on the

B. J. Jansen et al. (Eds.): *International Conference on Cognitive based Information Processing and Applications (CIPA 2021)*, LNDECT 85, pp. 784–789, 2022.
https://doi.org/10.1007/978-981-16-5854-9_101

teacher's classroom explanation. The teaching method is rigid, and there are some disadvantages: 1. Students can't get a complete image memory from the teacher's explanation; 2; 2. Some technical movements are dynamic, so it is difficult for students to distinguish the details of each movement; ③ The accuracy and standardization of the demonstration movement can not be guaranteed; 4. It is easy to be limited by the field equipment, which affects the teaching effect. Computer aided instruction can obtain the feedback information of students' learning through human-computer interaction activities in time, and can find and solve problems and strengthen the teaching content in time. The teaching process has also changed from one-way indoctrination by teachers to two-way communication between teachers and students, thus arousing students' interest and enthusiasm in learning. At present, the application of multimedia technology in the field of physical education is mainly manifested in the following aspects:

2.1 Development and Application of Sports Multimedia Courseware

Multimedia courseware is one of the most widely used forms of CAI in physical education. By using computer multimedia technology and combining with the characteristics of physical education teaching, we can research and make teaching courseware suitable for some subjects and learning courseware suitable for students' after class learning. Modern computer multimedia technology integrates text, graphics, sound, animation, video in one, with rich content, flexible forms and means of expression, making physical education more characteristic.

According to the different teaching contents, teaching objectives and teaching strategies, sports multimedia courseware can be roughly divided into four types: demonstration type, interactive type, intelligent type and comprehensive type. Sports multimedia learning courseware (CAL) is a new learning mode based on the interactive function of computer and modern network technology. The development and application of sports learning courseware make students' learning behavior change from passive to active, which can better reflect students' subjectivity, initiative and autonomy. In addition, we can also use computer multimedia technology for microteaching. Microteaching is a small-scale teaching activity that uses modern educational media to help students master certain skills.

2.2 Application in Physical Education Teaching Management

Office Management

Office software is widely used in daily physical education. Such as the use of software "word" to write teaching plans, the use of "excel" for data statistics. In addition, there are more practical application systems based on physical education developed by using office automation software, such as "physique monitoring system", "track and field games arrangement and performance management system", which greatly improve the work efficiency.

Applied to Sports Training

Using computer technology can accurately analyze and study the movement technology, and can also design high, difficult and new movements on the basis of studying the basic movements. Using multimedia technology to deal with sports skills and tactics can present diversification and multidimensional, make it vivid and clear at a glance, help students to establish the appearance of skills and tactics as soon as possible, improve the accuracy of action understanding, and achieve learning goals faster.

In sports training, we can also use multimedia technology to establish some simulation systems, such as "athlete injury first aid simulation system". This system is a kind of teaching system which can instruct students how to rescue correctly when there is a kind of sports injury in the simulated state. Such a system not only improves the students' understanding of the knowledge, but also exercises their ability to solve problems in emergency situations [2].

Applied to Sports Research

Sports scientific research is the top priority of physical education, which guides the development of sports. The introduction of computer and multimedia technology into sports scientific research is bound to play a huge role in promoting the smooth progress of sports scientific research. Multimedia technology can systematically display the topics and argumentation materials with sound, words, pictures and animation, which can most accurately express the author's ideas and achieve twice the result with half the effort. At the same time, the organic combination of computer and network technology has opened up a new field for the research of physical education teaching, making cross regional collaborative research possible, such as the comparative study of students' Physique in different regions.

3 The Realization of ICAI System in CAI

3.1 System Module Framework

The first mock exam is domain model: the contents of this model include the teaching knowledge, questions, ideal answers, and difficult questions. Generally speaking, the first mock exam is divided into two categories: one is the content of professional knowledge in physical education curriculum; The other is content related problems and solving knowledge points, which can also be called process knowledge.

Second, the first mock exam model: the role of this model is to evaluate whether students receive physical knowledge, and then analyze them in depth, and then understand the degree of students' knowledge of physical education and the causes of misbehavior in learning, and finally, they will be introduced into the student model in the form of current state.

Third, the first mock exam is the core technology of ICAI, which is used to understand students' real level of knowledge and analyze their learning ability, and to further achieve individual teaching and provide a theoretical basis for differentiated teaching strategies.

Fourth, the first mock exam of the first mock exam is to put forward the learning problem. The main task of this model is to ask students to learn questions. The main content and the requirements of the curriculum are mainly designed in the teaching of physical education. The corresponding knowledge points are designed to diagnose and supervise students' learning behavior. When students need further learning materials, this model can select corresponding contents for students.

3.2 System Model

The first mock exam is Overlay Mode: this model can show students' knowledge and knowledge. Such a model can make students know how to correct the behavior in the process of physical education learning.

Second, the first mock exam is (Baggy Model): this model can show students' right and wrong cognition of sports knowledge, and decide the professional deviation of students in this way by measuring the scale. This model can further find out the causes of students' wrong learning [3].

The purpose of ICAI computer-aided instruction system model can be analyzed from two aspects: one is to sort out students' information, and record students' names, gender and student numbers in the process of establishing the model; the other is to accurately grasp students' learning ability and level, so that the teaching system can effectively distinguish students' individual differences, So as to further realize the intelligent computer-aided teaching.

3.3 Advantages of ICA System

First, the system takes the cognitive theory as the core to design the physical education teaching content, so the formed courseware has certain flexibility and variability.

Second, to a certain extent, it can stimulate students' interest in learning, which can cultivate students' subjective initiative and provide students with autonomous learning atmosphere.

Third, to achieve the students as the main body of the difference in teaching, further reflect the new curriculum reform in teaching students in accordance with their aptitude.

Fourth, the classroom teaching effect and students' learning ability are effectively improved, as shown in Fig. 1.

Fig. 1. Students' learning ability

4 System Model

The system consists of a primary user receiving node Pu, a secondary user source node SS, a secondary user destination node SD and K relays {bkyk = 1} which can DF OFDM symbols transmitted by SS. All the above nodes have N orthogonal subcarriers and are configured with a single antenna and operate in half duplex mode. Each OFDM symbol is transmitted by N orthogonal subcarriers. The OFDM symbol transmitted by the transmitter can be expressed as x = [x (1) x (2), X (3), X (n). The relay R is configured with a data buffer BK with a finite length of L (and the length of OFDM symbol is taken as the unit length). OFDM symbols follow the principle of first in, first out in the buffer, and only one OFDM symbol can be transmitted in a time slot. U (BK) represents the number of OFDM symbols stored on BK, 0 ≤ U (BK) SL. Due to the severe path loss and shadow effect, the information transmission between SS and SD can not be completed directly, so it must be realized with the assistance of relay.

4.1 Working Principle and Signal Processing Flow of Two Hop OFDM-M System

At SS or R, NS subcarriers are selected from N orthogonal subcarriers according to the subcarrier group selection criteria. The set of all subcarrier group selection schemes can be represented by Q. The total number of possible selection schemes is given by CNSA, where CNS A is the binomial coefficient. B is the second choice, ∈ Q. N () represents the set of NS selected subcarriers in the scheme. Each OFDM group uses ns selected subcarriers to transmit information of Z bits. Among the NS selected subcarriers, only Na subcarriers are activated and transmit APM symbols as active subcarriers, while the remaining NSNA subcarriers remain silent. B-bit information is divided into index bits of zind length and ZM length by bit splitter. D modulation bit two parts. The index bit of B ndex is used to select the position (index) of the active Na subcarriers, and the modulation bit of ZMO is used for the traditional APM symbol modulation. The index selector will arrange the indexes of NS subcarriers in in ascending order and re label them as new indexes. For the convenience of distinguishing, the index of the selected subcarrier in nis called absolute index, and the new index marked by index selector is called relative index. Absolute index corresponds to relative index one by one.

4.2 The Simultaneous Interpreting Curves of Outage Probability and SNR of Two Hop Cognitive

OFDM-IM system under different transmission schemes are presented. The simulation parameters are k = 2, l = 2, (NSNA, m) = (3,2,2), a = 1, l0 = 20dbbpcu. P = 1 and P = 0.1 represent ideal CS and obsolete CS scenarios, respectively. It can be seen that the MABR / MSRs scheme can obtain lower system outage probability than the buk relay selection scheme without buffer assistance when the SNR is fixed in both ideal and outdated CS scenarios. This is because the buffer assistance relay does not rush to forward, but improves the transmission performance of the system by using the channel time-varying and selecting the best relay link for transmission. It can also be seen from the figure that the outage probability of the outdated CS scenario is higher than that of the ideal CS scenario. It is very difficult to get the ideal CS in the actual scene. There will be a certain delay between the relay selection time and the actual transmission time. The CSI between these two can change essentially, which will worsen the transmission performance of communication system [4].The main user interference threshold is different from the signal-to-noise ratio curve of ofdm-m system with two hops cognition. The simulation parameters are set as that the outage probability of two hop cognitive cache assisted relay ofdm-m system decreases with the increase of L in both ideal CS and obsolete CS scenarios. The reason is that the increase of L means that the primary user's receiver has a higher tolerance to interference, and the secondary user can transmit data with higher transmission power without affecting the primary user's QoS. The improvement of transmission power improves the transmission performance of the system. In addition, the outage performance tends to be stable at high SNR. With the increase of SNR, the outage probability converges to a fixed value.

5 Conclusion

In this paper, Cr technology is introduced into two hop buffer assisted relay ofdm-im system, and the transmission performance of two hop cognitive buffer assisted relay ofdm-im system under underlay frequency sharing mode is studied. Considering the obsolescence of CCS caused by feedback delay in practical application scenarios, a MABR/MSRs scheme influenced by time-dependent CS is proposed, and the closed form expression of system outage probability is derived. Through Monte Carlo simulation, the influence of different parameters on system interrupt performance is compared. The results show that the outage probability of the outdated CS scenario is higher than that of the ideal CSI scenario; When the primary user interference threshold is fixed, the outage probability will converge to a fixed value with the increase of SNR.

References

1. Peng, G.: Discussion on the application of multimedia technology in physical education. Sporting Goods Technol. 2012(03) (2012)
2. Shimin, L.: Computer aided Instruction Basic Course. Pudong Electronic Publishing House, Shanghai (2001)
3. He, K.: Theoretical basis of CAI and learning centered courseware design. J. Beijing Normal University 2000(1) (2000)
4. Basar, E., Aygolu, U., Erdal Panayirci, H., Poor, V.: Orthogonal Frequency division multiplexing with index modulation. IEEE Trans. Signal Process. **61**(22), 5536–5549 (2013). https://doi.org/10.1109/TSP.2013.2279771

Probability Analysis of China's Stock Market Rise and Fall Based on Data Mining Algorithm

Huan Xu(✉)

School of Statistics and Information, Shanghai
University of International Business and Economics, Shanghai 201620, China

Abstract. The stock market is called the economic barometer of a country. The stock price index can describe the general change of the price level in the stock market. It is a price average and can indicate the change of the stock market. In recent years, with the rapid development of China's economy, the research on stock index has become a hot topic. Based on technical analysis, this paper uses data mining algorithm and Poisson distribution to fit the dynamic process of stock index return rate, constructs a probability inference model, and predicts the rise and fall probability of stock index.

Keywords: Stock index · Data mining algorithm · Poisson distribution · Probability inference

1 Introduction

The price of a stock is determined by the value of the stock, and the price fluctuates around the value. Through the analysis of stock intrinsic value and its price form, we can find the fluctuation law of stock price. Stock intrinsic value analysis, namely fundamental analysis, mainly combines the business performance of the company and the development prospect of the industry to predict the future cash flow of the company, so as to determine the present value of the future value of the company. Buy when the price is below the intrinsic value and sell when the price is above the intrinsic value. The analysis of the form of price expression, namely technical analysis, is mainly to dig the historical price of the company's stock, trading volume and other technical indicators, to find out the law of stock price fluctuation. On the basis of the general law, the stock price trend is predicted, and then the current stock trading strategy is determined. Fundamental analysis is the analytical tool for long-term investment, while technical analysis is the analytical tool for short-term investment. This paper intends to infer the rise and fall probability of short-term stock indexes, so only technical analysis is considered [1].

Predicting the rise and fall of stock prices and stock indexes has been a matter of interest to academics and stock investors since the dawn of the stock market. For investors, the more accurate the stock price forecast is, the higher the investment return and the lower the investment risk. For the economic development of a country, the stock market is the barometer of the national economy, and the stock forecast also plays an

B. J. Jansen et al. (Eds.): *International Conference on Cognitive based Information Processing and Applications (CIPA 2021)*, LNDECT 85, pp. 790–795, 2022.
https://doi.org/10.1007/978-981-16-5854-9_102

important role. In view of this situation, this paper puts forward a new method to infer the probability of stock index rise or fall. Through data mining, Poisson distribution is used to fit the dynamic process of stock index return rate, so as to construct the inference model of stock index rise and fall probability, and predict the rise and fall probability of stock index [2].

2 Definition of Data Mining

Data mining is a kind of decision support method based on artificial intelligence, machine learning and statistical technology to mine and analyze data and draw conclusions. From the perspective of enterprises, data mining refers to the mining and statistical analysis of market and industry data related to enterprise operation, so as to discover the market conclusions hidden in the data, help enterprise decision makers to make correct decisions, dynamically adjust market strategies, and reduce the risk of enterprise operation. With the rapid growth of the company's database, especially the emergence of data warehouse, the original database tools have been unable to meet the needs of users. Users need not only general-purpose query and reporting tools, but also tools that can extract high-quality information from large amounts of data. The emergence and development of data mining conforms to this development trend [3].

Data mining is an important process that can extract effective, novel, and understandable patterns from a large number of incomplete and noisy data sets. The data mining process is shown in Fig. 1. A schema is an expression in natural language that can be used to describe a subset of a data set. As a pattern, expressions need to be simpler than enumerating subsets of data.The process is usually multi-stage processing in data mining, such as data cleaning, data integration, pattern search, pattern evaluation, knowledge representation, repeated modification and refinement, etc. The whole process requires a certain degree of intelligence and automaticity. Validity means that the new data mined by the existing mining mode is credible and valuable. Novelty requires the data mining model to have new adaptability for new projects. Comprehensibility means that the mined knowledge can be understood by users.

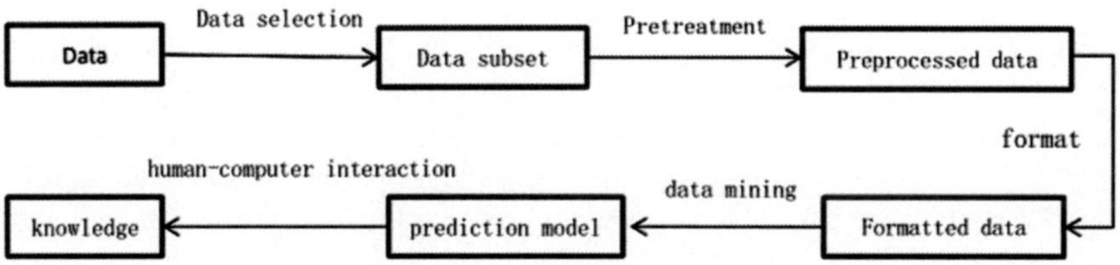

Fig. 1. Data mining process

3 Algorithm of Data Mining

Most data mining algorithms will use one or more objective functions and use some search methods, such as heuristic algorithm, gradient descent method, maximum and minimum method, network reasoning method, etc.

According to different mining methods, data mining algorithms can be divided into supervised learning and unsupervised learning.In supervised learning, an indication signal is first given, which can provide category markers and classification costs for each input sample in the training sample set, and find the fastest way to reduce the overall cost. There is no explicit instruction in the unsupervised learning algorithm, and the algorithm system automatically clustering the input samples [4].

Most data mining processes static non-real-time data, while some data mining algorithms need real-time or rapid mining and analysis performance. For example, data mining of stock market uses dynamic real-time data. Data mining involves many algorithms, as shown in Fig. 2. The specific data mining algorithm to be used depends on the specific application requirements. One algorithm can be adapted to many situations, and many situations can be solved by different algorithms.

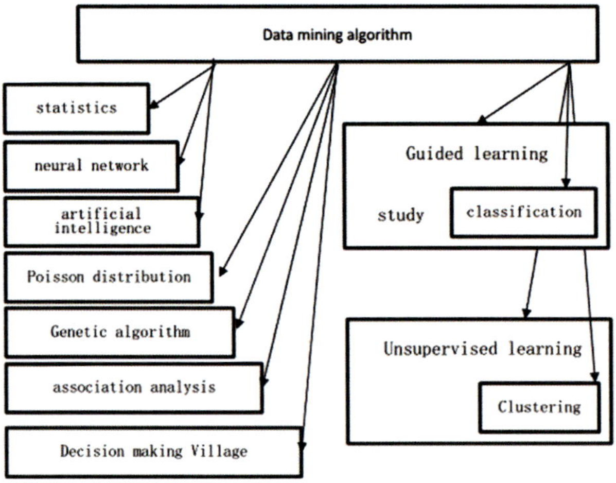

Fig. 2. Data mining algorithm

As can be seen from Fig. 2, data mining is essentially the establishment of mathematical models. Generally, mathematical models have generalization and adaptive abilities as well as autonomous or semi-autonomous learning functions. In this paper, the process of applying data mining algorithm to infer the rise and fall of stock index includes the following steps: mining technical indicators, descriptive statistical analysis, fitting Poisson distribution, constructing prediction algorithm, result analysis, and forming decisions. The classification algorithm with supervised learning is chosen here.

4 Inference Model of Stock Index Rise and Fall Probability Based on Data Mining

The fluctuation of stock index can be regarded as a discrete time series, and the daily closing price is a function of time. If we carry out technical analysis, we are actually predicting the stock index P_{t+1} on day t + 1 on the basis of knowing the information before day t of the stock index. P_{t+1} is a non-linear function. The recent technical analysis focuses on the dynamic process of return, mainly using neural network and survival model to predict the rise and fall of stock price and index. From the perspective of the dynamic process of yield, this paper also mines the historical data, adopts the Poisson distribution to fit the dynamic process of the rise and fall of the stock index, and based on the Poisson distribution, constructs the probability inference model of the rise and fall of the stock index, so as to judge whether the stock index on day $t + 1$ rises or falls.

5 Poisson Distribution Fitting Based on Historical Data

Before forecasting the rise and fall of the $t + 1$ day, the daily return of the stock index can be calculated. The yield of the stock index on day t is R. the yield is calculated by $R_t = (P_t - P_{t-1}) / P_{t-1}$, where P_t represents the closing price on the t day and P_{t-1} is the closing price on the t – 1 day [5].

There is a limit of 10% fluctuation in China stock market, and the daily stock yield fluctuates between –10% and 10%. The stock index is calculated by the method of the weighted composite price index, so its daily rise and fall are between –10% and 10%. The total yield interval [-10%,10%] is divided into N equal parts, and the length of each interval is 0.2/n. i is used to represent the interval number, i is from 1 to N, and the section I is represented as *sector* (i), then:

$$\text{Sector (i)} = [-0.1 + \frac{0.2(i-1)}{\text{n}}, -0.1 + \frac{0.2i}{\text{n}}], i = 1, 2, 3, ..., n \tag{1}$$

After the yield interval is divided into n equal parts, we can judge the stock index yield R on the t day. If R is in the interval $a(a \in \sec tor(i))$, in the historical data of the stock index, we can find the return rate on one day is R (R can be a small range centered on R), and the return rate on the next day is R' to form the a data set S, which is used to fit the Poisson distribution. If we use *count* (i) to denote the number of elements of stock index return in interval *sector* (i) of set S, and P (i) to denote the probability of stock index return in interval section (i) of set S, then:

$$P(i) = \frac{\text{count (i)}}{\sum_{\text{j=1}}^{\text{n}} \text{count (j)}}, i = 1,\ 2,\ 3,\ \ldots, n \tag{2}$$

5.1 Estimation of Distribution Parameters

Poisson distribution fitting is based on the division of yield interval into n-shares, and the best value of n is needed to be obtained, so that the return rate of stock index in set S is subject to Poisson distribution as much as possible. Let n's fit evaluation function be F (n), and now consider the elements that affect F (n). Firstly, the mean square deviation $\sqrt{\sigma_n^2/(n-1)}$ of the actual probability and the estimated probability is the main factor that affects the fitting degree of Poisson distribution. The smaller the mean square deviation, the smaller the deviation between the fitted Poisson distribution and the actual yield distribution, the smaller the difference between the return calculated from the fitted Poisson distribution and the actual return, the higher the fitting degree. So the mean square deviation is negatively correlated with $F(n)$. Secondly, after dividing the yield interval, we need to make more data available for Poisson distribution fitting in set S. Considering the extreme situation, if n is too large, that is, too many segments, so that the set S used to fit Poisson distribution contains too few elements, it is impossible to fit Poisson distribution; If n is 1, there is no segment in practice, then the set S contains all the historical yield data, and it cannot be predicted by this method.

Finally, we hope that the actual probability distribution of the return rate of stock index in set S is the curve as smooth as possible. Considering that the stock index return is divided into n intervals, the actual probability of the return rate of the set S stock index will appear abnormal point. As shown in Fig. 3, the outliers refer to the situation where the actual probability in a certain interval is lower than that of the adjacent interval on the left and right. In Fig. 3, the yield interval is divided into 15 intervals, and the vertical coordinate is the corresponding probability value, and point 4 is the abnormal point. The more the number of abnormal points $g(n)$, the more uneven the actual probability distribution. The number of outliers is negatively correlated with F (n).

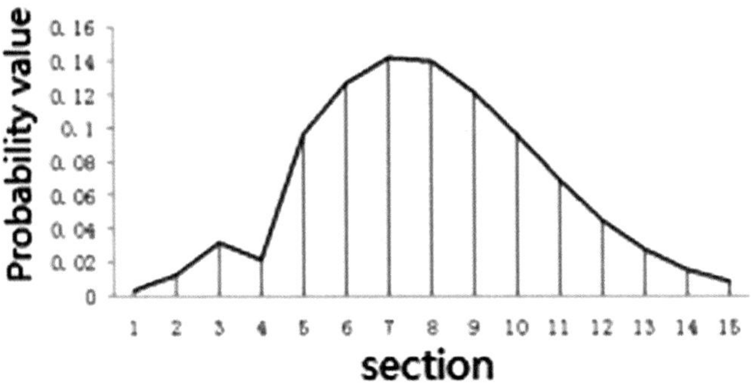

Fig. 3. Example of outliers

5.2 Construction of Up and Down Probability Inference Model

According to the expression of Poisson distribution, the probability distribution function of stock index return R_{t+1} in set S is constructed

$$P(t+1) = 1 - \sum_{i=1}^{n} B_i \times \frac{\lambda^i e^{-\lambda}}{i!} \tag{3}$$

The probability that R_{t+1} is in each interval follows the Poisson distribution, and the weight of return rate R_{t+1} in different intervals follows distribution B_i. If it is expressed by mathematical expression, there will be

$$B_i = \begin{cases} 1 & -0.1 + \frac{0.2i}{n} \leq 0 \\ 0 & -0.1 + \frac{0.2(i-1)}{n} \geq 0 \\ \frac{2}{n-2i+2}, & -0.1 + \frac{0.2(i-1)}{n} < 0 < -0.1 + \frac{0.2i}{n} \end{cases} \tag{4}$$

In this paper, we take $\{P(t+1) > 60\% | P(t+1) < 40\%\}$ as the criterion. When $P(t+1)$ is greater than 60%, we think that the probability of stock index rising on the $t+1$ day is larger; When the value of $P(t+1)$ is less than 40%, we think that the probability of stock index falling on the $t+1$ day is larger.

6 Conclusion

Based on data mining algorithm, this paper constructs a new model to infer the probability of stock index rise and fall. The Poisson distribution is used to fit the dynamic process of stock index return, and then the inference model of stock index rise and fall probability is constructed to predict the rise and fall probability of stock index, which has important reference value for investors to invest.

References

1. Li, Z.: Research on driving factors of stock market price deviation in China. Price Theory Pract. **10**(08), 104–108 (2020)
2. Wang, G., Chen, Y.: Data mining algorithm and its application in technical analysis of stock market. Comput. App. **24**(11), 104–106 (2004)
3. Deng, J., Tang, G.: An empirical analysis of the effectiveness of technical trading rules in China's stock market. East China Econ. Manage. **2009**(5), 15–17 (2009)
4. Qu, B.: A comparative study on the performance of fundamental analysis and technical analysis Evidence from China's Stock Market. Price. China. **18**(11), 77–80 (2019)
5. Fang, R.: Analysis and application of data mining algorithm based on association rules. Theory Algorithms. **01**(1), 36–39 (2016)

Research on Recommendation System of Internet Shopping Platform Based on Cognitive Internet of Things

Xuefeng Xu(✉)

School of Economics, Fuyang Normal University, Fuyang 236037, Anhui, China

Abstract. With the rapid development of Internet in China, online shopping has become a fast and common way of shopping in modern society. With the increasing variety of goods, the situation of "information overload" in online shopping is formed, which leads to more and more problems, such as consumers are difficult to find interesting goods, long search time and poor user experience. In this case, the user's demand for personalized product screening and push is more and more urgent. This paper will study the recommendation system of Internet shopping platform based on human behavior data in the process of consumption, to help consumers improve the efficiency of shopping, optimize the online shopping experience, help the network platform to achieve better service, and make a modest contribution to the development of China's Internet platform.

Keywords: Online shopping · Recommendation system · Internet

1 Introduction

In recent 20 years, the Internet has developed rapidly in China. So far, the Internet has been widely used in most urban and rural families in China. At the same time, online shopping has become one of the most common shopping methods in people's lives. In the development of online shopping platform, the online shopping platform has more and more commodities, more and more brands and quantity, which has formed the situation of "information overload". In this case, it is more and more difficult for people to search for the items they like on the shopping platform. Users need a "online shopping guide" to help them to screen and recommend products. Therefore, each online shopping platform has launched personalized customization recommendation system to solve the above problems such as "information overload" user experience. Alibaba e-commerce platform, as the largest e-commerce platform in the country, has its characteristic function "guess you like" and makes its shopping platform realize personalized customization function. Some scholars have made the following definition of personalized recommendation system: "recommendation system is a system for simulating store guide in e-commerce platform to provide commodity recommendation for consumers on the network platform to help consumers to better complete shopping" [1].

B. J. Jansen et al. (Eds.): *International Conference on Cognitive based Information Processing and Applications (CIPA 2021)*, LNDECT 85, pp. 796–801, 2022.
https://doi.org/10.1007/978-981-16-5854-9_103

Personalized recommendation system can reduce the user selection time and improve the efficiency of transaction, improve the user's use experience, and make the platform more competitive. In this context, a large number of research on personalized recommendation system is carried out on each network platform, which has become a hot research field in the society..

2 The Concept and Characteristics of Recommendation System

The so-called recommendation system refers to a tool to help users find useful information, without the need for users to provide clear requirements. The rec-ommendation system can analyze and mine the historical behavior information of users, construct corresponding learning models, infer the interests and preferences of users, etc. on this basis, users can get their own interested goods and services in time. In practice, the recommendation system has the advantages that traditional search engine technology can not compare. Such as diversity characteristics. Diversity is mainly used to measure the difference between different items in the user get list. Generally speaking, users have multiple different interests, and within a specific time range, they will be one of them. Therefore, the system will cover user interest in a certain proportion, so as to improve the user experience. As one of the main characteristics, real-time items that need to be recommended in many websites have strong timeliness characteristics, such as news, microblog, etc., and recommend them to users. In addition, the user's immediate behavior can reflect the user's interest. The recommendation items should be adjusted timely to enhance the user experience [2].

3 Recommendation System Based on Collaborative Filtering Algorithm

For the design of recommendation system, the most important part of the process is recommendation algorithm. Whether the algorithm selected is suitable directly affects the final recommendation accuracy. In order to improve the accuracy of recommendation, we need to extract data features, define the purpose of recom-mendation according to data characteristics and platform characteristics, select appropriate algorithms and improve them to achieve high-efficiency and accurate personalized recommendation.

Collaborative filtering algorithm is a common recommendation model. In short, it is a process of computing similar preferences according to the evaluation or preference of different users on the same platform. It is a recommendation algorithm using "group intelligence". Firstly, the system uses the user's historical data to judge the user's preference for the product, then calculates the similarity degree between each user based on the matrix formed by the user's interest. Finally, it finds out other users who are close to the user similarity, then calculates other evaluated goods by using it, infers the possibility of purchasing the product and completes the recommendation.

The cooperative filtering algorithm based on users is assumed as follows: the behavior preference between users is not completely independent but has certain similarity. The steps of implementing collaborative filtering recommendation based on users include the following steps: building user project evaluation ma-trix, calculating distance between users, calculating the purchase probability of the user for a project, and forming recommendation.

The following flow chart is used to show the basic principle and implementation process of user based collaborative filtering algorithm, as shown in Fig. 1:

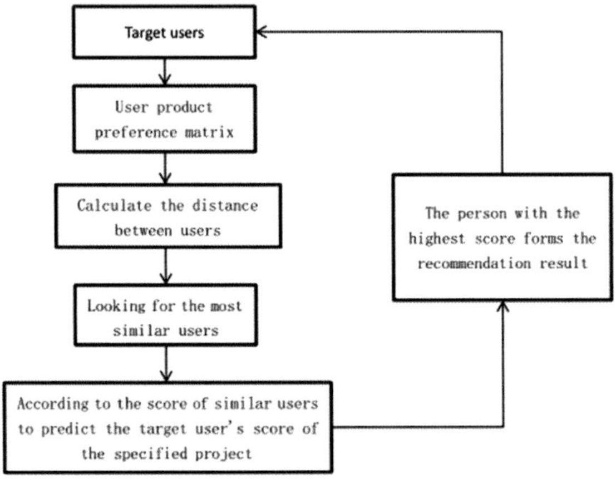

Fig. 1. Implementation process chart of collaborative filtering algorithm

Among them, Euclidean distance calculation method can be used to calculate the similarity between users, which shows the concept of "distance" more vividly. The calculated distance between two points in n-dimensional space can be ex-pressed by calculating the actual distance between them, that is, the score of N items given by two users can be regarded as two points in N-dimensional space, For the points that are closer to each other, they are more similar to each other. For the points that are farther away, they are less similar. The calculation formula is as follows:

$$w_{a,b} = \sqrt{\sum_{i=1}^{n} (r_{a,i} - r_{b,i})^2} \tag{1}$$

$r_{a,i}$ is the score of user a on item i, and $r_{b,i}$ is used Indicates the score of user B on the item.

4 Research and Application Platform of Internet Recommendation System

4.1 System Architecture

The recommender system consists of system interaction, batch processing and real-time processing. Among them, the interactive system mainly collects the user's request, and transmits the user's request to the application server, such as product browsing records, scoring and so on. In the second part, the user's behavior is incorporated into HDFS, and the recommendation matrix is obtained by batch processing[3]. Different parts have different functions. Suppose a scene, such as a book website, recommends similar products to users by using collaborative filtering method. Suppose users are interested in a Book unintentionally and find basic similar data continuously. If our recommendation system does not achieve real-time, then the user's recommendation will have a certain lag, unable to better meet the needs of users, to a large extent, dampen the user's loyalty to the website.

4.2 Key Technologies

Based on this paper, storm and Hadoop real-time recommendation system is proposed. Therefore, the key technologies mainly involve user behavior score simulation, collaborative filtering and other technologies. Sliding window technology, as a very important technology, because different nodes do not understand the network operation, and send data information at the same time, it is very likely to cause the middle node blocking, unable to achieve real-time data transmission and transmission. The introduction of this technology can effectively solve this problem. In practice, this technology can improve the throughput. In short, it allows the sender to receive any additional packets and ensures that the network environment runs smoothly [4].

4.3 Design and Implementation of System Module

First, user behavior score simulation. In the system design, the user rating is the evaluation of whether the user likes the goods or not. It is also an important basis for recommending other items to users. But in real life, users are not able to score all items, which will have a negative impact on the effective operation of the recommendation system. Based on this, in the design, we classify the users, such as browsing items, adding them to the shopping cart, and scoring the user behavior to get the highest scoring behavior.

Second, batch processing module. The batch processing module is composed of five modules, which use HDFS to store the historical data in a distributed way, and then send requests to interact with the next link to realize information sharing. In the system, we use MapReduce programming model to calculate and process data information. In practice, due to the large number of items, in the system, the user's historical data in collaborative filtering will be sorted to form a virtuous circle.

Third, real-time processing module. Due to the dynamic characteristics of user activities, it needs to be updated in real time to maintain the latest similarity matrix. There is a relationship between the similarity of items and the number of users who have scored items. On this basis, it can not only realize the storage of similarity, but also widely collect the relevant information of users, so as to improve the pertinence of recommendation.

Database is the most important part of recommender system, which includes user information table, browsing record information table and so on. The design of system database mainly follows two directions: on the one hand, the subordination between commodity categories; on the other hand, the subordination between commodity categories; On the other hand, the classification of the existing number of users is the classification and number of users. The database contains a large amount of information, which is an important guarantee and foundation for the realization of recommendation function. Therefore, in the process of recommender system design, we should pay more attention to the database technology, actively attract the technology, so that its positive role can be brought into full play, and give the recommender system more powerful recommendation function.

5 System Implementation

For the implementation of recommender system, we need to use Java, storm and other development tools to develop and design the system. The article recommendation system based on storm can generate user data information at any time, and generate simulation data after partial processing. Then import the history file into HDFS, start Hadoop, calculate the similarity of items, and start storm processing program. The user behavior is simulated for online module processing.

Real time item recommendation can only prove the availability of the system, but cannot judge the performance of the system. Therefore, for the item recommendation system, we mainly provide users with personalized recommendation list to improve the recommendation matching. For example, the ultimate goal of the development of input retrieval system and the design of recommendation system is to meet the actual needs of users. Therefore, in order to improve the quality and effectiveness of recommendation, technical personnel should select recommendation strategies scientifically according to the actual needs of users. For example, based on specific input method recommendation strategy, recommendation system provides users with relevant vocabulary information combined with relevant classification information. In the specific application of this strategy, users input words, and the system will make a chain reaction according to the user's needs, and recommend related words, and take the category words as the main category.

6 Conclusion

In the Internet era, recommender system is a new technology based on science and technology. Compared with traditional search engine technology, recom-mender system is more flexible and targeted. It can build a basic model according to users' behavior and provide targeted goods and services. This paper analyzes and studies the key technologies of recommender system based on storm technology, further understands the advantages of this technology, and discusses the implementation of recommender system combined with input retrieval, so as to help more people master this technology, so as to achieve the design goal of the whole personalized recommender system.

References

1. Zhang, D., Feng, X.: Research on a personalized mobile application integration framework and its implementation on android platform. Comput. Sci. **11**, 63–68 (2014)

2. Wang, B., Zhou, T., Liu, J.: Recommender system, information mining and internet based information physics. Complex Systems and Complexity Science. (z1), pp. 46–49 (2010)
3. Wu, Y., Shen, J.: Solving the problem of data sparsity in collaborative filtering recommendation system. Comput. App. Res. 35–41 (2007)
4. Peng, J., Feng, X., Zhou, W.: Optimization algorithm D for larege scale interney recommendations system Comput. Eng. Sci. **12**, 107–113 (2013)

Short Paper Session

The Trend of Elderly Service Industry Under Modern Information Technology

Xiu Sun(✉)

Dianchi College of Yunnan University, Kunming 650000, Yunnan, China

Abstract. With the deepening of population aging, China's elderly service industry needs to develop. The new generation of information technology based on big data, cloud computing, Internet of things, artificial intelligence, mobile Internet and so on, has brought unprecedented development opportunities to the elderly service industry, which can make the elderly service industry more intelligent and service-oriented, the elderly service mode more colourful, and the elderly life richer.

Keywords: Population aging · Modern information technology · Elderly service industry · Development trend

1 Introduction

With the development of medical and social economy, people's life expectancy is constantly improving, and the aging degree of population is deepening. In 1990, the aged 65 and above accounted for 5.57% of the total population. From 88.21 million in 2000 to 17.99 million in 2019, the corresponding elderly care ratio increased from 9.9% to 17.8%. With the deepening of population aging, the construction of home-based pension service mode, community pension service mode, public institutions pension service mode and medical care service system needs to be speeded up and improved, especially under the modern information technology. The next part of this paper will introduce the effects of modern information technology on the elderly service industry.

2 The Effects of Modern Information Technology

Modern information technology includes big data, cloud computing, logistics network, artificial intelligence, blockchain, 5G, VR/AR and other information technologies, which can realize the intelligent of industrial technology (see Table 1 for details). Among them, big data can store and calculate data as you like at any time on the computer. Cloud computing can up-and-download the available data conveniently and realize the sharing of network and services. The Internet of things can use intelligent identification and sensing devices to realize information exchange and link between things and things, or things and people. Artificial intelligence can simulate, extend and expand the theoretical

B. J. Jansen et al. (Eds.): *International Conference on Cognitive based Information Processing and Applications (CIPA 2021)*, LNDECT 85, pp. 805–809, 2022.
https://doi.org/10.1007/978-981-16-5854-9_104

methods and skills of human intelligence, and help human identify and process language, image and other relevant information. Blockchain is a distributed account book which cannot be tampered or forged by cryptography. It has many characteristics, such as multi-party writing, open account book, decentralization and tampering. 5G is the latest generation of cellular mobile communication technology, with theoretical transmission speed up to 10 GB per second. VR/AR is virtual reality (VR) and augmented reality (AR), which is a real world created by computer. Under the influence of virtual reality technology, what people see, hear and touch by hand can be seen as real. The latter is a technology that calculates the position and angle of camera image in real time and adds corresponding pictures, videos and 3D models. The goal of this technology is to stack virtual world on screen and interact with the real world [1].

Modern information technology has close relationship, can be integrated with each other, common development. The collection, sorting and calculation of big data cannot be separated from cloud computing. The real-time dynamic management of cloud computing needs the cooperation of Internet of things technology. The intelligent of Internet of things requires the use of artificial intelligence, cloud computing, big data and other intelligent technologies to realize monitoring and control the process of data and information. The development of artificial intelligence technology needs to be based on big data and the efficient processing ability of cloud computing. Blockchain builds a trust system, which is a part of cloud computing, which can reduce the credit cost and operation cost of big data and Internet of things, and promote the change of Internet data recording, dissemination and storage management [2]. 5G can improve the transmission speed of big data, and help artificial intelligence and blockchain to extract and respond to data [2]. VR/AR technology has made the carrier of information innovative, and the electronic screen is transferred from heavy electronic equipment such as TV and computer to small glasses, and the intelligent technology is portable.

Table 1. Main functions of modern information technology

Technologies	Major function
Big data	Store and calculate data
Cloud computing	Up-and-download the available data conveniently
Internet of things	Information exchange and link between things and people
Artificial intelligence	Simulate, extend and expand the theoretical methods and skills of human intelligence
Blockchain	Track records, cannot be tampered
5G	Communication faster
VR/AR	Stack virtual world on screen and interact with the real world

3 The Trend of Elderly Service Industry in Future

Modern information technology can obtain, process, store, spread and use the information of sound, image, text, digital and various sensing signals, which brings convenience and new opportunities for elderly service industry.

3.1 Bring Intelligent Medical Care

Low medical level, poor medical conditions and other factors often restrict the development of China's pension service industry. Integrating modern information technology into medical and health work to form intelligent medical care can not only bring convenience to medical staff and patients, but also improve the medical level with maximum effectiveness. Intelligent medical, also known as intelligent medical, mainly realizes the informatization and intellectualization of medical services through modern information technology, such as Internet of things, Internet, mobile computing, data fusion and other technologies. Patients and medical staff, medical institutions, medical equipment can synchronize information to achieve face-to-face interaction. Patients can register online, inquire online and enjoy treatment online. Medical staff can also receive patients online, check the patient's electronic cases, prescribe prescriptions, make wills, etc. At the same time, the statistics of chronic diseases, the exchange of medical information and the timely reporting of infectious diseases can be realized between regions. This will further improve the service efficiency and quality of medical diagnosis and treatment process, improve the comprehensive management level of the hospital, comprehensively change and solve the problems and difficulties of modern digital medical mode, intelligent medical and health management, hospital information system, etc., and greatly realize the high sharing of medical resources and reduce the public medical cost.

3.2 Provide Intelligent Services

The report on the loss of the elderly in China issued by the Research Institute of Social Assistance of the People, a research institute affiliated to the Ministry of civil affairs, is realistic. There are about 500000 people who have lost every year in the country, with an average of 1370 people walking every day. The global positioning system and corresponding electronic intelligent equipment in modern information technology can help to track, locate and find the elderly who are out of contact quickly, and ensure the safety of the elderly in the first time. In addition, the intelligent wearable equipment supported by modern information technology can monitor the body indexes such as heart rate, blood pressure, blood oxygen and so on, which suggests that the time of taking medicine and dietary advice cannot be replaced for the elderly health care. The newly developed elderly care robot has the functions of remote consultation, medication warning and combination of medical care, which can ensure the safety of elderly care at night, reduce the risk of various emergencies, make up for the intelligent room inspection scene of insufficient manpower in the night patrol by the elderly institutions, and is a friendly partner for the elderly to live in their home. These elderly robots can also monitor and analyze data for elderly with chronic disease, sleep disorder or other types of diseases, and form periodic health report, so as to realize early prediction, early prevention and accurate screening, and reduce the health risk of the elderly.

3.3 Innovation of the Elder Care Model

Various places such as elder care homes, welfare homes, old-age communities, old-age apartments, nursing homes, home-based care centers, day care centers, rehabilitation hospitals, nursing homes, etc. are the gathering places of the elderly care service industry [3]. With the support of modern information technology, the enterprise combines products and platforms organically, and takes the lead in proposing differentiated, personalized and systematic elder care solutions. It actively cooperates with government and industry customers, continuously understands the needs of the elderly, improves the products and platforms for the elderly, and provides the elderly care industry and individual users with a complete user experience and Internet features And services. “Smart community + elderly care”, “robot + elder care”, “new retail + elder care”, “medical service + elderly care + community service”, “family sickbed + home care”, “three assistance and one nursing (home-based elder care service with medical assistance, support, support + Remote Care)” New models of elderly care such as “school and local enterprises” and “system + service + elderly + terminal” are emerging [4].

3.4 Enrich Spiritual Life

The demand for elderly care services mainly includes elderly care, medical care, domestic service and leisure [4]. Through Internet of things, Internet, intelligent call, cloud technology, mobile Internet technology, GPS positioning technology, etc., the platform of “system + service + elderly + terminal” is built to assist the elderly in online diagnosis and treatment, and realize the information exchange between the elderly and their children, service centers and medical staff. The elderly does not have to live in the elderly care home to receive services passively, and can choose and enjoy professional elderly care services at home, such as the service items involving the combination of “medical care” such as life assistance, rehabilitation care, emergency assistance, day care, humanistic care, spiritual comfort, entertainment activities, legal aid, etc. The emergence of the Internet also enriches the spiritual and cultural life of the elderly. In addition to watching TV, listening to radio, reading and reading newspapers, planting flowers and growing grass and keeping pets, the elderly people have gradually become an important part of the cultural life of the elderly.

4 Conclusion

As a comprehensive industry, the elderly care service industry needs to meet a series of special needs of the elderly, such as living infrastructure, goods for the elderly, special services and so on, as well as the material, spiritual and cultural needs of the elderly, such as service, catering, housing, travel, entertainment, learning, medical care and so on. With the development of big data, cloud computing, Internet, Internet of things, artificial intelligence, blockchain, 5G, VR/AR and other modern information technologies, the pension service industry has obtained unprecedented development opportunities. In the future, the medical care in the elderly care service industry will be more intelligent, the service will be more comprehensive, the pension mode will be more abundant, and the life of the elderly will be more colorful. Therefore, China can speed up the integration of modern information technology into the pension service industry from the perspective of technology, talent, capital and system, and constantly improve the development level of pension service industry.

Acknowledgments. This work was financially supported by Yunnan Education Department Teacher Projects fund, "study on the sustainability of Chinese residents' pension under the integration of urban and rural areas" (2020J1308).

References

1. Zhao, X., et al.: Intelligent city intelligent medical sharing technology based on internet of things technology. Future Gener. Comput. Syst. **111**, 226–233 (2020). https://doi.org/10.1016/j.future.2020.04.016
2. Lian, W., Xue, T., Lu, Y., Wang, M., Deng, W.: Research on hierarchical data fusion of intelligent medical monitoring. IEEE Access **8**, 38355–38367 (2020). https://doi.org/10.1109/ACCESS.2019.2958854
3. Beijing Academy of Social Science: Analysis of the Development of Beijing 2019. Springer Science and Business Media, LLC (2021). https://www.sohu.com/a/345098404_762454
4. Mei, W., Jinghua, Z., Lixia, R.: Research on the development strategy of China's pension service industry from the perspective of demand. Hebei Enterprises **04**, 52–53 (2021)

Application of Computer Information Technology in Social Governance—Taking Big Data as an Example

Pei Zhang(✉) and Bei Ye

School of Literature, Law and Economics, Wuhan University of Science and Technology, Wuhan, Hubei, China

Abstract. The advent of the big data era has had a positive impact on social governance models, governance decisions, governance systems and governance processes. As a new starter of social governance, big data technology can not only eliminate the fragmentation of social governance and reduce the cost of social governance, but also meet the requirements of refined social governance, and truly improve the effectiveness of social governance and the accuracy of service targets. However, judging from many practices, the current big data technology has exposed many problems in the application of social governance. It is urgent to analyze these problems and seek corresponding measures based on the actual situation of social governance in my country to promote the smooth development of social governance innovation.

Keywords: Computer information technology · Big data · Social governance

1 Introduction

At present, the application of big data has brought about continuous changes in the social governance model, making the social governance mechanism face more opportunities for reform. However, the governance of society through big data also has various problems that need to be dealt with as soon as possible. At present, my country is in an important period of social transformation. The social structure and interest pattern are undergoing profound changes, and various social contradictions and conflicts of interest occur from time to time. There is an urgent need to strengthen and innovate social governance [1].

2 Big Data Technology Promotes Social Governance Changes

The rapid evolution and wide application of digital and intelligent technologies have left a deep and huge imprint on human society. Big data has the characteristics of large scale and capacity, diverse types and species, fast flow speed and low value density [2]. Benefiting from my country's high strategic importance to the diffusion and application of digital and intelligent technologies, my country has achieved leapfrog development in

B. J. Jansen et al. (Eds.): *International Conference on Cognitive based Information Processing and Applications (CIPA 2021)*, LNDECT 85, pp. 810–814, 2022.
https://doi.org/10.1007/978-981-16-5854-9_105

the fields of digital economy and digital governance in recent years, and to a certain extent plays the role of an explorer of government digital transformation in the international community. The rapid development of big data technology and smart technology and their multiple applications in the social governance system have promoted the innovation and transformation of my country's social governance.

2.1 Big Data Technology Promotes the Dynamics of Social Governance Models

For a long time, my country's traditional social governance model has adopted a static governance model based on the county system. This model shows strong vitality and superiority in maintaining social stability and national unity [3]. As human society enters the era of big data, the rich dynamic data information provided by big data technology for social governance not only promotes the development of the times, but also enables governance entities to grasp the changes in social governance data in a timely and comprehensive manner, and at any time according to the changes. Understanding and grasping the emergence of social problems and development trends, formulating countermeasures, effectively promoted the dynamics of social governance models [4].

2.2 Big Data Technology Promotes Scientific Decision-Making in Social Governance

Social governance involves all aspects of economic and social development and is related to everyone's vital interests. Therefore through scientific decision-, making, the formulation of targeted public policies has become the primary task of social governance. In the past, most social governance in our country relied on past work experience. However, with the continuous development of social economy, the empirical and social governance model no longer meets the needs of my country's current social development. The era of big data requires that social governance must use modern technology to dig out the two points of effective information needed for government decision-making from massive data to make decision-making more scientific and rational [5].

3 Difficulties Facing Social Governance in the Era of Big Data

3.1 Shortage of Big Data Technology Professionals

As the important role of big data technology in social governance has become increasingly prominent, the demand for talents with big data analysis capabilities is also increasing. At present, in addition to leading Internet companies such as Alibaba, Tencent, and Baidu, they have mastered and applied big data analysis technology. In addition to a group of big data analysis talents, on the whole, my country has a serious shortage of professionals who master data acquisition, data mining, data integration, and data use technologies. The application of big data in social governance is only in research and exploration. stage. At the same time, because the data analysis and processing process usually requires the cooperation of multiple professionals, this requires relevant personnel to not only be familiar with specific business knowledge, but also master big data technology. In reality, such compound talents are even more scarce. It directly affects the development and utilization of big data [6]. It can be seen that the lack of professional talents has become a prominent problem facing social governance in the era of big data.

3.2 Insufficient Sharing of Data Resources

In traditional social management, most departments handle their own affairs, and data resources cannot generally be shared [7]. Therefore, the first dilemma facing the big data era is the lack of data sharing theory. Most departments collect and analyze data according to their own governance needs. There is a lack of communication between various departments, and it is difficult to truly share information [8]. As a result, the comprehensive utilization rate of data is low. Although society has already entered the so-called network society, and many information resources have been shared on the Internet, my country has not yet truly realized the co-construction and sharing of data, the phenomenon of information islands still exists, and traditional social management concepts have been deeply rooted in the hearts of the people. Without realizing the effect of data, the government, enterprises, institutions, and civil society organizations are still at a break in information exchange, which makes the overall resource utilization rate of the society low and reduces the ability of social governance.

3.3 Data Security Issues Are Prominent

As a country with a large population, the amount of data generated by our country is bound to be huge. The collection of a large amount of data facilitates data analysis and processing, but it also inevitably increases security risks. One is the risk of data leakage. Data breaches became the most frequent security incident in 2018. For example, 240 million stay records of multiple hotel chains under China Lodging Hotel Group were suspected of being leaked [9]. According to data, in the first 9 months of 2018 alone, there were 3,676 data breaches in china [10]. As data leakage has brought huge damage to the government, market and the public, it has caused new social problems, legal disputes and even breeding of illegal and criminal activities, and has led to the formation of new social governance dilemmas. The second is the risk of data distortion. There is no lack of false and harmful information in the massive amount of information on the Internet, and the real and useful information is diluted, resulting in data distortion, which in turn leads to deviations in social governance decisions based on big data. The third is the risk of data monopoly. Because a large amount of data is usually in the hands of some enterprises or professional institutions, if these institutions manipulate the data collection, sorting, analysis, release, and utilization process for their own interests, it is easy to form a data monopoly (Fig. 1).

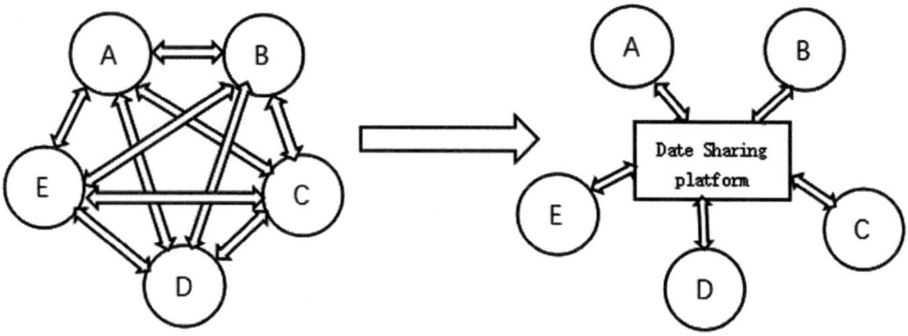

Fig. 1. Unified data resource management platform

4 Strategies for Innovating Social Governance in the Big Data Era

4.1 Cultivate Big Data Professional and Technical Personnel

Give full play to the advantages of universities and research institutes in cultivating talents, and use them as the main way and important base for cultivating professional talents in big data technology. Starting from the construction of disciplines, enhance the strength of the professional faculty, optimize the training plan, improve the teaching materials and curriculum system, reform the outdated teaching mode, strengthen the cooperation of industry, university and research, and cultivate a team of professionals with high theoretical level and strong practical ability. At the same time, the government should regularly carry out systematic training for professional data management workers so that they can keep pace with the times and improve the level of data information analysis, management, and application capabilities. The government can outsource social governance services that can open data to a professional technical team for processing by purchasing services from the society, thereby comprehensively improving the quality of data information processing.

4.2 Build a Data Co-construction and Sharing Platform

To create a social governance pattern of co-construction, co-governance and sharing, it emphasizes the multi-participation of governance bodies such as government, enterprises and society. Therefore, integrating resources and building a data platform has become an important part of the big data era to create a social governance pattern of co-construction, co-governance and sharing and social governance innovation. This requires the participation and cooperation of governance entities such as governments, enterprises, and society to strengthen data. Resource integration, realize the unification of data standards and formats, and build a unified and shared data platform.

4.3 Increase the Supervision of Cyber Security

In order to give full play to the advantages of big data in social governance, the Internet's security development strategy should be fully implemented, which can be achieved through comprehensive measures such as the protection of laws and regulations and the upgrading of security technologies. To ensure the security of national strategic information, people's personal consumption information, travel information, etc., while also dealing with various risks. The maintainers of each data platform should be responsible for information security and strengthen network supervision to avoid some large-scale data loss and illegal use incidents.

5 Conclusion

In summary, the advent of the era of big data has brought opportunities to social governance as well as new challenges. People do not need to be entangled in whether big data technology brings opportunities or challenges. It is this kind of debate that enables big data technology to develop better and contribute to social governance. The information and technology contained in big data have become an important part of my country's social governance. Therefore, my country must seize this development opportunity and use the rich data resources of the big data platform to provide services for my country's social governance, identify problems and formulate effective Response strategies to promote the orderly development of society.

Acknowledgements. The completion of this thesis is not my work alone. It is the result of the teaching, help and encouragement of all the teachers who have instructed me, helped my classmates and the family who have been caring and supporting me. I want to express my deep gratitude to them here!

References

1. Yue, S.: The precision of social governance requires big data thinking. Economic Daily, 15 December 2016
2. Tianguang, M., Ye, W.: Social governance innovation based on big data: local innovation and governance effectiveness. J. Guangxi Normal Univ. (Philos. Soc. Sci. Edn.) **57**(1), 111–123 (2021)
3. Jie, K.: The dilemma and path innovation of social governance in the era of big data. Admin. Law **12**, 60–66 (2019)
4. Pan, H.: Social governance innovation in the era of big data faces opportunities. China Economic Herald, 23 August 2014 (B07)
5. Yan, W.: Social governance dilemmas and countermeasures in the era of big data. Study Monthly **12**, 47–48 (2020)
6. Huihuang, Y.: On the status quo and development trend of computer science under the background of big data. Digital Commun. World **12**, 168 (2018)
7. Yuanyuan, P.: Difficulties and countermeasures of social governance in the era of big data. J. Huaihai Inst. Technol. (Human. Soc. Sci. Edn.) **15**(07), 113–116 (2017)
8. Ying, C.: An analysis of the social governance dilemma and innovative development in the era of big data. Xue Theory **34**, 87–88 (2015)
9. Chenjie, Q.: The top ten hot words of cybersecurity in 2018 gather industry hotspots to explore the development context. China Inf. Secur. **1**, 54–57 (2019)
10. Huangfeng, T., Jianwu, T.: China's national governance capacity building in the big data era. Explor. Content. **10**, 54–58 (2014)

Fetus Data Analysis System Based on Cloud Computing Platform

Jifeng Liang(✉)

School of Engineering and Technology, Xi'an Fanyi University, Xi'an, Shaanxi, China

Abstract. Combined with cloud computing, high-performance, easy extension, information sharing and storage characteristics of fetal data analysis system is designed, the fetus data analysis of the calculation of pressure is transferred to the cloud, implements fetus analysis of data in the cloud service model, users take no notice of pay attention to fetus analysis of data, boost the workpiece of analysis, lower user costs. besides, after analyzing the system, customers can use mobile devices such as mobile phones and tablets, mobile devices such as tablets allow data to be imaged and diagnosed anywhere, anytime. The fetus analysis of data on account of cloud computing platform supply an valid and fast diagnostic basis for doctors to estimate fetal development, and greatly increased doctors' operational capability and standard. besides, the put into effect of the system is of great sense to regional capital and remote consultation.

Keywords: Cloud computing · Fetus · Data analysis system

1 Introduction

CAT image data were used for functional tests, and the results were correct.

Cloud computing is a kind of supercomputing service mode in which large-scale network computing resources work together. With the help of virtualization technology, all the hardware and software will be combined through the network, make full use of and mobilize all the existing information resources, run the application required by the user on the server cluster, to meet the user's various application needs. At the moment, the cloud service is not only a distributed calculating, but also the result of the mixed evolve and computer technology rapid expansion such as distributed network memory computing, utility computing, parallel computing, load balancing, hot standby jumbled and virtualization [1]. With the help of the Internet and cloud computing, the cloud-based fetal data monitoring system records and analyzes all kinds of fetal monitoring data. The system not only provides the basis for doctors to evaluate and correctly diagnose the development of the fetus, but also improves the service level and capability of medical staff. besides, the system also plays a crucial part in telemedicine and cross-medical treatment.

B. J. Jansen et al. (Eds.): *International Conference on Cognitive based Information Processing and Applications (CIPA 2021)*, LNDECT 85, pp. 815–819, 2022.
https://doi.org/10.1007/978-981-16-5854-9_106

2 Cloud Computing Concept

The cloud computing concept is defined by the IEEE international network service conference (ICWS) in 2008.This definition is composed of three parts, as shown in Fig. 1 [2]:

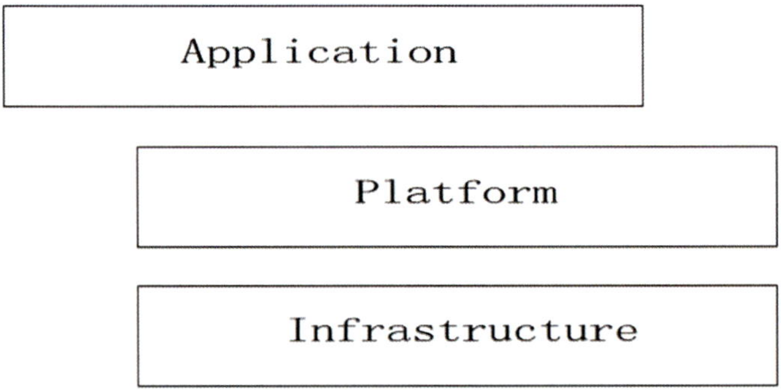

Fig. 1. The three-parts structure of cloud computing.

- The top part of Fig. 1 is about application, providing services to users through SaaS. This dimension is user-oriented, which means cloud computing is user-oriented information technology as a service (ITaaS). Centralized data centers can provide users with a wide range of scalable computing and applications.
- The middle part of Fig. 1 is the platform part, which provides the environment for the application or PaaS to run in. This part satisfies the needs of programmers and Web application developers. Cloud computing is a platform for software development and testing using the Internet network.
- The bottom part is infrastructure, very flexible, which can provide services from the Internet through the distributed data center. This part is for infrastructure developers and director that view IP infrastructure as a data center for cloud computing to connect to the Internet.

3 Features of Cloud Computing

Relatively with the traditional methods of calculation, cloud computing has the following property [3]:

- Multiple technology convergence: Cloud computing is the integration of various technologies joined efforts in developing. Cloud computing adopts use software, hardware and network resources as a computing resource.Because of its dynamism, extensibility, extensibility, and availability to virtualization, academia and the IT communion appellation IT the "cloud" [4].

- Large and secure data: Conduct propaganda from cloud computing promoters suggest that cloud computing can support the data centers for vocational management group. As a result, users are certain about security and a nigger in the woodpile.
- Handy: In the cloud computing environment, users so long as browser the terminal equipment to be acquired application services, without the need for highly configured computers or facilities [5].

4 Flow Structure Model of Foetal Data Analysis System

Foetal data analysis system based on cloud computing design idea is that by adopting a series of digital technology and equipment, continuous real-time fetal physiological and biochemical data, transmitted by the network technology to the cloud server, using the cloud platform super ability, such as storage, operation and logic combined with the expert team to assist service, carries on the analysis, diagnosis, prediction, tracking, evaluation and decision, ultimately provide data support for fetal medicine diagnosis. The deployment pattern is shown in Fig. 2, divided into two clouds and three clouds [6–9].

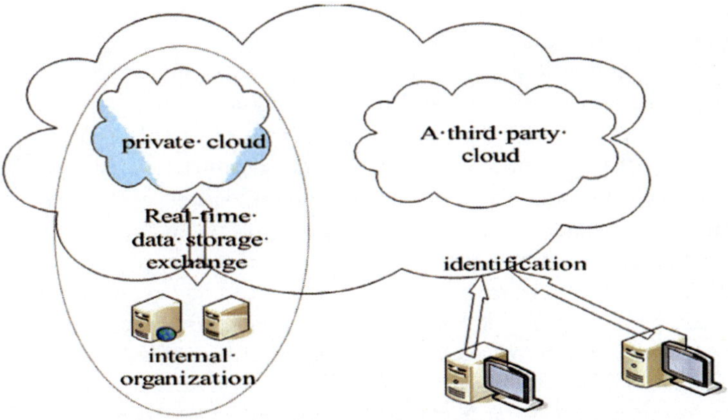

Fig. 2. Foetal data analysis system cloud deployment mode

This model takes advantage of powerful computing functions capacity of cloud computing platform to greatly improve the collection and processing of fetal data. The bottom part, Actual information resource one, makes full use of the acquisition of aboriginal data and supports hardware and virtual applications, Contains network middleware and memory. The second part is about supervise of cloud resources, which provides business logic control and an environment for the development, operation, supervise and control of cloud computing. The top part is the cloud data service one, consisting of self-service portal, business and where the application is executed, etc. (Fig. 3).

This architecture model is divided into three aspects: the first aspects is server sends requests and waits for replies; The second aspects is infrastructure and service application Terms; The third aspects is service creation, implementation, and monitoring. The

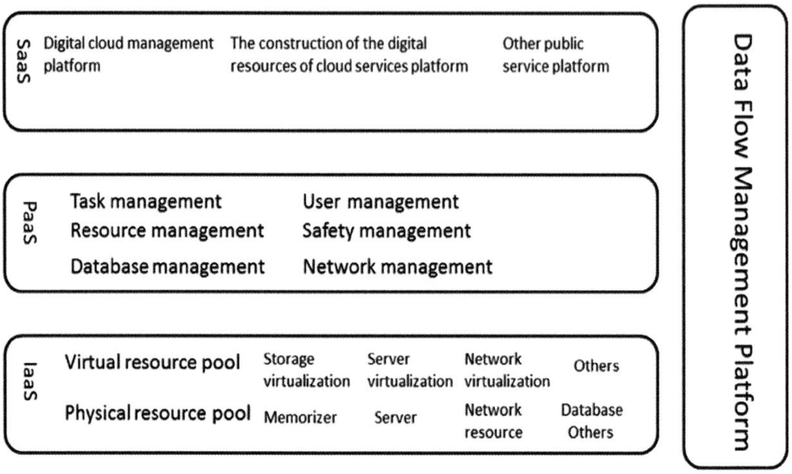

Fig. 3. Flow structure model of foetal data analysis system

core part is the second level, which is divided into three layers for three ways to run cloud computing: IaaS, PaaS and SaaS. The underlying physical information mainly include network information, server information, database information and software information. The virtual resource pool is a virtualized information base that combines the resemble type of information into isomorphism or closer isomorphism through different technologies. The information resource platform construction middleware is mainly responsible for the management of cloud computing resources, and the load balancing optimization and scheduling of numerous resources construction management, so as to provide reliable and safe services for the application of the upper layer. The management service construction layer based on multi-tenant technology mainly encapsulates the cloud computing capacity built by digital information resources into standard Web Service services through corresponding technologies, so that any user can customize the configuration of SaaS software due to their own needs without interrupting other users.

The Implement the general framework of systems for analysing vascular data in cloud environment includes essential picture processing functions and 3D stereoscopic visualization modules. System data flow analysis and function module division help realize the operating principle of medical picture analysis, multi-plane remould, three-dimensional stereoscopic be able to see and fetal data analysis server-side, as well as the fundamental function, overall arrangement, time limit, measurement tools and other assist functions of medical image processing in the client side. The server-side data interaction guarantees the working principle of data operation, plain picture conduct and be visible to in whole system framework.

5 Conclusion

This paper proposed a tentative idea of the development of fetal data analysis system, which theoretically shows that it is effective and feasible to apply cloud computing, an efficient and powerful new computing model, to fetal data analysis system. Cloud flow architecture model based on cloud computing is a form of development and utilization of fetal data analysis system, which solves the problems of fetal information resource management, monitoring and utilization. Its development and perfection will certainly bring more convenient and quick information service to people.

The image processing function of the method mentioned in the project in fetal detection has passed the test of a designated hospital, and has been implemented. With well operation effect and exhibition effect, it can astisfy the daily demand of doctors.

Acknowledgements. Project source: First-class major construction of electronic commerce in Xi'an Fanyi University.

References

1. NIST [EB]: http://csrc.nist.gov/groups/SNS/cloud-computing/index.html
2. Cloud Computing (2010). http://server.ctocio.com.cn/whatis/296/8046296.Shtml
3. Buyya, R., Yeo, C.S., Venugopal, S., Broberg, J., Brandic, I.: Cloud computing and emerging IT platforms: vision, hype, and reality for delivering computing as the 5th utility. Future Generat. Comput. Syst. **25**(6), 599–616 (2009). https://doi.org/10.1016/j.future.2008.12.001
4. Garfinkel, T., Matthews, J., Hoff, C., et al.: Virtual machine contracts for datacenter and cloud computing environments. In: Proceedings of the 1st Workshop on Automated Control for Datacenters and Clouds, pp. 25–30. ACM, New York (2009)
5. Armbrust, M., et al.: A View of Cloud Computing. Commun. ACM **53**(4), 50–58 (2010). https://doi.org/10.1145/1721654.1721672
6. Kantarcioglu, M., Clifton, C.: Privacy-preserving distributed mining of association rules on horizontally partitioned data. IEEE Trans. Knowl. Data Eng. **16**(9), 1026–1037 (2004). https://doi.org/10.1109/TKDE.2004.45
7. Bao, S.D., Garmen, C.Y.P., Shen, L.F., et al.: Authenticated symmetric-key establishment for medical boday sensor network. J. Electron. **24**(3), 421–427 (2007)
8. Zhang, Z., Min, H.: Analysis on the construction of personalized physical education teaching system based on a cloud computing platform. Wireless Commun. Mobile Comput. **2020**, 1–8 (2020). https://doi.org/10.1155/2020/8854811
9. Xiangqian, Y., Linxin, W., Chen Jianhua, S., Haijun, L.X.: Model design for big data clustering analysis of power dispatch automation system based on cloud computing platform. E3S Web Conf. **179**, 01028 (2020). https://doi.org/10.1051/e3sconf/202017901028

Application of Virtual Reality Technology in Digital Media Art Creation

Yi Li(✉)

Dalian Neusoft University of Information, Dalian 116023, Liaoning, China

Abstract. Both virtual reality technology and digital media art are modern new technologies, and the application of virtual reality technology in digital media art creation is already a relatively common situation. Through the integration and application of virtual reality technology, the form of digital media art creation is more Rich, creative works are more novel, and the quality level is higher. Therefore, this article expounds the connotation of virtual reality technology and digital media art, then analyzes the characteristics of the fusion creation of virtual reality technology and digital media art, and finally conducts research on its specific integration application.

Keywords: Virtual reality technology · Digital media art · Creative application

1 Introduction

Under the background of the overall improvement of the economic level, people pay more attention to the pursuit of artistic quality [1]. As the mainstream form of modern art creation, digital media art creation needs to be more innovative in creation. The application of virtual reality technology in digital media art creation can make the effects of digital media art creation more novel, and can better meet the needs of modern people for digital media art works [2]. Therefore, it is necessary to study how virtual reality technology can be better applied in digital media art creation.

2 The Connotation of Virtual Reality Technology and Digital Media Art

Nowadays, in the digital media art development market, competition is fierce. The development of modern industries is very fast. Digital media art creation needs to be able to keep up with the development of the times and constantly update and optimize the quality of art creation [3]. Occupying a favorable position in market competition, the connotation of virtual reality technology and digital media art will be explained below.

B. J. Jansen et al. (Eds.): *International Conference on Cognitive based Information Processing and Applications (CIPA 2021)*, LNDECT 85, pp. 820–825, 2022.
https://doi.org/10.1007/978-981-16-5854-9_107

2.1 Virtual Reality Technology

In simple terms, virtual reality technology is the use of technical means to simulate real scenes and things to form a virtual simulation world similar to the real world. Virtual reality technology is a modern new type of technology with comprehensive characteristics, and things built using virtual reality technology are restored [4]. The effect is excellent, it can make people immersed in the virtual environment, and it has applications in many contemporary fields. Virtual reality technology contains many modern advanced technology types, such as simulation technology, computer technology, network technology, sensor technology, and digital technology. Through virtual reality technology, people can have a more realistic sensory experience of the virtual world. However, it is still possible to communicate with the real world. Generally speaking, virtual reality technology has become popular in recent years. When people experience scenes constructed by virtual reality technology applications, they have good sensory and visual impact and a full sense of experience.

2.2 Digital Media Art

The content contained in digital media art is also quite complicated, and science fields such as nature, humanities, and society are all embodied in digital media art. Digital media art is both a technology and an art form. Digital media technology is also a modern new type of advanced technology. In simple terms, it combines ordinary art elements with digital media technology to form a new work of art. Or the activity is more modern and technological. In the creation of digital media art, there are a variety of artistic elements that can be applied [5]. Through the integration of digital image processing technology and art, and then the completion of artistic creation activities, the creation of digital media technology is more in line with the needs of modern art development, and has a great degree of artistic aesthetics and attainments. The level is relatively high. Digital media art creators have strong comprehensive quality capabilities, and are able to master theoretical knowledge and practical skills in computer, network, art, fine arts, design, etc.

3 The Characteristics of the Fusion of Virtual Reality Technology and Digital Media Art Creation

The fusion of virtual reality technology in digital media art is a very common phenomenon. In virtual reality technology, real scenes can be simulated and presented through data transmission, drawing, sensing, etc., to build a three-dimensional virtual scene environment, which can be restored in 360°Realistic scenes can also construct imaginary spaces and present them through three-dimensional spatial images, so that people's senses can accept a good, real and rich experience. The realization of virtual reality technology requires the combination of people and equipment to realize the transmission of sensory experience. The following is an analysis of the characteristics of its technology in digital media art fusion creation.

3.1 Immersion

The application of virtual reality technology is mainly to associate with people through special equipment, and then let people's senses enter the simulation environment constructed by virtual technology. Through the senses, people will feel really in the virtual environment and focus on the situation. Investment into it, so as to achieve immersion [6]. Immersive experience is very popular in modern times. It can make people feel like being on the scene. All sensory feelings will be affected by the virtual situation and environment. The application of virtual reality technology to realize the immersive experience is achieved through the use of virtual reality technology. The visual and auditory senses of the visual and auditory senses make the human's psychological and physical senses better feel the three-dimensional space constructed by virtual reality technology, making it feel the same in the virtual world as in the real world.

3.2 Interactivity

With continuous R&D innovation and development, virtual reality technology is now widely used and has significant value in various application fields. Virtual reality technology has the characteristics of interactivity. When people apply and run virtual reality technology, they need to wear special virtual displays, data gloves and other devices, and use these devices to connect people with the virtual scene space, so that people's bodies are in the environment. In the real world, the senses can use virtual interactive equipment to realize the interaction between people and the virtual world, and the interaction between humans and the virtual world is realized, and the interactive sensory experience is very comfortable and natural, making the interactive experience more authentic. Let people better feel the construction scene in the virtual three-dimensional world, and give people a novel experience. It is precisely because of the good interactivity of virtual reality technology that its own technological application innovation function is more significant. The integration and application in digital media artistic creation can promote the novelty of artistic creation and enhance the freshness of artistic creation to people. Promote.

3.3 Imagination

The application of virtual reality technology can build a diversified virtual world, not only limited to the restoration and simulation of real things and scenes, but also can build virtual Lenovo scenes through rich imagination [7]. Through technical means, the imaginary buildings, things, characters, scenes and other elements can be simulated and constructed. The virtual three-dimensional space constructed by virtual reality technology has strong openness and imaginative characteristics, which can be well satisfied. With the demands of modern art creation, virtual reality technology is unique in constructing imagination space, which has established advantages for the wide application and rapid development of virtual reality technology in modern times.

4 Specific Application of Virtual Reality Technology in Digital Media Art Creation

The application of virtual reality technology in digital media art creation has also experienced a relatively long time of practice, greatly enhancing the sense of science and technology of digital media art creation works, and further optimizing the integration and correlation between technology and art. In particular, it has changed the limitations of traditional digital media artistic creation expressions and low creative intensity. The scientific application of virtual technology has opened up a new development space for digital media artistic creation. Therefore, the following is a discussion of virtual reality technology in digital media artistic creation.

4.1 Application of Holo Studio Technology

Holo Studio technology is a holographic studio where virtual reality technology is used to construct a scene. In the holographic studio space, digital media art creators can use virtual reality technology to carry out creative activities at will. There is sufficient space for creation and realization of technology. Convenient. In the process of artistic creation, designers can directly use virtual reality technology equipment for practical operations based on design ideas to directly verify the effects that design ideas can present, and then they can also make corresponding adjustments by operating virtual reality technology equipment. It is possible to simulate the creative work completely according to the independent design idea, and to intuitively watch the artistic work formed under the creative idea, and then optimize and modify the creative method through the observation, analysis and feeling of the creative work. In the holographic studio of virtual reality technology, only the use of equipment can simply construct complex artistic creation effects, thereby enhancing the convenience of artistic creation.

4.2 Application of QuickVR Technology

The Quick Time VR technology has a wide range of modern applications, and the advancement and maturity of the technology have been significantly improved. The application value of this technology is relatively high, and the effects achieved through the application of the technology are relatively high-quality. Quick Time VR technology is to use computers and networks to build a virtual technology space construction platform to create omni-directional and multi-angle virtual images and virtual spaces. It can create a very realistic experience for people experiencing virtual scenes and a strong sense of atmosphere. Quick Time VR has powerful editing functions. It can also help creators process some captured images and sounds by connecting digital cameras and digital video cameras, so as to achieve effective control of the playback content and optimize human-computer interaction to a certain extent. Experience [8].

4.3 Application in the Protection of Urban Cultural Construction

Virtual reality technology can comprehensively present scenes, things, objects, buildings, information and other content. When people interact with virtual scenes, they can receive all the information in the scene. The types of information have the characteristics of diversification and novelty. People's acceptance and perception of information has also realized information dissemination. The application in digital media art creation can realize the publicity and dissemination of art work information and increase the mode of information dissemination. The information dissemination of this kind of art creation gives people a deeper impression, so the integration of virtual reality technology and digital media art creation there are also many applications in the protection of urban cultural construction. My country has a long history, each dynasty has a unique culture, and its architectural style is also deeply influenced by people's thoughts and aesthetics at that time [9].

My country has a vast territory, and all regions have relics left by historical development. Most of these ancient buildings have certain damages. These are the contents of great historical and cultural value. However, in the modern development, the protection of these ancient buildings Insufficient strength makes it difficult for many ancient cultures to be fully protected and inherited, which is not good for urban cultural construction. Therefore, in the protection of urban cultural construction, digital media art creation staff can use virtual reality technology to restore ancient buildings, and restore the appearance of ancient buildings in virtual scenes. Through the virtual interactive experience of technology, people can be more complete Perceive the landscape of ancient buildings, and have a clearer and richer understanding of the development of corresponding historical buildings, so as to promote the enhancement of the protection of urban cultural construction and establish the awareness of ancient building protection in people's hearts.

4.4 Application in the Entertainment Industry

In the creation of entertainment digital media art products, the application of virtual reality technology is also very extensive, and its specific application can promote the performance and innovation of entertainment products. Especially in the construction of modern Chinese cities, the entertainment industry accounts for a very large proportion. Video game towns and VR game experience halls are all forms of entertainment that modern people love very much. Through the application of virtual reality technology, entertainment products can give people a more realistic immersive entertainment experience. Virtual reality technology can be integrated with digital media art, allowing users to interact with the scene immersively after wearing a specific device. With the development of virtual reality technology, this degree of simulation will become more and more realistic [10].

5 Conclusion

The application of virtual reality technology in digital media art creation is to enhance the expressive power of art through technology, and make the expression of art more diversified. At the same time, art also allows technology to be better applied, providing users with more comprehensive and high-quality products.

References

1. Xiaofei, P.: Discussion on digital media art creation based on virtual reality technology. Sound Screen World **07** (2020)
2. Fu, M., Wang, N.: Research on digital media art creation based on virtual reality technology. Res. Commun. Power **35** (2019)
3. Peng, G.: Virtual reality technology and its application. Electron. Technol. Softw. Eng. **22** (2019)
4. Zheng, P.: Research on the application of virtual reality technology in the teaching of digital media art. Art Technol. **12** (2019)
5. Ye, Z.: Virtual reality technology changes the way of digital media interaction. Inf. Comput. (Theory Edition) **11** (2019)
6. Zhuqing, L.: Promote digital media art creation with the help of virtual reality technology. Electron. Technol. Softw. Eng. **01**, 104–105 (2020)
7. Mengyuan, F.: On the application research of digital media art to VR virtual reality technology. East West, South North **08**, 84 (2020)
8. Shangru, W.: Artificial intelligence application of virtual reality technology in digital media art creation. Inf. Record. Mater. **21**(01), 106–107 (2020)
9. Chengjing, Z.: Analysis of the commonality of the development mode of digital media art and virtual reality technology. Tomorrow's Fash. **22**, 195–196 (2020)
10. Tiantian, H.Y.: Research on the intersection and integration of Chinese digital media art works and virtual reality technology. Western Radio Telev. **08**, 13–14 (2020)

Flower "Dock"-Smart Flowerpot Based on Microcontroller

Shengyue Fan, Tiantian Wang, Bo Wang, Yuting Zhang, Yao Cai, and Baojun Liu(✉)

School of Electrical and Information Engineering, Jilin Agricultural Science and Technology University, Jilin, China

Abstract. The microcontroller smart planter project has an intelligent watering system which is based on a microcontroller programmed to adjust the upper and lower soil moisture limits, detection functions and so on. It also enables the acquisition, setting and comparison of data signals. In this paper, we discuss the microcontroller-based intelligent flowerpot project, the Flower "Dock", to understand the purpose, significance and basic conditions of the project, and focus on the design of the microcontroller-based intelligent flowerpot project.

Keywords: Microcontroller intelligent flowerpot · Flower "dock" · Basic conditions · Design scheme · System test

1 Introduction

If the flower "dock" microcontroller intelligent flower pot project is below the lower limit signal, its output command electric signal drive relay module can start the water pump work, it can realize the self-watering operation of plants; if its project is above the upper limit signal, its watering operation automatically stop, so that the intelligent flower pot project can realize the automatic watering process.

2 Purpose, Significance and Basis for the Implementation of the Smart Planter for Microcontroller Project

(1) Purpose

As far as modern life is concerned, people often live in a relatively dull, closed environment lacking warmth and care. This is why having a plant in the living space has become an important way for people to relieve stress. The reason for this is that people usually do not have the time to take care of plants, and also most people do not like the traditional flower-keeping model, so people are looking for a new model of flower-keeping that does not depart from the essential basis of flower-keeping, which is a creative and poetic smart planter project. This project has created a new intelligent system based on the definition of "automatic watering at regular intervals" and "automatic watering by sensing the external environment", thus protecting the plants more comprehensively [1].

B. J. Jansen et al. (Eds.): *International Conference on Cognitive based Information Processing and Applications (CIPA 2021)*, LNDECT 85, pp. 826–830, 2022.
https://doi.org/10.1007/978-981-16-5854-9_108

(2) Significance

In the "Dock" microcontroller project, it is important to clarify the significance of the project, which is to optimise the watering function of the flower pot by using the "automatic timer watering" function and the "automatic watering by sensing the external environment" function, based on a soil moisture detector sensing component that objectively reflects the watering level in conjunction with the actual soil conditions of the pot and feeds the watering data back to the microcontroller. The upper and lower watering limits are set on the basis of the microcontroller program and the values sent to the microcontroller are compared. If they are lower than the lower limit, the microcontroller outputs an electrical signal and transmits it to the pump via a relay module to start the watering activity [2].

(3) Basic Conditions

The "Flower Dock" microcontroller smart planter project uses a solar-powered system, an automatic/manual watering system and a voice system. The solar-powered system is an environmentally friendly concept that uses energy flow to convert solar energy directly into chemical and biological energy.

The second is the watering system, which enables regular automatic watering by means of a timer. In this process, the user presets the watering period and time using a software programme, and watering activities are carried out regularly in the form of fountains. The watering system also incorporates sensors which counteract the inevitable external environmental changes. It senses changes in soil temperature and humidity, for example, and it also creates a quantitatively triggered automatic watering system. This watering system consists of two types, an automatic watering system and a manual watering system [3].

The third is the voice system, which is pre-programmed with four voice packs in two styles, including different voices of men and women of different ages, with reference to the infrared sensor sense. When the sensor senses a nearby owner, the system establishes a connection with the owner and interacts with them through specific statements. If the sensor detects a lack of water in the pot, the system issues the appropriate statement and also includes a user-defined schedule reminder mode (alarm clock), while setting the need to provide a reminder service to alert the owner when the time is up via a preset language pack or music [4].

3 Smart Planter Project Design Scheme for Microcontroller

(1) Module Introduction

In the design of the intelligent flowerpot project of the Flower Dock microcontroller, it is necessary to first introduce the module, in which a voice recording chip APR9600 is placed inside the project flowerpot. The APR9600 chip does not require tape to record and play sound, in the recording process using the microphone to obtain a weak signal, after the microphone amplifier amplification by the filter for filtering, after the sampling circuit processing to achieve the analog flash memory FLASHROM optimization, which uses to analog/ulti-platform storage technology, effectively circumvent the ordinary digital recording circuit A/D, D/A conversion process due to quantization and compression in place to reduce distortion phenomena, to ensure high quality sound. Here, to analyse the oscillation resistance combined with the sound effect for stepless adjustment, making full use of the APR9600 chip's greatest advantage to achieve multi-segment control, through the ISD chip to achieve multi-segment control, this control process is quite complex. It requires the use of a large number of diode translators or microcontrollers to achieve functional assistance, using segmented recording to establish a microcomputer and programming machine technology system. In the APR9600 chip by setting the level of the 24, 25, 9 pin to achieve parallel two-band, four-band, eight-band control operation, to achieve parallel eight-band process, press M1–M8 any key to ensure that the chip recorded sound and can achieve sound replay [5]. In the process, to establish press the key not to release the chip to form a loop to play the wave, in parallel control state to achieve the optimization of the technical content of each time period. Here a set of double switch is added, if S1, S2 two switches are disconnected, it mainly uses the microphone circuit disconnection does not work, combined with RE terminal connected to low level value 0, at this time the chip into the recording state. At this point, press and hold M1 again to hear a "dri-" sound, then enter the first section of the recording, release the key and then hear a "dri-" sound to indicate the end of the recording, using the same method can achieve a total of 7 recording process.

In the system module is also set up an ultra-small relay, which mainly uses the microcontroller signal to pass to the pump to meet the practical requirements of the project, using the relay armature to turn on the contact control circuit to achieve smooth and cut off the control pump, so that a complete set of program control line is formed, and then realize the automatic operation. In the watering module part of the pump, the pump mainly processes the microcontroller information data and outputs through the relay module, if the soil is wet enough, at this point the upper limit the pump will stop watering. If the soil is dry enough to reach the lower limit the pump continues watering, so that it can be output normally as a load [6].

(2) Hardware Design and Assembly

The Flower Dock microcontroller smart planter project includes several hardware components, including a water pump, a moisture sensitive element, a soil temperature detector and a relay (including a normally open relay, a common relay and a normally closed relay), as shown in Fig. 1 [7].

As shown in Fig. 1, the smart planter project of the Flower "Dock" microcontroller can be connected to multiple device components and the project can be programmed to run properly [8].

(3) System Testing

In the process of testing the project system, it includes computer testing and testing of the modules. Firstly, computer testing, which enables the general framework of the program to be rewritten, and it enables modules to be written, which in general takes a lot of time to debug, so that the code can be modified directly to ensure that the program works properly. At this point it is necessary to use a computerised testing system to detect soil moisture, objectively show dryness and wetness and consider whether the relay can control the pump operation [9].

Next is a variety of module testing, which includes a moisture sensitive component test, which mainly uses a multimeter to measure the power supply positive and negative, which uses the 5v positive to analyse VCC, and connects the 5v negative directly to GND to ensure that the indicator above the moisture sensitive component is constantly on [10]. At this point reconnect the soil moisture detector to normally display the soil real-time conditions to avoid various problems with the monitor. In the soil sensor test process, the wet sensitive components need to be examined, the electricity is connected properly, and the soil detector is inserted into the water in the square board to ensure that the two indicators are fully lit, which indicates that the module is working as it should. The third is the water pump test, it will be the water pump to power, start the DC motor, turn on the power to work properly, to ensure that the normal work indicates that the module problems, reselect and disassemble the content of the components to find a normal working water pump. Finally, is the relay module test, you can relay VCC in the project directly to the power 5V positive pole, and GND to 5V negative pole, so that you can check the relay circuit, the microcontroller will be passed to the electrical signal small current control system contacts, to ensure that the module works properly [11].

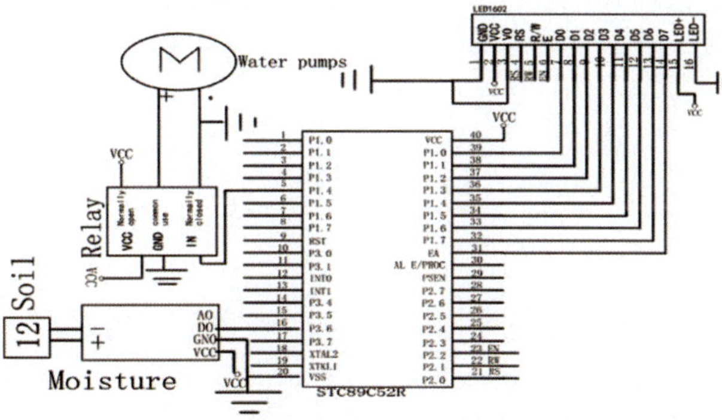

Fig. 1. Hardware Composition of the Flower Dock Microcontroller Smart Planter Project

4 Conclusion

In summary, the intelligent flowerpot project with the "Dock" microcontroller solves the problem of watering plants, but it also enables intelligent and fully automatic watering, using water resources for scientific and rational control of plant growth conditions. All of the above technologies provide better support for the smart planter project and promote the good development of the smart planter market in China.

Acknowledgments. This work was financially supported by China University student innovation and Entrepreneurship Competition fund.

References

1. Zhu, X., Hou, W., Hui, J.: Intelligent flowerpot design based on microcontroller. Inf. Technol. Inf. **2**, 172–173 (2020)
2. Wu, D., Kang, Y., Yang, H., et al.: Intelligent flower pot design based on microcontroller technology. Sci. Technol. Econ. J. **000**(002), 90 (2019)
3. Wang, Y., Xu, Y., Jin, G., et al.: Interactive intelligent potted plant design based on Arduino platform. Wireless Internet Technol. **180**(08), 64–65 (2020)
4. Liu, J., Yang, Z., Yang, C., et al.: IOT-based intelligent interconnection system for home landscape. Comput. Sci. Appl. **10**(9), 12 (2020)
5. Chen, Z., Chang, H.: Android mobile design for embedded smart flower pots. Fujian Comput. **35**(04), 36–39 (2019)
6. Lu, X., Ma, W., Sun, X.: Arduino-based smart watering planter. Electron. World **598**(16), 173–175 (2020)
7. Hou, C.: Intelligent flower farming system. Easy Learn Comput. **000**(019), 1 (2019)
8. Shen, L.: Making a "sunflower" smart planter. Radio **000**(001), 80–81 (2019)
9. Zhao, L., Yan, S., Wang, Y., et al.: Design and experimental study of solar-powered intelligent soilless cultivation flower pots. Agric. Mech. Res. **41**(12), 174–179 (2019)
10. Zhao, C., Li, C., Hou, X., et al.: Design and implementation of an intelligent ecological aquarium. Electron. Technol. Softw. Eng. **156**(10), 48–49 (2019)
11. Yin, X.: Design of intelligent flowerpot system for mobile terminals. J. Leshan Normal Coll. **8**, 13–17 (2020)

Application of Computer Data System in Intelligent Extraction of Fresh Rehmannia

Zhen Ren and Mingsan Miao(✉)

Department of Pharmacy, Henan University of Chinese Medicine, Zhengzhou, Henan, China

Abstract. Objective: To explore the rule of fresh rehmannia glutinosa using modern information technology, and to explore the application ideas and principles of high-frequency core drugs. Methods: with CNKI and ancient books as data sources, the clinical application of ancient and modern fresh rehmannia glutinosa was screened out, and the relevant Chinese medicine prescriptions or proprietary Chinese medicine preparations were screened out. The standard Chinese medicine was entered into excel 2013, and then SPSS modeler 14.1 was used to analyze the association rules to reveal the mechanism of high-frequency drugs. Results: among the 1 TCM prescriptions that meet the inclusion criteria, 5 association rules of drug pairs and drug groups were obtained. Conclusion: fresh rehmannia is often used in combination with ginger, sugarcane and Dendrobium.

Keywords: Fresh rehmannia glutinosa · Data mining · "Big Data" Era

1 Introduction

The research methods of artificial intelligence in the field of traditional Chinese medicine mainly include data mining, machine learning, deep learning, neural network and genetic algorithm. Data analysis is one of the basic application types of computer information processing technology, including human behavior and emotion analysis, time domain sequence analysis and correlation regression analysis [1].

Rehmannia glutinosa, as one of the bulk medicinal materials, has a large dosage every year and remarkable curative effect. Compared with raw Rehmannia glutinosa and cooked Rehmannia glutinosa, fresh rehmannia glutinosa is more cold and cool, but its clinical application is limited due to its difficult preservation and perishability. However, in many diseases, fresh rehmannia glutinosa can not be replaced by raw Rehmannia glutinosa and cooked Rehmannia glutinosa. Our laboratory has done a lot of work on the modern application and research of fresh medicine [2]. For example, the chemical composition, pharmacological action and clinical application of fresh rehmannia glutinosa and raw Rehmannia glutinosa were summarized and analyzed [3]. And observe the effect of fresh rehmannia powder on depression mice model, found that fresh rehmannia powder can significantly improve the tail suspension stress, forced swimming and reserpine induced blepharoptosis depression model in mice, and significantly improve the reserpine induced exercise depression model in mice [4]. This paper summarizes the

B. J. Jansen et al. (Eds.): *International Conference on Cognitive based Information Processing and Applications (CIPA 2021)*, LNDECT 85, pp. 831–835, 2022.
https://doi.org/10.1007/978-981-16-5854-9_109

application of Rehmannia glutinosa by summarizing the ancient records, modern experiments and clinical research, in order to expand the clinical application of Rehmannia glutinosa and develop new drugs of Rehmannia glutinosa.

Therefore, we use the relevant technology of computer big data to apply to the compatibility of traditional Chinese medicine, and get the common compatibility of fresh rehmannia from the clinical application of ancient and modern fresh rehmannia, so as to predict the future application direction of fresh rehmannia.

2 Ancient Records of Fresh Rehmannia Glutinosa

2.1 Record of the Efficacy of Fresh Rehmannia Glutinosa

In addition to the Chinese Pharmacopoeia 2020 edition, many ancient books also record the efficacy and indications of fresh rehmannia. See Table 1 for details. It can be found that the main functions of fresh rehmannia glutinosa are cooling blood and stopping bleeding, nourishing yin and tonifying kidney.

Table 1. Records of ancient books on functions and indications of fresh rehmannia glutinosa

Source	Record
《Dietotherapy herbs》	Main tooth pain, hematemesis, fracture
《Four tones of Materia Medica》	Black hair
《The origin of Medicine》	Cool blood, (moisten) dry skin, remove dampness. Soak in wine
《The image of Medicine》	Cooling blood, tonifying blood, tonifying kidney water, deficiency of true Yin

3 Analysis of High-Frequency Drug Association Rules

For high-frequency drugs, use SPSS Clementine 12.0 statistical software for association rule analysis, set support to 10%, confidence 80%, select the maximum number of antecedents to be 5 and other conditions, Apriori modeling and mining potential associations between Chinese medicines, total Obtained 12 groups of core Chinese medicine combinations, and the improvement of all the drug combinations was greater than 1, indicating that these drug combinations are statistically significant.Establish an associated network display of "data source-type-networked correlation diagram" See Table 2 and Fig. 1.

Table 2. Analysis of association rules between drugs

Consequent	Antecedent	Support %	Confidence level %
Rehmanniae Radix Praeparata = 1.0	Zingiber Officinale Roscoe = 1.0	13.04347826	100
Rehmanniae Radix Praeparata = 1.0	Ganzhe = 1.0	13.04347826	100
Rehmanniae Radix Praeparata = 1.0	Dendrobium nobile Lindl = 1.0	17.39130435	100
Rehmanniae Radix Praeparata = 1.0	Forsythiae Fructus = 1.0	13.04347826	100
Rehmanniae Radix Praeparata = 1.0	Ophiopogon japonicus(L.f.)KerGawl. = 1.0	21.73913043	100

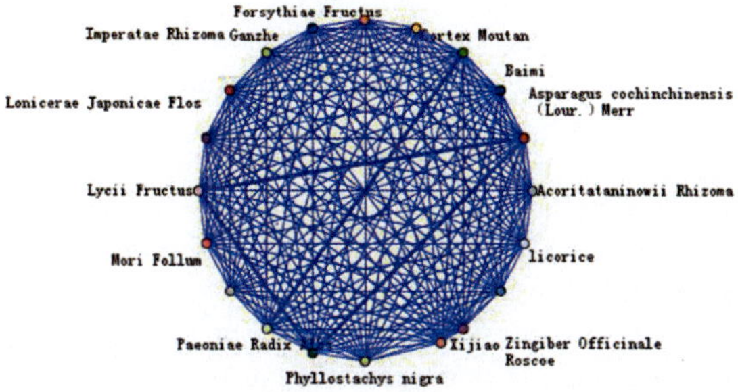

Fig. 1. Networked correlation diagram of high-frequency drugs

4 Conclusions

Clustering and classification analysis refers to the new data obtained on the basis of the induction and classification of the old data, predicting the later behavior of users based on reasonable structural assumptions, and classifying and grouping the possible limiting factors [5, 6]. Because the technology of data analysis is more careful, the conclusion is also more systematic, so it is often used in the related application of big data mining. Association computing refers to the statistical analysis of data to find the possible relevance between different types of data. This technology can be used in the field of artificial intelligence, so that the machine equipped with this technology has the ability of "learning". Through the continuous imitation of human behavior, its internal system will be more perfect. At present, computer related technology has begun to build in the direction of cloud computing, so that computer information processing technology should not only focus on the construction of hardware facilities, but also strengthen the connection between networks [7].

Fresh rehmannia glutinosa is a kind of seasonal traditional Chinese medicine, which has a clear harvest time. The fresh-keeping time of its deep-processing product (health preserving fresh rehmannia glutinosa) is only 180 days, which limits the application of fresh rehmannia glutinosa to a great extent. Through the excavation of ancient books and modern research, it is found that the application of fresh rehmannia glutinosa in ancient books is more extensive than that in modern research. The reason may be that fresh medicine is not easy to be preserved for a long time, and the preservation conditions are difficult to control. At present, the laboratory studies the use of freeze-drying technology to produce fresh rehmannia freeze-dried pieces, in order to maximize the retention of fresh rehmannia efficacy. Low temperature drying technology is applied to the drying of Chinese medicinal materials and preparations, which can avoid the destruction of thermosensitive active ingredients as far as possible, obtain higher product quality, and ensure the efficacy of drugs [8, 9]. Application of low temperature freeze-drying technology can effectively expand the scope of application of fresh medicine. At present, freeze drying technology mainly includes low temperature hot air drying, low temperature vacuum drying, vacuum freeze drying, low temperature adsorption drying and so on. Temperature is an important factor in the drying process of traditional Chinese medicine and its preparation, and it is also a key factor to determine its quality. Vacuum freeze drying is a process that makes the moisture in the material sublimate directly from the solid state to the gaseous state under the condition of low temperature and low pressure, and is carried away from the material by the vacuum system to achieve the purpose of drying. Freeze drying technology has many advantages in the application of traditional Chinese medicine, which is suitable for traditional Chinese medicine containing highly oxidizable and heat sensitive components, and the freeze-dried products can maintain good appearance, color, smell, complete dehydration, good rehydration and easy preservation [10].

Henan Province is the dominant producing area of Rehmannia glutinosa. It has a good source of fresh rehmannia glutinosa, and the preservation condition is poor, which not only affects the export of fresh rehmannia glutinosa, but also causes the waste of resources. Therefore, we should increase the research on the preservation direction of Rehmannia glutinosa, develop suitable fresh rehmannia glutinosa, ensure that fresh rehmannia glutinosa can play its pharmacological effect, and promote the development of genuine medicinal materials [2].

Acknowledgements. This work was financially supported by Standardization project of State Administration of traditional Chinese medicine (gzy-fjs-2020–219), major public welfare special project of Henan Province (201300310100), science and technology plan project of Henan Drug Administration (2020db050–55) fund.

References

1. Silva, A., Lee, B.Y., Clemens, D.L., et al.: Output-driven feedback system control platform optimizes combinatorial therapy of tuberculosis using a macrophage cell culture model. Proc. Natl. Acad. Sci. USA **113**(15), E2172–E2179 (2016)
2. Guo, L., Miao, M.: Characteristics of chemistry, pharmacology and application of raw (fresh) Rehmannia glutinosa. Acta Sinica Sinica **29**(03), 375–377 (2014). (in Chinese)
3. Kang, L., Chen, H., Liu, H., et al.: Effect of fresh Baihe Dihuang powder on behavior of depression model mice. J. Trad. Chinese Med. **33**(10), 1970–1973 (2018). (in Chinese)

4. Ding, X., Njus, Z., Kong, T., et al.: Effective drug combination for Caenorhabditis elegans nematodes discovered by output-driven feedback system control technique. Sci. Adv. **3**(10), eaao1254 (2017)
5. Huang, T., Zhong, L.L.D., Lin, C.Y., et al.: Approaches in studying the pharmacology of Chinese medicine formulas:bottom-up,top-down-and meeting in the middle.Chin. Med. (2018). https://doi.org/10.1186/s13020-018-0170-4
6. Li, S., Zhang, B.: Traditional Chinese medicine network pharmacology:theory,methodology and application. Chin. J. Nat. Med. **11**(2), 110–120 (2013)
7. Liu, J.: Protective effect and antioxidant mechanism of Catalpol on aorta of diabetic rats. Chin. J. Dual Pathophysiol. **30**(06), 1023–1028 (2014)
8. Liu, W., Li, Y.L., Feng, M.T., et al.: Application of feedback system control optimization technique in combined use of dual platelet therapy and herbal medicines. Front. Physiol. **9**, 491 (2018)
9. Wang, X., Wu, Z., Li, Y., Xiong, Y., Yang, M.: Application status and Prospect of low temperature drying technology in the field of traditional Chinese medicine. Chin. J. Pharmaceut. Indust. **50**(01), 42–47 (2019)
10. Han, R., Yang, Y., Li, X., et al.: Predicting oral disintegrating tablet formulations by neural network techniques. Asian J. Pharm. Sci. **13**(4), 336–342 (2018)

Innovation and Development of Higher Education Management Informatization Based on Big Data Technology

Guofang Miao(✉)

China West Normal University, Nanchong 637002, Sichuan, China

Abstract. As social life develops continuously, all aspects have undergone certain changes. People's living standards are constantly improving, and everyone's comprehensive quality is constantly improving. Big data, as the most novel technical concept at present, not only plays a crucial part in improving the country's economy and industry, but also has a certain impact on the education management of universities. The informatization of higher education management is mainly to integrate data generated in the process of education management. The information system was used for collection and integration. And use these data as the basis for the staff to make work decisions. To a certain extent, it promotes the development of higher education management, and improves the efficiency of management. However, there is no in-depth value development for the information and data generated in the education management process. With big data technology, this problem can be effectively solved. Therefore, research on the innovation and development of college education management information based on big data technology has positive practical significance.

Keywords: Big data technology · University education management · Informatization · Innovation · Development

Currently, the informatization development based on computer technology has brought about a broader space for technological progress, and new technologies have also emerged in an endless stream. In the context of the "big explosion" of technology, all aspects of society are making adaptive changes. Big data is a new technology developed based on computer technology. This is because in the current phase of network development, plenty of information data continue to appear. The related research staff discovered during the research work. These exponentially growing data are not just data on paper, but plenty of data can be analyzed to obtain more information we need. In this case, big data technology in use is born [1]. The impact of big data on our daily life, work, and so on is gradually deepening. For college education management, it will generate a lot of information and data in its daily operation. The current informatization of higher education management simply collects and organizes these data. No in-depth exploration of the value contained therein. This leads to a waste of resource value. Therefore, it is necessary to carry out the innovation and development of higher education management informatization with big data technology.

B. J. Jansen et al. (Eds.): *International Conference on Cognitive based Information Processing and Applications (CIPA 2021)*, LNDECT 85, pp. 836–840, 2022.
https://doi.org/10.1007/978-981-16-5854-9_110

1 An Overview of the Analysis of Big Data Related Content

For the definition of big data technology, a common explanation is that people take plenty of data as the basis and analyze these data through the high-speed computing mechanism of computer technology. The final process of obtaining some valuable information. Among the big data technology, the main thing is to dig out valuable information from plenty of data information. This information can make the development of related industries further. The characteristics of big data technology are mainly manifested in three aspects. The first aspect is that the spread of big data is very fast. This is mainly because of the popularity of Internet technology. The Internet covers almost every aspect of life. And under the influence of the openness of the Internet, there is almost no obstacle to data sharing between people. This accelerates the generation of data information [2]. In the traditional era, the transmission of information has always been restricted due to the influence of geographical or political reasons. The popularity of network technology has put an end to this situation. People use the Internet to freely obtain the information they need. In the current era, everyone is not only a producer of data information, but also a disseminator of data information.

The second aspect is the diversity of big data. Data collection is carried out by the Internet and equipment such as cameras installed daily. These are the collection channels of data information. These data information will be the basis for big data technology to function. In the big data time, people have a new understanding of data information. In different fields and different industries, more and more people have assumed the tasks of data producers and disseminators. This makes the data information not only more numerous, but also more colorful [3]. The third aspect is the usefulness of big data. This is mainly because big data has demonstrated its value in more and more ways. This is also why the big data technology develop rapidly. In terms of society, economy, scientific research, etc., big data has demonstrated its unmatched value. Big data technology can make people obtain more detailed and comprehensive data. Remedy the development deficiencies in many industries. People can have a deeper understanding of the law. The

Table 1. Statistics of professional education management Informatization in colleges in some provinces (unit: institute)

	Architecture major	Mechanics	Medicine major	Engineering major	Landscape major	Electronic major	Mathematic major
First category colleges	6	5	4	3	1	2	1
Second category colleges	5	3	5	2	1	6	2
The third category of colleges	4	4	6	1	2	3	3

practicability of big data helps the development of social life to be more stable. And gradually became the basis for the development of some new technologies. With the help of big data technology, every walk of life have made great progress [4] (Table 1).

2 The Development Status of Higher Education Management Informatization Under Big Data Technology

For the current stage of educational management informatization work in colleges, although there is a certain level of development, there are also some problems in the specific management details. For example, different majors all use the same educational management model. The difference of profession didn't start, and the detailed division of education management mode was carried out, which made education management work superficial. On the other hand, the security assurance of the informatization of higher education management should also be taken seriously. This mainly refers to the safe use of information and data and source security [5]. Safe use of information and data. It is the application of the distribution of relevant resource sciences to different majors during university education management. Information technology is used to formulate targeted education management plans. The security of information data source refers to the authenticity of its data. In the development of higher education management informatization currently, data resources security cannot be effectively guaranteed. Under the big data technology, the informatization of higher education management should have certain standards. This is mainly to ensure that education management can work under different educational technologies. This ensures the high efficiency and high accuracy of informatization. However, the current university education management work lacks the restrictions of standards and norms. This leads to deviations in the content of some related information resources [6].

3 The Path of Innovation and Development of College Education Management Informatization Under Big Data Technology

3.1 The Innovation and Development of College Education Management Informatization Should Be Based on the Network Platform

With big data technology, the innovation and development of education management informatization in colleges should take the network as the foundation. This is because the advantages of big data require a network. Including the daily information data generated by the educational management of colleges. Collect, organize, and analyze these data. With big data and information technology coming into use, the hidden value in these data will be unearthed. The relevant staff of university education management can use this as a basis to formulate education management strategies that are more in line with the actual conditions of the university. Moreover, under the thinking of big data, education management can collect and analyze information and data from the three points, including teaching experience, management methods, and student feedback, through systematic research and comparison. Then you can choose the implementation plan that best meets the educational requirements [7].

3.2 The Informatization of University Education Management Requires Further Improvement

Before, higher education management would have more one-sided or subjective decisions. This is mainly due to the imperfect informatization of higher education management. The information and data collected and integrated are not comprehensive and specific. On the basis of such data information as the basis for judgment, mistakes in decision-making will occur. Therefore, there is a need to further improve the informatization of higher education management [8]. To use big data as the technical support to improve the informatization of related higher education management, first of all, it is necessary to share information resources among colleges. This can broaden the channels for obtaining information and data. After the comparison of big data, university education management can have a more accurate reference basis; Secondly, we must conduct targeted information and data collection for different majors in colleges. Meanwhile, different majors should plan out different educational management content; Finally, there is a need to build a big data information application platform. The establishment of an independent big data application platform in universities can solve the problem of insufficient education and management information data resources [9].

3.3 Pay Attention to the Building of Network Security

The popularization of the informatization of higher education management has caused relevant information and data to be uploaded on the network. In this way, there is a need for network security. Colleges should attach importance to the all-round building of network security to prevent the leakage of information and data and cause unnecessary losses. There is a need to carry out real-time monitoring of network technology and timely repair the vulnerabilities in the system. There should be strict management authority and access authority settings for the network within the university. Moreover, the security of the internal network server must be strengthened, and at the same time, some network technical personnel can be recruited to carry out professional network security construction [10] (Fig. 1).

Fig. 1. Internet + education concept map

4 Conclusion

In general, the research on the innovation and development of higher education management information based on big data technology has a very important practical role. Related research has effectively promoted the progress of higher education management, and improved the technical level of higher education management. This makes the university education management system more perfect, and the corresponding work is more efficient.

References

1. Chao, W., Dejun, Q., Chao, C.: The path of innovation and development of college education management informatization in the era of big data. J. Jiamusi Vocat. College **37**(04), 113–114 (2021)
2. Xia, S.: Exploration of the innovation path of university education management informatization under big data technology. Road Success **02**, 10–11 (2021)
3. Meng, W.: Innovation and development of college education management informatization under big data technology. Employ. Secur. **23**, 155–156 (2020)
4. Wang, H.: Research on the innovation and development of university education management informatization based on big data technology. J. Shandong Inst. Agric. Eng. **37**(06), 157–158+188 (2020)
5. Yalin, X.: Research on the innovation and development of university education management informatization based on big data technology. Dig. World **04**, 71–72 (2020)
6. Ying, P.: Thoughts on the development of educational management informatization innovation in colleges and universities in the era of big data. Mod. Econ. Inf. **23**, 467 (2019)
7. Miao, C.: Research on the innovation path of university education management informatization under big data technology. J. Beijing Inst. Graph. Commun. **27**(11), 101–104 (2019)
8. Guimei, X.: Research on the innovation and development of university education management informatization based on big data technology. Manage. Technol. Small Med.-sized Enterp. (First issue) **09**, 83–85 (2018)
9. Li, N., Liu, J.: A preliminary study on the innovative development path of educational management informatization in colleges and universities in the era of big data. Educ. Modern. **5**(05), 295–296+305 (2018)
10. Li, X.: Research on the Innovation and Development of University Education Management Informatization Based on Big Data Technology. Xinyang Normal University (2017)

Smart Recognition Application of Android Mobile Garbage Classification from the Perspective of Big Data

Liping Liang, Chang Xing, Jiaai Zhang(✉), and Jie Hu

Jilin University of Agricultural Science and Technology, Jilin 132109, China

Abstract. This paper designs an App running on Android mobile phone system, which is designed to help urban residents with insufficient garbage classification knowledge to conduct accurate classification and treatment. Garbage classification is a reform of the traditional garbage collection and disposal and a scientific management method for effective disposal of garbage. Most people have the awareness of garbage classification, but the execution is not strong. On the one hand, because of poor awareness of environmental protection, I feel that garbage classification is of little significance to myself. On the other hand, there are a lot of household garbage and it is composed of different types of materials, which leads to low efficiency and time-consuming and laborious classification of ordinary garbage [1]. Based on the above reasons, this paper improves the ordinary garbage classification and treatment, and realizes the intelligent garbage classification and treatment.

Keywords: Garbage classification · Accurate and fast · Big data · Intelligent speech recognition

1 Introduction

1.1 Background of Big Garbage Sorting Platform

With the acceleration of the process of urban life in China, the production of domestic garbage has maintained a steady growth trend. In recent years, China has become one of the countries with the greatest pressure of urban domestic garbage treatment in the world. With the improvement of people's living standard, the society has also entered a stage of high consumption, and the problem of municipal solid waste has become increasingly prominent, hindering the harmonious development of the society [2]. In real life, garbage causes a lot of pollution to the environment. Furthermore, garbage classification has gradually attracted people's attention. In recent years, China has accelerated the implementation of garbage classification system, and the national garbage classification work has been gradually started from point to surface, with initial results. Through big data, valuable information can be quickly obtained from massive data and information, and intelligent and professional operations can be carried out to realize harmless disposal of waste, which will bring great benefits to society, economy and ecology [3].

B. J. Jansen et al. (Eds.): *International Conference on Cognitive based Information Processing and Applications (CIPA 2021)*, LNDECT 85, pp. 841–846, 2022.
https://doi.org/10.1007/978-981-16-5854-9_111

1.2 Domestic and Foreign Situation

In recent years, domestic also set off the upsurge of garbage classification. With the advent of big data, the problem has become more complex. As the first urban area to carry out garbage sorting in Shenzhen, Yantian District took the lead in implementing big data intelligent garbage sorting mode in 2015 [4]. Starting from the comprehensive coverage of intelligent collection and classification, classified collection and transportation, intelligent supervision, and base classification treatment, a large eco-friendly resource recycling system has been methodically formed. Foreign big data garbage governance and domestic governance is also different. Britain's Magpie system, for example, does this by placing unsorted rubbish on a conveyor belt, scanning it with infrared light, and sending the data back to the system's "brain". The system's "brain" activates high-pressure jets of air that blow similar types of garbage from the conveyor belt into a designated recycling bin. To date, practical knowledge and understanding of the benefits of waste separation from multiple performance perspectives remains limited [5]. But with the help of big data, garbage collection and transfer can not only effectively avoid the mixed transportation of garbage, but also effectively prevent the occurrence of such phenomena as garbage smuggling and illegal outbound transportation, and also ensure the standardized treatment mode.

1.3 Projects and Innovations

In this design, big data technology is applied in the treatment of municipal solid waste, which can not only realize the intelligent classification of garbage well, but also provide data support for the treatment of municipal solid waste – laying a good foundation for subsequent cleaning, disposal and recycling. At the same time, the current popular intelligent speech recognition technology can quickly obtain information and provide solutions for users. At the same time, we will push the "Wholesale" plan, which enables more people to participate in voluntary Marine services and jointly maintain the Marine ecosystem. The development of garbage classification software saves the cost of enterprises. Due to the high adaptability of big data itself, this APP will be more widely used after the arrival of 5G era, and help more urban areas to classify and treat household garbage intelligently. In recent years, China has accelerated the implementation of the garbage classification system, and the national garbage classification work has started from point to point and achieved initial results. Forty-six key cities have conducted trials first, making positive progress in garbage classification. Starting from 2019, domestic garbage classification will be fully launched in cities at and above the prefectural level. In the same year, Shanghai took the lead in issuing the "Regulations of Shanghai municipal Solid Waste Management", and garbage classification entered the compulsory era. With written laws and regulations, garbage classification has become a problem that residents must face in their daily life. So far, the benefits of practical knowledge and understanding of waste classification remain limited from an environmental performance perspective. With the advent of the era of big data, this problem has become more complex. Therefore, this design applies big data technology in the treatment of urban household garbage, which can not only achieve intelligent classification of garbage, but also provide data support for the treatment of urban garbage – laying a good foundation for subsequent cleaning, disposal and recycling.

2 Demand Analysis

2.1 Functions of Software Design

Big Data Analysis (BDA) has attracted a great deal of attention from academics and practitioners, and it is a rapidly growing field. [6] so the system has to do is based on the large mass knowledge atlas baidu search engine data, image recognition, speech recognition, text search can directly make the user more rapid, efficient to identify debris types (database for garbage classification standard will be released and of housing and urban-rural development of implementation with regular updates, to ensure the classification accurate, efficient and standard) [7]. At the same time, users can also sign up to participate in the voluntary "Whiletro" program pushed by our APP to regularly participate in the open sea, contributing their strength to the repair of the Marine circulation system [8]. In addition, we are also the main user upload garbage classification carefully relevant videos, that is, the use of experience, but also the dissemination of knowledge.

2.2 Main Problems Solved by Software

(1) Fast garbage classification.
(2) Increase residents' knowledge of garbage classification.
(3) Solve the problem of household garbage classification.

3 Overall System Design

3.1 Project Architecture

The project USES the MVC design architecture (can improve the reusability of the code, build good loosely coupled components, improve flexibility and controllability of the application), then system overall and detailed design as well as a variety of functions, including intelligent speech recognition, the design of the database, a user can query at any time by the Java language and the realization of the software of garbage classification. After the project runs, it can be downloaded and used by many people and the garbage classification is correct.

3.2 Architecture flow Chart of Garbage Classification Based on Big Data

Through data collection, the effect is presented [9, 10], as shown in the figure below (Fig. 1).

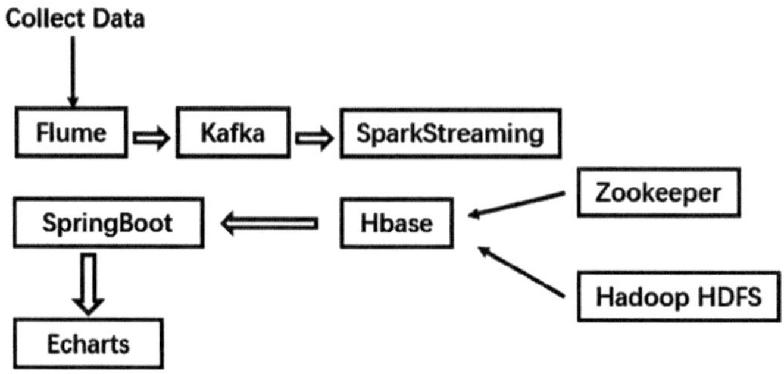

Fig. 1. The flow chart of real-time information analysis architecture of garbage classification based on big data

3.3 Development and Operation Environment

(1) Hardware environment

Processor: Intel®Core™ I5-8250U CPU @1.60ghz

(2) Virtual machine environment

Nexus 5XAPI 29 virtual Machine
Operating system: Android

(3) Software environment

Operating system: Windows10 64-bit operating system
Development language: Java
Development tools: Android Studio 2.2.2;The eclipse;JDK 1.8
Database: Navicat Premium
Server: Tomcat8.0

4 System Detailed Design

4.1 Software Subcontracting

BaseFragment and the creation of various sub-fragments.
Home page: HomeFragment.
Information: TypeFragment.
User: UserFragment.
Use FastJSON to parse the data.

4.2 Configuring the Networking Path (Creating the Constants Class)

```
public class Constants {
    public static  String BASE_URL="http://192.168.43.134:8080/ga/";
   // public static  String BASED_URL="http://192.168.43.134:8080/re

    public static  String HOME_URL=BASE_URL+"/json/HOME_URL.json";
   // public static  String IMAGE_URL=BASE_URL+"/img";
    public static  String BASE_URL_IMAGE=BASE_URL+"/img";

}
```

4.3 Setting up the Adapter for Banner Broadcasting (Using Banner Library)

```
//绑定数据
    @Override
    public void onBindViewHolder(RecyclerView.ViewHolder holder, int position) {
        if (getItemViewType(position) == BANNER) {
            BannerViewHolder bannerViewHolder = (BannerViewHolder) holder;
            bannerViewHolder.setData(resultBean.getBanner_info());
        }else if (getItemViewType(position) == CHANNEL){
            ChnnelViewHolder channelViewHolder=(ChnnelViewHolder)holder;
            channelViewHolder.setData(resultBean.getChannel_info());
        }else if (getItemViewType(position) == ACT){
           ActViewHolder actViewHolder=(ActViewHolder)holder;
            actViewHolder.setData(resultBean.getAct_info());
        }else if (getItemViewType(position) == RECOMMEND){
           RecommViewHolder recomViewHolder=(RecommViewHolder) holder;
           recomViewHolder.setData(resultBean.getRecommend_info());
        }


    }
```

5 Functions

5.1 Intelligent Speech Recognition

Use the SPEECH recognition SDK developed by IFLYtek Voice Cloud platform to download and configure according to the documentation. Gain access to the cellphone recorder, understand dictation recognition and other permissions, create the RecognizerDialog object, set the accent on Chinese, language on Mandarin, set the callback interface, and finally display the dialog receiving speech input.

5.2 Push "Jingluo" Program

Volunteer environmental protection activities will appear in the APP. When users click this item, they will see the information of volunteer activities closest to their location. At this time, users can participate in the registration, or set it to remind them to sign up for the next push. The feature uses the aurora to push messages. Use a long connection so that the connection to the server does not break. Long connection using XMPP protocol, suitable for message push, more streamlined, and without too much redundant network protocol socket based on TCP protocol to complete XMPP and long connection. Through this function can let more people together to protect the environment to make a contribution.

Acknowledgements. National Science and Technology Innovation and Entrepreneurship Training Program for College Students (202011439001).

References

1. Li, X., et al.: Study on the relationship between waste classification, combustion condition and dioxin emission from waste incineration. Waste Disp. Sustain. Energy **1**(2), 91–98 (2019). https://doi.org/10.1007/s42768-019-00009-9
2. Wen-jing, W., Xue-yi, Y.: Benefits analysis of classification of municipal solid waste based on system dynamics. J. Clean. Prod. **37**(29), 3–20 (2020). (in Chinese)
3. Zheng, J., Mingyuan, X., Cai, M., Wang, Z., Yang, M.: Modeling group behavior to study innovation diffusion based on cognition and network: an analysis for garbage classification system in Shanghai, China. Int. J. Environ. Res. Public Health **26**(18), 2–9 (2019). (in Chinese)
4. Sleeman IV, W.C., Krawczyk, B.: Multi-class imbalanced big data classification on Spark. Knowl.-Based Syst. **212**(54), 2–70 (2020)
5. Yuan, Y., Li, T., Zhai, Q.: Life cycle impact assessment of garbage-classification based municipal solid waste management systems: a comparative case study in China. Int. J. Environ. Res. Public Health **17**(15), 2–37 (2020). (in Chinese)
6. Raut, R.D., Yadav, V.S., Cheikhrouhou, N., Narwane, V.S., Narkhede, B.E.: Big data analytics: implementation challenges in Indian manufacturing supply chains. Comput. Ind. **125**(17), 17–43 (2020)
7. Ravikumaran, P., Vimala, K., Kartheeban, K., Narayanan, P.N.: Health data analytics: framework and review on tool and technology. Mater. Today: Proc. **23**(8), 1–7 (2020)
8. Yanan, L., Na, L., Xiao, Z., Yi, Q.: How wide is the application of genetic big data in biomedicine. Biomed. Pharmacother. **133**(48), 36–40 (2020). (in Chinese)
9. Alguliyev, R.M., Aliguliyev, R.M., Sukhostat, L.V.: Parallel batch k-means for Big data clustering. Comput. Indust. Eng. **152**(20), 2–30 (2020)
10. Rhee, C.K.: Chronic obstructive pulmonary disease research by using big data. Clin. Resp. J. **77**(23), 20–26 (2020)

Computer Games Based on Artificial Intelligence

Weiwen He(✉)

School of Information Engineering, Guangzhou Nanyang Polytechnic, Guangzhou 510000, Guangdong, China

Abstract. Chinese chess has a long history, but research on Chinese chess computer game lags behind that on International chess. In this paper, we apply Min-Max algorithm as a basis to implement artificial intelligence of chess. After generating moves, execution of alternate recursive search in finite depth and situation assessment in the searched leaf nodes is performed and execution of Alpha-Beta pruning judgment is obtained. Finally, we implement a chess game application program on the basis of application program framework of chess intelligent algorithm and Direct2D. The practice indicates that the program has better intelligent effect.

Keywords: Computer games · Chinese chess · Search algorithm · Game system

1 Introduction

The major breakthrough of international chess computer game attributes to two founding articles related to computer games published in 1950 by Claude Shannon: Programming A Computer for Playing Chess [1] and A Chess-playing Machine [2]. These two articles have provided trains of thoughts for implementing computer games. In 1958, three scientists Newell, Shaw and Simon had made in-depth research on Min-Max algorithm and found out that there was quite large proportion in the game tree that did not have search significance. A simple way named Alpha-Beta pruning algorithm could be used to cut off those branches.

The Taiwan scholars have began research on Chinese chess in the late 1970s and developed and published chess programs of Macintosh (6502) and IBM PC machine (8088) in the early 1980s. In 1983, Nankai University developed endgame program and global program in later few years. In the middle of 1980s, Hongkong published chess program. There were some specialists and scholars hoping to transplant international chess computer technology to Chinese chess to drive faster development of Chinese chess computer [3]. Meanwhile, Chinese chess has entered to theoretical research from technological research, and relevant theories of opening, middle game and endgame and chess solution are set up successively, which have offered theoretical basis for Chinese chess computer game.

B. J. Jansen et al. (Eds.): *International Conference on Cognitive based Information Processing and Applications (CIPA 2021)*, LNDECT 85, pp. 847–851, 2022.
https://doi.org/10.1007/978-981-16-5854-9_112

The paper has implemented Direct2D application program framework, Min-Max algorithm, Alpha-Beta pruning algorithm, chess artificial intelligent algorithm. It also includes the algorithm implementation of the whole chess program and offers logical control, status switching, move generation, move assessment and move suggestion, and the application program can be interactively interfaced.

2 Data Representation

2.1 Pieces

Both parties have 14 kinds and 32 pieces in chess. It can use different kinds of chess to represent the category encoding from 0 to 13 to represent these 14 kinds of chess, such category encoding can be used as array index, and 0–6 refer to the piece type of the red, 7–13 refer to the piece type of the black. In this way, only to judge whether it is less than 7, then you will know it is the red or the black.

The 32 pieces also need encoding, because it is necessary to differentiate specific pieces to specifically analyze the situation of a certain piece. Here you can encode the red as from 0 to 15 and black from 16 to 31. In this way, the pieces belonging to which party can be judged and used as subscripts through the fifth place in binary system. Besides, there is no pieces in some places in chessboard, thus "empty piece" needs to be defined.

2.2 Chessboard

Chessboard is used for containing the pieces. As for the representation of chessboard, the most direct way is the two-dimensional array. But the weakness of two-dimensional array is that it has lower efficiency and inconvenient in logical control when some operations are proceeded. Here use a one-dimensional array of 144 elements to represent chessboard and save the serial number of pieces in each place [4]. The chessboard that each element specifically represents is shown as Fig. 1, in which the number encoding is of hexadecimal.

One advantage of representing chessboard in this way is to calculate relative location in a more convenient way. For example, the index of top-most place of x place is x – 1, and the index of right place is x + 16. The most convenient is the judgment of horse vaulting, which can be conveniently calculated by the different value of relative location.

3 Algorithm Implementation

3.1 Move Generation

Different piece categories have different movement rules. Take the red as an example, the rules are: pawn moves one standard at a time; before crossing the river, it only moves forward; after passing over the river, it can move left and right or forward. If the movement goal is the empty piece (namely no piece is captured), then the move is general move, or it is attacking move.

In the process of generating move, if there is no effective move, it means it is being caught, and the game is over; if the opposite king can be directly captured, the game

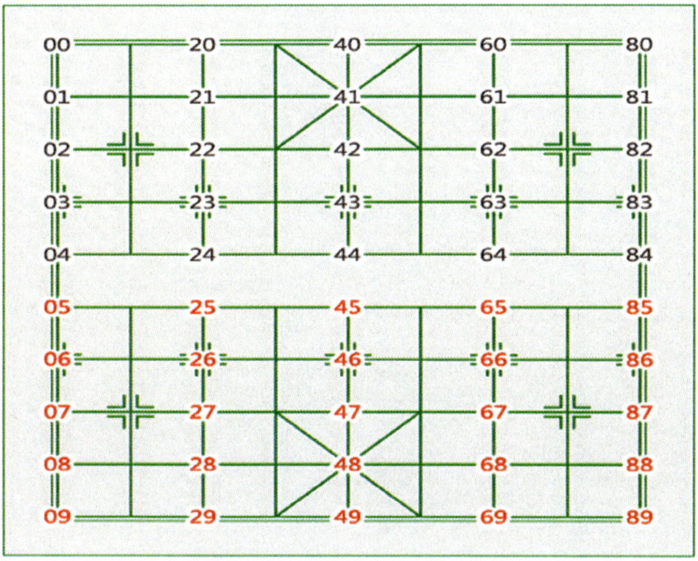

Fig. 1. Encoding of chessboard place.

is over as well. Besides, when the kings of each party are opposite to each other, then the movement party wins according to the rules, the game is over. In the paper, when traverse move, we regard this kind of situation as one of the effective moves. But in fact, it can not execute the move forever.

3.2 Situation Assessment

Situation assessment is a very important part of artificial intelligence in chess, and also a very wild part. The programming of this part pays more attention to the balance of knowledge and efficiency [5]. The adopted assessment strategy in the paper is: use the sum of piece values as basic scores, analyze the difference of the times of being protected and threatened of opposite each piece. If certain piece of the opposite is being threatened as a whole, then minus the value of the piece from basic score.

3.3 Search Algorithm

The use of search algorithm of Alpha-Beta must be an in-depth and preferential search algorithm and needs to be implemented by recursion [6, 7]. Because sides changing movement occurs with every further step forward in the searching process, it is an alternate recursion. Take the red movement as an example, the search procedure is shown as Fig. 2.

At the beginning, Alpha is –MAXSCORE, Beta is +MAXSCORE, namely the highest score is the minimum value is thought to be found [8]. The already found lowest score is the maximum value, thus their values in the searching process can be continuously close to the optimal value; the end of recursion is game-over, or reaches definitive depth.

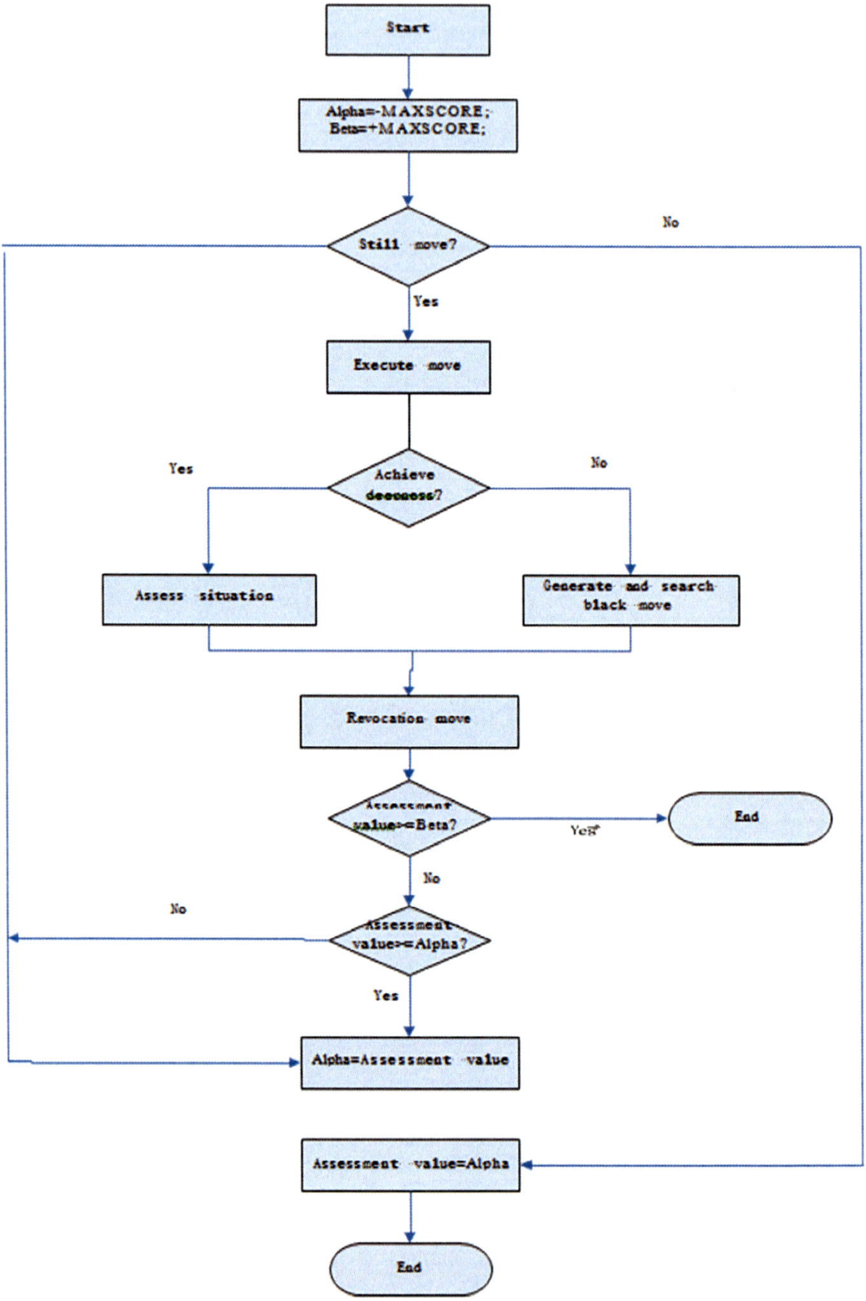

Fig. 2. Flow chart of search algorithm

Both assessing situation or searching opposite move will generate assessment value. When the assessment value is higher than Beta, Beta pruning occurs, and the search is terminated [9]. Or until all the move searches are over, the found maximum value will be regarded as ultimate assessment score. In the process of searching, the attacking move is preferentially searched, and then searching the general move, which will make pruning easily occur as soon as possible [10].

Due to the fact that the searching process of artificial intelligence is generally longer, thinking too long in the user interface thread will lead to losing response of user interface. Therefore, establishment of another thinking thread is needed. If it is artificial intelligence's turn to think, notify the thread to execute thinking.

4 Conclusions

Through testing artificial intelligence, the paper has found out: in the earlier stage, computer sometimes occupies initiative position by virtue of itself and temporarily gives up capturing pieces to expand its threatening situation. In this way, it will lose its initiative position for constant change of situation; in the middle stage, the response of artificial intelligence is rapid and tactful, it often moves good pieces and excellent pieces for overall consideration, which is contributed to the feature of computer's mechanical thinking. By contrast, human being generally focuses so much on the part that they lose the overall consideration; the later performance is more rapid and tactful. The paper uses Min-Max algorithm as basis to implement artificial intelligence of Chinese chess. After generating move, it executes alternate recursion search in limited depth and executes situation assessment in the searched leaf nodes. After that, it executes Alpha-Beta pruning judgment and finally launches an intelligent chess software.

References

1. Shannon, C.E.: Programming a computer for playing chess. Philos. Mag. **41**, 256–275 (1950)
2. Shannon, A.: Chess-playing Machine. Sci. Am. **2**, 48–51 (1950)
3. Ban, Y., Zhao, Y.: Algorithm research on online game of Chinese chess. In: Proceedings of the 2012 International Conference on Computer Science and Service System, pp. 1659–1661 (2012)
4. Encyclopedia of Chinese Chess. Computer Game. http://www.xqbase.com/computer.htm
5. Yen, S.-J., Yang, J.-K.: Two-stage Monte Carlo tree search for connect 6. IEEE Trans. Comput. Intell. AI Games **3**(2), 100–118 (2011)
6. Chen, B.-N., Liu, P.-F., Hsu, S.-C.: Aggregating consistent endgame knowledge in Chinese Chess. Knowl.-Based Syst. **34**(SI), 34–42 (2012)
7. Gui, W., Jun, T.: Chinese chess recognition algorithm based on computer vision. In: The Processing of CCDC 2014, vol. 5, no. 1, pp. 3804–3809 (2014)
8. Microsoft Corporation. MSDN Resource Library. http://msdn.microsoft.com/library/
9. Kerr, K.: Layered Window with Direct2D. http://msdn.microsoft.com/en-us/magazine/ee819134.aspx
10. Bin, Z., Xu, Y.: Adaptive Genetic Algorithm Application in the Chess Game System. Computer Program Skill and Maintenance, pp. 122–124 (2012)

Construction and Practice of Professional Ability Embedded in Clothing Major Under Cloud Data Service

Qi Guo(✉) and Jiatong Yan

Jilin Animation Institute, Changchun, Jilin, China
guoq911@nenu.edu.com

Abstract. As modern clothing majors continue to use the advancement of technology to improve the training of professional talents, vocational ability embedded education is the main direction for future education reform and innovation. Vocational ability embedded education mainly uses the school-enterprise cooperation model to tap the industry's potential and guides students' learning at school with employment as the orientation. At the same time, it uses cloud data services to realize information sharing between schools and enterprises through the cloud platform, which can better set the development direction for students and improve the practical problem of students' difficulty in obtaining employment after graduation. This paper mainly explores the construction and practice of professional ability embedded in the clothing major under the cloud data service. In addition, this paper analyzes the current development in the form of questionnaire surveys, and it also explores practical methods and future development trends.

Keywords: Cloud data · Professional ability · Embedded training · Clothing major

1 Introduction

With the advancement of technology and the acceleration of the pace of modern life, every industry is seeking an efficient and favorable development model, and continuously adjusting and upgrading the industry. This leads to the problems of "difficulties in employment" and "labor shortage", which reflects the contradiction in China's talent structure [1]. Through the joint development of school-enterprise connection, this paper uses the service of cloud data to explore the construction and practice of professional ability embedded in clothing majors, so as to solve the current problems.

2 Current Construction of Professional Ability Embedded in Clothing Major Under Cloud Data Service

In recent years, cloud data has become a hot spot for current development, and cloud data services are the integration of new technologies such as virtualization, automation,

B. J. Jansen et al. (Eds.): *International Conference on Cognitive based Information Processing and Applications (CIPA 2021)*, LNDECT 85, pp. 852–857, 2022.
https://doi.org/10.1007/978-981-16-5854-9_113

parallel computing, and security strategies. It is now one of the key infrastructures in most industries [2]. The clothing major under the cloud data service, through the collection of school information, can directly correspond to a company's job selection before working, which improves the efficiency of talent selection.

2.1 Survey of Satisfaction of Schools and Enterprises with Cloud Data Services

In order to understand the development satisfaction of cloud data services to help the construction of the clothing profession, we conducted a detailed survey of school and related enterprise personnel through questionnaire surveys. Among them, there are 300 valid questionnaires, and the specific survey data is shown in Table 1.

Table 1. Satisfaction of schools and enterprises with cloud data services

Satisfaction of schools and enterprises with cloud data services				
Very satisfied	Relatively satisfied	General	Not so satisfied	Unsatisfied
52%	23%	17%	8%	0%

It can be seen from the above table that schools and enterprises are highly satisfied with the talents cultivated through the clothing major under the cloud data service. Very satisfied answers accounted for more than half of the total number of people. Relatively satisfied and general adaptation accounted for 40% of the total number of surveys, and only 8% were less satisfied. There was no dissatisfaction data among the people who received the questionnaire. Preliminary conclusions can be drawn from the data. Cloud data service has become a very common service technology and is recognized by most people. It is not only the choice of clothing professional education in colleges and universities or the talent selection of enterprises in the society, but also the professional construction of professional ability embedded clothing professional under the cloud data service to meet the needs of both. For colleges and universities, with the innovation of education mode, they can better cultivate talents needed by society, and enterprises can also find talents suitable for their own needs through the sharing of cloud data.

3 Construction of Professional Ability Embedded in Clothing Major Under Cloud Data Service

3.1 Purpose of Education for Embedded Clothing Major

The ultimate goal of choosing embedded teaching for clothing majors is to ensure that college students can find suitable jobs and complete their employment plans in a short time after graduation. In the western education system, embedded teaching has been very mature and perfect, but it is still in the early stage of teaching reform in China [3].

The embedded education construction of professional vocational ability for clothing majors is to let students improve their knowledge and skills together and apply the knowledge they have learned in practice. Through the connection with enterprises, students can have the opportunity to enter the enterprise for time learning during the school period and conduct independent communication and interaction with seniors with work experience. Students can understand the composition of their major in advance, and after each task is completed, they can summarize and upload to the cloud platform to summarize and accumulate their own experience. Students gain a sense of accomplishment through the guidance of the project, which can improve their motivation and enthusiasm for learning [4].

3.2 Investigation on the Adaptability of Embedded Clothing Major Education to Work Posts

In order to understand the progress made in the construction and practice of the professional ability embedded clothing major under the cloud data service, we conducted a detailed survey on the students receiving the ability embedded teaching and related enterprise personnel through the form of questionnaire survey. Among them, there are 300 valid questionnaires, and the specific survey data are shown in Table 2.

Table 2. Adaptability of embedded clothing major education to work posts

Adaptability of embedded clothing major education to work posts				
Very adaptable	Relatively adaptable	General	Not so adaptable	Unadaptable
49%	23%	15%	10%	3%

It can be seen from the above table that students studying in the clothing major are more adaptable to the professional ability to accept embedded teaching and work positions. Very satisfied answers accounted for 49% of the total number of people who received the questionnaire survey, relatively satisfactory and general adaptation accounted for 38% of the total number of surveys, and less adapted and unsuitable accounted for only 13%. It can be seen from the data that the construction and practice of the professional ability embedded clothing major has achieved initial results. According to the survey of clothing companies, the post personnel can adapt quickly and carry out corresponding work well, reducing the time for induction training. And with the help of cloud data, it is possible to find graduates who are suitable for recruitment positions faster and more accurately. This school-based education model can promote the efficient and sound development of enterprises after students enter and leave the society.

4 Professional Practice and Development of Professional Ability Embedded Clothing Under Cloud Data Service

Vocational ability embedded clothing teaching is to comprehensively cultivate and improve students' abilities, combined with the practical teaching experience of embedded system courses, based on practicality, application and innovation [5]. Through cloud data, this research collects and predicts the strengths of students, and assigns tasks for practical courses and still provides employment guidance when they are employed in the future. Therefore, in the clothing major, the practice of vocational ability embedded teaching can improve the quality of running a school and improve the employment competitiveness of students, which is the main reform direction of future teaching. It can improve students' comprehensive quality and skills [6].

4.1 Change the Status of Practical Teaching

In the past, the teaching was mainly based on the content of the books, and the teacher's explanation became the acquisition of professional knowledge, ignoring the importance of practice. This leads to students' lack of hands-on ability, and their knowledge of the clothing profession only remains on the surface. After graduation, students will be confused, do not know what kind of work they should do or start a job that is different from what they think, resulting in wasting a lot of time in finding and re-planning their career direction.

Embedded courses increase many opportunities for teaching, adhering to the principle of equal emphasis on learning and practice, allowing students to create innovation groups when they are at school, and develop and complete innovative projects through joint training between the school and related companies [7]. During this period, all the processes that students participate in will be uploaded and recorded through the cloud platform. A large amount of cloud data storage can provide a reference for the school to revise the syllabus and curriculum development in the future, make up for the shortcomings, and continue to play to its strengths.

Students build a benign, stable and sustainable school-enterprise interactive environment under the embedded practice and development education mode of vocational competence under the cloud data service [8]. The introduction of the Internet has also continuously deepened education. The cloud data platform provides convenient conditions for our study and future work.

4.2 Seamless Connection with Business Development

The Internet is the mainstream technology of modern development and the core tool that we can use in our study, work and life. With the continuous development of cloud data services, the construction and practice of embedded clothing majors in the professional abilities of clothing is also continuously deepened [9]. With the help of cloud platforms, sharing data has become something that can be done anytime, anywhere.

In the mode of school-enterprise cooperation, maximizing the role of professional ability embedded clothing professional construction and practice can enable both the school and the enterprise to achieve a "win-win" result [10]. In the case of selective sharing of cloud data, students can apply for an internship to the company when they are doing functional practice in the company and wait for the approval of the company. During the holiday, they can carry out short-term internships to accumulate experience for themselves.

Not only can students apply on their own initiative, but also during the school-enterprise linkage teaching period, the company can also conduct professional aptitude tests on students under cloud data analysis. Those who meet the corporate talent standards can provide an offer to students in advance. In addition, schools can also use cloud data services to learn more about students' specialties through practical teaching of professional ability embedded clothing professional, so as to transport and train more talents for enterprises. Moreover, through such a functional embedded teaching model, students can have a basic understanding of some positions in advance, which can greatly shorten the practice of corporate training and try to be directly on the job. School-enterprise cooperation and collaborative education is an important way and mode for higher education to integrate into modern social production practices. The use of cloud data services is also an important factor for better completion of the embedded teaching construction and practice in the apparel industry, allowing students to seamlessly connect with corporate jobs after graduation, and grow rapidly.

5 Conclusion

In the era of rapid development, in order not to be abandoned by the torrent of the times, it is necessary for ours to continuously improve the teaching mode in schools. Integrating development with technological innovation, cultivating innovative and practical talents, who can quickly become their own development capabilities are the needs of society and the talents most companies want. From the perspective of students, students' knowledge is explained through the teacher's professional ability embedded teaching, and they understand the basic theories of the profession. Through cloud data matching companies and participating in school-enterprise joint practice activities by themselves, students can directly improve their practical experience. The construction and practice of the professional ability embedded clothing specialty under the cloud data service is an innovative development model of the future college education model. The help of cloud data will also provide enterprises with more comprehensive and high-quality talents.

Acknowledgements. This article is the phased research result of the higher education teaching research project of Jilin Animation Institute (Z1805) "Construction and Practice of the "Embedded Professional Ability" Curriculum System of Film and Television Clothing under the Background of Transformation".

References

1. Bao, Z.: Exploration of multi-dimensional "embedded" talent training model. J. Wuhan Polytech. **16**(06), 34–38 (2017)
2. Qiong, Y., He, B., Shuang, L., Yue, T., Long, Q., Jiasheng, Y.: Research on cloud data center and cloud platform technology application. Radio Telev. Technol. **42**(12), 74–87 (2015)
3. Fei, Z., Quan, L.: Research and exploration of practice embedded teaching reform. Univ. Educ. **11**, 62–63 (2015)
4. Chen, W.: Teaching thinking of ability-based embedded system specialty in higher vocational colleges. Talent **15**, 28 (2013)

5. He, J., Ye, Z., He, Y., Lu, L., Xu, W.: Exploration of the innovative talent training model in the transformation pilot of applied undergraduate colleges and universities from the perspective of interdisciplinary: taking the practice teaching reform of embedded system courses as an example. Vocat. Educ. Forum **33**, 42–44 (2016)
6. Jia, B., Xu, H., Gong, B.: Exploration and application of experimental teaching method reform for the Internet of Things engineering specialty oriented to cultivate practical ability. Heilongjiang Educ. (Theory Pract.) **05**, 73–74 (2020)
7. Zhang, M., Zhang, B., Song, J., Huang, Q.: Exploration and practice of the teaching reform of embedded system development course. Sci. Technol. Innov. Her. **17**(04), 213–214 (2020)
8. Xu, D.: Research on the improvement of teachers' practical teaching ability based on "embedded" school-enterprise cooperation: taking Jiangyin Vocational and Technical College International Trade Practice Major as an example. J. Changchun Inst. Educ. **37**(01), 52–56 (2021)
9. Yao, X., Xiang, L.: The application of network interactive teaching mode in the reform of modern ideological and political curriculum education: comment on "Innovative exploration of college students' ideological and political education from the perspective of new media". Chem. Reagents. **42**(04), 475 (2020)
10. Zhang, S.: The exploration and practice of the talent training model of school-enterprise cooperation: taking Changchun University of Science and Technology as an example. J. Changchun Univ. Sci. Technol. (Soc. Sci. Edn.). **34**(03), 131–134 (2021)

The Evolution Trend of "Digital Economy" and Its Enlightenment to Energy Enterprises

Guang Chen[1](✉), Hongda Gao[1], Guozhen Ma[2], Jianye Liu[1], and Wanting Yin[1]

[1] State Grid Energy Research Institute Co. Ltd., Beijing 102209, China
chenguang@sgeri.sgcc.com.cn

[2] Economic and Technology Research Institute, State Grid Hebei Electric Power Company, Shijiazhuang, Hebei, China

Abstract. Firstly, the 2021 Government Work Report and the 14th Five - Year Plan are reviewed, then analyze the evolution trend of "digital economy" from the change of status, the new policy opportunities of "digital economy", the attention, data opening and sharing, development and utilization, and finally propose corresponding enlightenment and suggestions for energy enterprises.

Keywords: The digital economy · Evolving trend · Data

"Digital economy" are the hot topics in the Government Work Report of 2021 and the Outline of the Fourteenth Five - Year Plan of the People's Republic of China and the National Economic and Social Development of the 2035 (hereinafter referred to as the Plan Outline) [1].

"Digital economy" is ushered in a new round of policy dividends, further extending from the economy to people's livelihood [2], government affairs, ecology and other fields, and putting forward new requirements and expectations for energy enterprises in improving the efficiency of government decision - making, standardizing internal platform management, promoting the development of power data, and serving the construction of smart city and digital countryside (Fig. 1).

Fig. 1. Big data

1 Policy Interpretation

In the 2021 Government Work Report, it is proposed to "accelerate the development of digital development, create new advantages of digital economy, jointly promote digital

B. J. Jansen et al. (Eds.): *International Conference on Cognitive based Information Processing and Applications (CIPA 2021)*, LNDECT 85, pp. 858–863, 2022.
https://doi.org/10.1007/978-981-16-5854-9_114

industrialization and industrial digital transformation [3], accelerate the pace of digital society construction, improve the level of digital government construction, create a good digital ecology, and build a digital China". The Planning Outline makes the "digital economy" alone. This paper mainly analyzes the content related to energy enterprises [4].

First, we will promote the deep integration of digital technology with the real economy, and accelerate the building of new advantages in the digital economy. Mainly include: strengthening the innovative application of key digital technologies. We will focus on key areas such as high - end chips, operating systems, key artificial intelligence algorithms and sensors, and accelerate the research and development, breakthroughs and iterative application of basic theories, basic algorithms, equipment and materials [5]. We will speed up digital industrialization. Cultivate and strengthen emerging digital industries such as artificial intelligence, big data, blockchain and cloud computing; carry out pilot and demonstration in key areas such as smart energy; encourage enterprises to open up search, e - commerce and social data, and develop the third - party big data service industry. We will promote digital industrial transformation. Build international industrial Internet platforms and digital transformation promotion centers in key industries and regions; promote digital transformation of the service industry [6].

Second, we will promote innovation in public services and social operations, and accelerate the construction of a digital society. Mainly include: providing intelligent and convenient public services. Focus on key areas such as education, medical care, promote the inclusive application of digital services; encourage social forces to participate in "Internet public services". Build a smart city and a digital countryside. Promote urban and rural development and governance mode innovation; build urban data resource system, promote the construction of urban data brain; explore the construction of digital twin city; and promote the digitalization of rural management services. Build a new picture of a better digital life. Promote the digitization of shopping consumption, home life, tourism and leisure, transportation, and other scenarios; strengthen the education and training of digital skills, popularize and improve citizens' digital literacy; accelerate the construction of information barrier - free, and help the elderly and the disabled share digital life [7].

Third, digital technology will be widely used in government management services to improve the level of digital government construction. Mainly include: strengthening the open sharing of public data [8]. Deepen the sharing and utilization of basic information resources such as national population, legal persons, and spatial geography, give priority to the opening of high - value data sets such as enterprise registration and supervision, health, transportation, and meteorology to the public, and encourage third parties to deepen the mining and utilization of public data. We will improve the efficiency of digital government services. Accelerate the construction of digital technology auxiliary government decision - making mechanism, improve the level of accurate dynamic monitoring, prediction and early warning based on high - frequency big data, strengthen the application of digital technology in public health, natural disasters, natural disasters, accident disasters, social security and other public emergencies, and comprehensively improve the ability of early warning and emergency response [9].

Fourth, build a digital rule system and create an open, healthy and safe digital ecology. Mainly include: establish and improve the data factor market rules. Accelerate the establishment of basic systems and standards and norms for data resource property rights, transaction circulation, cross - border transmission and security protection, accelerate the basic legislation in data security and personal information protection, and improve the data classification and grading protection system applicable to the big data environment. We will create a standardized and orderly policy environment. Build a policy and regulations system corresponding to the development of digital economy; improve the management norms of sharing economy, platform economy and new individual economy; and strengthen the supervision of Internet platform economy according to law and regulations. We will promote the building of a community with a shared future in cyberspace. Actively participate in the formulation of international rules and digital technology standards such as data security, digital currency and digital tax.

2 Evolving Trend

First, the digital economy has become a new momentum of China's economic development, and has played an increasingly important position in the economic system. Under the "two - wheel drive" of policy and market, China's digital economy has developed rapidly and has become the world' s second largest digital economy. According to data from the China Institute of 2019, by the end of 2019, China's digital economy reached 35.8 trillion yuan, accounting for 36. 2% of GDP. The scale ratio of the two parts of "digital industrialization" and "industrial digital" is about 1: 4. The Outline of the Plan first proposed the new economic index of "the added value of core industries of digital economy occupies GDP", which requires the added value of core industries of Chinese digital economy to increase from 7. 8% in 2020 to 10% in 2025. We can simply calculate that China's digital economy will be over 50% of GDP by 2025.

Second, the development of the "digital economy" has ushered in a new round of policy dividends. After the NPC and CPPCC sessions and the NPC and CPPCC sessions, with incomplete statistics, at least 18 provinces (cities), including Beijing, Shanghai and Zhejiang, have clearly proposed to vigorously develop the "digital economy" and make detailed plans for the related work of the digital economy. The high - frequency vocabulary for the development of digital economy in 2021 are: industrial Internet, 5G, big data center construction, smart city, smart government, and digital industrialization, etc.

Third, the negative effects caused by the digital economy while generating many new technologies, new forms of business and new models have also attracted attention from all parties. While promoting economic and social development, digital technologies such as big data and artificial intelligence, as well as new business forms and models such as the platform economy and the sharing economy, have also brought risks and challenges. For example, artificial intelligence finds and summarizes people's behavior rules through big data, producing "big data killing" problem; platform enterprises force businesses to "choose one" station, causing monopoly and unfair competition; unbalanced distribution between different regions, urban and rural areas and age groups, resulting in "digital gap" in access, use, ability; fuzzy personal data concept and boundaries, imperfect personal security supervision system, leading to excessive collection, use and leakage of user personal information data.

Fourth, promoting data open sharing, development and utilization gradually has practical conditions. "establish and improve data elements market rules" "speed up the establishment of data resource property rights, transaction circulation, cross - border transmission and security protection and other basic system and standards" "development of data assets evaluation, registration and settlement, transaction matching, dispute arbitration and other market operation system" "accelerate the data security, personal information protection basic legislation" "perfect the data classification protection system applicable to big data environment" and other content, etc., To promote the opening, sharing, development and utilization of data for all parties to solve the potential risks of law, security, system and other aspects, Open data sharing and development and utilization will drive into the "fast lane".

3 Relevant Enlightenment

First, with the Planning Outline as the guidance and direction, actively carry out the research, development and application of digital technology. It is suggested that energy enterprises should take the guidance of the key digital technology and application scenarios clearly defined in the Planning Outline, increase investment in infrastructure construction and technology research and development in relevant fields, and vigorously carry out scene innovation, business form innovation and business model innovation with the application of digital technology as the goal [10].

Second, while sharing the "dividend" of the rapid development of the digital economy, we also pay high attention to the negative effects derived from digital technology. Energy enterprises should not only enhance the awareness of opportunity, but also to establish a risk consciousness, attach importance to the development of new contradictions and new problems in the process of developing digital economy: first, constantly strengthen the social responsibility of enterprises, strengthen the internal review of various digital platform compliance, to ensure that the platform development strategy in line with the country's policy orientation, avoid monopoly, unfair competition and other policy risks. Secondly, in the design of digital products (such as APP), we should fully consider the "digital divide", strengthen the "digital care", and increase the product functional modules for the less developed areas, rural areas, the elderly, the disabled and other vulnerable areas and groups, such as "care mode".

Third, actively strive for the pilot and demonstration policies issued by the state. According to the Plan Outline, a number of international industrial Internet platforms and digital transformation promotion centers will be built in key industries and regions. At present, energy enterprises are accelerating the digital transformation based on the industrial Internet and other digital technologies, and it is suggested to actively apply to the state for the demonstration and pilot policies of the construction of the industrial Internet platform and the digital transformation promotion center.

Fourth, accelerate the development and utilization of electric power big data to fully release the value of data. First, the external demand for power big data constantly expands. The state proposed the construction of a "digital government", requiring government departments to "accelerate the construction of a digital technology - assisted government decision - making mechanism, and improve the level of accurate dynamic monitoring, prediction and early warning based on high - frequency big data". Due to the characteristics of power big data with high authenticity, high timeliness, fine granularity and high statistical frequency, the dependence of government departments at all levels on power big data will be further enhanced in the future. In addition to the previous power data analysis needs in economy, people's livelihood and environmental protection, the government will also add data analysis and data decision - making needs for public emergency response of public health, natural disasters, accidents and social security. It is suggested that energy enterprises strengthen the design and development of power big data analysis products around the above new fields, further give full play to the value of power data, and serve the modernization construction of national governance. Secondly, the open sharing of public data is conducive to the power big data to play the huge value of multi - dimensional data integration. The government actively encourages and supports the opening - up and sharing of high - value public data, including government affairs, population, legal persons, spatial geography, geography, enterprise registration supervision, health, transportation, meteorology, to the whole society, providing possibilities for power big data to give full play to data value on the basis of the integration with multi - dimensional data. Thirdly, the government will accelerate the establishment and improvement of market rules and legal systems in the aspects of data resource confirmation confirmation, data asset evaluation, data transaction and circulation, and data security protection will clear the obstacles for the opening and sharing of power data, and help energy enterprises promote the development, utilization and value improvement of power big data. Finally, it is suggested that energy enterprises should gradually explore the non - core power data such as e - commerce data to the society in an orderly manner under the premise of ensuring data security. According to the Outline of the Plan, enterprises are encouraged to open up search, e - commerce and social data, and energy enterprises should carefully study and deal with them.

Fifth, we will fully support the state in promoting the building of a digital society. In terms of supporting the construction of smart cities, on the basis of electric power big data, we will support the state to build an urban data resource system, urban data brain and digital twin city, and constantly improve the level of urban governance. In terms of supporting the construction of digital rural areas, we will accelerate the research and development of applicable rural power infrastructure in view of the characteristics of low rural population density and weak farmers economic affordability.

Acknowledgments. This work was supported by the science and technology project of State Grid Corporation of China, called "Research on bidirectional data value mining technology and typical application of "power economy" in SGCC" (Project Code: 1400-202157207A-0-0-00)".

References

1. Tea, H.: The development of China. China Econ. Times **05**(004), 24 (2021)
2. Wenfang, Qin, H.: Grasp the six trends of digital economy development accurately. Anhui Daily **05**(006), 18 (2021)
3. Huang, W.: Grasp the new trend of digital economy. Henan Daily **05**(007), 12 (2021)

4. Bing: Promoting digital economy legislation. China J. Soc. Sci. **05**(008), 11 (2021)
5. Wang, C.: Accelerate digital development and build digital China. Oper. Manag. **05**, 1 (2021)
6. Chen, G., Zheng, H., Yin, G.: To achieve carbon peak and neutral goals, energy digital economy. China Energy News **05**(028), 10 (2021)
7. Zang, Z.: Data mining tool WEKA and its application research. Enterp. Sci. Technol. Dev. **09**, 38–39 (2018)
8. Zhan, P., Xie, X.: Summary of big data systems and key technologies and tools. Netw. Secur. Technol. Appl. **08**, 50–52 (2018)
9. Wang, W., Chen, X.: Data mining processing and application research based on excel. Chin. Foreign Entrep. **09**, 44–45 (2018)
10. Xianghaihua: A summary of the development of database technology. Mod. Inf. **12**, 31–33 (2003).

Check for updates

Virtual Interactive System of Manipulator Based on Digital Twin

Yawei Li(✉), Guofu Luo, Ningbo Zhang, Junhao Zhang, and Xingbo Yang

College of Mechanical and Electrical Engineering, Zhengzhou University of Light Industry, Zhengzhou, Henan, China

Abstract. Starting from the concept of digital twins, the virtual interactive system of the robotic arm was studied. Firstly, the overall framework of the system was determined, then the functional modules of the system were divided, the system network architecture was designed, and the development tools used to build the system development environment were finally given. The platform and open source program library lay the theoretical foundation for the development of the virtual interactive system of the robotic arm based on the digital twin.

Keywords: Digital twin · Robotic arm · Virtual interaction · Remote control

1 Introduction

With the "Made in China 2025", "Industry 4.0" and other plans proposed, the manufacturing industry is developing towards intelligence [1]. Digital twin technology realizes the interconnection and integration of the physical world and the information world, providing strong technical support for the intelligent transformation of traditional manufacturing [2]. Manipulator is an indispensable equipment in modern manufacturing industry. This article takes manipulator as the research object, combined with the connotation of digital twin "virtual and real integration" and "virtual control real", and gives the overall design scheme of the virtual interactive system of the manipulator [3].

2 Overall Framework Design of Virtual Interactive System

2.1 Designed With Object-Oriented Thinking

The development of virtual simulation system of manipulator mainly uses object-oriented design ideas [4]. The purpose is to encapsulate object information, hide internal details, copy reusable function codes, and expand corresponding interfaces. A series of configurations and responses of the system are realized through the message response mechanism. Classification of these messages can minimize the degree of association between different objects in the system [5]. The overall frame design of the virtual interactive system is shown in Fig. 1:

B. J. Jansen et al. (Eds.): *International Conference on Cognitive based Information Processing and Applications (CIPA 2021)*, LNDECT 85, pp. 864–867, 2022.
https://doi.org/10.1007/978-981-16-5854-9_115

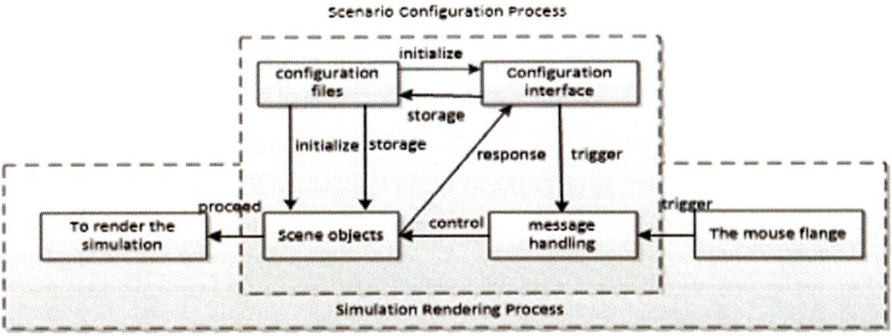

Fig. 1. The overall frame diagram of the object-oriented virtual interactive system

During the process of the scene configuration, the configuration file can initialize the scene object. After the modification operation, the modified information can be saved to the corresponding configuration file. The configuration file stores the initialization information of all customizable interfaces of the scene object. The configuration file can also initialize the configuration interface, and rewrite the corresponding configuration file after the interface information is changed [6]. When the function button of the configuration interface is clicked, a message will be triggered, and the scene object will respond to the change of the configuration interface, which is convenient for the operator to observe and adjust the state of the scene.

In the process of the simulation rendering, the operator can use the mouse and keyboard to trigger a message, and then control the state change of the scene object. The scene objects must be continuously redrawn, rendered, and refreshed in order to present the motion simulation process of the model in the three-dimensional view [7].

2.2 System Function Module Design

Using modular design ideas, the robotic virtual interactive system is divided into different modules according to functions [8], including: the initialization module is used to initialize the scene configuration when the virtual interactive system starts, including scene initialization, data initialization and twin model initialization; real-time the data communication module has the functions of data collection, transmission, and transmission of commands. The visual monitoring module can display the operating status information of the robot arm to realize the motion simulation of the robot arm; the remote control module can send the start and stop signals to the physical robot arm through the twin robot arm to realize remote start and stop or manual remote operation; The system interaction module can pick up the model and modify and adjust the model [9]. The adjustment of the viewing angle is convenient for the system operator to observe better.

2.3 The Overall Network Architecture Design of the System

According to the functional modules of the system, the overall network architecture of the virtual interactive system is designed, which is divided into the device layer, the middle layer, and the client layer from top to bottom. The equipment layer is the bottom layer of the system, which consists of robotic arms, industrial Ethernet, and production line servers [10]. The robotic arm and the production line server are connected through industrial Ethernet and switches. The data acquisition server and the remote control server together constitute the production line server [4]. The middle layer includes message middleware and database. The database is used to store the operating data of the robotic arm during its work. The message middleware plays a role in connecting the client layer and the device layer. To use this virtual interactive system, a designated virtual interactive system client must be installed. The specific network architecture of the robotic arm virtual interaction system is shown in Fig. 2.

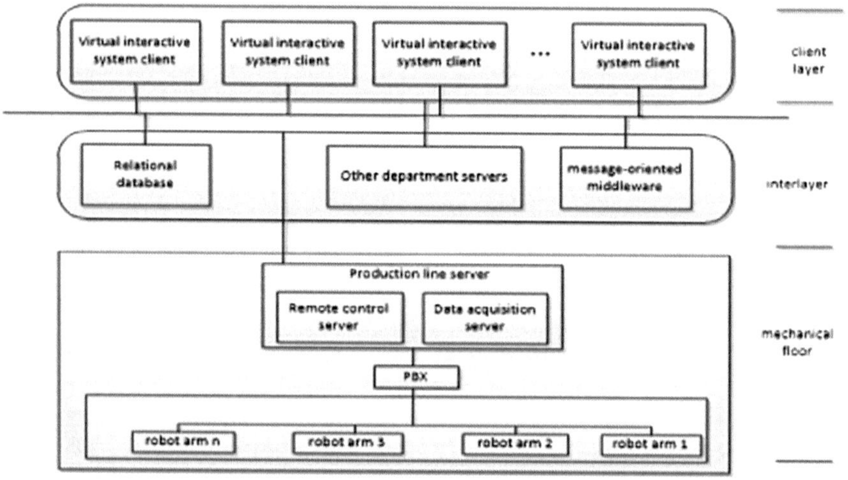

Fig. 2. Virtual interactive system network architecture

3 Construction of Development Environment for Virtual Interactive System

3.1 Selection of Development Tools

There are many tools that can be used to develop virtual interactive systems, such as Cult3d technology, Unity3D, etc., but there are some problems of flexibility and poor compatibility [11]. The OpenGL programmable pipeline can flexibly use GPU acceleration, improving the graphics rendering efficiency of the system client, and ensuring that Stable operation and good compatibility and scalability. Therefore, the development tool of the virtual interactive system is selected as the OpenGL programmable pipeline.

3.2 Construction of Development Environment

After the basic development tools were selected, the Windows 10 operating system is selected as the development platform, the integrated development environment is Visual

Studio 2015, and the programming language is C++ to develop the system. However, there will inevitably be problems with large amounts of code and complex programs in the process of developing the system. You can use free open source tool libraries, such as glew32.lib, glfw3.lib, Assimp3.lib, SOIL and other new and more powerful tool libraries to realize the rapid development of the system.

4 Conclusions

Digital twin is one of the important ways to realize intelligent manufacturing, and it puts forward higher requirements for virtual reality intelligent interaction in intelligent manufacturing system. Based on the concept of digital twins, this paper proposes a virtual interactive system of robotic arms based on digital twins, and completes the framework of the system. With object-oriented design, the functional modules of the system are divided according to the idea of modularization. According to the requirements of each functional module, the overall network architecture of the system is designed. According to the system requirements, the basic development tools of the virtual interactive system were selected, and the platform, environment and open source tools needed to build the system development environment were given. It has laid a theoretical foundation for the landing application of the virtual interactive system of the robotic arm.

References

1. Tao, F., Liu, W., Liu, J., et al.: Digital twin and its potential application exploration. Comput. Integr. Manuf. Syst. **24**(01), 4–21 (2018). (in Chinese)
2. Wang, P., Yang, M., Zhu, J., et al.: Dynamic data driven modeling and simulation method for digital twin. Syst. Eng. Electron. **42**(12), 2779–2786 (2020)
3. Wang, M.: Key Technologies of Virtual Remote Control of Quadcopter UAV in Underground Coal Mine. University of Science and Technology of Xian (2020)
4. He, L.: Design and implementation of virtual interactive system for CNC machine tools based on digital twin. University of Electronic Science and Technology of China (2019)
5. Hong, H., Zheng, L., Xie, W.: Simulation research of industrial robot based on network. Inf. Technol. Netw. Secur. **36**(11), 5–7+11 (2017)
6. Zeng, X., Song, A., Zhou, Y.: Design of a virtual human-machine system for manipulator teleoperation. Manned Spacefl. **26**(03), 353–360 (2020)
7. Cai, L., Li, Q., Li, Q.: HMI system for manipulator based on VR technology. Machinery **57**(03), 1–3+15 (2019)
8. Cui, J., Li, Z., Liu, W.: Technology of glazing robot virtual modeling and interaction. J. North China Univ. Sci. Technol. (Nat. Sci. Ed.) **38**(02), 92–99 (2016)
9. Liu, X.: Design of virtual interaction of ship cabin based on Unity3D model. Ship Sci. Technol. **43**(04), 4–6 (2021)
10. Zong, A., He, L., Wang, W.: Research on virtual excavator operating system based on virtual interaction. Mach. Des. Res. **36**(02), 175–180 (2020)
11. Li, J., Zheng, X., Sun, Y.: Virtual assembly interaction system of hydraulic support based on autonomous learning. Coal Mine Mach. **41**(03), 167–169 (2020)

Discussion on the Application of the Internet of Things Technology in English Teaching

Xiaowen Du(✉)

College of Foreign Languages, Xinjiang University, Urumqi 830046, Xinjiang, China

Abstract. This paper puts forward a new idea for the present situation of English procedural evaluation based on the Internet of Things. With the help of this technology, the intelligent data can be obtained and the flexibility of teaching can be enhanced. Through the data mining and the data analysis in the process of procedural evaluation, the procedural evaluation of teachers in English teaching is inferred. English teaching is linked with the Internet of Things ecosystem to improve students' enthusiasm, interaction and participation in English learning. English teaching is bound to rely more and more on technology, but policymakers need to actively consider digital technology as a teaching tool and work extensively with educators and technical experts to develop innovative educational strategies.

Keywords: Computer · Internet of Things Technology · English teaching · Educational strategy

1 Introduction

In educational activities, language teachers can use mobile devices to prepare teaching and learning activities. Given that all students have mobile devices, this has become an additional motivation for students to use them in English classes. And teachers have been developed because students use these devices as learning tools [1]. The use of mobile devices as a learning tool can be achieved in a number of ways, depending on the compatibility of each device with many existing applications and the operating system of the device. In today's globalized environment, the importance of English for specific purposes cannot be underestimated because it is widely used by users of other international languages [2]. Just as an economist needs to learn economic terminology, a practitioner in the ICT industry needs a language to communicate with foreign businessmen. Companies are always implementing communication strategies because they need to make quick decisions, and miscommunication with others can have a negative impact on the speed of business collaboration, business cooperation, or the introduction of new products on the market. Globalization has accelerated the application of English in the business environment and cross-cultural communication. Researchers have studied the use of business English in other languages for decades [3]. Advances in emerging technologies provide educators with the opportunity to transcend some of the traditional challenges, which not only liberates teachers from the physical screen, but also promotes

B. J. Jansen et al. (Eds.): *International Conference on Cognitive based Information Processing and Applications (CIPA 2021)*, LNDECT 85, pp. 868–872, 2022.
https://doi.org/10.1007/978-981-16-5854-9_116

task management. Many technologies are still a link in the classroom, and the connecting devices can improve the core skills of teachers and may even herald a different and exciting educational technology. In addition, we would also like to mention the role of the platform in integrating English learning activities and student assessment. For example, diversifying the forms of learning tools, setting changes in two to four answers for practice, and managing them as a competition to enhance the spirit of competition among students [4]. In addition, mobile devices can be used to exchange information and respond quickly to enhance listening activity. Anyway, mobile devices promote collaborative learning, creativity and self-assessment. In fact, the use of these devices is not as a simple language teacher. Project training is important because it can help teachers understand the use of mobile devices, effectively create courses and teaching applications, and stimulate students' innovation and creativity.

2 English Teaching Based on the Internet of Things

Different from the traditional English teaching method, English teaching based on the Internet of Things technology is of an innovative teaching method, which has been successfully applied and studied scientifically in the world for decades [5]. In a world dominated by cultural and occupational mobility, teachers need to prepare their students for a world of amazing linguistic diversity. English teaching based on the Internet of Things is a kind of learner-centered English teaching, which applies English as a foreign language to specific research or professional fields. For the computer major students majoring in economic informatics, the particularity of this innovative method lies in the use of real materials (professional magazines, reports, conference materials, expert interviews, etc.). Teachers will plan and develop courses for students according to their needs and expectations. The acquired knowledge and technological innovation in information and communication technologies can be applied to current and future tasks. Our course plan aims to develop professional language skills, focusing on professional examples in the field of the Internet of Things. The main goal is to acquire and use specific concepts, strategies and working methods in this field with the support of English language, with emphasis on the acquisition of various professional vocabulary [6]. As a result, students will have the opportunity to understand the format, structure and practices of specific professional schools in the field and to develop their ability to use the English language structures specific to the above fields. The importance of technological progress is reflected in the participation of computer literacy in teaching and the in-depth analysis of the various effects on educational methods. The real impact of technology on the English learning process can be understood by clearly establishing the roles of teachers and students in the teaching and learning, and clearly setting educational objectives and educational significance through information and communication technologies. The use of modern educational technology is a financial, social and political issue that should occupy a central place among the concerns of governments and international organizations. Since national and international political regions are determined by government policies, education is less open to commercialism that promotes successful innovation in other industries. As a result of this political environment, the management structure of educational institutions is often hierarchical and lacks entrepreneurship and the capacity to adapt to change. Teachers are rarely involved in policy decisions, so they have a weak ability to reform the teaching and learning processes required by new technologies and methods. Teaching acts as an important role in providing quality education to all. In the field of English teaching based on the Internet of Things, systematic teaching includes both the overall teaching model and the cognitive teaching model. The overall model of teaching places learners at the centre of the teaching process, which occurs at the following stages: procedural aspects (planning, curriculum design, classroom dynamics, teaching techniques, assessment of learning and implementation of national standards); learning perspective (behavior and cognition); attitude (active learning, supportive environment). On the other hand, students' cognitive model provides enough experience learning and teaching. Obviously, students need to be exposed to knowledge, know how to organize knowledge, and have the self-motivation to learn. The principle of cognitive teaching mode is: students take the initiative to learn and solve problems, students use a variety of learning strategies, teachers allocate students time to apply new skills, the responsibility for learning and solving problems is transferred from teachers to students, the learning strategies are clearly defined for students, student's learning efficiency is up to the teacher to decide, and the teacher is responsible for teaching decisions [7].

3 The Application of Connecting English Teaching to the Internet of Things

The Internet of Things is regarded as a catalyst. Skills in the 21st century fall into the following four categories: thinking mode (creativity and innovation, critical thinking and problem-solving, decision-making and learning); working mode (communication, collaboration and teamwork); working tools (information literacy, digital literacy); life skills (life and career management, personal and social responsibility). The reason for the failure of educational technology is not technology, but the lack of digital education. Some teachers tend to put technology before education and let technology drive teaching. Teachers need to be trained and show students how to use the skills in the classroom. Similarly, students need to show how to use technology to support their learning. When a teacher plans a class, what he thinks is that students must acquire a professional skill. This can be understood as a set of professional knowledge in relatively determined fields such as science and technology, which enables students to engage in specific occupations at a reasonable level of competence [8]. It includes expertise, specific skills and an understanding of yourself and the surrounding environment. Learners know the terminology in their mother tongue. Students should focus on the vocabulary used in the selected field of activity and develop communication skills related to the selected field (Fig. 1).

Our course plan aims to develop professional language skills, focusing on professional examples in the Internet of Things ecosystem. The main goal is to acquire and use specific concepts, strategies and working methods in this ecosystem (supported by English language), with a focus on the acquisition of various professional vocabulary. The first phase deals with text searching and subject areas about students' concern. Therefore, students will have the opportunity to understand the format. Aiming at the professional genres in this field, they should be trained to use English language structure

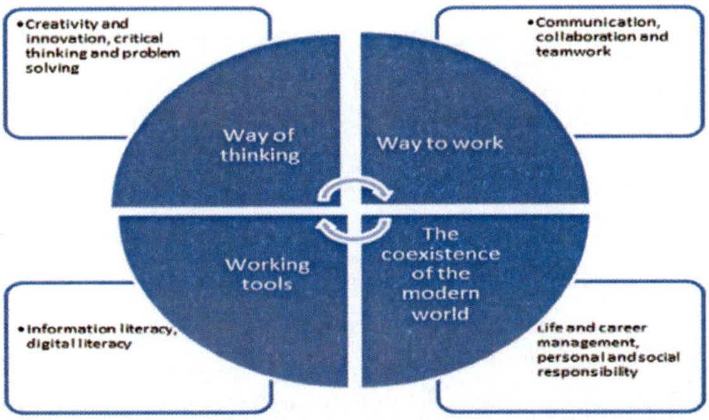

Fig. 1. Skills for connecting English teaching to the Internet of Things fall into broad categories

to practice [9]. Learning English for a specific purpose can be described as a realistic construction process, involving the network, the existing cognitive schema and the favorable environment. If the absorbed information has nothing to do with the current interests and needs of the learner, it will be shelved. The development of technology is affecting research and language teaching in many ways, which is related to teachers and researchers. In addition, they stressed that information technology plays an important role when the internet becomes the center of student activities and teacher support. The point of view is based on the premise that, in a changing world, cross-cultural and professional communication, as well as their participation in decision-making processes, are essential, and there is a growing need for permanent links with more sources of information. The English course of the School of Business Management computer Science focuses on describing communication and the use of language in economic conditions and communication technologies. Practical courses are tailored to provide discourse descriptions from empirical surveys so that students can better understand communication and language use in the above areas of specialization.

4 Conclusion

This paper analyzes the management structure of universities. Especially, the policy makers must establish a new educational paradigm. Through open and intensive training projects and mutual learning exercises, all educators and students are connected to the Internet of Things. English teaching based on the Internet of Things can respond to the needs of students, bring materials related to their professional fields, and manage to connect the classroom to the Internet of Things through a comprehensive curriculum plan of digital technology [10].

References

1. Ma, S.: A study on college oral English teaching based on real-time dialogue on the internet. Overseas Engl. **04**, 136–137 (2021)
2. Zhang, Y.: College English teaching reform under the internet model. Overseas Engl. **2021**(04), 158–159+171 (2021)

3. Zhu, Y.: A probe into the transformation of college English listening teaching model under the mixed teaching concept. Overseas Engl. **04**, 170–171 (2021)
4. He, Y., Li, C.: An analysis of the application of online and offline mixed teaching model in college English teaching. Engl. Square **06**, 112–114 (2021)
5. Shi, L.: A probe into the construction of English translation teaching model based on internet. Overseas Engl. **2021**(03), 59–60+70 (2021)
6. Gao, Y.: Discussion on the application of computer internet of things technology in English teaching in secondary vocational schools. J. Jiamusi Vocat. Coll. **2**, 356 (2018)
7. He, L.: Discussion on the application of computer internet of things technology in English teaching in secondary vocational schools. Campus Engl. **6**, 111–112 (2020)
8. Wen, Y.: On the application research of multimedia technology in computer English teaching in secondary vocational schools. Mod. Vocat. Educ. **2018**(9), 10 (2018)
9. Yu, J.: Research on the application of modern educational technology in English teaching in secondary vocational schools. Comput. Knowl. Technol. **10**, 170–172 (2020)
10. Chen, J.: The theoretical connotation and application analysis of college English network teaching. Foreign Lang. Teach. **6**, 46 (2004)

The Creation and Presentation of Sculpture Art in the Context of Digitization

Zizhao Han(✉)

Xi'an Jiaotong University City College, Xi'an 710018, Shaanxi, China

Abstract. The rapid growth of digital information technology has brought intensive development opportunities to the traditional sculpture industry, but these opportunities come with many challenges. Computer and digital sculpture technology are now being widely used in the sculpture industry. The advantages of quickness and accuracy have been favored by more and more sculptors. Digital technology enriches the creative techniques of sculpture art, while bringing infinite possibilities to its presentation.

Keywords: Digital context · Sculpture art · Creation and presentation

1 Introduction

The development of science and technology allows digital technology to be able to benefit all walks of life, while sculpture industry can also feel its strong presence. The use of digital technology has injected new vitality into sculpture creation and realized the organic combination of materialization and virtualization of sculpture art.

2 Features of Digital Sculpture

As an ancient art form, what differs it from painting is its three-dimensional look. The carrier form of sculpture art also has diverse characteristics, such as clay, metal, stone, wood. In its long development journey, the sculpture has gradually developed its mature creative process and artistic style. With the rise of digital computer technology and various 3D simulation technologies, computer modeling technology and digital sculpture technology are gradually applied to the sculpture industry. More and more sculpture artists rely on digital information technology for creation. Compared with traditional sculpture, the digital sculpture has the following features:

2.1 Non-substantiality

The non-substantiality of the design and creation process is the most important feature of digital sculpture technology. Traditional sculpture art requires physical carriers for design and creation, such as drawing on paper for design. Clay, metal, wood and other materials

B. J. Jansen et al. (Eds.): *International Conference on Cognitive based Information Processing and Applications (CIPA 2021)*, LNDECT 85, pp. 873–877, 2022.
https://doi.org/10.1007/978-981-16-5854-9_117

are required for creation. Traditional sculpture needs to utilize physical materials to present the artistic shape of the sculpture and the author's emotion and creative concept [1]. In contrast, the design and creation process of digital sculpture is virtual and relies on the rendering techniques of computer modeling software; the artist can also conduct modifications at will to the 3D sculpture model during the creation process. Therefore, digital sculpture technology is more virtual and convenient.

2.2 Digitization

The creation of digital sculpture is no longer limited by the size of the traditional sculpture carrier. The design of points, lines, and surfaces of the sculpture can be realized with the computer. In addition, the size of the sculpture is no longer determined by the size of the physical carrier but the storage capacity of the computer. When a design changes in the creation process, traditional sculpture can only be re-created on a new physical carrier, but digital sculpture can be manipulated in the computer software for the established digital model. The sculpture will be created at once after the final design is determined, with no unnecessary material waste.

3 Differences in the Digital Creation Process

There is a huge difference between digital sculpture and traditional sculpture in terms of the creation process. Although digital sculpture converts the traditionally substantial sculpture creation to a design on software, there are differences in the essence of the two creation processes due to the difference in techniques and sequence.

3.1 Different Observation Perspectives When Creating Digital Sculpture and Traditional Sculpture

To begin with, data application plays a role in digital creation, which can present the angle, scale, and size of the sculpture with precise data, while traditional sculpture needs to be observed by the naked eye. Also, light and distance will have an impact on the observation. In addition, the creators' observation is not limited by the volume and space of sculpture in digital creation, so it effectively avoids the impact of traditional large sculptures caused by the space angle and observation distance to the creator.

3.2 Differences Between the Creation Process of Digital Sculpture and Traditional Sculpture

Traditional sculpture creation requires sculptors to use their hands and tools to shape the sculpture. In digital creation, sculptors can simulate various sculpting tools using computer software to cut, carve, polish, and other tedious operational steps for the basic outline of the sculpture, therefore greatly improving the convenience of sculpture art creation [2]. Also, digital sculpture means allow sculptors to save their strength in the creation process so that sculptors have more energy to think. The sculptor can design the geometry of the sculpture precisely without the limitation of site, material, and space in the process of creation. Making unlimited modifications to the sculpture is made possible. Meanwhile, as the digital modeling software becomes more and more advanced, various fine textures, lines, and patterns can be designed, making the digital sculpture more realistic to the audience.

3.3 Digital Sculpture Requires a Combination of Various Software and Hardware

In digital creation, modeling software and hardware are equally essential. Hardware is also the main creative tool of digital sculpture, such as mouse and graphics tablet. 3D scanning technology provides a variety of technical means for the creation of digital sculpture, which can establish its digital model by scanning the physical material, allowing direct design and modification on the digital model during creation. The emergence of 3D scanning technique has broken through the traditional boundary between the solid and the virtual, realizing the free conversion between virtuality and reality. In the physical conversion of digital sculptures, various hardware is also applied, such as laser engraving machines, water-ripping machines, 3D printers, etc. The use of these devices allows for greater accuracy, more convenience, and lower cost, financial wise and labor wise, in the conversion of digital models to physical finished products [3].

4 The Presentation of Sculpture Art in the Context of Digitization

4.1 The Transformation of Sculptural Expression Brought by Mobile Internet

The digital storage and presentation forms of sculpture art do not have the time and space limitations as traditional sculpture works do. When traditional sculpture works are presented to the audience, the distance between the audience and the work and the light of the exhibition environment will influence the artistic tension around the works. Sculpture artists can make full use of digital tools and mobile Internet to achieve flexible presentation of sculpture works. In addition to being applied to the creation process, 3D scanning technology can also be used in the presentation of sculptures to achieve a comprehensive rendering of sculptures [4].

Although digital sculpture is not limited to physical carriers, this does not mean that it is defective in terms of visual expression. With the rapid development of electronic products in recent years, various rendering software and hardware have been upgraded and improved, and the rendering techniques of the software have been able to achieve the level of extreme closeness to reality. PBR material based on physical reflection is a technique that imitates real objects [5]. The PBR material can not only simulate the texture of sculpture materials, but also enhance the overall visual effect, giving the audience an immersive and realistic feeling. Today, the most advanced rendering technique can be applied to dynamic rendering of sculptures, presenting them to the audience from a complete piece of metal, stone, wood, etc. to a beautiful sculpture. presenting people the process of transforming a mere piece of metal, stone, or wood to an exquisite sculpture. The application of digital technology has brought about a radical change in the presentation of sculptures and presented has the audience an exquisite visual feast.

4.2 The Transformation of the Sculpture Function Brought by the Digital Context

In the digital age, sculpture is no longer a display or decorative artifact, but an important material in the digital animation industry. Sculptures can be scanned and post-produced by digital means to become a true-to-life character in animation and game works. Digital technology has transformed the function and role of sculpture, making it no longer just a humanistic artwork, but also a digital art product full of vigor that enriches the entertainment life of the public. In art museums and galleries, digitization can also bring new vitality to sculptures. Many sculpture art museums have already combined digital information means to realize dynamic display of sculptures, using 3D model technology to display the creation process of sculptures made by artists, the historical background of sculptures, 3D restoration of damaged sculpture relics, etc. Combined with AR technology, VR technology and other technologies, the audience can have an authentic and intuitive visiting experience [6, 7].

4.3 Digital Background Propels the Development of Sculpture Industry

The development of the Internet and the upgrading of mobile devices have created favorable conditions for the development of sculpture art in the digital context. On one hand, Internet computer technology can be used as a creative tool for sculpture artists, for that the efficient and fast modeling software can speed up the creation process. On the other hand, Internet computer technology can be used as a promotional tool for sculpture artists. Both digital and physical sculpture works can be disseminated and promoted on the Internet [8]. Also, sculpture artists can publish their own works on the Internet to let more people enjoy their works at home, thus boosting artists' popularity. Sculpture artists can take in the comments and suggestions from the audience in terms of their works on the Internet, so as to gain new creative ideas and motivate themselves to innovate. In addition, as more and more sculpture artists display their physical and digital sculptures on the Internet, artists can browse their peers' works on the Internet, interacting and learning from each other, which will fuel a continuous growth of their skillset and strength [9, 10].

However, digitization is a double-edged sword. The development of digital information technology and the Internet also gives access to imitating and plagiarizing by unscrupulous people. Driven by self-benefiting interests, plagiarizing is becoming ubiquitous. The digital information technology and the Internet technology, while providing sculpture artists with more convenient creative conditions, should also go the extra mile to protect the intellectual property of the sculpture artists.

5 Conclusion

As an inevitable trend for the future sculpture industry, digital technology does bring certain convenience to the creation and presentation of sculpture. However, as an ancient art form, the most important thing in sculpture art is the realization of the author's "humanity" and the integration of ideas. Although the digital means seem to be precise and swift, the creation of sculpture cannot rely entirely on the machine. What's more important is the participation and creation of human, who is the only shaper of the ideas and emotions of sculpture. If lacking human participation, the sculpture is just a cold artifact instead of a humanistic artwork.

References

1. Xi, B.: On the creation and presentation of sculpture art on the digital context. Art Sci. Technol. **033**(010), 87–90 (2020)
2. Bin, Q.: Study of digital sculpture modeling methods in the context of 5G. Art Mark. **06**, 92–93 (2020)
3. Ji, X.: The impact of digital sculpture and 3D printing on sculpture art. Hubei Institute of Fine Arts (2019)
4. Gao, H.: The construction of light on the visual presentation of sculpture. Shenyang University (2019)
5. Chen, W., Zhang, H.: A brief analysis on the digitization of mobile terminal of large wood carving art – taking Antian ancient residential buildings in Dongyang as an example. Arch. Cult. **05**, 263–265 (2021)
6. Duan, L.: The art development of Wuhu iron painting integrating 3D digital design. J. Shandong Agric. Eng. Univ. **36**(10), 79–80 (2019)
7. Zhang, X.: Three-dimensional digital protection of ancient pagoda sculpture art – taking Beijing Tianning Temple Pagoda as an example. Grand View Fine Arts **09**, 96–97 (2016)
8. Sun, F., Shu, W.: Research on the application of 3D scanning technology in the digital protection of "Huizhou Three Carving" – taking the digital model collection of Huizhou Building wood carving component as an example. J. Jingdezhen Univ. **31**(03), 72–75 (2016)
9. Wang, L., Wang, J., Wang, Y., Bo, B.: The positive significance of digital model carving to the model making of Chinese ancient architecture. Shanxi Arch. **44**(36), 33–34 (2018)
10. Wu, Q., Kuang, Y.: Protection and development of wood carving intangible cultural heritage in the digital era. J. Guangzhou Inst. Social. **11**(02), 81–84 (2013)

Prevention and Control of Systemic Financial Risks from the Perspective of Digital Finance

Jinjie Xiao(✉) and Shengbo Zeng

School of Economics, Sichuan University, Chengdu, Sichuan, China
xiaojinjie93@keyaninfo.com

Abstract. In recent years, due to the vigorous development of digital finance, the impact of digital finance on systemic financial risks has entered the research horizon of experts and scholars. This paper innovated the research path of the impact of digital finance on systemic financial risks, explored its impact on systemic financial risks from three dimensions: the coverage breadth index of digital finance, the depth index of digital finance use, and the digital degree index, and selected the data from 2011 to 2018 as relevant variables to analyze deeply by using the VAR model, and the impulse response method. Research shows that the above three dimensions of digital finance have a differential impact on systemic financial risks.

Keywords: Digital finance · Systematic financial risks · Risk prevention

1 Introduction

With the rapid development of digital technologies such as big data, cloud computing, Internet of Things, and blockchain have rapidly spread to all areas of the traditional financial industry. Digital finance has emerged at the historic moment. Digital finance, which integrates payment, financing, investment, insurance and other functions, not only has the three dimensions of coverage, depth of use, and digitization, but also has features like a wide range of services, popularization of customer groups, digital risk management, and low transaction costs. This new generation of financial service model that organically integrates traditional financial institutions and Internet companies has had an important impact on promoting economic development, facilitating the lives of residents, alleviating corporate financing, alleviating relative poverty, and promoting financial institutions and ecological changes. However, the financial innovation brought about by digital technology also brings new challenges to the security of financial services. Because the maturity and stability of digital technology are not advanced enough, its disaster recovery system is still not perfect. Once it is attacked by external attacks or other damages, its digital system can be easily paralyzed, result in stagnant financial services. Financial entities will face a series of losses such as capital loss and data leakage, which are likely to cause financial panic and systemic financial risks, increasing the threat of potential systemic financial risks. Therefore, it has important theoretical and practical significance to explore the impact of digital finance on systemic financial risks.

B. J. Jansen et al. (Eds.): *International Conference on Cognitive based Information Processing and Applications (CIPA 2021)*, LNDECT 85, pp. 878–883, 2022.
https://doi.org/10.1007/978-981-16-5854-9_118

2 Current Status of Digital Finance

Institute of Digital Finance Peking University and Ant Financial Services Group jointly compiled the Peking University Digital Inclusive Financial Index based on the big data of Ant Financial's trading accounts. The index has good representativeness and reliability [1]. China's digital finance index can effectively reflect the degree of development of digital financial inclusion in various regions [2].

As shown in Fig. 1, from the perspective of the overall level of digital finance in the country, the overall development level of China's digital finance from 2011 to 2020 was continuously improving, showing a trend of year-on-year growth. The digital finance overall average index rose from 40 in 2011 to 341.22 in 2020, indicating that China's digital finance is developing rapidly. It is mainly due to the rapid development of Internet technology in recent years, which provides a good foundation for its rapid development. From the perspective of the three secondary indicators of digital finance, the development trends of different dimensions of digital finance have their own characteristics. The digitization degree index is developing in volatility. It showed an upward development trend from 2011 to 2015, but declined after 2015. It began to rise in 2017 and maintain a continuous growth trend. However, in general, the digitization degree index is much higher than the coverage breadth index and the depth of use index. The coverage breadth index showed a continuous and stable growth, from 30.48 in 2011 to 326.44 in 2020. It also maintains a relatively consistent development with the overall digital finance index. With the use of wealth management platforms such as Yu Ebao, the depth of use index began to increase in 2014. On the whole, China's digital finance has developed relatively rapidly, with outstanding results.

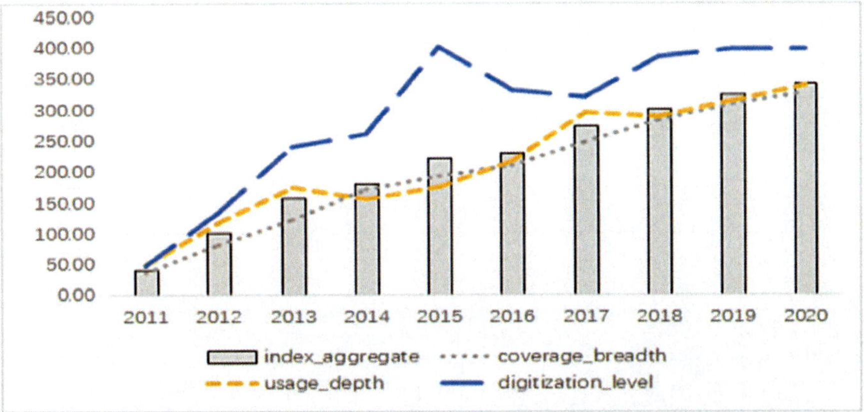

Fig. 1. 2011–2020 China digital finance index and 3 dimensional indexes

3 Measurement of Systemic Financial Risks

The systemic financial risk in this paper is calculated based on the financial stress index method weighted by the EWMA correlation coefficient. With reference to the related studies of China CISS (Ma et al. 2019) [3] and Li and Zhang (2020) [4], which were constructed by the National Institute of Finance of Tsinghua University, a total of eight indicators in the four dimensions (currency and bond market, stock market, financial market and foreign exchange market) were selected based on the needs of this paper to re-construct China's systemic financial risk comprehensive index. Specific indexes are shown in Table 1 and Fig. 2:

Table 1. Four markets and corresponding indexes

Market	Index
Currency and bond market	FR007 rate swap (three months)
	5-year government bond yield
	The interest margin between 5-year AA grade corporate bond and 5-year government bond
Stock market	The ratio of the total market value of listed A-share companies to GDP
Financial market	Financial industry stock index volatility
Foreign exchange market	Fluctuations in the exchange rate of US dollar against RMB
	Fluctuations in the exchange rate of the euro against RMB
	NDF volatility

In view of the availability of data, this paper selected the data time span from 2011 to 2018. According to the method of Hollo et al. (2012), the empirical cumulative distribution function was used to standardize the indicators, thereby reducing the value range of all indicators to (0, 1). The empirical cumulative distribution does not need to set the distribution in advance, which is more in line with the actual structure of the data.

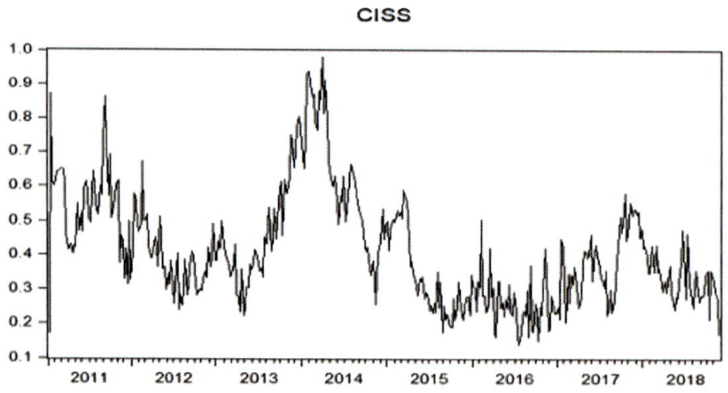

Fig. 2. 2011–2018 China systemic financial risk composite index

This paper used correlation coefficients to form a weight matrix, which better reflected the "systematicness" of risk, and assigned weights according to the degree of risk infection. The index to some extent heralded the outbreak of systemic pressure on China's financial system.

4 Conclusion

First, the coverage of digital finance has a direct inhibitory effect on systemic financial risks. The breadth of digital financial coverage is mainly reflected in the number of electronic accounts (the number of Alipay accounts owned and the Alipay accounts bound to bank cards) [5]. With the expansion of digital finance coverage and the increase in the number of users, it can effectively reduce the level of systemic financial risks. This shows that the development of digital finance has the characteristics of reducing systemic financial risks.

Second, the depth of use of digital finance will respond to changes in the level of systemic financial risk, that is, when the level of systemic financial risk is high, the depth of use of digital finance will be limited, the scope of business will be rigid. The specific performance will be the stagnation in growth of the number of users and transaction amount of payment, cargo base, credit, insurance, investment, and credit related businesses, et.al. The depth of use of digital finance is mainly based on the actual use of Internet (Alipay) financial services. The actual number of users, the number of transactions per capita and the transaction amount per capita are used to measure the usage [3]. It mainly includes payment business, money fund business, credit business, insurance business, investment business and credit business. The empirical results show that changes in the systemic financial risk itself will affect the depth of changes in digital finance and have a negative impact on it. This shows that on the contrary, systemic financial risks have an impact on the depth of use of digital finance. That is, when the systemic financial risk increases, the amplitude of the digital financial depth decreases, and the change of the amplitude of the digital financial depth index will not affect the systemic financial risk. This is due to the rapid transfer, spread, and spread effects of systemic financial risks.

Third, the degree of digitization of digital finance has no effect on systemic financial risks, that is, the characteristics of the digital finance (technology neutral) platform have nothing to do with systemic financial risks. This shows that digital financial technology is a neutral tool and method, and the key depends on how to use it. The degree of digitalization refers to the degree of digital support services, which mainly refers to the mobility, affordability, creditability and convenience of services. Among them, convenience and affordability are the main factors that affect users' use of the digital financial service platform (Alipay). The specific manifestation may be that the nature of emerging digital financial instruments such as mobile payment and Ant Credit Pay does not have much impact on systemic financial risks.

Fourth, the rise of the overall digital finance index has a time lag in the impact of systemic financial risks and will reduce systemic financial risks. Therefore, the continuous development of digital finance under effective financial supervision has a positive inhibitory effect on the prevention and control of systemic financial risks. This shows that digital finance has a direct inhibitory effect on systemic financial risks, and plays an active role in the prevention and control of systemic financial risks. However, with the development of digital finance, the intervention of regulatory authorities should be required to further strengthen supervision, effectively limit the scope of risks, and balance the relationship between financial innovation and financial security [6].

5 Enlightenments on the Study

First, it is necessary to further expand the geographical and spatial scope of the use of digital finance and expand the scale of digital finance user groups to increase the breadth of digital finance coverage. Although digital finance is an immature and irregular stage in the early stage of its development, and it hides the threat of potential systemic financial risks, with the standardized management and scientific development of digital finance, it will have a negative impact on systemic financial risks [7], and this negative impact will continue to decay over time. Therefore, we must accelerate the pace of digital finance development.

Second, we must further strengthen the prevention of systemic financial risks. This is because from historical and international experience, systemic financial risks have four important essential characteristics: wholeness, contagiosity, endogeneity and probability. Therefore, when the level of systemic financial risks is high, it will affect the depth of use of digital finance. Thus, the depth of use of digital finance should be increased.

Third, we must further promote the in-depth integration of digital technology and traditional financial services. The potential for digital finance to cause systemic financial risks is related to the level of digital financial technology development [8]. In addition, China still has problems such as insufficient digital technology conversion rate, imperfect digital infrastructure construction, and lack of digital talents. Therefore, it is necessary to further improve the level of digital technology research and development, strengthen the organic integration of digital technology and traditional financial services, and be more conducive to the control and prevention of systemic financial risks.

Fourth, the supervision of digital finance must be further strengthened.

Specifically, on the one hand, in China, it is necessary to further improve the digital financial risk monitoring, early warning, and disposal mechanisms, strengthen the construction of China's digital financial supervision system, and build an inclusive and prudent digital financial new regulatory framework. On the other hand, in the international arena, China has also taken the initiative to strengthen the coordination of the international supervision of digital finance, promote the reaching of a consensus on the international supervision of digital finance, and build a unified standard and norm for the international supervision of digital finance.

References

1. Liu, Z., Lai, M.: Virtual collateral, soft information constraint and online peer-to-peer lending. China Soft Sci. **01**, 35–46 (2015)
2. Liang, H., Zhang, Y.: The development of digital finance, financial supervision and the risks of China's commercial banks. Finance Econ. **01**, 30–39 (2021)

3. Song, X.: Empirical analysis of digital inclusive finance bridging the urban-rural residents' income gap. Finance Econ. **06**, 14–25 (2017)
4. Wang, Y., Ye, X., Xu, L.: Can digital finance boost the real economy? Finance Econ. **03**, 01–13 (2020)
5. Teng, L., Ma, D.: Can digital finance promote high-quality development? Stat. Res. **37**(11), 80–92 (2020)
6. Lin, W., Li, S.: Spatial externalities in digital financial supervision and the choice of central and local decentralization. Zhejiang Soc. Sci. **01**, 31–39+156 (2020)
7. Cai, Y.: Analysis of the legal management of China's digital finance—taking ant credit pay as an example. Theory Observe **03**, 89–91 (2019)
8. Wu, Y.: Analysis of the rise of digital economy and the development of digital finance. Mod. Bus. **07**, 119–121 (2021)

Research and Development of Learning Early Warning System Based on Artificial Intelligence Technology

Shaohua Fu(✉)

Big Data and Internet of Things School, Chongqing Vocational Institute of Engineering, Chongqing, China
fsh1108@cqvie.edu.cn

Abstract. Aiming at the current situation of staff shortage and low efficiency in learning quality assessment, this research conducted in-depth research on artificial intelligence and data analysis technology, and designed an intelligent and automated learning quality analysis and early warning system. The system is divided into four functional modules: evaluation index management, evaluation management, result management and system management, which can complete the questionnaire design intelligently and quickly according to the evaluation index. After the user completes the entry of the evaluation results, the system database can quickly generate data tables, use the embedded analytic hierarchy process algorithm module to sort out the evaluation index system, and finally use the artificial neural network model to get the evaluation results. The system has complete functions and simple interaction, which can greatly improve the efficiency of learning quality evaluation.

Keywords: Artificial intelligence · Data analysis · Learning early warning

1 Introduction

With the continuous expansion of the scale of education and teaching, the energy and resources consumed by various colleges and universities in the work of student affairs are also increasing, and the trend is gradually increasing [1]. Learning quality evaluation is an important means to measure the effect of student learning, standardize teacher management team, and improve teaching level. It is also one of the important educational tasks of college student management [2]. However, due to the limited number of academic staff involved in teaching management in institutions of higher learning, relevant counselors are under tremendous work pressure. With the development of computer and big data technology, artificial intelligence technology has been widely used in different scenarios [3]. Introducing artificial intelligence technology into the learning early warning work, collecting the multi-dimensional evaluation results of the evaluation objects in the computer software system, and analyzing the data, you can quickly get the learning quality evaluation results [4]. At the same time, a threshold value is set in

B. J. Jansen et al. (Eds.): *International Conference on Cognitive based Information Processing and Applications (CIPA 2021)*, LNDECT 85, pp. 884–889, 2022.
https://doi.org/10.1007/978-981-16-5854-9_119

the system. When the learning quality drops to the threshold value, an early warning is issued to the evaluation object and can be urged to adjust the learning strategy in time. In the above application scenarios, automated and intelligent evaluation of learning quality can be realized, greatly reducing human resource consumption and improving evaluation efficiency.

Based on the above analysis, this article designs a learning quality evaluation system under the guidance of software engineering theory [5]. When designing the system, detailed requirements analysis should be carried out to ensure the completeness of system functions. When implementing the system, based on the existing evaluation index system, the analytic hierarchy model is introduced to assign unequal weights to different evaluation indexes to ensure the fairness of evaluation [6]. When the system performs early warning of learning quality, it uses the neural network model in artificial intelligence technology to ensure the rapidity, accuracy and real-time performance of recognition [7].

2 System Design

2.1 Demand Analysis

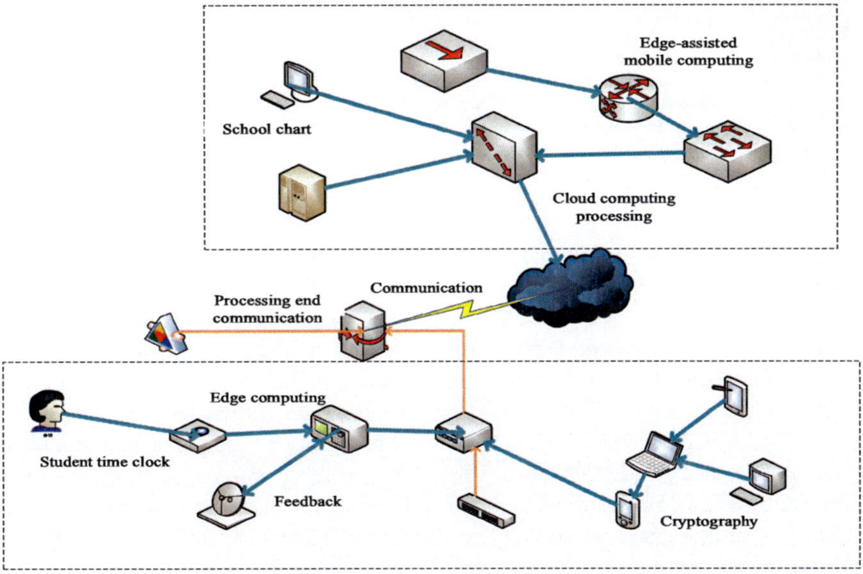

Fig. 1. System user use case

In order to relieve the work pressure of teaching assistants and improve work efficiency, this article uses artificial intelligence technology to empower learning quality assessment [8]. Through planning the information system, establishing a learning evaluation index system, and using relevant machine learning and data analysis methods to automatically realize the evaluation and scoring of student learning [9]. At the same time, according to the evaluation results, dynamic monitoring and early warning of students with low learning quality will be carried out to ensure the steady improvement of teaching quality [10]. In order to achieve the above goals, this article takes software engineering theory as a guide. In software engineering theory, demand analysis is an important part of system design [11]. Through demand analysis, it is clear that the system users and the goals they want to achieve, and then the corresponding functional modules are designed based on the currently achievable computer software technology [12]. In the teaching evaluation system designed in this paper, the user use cases and user activities shown in Fig. 1 are involved.

2.2 Function Module Design

According to the relevant content of the above demand analysis, sort out the flow chart of each user's activity. Then design the functional modules of the system according to the business process, as shown in Fig. 2.

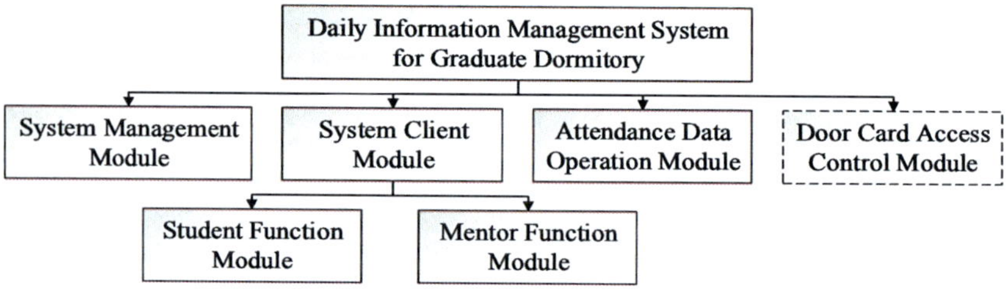

Fig. 2. System function module

It can be seen from Fig. 2 that the modules of the system are mainly divided into two levels: the first-level sub-module and the second-level sub-module. Among them, the first-level sub-modules are index management, evaluation management, data analysis and system management. Different first-level sub-modules correspond to different second-level sub-modules, and the functions under each module are detailed as follows:

Indicator management: The learning quality evaluation system designed in this article is the same as the method used in the current university evaluation and management work, that is, based on questionnaires. The questionnaire can ensure the authenticity and comprehensiveness of the evaluation, and different evaluation indicators correspond to different questionnaire questions. The teaching indicators are issued by leaders and experts, and the questionnaires are made by system administrators. The specific questionnaire types, parameters and generation are realized through corresponding functional modules.

Evaluation management: Because different users have different levels of awareness of learning quality, their understanding of learning evaluation indicators is also different. Therefore, all the teaching evaluation results cannot be confused. Students, schools, and experts are required to participate in learning assessment. Evaluation management mainly issues different evaluation methods according to different user groups:

that is, different users carry out evaluation feedback of learning quality within the system permissions according to their own role permissions.

Data analysis: As this article needs to realize intelligent and automated learning quality evaluation, it is necessary to summarize all the collected evaluation results and effectively analyze the results. In the result analysis, this paper uses the analytic hierarchy process to analyze the data, and then uses the neural network algorithm in artificial intelligence to intelligently score all the evaluation results, and finally gives the evaluation results of all the participants. In the data analysis module, the system can also visually display the evaluation results through charts; in addition, the system sets thresholds to alert participants who have unsatisfactory evaluation results to ensure the quality of learning.

System management: For a huge software system such as learning, evaluation and management, it is also necessary to set up modules for system operation and maintenance personnel to perform system management. System management specifically includes: system personnel authority management, database management, and operation and maintenance management. Through the setting of the system management module, the stable and reliable operation of the system is ensured.

2.3 System Architecture Design

In addition to meeting related functional requirements, the realization of the system also needs to meet related non-functional requirements in order to ensure the smooth and reliable operation of the system. The non-functional requirements that this system needs to meet mainly include: First, stability. This system is oriented to colleges and universities. During the design process, it is necessary to ensure that the system can run smoothly after it is put into operation to avoid software jams and other phenomena. The second is ease of use. The personnel involved in the learning evaluation are not professional software engineers. The computer skills of the users must be considered during the design of the system to ensure the friendly interface and convenient operation. Once again, it is security. The work of learning evaluation is an internal work for colleges and universities, involving the privacy of the participants and the relevant learning data of the learners. During the operation of the system, data security and privacy must be guaranteed. The last is scalability. Taking into account the continuous changes of learning evaluation indicators and the continuous expansion of teaching scale, the system must ensure good scalability in terms of functions and data storage capabilities.

In order to meet the above non-functional requirements, this article needs to use mature software architecture. Based on the above analysis, the layered architecture used in this article is shown in Fig. 3.

As can be seen from Fig. 3, in terms of software architecture, the system is divided into 4 layers: presentation layer, business logic layer, persistence layer and data layer. The system uses JSP technology in the presentation layer, and the business logic layer uses the Struts framework in the J2EE framework; the data layer and the persistence layer use the Hibernate framework in J2EE. Among them, Hibernate maps the classes in the application to the database tables, and directly performs database operations. The business logic of the system relies on the Struts framework. When the user issues an

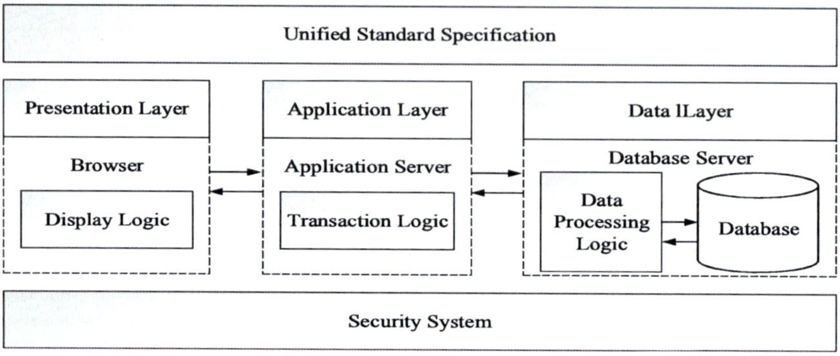

Fig. 3. System composition

instruction from the presentation layer, JSP will pass the request from the presentation layer to Struts, and then complete the business operations in the system.

3 Conclusion

In modern teaching management, the evaluation of learning quality is an important feedback method to improve the learning effect of students. In order to improve the efficiency of learning quality evaluation, this paper designs a learning quality evaluation early warning system. This system is based on artificial intelligence and data analysis technology, with the help of neural networks to complete the automatic evaluation of learning quality. The system designed in the article has complete functions and stable operation, which has important reference significance for the automation and intelligence of teaching evaluation in the future.

References

1. Kim, H.-W., Jeong, Y.-S.: Secure authentication-management human-centric scheme for trusting personal resource information on mobile cloud computing with blockchain. HCIS **8**(1), 1–13 (2018). https://doi.org/10.1186/s13673-018-0136-7
2. Gaaloul, W., Zhou, Z., Panetto, H., et al.: Special issue on fog and cloud computing for cooperative information system management: challenges and opportunities. Future Gener. Comput. Syst. **109**, 704–705 (2020)
3. de Bruin, B., Floridi, L.: The ethics of cloud computing. Sci. Eng. Ethics **23**(1), 21–39 (2016). https://doi.org/10.1007/s11948-016-9759-0
4. Eivy, A.: Be wary of the economics of "serverless" cloud computing. Cloud Comput. IEEE **4**(2), 6–12 (2017)
5. Asadi, S., Nilashi, M., Husin, A.R.C., Yadegaridehkordi, E.: Customers perspectives on adoption of cloud computing in banking sector. Inf. Technol. Manag. **18**(4), 305–330 (2016). https://doi.org/10.1007/s10799-016-0270-8
6. Almarabeh, T., Majdalawi, Y.K.: Cloud computing of e-learning. Mod. Appl. Sci. **12**(8), 85–89 (2018)

7. Taylor, R.P., Cordeiro, C., Giordano, D., et al.: Consolidation of cloud computing in ATLAS. J. Phys. Conf. Ser. **898**(5), 52–58 (2017)
8. Abdel-Basset, M., Mai, M., Chang, V.: NMCDA: a framework for evaluating cloud computing services. Future Gener. Comput. Syst. Int. J. Esci. **86**(9), 12–29 (2018)
9. Nawrocki, P., Reszelewski, W.: Resource usage optimization in mobile cloud computing. Comput. Commun. **99**(4), 1–12 (2017)
10. Baldassarre, M.T., Caivano, D., Dimauro, G., et al.: Cloud computing for education: a systematic mapping study. IEEE Trans. Educ. **61**(3), 234–244 (2018)
11. Senyo, P.K., Addae, E., Boateng, R.: Cloud computing research: a review of research themes, frameworks, methods and future research directions. Int. J. Inf. Manag. **38**(1), 128–139 (2018)
12. Shojaiemehr, B., Rahmani, A.M., Qader, N.N.: Cloud computing service negotiation: a systematic review. Comput. Stand. Interfaces **55**(1), 196–206 (2017)

New Application of WeChat Marketing in the Context of Internet

He Huang(✉)

Hunan Vocational and Technical College of Finance and Economics, Hengyang 421000, Hunan, China

Abstract. With the rapid development of the Internet, China's economy is still developing rapidly. Although the performance growth of China's auto sales industry has been continuously increasing, the competition in the traditional auto sales industry is fierce under the background of the rapid development of the auto industry. In such a competitive sales environment, how to keep the market and tap more customers? By virtue of the convenience of communication brought by social media on mobile terminals, automobile sales can be carried out with twice the result with half the effort.

Keywords: Internet background automobile sales · WeChat marketing · O2O

According to statistics from the China Association of Automobile Manufacturers, China completed the sales of 9,979,900 passenger vehicles of Chinese brands in 2018, and the production of 13,864,100 passenger vehicles of Chinese brands, creating the best sales record of Passenger vehicles of Chinese brands. In the context of the current critical period of building a moderately prosperous society in all respects, the automobile sales market in China's big cities is gradually shrinking, and the demand for automobiles in big cities is constantly saturated [1, 2]. However, the demand for automobiles in China's second-tier cities, third-tier cities and the vast rural markets is still strong. Therefore, in order to maintain their competitive position in the market and expand their potential customer base, enterprises in the automobile sales industry need to innovate their automobile marketing methods, change their automobile marketing ideas, and find more suitable sales methods and methods for second-tier cities, third-tier cities and rural markets. With the rapid development of Internet economy, WeChat marketing and automobile sales collide together to ensure the smooth transformation of marketing methods of automobile sales companies [3, 4].

1 WeChat and WeChat Marketing Overview

Tencent, an Internet giant, launched a new chat and dating APP called WeChat in 2011, which is different from QQ. This APP is a free mobile APP. The core of this APP is multimedia information communication [5, 6]. Since the launch of the WeChat APP, the number of users has skyrocketed. Especially with the promotion and development

B. J. Jansen et al. (Eds.): *International Conference on Cognitive based Information Processing and Applications (CIPA 2021)*, LNDECT 85, pp. 890–894, 2022.
https://doi.org/10.1007/978-981-16-5854-9_120

of smart phones and free WIFI networks, as well as the popularization of 4G networks, the current society can be said to have directly entered the WeChat era [7, 8]. In today's social life, it is easy to find people in a hurry or leisure holding a mobile phone to send voice messages or swipe the screen to swipe the circle of friends.

Under the background of the Internet era, WeChat, an important tool for Internet information dissemination, innovatively creates WeChat marketing methods. WeChat marketing refers to under the background of Internet economy, relying on WeChat APP this Internet platform, to focus on the company's WeChat users to promote their product information or their own enterprise image, with this kind of promotion way to promote the enterprise products and brand image, with the aid of WeChat platform to communicate with the customers at any time a new marketing mode of the Internet era.

2 Feasibility Analysis of WeChat Marketing Application in Automobile Sales

(1) The fierce competition in the automobile market promotes the generation of WeChat marketing
Although there are a lot of marketing modes in the current automobile enterprises, these marketing modes include entertainment marketing, experiential marketing and other marketing modes.
However, with the continuous development of mobile Internet technology, it has been penetrating into People's Daily life. And in the entertainment marketing, etc. These traditional marketing model, the seller can't instant communication and communication with potential customers, especially in some big cities and major cities in the big cities such as car lottery restrictions, the main traditional automobile sales market is saturated, the car sales market competition greatly. Fierce competition of auto sales market and the existence of communication time delay traditional automobile marketing mode, to promote the production of the emerging Internet automobile marketing mode, under this big background, WeChat using the highest rate in the vast consumers, and helps sellers and potential purchasers, instant communication, promote the instant work of sales, WeChat marketing model then arises at the historic moment under this kind of background.

(2) Intersection between automobile sales potential customer groups and WeChat user groups
With the rapid development of Internet technology, the function of traditional media such as newspapers, magazines and other publicity channels to obtain information appears to be slightly backward and limited. In the era of information explosion, it cannot meet people's demand for information at all.WeChat APP generated under mobile Internet technology can not only better protect users' private information, but also provide users with social tools with media functional services such as instant communication. With the popularity of credit industry, automobile sales potential customer group of younger, these after 80, after 90, and even after 00, are growing up in the Internet age, the way they get information are to use the Internet, when they produce shopping goal in buying a car, they will often use the Internet in the network to capture information sold the car, and compared with the traditional

mode of mobile communication, WeChat can provide free voice communication but also young people's favorite. Accordingly, with the help of WeChat this mobile Internet communication software, to help the car sales work.

(3) WeChat marketing has more advantages in automobile sales
Compared with traditional advertising methods relying on traditional media, WeChat marketing has the advantages of low cost and high transmission rate. First of all, WeChat users are free at the time of registration and are free as they go along. There are no charges other than the cost of mobile data traffic. Automobile sales enterprises can use WeChat public number platform to set WeChat public number for automobile sales, and set questions that can be answered automatically according to classification standards such as models, as well as reduce labor costs. Compared with traditional marketing, WeChat marketing can win the trust of potential consumers more. WeChat'sfriend-making mode is more inclined to friends and relatives or familiar people, so WeChat marketing mode can win the trust of potential users more.

3 Optimize the Application of WeChat Marketing in Automobile Sales

Car sales companies in the era of mobile Internet, ways to get a better development, essential WeChat marketing this way, using WeChat WeChat tools such as number of public platform to carry out the auto marketing activities, can from WeChat marketing combined with O2O business model, combining WeChat marketing and precision marketing, WeChat combined marketing and relationship marketing direction, car sales WeChat marketing activities.

(1) Combination of WeChat marketing and O2O business model
As the current business model is new O2O, the innovative business model, the Internet can be a combination of online and offline service innovation, in combination with WeChat APP precision pushed to the potential customers under the characteristics of the car sales work of WeChat marketing combined with O2O business model, create a car sales WeChat O2O business model of marketing platform. In this innovative commercial marketing platform, automobile sales enterprises can effectively combine Internet online publicity and promotion work with Internet offline automobile sales work, so as to fully improve the sales efficiency and quality of automobile sales enterprises. Working in the automobile sales company sales brand propaganda work, car sales enterprise can be sold in or on the open to booking a car models and preferential activities such as marketing promotion in the automobile sales companies for information WeChat in public, potential customers by WeChat public attention, can be directly in the field of automobile sales companies outside the choose and buy or purchase in advance to their own cars, reduce the potential customers in the choice of purchase decision stage paid to pay on time and money. WeChat O2O business model is applied in the process of automobile sales, which can not only reduce the amount of money that automobile sales enterprises need to pay for advertising promotion, but also retain potential customers attracted by the automobile sales company to the maximum extent.

(2) Combination of WeChat marketing and precision marketing
Precision marketing is the most popular marketing model among sellers. WeChat marketing is also a kind of precision marketing because most of its targets are potential customers who buy cars.
Automobile sales companies in the company brochure printed and distributed by the printing enterprise WeChat public good qr code, in auto sales company staff business card printing enterprise WeChat public qr code, in the car needs of customers after adding the WeChat public, categorizing these potential customers according to the buyers demand, provide accurate car demand analysis and models recommended, in this kind of WeChat marketing campaigns, can make the car sales enterprise propaganda work to get a better return, do more precise marketing. Automobile sales companies can also use the "like" activity in WeChat APP to plan thumb up to exchange coupons and other marketing activities of the automobile sales company, which can not only dig out potential customers to buy automobiles, but also virtually enhance the popularity of automobile sales companies. Based on WeChat friends of the user through the user's own mobile phone address book, or another dating chat software QQ is from tencent, add in WeChat marketing activities, can add a automobile marketing enterprise number WeChat public platform for the customer as the forward by friends or thumb up to the user's WeChat friends together to interact, enlarge the propaganda of automobile sales companies, reduce car sales of the company's propaganda work difficulty, alleviate the pressure of the automobile sales company's sales, reduce automobile sales company's sales staff work harder.
(3) Combination of WeChat marketing and relationship marketing
Relationship marketing is a process in which marketing activities are regarded as an interactive process between enterprises and consumers, suppliers, distributors, competitors, government agencies and other public relations. In the process of marketing activities, the marketer needs to create an environment in which there is a good relationship between various micro-environmental factors, so as to ensure a good competitive relationship between marketers and consumers.

In WeChat APP, the basic point of marketing activities established by using the function of moments is the relationship of trust between marketers and consumers.

To use the function of moments to carry out automobile sales and marketing activities, automobile sales staff as marketers need to invest more emotions in this marketing activity than other marketing activities. This kind of "point-to-point" communication method can effectively help the automobile sales staff of the automobile sales company to build up the relationship and strengthen the relationship with the potential automobile buyers in the circle of friends. At the same time, relationship marketing can also develop new customers on the basis of maintaining old customers.

4 Conclusion

As car sales industry increasingly fierce competition, car sales enterprise propaganda needs from the traditional newspapers, magazines, television advertising to online advertising, and to expand into new media channels, such as weibo, WeChat sales directer also play Internet line through Internet, online car sales enterprises should fully tap WeChat powerful function, make full use of and develop WeChat marketing, enhance brand force and marketing force.

References

1. Liu, Y., Xiao, J.: Research on interactive automobile marketing model of "internet +." J. Hubei Inst. Automob. Ind. **29**(4), 76–80 (2015)
2. Zhang, M.: Research on the advantages of "internet +" interactive automobile marketing model. Comput. Nerd **3**, 97 (2018)
3. Zhang, B.: "Internet +" interactive automobile marketing model thinking. China Mark. **27**, 133 (2018)
4. Gong, W.: Research on O2O operation mechanism and three-dimensional marketing of e-commerce. Nanchang Univ. (2014)
5. Zhao, B., Sun, A.: Thinking on distribution channel management in automobile industry. Mod. Bus. Trade Ind. **2013**(01) (2013)
6. Zuo, P.: Analysis of existing problems in WeChat marketing. China Commer. Trade **2013**(36) (2013)
7. Long, Y.:WeChat marketing – SoLoMo based application. Hopping Mall Modern. **2013**(22) (2013)
8. Yu, Q.: Analysis on the advantages of WeChat marketing in retail industry. Inf. Comput. **2014**(21) (2014)

Mobile Game Design in the Context of the Mobile Network

Jianlong Jiang and Gang Liu(✉)

School of Arts, Anhui Xinhua University, Hefei 230088, Anhui, China

Abstract. With the upgrading of mobile network and the rapid development of smart phones, mobile phone hardware devices running mobile games are getting better and better, which promotes the in-depth development of mobile game design. With the diversification and simplification of mobile game development platform, more mobile game developers join in, inject more vitality into mobile games, accelerate the innovation of games and meet the needs of people for games. This paper analyzes the development of mobile games, the common platforms of mobile games, the current situation and development prospects of mobile games.

Keywords: Mobile networks · Mobile games · Game design · 5G

1 Introduction

With the popularity of smart phones, mobile games have become the main way of mobile electronic entertainment, and more and more people have joined in the design and development of mobile games. The development of mobile games depends on the development of mobile network and mobile hardware devices, and mobile games are always restricted by mobile hardware technology. However, with the improvement of smartphone hardware, the technical limitations of mobile games are becoming less and less, and mobile games can also have gorgeous art design and complex and complete story plots. Mobile games also have better development prospects due to their inherent portability and network. Games have a long history, with the release of the mobile game Space Wars in 1962, which heralds the birth of video games. Nowadays, as an important part of electronic games, mobile games have only developed and matured in the last ten years. Compared with TV games and PC games, mobile games started late [1]. In today's technology rapid development, mobile phone games already showed a huge developing space and prospect, mobile game development cycle is short, the cost is low, at the same time mobile game development technology is easy to be mastered developers, for you can't do a big game development lovers, choose to develop mobile phone game is help you achieve the dream of a shortcut.

Human-computer interaction is the foundation of video games. The most important way for mobile games to get input from players is the buttons or touch screen, which are just like the keyboard and mouse of a computer. At the same time, players can get the game content output from the phone through the screen and speakers of the phone.

B. J. Jansen et al. (Eds.): *International Conference on Cognitive based Information Processing and Applications (CIPA 2021)*, LNDECT 85, pp. 895–900, 2022.
https://doi.org/10.1007/978-981-16-5854-9_121

The input of players and the output of mobile phones constitute the human-computer interaction of mobile games. The development of mobile phone display ability promotes the presentation of game content, while the development of mobile phone input devices enriches the gameplay of mobile games, allowing more excellent ideas to be realized on the mobile platform. Nokia mobile phones in the late 20th century provided the first mobile phone game, the game is called "the snake", believe that most mobile phone users have played this a classic mobile phone game, the game at the start of a few small snake around obstacles, try to eat small pill then grew up, as the growth of the length of a serpent and complex degree of change, The player becomes more and more difficult as the game becomes more and more manageable, so that the simple game itself does not get the player stuck, and the game can be easily restarted. The processing power of today's mobile phones is enough to develop more complex and massive games, and the release of the 3D version will retain the original gameplay while improving the visual aesthetics to attract more players. The simple game controls, intuitive graphical display, and interesting game content can attract players' attention in a short time without causing excessive addiction are all the reasons for the success of this game on the mobile platform.

2 A Common Platform for Mobile Games

There are many mobile phone platforms, developers need to choose the appropriate platform for development, according to the developer has a professional foundation, choose the appropriate platform, the following introduction of iPhone, Android, J2ME and other common mobile phone system platform.

2.1 iPhone Platform

Apple CEO Steve Jobs announced the launch of the iPhone, which combines the functions of a mobile phone, a touch screen and web browsing with a communications device. IOS is the operating system developed by Apple for the iPhone. It is mainly used on the iPhone, iPodtouch and iPad. Like the MacOSX operating system on which it is based, it is based on Darwin. The system architecture of IOS is divided into four levels: the core operating system layer, the core service layer, the media layer and the touch layer. System operations take up about 240 MB of memory space. Like other mobile platforms, iOS also has an SDK for developers to create apps, but in order to distribute the software better, developers must join the iPhone program. After joining the program, developers will get a license to distribute their software to Apple's App Store. The iPhone emulator included with the SDK development tool is used to simulate the look and feel of the iPhone on a developer's computer. However, the SDK requires a Mac with an Intel processor running MacOSXLeopard. Other operating systems, including Microsoft's Windows operating system and older versions of MacOSX, are not supported [2]. Compared with Symbian, Android and J2ME, which will be discussed later, it is obvious that the basic requirements of developing games for iPhone are more, but correspondingly, iPhone's excellent graphic display ability, multi-touch interactive mode and excellent gravity sensing give game developers more possibilities to realize their own ideas. Compared with other mobile game development platforms, the development of iPhone is much easier. Although new versions of iPhone are introduced due to continuous technological updates, they are all the same model, so developers can only design games for the real device once, and the development cycle of iPhone games is shorter. IOS comes with a nice package, and the iPhone uses a great 3D game engine like Unity to develop games. The development of the iPhone won't be covered in this book, but for those who want to learn more about mobile games, the iPhone is a platform that you can't miss.

2.2 Android Platform

The symbol of the Android operating system – a small robot, as shown by the word Android – means "robot" in its original sense. Android was founded by Silicon Valley geek Andy Rubin, who created the platform shortly after he left Danger Mobile Computing. Andy has long dreamed of making Android a fully open mobile platform. Android was later acquired by Google. In 2007, Google officially announced the formation of Open Mobile Phone Alliance with other 33 mobile phone manufacturers, mobile phone chip suppliers and mobile operators, and released the open mobile phone software platform named Android, hoping to establish a standardized and open mobile phone software platform and form an open ecosystem in the mobile industry [1]. Since then, Google, known as the world's largest search service provider, has aggressively entered the mobile communication market. After several years of development, Android has become an important part of Google's corporate strategy, promoting the realization of the corporate goal of "providing information to everyone at any time and anywhere".

(1) Linux kernel. Android provides core system services based on Linux 2.6. For mobile application developers, there is no need to consider the underlying hardware details in the development process, while for mobile phone manufacturers, the Android platform can be run on their own hardware platform after certain modifications to this layer [3].
(2) Android runs. Each Android application is in the Dalvik virtual machine, running in their own process. The Dalvik virtual machine executable file format is.dex, which is a compressed format specially designed for Dalvik and suitable for systems with limited memory and processor speed [4].
(3) Application framework layer. The design of Android's application framework simplifies the reuse of components, in which any application can publish its own functions for other applications to call [5], which makes it convenient for users to replace components of the program without affecting the use of other modules, but such replacement needs to follow the security restrictions of the framework.

2.3 J2ME

Java ME, formerly known as J2ME, is a Java language platform for embedded consumer electronic devices such as set-top boxes, mobile phones, and PDAs. It includes a virtual machine, KVM, and a set of standardized Java APIs [6].

J2ME, J2SE and J2EE together constitute the three major versions of Java2 technology. In order to solve the contradiction of the diversification of wireless devices, Sun divides the J2ME technology architecture into three layers of Java Virtual Machine,

(Coniguration) and Profile [7] according to the resource characteristics of various devices, and then further subdivides it. This enables J2ME to work within the limitations of each type of device while providing a minimum of Java language functionality.

J2ME is the most common and mature development method in mobile game development at present. Despite the rapid development of iPhone and Android in recent years, J2ME development is still the most populous and most important part in the company that makes mobile game. To this day, most mobile phone users still support J2ME on their phones, and the people who master J2ME development are obviously the largest part of mobile game developers. Learning J2ME can help students better master the Java programming language and lay a foundation for the development of Androld which also uses Java language [8]. Of course, the choice of J2ME for teaching, its low investment, easy to use is also a factor we must consider.

3 The State and Future of Mobile Gaming

Mobile phone platform has developed rapidly in recent years, which has promoted the progress of mobile phone games. Similar to the development direction of man-machine interaction mode of stand-alone games, mobile phone games are also increasingly pursuing realistic man-machine interaction. The portability of mobile phone makes mobile phone itself the main carrier of interaction. With the widespread use of 5G network, the mobile advantages of mobile phones are further reflected. Online games on mobile phones enable activities that must be carried out in front of the computer to be carried out wherever you go. This trend of entertainment accompanying development is obviously complementary to today's information accompanying. If the popularity of WeChat is to share everyone's thoughts and current situation with others, the development trend of mobile online games in the future must be to share one's happiness with others anytime and anywhere, and the sharing method of the latter is closer to the real interaction between people.

The growth of the mobile gaming market shows that mobile gaming can be a profitable business, but it should also be noted that most of the games installed on users' phones are developed by foreign companies, with only mobile versions of online PC games. Faced with broad prospects, more investment will appear in the mobile gaming industry, while for China mobile game development pattern, also should draw lessons from the development of China's PC game, PC game development began in single-player role-playing game, can truly becoming profitable industry because of the network game, spend money to buy equipment is one of China's online game unique profit model, Can use their own wisdom to develop their own national conditions of the characteristics of the model, can also apply the same successful experience in the mobile game industry. At the moment, low-income people account for the vast majority of mobile game users. It's hard for these players to buy a game as a whole, so the fragmented approach is still something that mobile game operators can learn from. Paying per level and spending money on equipment are both good choices [9].

Due to product quality, incompatible operating platforms, and limited payment channels, mobile game users are not spending much money on mobile games. While there are a lot of mobile games out there, there are very few high-quality mobile games out there; Since there are several operating platforms on mobile phones, games between different platforms cannot be compatible, so making a game cannot be shared by all users; At present, the consumption of mobile game market in China is mainly charged by telecom operators, and users have very limited choice of payment channels. Users are often unable to get the mobile games they desire in the most convenient way. Currently, mobile games are still mainly free games, and many people do not have the habit of buying games. The games that come with the purchase of a mobile phone are certainly the first choice for mobile users. The consumption habits of mobile games still need to be cultivated, but there is no doubt that the consumption potential of mobile games is huge and worth tapping.

In the future, with the gradual development of mobile technology, mobile phones will be further developed in terms of display ability and operability, and more types of games will be better presented on mobile phones. It is believed that strategy games, which have high requirements for operation and are highly respected on PC platform, can also be perfectly presented on mobile phones in the future. We can only imagine what the future of mobile games will look like, but we will have to experience it first hand. First, the combination of virtual and real is still the main trend of mobile game interaction. Second, the genre of mobile games will be much broader, and strategy games that haven't been a huge success on mobile will be potential players. Third, charging is certainly the main means for mobile games to gain revenue. Whether it is the electronic market model launched by iPhone and Android, or the carrier charging model, it may be integrated in the future to form a systematic and complete payment model. Mobile has the power to combine disparate and unrelated technologies in new and creative ways to explore and create an ever-changing world of mobile gaming.

Acknowledgements. This work was supported by the animation Research Institute of AnhuiXinhua University (yjs201708), High level teaching team of quality engineering project of Anhui Xinhua University (2019jxtdx05), Key projects of Humanities and Social Sciences in Anhui Universities (SK2019A0744), Quality Education Research Center project of Anhui Xinhua University (IFQEZD201909).

References

1. Sun, Y.: Research on the Optimization of Mobile Game Interaction Interface Design. Beijing Jiaotong University, Beijing (2017). (in Chinese)
2. Cai, W.: Design and Implementation of Android Apps Based on Privacy Protection. Beijing Jiaotong University, Beijing (2018). (in Chinese)
3. Su, Y., et al.: Development and application of "office worker healthy life housekeeper APP." Sci. Inf. Technol. **31**, 139–141 (2019). (in Chinese)
4. Meng, H.: Preliminary study on common technologies of android development. Sci. Technol. Innov. Appl. **6**, 26–28 (2017). (in Chinese)
5. Zhang, J.: Application development and research based on android environment. Silicon Valley **2**, 86 (2019). (in Chinese)
6. Wei, Y., et al.: Application research of J2ME technology in embedded field. Comput. Knowl. Technol. **9**, 129 (2015). (in Chinese)

7. Yan, Y.: Interpretation of BREW and J2ME warring puzzle. Commun. World **7**, 216–218 (2018). (in Chinese)
8. Yuan, S.: Analysis of mobile game market. Heilongjiang Sci. **4**, 66–67 (2018). (in Chinese)
9. Li, Y.: Analysis on the potential of college students' mobile game consumption. China Mark. **3**, 156 (2017). (in Chinese)

Structural Modification Design and Application of Back Support Device for Coal Mine Drilling Rig

Xin Gao(✉)

Xi'an Research Institute of China Coal Technology and Engineering Group Corp., Xi'an, Shaanxi, China

Abstract. Under the environment of the structure optimization of the coal mine crawler full hydraulic drilling rig, aiming at the current structure of the stabilizing device behind the drilling rig as the research object, combined with the actual operation and application in coal mine, the shortcomings and defects are analyzed. The mechanical structure of the device is optimized and improved, and compared with the previous structure. After the large area configuration application and popularization of the field prototype in the coal mine, it is verified that the optimized post stabilizing device can effectively adapt to the complex situation of the coal mine laneway. It improves the reliability and stability of underground drilling in daily use, and is of great significance for the safety of mining and construction.

Keywords: Crawler drilling rig · Back support device · Optimization · Structure design · Application · Safety

1 Introduction

In the daily drilling construction of crawler full hydraulic tunnel drilling rig in coal mine, it is necessary to stabilize the feed body to ensure that the drilling rig can still maintain a stable working state under the working environment of large torque of rotary device [1]. Otherwise, it may cause the movement of the body, displacement and dropping of drilling tools in the hole, or it may cause the tragedy of related casualties [2, 3]. At present, all types of drilling rigs in coal mines adopt the form of front and back up and down rigid stabilizing devices [4]. The main working principle is to extend the hydraulic cylinder installed in the front and back up and down stabilizing devices [5], and install an appropriate number of connecting rods at the end of the cylinder, so that the terminal grounding device can contact the roadway wall, and ensure the stability of the fuselage through the expansion of the cylinder [6], So that the drilling rig can achieve the purpose of stable drilling operation [7]. In general, as long as the drilling rig can be placed in the roadway, there is space for top support when it is stable up and down and in front [8], but only the rear stabilizing device. After adjusting the dip angle and azimuth angle before drilling, there are many restrictions on the installation space of this single

B. J. Jansen et al. (Eds.): *International Conference on Cognitive based Information Processing and Applications (CIPA 2021)*, LNDECT 85, pp. 901–905, 2022.
https://doi.org/10.1007/978-981-16-5854-9_122

direction rear stabilizing method [9], Although it can meet the needs of drilling rig for post stable support under the conditions of some underground roadways in coal mines, for some underground roadways with complex environment, there are some objective facts such as narrow space and limitations. The post stable device parallel to the single direction of the fuselage is unable to carry out the jacking operation due to the limited space, it can't meet the needs of stable support at the back of the drilling rig under complex environment conditions, and the environmental adaptability is weak, and the structural design is not flexible [10]. Therefore, in order to solve the above problems of space adaptability, this paper optimizes and modifies the rear stabilizing device used in the current crawler full hydraulic tunnel drilling rig in coal mine, so as to achieve the purpose of flexible auxiliary support and stability.

2 Overall Design

In the daily construction of crawler full hydraulic tunnel drilling rig in coal mine, there is a demand for stable support in the front and back up and down directions. The optimized and modified rear stabilizing device is a space adaptive structure, which can be supported for different positions, and can be stabilized adaptively according to the conditions of underground roadway in coal mine [11]. The structural layout of the rig assembly stabilizer is shown in Fig. 1.

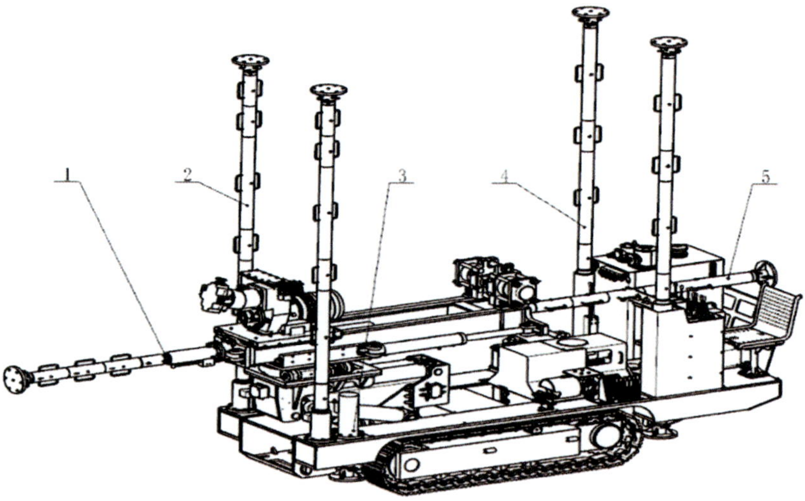

Fig. 1. Layout of stabilizers for crawler drill 1. cardan shaft type rear stabilizer 2. plug in oil cylinder upper stabilizer 3. lower stabilizer 4. plug in oil cylinder upper and lower stabilizer 5. plug in front stabilizer

3 Design of the Rear Stabilizer

(1) Before modification
Before the modification of the rear stabilizer, the structure of the standard plug rod oil cylinder is in a single direction. The screw connection is installed on the body. The upper grounding device contacts the roadway wall through the expansion rod and the extension of the cylinder for post stabilization operation. The structure of the front and rear stabilizer is shown in Fig. 2.

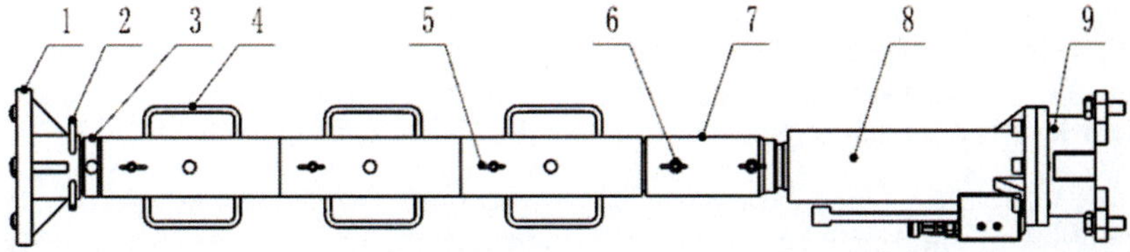

Fig. 2. Overall structure of rear stabilizer 1. upper grounding device 2. U-pin 3. ejector 4. telescopic rod 400 5. split pin 6. inserting rod 7. plug and change diameter 8. rear top oil cylinder 9. connecting plate of rear jacking oil cylinder

(2) After modification
After modification, a connecting component of cardan shaft structure is added on the basis of the plug-in rod refueling cylinder, that is, a cylinder lug seat and a connecting plate are added between the connecting plate of the rear top oil cylinder and the rear top oil cylinder, so that the rear stabilizing device can be stably operated at different positions and contact with any position of the underground roadway in the coal mine, To achieve the purpose of drilling rig fuselage in all directions. The structure is shown in Fig. 3.

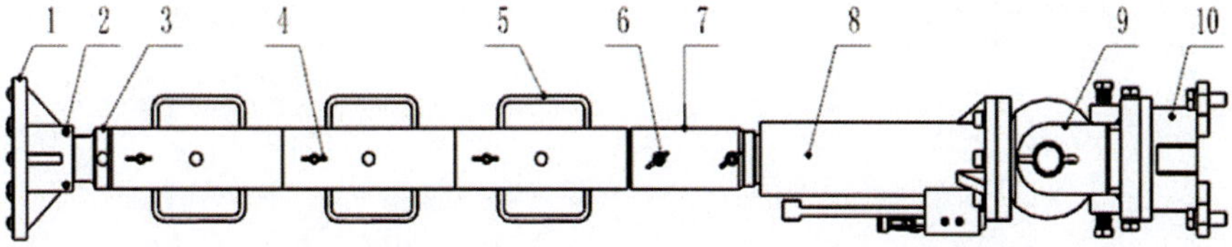

Fig. 3. Overall structure of cardan shaft type rear stabilizer 1. upper grounding device 2. U-pin 3. ejector 4. telescopic rod 400 5. cotter pin 6. inserting rod 7. inserting variable diameter 8. rear top oil cylinder 9. oil cylinder ear seat 10. rear top oil cylinder connecting plate

4 Structure Design of Cardan Shaft

In this paper, the modified universal shaft structure is mainly composed of two parts, which are the stable adjusting mechanism and the stable locking mechanism. The overall structure of cardan shaft is shown in Fig. 4.

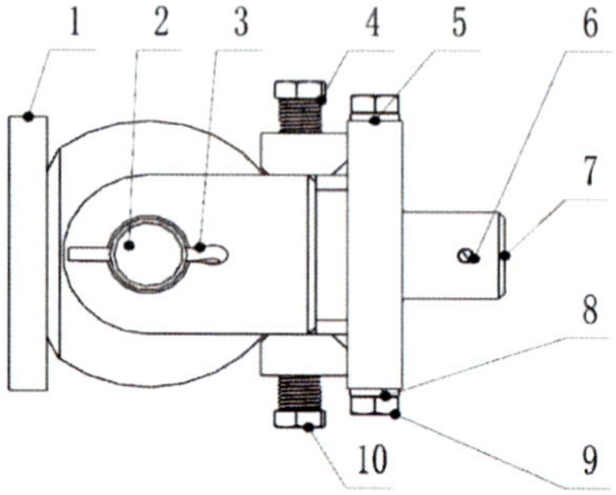

Fig. 4. Overall structure of cardan shaft 1. cylinder adapter plate 2. pin shaft 3. short split pin 4. locking spring 5. connecting base plate 6. long split pin 7. oil cylinder ear seat 8. spring washer 9. hexagon head bolt 10. special bolt

The stable adjusting mechanism is the main structure of the device, which is composed of radial adjusting mechanism and axial adjusting mechanism. The radial adjusting mechanism is the flat ear end of the oil cylinder ear seat, which is connected with the hydraulic cylinder adapter plate through a pin shaft; The axial adjustment mechanism is the pin shaft end of the oil cylinder ear seat, which is inserted into the connecting plate of the rear top oil cylinder, and the opening pin is added to limit the position (Fig. 5).

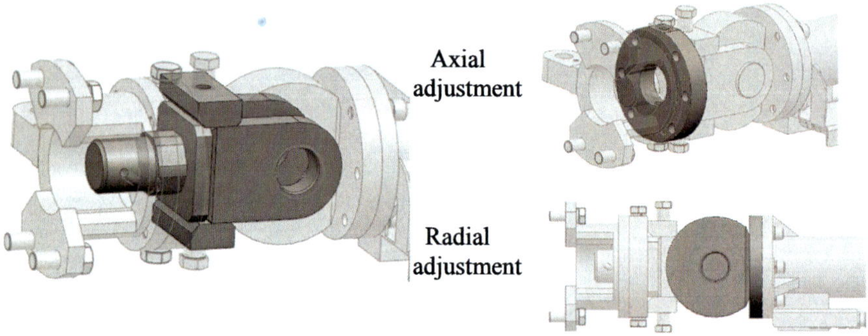

Fig. 5. Schematic diagram of important parts of regulating mechanism

The locking mechanism is composed of radial locking and axial locking. Radial locking is the locking operation between radial positions by locking spring of cylinder lug and special bolt in the form of wedge plug; Axial locking is to screw the hexagon head bolt on the bottom plate, so as to lock the end of the pin shaft end of the cylinder lug seat in axial position.

5 Modification Effect and Field Application

The modified and optimized universal shaft type rear stabilizer has a good effect, which can meet the stability requirements of different positions for the feed body needle of crawler drill in underground coal mine. It has a wider scope of application, convenient use and good fixation effect, which opens the deadlock of single and solid form before, So that the drill can deal with the complex working conditions of coal mine roadway environment for the stable operation of feeding device. After the modification design, the device has been produced, assembled and mass produced on the rig body. After a long time of practical application in the coal mine, the feedback is good, and the customer's praise is obtained. In the normal construction operation of the coal mine drilling rig, the target of stable body can be achieved excellently, which ensures the safety and stability of drilling operation, and achieves the expected design purpose perfectly.

6 Conclusions

According to the characteristics of complex and limited coal mine environment, and considering the compact structure of the current coal mine drilling rig, there are certain requirements for the width and height, so a certain optimization design and modification of the rear stabilizing device are carried out. Through the application and experiment of the prototype, the characteristics of the post stabilizing device are demonstrated, which can basically meet the needs of the drilling machine for post stabilizing support at any position in the underground roadway of the coal mine, and has strong adaptability, which further improves the stability and reliability of the crawler full hydraulic tunnel drilling machine in normal construction, It also provides a new space for the optimization of the whole rig.

References

1. Fang, P.: Research on key components and dynamic characteristics of 12000N·m fully hydraulic durectional drilling rig in coal mine. China Coal Res. Inst. (2019)
2. Shi, L., Song, H., Sha, C., et al.: Development and application of all hydraulic crawler drilling rig in fully mechanized roadway. Coal Mine Mach. **40**(06), 129–131 (2019)
3. Li, X., Han, J., Yao, Y.: Narrow-body directional crawler drilling rig used in coal mine. Saf. Coal Mines **51**(11), 142–145 (2020)
4. Yao, K., Sun, B.,Yin, X., et al.: Development and application of ZDY series crawler mode directional drilling rig. Coal Sci. Technol. **41**(03), 28–31+54 (2013)
5. Wang, Q., Chen, H.: Development and prospect on intelligent drilling technology and equipment for gas drainage. Ind. Mine Autom. **44**(11), 18–24 (2018)
6. Bai, W.,Yao, N., Song, H., et al.: Key technology for track-mounted drill of full section bed plate tunnel. Saf. Coal Mines **48**(01), 113–115+119 (2017)
7. Yao, Y., Yao, N., Peng, T.: Development situation and prospect of luffing mechanism of drilling rig in underground coal mine. Coal Geol. Explor. **48**(02), 14–19 (2020)
8. Han, G.: China Coal Industry Drilling Engineering. Coal Industry Press (2000)
9. Feng, D.: Rig Design. China University of Geosciences Press, Wuhan (1992)
10. Chugh, C.P., Sharma, V.M., Steele, K.: Design Criteria for Drill Rigs. CRC Press, Boca Raton (2020)
11. Wu, D., Liu, Z., Wang, S.: Modular design of crossing boreholes drilling rig in underground coal mine. Saf. Coal Mines **52**(02), 126–130 (2021)

Design of the Aging of the Residential Area Environment by Using Computer Three-Dimensional Drawing Technology Under the Home Care Mode

Jie Gao(✉)

School of Life Sciences, Yan'an University, Yan'an, Shaanxi, China

Abstract. With the deepening of aging, the landscape environment design of residential areas related to the elderly needs to be further explored. The purpose of this paper is to study the aging design of residential area environment based on computer three-dimensional drawing technology. First of all, it sorts out the research theory of the aging landscape environmental design principles of residential areas, and provides a theoretical basis for the aging design of residential areas. Through on-site investigation and other methods, the problems of the elderly in old residential areas are found. Then the computer three-dimensional drawing technology is analyzed in detail, and the technology is applied to the aging design of residential area environment. Finally, this article conducted a questionnaire survey on the elderly in the residential area. The results of the survey showed that 62% of the elderly have the desire to live independently, but they are close to their children and hope to provide convenient services and facilities for the elderly with the help of the community.

Keywords: Aging · Residential environment · Computer three-dimensional mapping technology · Questionnaire survey

1 Introduction

Facing the trend of rapid aging population and various severe problems brought about by aging population, it is urgent to solve the related problems of clothing, food, housing and transportation for the elderly [1, 2]. People's requirements for the living level have not only stayed in solving the housing problem, but also gradually attached importance to the importance of the landscape environment in the residential area, and the targeted design of the residential environment has also been gradually concerned, whether in terms of material or spiritual culture [3, 4]. The elderly spend most of their time in residential areas, which should be paid more attention [5]. Whether in terms of physiology, psychology or behavior, the elderly have great differences with young people, and have higher requirements for environmental quality [6, 7].

China's family tends to be miniaturized, the pension resources are scarce, and the function of family pension is gradually weakening. However, the existing social and

B. J. Jansen et al. (Eds.): *International Conference on Cognitive based Information Processing and Applications (CIPA 2021)*, LNDECT 85, pp. 906–911, 2022.
https://doi.org/10.1007/978-981-16-5854-9_123

economic level cannot meet the requirements of realizing social pension in an all-round way, the key to solve this problem is to find a pension model suitable for the current situation and in line with China's social and economic development level [8, 9]. Obviously, to promote the development of home-based care model can better adapt to the existing economic and productivity level of China, and complete the complementary advantages of the pension model [10].

The innovation points of this paper are as follows: first, at present, China pays more attention to the landscape design suitable for the elderly, focusing on the elderly apartment, rarely applied in the residential area, even in the design of the elderly apartment, there are many deficiencies in the design. This paper explores the residential landscape design suitable for the elderly from the basic characteristics and needs of the elderly, to meet the daily needs of the elderly. Secondly, from the perspective of human perception and cultural connotation, this paper attempts to analyze the promoting effect of green plant design on the physical and mental health of the elderly in the residential area, which enriches the design research on the material and spiritual civilization of the elderly in the landscape part of the residential area.

2 Proposed Method

2.1 Aging Design

"Design for aging" is simply understood as the design to adapt to the aging. It adheres to the design concept of taking the elderly as the basis. It is not only the category of barrier free design, but also from the perspective of the elderly, paying close attention to the physical and psychological changes and needs of the elderly. Whether the interior design and supporting functional space design, facilities and equipment configuration, furniture selection and detail treatment are in place, which have a direct impact on the safety, convenience, comfort and economic beauty of the elderly home care.

2.2 Design Principles of Landscape Environment for Aging Residential Areas

(1) Social principle
The design of outdoor environment should be easy for people to contact and communicate with each other more frequently, so as to increase people's interaction opportunities. For example, the outdoor rest area can be set near the hall, restaurant or indoor activity area, and the seats at the main entrance can be arranged in a U-shaped shape to increase the opportunity for people to meet and greet each other; the seats should be placed to avoid embarrassment caused by direct eye contact; some game places and children's activity areas should provide seats for the elderly to increase the chance of chatting among the elderly.

(2) Humanization principle
In view of the different needs of the elderly, the landscape design of residential areas should be considered comprehensively, providing a variety of space, facilities and activities, leaving a choice, so as to meet the participation of different needs of the elderly. For example, group communication space can provide a place for

the elderly to participate in collective activities to meet their sense of belonging; small scale communication space can be used for chess, card, song, dance and other entertainment activities to meet the needs of the elderly's sense of neighborhood; the elderly can bask in the sun, breathe fresh air, raise flowers or feed birds in the rest space. Each type of space has different location selection, different design forms and ancillary facilities configuration.

3 Experiments

3.1 Brief Introduction of Residential Area

(1) Community A
The transportation is convenient, and the surrounding public facilities are relatively complete. The overall greening of the residential area is good, and the pedestrian and vehicle entrances and exits are set up in each plot of the residential area. The composition of the people living here is complex, basically ordinary workers.
(2) Community B
The allocation room used to be provided by the unit to the staff and workers has certain supporting facilities, but the greening environment is poor, and the greening and activity venues are lacking; there are many vendors and noisy living environment. Most of the residents living in the community are technicians with good income and stable income.

3.2 Investigation Contents

Survey object: mainly people over 50 years old living in the community. The purpose of the survey is to further understand the use needs of the elderly through public participation, and to explore the ideal state and composition of outdoor living environment. Nearly 100 residents who are old or about to step into the ranks of the elderly were interviewed or investigated by questionnaire.

4 Discussion

4.1 Residents' Satisfaction of Community A

45% of the people said that they lived apart from their children, but they wanted to live close to their children; 45% said they were old and lived with their children; 17% said that they were old and went to nursing homes, as shown in Fig. 1.

Basically, all the respondents are satisfied with the environment, location and supporting facilities of the community. The degree of satisfaction and demand is as follows: there are fitness facilities and activity places, small squares for entertainment, walking roads, rest seats, chess and card tables, good greening, parks and convenient transportation. The order of demand for supporting facilities is as follows: provide medical facilities that can provide y, having activity centers for the elderly, such as playing chess, playing

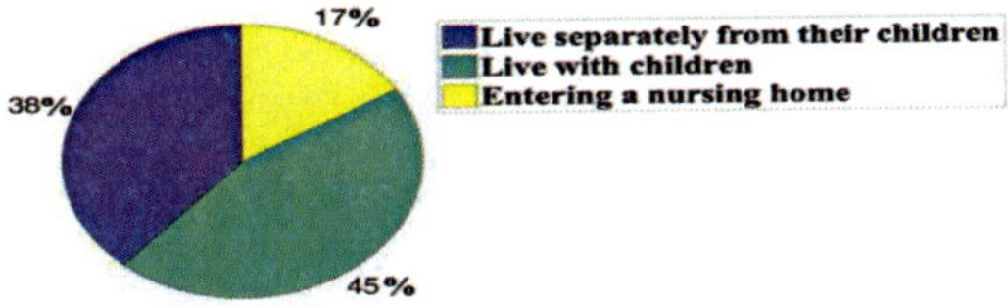

Fig. 1. Willingness of elderly residents in community A

cards, drinking tea, reading newspapers, etc., and having convenient commercial outlets, such as restaurants, daily necessities stores, clothing processing stores, etc., public toilets, home economics and nursing for door-to-door services are shown in Table 1. But the interest in fountains, sculptures and flower beds is low. Dissatisfaction is mainly reflected in: too many cars, disorderly parking, unsafe.

Table 1. Demand ranking of supporting facilities

High to low demand	Supporting facilities
1	Medical care
2	Activity center
3	Commercial outlets
4	Sketch

4.2 Residents' Satisfaction of Community B

62% of the elderly have the desire to live independently, but they are closer to their children, hoping to provide convenient services and facilities for the elderly in the community; about 20% of the elderly hope to live with their children; 10% of the elderly express that they want to go to the elderly apartments or pension institutions, but their children can visit them often; about 8% of them are willing to live in the elderly community, as shown in Fig. 2.

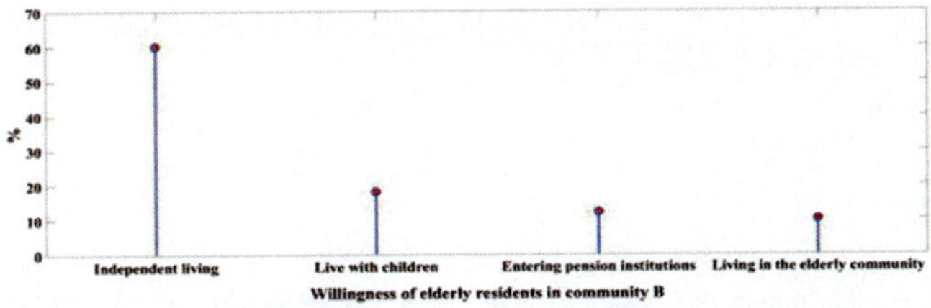

Fig. 2. Willingness of elderly residents in community B

The respondents were basically satisfied with the supporting facilities, and most of them were not satisfied with the environmental construction, which were mainly reflected in the following aspects: poor greening, noisy and unsafe environment, lack of lax management in the community, serious building phenomenon, lack of activity venues, insufficient facilities, serious lack of space for elderly people to chat, and lack of small pieces and facilities such as sitting, playing chess and playing cards, the external environment is difficult to meet the needs of people. The most demanding are: medical and health services, sports venues, good green environment, benches and public toilets, lighting facilities, cultural facilities, quiet and safe. It can be seen from the above: the choice of pension mode and living mode is closely related to income, education level and occupation. The elderly with good economic conditions, education level and high social status before retirement are easy to accept the new pension mode, prefer to live on their own, and have low dependence on their children. But the requirements of outdoor environment facilities, content, style and sketch are basically consistent.

5 Conclusions

At present, China's pension system, social security system is not perfect, pension facilities are not perfect, in the face of the rapid aging of the population, the urgent need to improve the outdoor living environment of the elderly, this paper puts forward the landscape design research of urban residential area which meets the physical and mental needs of the elderly. This paper summarizes the physiological, psychological, behavioral characteristics and needs of the elderly by studying the relevant literature at home and abroad. Combined with the theories of geriatric psychology and environmental behavior, this paper analyzes the landscape design principles of residential areas in aging society, studies the design methods of residential landscape under the background of aging society, and investigates and analyzes the two typical cases of community A and B, Put forward suggestions and prospects for the future residential landscape design, further enrich the relevant theories of residential landscape design, and provide theoretical support and practical methods for the design of living environment suitable for the elderly.

Acknowledgment. Scientific Research Project of Education Department of Shaanxi Province "The Design Research of Elderly-Oriented Residential Environment under the Home-Based Care Model" (Project Number: 18JK0882).

Science and Technology Research Project of Yan'an Science and Technology Bureau of Shaanxi Province "The Research on Outdoor Environment Design of Yan'an Pension Service Institutions" (Project Number: 2017KS-09).

References

1. Juang, L.W., Kollmeyer, P.J., Anders, A.E., et al.: Investigation of the influence of superimposed AC current on lithium-ion battery aging using statistical design of experiments. J. Energy Storage **11**(June), 93–103 (2017)
2. Shao, Q., Gou, X., Huang, T., Yang, S.: Anti-aging analysis for software reliability design modes in the context of single-event effect. Softw. Qual. J. **28**(1), 221–243 (2019). https://doi.org/10.1007/s11219-019-09464-3

3. Luceno, J., Martin, M.: Optimal design of aging systems: a-frame coolers design under fouling. Comput. Chem. Eng. **122**(Mar. 4), 47–58 (2018)
4. Silbaugh, K.B.: Aging policy design: building from Anne Alstott. Boston Univ. Law Rev. Boston Univ. Sch. Law **97**(5), 1943–1945 (2017)
5. Raviteja, P.B.V., Durga, D.: Efficient adaptive hold logic aging-aware reliable multiplier design using verilog HDL. Int. J. Comput. Appl. **179**(26), 5–12 (2018)
6. Yuejun, Z., Zhicun, L., Dailu, D., et al.: Design of SH aging sensor for real time and application in sensing network. Can. J. Electr. Comput. Eng. **43**(2), 73–82 (2020)
7. Dey, S., Nandi, S., Trivedi, G.: Machine learning approach for fast electromigration aware aging prediction in incremental design of large scale on-chip power grid network. ACM Trans. Des. Autom. Electron. Syst. **25**(5), 1–29 (2020)
8. Rossi, D., Tenentes, V., Reddy, S.M., et al.: Exploiting aging benefits for the design of reliable drowsy cache memories. IEEE Trans. Comput. Aided Des. Integr. Circuits Syst. **37**(7), 1345–1357 (2018)
9. Singh, J., Shahi, A.S.: Weld joint design and thermal aging influence on the metallurgical, sensitization and pitting corrosion behavior of AISI 304L stainless steel welds. J. Manuf. Process. **33**(June), 126–135 (2018)
10. Peruzzini, M., Pellicciari, M.: A framework to design a human-centred adaptive manufacturing system for aging workers. Adv. Eng. Inform. **33**(Aug.), 330–349 (2017)

Translation Quality Evaluation Method and Its Application in Computer Translation Evaluation System—From the Perspective of Translation Criticism

Shengliang Xie(✉)

Guangdong Mechanical and Electrical Polytechnic, Guangzhou, Guangdong, China

Abstract. Traditional translation criticism mainly comes from subjective introspection and impression judgment, lacking convincing scientific data and statistics. The application of computer translation evaluation system makes objective and scientific translation criticism come true. This paper reviews the research background and current situation of translation criticism, expounds the concept and connotation of computer-aided translation criticism, and puts forward the method and application of translation quality evaluation based on computer translation evaluation system, which provides a new paradigm for the form of translation criticism.

Keywords: Computer-aided · Translation criticism · Method

1 Introduction

"Translation criticism is an essential link between translation theory and its practice" (Newmark 1988:184).Translation criticism has long been regarded as "the weakest link in translation studies" (Zhou and Luo 1999:159). The traditional methods of translation criticism are mainly empirical and introspective, which often appear weak and lack of scientific criticism principles and good atmosphere of overall criticism (Du and Wen 1994:774). Moreover, the concepts of translation standards and translation criticism standards are confused, with unclear academic theories and narrow perspectives. They always adhere to literal translation, free translation, faithfulness and equivalence, which makes the translation criticism view the same (Li and Pan 2011: 126–131).

In fact, translation criticism is a mirror to examine translation activities. Correct and scientific translation criticism can promote translators' translation level, thus improving translation quality and helping readers better read and understand translation (Yang 2005:21). With the development of computer and other modern information technology, scientific translation criticism has gradually become possible.

B. J. Jansen et al. (Eds.): *International Conference on Cognitive based Information Processing and Applications (CIPA 2021)*, LNDECT 85, pp. 912–916, 2022.
https://doi.org/10.1007/978-981-16-5854-9_124

2 Research Background and Current Situation

Since the 1980s and 1990s, translation criticism has made great progress in China. Especially since the 21st century, with the rapid popularization and development of computer and Internet, under the guidance of new translation theories and ideas, there has been a re-examination of translation criticism and a special study of translation quality assessment [1–3]. With the quantitative means of electronic corpus and the concept of Meta translation criticism, the academic activities related to translation criticism are also developing vigorously. For example, in April 2011, the Symposium on translation criticism from a multidimensional perspective was held in Hunan University of Science and Technology. More than 80 seminars have been held on the current situation of translation studies in China and abroad, and the problems of translation research in China and abroad have been discussed. The success of the seminar shows that with the vigorous development of translation studies, the importance of translation criticism has been further valued by the translation community [4–6]. Compared with the tradition, the theoretical consciousness of translation criticism in the new era has been enhanced, and translation criticism is no longer limited to the subjective impression. However, translation criticism still has many problems in theory and practice.

From the current situation, the development of translation criticism in China is still relatively slow, which is the field that translation studies should be particularly strengthened (Zhang 2011:110–111). Although there are a large number of research papers, multi-dimensional perspectives and novel viewpoints, literary translation criticism is still the main research object, and there are few monographs on translation criticism [7–10]. In particular, the study of translation criticism combined with computer and network technology is obviously backward, which is not commensurate with the rapid development of computer-aided translation research. For example, corpus based translation studies have made great progress, but there are only a few papers on translation criticism based on it, and few studies on translation criticism combined with the Internet.

3 The Concept and Connotation of Computer-Aided Translation Criticism

Computer-aided translation criticism should refer to all the studies of translation criticism carried out with the help of modern technologies such as electronic corpora, Internet, various translation software, and translation research software. But at present, corpus is mainly used to carry out translation studies, and the results are also rich. Therefore, we try to use network technology and computer software to carry out a series of research on translation criticism, and conduct scientific research on translation criticism in a new way, which is consistent with Professor Xu Jun's advocacy of "establishing a scientific critical spirit" (Xu 2003:409) [11–14]. Translation criticism is a core concept in translation studies, but the definition of translation criticism in academia is different. For example, Yang (2005:3) believes that translation criticism refers to the analysis, comment and evaluation of a translation according to certain translation standards and certain argumentation methods, or comments on certain phenomena in translation by comparing different versions of a work. Wen (2007:38) [15, 16] pointed out that translation criticism is a scientific cognitive activity to analyze, interpret and comment on various translation phenomena, translated works and translation thoughts based on certain translation theories and translation criticism theories. In addition, many well-known scholars, such as Xu Jun and Yang Zijian, have also expressed their views on the definition of translation criticism.

According to the above definition of translation criticism, we can realize that the essence of translation criticism is to analyze and comment on translation works or phenomena in a certain way. Therefore, translation criticism has both subjectivity and objectivity (Xu 2006:56). But the essence of scientific research is to emphasize the objective and realistic description of things and phenomena, that is, objectivity, and translation criticism is no exception. Perhaps due to the limitations of the times, traditional translation criticism is evaluated by introspection or subjective impression. In the new era of rapid development of science and technology, with the mature development of computer and network and other electronic technologies, it is inevitable to integrate the achievements of high technology into translation studies. The rise of corpus translatology is the proof of this trend. Therefore, computer-aided translation criticism is not only the inevitable outcome of the development of computer information technology, but also the urgent need for the further development of translation studies (Fig. 1).

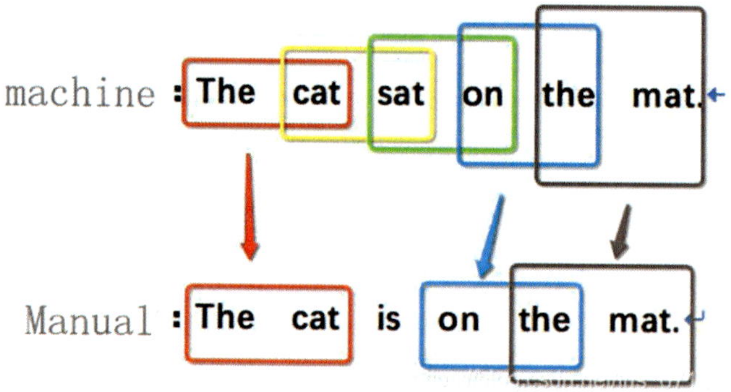

Fig. 1. Manual translation and machine translation

4 Translation Quality Evaluation Method and Its Application Based on Computer Translation Evaluation System

With the help of computer and network technology, the establishment of computer-based translation evaluation system is actually a powerful tool for computer-aided translation criticism. It can carry out translation quality evaluation and carry out multi-perspective scientific research on translation criticism, including criticism mode, and standard, etc. Generally speaking, we can explore from the following two perspectives.

First of all, we use network technology to establish a different mode of network translation criticism, which is different from the traditional one. We carry out translation criticism on the platform of computer and Internet. With its dynamic, timely and interactive advantages, it provides a new modern platform for translation criticism, including literary translation criticism. At this time, the elements of translation criticism, such as the subject, the form and the object of criticism, have changed greatly. The limitations of network criticism, such as conformity, vulgarity and duplication, bring new thinking to the development of translation criticism theory.(Lin 2010).Therefore, the method of network translation criticism model has opened up a new space for translation criticism, injected new vitality, and made translation criticism develop in a healthy direction.

Secondly, we explore the translation methods from the perspective of corpus. The criterion of translation criticism and the criterion of translation are two different concepts. There is no fixed standard for translation criticism, because the diversity of the identity of the critic and the diversity of the translated text determine the scope and factors that translation criticism should involve. Therefore, the criteria of translation criticism should be diversified and multi perspective (Mu 2006).Therefore, we can establish parallel corpora through the translation of scientific and technological texts, and use corpus analysis software such as Wordsmith to explore the effective factors of translation criticism standards, such as word frequency and sentence length, and scientifically discuss the translation criticism standards. At the same time, we can summarize the feasible methods of translation quality testing through the corpus, so as to find out the objective and effective translation quality testing methods.

5 Epilogue

For a long time, traditional translation criticism mainly focuses on whether the translation and the original text are faithful and equivalent, and the conclusions come from subjective introspection and impressionistic judgment, lacking convincing scientific data and statistics. The method of translation quality evaluation by computer translation evaluation system makes the objective and scientific translation criticism come true and puts forward new research ideas for translation criticism. Computer aided translation criticism, especially corpus based translation criticism, can make us try to analyze the real features and nature of the translated works based on a large number of real corpus, and reveal the characteristics that may exist in translation, such as the translators' preference for words, the characteristics of sentence structure, the degree of explicitness and normalization, etc.. In addition, with the application of network technology in translation criticism, readers can communicate directly with publishers, translators and even the original author, thus giving full play to the functions of supervision, value judgment, interpretation and coordination of translation criticism. At the same time, the universality, spontaneity and sharpness of network criticism also open up a new space for translation criticism, which helps to improve the quality of translation and provides a new paradigm for the form of translation criticism.

In a word, the evaluation of translation quality based on computer and network technology provides many new possibilities for translation criticism. It not only provides new tools for translation studies, but also expands the scope of translation studies. It makes the "aesthetic evaluation" and "scientific judgment" of translation criticism jump out of the shackles of introspection of experience and intuition (ibid: 38), and achieves the organic unity of subjectivity and objectivity, thus promoting the historical turn from prescriptive to descriptive in translation studies.

Acknowledgments. This work was financially supported by Guangdong Education Bureau's special projects in key areas of "Serving Rural Revitalization Plan" of Guangdong Province's colleges and universities in 2019 fund (Project No. 2019KZDZX2047).

References

1. Newmark, P.:A Textbook of Translation. Prentice Hall, New York (1988)
2. Du, C., Wen, J.: A Hundred Studies on Contemporary Translation in China. Chongqing University Press, Chongqing (1994)
3. Guo, J., Wang, Y.: Translation teaching and testing under the multimedia network environment. Res. Audio-Vis. Educ. Kong Lingcui **33**, 115–116 (2008)
4. Kong, L.: Application of modern information technology in translation. China Sci. Technol. Inf. **2**, 36 (2005)
5. Li, Y., Pan, X.: A new perspective of translation criticism: the theory of translation criticism parameters based on historical contextualization. J. Huaqiao Univ. **1**, 126–131 (2011)
6. Lin, Z.: Translation criticism research in the network environment. Contemp. Lit. **2010**(2), 69 (2010)
7. Lv, J., Hou, X.: Introduction to Translation Criticism. Shanghai Foreign Language Education Press, Shanghai (2009)
8. Mu, L.: Translation criticism and translation standards. Foreign Lang. Foreign Lang. Teach. **4**, 45–47 (2006)
9. Wang, L., Wang, J.: Computer aided translation research methods and their application. Foreign Lang. Foreign Lang. Teach. **5**, 41–44 (2008)
10. Wen, J., Liu, P.: Fifty years of Chinese translation criticism: retrospect and prospect. Gansu Soc. Sci. **2**, 38–41 (2006)
11. Wen, X.: On the subject of translation criticism. J. Nankai, Xu Jun **2007**(3), 38 (2007)
12. Xu, J.: Translation Theory. Hubei Education Press, Wuhan (2003)
13. Xu, W.: The application of parallel corpus in translation criticism: a case study of different versions of bacon of studies. Foreign Lang. Stud. **2**, 54–59 (2006)
14. Yang, X.: Introduction to Translation Criticism. China Translation and Publishing Company, Beijing (2005)
15. Zhang, J.: A summary of the first national symposium on translation criticism. Foreign Lang. Stud. **2**, 110–111 (2011)
16. Zhou, Y., Luo, P.: Translation and Criticism. Hubei Education Press, Wuhan (1999)

Design and Implementation of Chinese Pronunciation Automatic Scoring System Based on Speech Recognition

Chao Wang(✉)

School of International Education,
Dalian Polytechnic University, Dalian 116034, Liaoning, China
wangchao@dlpu.edu.cn

Abstract. The development of speech recognition technology makes the interaction between human and computer possible. In view of the deficiency of pronunciation teaching in TCFL, this paper proposes the design of Chinese automatic pronunciation level evaluation system in TCFL, and describes the structure, function and process of the system in detail. This paper introduces the key technologies and steps in the implementation of the system: dynamic time warping algorithm, corpus building, phonological segmentation technology and evaluation and grading standards. Through a small range of experiments, it shows that the system has a certain reference value for the test of foreign students' Chinese pronunciation level.

Keywords: Speech recognition · TCSL · Automatic scoring · Dynamic time warping · Phoneme segmentation

1 Introduction

Language is the main tool for people to communicate with each other. With the continuous development of international communication, many people are not satisfied with mastering their own mother tongue, and have begun to learn a second language, some even more. Therefore, language learning has become a hot spot in the field of education, and the traditional classroom teaching is far from enough for the second language learning. A technology called computer aided language learning (call) should be born. However, call and classroom teaching have a common shortcoming, that is, they put too much emphasis on language reading, writing and listening comprehension, and neglect the training of speaking ability, which is an important part of language communication. As a result, "dumb English" and "dumb Chinese" appear. At present, the "Chinese fever" in the world is heating up, and more and more people are learning Chinese. Not only in China, but also in many countries in the world, there is an upsurge of learning Chinese. However, there is a lack of professional teachers resources. As we all know, feedback is an important part in language learning, especially in pronunciation learning. Timely, accurate and objective feedback can help learners find out the gap between their own pronunciation and standard pronunciation and correct pronunciation errors.

B. J. Jansen et al. (Eds.): *International Conference on Cognitive based Information Processing and Applications (CIPA 2021)*, LNDECT 85, pp. 917–921, 2022.
https://doi.org/10.1007/978-981-16-5854-9_125

2 Speech Recognition Technology

Speech recognition is essentially a process of pattern recognition, and its basic principle block diagram is shown in Fig. 1. It mainly includes speech signal preprocessing, feature extraction, feature modeling, similarity measurement and post-processing.

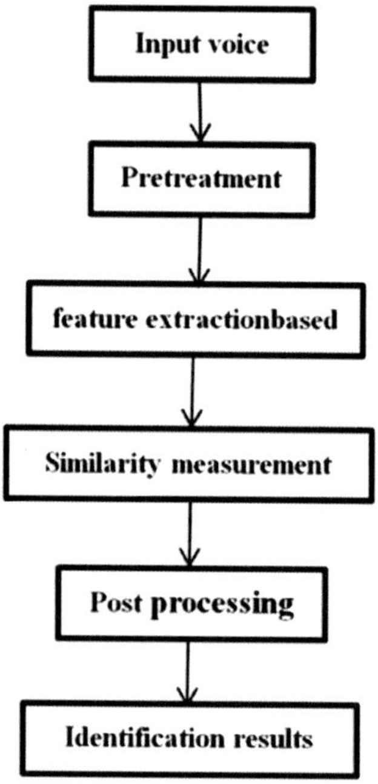

Fig. 1. Principle of speech recognition

A speech recognition system consists of two stages: training and recognition. Whether training or recognition, it is necessary to preprocess the input original speech and extract the features [1]. The feature extraction module is responsible for calculating the acoustic parameters of speech and calculating the features, so as to extract the key feature parameters reflecting the signal features, so as to reduce the dimension and facilitate the subsequent processing. short-term spectrum, formant frequency, Mel frequency cepstrum coefficient (MFCC), etc. Feature selection and extraction is the key to system construction.

3 Automatic Scoring System for Chinese Pronunciation

Considering the law of language learning and the characteristics of Chinese pronunciation, learners should follow the process from easy to difficult, from simple to complex. It is usually divided into the following three stages: the first stage mainly studies the basic pronunciation units of Chinese, including the initial consonants and finals in Chinese; the second stage mainly studies the pronunciation of Chinese characters and words, including the combination of consonants and finals of various Chinese characters and Chinese tones; the third stage mainly studies the pronunciation of phrases and sentences in Chinese, It includes the co pronunciation of words in phrases and sentences, the intonation of sentences, the speed and rhythm of sentences.

3.1 Basic Framework of the System

The system is divided into four relatively independent modules: (1) the standard model training module is mainly responsible for the establishment of standard corpus. The Chinese automatic scoring system mainly compares learners' pronunciation characteristics with the pronunciation in the standard corpus, and scores the learners' pronunciation according to the similarity. Therefore, for the whole system, the training of the standard model is the most basic part. Generally, the standard model is selected to train the standard corpus, In order to achieve the expected standard effect. (2) The expert grading module enables experts to grade learners' pronunciation. Because the pronunciation score directly given by the automatic scoring system may be obscure and can not reflect the actual pronunciation level of learners, experts who use Chinese pronunciation give a grading standard for pronunciation in non-standard corpus, and then combine the pronunciation score given by the system to show the learners' pronunciation level intuitively and clearly. (3) The automatic scoring module is responsible for comparing learners' pronunciation with those in the standard corpus. According to the principle of speech recognition, a similarity algorithm is used to give the quantitative index of pronunciation. This is the most important part of the whole system. After the automatic scoring module collects the learners' pronunciation, after preprocessing and feature extraction, according to the unit model trained in the standard corpus, it adopts forced alignment to segment the pronunciation into small units needed to calculate the pronunciation quality measurement. According to the characteristics of Chinese pronunciation, small units are usually phonemes. Finally, a certain similarity measurement method is used to get the score of learners' pronunciation. (4) The error analysis module is responsible for giving learners' pronunciation errors and corresponding correction suggestions based on the expert knowledge base and the collected features of learners' pronunciation. This module is very helpful for learners to correct their pronunciation, and can give learners the most direct suggestions. However, due to the related professional knowledge, the specific work will be carried out in the stage of system improvement and improvement.

3.2 Key Technologies of System Implementation

In speech recognition, there are many disadvantages in simply comparing the input template with the corresponding reference template. Because of the great randomness of speech signal, even if the same person at different times of the same speech, also can not have the same length of time, so time normalization processing is essential [2]. Dynamic time warping (DTW) uses dynamic programming (DP) to transform a complex global optimization problem into many local optimizations to make decisions step by step. Suppose that the feature vector sequence of the reference template is = {1, 2,}, and the input speech feature vector sequence is = {1, 2,},. Let the time normalization function = {1, 2,}, where n is the path length; =, indicating that the first matching point pair is a matching point pair composed of the first eigenvector of the reference template and the first eigenvector of the template to be tested. The distance (or distortion) between them is called local matching distance. DTW algorithm is to minimize the sum of weighted distances by local optimization:

$$= \min \frac{\sum_{n=1}^{1} d(x_{i(n)}, y_{i(n)})}{\sum_{n=1}^{N} W_n} \tag{1}$$

3.3 Phonological Segmentation Algorithm

In a sense, the overall difference between two speech segments in the feature parameter space is called the distance between the two speech segments, which is referred to as the inter segment distance. The method of calculating distance and distance between rules is called segment. Since the speech feature vector is usually calculated according to the time frame of a certain length, a certain number of continuous speech frames can be taken to form a speech segment when calculating the inter segment distance [3]. A segment distance measure based on the average distance of the frame feature vectors between speech segments is used in the experiment. Let and denote two segments of the same continuous speech, each of which contains, time frames. The feature vectors of speech segment and time frame are respectively $1 \leq \leq, 1 \leq 5$. Let the Euclidean distance between and be, then the distance between segments of you and segment is

$$= \frac{1}{N_a N_b} \sum_{i=1}^{N_a} \sum_{j=1}^{N_b} D_{ij} \tag{2}$$

3.4 Classification Standard

Chinese pronunciation scoring system shows the level of learners' pronunciation by giving certain feedback to learners. One method is to evaluate the similarity between learners' pronunciation and standard pronunciation, such as DTW distortion distance, by calculating the score of pronunciation quality measure. However, the disadvantages of this method are: the scoring is absolute, unstable, vulnerable to the influence of individual differences of learners, and the expression is not intuitive, the meaning is abstract, and is not consistent with people's perception; the other method is to take the expert's judgment as the standard, and convert the calculated pronunciation score value into the expert's score through mapping. Expert's grading is a kind of fuzzy classification, which is generally combined with professional knowledge and subjective feeling to divide the pronunciation quality into several grades, such as very good, good, general, poor, very poor, etc. [4]. Although this kind of classification has a certain subjectivity and is relative, it has a clear meaning, more in line with human perception habits, and has certain stability. In addition, articulation, naturalness or fluency can also be used as a measure of expert grading.

4 Conclusion

In order to verify the reliability of the system and the consistency with expert's narration, 20 foreign students with different Chinese proficiency were randomly selected to test the system. The test materials are mainly from foreign students' elementary and intermediate textbooks of Chinese as a foreign language, and the common words covering Chinese pronunciation syllables and tones are selected. The test site is arranged in a professional phonetic classroom. Each foreign student reads 20 Chinese characters and 20 words or phrases randomly, and the system records and scores them. At the same time, five TCSL teachers also grade the students' pronunciation according to the recording. Due to operability, the two scoring mechanisms are divided into five levels (1 worst; 5 best). After the analysis of the data, we can see that the system can accurately score the Chinese pronunciation of foreign students of different levels, basically reflects the learners' Chinese pronunciation level, and also can maintain good consistency with the manual score, which is helpful for training Chinese pronunciation level.

References

1. Yi, K., Tian, B., Fu, Q.: Speech Signal Processing, pp. 154–233. National Defense Industry Press, Beijing (2000)
2. Han, J., Zhang, L., Zheng, T.: Speech Signal Processing, pp. 191–258. Tsinghua University Press, Beijing (2004)
3. Han, J., Wang, H., Li, H., et al.: Pronunciation learning technology based on speech recognition. Electroacoust. Technol. **9**, 47–51 (2004)
4. Zhang, J., Yang, H., Lin, L., et al.: An automatic speech segmentation method based on interval distance measure. J. Shanghai Jiaotong Univ. (Chin. Ed.) **9**, 1362–1365 (2001)

Analysis of Computer Network Security Problems and Countermeasures Based on Cognitive Big Data

Jie Dong(✉), Peng Wang, Haihong Chen, and Yazhen Gu

College of Mathematics and Computer Science, Chifeng University, Chifeng 024000, Inner Mongolia, China

Abstract. Big data provides convenience for social development and work, and computer network has been widely used for its convenience and efficiency. However, nowadays, the problem of computer network security is becoming increasingly prominent. How to find feasible strategies to ensure the security of computer network has become the main research content, which requires us to ensure the stable operation of computer network with high standard security.

Keywords: Big data environment · Computer network · Security issues · Protection methods

1 Big Data Background

1.1 Big Data Concept

Through the exchange, integration and analysis of large-scale data, we can dig out useful information, create new value, and bring "big knowledge", "big technology", "big profit" and "great development" [1]. Big data is not a new thing. In recent years, the rapid development of cheap storage, sensor and data acquisition technology, the increase of information links through cloud and virtualization devices, and the application of innovative software and analysis tools are driving big data and making data generation fast.

1.2 Big Data Features

(1) Volume: today's amount of data storage is exploding, and we're stuck in data. The storage unit of big data has changed from TB level to Pb level, and even to ZB level. (2) Diversity: the amount of big data associated with the big data phenomenon has brought new challenges to the data processing centers that deal with it: its variety. With the proliferation of sensors, intelligent devices and social technology, data not only contains traditional structured data, but also includes (3) Velocity: the high speed of big data means that the speed of data generation and processing is extremely fast. Therefore,

B. J. Jansen et al. (Eds.): *International Conference on Cognitive based Information Processing and Applications (CIPA 2021)*, LNDECT 85, pp. 922–926, 2022.
https://doi.org/10.1007/978-981-16-5854-9_126

effective processing of big data needs to analyze the number and types of big data in the changing process in a specific time and space. (4) Value: there is no doubt that big data itself contains great value, especially business value. Big data can bring direct value to the core business of enterprises, which is the main reason for the popularity of big data. The value of big data is sparse. Mining big data is like scouring sand in waves, but it is precious.

1.3 Big Data Based on Data Mining Algorithm

K-means clustering algorithm is a classical clustering algorithm based on partition. Its goal is to divide the data set into K clusters, so that the difference degree between the data in each cluster is as small as possible, and the difference between the data between clusters is as large as possible. Because the idea of K-means clustering algorithm is relatively simple [2]. K-means clustering algorithm has been widely used.

D-dimensional data set $X = \{x_1, \ldots, x_i, \ldots x_n\}$ With n points, where $x_i \in R^d$ and $S = \{S_1, \ldots, S_i, \ldots S_k\}$.

The formula for calculating the cluster center μ_i of each cluster is as follows

$$\mu_i = \frac{1}{\|s_i\|} \sum_{x_j \in s_j} x_j \tag{1}$$

The value range of i is an integer from $X = \{x_1, \ldots, x_i, \ldots x_n\}$, and the cluster center set $\mu = \{\mu_1, \ldots, \mu_i, \ldots \mu_n\}$.

The clustering objective function is defined as follows

$$v = \sum_{i=1}^{k} \sum_{x_j \in s_j} (x_j - \mu_i) \tag{2}$$

Randomly select k data points from the data set X as the initial clustering center; use Euclidean distance to calculate the degree of difference between data points, and divide the remaining points into the cluster center with the least dissimilarity in the cluster represented by.

2 Influence Factors of Computer Network Security Under Big Data

The Internet has led people into a new era, but the problem of network security has gradually emerged. When the computer began to rise, people did not pay attention to the mechanism of computer network information system, making the computer network in the future development of security risks. Network security incidents have increased year by year in recent years, which threaten the security of information management.

2.1 Hacker Attack

Under the background of big data, hacker attack technology is more and more high, and it is not easy to be found, which poses a serious threat to the computer network.

At present, there are several ways to attack hackers: first, web applications. Hackers seize web vulnerabilities and steal user information. Second, backdoor procedures. In the process of programming, many programmers leave backdoor programs for the convenience of subsequent program operation and maintenance [3]. Although convenient, it also gives hackers an opportunity to take advantage of it, which becomes an attack entrance. As shown in Fig. 1. Third, hackers send junk information to the computer system, which increases the load of the system. Finally, the system can not carry a large amount of information, resulting in system paralysis. Fourth, hackers collect network information and steal user information by setting monitoring program. Fifthly, through the form of distributed denial of service attack, the computer network host can not collect information.

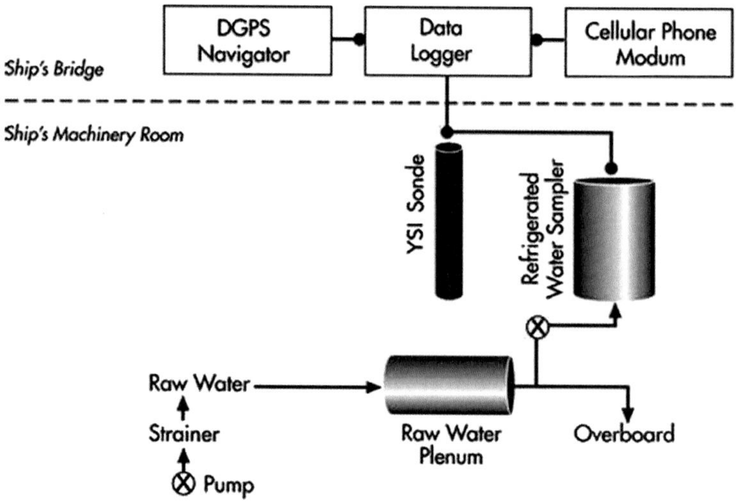

Fig. 1. Setting monitoring program

2.2 Virus

Although the current computer system has become mature, there are some loopholes in any kind of computer operating system. Especially under the background of big data, network data sharing and opening provide convenience, but also leave security risks. Operation error, system not updated for a long time, etc. may lead to virus invasion of computer network, software and hardware damage, data damage, and bring serious losses to users. Under the background of cloud computing big data, there are many security problems in cloud computing, such as: API is attacked by hackers, information exposure, identity information disclosure, DoS attack, etc. Cloud platform infrastructure is distributed data storage technology, parallel technology, which increases the cloud platform physical security risk, host risk, network security risk. In addition, the cloud platform stores huge amount of information, and the data management rights and ownership are different, which also increases the difficulty of cloud platform security management.

3 Computer Network Security Scheme Strategy Under the Background of Big Data

3.1 Install Firewall

Install firewall technology, deploy web protection system, and create security situation awareness management platform to achieve the purpose of computer security protection. The firewall has the functions of network information screening, monitoring network login, preventing network attacks and intercepting harassing information. As long as users choose reliable firewalls, they can ensure that their protection ability extends to deep-seated functions. Firewall can quickly monitor different forms of DDoS attacks, compatible with security management, SOP security virtualization, and strengthen access management [4]. In addition, the firewall can also identify and manage the application layer traffic, use the first intrusion detection engine to monitor the whole process, and quickly find malicious code.

3.2 Vulnerability Scanning Technology

Vulnerability scanning technology can realize the whole process scanning, including servers, firewalls, hosts, etc. it can intelligently analyze the causes of vulnerabilities and ensure the security of network system. In addition, attention should be paid to operation safety management. The selection of Intel instruction system, microprocessor and virtual host IP address of computer operating system can also improve the stability of the system and discover the existing risks in time.The installation of intrusion detection system can improve the security of computer network and prevent hacker attacks.After installation, the system can monitor whether there are attacks inside or outside the firewall comprehensively to ensure the stability and security of firewall information. In addition, the installation of encryption authentication system and isolation device can prevent hackers from stealing user information. If the user information is stolen by hackers, the system will immediately isolate the program and repair the damaged data.

4 Conclusion

Comprehensive analysis, the computer network security protection needs us to adhere to maintenance for a long time, constantly optimize and improve, integrate various influencing factors, and supervise and protect the whole process. Under the background of big data, the use of firewall, vulnerability scanning, physical isolation and other technologies can ensure computer network security, avoid hacker attacks and virus invasion, and improve the level of computer security protection.

References

1. Gao, Y.: Research on library health information service mode in big data era. Henan Libr. Sci. J. **41**(05), 106–107 (2021)
2. Yuan, S., Cai, D., Su, X.: A study on the cognitive differences between Chinese and foreign tea brands based on big data and bell model: a case study of Wu Yutai and Lipton. Pract. Underst. Math. **51**(10), 90–105 (2021)
3. Liu, K., Li, Y., Zhang, X.: Comparative empirical study on data intelligence and expert knowledge from the perspective of trust [J/OL]. Library and Information Work, pp. 1–8, 21 June 2021. https://doi.org/10.13266/j.issn.0252-3116.2021.06.012
4. Zhang, A., Wang, F.: Political security risks of "big data maturity." Future Commun. **28**(02), 46-51+124-125 (2021)

Application of Artificial Intelligence Aided Cognitive Computing Method in Building System

Gu Yue(✉)

Panjin Vocational and Technical College, Panjin 124010, Liaoning, China

Abstract. Teaching quality evaluation is a process in which a specific evaluation institution analyzes the evaluation information and makes a conclusion according to the activity information of the evaluation object, and finally feeds back the evaluation results to the school management department. Data analysis is a process of identifying effective, potential and understandable patterns from sample data sets by data mining technology. The information system based on Web technology has the characteristics of low cost, good scalability, easy maintenance and management, which is the best way to realize information management. This paper attempts to use rich Internet application (RIA), the most advanced rich client technology, and Java2 Platform Enterprise Edition (J2EE), which is the most widely used enterprise server-side development technology, to realize a teaching quality evaluation system based on the integration of Flex technology and J2EE technology.

Keywords: Data mining · Teaching evaluation · Clustering

1 Introduction

With the continuous enrollment expansion of colleges and universities, the quality of higher education has gradually become the focus of attention of the general public, and the teaching reform is more prominent [1]. At present, the domestic colleges and universities basically use the educational administration management system based on the network to carry on the teaching evaluation. Although the advanced technical means are used, the evaluation results are always difficult to have enough persuasion. As a kind of management means, the basic requirement of teaching evaluation is that "the result of teaching evaluation should be able to objectively explain the level of teachers' teaching quality". But in practice, many teachers, including teaching administrators, are skeptical about the teaching evaluation results in hand. One of the core problems is that the evaluation results based on simple mathematical statistics seem to be unified, but it does not take into account the actual situation of various objective existence, such as student factors, curriculum factors, teaching arrangement factors, etc. it is precisely because these factors are not taken into account, so the evaluation results are not objective, so it is difficult to convince the public.

B. J. Jansen et al. (Eds.): *International Conference on Cognitive based Information Processing and Applications (CIPA 2021)*, LNDECT 85, pp. 927–931, 2022.
https://doi.org/10.1007/978-981-16-5854-9_127

This system attempts to use advanced data mining technology, through the mining of historical data, the objective factors that affect the teaching evaluation results are mined out, and their influence ability is evaluated to form an "influence factor". Then the "influence factor" is applied to the teaching evaluation system to shield the influence of various objective factors on the teaching evaluation results, A relatively objective evaluation result is formed.

2 Overview of Data Mining

2.1 The Concept of Data Mining

In the past 30 years, the rapid and lasting development of computer technology has made a large supply of data collection and storage equipment with stable performance and low price. These technologies greatly promote the development of information industry and related industries, making a large number of databases and information repositories used for things management, information retrieval and data analysis [2].

Huge amounts of data are stored in a large number of data repositories. Due to the lack of effective data processing and analysis tools, people can not obtain useful knowledge from the huge database, which is the phenomenon of "rich data, but lack of information". People need an automatic, efficient and reliable data analysis tool to transform the data grave into a knowledge "gold nugget".

In this context, data mining technology emerges as the times require. Data mining refers to the process of discovering valuable knowledge from a large number of data stacked in the database or other storage file libraries. In short, it is "mining knowledge from data". Through data mining, we can find some unknown interesting knowledge or rules from a large number of data. The discovered knowledge or rules can be used for data analysis, decision-making basis, guidance and improvement of process control, assistance in information management and data processing. Data mining is a comprehensive subject, involving database technology, artificial intelligence, statistics, probability theory, pattern recognition, neural network, information retrieval and other fields of knowledge. Therefore, data mining is one of the most important frontiers of information and database system and one of the interdisciplinary subjects with the fastest development of information technology.

2.2 Steps of Data Mining

In the early research, data mining is regarded as a basic step of KDD in data. The steps of KDD are generally composed of seven steps: data cleaning, data integration, data selection, data transformation, data mining, pattern evaluation and knowledge representation. The specific process is shown in Fig. 1.

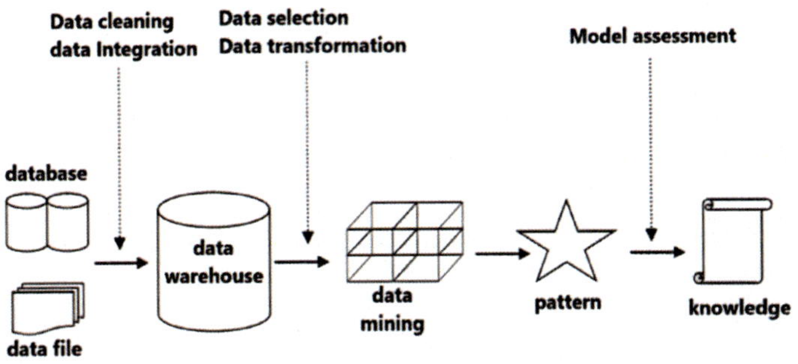

Fig. 1. General steps of data mining

3 Data Types in Cluster Analysis

When the variable type is a continuous scale value, the degree of difference between objects is usually calculated based on the distance between each pair of objects. The most commonly used distance measures are Euclidean distance and Manhattan distance.

Euclidean distance is defined as follows:

$$d(i,j) = \sqrt{(x_{i1} - x_{j1})^2 + (x_{i2} - x_{j2})^2 + \cdots + (x_{in} - x_{jn})^2} \quad (1)$$

Manhattan distance is defined as follows:

$$d(i,j) = |x_{i1} - x_{j1}| + |x_{i2} - x_{j2}| + \cdots + |x_{in} - x_{jn}| \quad (2)$$

4 Classification of Data Mining System

Data mining is an interdisciplinary field, which is influenced by many disciplines (as shown in Fig. 2), including database technology, statistics, machine learning, information science and so on [3]. It is helpful for users to distinguish different data mining systems and choose the most suitable mining tools. According to different classification standards, data mining system can be divided into the following categories.

(1) Classification according to the types of databases mined: the database system itself can be classified according to different data models and data types. For example, according to the data model, there are relational databases, thing databases, object relational databases, etc. according to the data types, there are space, time series, text, streaming data, multimedia data, etc.
(2) Classification according to the knowledge type of Mining: according to the knowledge type classification of data mining, that is, classification according to the function of data mining, such as characterization, data differentiation, association analysis, classification and prediction, clustering, outlier analysis and evolution analysis.

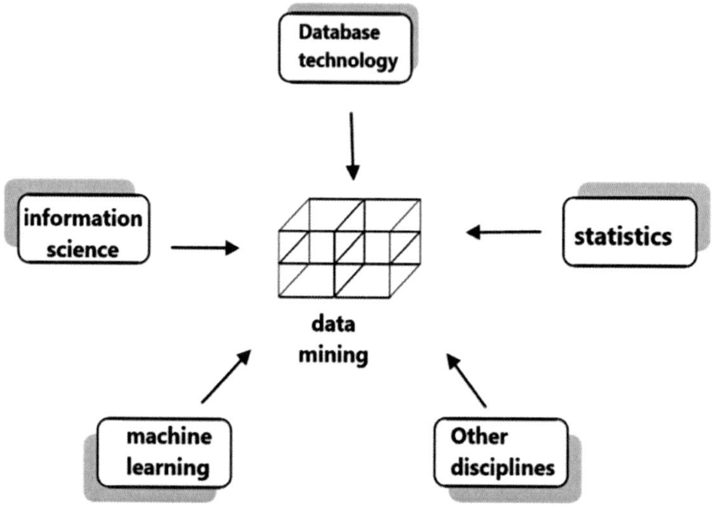

Fig. 2. Data mining is an interdisciplinary field

(3) According to application classification: different application fields generally have specific data mining tasks, such as finance, telecommunications, biological science, stock market, etc.

According to the general law of software development, this chapter starts from the system requirements analysis, analyzes the overall requirements of the system and the functional requirements of each module in-depth; on the basis of the requirements analysis, the system activity flow is analyzed through the use case diagram; after all the processes of each module are determined, the database is analyzed and designed; Finally, the multi-layer architecture model of J2EE is used to implement the system [4].

5 Conclusion

This paper summarizes the application and research progress of data mining technology in teaching quality evaluation system in detail, studies the design and implementation of teaching quality evaluation system based on multi-layer framework system integrating Flex technology and J2EE technology, and analyzes and discusses the influence of curriculum factors on teaching evaluation results. J2EE is an enterprise level platform based on standard. J2EE provides a platform independent, portable, multi-user and safe enterprise platform, which simplifies the development, management and deployment of system applications. The data persistence layer adopts hibernate framework, abandons the idea of database center, completely object-oriented programming idea, and realizes the encapsulation of database operation, object/relational database mapping, transaction management, caching mechanism and so on. Spring framework is adopted in business logic layer. Spring provides a method to manage objects, which can organize intermediate objects effectively, make object management clear, and help programmers form good habits of interface oriented programming.

References

1. Liu, J., Fang, W., Guo, D.: Report on 2012 annual report on talent cultivation quality of China's higher vocational education. Vocat. Techn. Educ. **21**, 2–12 (2012)
2. Lu, J.: Teaching evaluation system based on data mining. Master's thesis of Hohai University (2007)
3. Xie, C., Chen, Y., Wang, Y.: Application of Apriori algorithm in data mining of user behavior monitoring in AC V is. Acta Sin. Sin. **2010**, 50–54 (2010)
4. Wang, L.Z., Cao, C.: Decision forest data mining model based on generalized information theory. J. Jilin. Univ. **2010**, 89–92 (2010)

Data Mining Algorithm and the Role of Business Administration in Promoting Economic Development

Yanping Tan(✉)

Guangdong University of Science and Technology, Dongguan 523000, Guangdong, China

Abstract. In many business administration personalized recommendation technologies, as the most widely used and most successful recommendation strategy, it improves the performance of SVM algorithm in e-commerce data mining, and verifies that the application of data mining algorithm to business management research is feasible and has high accuracy. The combination of data mining and business management is based on the development of modern industrial production and business contacts, and constantly studies and formulates the actual situation, and discusses the positive role of business management in economic development. Strategy, business management plays a standardized role, can promote the healthy development of the market, also has an important role in promoting the overall development of the economy. Based on this, this paper concludes.

Keywords: SVM · Clustering · Data mining · Business administration · Economic development · Driving effect research

1 Key Technologies of Data Mining

According to the characteristics of the original data, the technology makes contributions to various fields. The following are the key technologies in data mining. Data mining is a new technology with broad prospects. Facing the huge and complex information in the information age, data mining is an effective means of processing. Data mining is not a monotonous technology. It covers a variety of modern technologies including computer, statistics and so on. When it is necessary to analyze and process complex data, various types of.

1.1 Application of Association Rules

Finally, we find the financial risk in management accounting by mining hierarchical data. According to the data mining process, firstly, the risk analysis model of management accounting is constructed, which can be divided into three levels: the first level is the root node of management accounting; the second level includes four modules of operation ability, profitability, solvency and growth ability; the third level is the detailed contents

B. J. Jansen et al. (Eds.): *International Conference on Cognitive based Information Processing and Applications (CIPA 2021)*, LNDECT 85, pp. 932–935, 2022.
https://doi.org/10.1007/978-981-16-5854-9_128

of each of the four levels, thus forming a tree hierarchy structure. In the analysis of financial risk in management accounting, we need to mine from the bottom data, and we can also mine at the same time [1]. By mining the bottom data, we can find the financial risk of public business management.

1.2 Association Rules

It is universal in practical application. Association rule mining is an important part of data mining. Its purpose is to discover the relationship between different elements in database. These association rules can help users find the behavior characteristics needed for decision-making. For example, when analyzing the shopping basket of customers in a store, we can get the information that "many customers will buy salad dressing at the same time when they buy bread". Therefore, the sales of bread and salad can be increased through discount promotion of bread. At present, the research on the application of association rules focuses on proposing different kinds of rules mining algorithms, improving the efficiency of existing algorithms and applying them to new fields [2]. In all kinds of fields, it is very important to find the relevant things and find their laws, which determines the association rules.

The big data correlation formula is as follows:

$$r = \frac{\sum_{i=1}^{n}(x_i - x)(y_i - y)}{\sqrt{(\sum_{i=1}^{n}((x_i - x)^2)(\sum_{i=1}^{n}(y_i - y)^2)}} \quad (1)$$

Note: R is the correlation coefficient of X and y, when r = 1, it means that there is a positive correlation between X and Y; when $r = -1$, it means that there is a negative correlation between X and Y; when r = 0, it means that there is no correlation between X and Y. Generally $-1 \leq r \leq 1$ indicates that there is a linear relationship between variables X and y.

1.3 Decision Tree

Decision tree is one of the most commonly used algorithms in data classification. Its representation is relatively simple, and the analysis objects are divided into different categories in a step-by-step manner. Building decisions.Trees can be simply made for three steps (Fig. 1).

1.4 Apriori TID Algorithm

The advantages of Apriori TID algorithm in searching frequent itemsets are as follows: transaction database d is used to calculate the support of candidate frequent itemsets only in the first scan, and candidate transaction database d generated in the last scan is used to calculate the support of candidate frequent itemsets in other scans. In this way, the number of scanning database will be reduced, and the candidate frequent itemsets can be reduced rapidly under certain circumstances. Even though it is optimized, some

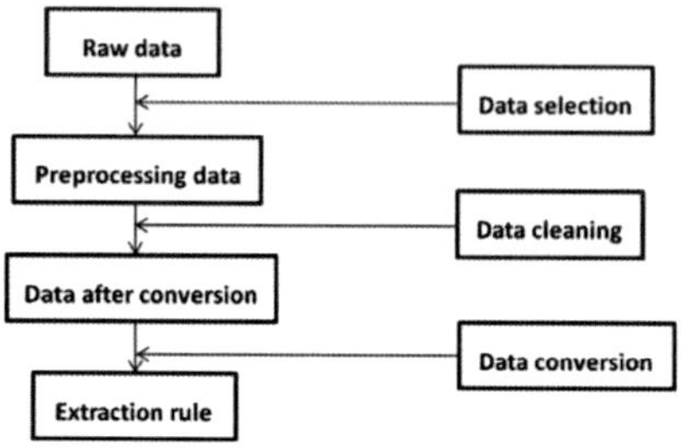

Fig. 1. Decision tree

inherent defects of AprioriTid method can not be overcome, which may produce a large number of candidate sets. When there are 10000 frequency sets with length I, the number of candidate sets with length 2 will exceed 10m. In addition, if you want to generate a very long rule, the number of intermediate elements to be generated is also huge. You may need to repeatedly scan the database to check a large candidate set through pattern matching. It is impossible to analyze rare information.

2 The Particularity of Business

Supervise the trading behavior in the market. On this basis, the behavior that does not conform to the standard should be dealt with in time. Compared with other management means, the particularity of industrial and commercial administration lies in its wide management scope, numerous management affairs and different application methods. In the scope of management, the most important function of business administration is to regulate the market and production. Market behavior is spontaneous. There are many participants in the market with different qualities, which will affect the overall stability of the market. For example, the most common ones include dishonest transactions, which can be restricted according to the requirements of the specification through the means of industrial and commercial administration. In the process of business administration, first of all, we should apply certain norms to restrict market behavior, and secondly.

3 The Role of Business Administration In Promoting Economic Development

3.1 The Mutual Promotion of Business Administration and Economic Development

Normative and healthy. Therefore, there is a fundamental development link between the two. The level of business administration is affected by the regional economic development. The basic idea of industrial and commercial management is to regulate the behavior of industry and commerce according to the relevant regulations of the month. It requires both parties to have a normative understanding and the spirit of buying a contract. The default content of the regulation is correct [3]. The generation of this idea needs the basis of industrial and commercial development in the region. The interaction between people is not due to the basic The community element is due to the commercial factors, in which the formulation of contract and the blue keeping become the important matters restricting this behavior. With this word, the first strong Xiang I rule will be gradually improved with the progress of regional economic development, and continue to play a role in economic development. This role needs to be carried out through the overall restriction of production and commercial activities. This overall way of temperature management can ensure that the festival field can operate in each festival.

3.2 The Normative Function of Business Administration on Production

In the process of business administration related work, the application of management strategy needs to be based on the specific management affairs and correctly selected. Quantity, stable market. First of all, in the production process, the relevant departments of industrial and commercial administration can apply industrial and commercial management laws to restrict and supervise production behaviors. In the current market demand, people need more kinds of materials for life and production. Therefore, the number of relevant production enterprises and producers is also unclear. In this case, the production environment is observed as a whole,. You can discover the complexity. The industrial and commercial administration department can supervise the production of the enterprise, and use the production requirement rules of standard temperature to estimate the production behavior of the enterprise [4]. Only when the standard is met can the production be continued. Secondly, industrial and commercial management has a certain regulatory effect on the production of materials in enterprises.

4 Conclusion

Based on the data mining algorithm, industrial and commercial management plays an important role in promoting the development of China's market economy. Industrial and commercial management can create a fair and fair competition environment for enterprises by strictly regulating market subject access system, strictly controlling marketing behavior, and improving various market systems. At the same time, it also provides a transparent consumption environment for consumers, effectively protecting the legitimate rights and interests of both enterprises and consumers, so as to promote the good development of market economy.

References

1. Wang, T.: Exploring the role of business administration in promoting economic development. Value Eng. **39**(3), 142–143 (2020)
2. Hao, X.: Research on the role of business administration in promoting economic development. Bohai RIM Econ. Outlook **2018**(3), 40 (2018)
3. Yu, H.: Research on the promotion effect of business administration on economic development. Shopping Mall. Modern. **2015**(28), 212–213 (2015)
4. Xing, C., Wan, W., Wang, X.: Improved algorithm for mining frequent itemsets in vertical data format. Comput. Eng. Sci. (07) (2017)

Research on Real Time Detection of Human Fitness Based on Artificial Intelligence Assisted Cognitive Computing

Chenqi Xiang(✉)

School of Public Administration,
Chongqing Vocational College of Transportation, Chongqing 402247, China

Abstract. In recent years, China's achievements in the field of artificial intelligence belong to the world. Artificial intelligence technology has penetrated into all walks of life. In particular, the deep integration of artificial intelligence and health management has made remarkable progress and achievements, improved people's quality of life, and provided more high-quality, customized and personalized health management services. Under this background, this paper puts forward the definition of personalized sports health management, analyzes the necessity of personalized sports health management service in the era of artificial intelligence, and discusses the design idea, system operation process and management stage design of "artificial intelligence + personalized sports management service".

Keywords: National fitness · National health · Physical health test · Fitness assessment · Fitness services · Exercise intervention prescription

1 Introduction

In the past 20 years, aging population, younger cancer and sub-health have become the focus of public health problems. The elderly have long been plagued by osteoporosis and cardiovascular and cerebrovascular diseases. Lymphoma, gastric pox and other cancer patients are more common in young adults, most of them are found in the middle and late stage, and the mortality is increasing year by year. The distribution of population health status in China shows that only 5% of the population is healthy and 80% is sub-healthy. Among them, about 300 million obese people are in sub-health status for a long time, leading to chronic diseases. In 2018, the number of chronic diseases in China has exceeded 2.6 billion, and the number of deaths caused by chronic diseases has accounted for 866% of the total deaths in China, and the burden of chronic diseases has accounted for nearly 70% of the total disease burden [1]. Chronic disease has become a health crisis that can not be ignored. These problems have attracted the attention of our country. The State Council issued the "healthy China 2030° planning outline", which proposed that people need to pay attention to their own health, emphasize prevention rather than treatment, and advocate self-health management. Practice has proved that

B. J. Jansen et al. (Eds.): *International Conference on Cognitive based Information Processing and Applications (CIPA 2021)*, LNDECT 85, pp. 936–940, 2022.
https://doi.org/10.1007/978-981-16-5854-9_129

the use of sports behavior intervention and control can effectively reduce the proportion of sexually transmitted diseases, promote the physical and mental health of the whole people, prolong life and improve the quality of life. Therefore, sports health management service is also an indispensable social service in the future.

2 Definition of Individualized Sports Health Management

In 2003, foreign scholars began to study personalized health management, and then steadily increased, which has been in the exploratory stage. At present, 88 countries or regions with the United States and the Netherlands as the core participate in the field of personalized health management, involving the research fields of medicine, computer science and sociology, The research content mainly includes disease health management, mental health management, physical health management, lifestyle management, catastrophic disease and injury management, disability management and so on. The population involved includes young people, middle-aged people, elderly people, women and disabled people. Considering the application of Internet and big data, the future research focus is more on the realization of personalized health services and electronic health records. In 2010, the Vice Minister of the Ministry of health proposed to implement the "preventive health project", build a preventive health care service system, and improve the service mode and service specification. Therefore, personalized health management in China is challenging and has great research space, and we need to increase human and material resources investment in this service field.

From the definition of personalized health management, scholars at home and abroad have no accurate definition [2]. Wu Zhijie (2014) introduced the occupational skill training textbook "health manager" published by the Ministry of health in 2013 to define health management, which is guided by the concept of modern health and the thought of "preventive treatment of disease" of traditional Chinese medicine, and uses the theories, techniques and methods of medical management and other related disciplines to comprehensively and continuously detect, evaluate and intervene the health status of individuals or groups and the risk factors affecting health, so as to achieve the goal of health management To promote the health of all as the goal of the new medical service process. Cai Feng (2009) proposed that the definition of health management in the United States is a process of comprehensive monitoring, analysis and evaluation of individual and group health, provision of health consultation and guidance, and intervention of health risk factors. Gu Dongxiao (2018) proposed that personalized health management is a process of comprehensive assessment, analysis and management of residents' health status, which aims to improve their health status through effective personalized health guidance and behavior intervention.

3 The Necessity of Personalized Sports Health Management Service in The Era of Artificial Intelligence

3.1 Intelligent Products Provide Scientific Data For Personalized Sports Health Management Services

In recent years, China's artificial intelligence industry has developed rapidly. The artificial intelligence market or the company's "blowout growth" is estimated that the scale of

China's artificial intelligence industry will reach 335.6 billion yuan in 2022. Intelligent technology has been applied to nearly 20 industries, such as shopping malls, hotels, banks, museums, hospitals, teaching and so on. In recent years, it will be applied in various fields of society. At present, wearable intelligent devices such as smart bracelet, smart sports shoes, smart glasses and smart shirt in the sports fitness industry have become indispensable sports products for residents' fitness. It can monitor the health data of users in real time and feed back the sports effect in time, so that residents can accurately understand their own health status. Wuxi City uses the intelligent physique test system to complete the students' physique test files, sports risk prediction, scientific fitness new method training and fitness effect monitoring and other technical means, so as to provide the basis and service for the prevention, intervention and improvement of students' health. There are also keep, hot fitness, fittime and wake in the field of yoga for vertical fitness. App software is used to monitor the speed, heat and step frequency in the process of fitness exercise, and make the most appropriate fitness management service scheme for exercise intensity in time [3]. These intelligent products will inevitably lead the sports health management service to personalized, and provide convenient conditions to meet the individual needs. The maximum entropy model 15 is based on the concept of entropy. The definition of entropy comes from the thermodynamics of physics, which is used to describe the uncertainty of matter in molecular state. There are several types of Di, the most commonly used one is Shannon's information entropy, and di mentioned in artificial intelligence algorithm is also "Shannon's information entropy", as shown in Eq. (1).

$$P_{ij} = \frac{\tau_{ij}(k-1)}{\sum\limits_{l \in T} \tau_{ij}(k-1)}, j \in T \tag{1}$$

$$L(s) = L(s) \cup \{j\}, i := j \tag{2}$$

3.2 Intelligent Stadiums Provide Efficient and Convenient Sports Environment For Personalized Sports Health Management Services

Smart stadiums and gymnasiums, which are the perfect combination of intelligent technology and sports industry, provide personalized sports environment for sports health management service. For example, in September 2016, Foshan launched an intelligent basketball court, which is a business model of all self-help without any management personnel. In the past two years, self-help fitness optical pigsty king, super orangutan, etc. combined with app booking venue courses, give users a more perfect intelligent experience, and let users experience more freely in member registration, class appointment, payment and other links. In 2018, beiti group and Zhimei sports jointly launched the online and offline intelligent stadium sports health platform. The first batch of intelligent standardized stadiums of "thousand stadiums" plan will be invested in Beijing, Zhejiang, Shanxi and other provinces and cities. In April 2019, the newly developed intelligent unattended venues in Hangzhou will be put into operation.

3.3 Sports Health Manager is a Talent Reserve to Adapt to the Personalized Sports Health Management Service in the Era of Artificial Intelligence

In recent years, with the increase of hypertension, obesity, diabetes and other infectious diseases in China, the government began to pay attention to this problem, and proposed that in the future, all medical institutions, schools, communities, restaurants, food and health products production or sales institutions must be equipped with health managers. According to the statistics of the National Health China 2030 planning outline, the number of people enjoying health management services in China only accounts for 2/10000 of the total number. According to the relevant international standards, the demand for health managers in China is more than 10 million. It can be imagined that sports health managers are even fewer, but they are in urgent need of service personnel in the future. From the requirements of the development of our times, sports health manager should be a new occupation in the era of artificial intelligence. In the near future, intelligent customer service (shopping guide, medical guide), intelligent medical diagnosis, intelligent teachers, intelligent logistics, intelligent financial system are expected to widely appear in our lives. It is the need of the development of artificial intelligence society and is moving towards the direction of scientific and personalized sports health guidance. With the rise of artificial intelligence revolution, the mode of human production is about to have a revolutionary change, and many traditional occupations are about to disappear [4]. For example, teachers, doctors, journalists, lawyers, designers and so on will also be deeply influenced by AI, and some work functions will also be replaced by AI.

4 Concluding Remarks

In terms of classification service, the artificial intelligence algorithm analyzes the health status of the population through cloud service data, such as health, sub-health and disease, and then subdivides the health population into three health status, namely children health care, maternal health care and elderly health care. With the help of artificial intelligence equipment or cloud data technology, personalized health guidance scheme can be formulated according to sports health report. For sub-health or sick people, the combination of medical and physical education is implemented, and personalized sports health guidance plan is formulated through online guidance, offline medical and physical therapy prescription. After the comprehensive evaluation of the health status of users or residents, we enter into the implementation process of intelligent personalized sports health intervention, and finally achieve the goal of sports health promotion and physical and mental health management, so as to complete the operation of personalized sports management service system in the era of human intelligence.

References

1. Tang, L., Ying, C., Tang, L.: Review of sports health promotion research in China. J. Tianjin Inst. Phys. Educ. **25**(3), 201–205 (2010)
2. Liang, L.: Research on the collaborative development mode of community health care and sports health industry under the background of sports medicine integration. Econ. Res. Guide. **2017**(30), 54–55 (2017)

3. Chen, W., Yang, J.: Building personalized health management system for enterprise knowledge workers. China. Human. Resour. Dev. **17**(9), 35–40 (2010)
4. Wu, Z., Guo, Q.: Research on the development strategy of China's health management industry in the era of big data. Health. Econ. Res. **326**(6), 14–16 (2014)

Application Analysis of Cognitive Internet of Things Technology on Permeability in Mining

Jing Meng(✉)

Oil Production Plant of Daqing Oilfield Company Ltd.,
Geological brigade in No.7, Daqing 163458, China
mengjsy@petrochina.com.cn

Abstract. Computer simulation technology is a comprehensive technology, which realizes the simulation of the research object through the effective combination of computer software and control system and display system. In the process of coal mine production, the efficiency and safety of mining has always been a problem that coal mining enterprises pay more attention to. Through the application of computer simulation system in the process of coal mining, this paper mainly analyzes the simulation efficiency of different mining methods, and simulates the underground operation, so as to find the potential safety hazards and avoid the safety problems of mining. This paper focuses on the application of computer simulation system in the process of coal mining.

Keywords: Computer · Simulation system · Coal mine · Mining

1 Introduction

With the continuous development of science and technology, the combination and application of computer technology and all walks of life are gradually promoted. In software technology, the formation of simulation system realizes the simulation of real scenes for many industries. In the process of coal mining, coal mining equipment and construction personnel are generally working underground. Coal mine production needs to achieve high efficiency and high output, and the most important one is safety production. Coal mining is the most important production link in coal industry [1]. The computer simulation system is used to comprehensively analyze the underground parameters and detection results; through the simulation software system, the underground environment is simulated, so as to realize the simulation of the real mining scene.

2 Analysis of Computer Simulation System

With the rapid development of computer technology, it is one of the most widely used technologies in the world. For the development direction of computer technology, generally can be divided into software and hardware two main directions. The hardware direction is mainly single chip microcomputer, and the application of software is very broad. Application software in different fields has different practical functions, which brings great convenience to human production and life.

B. J. Jansen et al. (Eds.): *International Conference on Cognitive based Information Processing and Applications (CIPA 2021)*, LNDECT 85, pp. 941–945, 2022.
https://doi.org/10.1007/978-981-16-5854-9_130

2.1 Understanding of Computer Simulation Technology

With the continuous improvement of people's requirements for computer technology, some industries began to put forward the demand for imitation. The so-called simulation is to set the parameters of the actual object through computer software technology, and realize the simulation and Simulation of the actual object through computer imaging and software analysis technology.

The meaning of computer simulation.Computer simulation is the abstract simulation of a certain level of a specific object through computer software technology and three-dimensional imaging technology. Through the computer simulation technology, researchers can dynamically imitate the actual situation, and use the theory and method of model experiment to get the required information from the simulated object, so as to help researchers find solutions to the problems they study. From the technical point of view, simulation is a relative concept and theory. Simulation is an infinite close to the real mode, but it can never be completely consistent with the real situation. Therefore, the computer simulation technology is actually an infinite approach to the research object.

Realization and composition of computer simulation system.In the actual process of computer simulation, we need to use a number of computer technology to achieve efficient simulation [2]. The system composed of many computer technologies is called computer simulation system. Among them, an effective computer simulation system must go through several steps, such as the establishment of the model, the data comparison of the simulation experiment, the processing of the simulation data and the later analysis and verification, in order to realize the basic functions of the computer simulation system. In order to build a more powerful computer simulation system, it also needs the assistance of control system, display system and other technologies.

2.2 Advantages of Computer Simulation Technology

The advantage of computer simulation technology is that it can use the way of dynamic experiment to analyze and verify the data in the later stage, and get the data closest to the actual research object, so as to realize the "real" presentation of the research object. The research object can be truly presented, and the results of different situations can be presented by changing the relevant data and parameters. In general, highly accurate computer simulation technology, It can realize the effect of cost saving and safety production, and ensure the project to be completed on time and with quality to the maximum extent.

Binary exponential backward algorithm

This paper describes the binary exponential back-off algorithm (beb), which is an algorithm model of binary exponential relationship formed by the number of retransmissions of conflicting terminals. That is, with the increase of the number of retransmissions of conflicting terminals, the value range of backoff delay increases by 2. The algorithm idea is as follows: when the system is in conflict, The conflicting terminal must randomly select a value within the time slot window provided by the algorithm each time.

$$\begin{cases} \varsigma = random \\ t_\varsigma = \varsigma \cdot \tau \end{cases} \tag{1}$$

In the improved algorithm, the setting of parameter 3 is obtained through analysis and computer simulation comparison. Because the terminal conflict in the network is generally very small, setting the parameter to 3 after comparison in the above table can give full play to the advantages of the improved algorithm. Computer simulation experiments show that the improved algorithm of binary exponential backward decomposition can significantly reduce the total delay required to decompose the conflict, The average data of binary exponential backoff algorithm and improved binary exponential backoff algorithm obtained from 100 computer simulation experiments on s conflict terminals. In addition, considering different arrival rates (g), the calculation formula of throughput (s) is

$$S = \frac{Ge^G}{1 + G + \sum_{k=1}^{\infty} E(L_k)} \tag{2}$$

Where $E(L_K)$ represents the average maximum time slot value required for successful conflict resolution in K conflict terminals. The throughput characteristic curves of the basic algorithm and the improved algorithm are shown in Fig. 1.

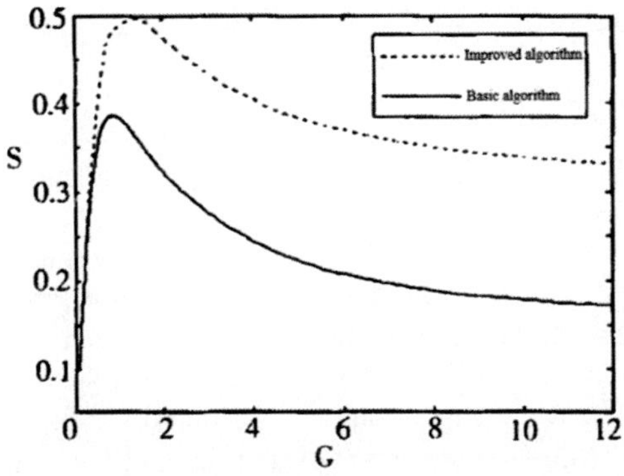

Fig. 1. Discharge curve of the system.

As can be seen from Fig. 1, with the increase of load (g), compared with the basic algorithm, the system throughput of the improved algorithm tends to be stable, and its value is about 0.33, which is significantly higher than the stable value of 0.18 of the basic algorithm. This shows that the improved algorithm proposed in this paper is stable and effective.

3 Analysis of Coal Mine Production and Mining

Coal mining industry is not only the main energy industry in China, but also one of the pillar industries of China's energy industry. With the continuous improvement of people's living standards, people's demand for energy is also increasing. For coal industry, high efficiency and high yield are its main development goals. From the analysis of coal mine construction site, coal mine production links are mainly coal mining and coal mine transportation. Among them, coal mining plays an important role. Therefore, coal mining operation is the key link to achieve high yield and high efficiency. In addition, the most important idea of coal mine production is to achieve safe production. Compared with other industries, coal mine industry is a high-risk industry [3]. As coal mining generally adopts underground operation, it has a certain threat to the life safety of coal miners. Therefore, for safety production, we should pay close attention to the mining link to ensure the safety of coal mining.

4 Application of Computer Simulation System in Coal Mining

Through the analysis of computer simulation technology and simulation system, it is clear that the computer simulation system can realize the simulation of the research object, and simulate the possible phenomena and results in different situations. The way of dynamic verification makes the computer simulation technology have a great application in the mining production of coal mine.

4.1 Computer Simulation System for Increasing Coal Mining Output and Efficiency

The computer simulation system is also very important for the simulation effect in the process of coal mining. For the simulation system, the geographical environment of the mining area can be simulated. For example, first of all, the soil and geographical environment of the mining site are detected to understand the soil distribution and geographical environment. For the form of mining mine, simulation analysis should also be carried out. Efficient production and efficient mining is based on whether the natural environment of mining can be simulated through computer simulation technology and simulation system, so as to get close to the actual situation of the mine to the maximum extent. Then, through different mining methods to simulate the mining efficiency, using two different methods, and the results of comparative analysis, get the most suitable for efficient mining method. In fact, it can also simulate many factors, such as personnel allocation and the use of mining tools, so as to clarify the mining strategy.

4.2 Application of Computer Simulation System to Improve Coal Mining Safety

In the process of coal mine production, mining is the most important production link. Because the working environment of the coal mine can not really understand and master, it is impossible to predict what kind of accidents will be encountered in the mining process, resulting in mining operations with high risk. The computer simulation system can simulate the actual environment through the data of underground soil and stress points [4]. During the stage mining, the detected data can be verified one by one through the computer simulation system. When mining for a period of time, the construction personnel need to carry out manual mining. So, we still need to use the computer simulation system to simulate the position and density of personnel work, so as to ensure the safety of personnel work. In addition, the simulation and Simulation of mine disaster can also be carried out. Through positioning and Simulation of the underground position, once a collapse event occurs, the computer simulation system can simulate the collapse site, and judge the geographical location of the construction personnel waiting for rescue, so as to greatly improve the safety in the process of coal mining.

5 Conclusion

In the process of coal mining, is the most prone to security risks, mining machinery and equipment and construction personnel need to work under the unknown mine, so in terms of mining efficiency and mining safety, can not be effectively guaranteed. The use of computer simulation technology, can simulate the unknown environment, so that when the environment changes, simulation and prediction. For coal mining, the computer simulation system is of great help. In the unknown geographical environment, through exploration and measurement, the data and parameters are verified and processed. By using the simulation system, the real reproduction of underground operation can be realized, and the targeted mining work can be carried out according to the simulation data. So as to maximize production efficiency and ensure production safety.

References

1. Song, B.: Gas seepage law and determination method of high level drilling horizon in sequential mining of large mining height working face. China University of Mining and Technology (2020)
2. Zhang, L.X.: Study on microscopic characteristics of gas hydrate phase transition and gas displacement mechanism. Dalian University of Technology (2019)
3. Dong, M.: Driving conditions of crude oil in tight low permeability reservoir and its influence on production effect. China University of Petroleum (Beijing) (2019)
4. Zhang, Z.: Theory and application of fluid induced fracturing and permeability increasing coal and rock mass. Chongqing University (2019)

Monitoring Method of Hydraulic Long Auger Drilling Process Based on Cognitive Interaction

Haifeng Wu(✉)

Zhangjiajie Institute of Aeronautical Engineering, Zhangjiajie, China

Abstract. Long screw drilling machine is a kind of pile machinery products. It is widely used because of its low cost, wide application range, good quality of pile foundation, high bearing capacity of single pile, less noise, no mud pollution, less steel consumption and other advantages. However, there is still a problem of low drilling efficiency in the long screw drill, which affects the efficiency of drilling and piling and the performance of the whole machine. In order to improve the drilling efficiency, this paper makes a simulation study on the drilling efficiency of the long screw drill bit, and designs a twist bit with high drilling efficiency.

Keywords: Drilling efficiency · Mechanical analysis · Twist bit · Drilling simulation · Structure optimization

1 Introduction

In recent years, with the development of various pile forming technology, all kinds of piling equipment are constantly upgrading. From the original hand hammer to the free falling hammer, and then to the steam driven, diesel driven, compressed air driven, hydraulic driven and motor driven pile driver. Typical products include diesel hammer pile, vibrating hammer pile, hydraulic grab, three-axis drill, rotary drill, static pile driver, pile frame, rotary drill and long screw drill. With the development of construction technology of cast-in-place pile in our country, the long screw drilling machine is developed [1]. It is the main type of bored pile construction machinery, because it has the advantages of low equipment cost, wide applicability, good pile quality, high single pile bearing capacity, less noise, no mud pollution, less steel consumption and so on, it is widely used. The long auger drill first appeared before World War II, and it came out in Calvert company of the United States. After World War II, it developed in Europe: in 1948, it was first developed by Maite company of Italy, and then developed in Italy and Germany. In the 1970s and 1980s, it developed rapidly in Japan. At that time, Japan called the long screw drill rotary bucket pile. In the early 1980s, China imported this kind of working device from Japan and installed it on kh-125 crawler crane; in 1984, Tianjin exploration machinery factory imported the long spiral cobalt machine from RD company of the United States and digested and absorbed it.

B. J. Jansen et al. (Eds.): *International Conference on Cognitive based Information Processing and Applications (CIPA 2021)*, LNDECT 85, pp. 946–950, 2022.
https://doi.org/10.1007/978-981-16-5854-9_131

2 Bit design and Mechanical Analysis of Long Auger Drill

As the core part of the long screw drill, the bit of the long screw drill directly affects the drilling efficiency of the long screw drill. In view of this problem, this chapter analyzes the main factors that affect the drilling efficiency, designs a twist bit of long screw drill which is conducive to cutting soil, and makes mechanical analysis on the drilling process of the bit [2].

2.1 Size Design of Twist Bit

The main parameters of twist bit are helix angle, vertex angle, main deflection angle, rake angle, secondary deflection angle, secondary back angle and transverse blade length.

Helix angle β, which mainly reflects the chip removal of bit. The larger the helix angle is, the larger the feed rake angle of the bit is, and the sharper the bit is, which is also conducive to chip removal. However, if the helix angle is too large, the strength and heat dissipation condition of the bit will be weakened, and the wear of the bit will be intensified. Because the length of the drill tip is short, the helix angle is designed to process two main cutting edges. There are not many functions in chip removal. Therefore, the helix angle can be set in size optimization.

The calculation formula of helix angle is as follows:

$$\tan \beta = \pi d_0 / p_h \tag{1}$$

At the same time, the best main deflection angle of the blade is found. The calculation formula of main deflection angle is as follows:

$$\tan k_r = \tan \varphi \cos \lambda_{stm} \tag{2}$$

After the helix angle is given, the optimal rake angle of the blade is found while the vertex angle is optimized. The formula for calculating the position of any point m at the rake angle of the main cutting edge is as follows:

$$\tan \gamma_{om} = (\tan \beta_m / \sin k_{rm}) + \tan \lambda_{stm} \cos k_{rm} \tag{3}$$

In the process of drilling, the torque balance of the long screw drill is as follows:

$$T = M_1 + M_2 + M_3 \tag{4}$$

By analyzing the twist bit of long auger, we can see that it combines the advantages of twist drilling, and uses the cutting ability of the cutting edge of twist drill to break up the soil near the cutter teeth. At the same time, the cutter teeth carry out the second cutting to break up the soil. Therefore, the cutter teeth are mainly subject to two parts of force in the process of drilling and squeezing soil, one part is the loose soil after the twist drilling, and the other part is the soil that has not been drilled.

2.2 Summary of this Chapter

This paper expounds the process of long spiral drilling, analyzes the main factors affecting the drilling efficiency, and sums up the necessary conditions for the design of a new drill bit: there is a large enough area s between the soil and the drill tip to produce extrusion contact force, which can overcome or reduce the drilling deflection force to the greatest extent; more cutter teeth can cut the soil in the same time, and the soil after drilling can be discharged into the blade in time to bring out the hole; The blade has high strength and blade root stiffness [3].

The twist bit of long auger drill is designed, which is a long auger bit with twist drill tip structure. It mainly uses the accurate centering of twist drill without deflection or hole channeling, and takes the twist drill bit with twist drill structure as the long auger drill tip. Therefore, the drill point has the cutting ability to cut the soil at one time, which is equivalent to increasing the number of cutting teeth of the drill bit; at the same time, the twist drill point will squeeze the cut soil into the spiral guide groove, so that the soil and the drill point have a large enough area s to have extrusion contact, so as to generate anti deflection force, so that the drill bit will not deflect or channeling; Secondly, the two auxiliary cutting edges of twist drill also assist in centering when drilling soil, so as to make the bit centering accurate.

3 Drilling Process Simulation and Efficiency Analysis of Long Auger bit

This paper mainly designs a twist drill bit for long auger drill, and analyzes its mechanics. This chapter focuses on the drilling simulation of the long screw drill, and uses ANSYS/s-dyna software to simulate the drilling process of two kinds of long screw drill. The drilling process and soil deformation of long auger are obtained by simulation, and the drilling efficiency of twist bit and traditional centering long auger bit is compared by simulation.

3.1 Constitutive Model of Soil

The constitutive relation is the mathematical expression of the mechanical properties of the material. It is mainly composed of liquid, gas and solid mixture. In addition, it is affected by the stress history, so its mechanical properties become more complex. Firstly, the deformation of soil has obvious nonlinear characteristics, and its stress-strain curve shows nonlinear characteristics from the beginning of bearing load. Secondly, the stress path, stress history, consolidation degree and principal stress all have great influence on the deformation of soil. In addition, the soil has some special properties such as dilatancy and anisotropy. It is quite different from metal and other materials. The research on soil model has a history of several decades. Scholars have also developed more than a dozen soil models. However, due to the great difference of soil properties, it is difficult to summarize all the properties by one model. For the occasions with high requirements, the test models that are more suitable for specific areas, projects and soil properties are used.

3.2 Finite Element Analysis of Drilling

In the process of drilling, the depth changes all the time, and there is a certain initial stress in the soil before drilling. After drilling out, not only the geometric dimension of the drilling object changes, but also the stress and deformation of the remaining structure changes. Therefore, the stress and deformation in the finite element analysis of drilling process is special compared with the gravity and deformation in the finite element analysis of general structure. The key problem is how to use the finite element to simulate the characteristics of different stages in the drilling process, so that the stress on the drilling surface can be completely relieved and become a stress free surface. It is an effective method to analyze the character of drilling hole with finite element method, but it involves many factors, such as the stress-strain character of soil after unloading, the selection of model and the determination of parameters, the consideration of residual stress in the soil below the cutting surface after unloading [4]. This chapter mainly uses Lagrange display algorithm. In the Lagrange display algorithm grid, with the cutting process, the soil element deforms under the drilling brake of the bit. The reason for the deformation is the extrusion failure of the cutter teeth on the soil element. The deformed element extrudes another soil element, forming a continuous deformation process. Although the Lagrange algorithm can not obtain the continuous process of soil cutting debris like ALE algorithm, it is completely consistent with the failure mechanism and the stress and deformation in the finite element analysis of drilling process.

3.3 Drilling Result Analysis and Efficiency Comparison

The analysis of drilling results is mainly divided into two parts: first, the dynamic characteristics of bit are analyzed by studying the change of contact force, so as to prepare for the mechanical analysis in Chapter 4, and also provide theoretical basis for further research in the future; the efficiency of two kinds of long spiral bit is compared by comparing the simulated cobalt head after drilling for lmin. There are two kinds of comparison methods: one is to compare the drilling efficiency of twist bit and traditional centering long screw bit by studying the total deformation of soil in the drilling process of two kinds of long screw bits and comparing the total deformation of soil in the same time; The other is to compare the drilling efficiency by comparing the vertical displacement of twist bit and traditional centering long screw bit.

4 Conclusion

Long auger drill is widely used because of its low cost, wide applicability, good quality of pile foundation, high bearing capacity of single pile, less noise, no mud pollution and less steel consumption. However, there are still a series of problems with the increasingly mature long screw drill, such as the low drilling efficiency of the long screw drill, the deviation or channeling of the drill, the excessive deformation of the drill bit, the fracture of the root of the cutter teeth, etc. The problem that affects the market prospect of long auger bit and has no competitive advantage compared with other kinds of piling machinery products is that the drilling efficiency of long auger bit is not high. The existence of this problem seriously affects the drilling speed of the long screw drill. In order to improve the drilling efficiency and the performance of the drilling machine, this paper designs a twist bit of the long screw drilling machine, which improves the drilling efficiency.

Acknowledgement. Hunan Natural Science Foundation Science and Education Joint Project, 2019 JJ70037, Monitoring Method for Drilling Process of Hydraulic Long Spiral Rig.

References

1. Zhang, L.: Status and development of foreign automatic drilling equipment. Foreign Petrol. Mach. **7**(1), 1–8 (1996)
2. Wang, Y.: Calculation of main parameters of screw drill. Road Constr. Mach. Constr. Mech. **15**(6), 8–10 (1998)
3. Yang, G.: Fault Diagnosis and Elimination of Rock Drill, Drilling Rig, Pile DriveRoad header and Crusher. China Machine Press, Beijing (2009)
4. Li, S., Liu, X., Yang, J.: Construction technology of bored pile. Geol. Publ. House **1990**, 102–108 (1990)

Application of Cognitive Internet of Things in Well Test Data Stimulation Measures

LiLi Zhang(✉)

Geological brigade in No.7 Oil Production Plant
of Daqing Oilfield Company Ltd., Daqing 163458, China
zhang_lili@petrochina.com.cn

Abstract. In order to slow down the decline of crude oil production, it is necessary to continuously implement stimulation measures for oil wells when the production of oil and gas fields enters the middle and late stage of development. Choosing the right time to carry out measures is one of the important means to control costs and improve benefits. If the cost of measures is regarded as an investment, the investment recovery is the basis to judge whether the measures are feasible. This paper analyzes the cost and benefit of stimulation measures from the perspective of economics, and gives the best time to implement stimulation measures and the model of end point production. The results show that the best time of stimulation is when the income and cost of oil well are equal. The end point production model can be used to judge whether the stimulation has economic value, so as to provide a theoretical basis for single well stimulation decision.

Keywords: Single well · Stimulation measure · Measure production decision model

1 Introduction

Most of the oil field enterprises in the later stage of development in China are facing the pressure of maintaining production and controlling cost. According to the general law, during the whole economic life of oil wells, the oil production decreases year by year and the water content increases continuously until there is no economic value. In order to slow down the decline of crude oil production and control the rise of water content, all oilfield enterprises take various measures to increase production, but the cost of the measures is a large investment. Therefore, choosing the right time to carry out the measures is one of the important means to control the cost and improve the efficiency [1]. For an oil well, it is impossible to implement stimulation measures to increase its production indefinitely. If the cost of stimulation measures is regarded as an investment, investment recovery is the basis to judge whether stimulation measures are feasible. In the past, the research on single well stimulation measures only focused on judging whether there is economic production. From the perspective of economics, the author makes a comprehensive and systematic study on the timing and end point of

B. J. Jansen et al. (Eds.): *International Conference on Cognitive based Information Processing and Applications (CIPA 2021)*, LNDECT 85, pp. 951–955, 2022.
https://doi.org/10.1007/978-981-16-5854-9_132

single well stimulation measures, and puts forward a decision-making model of single well stimulation measures, so as to provide a theoretical basis for decision-makers to make decisions on single well stimulation measures.

2 Application of Computer Technology Under the Background of Big Data

In the application of computer technology, people are more and more aware of computer technology and the potential of computing technology. Under this background, many enterprises begin to attach importance to the research and development of computer software, and at the same time, they also put the well researched computer technology into practical use. Today, computer applications are more and more widely used, More and more people pay more attention to the innovation of computer technology.

2.1 Application Scope of Computer Technology in Big Data

First of all, computer technology in the business field, with the emergence of computer technology, the whole industry management has brought a comprehensive update, which has played a good role in improving the economic benefits of enterprises. At this stage, the industry effectively improves its management level by using computer technology, such as the support of RP software, which can effectively improve the level of management. With the support of computer technology, the enterprise data processing does not use manual operation, which can effectively reduce errors, make data processing more efficient and more accurate. With the support of computer software technology, the speed of data processing in the enterprise is gradually strengthened, and the pace of modern development is gradually improved [2]. The analysis of all kinds of data of the content is enough to grasp the needs of customers more effectively, so that they have higher market competitiveness and achieve more long-term development. Secondly, in the field of teaching, computer technology also has a very important significance. In the comprehensive innovation of enterprise management, computer technology will also effectively improve the level of teaching practice. Digital teaching method makes it more interesting, students' interest in learning has been effectively improved, and students' active learning builds a more efficient platform.

2.2 Problems in the Application of Computer Software Technology in Big Data Environment

Considering the domestic development situation, big data has achieved higher results in a short period of time, and the social support is getting higher and higher. Based on this background, the computer software technology must be reformed, aiming at the effective improvement of technology, in order to have more development opportunities. The background of big data plays an endless role in the society, which makes the transparency of data information become more prominent through the support of computer technology, and the data has higher authenticity. At the same time, accurate and real information and data can be used to judge the future development of the industry, formulate good development plans, and achieve the goal of sustainability.

3 End Point Analysis of Oil Well Stimulation Measures

3.1 The Marginal Production of Increasing Production Measures is a Decreasing Law

According to the natural law of oil and gas exploitation, thick pay zones are put into development in the early stage of oil and gas development. These pay zones have good permeability and high oil saturation. With the increase of water cut and the deepening of development degree, only some low permeability, low oil saturation, high water cut, low grade and some oil increasing effect per unit well can be used to release production capacity [3]. For the water injection oilfield, if the liquid extraction measures are adopted, the amount of oil added by each liquid extraction will also decrease with the increase of water cut. Therefore, the increased production per unit well is bound to show a decreasing trend.

3.2 Increasing Law of Marginal Cost of Increasing Production Measures

With the development of oil well operation technology and the maturity and matching of technology, the operation cost per unit well should be on the decline, but according to the relevant statistical data, the cost has an increasing trend. With the deepening of oilfield development, the difficulty of operation will also increase, so that the operation cost has an upward trend. From another point of view, because of the law of diminishing marginal production, the cost of measures for increasing 1 ton of crude oil is bound to rise.

3.3 Incremental Law of Marginal Cost of Oil Increase by Measures

In addition to the direct cost of measures, the marginal input of measures to increase oil production also includes the direct consumption of the measures to increase oil production, such as raw materials, gathering and transportation treatment, water and gas injection, etc., that is, the marginal cost of measures is the sum of the measure cost per ton of oil and the variable cost per ton of oil.

With the increase of water content, the marginal cost and the variable cost per ton of oil increase, so the marginal cost of oil increase is bound to increase. Under the above rule, the application of oil well stimulation measures is not unlimited [4]. With the decrease of marginal production and the increase of marginal cost of stimulation measures, after reaching a certain extent, the marginal benefit of the measures will be less than the marginal cost, At this time, it is uneconomic to implement stimulation measures again, and the decision-maker should stop the operation of stimulation measures for the well.

An oil well can continuously implement measures for many times, but when a certain oil well can not obtain economic benefits by implementing stimulation measures, it can be concluded that this oil well is the economic limit point, as shown in Fig. 1.

When a is in the balance of input and output, the output can be increased to B. if the measure is effective, the input-output equilibrium point C can be obtained. After C, the D can be obtained, which is the income that can be obtained by input. That is to say, at this point, the economic limit measure model can be obtained.

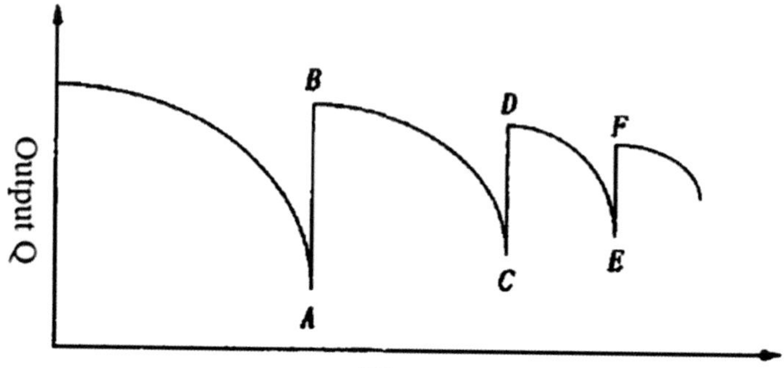

Fig. 1. Production curve of measure well

4 Calculation Method of the Limit of Stimulation Measures

(1) If the oil increment of the measures is ΔQ, the cost of the measures is regarded as reinvestment L, the variable cost of the oil increment after the measures is Δ CV, and the tax is Δ T_X, then

$$\Delta C_v = \Delta Q[A_u + B_u F/(1 - F)] \tag{1}$$

$$\Delta Tx = \Delta QKTx \tag{2}$$

Where, C_1 is the oil and gas treatment cost per ton of liquid production; C_2 is the water injection cost per cubic meter of water; C_3 is the paid use cost of reserves per ton of oil extraction; m is the injection production ratio; B is the crude oil volume conversion coefficient; β is the crude oil volume coefficient; γ is the relative density of crude oil; is the target profit rate,%; is the crude oil price, yuan; τ x is the tax, yuan; factory is the comprehensive water cut of oil well; K is the crude oil commodity rate.

5 Conclusion and Suggestion

(1) The best time to increase production is when the benefit and cost of oil well are equal.
(2) The model of measure end-point yield is established, which can be used to judge whether the measures of increasing production have economic value.
(3) In the actual production process, oil production enterprises can use relevant software to monitor the cost and income of oil wells at any time according to the production data, and make decisions on whether to implement the production increase measures and when to stop using the production increase measures, so as to overcome the blindness of decision-making.

References

1. Yue, L., Yue, D.: Economic evaluation method of recoverable reserves and stimulation measures in high water cut stage of old oilfield. Acta Petrologica Sinica **21**(5), 39–44 (2000)
2. Ji, C.: Discussion on the development and application of computer software technology in the era of big data. Sci. Technol. Innov. (24), 158–159 (2018)
3. Cai, J.: Application of computer software technology in the era of big data. Electron. Technol. Softw. Eng. (07), 50–51 (2019)
4. Li, B., Zhang, G., Liu, W., et al.: Study on the method of annual oil production allocation. Acta Petrologica Sinica **22**(2) (2001)

Analysis of Big Data Computing Logistics Supply Chain Management Mode Inspired by Cognition

Yinghuan Liu(✉)

Software Engineering Institute of Guangzhou, Guangzhou, Guangdong, China

Abstract. At present, the research of supply chain management is more and more inclined to the research of the overall method, and the research methods also emerge in an endless stream. Among them, operational research method is the most widely used in the research of supply chain management. However, in the research process of operations research, some nonlinear behaviors in the supply chain are often ignored, resulting in some conclusions that can not meet the high-level decision-making of logistics managers. In this paper, the author uses the system dynamics method to study the supply chain management. By applying system dynamics to the field of logistics system, we can deeply analyze the dynamic development and operation mechanism of logistics activities from qualitative and quantitative perspectives, and then provide assistance and reference for logistics decision-making.

Keywords: System dynamics · Supply chain management · Order system model · Sales system model

1 Introduction

In recent years, with the diversification of customer demand, the intensification of enterprise competition, the development of global manufacturing concept and China's entry into WTO, logistics has attracted global attention, and our government also regards it as a new economic hot spot. As an economic activity related to the spatial displacement of commodity entities, logistics plays an important role in the production and circulation of materials. Logistics support our industry and life, social development is inseparable from logistics. In the past, people focused on qualitative analysis in the study of logistics system, because it is difficult to describe the logistics system quantitatively. Later, operational research method has been widely used in the research of logistics system, but it still can not meet the high-level decision-making of logistics managers. Because it is difficult to find the optimal solution by using operational research method. Sometimes in order to find the "optimal solution" and simplify the problem, some nonlinear behaviors in the system are often ignored, so that the final result is meaningless to the actual problem [1].

B. J. Jansen et al. (Eds.): *International Conference on Cognitive based Information Processing and Applications (CIPA 2021)*, LNDECT 85, pp. 956–960, 2022.
https://doi.org/10.1007/978-981-16-5854-9_133

This paper analyzes and thinks about the behavior and results of each link in the logistics system from the perspective of system, and examines the operation mechanism of the logistics system from the perspective of overall situation, rather than limited to a single link or element of the logistics center. This way of thinking has a very positive practical significance for solving the problems existing in the logistics system, making relevant logistics decisions, improving the overall operation efficiency of the logistics system, and then improving the overall service level of China's logistics industry.

2 Basic Theory of Logistics

2.1 Definition of Logistics

Logistics is composed of two basic elements: material and flow. The "thing" in logistics refers to the part of material data in the world of material data that has the characteristics material entity and can carry out physical displacement, such as raw materials, semi-finished products, finished products, etc. As shown in Fig. 1. The "flow" in logistics mainly refers to the physical movement relative to the earth, which is also called "displacement", such as the change of spatial position and time. Logistics is not a simple combination of "material" and "flow", but a unity of economic, political, social and physical movement, with its own movement law.

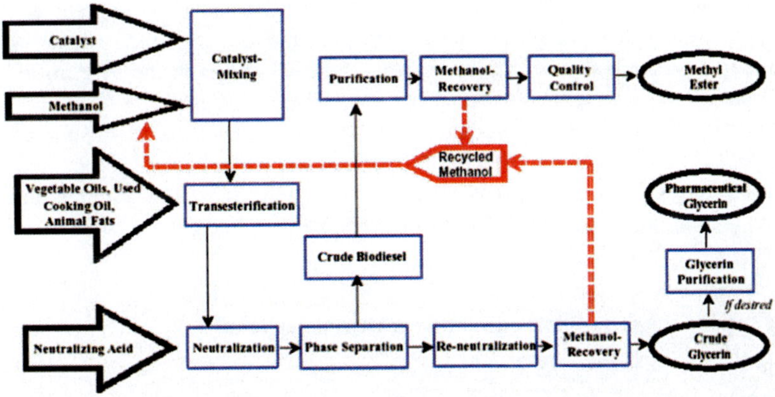

Fig. 1. Simple combination of "matter" and "flow"

2.2 Logistics Development Stage

According to the characteristics of logistics practice, such as content, application technology, implementation means and methods, the development of logistics activities can be divided into two stages, and the mid-1980s can be regarded as a time mark of development [2].

(1) Primary stage (traditional logistics stage): mainly manual operation and mechanical operation, paying attention to various functions of logistics. From the analysis of the overall development process of logistics, the primary logistics period is from the 1940s to the mid-1980s; the main feature is the development of specialization and mechanization, focusing on improving the efficiency and benefit of transportation, warehousing, distribution, outsourcing and other logistics links.
(2) Advanced stage (Modern Logistics stage): Based on electronic information technology, focusing on the comprehensive integration of service, personnel, technology, information and management, is the embodiment of the combination of modern production mode, modern management means and electronic information technology in the field of logistics. With the help of electronic information technology, the operation entities of the same logistics channel can effectively carry out logistics activities within enterprises, among enterprises, regionally, nationally and internationally.

3 Basic Theory of System Dynamics

3.1 Overview of System Dynamics

System dynamics (SD) is a system analysis method based on feedback control theory and combining qualitative analysis with quantitative research to study social economic management system. In essence, it is a series of first order differential equations with time delay, and its model is based on causality diagram and flow diagram. Its main ideological basis is the causal relationship and system structure of the system. It thinks that there is a universal causal relationship among all things in the system, and any system has a certain structure and functions shown by this structure.

The uniqueness of solving problems in system dynamics is to establish a standard mathematical model. Starting from the micro structure of the system, the internal relationship between the structure function and dynamic behavior of the system is analyzed with the help of computer simulation technology, so as to find out the countermeasures to solve the problems.

3.2 Vensim Simulation Software and Its Characteristics

Vensim software has the following features:

(1) Modeling by graphic programming. In Vensim, "programming" actually does not exist, only the concept of modeling. In the main window after starting the Vensim system, draw the simplified flow rate into the tree or flow chart according to the operation button, and then input the equations and parameters through the equation editor to directly simulate the use [3]. In Vensim, equations and variables do not have time scales, and the model is built around the causal relationship between variables.
(2) It runs under Windows operating system and uses a variety of analysis methods, which makes the output information of Vensim very rich. Output compatibility, general simulation results, in addition to instant display, but also provides a save to file and copy to the clipboard and other methods of output.

(3) Various analysis methods are provided for the model. Vensim can analyze the model structure and data set. Structural analysis includes cause tree analysis, result tree analysis and feedback loop list analysis. Data set analysis includes data value and curve analysis of variables changing with time.

3.3 Test Input

The typical retailer's sales level generally fluctuates around the average sales value [4]. During this period, there are some seasonal changes and the overall development trend, which may be rising year by year. In this way, we need to complete a complex test input function to describe the real retail sales level. As shown in Fig. 2.

The step function (PQ) is expressed by the following equation:

$$STEP(P, Q) = \begin{cases} 0, \text{ Time } < \text{ Q} \\ \text{P, Time} \geq \text{Q} \end{cases} \tag{1}$$

Where: P is the step amplitude.

The simulation results of retailer order system model 1.

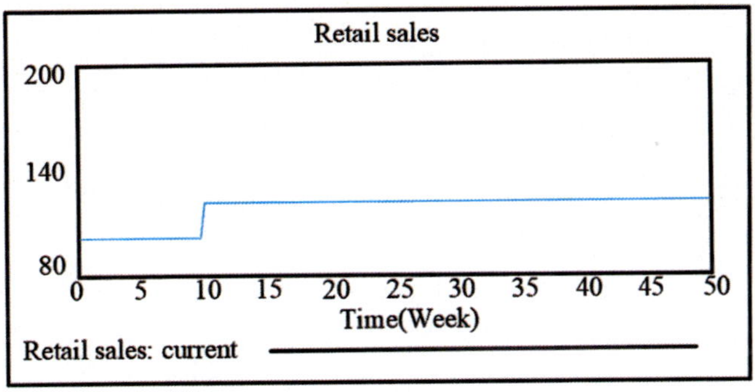

Fig. 2. Simulation results of retailer order system model 1

If the inventory change rate at time t calculated in step is recorded as "rate (T)". Then the inventory of time $T + T_1$ can be expressed by the following equation:

$$T + T_1 = T + T_1 \times T \tag{2}$$

From the above process, it can be seen that the accuracy of Euler method mainly depends on that the "time step" of the model constant is less than 1/3 of the minimum time constant in the model.

4 Conclusion

Supply chain management is not only a hot research topic in today's society, but also a complex research topic. Supply chain management includes a wide range of contents, including the flow process of goods from suppliers to final customers through manufacturers and distributors. In the past, most of the researches on supply chain management used qualitative analysis and operational research methods, and a small number of researches used system dynamics methods, but most of them focused on transportation and inventory.

Acknowledgements. Logistics system control, school level research team, number: ST202003.

References

1. Zhang, Q.: Analysis of new energy supply chain management model – taking photovoltaic industry as an example. Appl. Energy Technol. (01), 1–3 (2021)
2. Xu, S., Wang, D., Hu, Y.: Analysis of aerospace model material management mode under supply chain management thinking. Aerosp. Ind. Manag. (01), 46–49 (2021)
3. He, X.: Logistics management status and countermeasures of feed enterprises under supply chain management. China Feed (01), 111–114 (2021)
4. Ouyang, Y., Yang, Z.F.A.: Analysis on collaborative management mode of automobile supply chain with big data suppliers. Small Medium Sized Enterp. Manag. Technol. (Zhongxunjiao) (11), 20–21 (2020)

Design and Implementation of New Media Resource Search System Based on Cognitive Heuristic

Qing Liu[1](✉) and Jinghua Zhao[2]

[1] Hunan International Economics University, Changsha 410000, Hunan, China
[2] University of Shanghai for Science and Technology, Shanghai, China

Abstract. The rapid development of network promotes the development of multimedia teaching resources, which contains a large number of multimedia teaching resources. In order to effectively use these multimedia teaching resources, the importance of retrieval tools is highlighted. This paper designs and implements a teaching resource search system, which is used to search new media teaching resources related to education.

Keywords: Multimedia teaching resources · Algorithm · Search system

1 Introduction

With the development of network technology and multimedia technology, as the main form of Web information, multimedia resources, such as image, sound, video, animation, text and so on, play an increasingly important role in teaching. How to effectively use these teaching resources and make them play their due role in the teaching process has become an urgent problem for educators. According to a certain topic, according to the semantic, content, visual characteristics of media and the space-time relationship between objects, we make full use of multimedia retrieval technology to retrieve the multimedia database, so as to obtain a large number of multimedia data. But this kind of retrieval technology is mostly content-based retrieval technology, which needs a multimedia database, In addition, the retrieval of multimedia data in the database is characterized by content, semantics, visual features and so on. The process is more complex, time-consuming, low efficiency, and requires a special agency. At present, the major search engines mainly use the relevant text in the web page to extract the keywords describing multimedia information for multimedia information retrieval. This kind of search engine can directly and quickly find multimedia resources from the web. But this kind of retrieval also has disadvantages, that is, sometimes the content of multimedia information can not be described by several keywords, and the extracted keywords may not match the content, which leads to the retrieval of multimedia information content has nothing to do with the teaching theme [1].

Based on the application of topic search in the field of teaching, this paper studies and improves the search algorithm, process and topic. We design and implement a network

B. J. Jansen et al. (Eds.): *International Conference on Cognitive based Information Processing and Applications (CIPA 2021)*, LNDECT 85, pp. 961–965, 2022.
https://doi.org/10.1007/978-981-16-5854-9_134

multimedia resource search system, which is used to search the multimedia resources in the network. This search system first establishes the subject word set of teaching resources in multiple disciplines according to the needs of education and teaching, and then automatically analyzes the structure of the seed website Finally, it analyzes the next web page according to the obtained web page links, and finally realizes the analysis and judgment of all the pages of the whole seed website, so as to find the multimedia teaching resources contained in this website. The following is a detailed introduction of the architecture and working principle of the subject search system based on network multimedia teaching resources, the realization of the key technologies in the system and the test results of the system.

2 Architecture and Working Principle of Search System

"Multimedia teaching resources subject search system" is designed to search the multimedia resources existing in the network, and its architecture is the same as that of the conventional subject searcher. In this architecture, the web spider is the core of the whole search system. Firstly, the teaching word set of a certain subject at a certain stage is established, which is related to the content of each subject and linked to the database engine, The objects it deals with include: static web pages (text, multimedia information, etc.), web development dynamic web pages, the internal structure of web pages, the topic searcher uses the close relationship of the three to extract the content representation information of multimedia teaching resources, and the text in the web page, as the basic content information representation of multimedia web pages, can be used as the agent of multimedia content. The whole searcher is divided into online part and offline part, as shown in Fig. 1. The online training is mainly responsible for web information extraction and URL selection in the queue to be crawled; Offline training is mainly responsible for URL rule training and webpage correlation calculation [2].

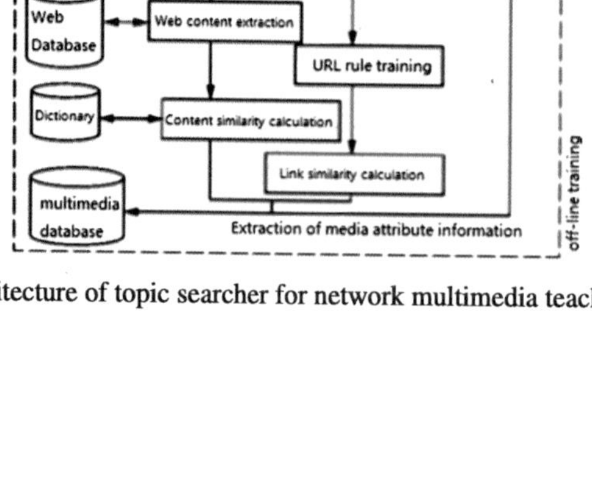

Fig. 1. Architecture of topic searcher for network multimedia teaching resources

"Theme spider" grabs a web page from the Internet and extracts two parts of the information of the web page. One is the content of the web page, so as to determine the relevance between the web page and the query multimedia theme; The second is the webpage link, which determines the crawling page of the theme spider "The theme spider obtains the text information of the web page through the HTML parser, and stores the link with the web page into the web page database. The similarity calculation of the web page content is used to judge the relevance between the web page and the query multimedia theme" The theme spider extracts the URL of the web page at the same time, and then judges whether the web page contains multimedia. If it contains multimedia, it carries out "URL rule training" and uses it for "link similarity calculation". After the content and link similarity calculation, the web page links are weighted with the web page links in the "URL database", so as to determine the next web page to crawl, The crawling direction of "web spider" is determined by "URL link filtering"; If it does not contain multimedia, the extracted Web links will be directly stored in the "URL database", waiting for the "web link weighting" to assign the weights. The multimedia links obtained by the final search are stored in the "multimedia database" together with the information representing the multimedia content after the "media attribute information extraction". At the same time, they are also stored in the "URL database" to indicate the next crawling direction of the "theme spider".

3 The Realization of Key Technologies of Subject Search Engine

According to the technical specification for the construction of modern distance teaching resources, the paper describes the attribute of multimedia teaching resources used in distance education, and the attribute of multimedia resources includes file attribute and content feature. File attribute refers to the meta information related to the file, such as file name, file format, file size, file establishment time, URL, etc.: content feature refers to the audio-visual features and structure and theme information contained in multimedia itself, such as name, composition, layout, picture color, object shape, sound intensity, tone, etc. Considering the characteristics of web multimedia resources and the computer intelligent processing ability, we extract the description information about multimedia contained in the web page. The web page is usually transmitted in the form of HTML semi-structured through c/s, and the information that can describe multimedia content in the web page: Web page URL, web page title, file name, hyperlink text, file format, etc. [3].

Web page text is written in HTML, and it is an identity language. A large number of tags are used to mark information categories, such as title of text and <p> of natural segment segmentation. By analyzing these tags, the computer can get all kinds of information contained in the text of web page. In the web multimedia search engine, we extract the following information to represent the theme of Multimedia:

① URL of the web page; ② The text content of the tag of the web page <it> ③ anchor (anchor text) of multimedia link in the web page; ④ URL of multimedia link.

4 Search Strategy of Searcher

Usually, the searcher starts from a seed, requests to download network resources through HTTP network protocol, analyzes resources and extracts links, and then accesses the network by a certain search strategy through cyclic iteration. The search strategy of searcher directly affects the crawling direction of web spider and the recall rate of multimedia web pages. So first, we should choose the suitable seed website as the starting seed, and then we should have good search strategy. PageRank algorithm "proposed by brin and page is the first algorithm to use hyperlink information for Web mining", and also the most successful algorithm in commercial applications, which is adopted by Google search engine [4]. The basic idea of this algorithm is as follows: set the chain in set of page i as {T1, T2,..., Tn}, that is, every page in {T1,..., Tn} is linked to page i, C(i) is the number of pages chained out of page i, then the rank value PR(i) of page i can be calculated by two steps: (1) randomly take Web. any page by probability e (2) Randomly take any link of the current page to get the page with probability $1 - e$.

$$PR(i) = 1 - e + e * (PR(T1)/C(T1) + \ldots + PR(Tn)/C(Tn)) \tag{1}$$

Obviously, the larger the pr(I) value is, the more authoritative the page is. This algorithm has nothing to do with the user's query conditions, but only gives the rank value of each page as a reference value for the ranking of search engine search results. The higher the rank, the higher the ranking. We use the improved PageRank algorithm to add the click through rate and the amount of multimedia information contained in the webpage to the PageRank algorithm, which ensures the relevance between the crawling webpage and the topic, and solves the problems of the authority and coverage of the webpage search. The improved PageRank algorithm is as follows:

$$PR(P) = \frac{1-d}{N} * W(T_i) + d * \sum_{i=1}^{n} PR(T_i) * P(T_i, p) \tag{2}$$

PR(p) represents the PageRank value of a web page; $PR(T_i)$ represents the PageRank value of the web page, where the web page T: points to the web page p: $P(T_i.P)$ is the damping coefficient, and the actual value is 0.85; $P(T_i, P)$ is the probability from page T_i to page P. the calculation method is as follows: n is the number of pages that have been downloaded to the queue to be crawled and related to the topic: n is the number of pages linked to page P: $w(T_i)$ It represents the number of T_i embedded multimedia in web pages.

5 Conclusion

This paper introduces each link of the whole network multimedia teaching resources subject search system in detail, and the experimental effect is good. But the performance of the system still needs to be improved in the following aspects: first, expand the subject word set of basic education, expand the scope of search, and immediately update the records in the "multimedia database". Second, continue to improve the efficiency of multimedia topic search algorithm, focusing on optimizing the parameters of the experimental algorithm. Third, consider the acquisition of multimedia resources in the database (dynamic web page) of multimedia network.

References

1. Zhang, Y.: The construction of learning community network based on CRP concept. J. Jiangsu Radio Telev. Univ. **12**(4), 3234 (2010)
2. Meng, X.: Content feature extraction and search of multimedia network teaching resources. Res. Educ. Electrification **3**(2), 33–37 (2007)
3. Liu, Z.Y.: Design and Implementation of web Teaching resources subject retrieval system. J. Northeast Dianli Univ. (Nat. Sci. Ed.) (2008)
4. Yang, R., Meng, X.: A topic search algorithm based on Web content and link analysis. Intell. J. **6**, 64–66 (2008)

Airline Slot Allocation Optimization Decision Under CDM Mechanism

Wenya Xiao[1](✉) and Jinghua Zhao[2]

[1] ChongQing City Management College, Chongqing 401331, China
[2] University of Shanghai for Science and Technology, Shanghai, China

Abstract. Air transportation system involves three aspects: air traffic control, airlines and airports, but at present, the solution to abnormal flights only depends on air traffic control. This paper studies the second time slot assignment problem under the airline CDM mechanism. Firstly, RBS algorithm is used to allocate flights for the first time, then a new algorithm is proposed to allocate time slots for the second time by combining compression algorithm and fixed priority TTC algorithm, and this algorithm and genetic algorithm are used to solve the model at the same time. Finally, the actual data of a domestic airport is used for simulation verification. The results show that compared with RBS algorithm and genetic algorithm, the proposed algorithm can not only reduce the delay cost, but also balance the average delay cost of different airlines.

Keywords: Time slot allocation · Collaborative decision making · Transport aviation

1 Introduction

Nowadays, with the increasing improvement of people's living standards, people's travel demand for comfort and speed is growing, so that with the rapid development of civil aviation transportation industry, the supply of airspace airport capacity is not enough. Especially in the case of bad weather, the capacity of the airport is greatly reduced, the slot resources are tight, most flights can not land on time, they can only hover in the air and wait, which not only increases the risk, but also consumes huge costs. Nowadays, the main way to solve this problem is to use the ground waiting program (GDP) to let the flights that do not take off wait at the airport, because the cost of ground waiting is far lower than that of air waiting, and the strategy of ground waiting instead of air waiting can reduce the cost and the risk factor at the same time. It is related to the allocation of time slots to change the take-off time of the aircraft through the ground waiting strategy, so as to avoid the congestion period. The allocation of time slots is usually decided by the air traffic control department, which in fact ignores the actual decision-making objectives of airlines. This paper mainly studies the secondary allocation of time slots based on CDM (cooperative decision mechanism) under the joint decision of airlines and air traffic control bureau according to their respective decision objectives, so as to improve the utilization rate of time slots [1].

B. J. Jansen et al. (Eds.): *International Conference on Cognitive based Information Processing and Applications (CIPA 2021)*, LNDECT 85, pp. 966–970, 2022.
https://doi.org/10.1007/978-981-16-5854-9_135

Since the 1970s, there has been an upsurge in the research on abnormal flight recovery. The research on abnormal flight recovery can be divided into two categories. One is to design an optimized flight recovery scheme to reduce flight delay from the perspective of airlines; The other is to design an optimized time slot assignment scheme to achieve the same goal from the perspective of air traffic control. Due to the commonality of time slot resources, time slot assignment also involves the fairness of the optimization scheme. However, there are some defects in these two methods; The second one ignores the decision-making status of airlines, resulting in the optimized slot allocation scheme does not meet the requirements of airline production and operation organization.

2 Optimization Model

2.1 Question Hypothesis

The shortage of time slot resources caused by weather and flow control mainly occurs in large hub airports, so this study mainly considers the time slot arrangement of hub airports. In airlines has the characteristics of flight wave. The so-called flight wave refers to a group of flights arriving (or departing) in an adjacent period of time, and these flights will be processed by the crew, the airline and the airport Due to the existence of flight wave, flight delay is not only the delay of single flight, but also the delay of flight wave [2].

In order to simplify the analysis and processing of the model, this paper ignores the specific number of passenger transfer and the mutual transfer between flight waves, and only considers the arrival slot allocation in the time slot allocation.

Due to the importance of hub airports, general airlines will make full use of the time slot resources, so the possibility of autonomous flight cancellation is very small. The airlines cancel some flights in order to meet the objectives and requirements of the overall recovery plan, so that the number of time slots used is less than the number allocated by ATC.

In order to focus on the essential characteristics of airlines' autonomous decision-making of one time slot, this paper does not consider the cancellation of flights.

According to the actual situation, the following assumptions are abstracted.

(1) The total number of time slots allocated to airlines by the time management department for weather reasons and flow control remains unchanged, but the number in each time zone is less than the original number; This shows that the number of time zones has increased. For example, the original 800–1000 allocates four time slots, but now 800–1000 allocates two time slots, and 1000–1200 allocates two time slots.
(2) Airlines are free to decide the time slot to be used by their own flights, but the time slot used by flights should not be earlier than the scheduled arrival time of flights.
(3) The number of arrival flights arranged by airlines is equal to the number of time slots obtained.

2.2 Establish Mathematical Model

Variable definition

The following are explained according to the above explanation and the parameters related to the following model:

(1) S: the schedule of time slot flights arranged for the first time;
(2) P^F_{fa}: indicates the priority of the flights in airline A in all flights;
(3) P^F_{fO}: indicates the priority of the flights in airline B in all flights;
(4) F_a : a: the number of airline flights;
(5) F: the number of flight sets of all airlines;
(6) OTA_f/OTD_f: Flight f initial entry/exit time;
(7) CAT_f/CTD_f: Flight f planned time to enter/exit;
(8) x_g: When f_i light acquisition time slot s_i, $x_{ij} = 1$; Otherwise $x_{ij} = 0$;
(9) $g_{fi}(t)$: the cost of delay corresponding to the flight f delay time t.

Mathematical model

In this paper, the airlines can assign time slots freely under CDM mechanism, which can make the cost balance of each airline delay, which can improve the enthusiasm of airlines [3].

The objective function is the absolute value of the difference between the average delay cost of each flight of each airline and the average delay cost of each flight of all airlines:

$$min \sum\nolimits_{S_j \in S} \left| \frac{\sum_{f_i \in F} \sum_{s_i \in s} gf_i(t)xij}{F_a} - \frac{\sum_{f_i \in F} \sum_{s_i \in s} gf_i(t)xij}{F} \right| \tag{1}$$

Constraints:

$$\sum\nolimits_{s_i \in s} is_j \geq OTA_f/OTD_f x_{ij} = 1, f_i \in F \tag{2}$$

$$\sum\nolimits_{A \in F} x_{ij} \leq 1, s_j \in S \tag{3}$$

$$x_{ij} \in \{0, 1\} \tag{4}$$

(2) Formula (3) indicates that each flight can only be allocated to one time slot at most.

3 Example Verification

(1) Assumptions

Firstly, the following assumptions are made according to the model

(1) The time slots provided by the air traffic control department are all known;
(2) All airlines are free to allocate time slots;
(3) The passenger delay time is the difference between the planned arrival and departure time and the initial arrival and departure time.
(2) Case analysis

According to the actual data of Guangzhou airport from 4:00 to 5:00 in 2008, this paper uses the genetic algorithm to process the actual data. On this basis, this paper uses this algorithm to process the actual data. The results show that this algorithm can get a more optimized average delay cost than genetic algorithm.

First of all, the actual data involved 10 aircrafts, of which 5 belonged to CAG airlines and the other 5 belonged to CES airlines. These 10 aircrafts participated in the ground waiting process (GDP) in the afternoon of that day, and the time of ground waiting process was from 4:00 pm to 5:48 PM. The two airlines allocate time slots twice based on CDM mechanism. Due to the different operating costs of different aircrafts, the flight with higher operating costs should be considered first when selecting the time slot of the flight. See Table 1 for the delay operating costs of various aircrafts. In the first assignment of flight schedule, RBS algorithm is used for the first assignment, and this algorithm and genetic algorithm are used for the second assignment.

Table 1. Operation cost of various types of aircraft delay

Model	Representative symbol	Maximum takeoff weight/T	Wake level	Delayed operation cost yuan/h
Heavy machine	H	>136	Heavy	4200
Medium model	M	7–136	Medium-sized	3000
Light model	L	≤7	Light	240

According to the model and algorithm, we can get the optimized flight slot assignment scheme. The delay time of the optimized scheme is 1850 min. If the traditional first come first served (FCFS) method is used to arrange the flight slot, the delay time of the flight slot is 2 030 min, The flight slot assignment can reduce the total delay time by 180 min by using the optimization method. The flights assigned in Table 2 are exchanged, and this exchange does not affect the flight wave delay, but can improve the normal rate. In these 12 flights, there will be 3 normal flights [4], but none normal flights according to the FCFS mode, which indicates that when the airlines have the independent decision-making power, the flight slot assignment can reduce the total delay time by 180 min, It can not only reduce the total flight delay, but also improve the flight punctuality.

Table 2. Optimized flight slot assignment scheme

Time slot number	Flight number	Flight wave	Time slot number	Flight number	Flight wave
1	4	3	7	12	4
2	7	4	8	1	1
3	8	3	9	3	1
4	9	3	10	5	1
5	10	4	11	2	2

4 Conclusion

By using the assignment model and parameter heuristic algorithm, airlines can effectively reduce the total passenger delay time. Because the recovery of abnormal flights involves three parties, namely, air traffic control, airlines and airports, and often air traffic control is in an absolute dominant position. Although the time slot allocation scheme of airlines determined by air traffic control from the perspective of efficiency and fairness may be optimized, it does not mean that the participation of Airlines is unnecessary. This paper fully shows that under the collaborative decision-making mechanism, the active participation of airlines is needed to better solve the problem of abnormal flights. If multiple airlines set up dynamic time slot transactions, it will maximize the role of airlines in collaborative decision-making.

References

1. Zhang, R., Zhou: Research on fairness of initial slot allocation in CDM. Command Inf. Syst. Technol. (2), 51–55 (2010). Decimal
2. Zhang, R., Hu, M., Zhang, H.: Research on time slot fair allocation in cooperative ground waiting program. Traffic Comput. **26**(4), 62–65 (2008)
3. Zhou, X., Zhang, X., Liu, C.: Application of slot allocation algorithm in CDM GDP program. Coll. Beijing Univ. Aeronaut. Astronaut. **32**(9), 1043–1045 (2006)
4. Xu, X., Li, X.: Analysis and simulation of delay cost in flight ground waiting model. J. Nanjing Univ. Aeronaut. Astronaut. **38**(1), 115–120 (2006)

On the Combination of College Translation Curriculum System and Information Technology

Changhong Zhai(✉)

School of Foreign Languages, Wuhan Polytechnic University, Wuhan 430028, Hubei, China

Abstract. In recent years, the continuous interaction and integration of information technology and college translation curriculum system have challenged the traditional college translation teaching model and curriculum system, and a new technological paradigm is being formed. This paper aims to study the application of information technology in foreign language teaching, and further use information technology to provide high-quality and efficient network information resources and services for college translation elective courses, which is conducive to providing suggestions and strategies for the teaching of college translation elective courses.

Keywords: Information technology · College translation · Curriculum system

1 Introduction

Over the last years, the traditional teaching model and curriculum system in college translation education have been facing with challenges and a new technological paradigm is coming in shape due to the continuous interaction and integration between the information technology and college translation curriculum system [1]. Breaking the limitation of the teaching mode and teaching method in the traditional foreign language teaching and curriculum designs that depends on a chalk, a task book, a mouth, this new technological paradigm is more than just introducing computers and internet as the aiding study tools in the foreign language teaching and foreign language curriculum. The relationship between the information technology and foreign language curriculum is neither about dominance nor separate equivalence, but a relationship of harmony and interdependent. Information technology is not only serves as the means but the aim and content of the foreign language education. With a reference to what the internet information management summarize from the information technology content, and what features possessed by the foreign language teaching, the foreign language information technology is supposed to include the building of the shared internet platform; the storage and transmission of the foreign language resource; the building of the service cooperation platform for foreign language resources; various software and technologies and so on. All kinds of internet information platforms in foreign language education require good user interface as well as diversified functions for data transmission and satisfying requirements that varied form client groups. The internet information platforms offers the foundation

B. J. Jansen et al. (Eds.): *International Conference on Cognitive based Information Processing and Applications (CIPA 2021)*, LNDECT 85, pp. 971–976, 2022.
https://doi.org/10.1007/978-981-16-5854-9_136

and guarantee for all kinds of foreign language resources (including E-books, internet resources and multimedia resources and so on) and foreign language information services (including the training for teachers and students, communication between teachers and students, homework and examination check and so on). The idea of integrating the internet information of foreign language information service with the foreign language education originates from the popularization and application of the internet information technology in the teaching field, which produces some fruitful studies. And the studies provides some typically cases, relatively systematic theories as well as data for the explorations of the integration between the internet information and the curriculum [2]. The internet information technology that focuses on multimedia and internet, is getting into every field and part of education at an increasingly high speed, continuously upgrading the teaching means and enriching the information resources. As the internet information technology popularizes, the advantage of integrating this technology with the college translation curriculum will surely more observable. And the college translation resources will transform from the traditional media that mainly represented by paper to the computer and the Internet. The internet information technology will have a closer relationship with the college translation teaching since this technology is powerful in imparting knowledge. It is thought that information technology are not only serving as an assistance tool in the college translation teaching, but becoming very important part of the curriculum system as a whole. Every part of the English curriculum, including the teaching subject, teaching plan, textbook compiling, course design, teaching processes, teaching model, teaching management, teaching evaluation and some others in the system, can offer information services through the integrating with the internet technology software or hardware and being transformed and transmitted in different forms like sound, light and color [3].

2 The Building of Connecting Information Technology with the College Translation Elective Curriculum System

With the context of the rich information resources and advanced technology, accelerated transmission speed and multiple information services, "guide students based on their characteristics and abilities" is the prerequisite and the guideline of building the college translation curriculum system. Information technology can offer help and support for the establishment of college translation curriculum in an all-round way [4]. Theoretically, some foreign language educational theories like Learning Theory of the Constructionism, Teaching Theory of Cognitive Linguistics, the theory of multiple intelligence, the teacher-led-learner-centered teaching theory are the knowledge basis for building the college translation curriculum system. The arrival of the information technology enriches the content of the original theories. As shown in the following figure:

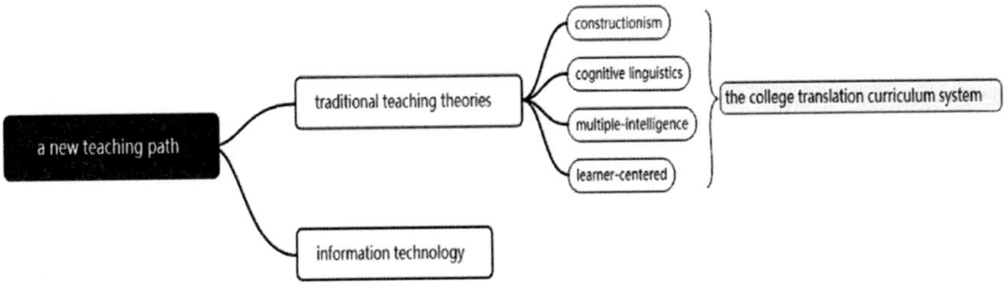

For instance, situational learning are advocated in Constructionism. In the traditional college translation elective course system, teachers and students mainly depend on the traditional classroom, class, and concrete teaching tool to obtain new knowledge and experience. However, in the modern curriculum system, teachers and students can depend more on the virtual platform for interaction. The aim of building the college translation elective curriculum system has everything to do with the information technology [5]. Training brained youths with clear understanding of the international rules and international views is the aim of the college translation curriculum system even the whole college translation teaching. Language objective is also the directive teaching objective. But that students subconsciously master the skill of using the information technology should be regarded as an independent part of building the curriculum system. In addition, information technology exerts increasing influence on the college translation curriculum system. For one thing, the information technology gradually updates the online elective course system, for another, information technology can classy students' choices for analysis and gaining the necessary data to better satisfy students' demands of different kinds. The college translation elective course system can be base on this and make correspond adjustment to dynamically reflect and meet students' requirement. At the same time, the function of perpetration and importance of the information technology can be demonstrated in the foreign language teaching resources, teaching designation and teaching process, including the 3D -hologram text books, online classroom, flipped classroom, online teacher-with-student interaction, homework submission system, curriculum evaluation system and evaluation system about teachers and students.

3 Suggestions on Optimizing the Curriculum System and Improving the Internet Information Services

When optimizing the college translation curriculum system, we must take full consideration into many different factors which contain internal and external environment, teachers and students, policies and motivations, and some others. Attention should be paid to the followings are the aspects in the process of integrating information technology and the college translation curriculum system:

3.1 Materializing the Curriculum Sittings Logically and Diversely

Using the advanced internet software and technology to classy the elective courses and materialize the curriculum settings so as to meet the individual study demands. Meanwhile, the information technology can accurately divide students into different groups, and provide different classes of even the same curriculum according to the students' various language level and ability. In this way, students' demands are satisfied on the one hand, and the teaching content is more targeted on the other hand. What's more, it helps to optimize teaching resources and enhance the teaching effect.

3.2 Motivating Students and Optimizing the Curriculum through the Cyclical Feedback Information

To encourage students to autonomously take class in elective course through the bouquet classes offered on the school internet. Teachers can put what they impart and the relative study resources on the internet as well as design interaction space for students. If so, students can get the opportunities to choose what they like on the courses, class time and teachers based on their own reality; more opportunities are created for communications between teachers and students, and between students, which makes students actually become the subject of the curriculum. With the convenience of the interaction platform, students can sent the curriculum suggestion to teachers, and teachers can adjust the teaching content timely or ask about more students' opinions, realizing the optimization of the dynamic class [6].

3.3 Making Better Use of the Educational Resources by Introducing the MOOCs

Schools that offer college translation elective courses are supposed to strengthen cooperation and communication with the university network center and other functional departments, properly make use of foreign language resources and give reasonable suggestion on introducing investment in school and information resources. Strengthening the building of the MOOCs for the college translation elective course, making full use of MOOCs resources from other colleges. MOOCs, which is a new online development mode, is free from the time limitation to educate students, making it possible for students to arrange their student flexibly. In addition, students can take class in the first-class universities from home and abroad, which indeed expend students' study platform. MOOCs are conducive to students' ability in autonomous learning and popularizing the common share of the qualified teaching resources [7]. That is to say, all universities should actively join in the curriculum union to help students to take class of the in other universities' general education on common elective courses. In this way, the mixed teaching mode that students become the subject in the study through the autonomous learning on line, can be established, providing solutions to the shortage of educators and inadequate classes for students to select and meeting the individual demand perfectly.

3.4 Enhancing Teachers' Ability on Information Technology to Make Full Use of the Modern Teaching Means

College translation teachers that only master foreign language can not keep pace with the developing age. And how to greatly integrate the advance information technology to reach the teaching aim has become the basic quality for teachers. The measurements for expanding the curriculum platform should include the followings: encouraging teachers to have further study in projects about foreign language information resources and information serves so as to apply what they learn to create his own characteristics and give creative classes; being optimistic to exploit and use the international resources; strengthening the team work cooperation of the teachers from home and abroad, such as asking foreign teachers to give selected class. Also, it is good to drawing on the advantage of cooperation of the teachers and adding more colorful curriculum modes like lecture, forum and working groups. Modern teaching means should be making full used of to achieve 3D teaching.

3.5 Making Reasonable Rules for the Online Teaching Management

On the one hand, online qualified teaching files, including the syllabus, curriculum description, teaching arrangement, teaching schedule, should be established by teachers; on the other hand, teachers should regularly get students' opinions on the class, encouraging teachers to draw on advantages and avoid disadvantages from communicating with each other [8]. Information about the booklet Guideline on elective course, curriculum description and other suggestions about the elective courses should be announced and online-updated so that students can get reference on what class to take, and evaluate the optional courses.

4 Summary

Information technology not only bring the advantage teaching means, rich study resources and multitude learning modes to the college foreign language education, but also exert profound influence on the teaching concept of both teachers' and students', tremendously strengthening teachers' ability on imparting knowledge and students' ability on acquire foreign language [9]. Pragmatical researches have been given to this study on the college translation elective curriculum, and information service, and the questionnaire have been tested and received some relative experts' advises and given serious tested. Through the date analysis, it can be found that integrating the information technology with the college translation elective curriculum is an inevitable trend as well as it will become the core of the modernized and informationized English education. Though the study parameter involved are not very much, this study can be the fundamental for the deep study about integrating the information technology with the college translation elective curriculum, given the research on foreign language education is still at the beginning [10].

Acknowledgement. This research was supported by 2021 research project of Wuhan Polytechnic University: Reinterpretation of "Faithfulness" in Translation from the Perspective of Translator's Behavior (Project Number: 2021Y70). It was also supported by 2020 teaching research project of Wuhan Polytechnic University: A four-dimensional model construction of case base for translation courses (Project Number: XZ2020011).

References

1. Hu, J., Chen, J.: An outline study on e-educology of foreign languages. Comput.-Assoc. Foreign Lang. Educ. **3**, 3–12 (2013)
2. Mu, Y.: Construction of the Library Information Resources shared platform in college and universities library work in colleges and universities. Libr. Work Coll. Univ. **2**, 55–56 (2014)
3. Ma, C., Chen, J.: VILL@GE: virtual language learning through edutainment activities -project analysis and elicitation. China Educ. Technol. **2**, 121–125 (2013)
4. Cai, L., Wu, W.: On taking information communication technology as an ability construct for modern foreign language teachers. Comput.-Assoc. Foreign Lang. Educ. **1**, 45–53 (2014)
5. Ma, W., Hu, J.: On the impact of international MOOCs and redesign of college English curriculum. Comput.-Assoc. Foreign Lang. Educ. **03** (2014)
6. Wang, H.: Translation technology teaching practice in the background of information age. Chin. Transl. J. **03** (2012)
7. Cui, Q.: Teaching practice of MTI translation and localization course in colleges and universities. Chin. Transl. J. **01** (2012)
8. Yu, J., Wang, H.: Research on the teaching of master's degree in computer-aided translation. Chin. Transl. J. **03** (2010)
9. Xu, B.: CAT and translation studies and teaching. Shanghai Transl. **04** (2006)
10. Yang, L.: The prospect of information-based translation teaching. Foreign Lang. Foreign Lang. Teach. **11** (2005)

An Exploration of College Students' Second Language Mental Lexicon Storage Model Based on WordNet Computer Data Analysis

Chunxia Zhang(✉)

Foreign Languages School, Jingchu University of Technology, Jingmen 448000, Hubei, China

Abstract. Vocabulary plays an extremely important role in language learning. Because of its importance, more and more researchers use lexical relevance test to study psychological vocabulary. In the past half century, there have been dozens of researches on the representation structure of bilingual mental vocabulary. Through the analysis of WordNet computer data, this study analyzes the storage mode of college students' second language psychological vocabulary. This research is carried out from the aspects of research objects, research methods, research results and research significance. It can be seen from the statistical data that semantic responses account for the largest proportion of these two types of responses, followed by formal responses and meaningless responses, which indicates that the second language research objects are mainly stored in the semantic association of psychological vocabulary. This provides support for the semantic theory of psychological vocabulary organization model.

Keywords: Computer data analysis · Mental lexicon · Storage model · Second language · Word association · WordNet

1 Introduction

Vocabulary acquisition is the central task of second language acquisition [1]. There is a mental dictionary in everyone's brain, which refers to the connected storage state of lexical knowledge in the brain. The mental lexicon is quite different from the common one. The words in the dictionary are arranged in alphabetical order and have strong regularity. Mental dictionaries vary from person to person. Mental lexicon is connected by phonetic, morphological or semantic networks that exist in the brain. In everyone's brain, frequently used words are placed at the top of the Internet for easy retrieval, and mental words store more non-linguistic information than ordinary dictionaries. With the further development of psycholinguistics, the study of second language mental vocabulary has been gradually developed.

Researchers in China and abroad have done considerable researches in this field. There are many empirical studies on L2 Mental Lexicon Organization model in China. For example, Zhang Shujing found that most of the students' lexical response is a response based on pronunciation when they take a vocabulary free association test;

B. J. Jansen et al. (Eds.): *International Conference on Cognitive based Information Processing and Applications (CIPA 2021)*, LNDECT 85, pp. 977–981, 2022.
https://doi.org/10.1007/978-981-16-5854-9_137

the test subjects' acquired vocabulary had not yet formed meaningful connections in their mental lexicon, which made it impossible for the students to use the stored mental vocabulary correctly [8]. The results of Zhang Shujing's experiment show that the way words are connected in the brain is related to the degree of familiarity of words. The low-and medium-familiarity words of learners mainly depend on phonological connection, while the highly-familiarity words depend on semantic connection [9]. Zhang Shanshan's research shows that for second language subjects, the response to different parts of speech of English words is very different [7]. Zhang Ping's research shows that semantic connectedness plays a more important role in the mental lexicon network of learners with higher language proficiency [5].

There are also many discussions and researches on learners' mental lexicon construction model. Zhang Ping defined the concreteness of different parts of speech for the first time from the perspectives of spatial concept and sensory experience, and tested the concreteness of the selected words with a scale, however, cross-combination knowledge grows slowly, and the proportion of aggregated knowledge is much higher than cross-combination knowledge [6]. Some researchers found that the high-level group showed significant advantages in the semantic types of vertical aggregation (such as synonyms, Antonyms, etc.), and the high-level group significantly higher than the low-level group in the proportion of cross-combination conjunctions [3]. Yu Cuihong and Cai jinting examined Chinese English learners' mental lexicon and their depth of vocabulary knowledge by means of Word Association and Vocabulary Ability Tests, semantic connection is dominant, network construction of nouns and adjectives is superior to that of verbs, the development of vertical and aggregate knowledge is remarkable, and the development of horizontal and horizontal combination knowledge is slow, and it has an important influence on the depth of vocabulary knowledge. Different mental lexicon has different influence on the depth of vocabulary knowledge [4]. Li xiaosa and Wang Wenyu explored the association model of Chinese efl learners' mental lexicon by means of Word Association test. The results showed that the proportion of aggregative Lexical Association of high-level learners was significantly higher than that of middle-level learners, however, there was no significant difference in cross-combination between the two groups [2].

2 The Research Methods

Table 1. The subjects of study

	Experimental grouping				
Vocabulary	Level 1	Level 2	Level 3	Lever 4	Total
	<2500	2500–3500	3500–4500	>4500	
Subjects	5	5	5	5	20

2.1 The Subject of Study

All of the 20 English majors from 2 classes in Jingchu University of Technology were chosen to take part in the study. Before the experiment, the vocabulary level of the subjects was tested by vocabulary level test. According to test results (as shown in Table 1) were divided into four groups, the lowest level 1 subjects vocabulary, grasp the vocabulary by less than 2500 words, level 2 participants grasp the vocabulary between 2500 to 3500 words, which is above the average of level 3, who can reach 3500 to 4500 words. The final group, level 4 subjects, whose vocabulary is the largest, more than 4500 words. Five students from each group of pretests were randomly selected as subjects of this study.

2.2 Research Methods and Means

From the requirements of college English teaching, 25 high-frequency words are selected and the stimulus words are randomly arranged. Participants were given an association test in which they were exposed to a mental word.

A "look - write" free association test was used before and after the test. The subjects looked at the words that appeared on the slide, at the same time and wrote down the words that could come to mind in 5 s, which were the subjects' response words. Finally, these response words were statistically analyzed to answer the question posed by this study: What is the mental lexical connection pattern of L2 learners? (1) Semantic collection or formal collection? 2. What kind of semantic or formal connection are they?

The experimental tools used in this study is English word association test.

This research attempts to use WordNet an BNC to identify the semantic relationships between words. WordNet is a huge vocabulary designed by psychologists, linguists and computer engineers at Princeton University. Its characteristic is that it is consistent with words in word meanings, rather than organizing vocabulary information. In WordNet, words are organized in accordance with the organization of synonym sets. Each synonym set represents a basic vocabulary concept, and a synonymous relationship (synonym, antonym relationship (antonym), relationship (upper meaning) is established between these concepts). BNC is used to identify other reaction words from the aspect of forms, i.e. phonological and inflection, which is convenient and quick.

A prediction test was conducted before the formal word association test to predict the time required to complete the test and to improve the experimental process and deixis. Before the word association test began, the test requirements were clearly explained in English and Chinese and examples were given to each participant to show them how they should complete the vocabulary test.

After receiving the association test paper, the participants were asked to provide an unlimited number of answers for each stimulus word as soon as possible, and not to modify the writing. The test should be finished within 30 min. After the completion of the word association test, data would be analyzed through computer, that is, the researchers fed the raw data into a table and classified the reaction words to analyze the subjects' scores (Table 2).

Table 2. The vocabulary test

		Level 1 (n = 5)		Level 2 (n = 5)		Level 3 (n = 5)		Lever 4 (n = 5)	
		Num.	Percent	Num.	Percent	Num.	Percent	Num.	Percent
Semantics	Synonymy	28	16%	42	19%	54	21%	58	23%
	Antonymy	31	18%	45	20%	52	20%	52	21%
	Hypernym	39	22%	37	17%	43	17%	44	18%
	Meronym	28	16%	31	14%	32	12%	32	13%
Forms	Phonological	25	14%	28	13%	46	18%	29	12%
	Inflection	11	6%	27	12%	21	8%	27	11%
Meaningless response		15	8%	13	6%	9	4%	5	2%
Total		177	100%	223	100%	257	100%	247	100%

3 Discussion and Analysis

First, the author collected the total number of reaction words. As can be seen from the statistics, in the two types of reaction words, the proportion of semantic response is about 67%, followed by the formal reaction, i.e. 18.3%, and meaningless response accounts for 5%, which suggests that the subjects of the second language vocabulary mainly stored by means of semantic association in the mental lexicon, which support the mental lexicon semantic theory of organization.

According to the collected data, the reaction words in synonyms, antonyms, hypersemous words and meronym are 182,180,163,123 in semantic reaction words, respectively, accounting for 19.8%, 19.7%, 18.3%, and 13.8%, respectively. The phonological response (16%) and the inflection response (9%) were significantly higher than the meaningless response (5%). In English vocabulary teaching, the author found that in traditional vocabulary learning, learners mainly rely on Chinese word meanings to check up and memorize new words. Most Chinese learners mainly care about the Chinese meanings of new words when learning new English words, and seldom pay attention to the English interpretation and collocation of new words. Memorizing new words in one's the native language can help learners master the new words in the second language for a short time. However, this way of vocabulary acquisition is that it reduces the learner's chances of exposure totally to a second language. Learners doesn't obtain the meaning of words through context, which makes them different to establish connections between words in the second language. Learners of l2 mental lexicon many words related to the concept of mother tongue or semantic knowledge, caused many unrelated words or wrong response, weaken the semantic relations of the second mental lexicon, which directly cause the learners to identify the basic meaning of vocabulary, but cannot extracted them from vocabulary accurately and rapidly, thus influencing the accuracy of the second language use and fluency. The translation of the mother tongue can only be used in the initial stage of second language learning. For English majors, they need to access a lot of language materials through reading, listening, speaking, etc., and rely on context to acquire word collocations, fixed phrases, grammatical features, etc., so as to accumulate language knowledge and establish vocabulary connections.

4 Conclusion

It is of theoretical and practical significance for researches and educators to study psychological vocabulary. Correctly and effectively guiding students to establish their own L2 vocabulary association network system in the process of vocabulary learning can effectively help students master more words. However, based on the individual differences of learners in learning, teachers should teach students in accordance with their aptitude, so that learners of different levels can maximize their potential and expand their vocabulary.

Acknowledgments. This work is supported by Humanities and Social Sciences Project of Hubei Education Department (Project No.19G056).

References

1. Lewis, M.: The Lexical Approach. Language Teaching Publications, Hove/London (1993)
2. Li, X., Wang, W.: WordNet and the Second Language Mental Lexicon under the Intervention of WordNet: Language Science (2016). (in Chinese)
3. Liu, S., Fu, B., Hu, A.: An empirical study on the vertical and horizontal network of mental lexicon representation of people with different second language proficiency. J. PLA Coll. Foreign Lang. **2**, 57–70 (2012). (in Chinese)
4. Yu, C., Cai, J.: The relationship between psychological vocabulary, organizational pattern and vocabulary knowledge depth of Chinese English learners. Foreign Languages in China (2014). (in Chinese)
5. Zhang, P.: Rethinking the word selection and reaction classification criteria of word association test. J. PLA Foreign Lang. Coll. (2010). (in Chinese)
6. Ping, Z.: A comparative study on the part of speech effect of learners' mental lexicon construction between English and Chinese. J. PLA Foreign Lang. Coll. **6**, 68–87 (2011). (in Chinese)
7. Zhang, S.: Exploring L2 lexical structure through word association experiment. Modern Foreign Lang. (2006). (in Chinese)
8. Zhang, S.: Vocabulary acquisition from the perspective of reaction types. Foreign Lang. Teach. Res. (2003). (in Chinese)
9. Zhang, S.: The relationship between L2 mental lexicon from the perspective of lexical association. J. PLA Foreign Lang. Inst. (2005). (in Chinese)

Information Technology-Based Picture Book Design of Children's Intellectual and Educational Anecdotes

Gang Liu and Lei Zhu(✉)

School of Arts, Anhui Xinhua University, Hefei 230088, Anhui, China

Abstract. With concise and vivid language and delicate and beautiful pictures, picture books have become a fashion for children's reading training in the 21st century. In addition, with the improvement of people's living standards, people focus on the spiritual level, cultural improvement, more and more attention to children's reading education. The development of picture books based on information technology has a broad development prospect. The following through the picture book introduction, picture book creation, development prospects and other contents are discussed.

Keywords: Information technology · Children's fun · Picture book design

1 Introduction

Picture books are picture books, which mainly use fine arts as the main body according to the plot, sometimes with a small amount of text. The picture book is not a combination of pictures and words written for children in the general sense. The pictures and words used in the picture book represent the plot of the story, which is a picture formed after careful design, planning and modification for many times. In the picture book, fine art painting is no longer the role of ornament, but the main content of story plot expression, which can resonate with children. The modeling elements and color expression of fine art painting are carefully designed. Picture books emphasize the emotional interaction of the story and the connection of the resonance of the theme content, and then form a continuous visual image of the plot [1], The creators of picture books are equivalent to the directors of animation films and television, who need to present the content within a limited time to be both beautiful and clear, and resonate with the audience. An excellent picture book can let children who cannot read read out the meaning of it. Another picture book is more beautiful, and the layout is exquisite and unique, so that it forms a nearly perfect whole [2].

Read this is the meaning of children through to draw this picture of ornamental understanding, interpretation to understand the painting is the beauty of the language rich in vocabulary, make children acquire knowledge and enjoy the pleasure of picture books reading and interesting, causing the child's thinking and asking questions, express their

B. J. Jansen et al. (Eds.): *International Conference on Cognitive based Information Processing and Applications (CIPA 2021)*, LNDECT 85, pp. 982–985, 2022.
https://doi.org/10.1007/978-981-16-5854-9_138

questions in inner thoughts constantly, can also help children establish comprehensive mental world [3]. Children's parents can also communicate with their children in the rich language of fine arts and paintings, inadvertently promoting the cultivation of children's language expression and communication skills. Children in the life experience is very limited, but through the painting instinct rich reflects the child's life experience. Through the colorful contents of picture books, children can find and dig out the fun of life. There are many life experiences that cannot be directly contacted in life. By expressing them through the media of picture books, children can understand and experience life directly or indirectly, which can broaden children's horizons intentionally or unintentionally [4]. As a friend of books, an excellent picture book is not only a fine representation of painting, but also the composition and color changes can give readers a good visual impact and make them feel happy. Through language, emotional communication and unreserved transmission of thoughts to children, the great interest in reading experience can be enhanced [5]. The fine arts in picture books can better replace the interpretation of language, so that children have a good bridge in the cognitive world. Picture books are mostly books specially drawn by independent painting authors, mainly based on painting itself. For children, fine art pictures can make them more interested than words. During the cognitive period of picture books, colors will be more rich, with high saturation, which is more acceptable and interesting for children. Relevant institutions of child psychology research believe that children's ability to recognize fine arts and draw graphics starts at a very young age [6]. If children's parents can consciously guide and create a warm reading environment to explain graphics and read words for children, it will help children better understand [7]. Then from the beginning of the high level of pictures and words, children can taste the expressive force of painting art in listening to stories, and understand words and familiar with literature knowledge in reading and learning fine arts and paintings. Compared with the fast-food network cultural languages popular in the society, which can only bring temporary pleasure, appreciating picture books is undoubtedly a wonderful and good experience that can make people's eyes enjoy the beauty, make their hearts happy and elevate their spirits [8]. Of course, picture books can't immediately realize all our expectations for children, but they can highlight the charm of high-quality fine art paintings and words in picture books, which can help foster children's cognitive ability, observation ability, communication ability, imagination and creativity. And emotional development and so on has an immeasurable subtle impact [9].

2 Picture Book Story

The theme of this picture book creation is the picture book series about children's intellectual education. The plot narration of the story is used to analyze and study the characters, In this way, children who are interested in this kind of picture books can have a deeper understanding of picture book stories, so as to stimulate their interest in reading, and make the market of children's picture books more and more rich, so that they can have a better development [10]. The following is the drawing of the picture book named "Boo Boo's growth story" for elaboration.

The plot of the picture book is a cute duckling named Bubu, who likes to wear red wellies, and his sister and good friend Carla. Bob likes to wear red wellies. One day,

Bob's red wellies are missing. He is very worried to find his wellies and finds that he is wearing Carla's feet. They all like to wear rain. Carla was a little yellow baby duck who liked to wear a blue lintel. Later, Carla also has her own boots, they walk in the grass, play together, became the best of friends. Weeks later, Boo Boo and Carla cared about little Boo. Because my sister is on the way. Boo poked the shell with his mouth. Carla listened to the sound inside with her ears. They sat on the shell together, waiting for Boo's baby sister to be born. Little sister after birth to see cloth cloth and Carla are wearing wellies, it also wants to wear, they gave cloth little sister one, the letter to play in the grass, happy to spend a good day. On Boo Boo's birthday, they give him a surprise birthday cake and celebrate his birthday together. Boo Boo may have a birthday wish of his own, hoping to grow up to be a kind and brave duck who will learn to share.

3 The Process of Creating Picture Books

First, a large amount of accumulation should be prepared in the early stage of painting. Before design, refer to online information and collect various materials to make full preparations for the role modeling and scene modeling involved in the creation process. I save pictures that I am interested in, and occasionally record every bit of things in life with my brush, which provides rich materials for the creation of picture books.

Second, draw a sketch of the plot according to the script, and show the plot described by the script, thus forming a complete structural system. When determining the style of a picture book, there are two major aspects to consider. First, I involved in the content of the work to be determined according to the story plot and need factors. The second is to highlight their own painting style in the creation. Different paintings can reflect the artist's own painting style. The expression method of lines in the picture, the coloring method of the picture and the processing of the relationship between light and shade are different from person to person.

Third, design specific characters and scenes, and make everything in detail. After the initial establishment of the design scheme, I began to look for the available conditions for creation, hoping to find a suitable creation concept from the materials.

Fourth, draw the final draft of several different angles of each shape to complete the production of characters. Color not only has a strong expressive force in reproducing the role, but also different colors can cause people's different psychological feelings. According to the steps from the composition, color, spread large color block to make the basic style of the picture roughly determined and then carry out in-depth description processing. In terms of the use of software, we chose different brushes to produce different effects. Finally, we checked the coherence of the story and the design of characters as a whole.

4 Development Prospect of Picture Books

Picture books, or simply picture books, have become a trend in parent-child reading in recent years. While the sales volume of picture books in the publishing industry is increasing year by year, a lot of capital is focusing on the derivative market related to "picture book reading". Among them, picture book stores are a platform integrating picture book sales, borrowing, reading guidance and picture book extension activities, springing up like mushrooms after a spring rain.

The development of picture books is very tortuous. In the past, most picture book libraries were not profitable or even in the state of being the grandchild. However, with the increasing number of post-80s and post-90s parents, they have absorbed a wider range of foreign educational concepts, and picture book education is essential in foreign countries, especially in Japan. After investigation, it is not difficult to find that children who love reading since childhood, and children who have never been exposed to books, usually have very different behavior habits. In fact, children's reading habits are inseparable from their parents' daily habits. Most children who love reading have parents who are intellectuals. Therefore, as more and more talents can be accepted by higher education, more and more literate intellectuals in the society have also become literate. In addition, with the improvement of people's living standards, we should focus on the spiritual level and cultural accomplishment. With the improvement of parents' cultural background, they are bound to pay attention to children's quality-oriented education since childhood. Picture books, as a form of picture reading, have been attached great importance and become a fashion for children to read in the 21st century.

Acknowledgements. This work was supported by The animation Research Institute of AnhuiXinhua University (yjs201708), High level teaching team of quality engineering project of Anhui Xinhua University (2019jxtdx05), Key projects of Humanities and Social Sciences in Anhui Universities (SK2019A0744), Quality Education Research Center project of Anhui Xinhua University (IFQEZD201909).

References

1. Tian, Y.: Picture book design in the digital age. Comput. Knowl. Technol. **8**, 76–78 (2019). (in Chinese)
2. He, C.: Preliminary attempts of group text reading. Educ. Mod. **12**, 126 (2019). (in Chinese)
3. Quan, X.: Making full use of picture book reading to carry out language teaching for children. Curric. Educ. Res. **19**, 93–94 (2018). (in Chinese)
4. Zhao, J.: Strategies of independent reading guidance for lower graders in primary schools. Yangtze River Ser. **25**, 32–35 (2019). (in Chinese)
5. Hu, J.: A Practical Study on the Influence of Picture Books in Basic Art Areas on Children's Art Creation in Middle Class. Shanghai Normal University, Shanghai (2018). (in Chinese)
6. Hu, X.: Discussion on the role of picture book reading in family education. Child. Big World **11**, 58 (2019). (in Chinese)
7. Ren, S.: Micro curriculum helps children's speech learning. Jiangsu Educ. **10**, 86–87 (2019). (in Chinese)
8. Ma, Z.: Analysis of the application of picture book reading in Chinese writing teaching in primary schools. Exam. Wkly. **5**, 89–90 (2020). (in Chinese)
9. Lan, F.: Painting the future with colors – a brief discussion on the application of children's picture books in art teaching in primary schools. Questing Knowl. Guide **8**, 65–68 (2019). (in Chinese)
10. Zhu, J.: Study on the visual art of folklore – the legend of Fengcheng as an example. Shanxi Youth **8**, 96–97 (2018). (in Chinese)

Human Effectiveness Analysis Based on Big Data Thinking – a Case Study of Human Resource Strategic Management Course

Lei Tan(✉) and Yuelin Yin

Dalian Neusoft Information Institute, 8 Software Park Rd., Dalian, Liaoning, China
tanlei@neusoft.edu.cn

Abstract. In recent years, the rapid development of big data technology has brought far-reaching impact to various industries. In the field of human resources strategic management, big data technology and thinking also affect the process and mode of human resource management more and more. In the field of human resource strategic management, big data technology and thinking have more and more influenced the process and mode of human resource management. Based on big data thinking, this paper selects the human effect analysis module which is closely combined with big data in the course of human resource strategic management. Multidimensional analysis is carried out from three aspects: data collection and analysis, data visualization, data security and privacy, in order to guide enterprises to change in time to deal with this wave of information.

Keywords: Big data thinking · Human-effect analysis · Human resources strategic management course

1 Introduction

In the 21st century, human resources management has experienced unprecedented scientific and technological progress, global integration, knowledge and innovation, organizational change, international market competition and production division of labor and other forces to promote and impact, the strategic position of human resources management is increasingly prominent, human resources strategic theory and practice continue to obtain new development. With the rapid changes of the internal and external environment of the enterprise and the new requirements of the enterprise strategic management to the human resource management, the human resource planning is gradually integrated with the human resource strategy. As the source of human resource management activities, human resource strategy and planning have increasingly become the core issues of management [1]. With the development of the times, big data technology and thinking also affect the process and mode of human resource management more and more. Based on big data thinking, this paper selects the human effect analysis module which is closely combined with big data in the course of human resource strategic management.

B. J. Jansen et al. (Eds.): *International Conference on Cognitive based Information Processing and Applications (CIPA 2021)*, LNDECT 85, pp. 986–991, 2022.
https://doi.org/10.1007/978-981-16-5854-9_139

From the data collection, mining and visualization and other big data technology applications. Using big data technology to improve the ability of human resource quantitative management is an innovation in human resource management 2.0 context [2].

2 Expected Learning Outcomes and Difficulties Based on Curriculum

In the human resource management planning module of human resource strategic management course, the related knowledge of human resource cost and human efficiency is involved. Before the analysis method of big data was not used, the curriculum was reformed in order to make students understand the concept and process of human cost and human effect data analysis. Using scientific human cost analysis and prediction method, using Excel to carry out quantitative analysis of human cost and human efficiency, to provide reference for A company's strategic decision and development.

This section addresses the expected learning outcomes as follows (Table 1):

Table 1. Expected learning outcomes

Project content	Expected learning outcomes	Difficulties
Integrity of manpower cost and manpower efficiency design	Expected learning effect 1: understand organizational human resources strategy and planning, organization and position management, recruitment and talent evaluation, compensation and benefits, performance management, training and development, employee relationship management and other professional knowledge	KT6 human cost and effectiveness
Analysis of rationality	Expected learning effect 3: can use the knowledge to analyze the current situation of enterprise human resource management and information human resource management	KT9 data analysis, drawing visual charts

(continued)

Table 1. (*continued*)

Project content	Expected learning outcomes	Difficulties
Solution feasibility	Expected learning effect 5: applying human resource tool method and information management tool, combining the problems found and analyzing the causes, this paper puts forward the solution of human resource management and information human resource management	KT11 according to the business of the enterprise, put forward the optimization suggestion
Project content	Expected learning outcomes	Difficulties
Accuracy of use of Excel tools	Expected learning effect 6: basic document processing ability, access to information	KT7 data collection

3 Data Collection

The multi-dimensional collection of data determines the overall effect of quantitative management of human resources. It is the most basic part of the application of big data technology to collect complex data information through various data collection and analysis and data visualization methods. Data multidimensional collection refers to broadening the multi-modal data sources and channels needed for quantitative management of human resources in data analysis, obtaining the collection and monitoring of real-time data of human resources, and ensuring the timeliness, continuity and comprehensiveness of relevant data [4].

Multidimensional data collection is the initial stage of data processing in human resource quantitative management. This procedure requires human resource management departments to change the conventional "sample" thinking and static "report data" behavior. It becomes the thinking and action of "source data" and "whole picture" to realize the principle of data collection. Not only focus on the basic factual data representing the current situation of human resources, such as human resources basic data and other structured data [3]. Also pay attention to web pages, surveillance video, pictures and other semi-structured and unstructured data, such as employees WeChat, Weibo chat keywords, can extract their psychological attitude and work collaboration, and so on. Under the background of big data, the data collection tools that human resource management can be used effectively include search engine classification tools and distributed computing systems. In particular, the popularity of the Internet, through the Internet in business information websites and competitors' websites to collect information is conducive to strengthening the scope and accuracy of human resources dynamic data collection [5, 6]. The data content should be multi-dimensional: first, the dynamic data representing the vitality of human resources, such as the relationship between the recruitment cycle of core posts and the market market; second, the analysis data representing the quality of human resources; third, the data related to the return on investment of human resources projects; fourth, the temporal and spatial sequence data of human resources management activities, that is, cross-regional, cross-annual correlation data, through which the evolution of human resources management activities can be obtained [8].

For specific A companies for human effectiveness analysis, we can analyze from the following aspects.

(1) Calculate the efficiency of human cost profit (efficiency of human cost profit = net profit/total human cost),
(2) Calculation of human cost efficiency by filling in net profit and total human cost data based on known data (human cost efficiency = operating income/total human cost)
(3) Calculating per capita human cost based on known data on the number of people on the job and total human cost (per capita human cost = total human cost/total number of people on the job)
(4) Total cost and labor cost data based on known data, calculate labor cost content (labor cost content = total cost/total labor cost)
(5) Draw a comparative picture of the cost growth of A companies based on known data [7].

4 Data Visualization, Giving Predictions and Recommendations

Data visualization "refers to the results of data mining analysis, which are presented in visual form with the help of certain tools, as the basic basis of related management decision-making work. The most important concern of human resources managers is not the details and processes of data mining, but the display and interpretation of data analysis results, and the current situation, trend judgment and related management suggestions of human resources quantitative management [10]. On the basis of human data multidimensional collection and data integration mining program, human resource quantitative management should also use "data visualization technology", which is the most intuitive presentation of data mining results. It can directly reveal the hidden law and value of big data and facilitate the understanding and recognition of big data analysis results. Visualization technology generally includes collection technology, icon technology, pixel technology and distributed technology.

Through visualization technology, it can directly and effectively assist enterprise managers and human resources management departments to carry out human resources management and planning. In the interpretation of data visualization results by using big data tools, we should also encourage multi-department participation in the information cooperation mechanism, and encourage professionals to carry out objective argumentation on the results of data analysis. Enhance the objectivity and accuracy of data analysis [9].

For specific companies, we can forecast the corresponding labor costs and give suggestions for improving human efficiency.

(1) Human cost forecast for the next year: according to the scientific human cost forecast method (it is suggested to choose the backward method to forecast, the concrete analysis can refer to the video of Mu class), considering the enterprise strategy and internal and external environment of the A company, make the manpower cost forecast for the next year;
(2) Propose proposals for improving human efficiency: in view of the actual situation and data analysis of enterprises, put forward proposals for improving human efficiency.

5 Form Analysis Report and Complete Report

(1) Complete the analysis report as required;
(2) Complete manpower cost and effectiveness analysis form and manpower cost analysis process chart.

Human resources quantitative management has been a difficult problem. Under the background of big data era, the use of big data technology can promote the quantitative management of human resources, make them create greater strategic value and form strategic support to business departments. For this reason, it is necessary to change the traditional human resource management system, explore the "big data technology +" human resource quantitative management system, and help organize human resources departments to carry out "data multidimensional collection", "data mining analysis", "data visualization". Through the data to find the relationship between talent management and labor productivity, to promote the transformation of enterprise human resources management model.

Acknowledgments. Liaoning Province Science and Technology Talents Innovation and Entrepreneurship Ecological Environment, project number LNKX2018-2019C37.

Research on the coordinated development of scientific and technological innovation talents and regional economy under the labor supply side reform, project number L19BRK002.

Research on the training and incentive mechanism of scientific and technological innovation talents in Dalian, project number 2020dlskzd322.

References

1. Wang, M.: Research prospect and multidimensional analysis of "big data technology +" and human resource quantitative management. J. Tianshui Inst. Adm. **17**(02), 31–35. (in Chinese)
2. A Study on Enterprise Training and Development Based on Quantitative Management of Human Capital and Human Resources. Ocean University of China (2008). (in Chinese)
3. Na, J.: Human resource quantitative management. Embrac. Big Data Age Sci. Technol. Econ. Mark. **01**, 103–105 (2017). (in Chinese)
4. Wang, Q., Zhu, X.: On the Innovation of Enterprise Human Resource Management in Big Data Age. J. Shenyang Univ. Technol. (Soc. Sci. Ed.) **3** (2015). (in Chinese)
5. Li, L.: Innovative Management of Human Resources in the Age of Data Leadership Science **10** (2014). (in Chinese)

6. Wang, S.: The influence of big data age on human resource management .China Manag. Inf. **2** (2015). (in Chinese)
7. Du, J.: Reflections on the reform of enterprise human resource management in the age. Data China Manag. Inf. **5** (2015). (in Chinese)
8. Heyun, A., Xue, J.: Reflections on the reform of enterprise human resource management in the data age. Econ. Res. Ref. **11** (2014). (in Chinese)
9. Li, Z.: Quantitative Value Management: Activating Human Resource Efficiency of Grass-Roots Organizations China Human Resources Development **11** (2013). (in Chinese)
10. Tu, Z.: Big Data: The Coming Data Revolution, pp. 349–351. Guangxi Normal University Press, Guilin. (in Chinese)

Portrait of College Students and Personalized Teaching Mode Under the Blended Learning Mode Based on Big Data Technology

Yuelin Yin(✉) and Xiaomu Yu

Department of Information and Business Management, Dalian Neusoft University of Information, Dalian, Liaoning, China
yinyuelin@neusoft.edu.cn

Abstract. With the advent of the information age, information technology education has become an inevitable trend in higher education reform. Targeted individualized teaching can help teachers develop more effective teaching models. This article mainly studies the collection and sorting of large amounts of data generated through online learning platforms based on a blended teaching model. The effective data generated in the learning process of students is aggregated and classified, and student portrait tag libraries of different dimensions are developed according to the specific influencing factors reflected by the data. Use portrait labels and statistics to perform variance analysis to understand the association between each label and the result attribute scores, so that the students represented by the portrait labels have clearer problems in the learning process. The design of teaching and the formulation of learning plans for each portrait label help students to learn more efficient professional courses.

Keywords: Blended teaching model · Student portrait · Individualized teaching

1 Introduction

With the advent of the information age, informatized online learning tools and platforms have become a popular way of autonomous learning in higher education and adult education. Self-learning using fragmented time has freed the fixed learning location and time. Online learning not only improves the overall quality of the people, but also provides a development opportunity for the university education industry. How to make good use of online learning to carry out personalized education and evaluation based on blended teaching is a challenge for high-calibration. The emergence of the seventh information revolution brought the world into the 5G era of intelligent Internet, using big data and related tools to collect information network data, and fully tap the value of big data. This method can not only be used for commercial big data analysis to make business decisions, but also can be used in education and teaching. In recent years, especially affected by the global epidemic in 2020, college teachers have used modern technology to achieve high-efficiency and high-quality teaching based on a blended teaching model.

B. J. Jansen et al. (Eds.): *International Conference on Cognitive based Information Processing and Applications (CIPA 2021)*, LNDECT 85, pp. 992–997, 2022.
https://doi.org/10.1007/978-981-16-5854-9_140

Therefore, using big data tools and big data thinking mode to analyze the teaching process under the blended teaching mode, and formulating a personalized teaching design based on the analysis of student portraits may provide a better way for the personalization and evaluation of teaching s solution [1–3].

2 Necessity of Analyzing Student Portraits Based on Big Data

The mode of talent training and teaching process in my country's universities is relatively fixed. Teaching evaluation often only focuses on the results of summative assessment [4]. It is difficult to evaluate the actual problems of student learning in the teaching process, and it is difficult to find problems in the formation process of students' learning, and conduct timely Targeted teaching adjustment and design.

In the teaching of colleges and universities, based on the blended teaching mode, through the construction and design of online resources, it helps students to formulate their own learning plans before and after class, and teaches in accordance with the teaching content of the classroom. Blended teaching, that is, the blend of online and offline, that is, the blended mode of "E (e-Learning) + C (Classroom)". The concept of blended learning was first proposed to refer to the blending of online and offline. -The combination of learning and traditional classroom learning has achieved good results through practice. However, not all courses have a fixed teaching model, and not all students have the same learning methods and paths. Then, it is necessary to carry out advanced curriculum teaching design for students of different types of portraits [5–7].

The advantage of the blended teaching model is that it pays more attention to the learning process of students. Through the online platform for learning ability training, through the assessment software to test, get the data. Use big data technology to evaluate and analyze the learning process. Big data not only refers to the huge amount of data, but also includes complex data types. The data that can be exported such as the online learning system platform of the blended teaching, the questionnaire star test, and the course assessment scores have such characteristics. The data includes structured data, unstructured data, text data, pictures, and other complex data [8–10].

3 Construction of Student Portrait Model

The construction of the student portrait model is carried out through the following 4 steps, as shown in Fig. 1.

3.1 Data Collection

Compared with the student learning data of traditional teaching, the data types of the blended teaching model are easier to collect, but there are more types of data. Such as online learning system platform, questionnaire star test, course assessment results, etc. The data includes structured data, unstructured data, text data, pictures and other complex explicit data and potential data.

Collect explicit data on the platform during the learning process of students, including the time spent watching MOOCs, the number of logins, the number of downloads,

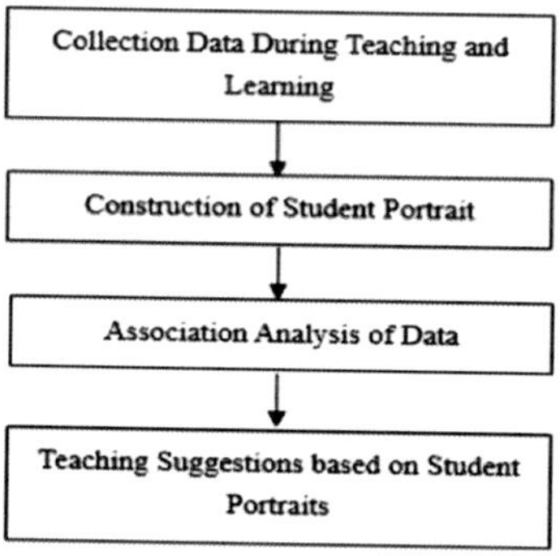

Fig. 1. Construction and analysis steps of student portrait model

and the download rate of students on the blended teaching resource platform. The purpose is to understand students' preferences and usage rates of resources; Collect the preparatory test and in-class test results of the questionnaire star platform, the purpose is to understand the students' self-learning before class and the acceptance of classroom learning, and grasp the key and difficult points to achieve satisfaction in the teaching; Through practical homework in class and the completion of level three projects, sort out student performance, mark questions to form text data and structured data; Analyze students' classroom participation and learning activity through students' classroom performance and the results of answering questions, as well as the potential data of students' speeches in the class online discussion area; The above data is summarized and converted to analyze whether there is a significant relationship between the summative results.

3.2 Construction of Student Portrait Model

(1) Analysis of the factors of student learning attributes

The data obtained is divided into behavior attribute and result attribute data through cluster analysis, and a student portrait tag library is formed through 4 dimensions. Behavioral attributes refer to the learning process, including online learning platform resource usage, classroom participation and so on. Result attributes refer to students' autonomous learning before class and the results obtained after class learning, mainly including homework Results, assessment results, and online test results. Achievement is the most direct criterion for judging students' learning results. Therefore, in the subsequent analysis, this variable will be used as the dependent variable and the correlation analysis between the data of each learning process.

(2) Student portrait model design

Based on the above student behavior and result attribute data, 4 dimensions are selected as student portrait tags to form a portrait tag library; According to homework, after-class quizzes, and three levels of three-level projects, students' classroom acceptance capabilities are divided into three types, applied, comprehensive, and beneficial; According to classroom performance and answering questions, consider students' classroom activity to analyze student participation, and divide students' learning behaviors into 3 types, namely active, occasionally active and dull; According to the number of visits, duration, download volume, and playback volume of the resource platform, students are divided into 3 types, namely resource collection enthusiasts, that is, the learning behavior of all resources are browsed and downloaded; MOOC enthusiasts are targeted The attention of MOOC videos is greater than the learning behavior of document-type learning materials; document lovers, that is, for extracurricular readings of literature, the attention of courseware PPT is greater than the learning behavior of MOOC videos; Reflect the students' autonomous learning ability through pre-class tests, and divide students into two types, namely, those capable of autonomous learning and those not capable of autonomous learning.

3.3 Statistical Analysis

In the last round of mixed teaching, the learning data of 149 students in 2 teaching classes were tracked. First, analyze whether there is a significant relationship between resource platform usage, autonomous learning, and formative performance, as shown in Tables 1 and 2.

Table 1. Test for homogeneity of variance

	Sum of squares	df	Mean square	F	Sig
Calibration model	2818.823[a]	3	939.608	19.663	.000
Intercept	189543.672	1	189543.672	3966.499	.000
Self-learning ability	630.988	1	630.988	13.204	.000
Resource preference	399.837	2	199.919	4.184	.017
Error	6928.989	145	47.786		
Total	216476.000	149			
Corrected total	9747.812	148			

Table 1 shows that the model is significant based on the effect of resource platform usage preference, autonomous learning ability and formative performance. Since the significance level of the two factor tests is close to 0, it indicates that resource use and autonomous learning ability influence formative performance. All have a significant impact.

Table 2 shows the parameter estimates of the analysis of variance model. In the table, the self-learning ability is weak and the document enthusiasts are the reference levels. The strong self-learning ability significantly improves the result of formative performance. Resource collection enthusiasts and MOOC hobbies the result of formative performance

Table 2. Parameter estimation

Parameter	B	Error	t	Sig	95% confidence interval	
					Upper	Lower
Intercept	32.509	1.043	31.177	.000	30.448	34.570
Strong self-learning ability	5.597	1.540	3.634	.000	2.553	8.641
Weak self-learning ability	0^a	-	-	-	-	-
Resource collector	5.605	1.948	2.877	.005	1.755	9.456
Mooc collector	2.720	1.418	1.918	.057	−.083	5.523
Document collector	0^a	-	-	-	-	-

is also improved to varying degrees. It can be found that the use of resources and the cultivation of self-learning ability are of great help to the formation of usual grades.

Using the same method, the correlation analysis of the four-dimensional factors and the formative assessment results and the summative assessment results were carried out respectively. From the analysis results, students with strong independent learning ability have an advantage in the final scores, while those with a high download rate of documents have worse final scores than those with a high rate of watching MOOC. Only 60% of the final scores of resource collection enthusiasts It is directly proportional to the formative performance, indicating that the self-learning ability of some of the resource enthusiasts needs to be improved; and among the students with high activity and class participation, only 45% of the students have good grades, so they cannot be used as an assessment. The main basis: students who have completed well in class assignments, in-class experiments, and three-level projects have better summative and formative results.

3.4 Suggestions for Teaching Design Ideas Based on Student Portraits

Realize early warning during the learning process. Early warning interventions in the learning process are divided into groups of people according to portrait tags, including collective intervention and individual intervention. Personal interventions include students who are resource lovers or document lovers who are dull in the classroom, dependent on homework, and poor in class tests. For MOOC enthusiasts, collective intervention is enough.

Develop a study plan and preview the stepwise setting of difficulty. Based on different types of portrait tags, develop an advanced learning plan, and reflect the advanced nature of the difficulty of the question in the pre-class test. Appropriately consider the progress of students, and conduct formative performance evaluation.

Record the video of the micro lesson. The recording time of micro-videos with key knowledge points should not exceed 3 min. The content needs to be combined with current hot topics, and the theoretical content is assisted by cases or sample questions.

The teaching evaluation system should consider students' progress and the application of theoretical knowledge. Appropriate consideration of grade changes in the learning process, focusing on the improvement of grades, the cultivation of learning ability, and the flexibility of applying theoretical methods.

References

1. Wei, S., Han, Y., Wang, L.: Research on online teaching reflection based on data mining and analysis of learning process. Mod. Educ. Technol. **25**, 89–95 (2015). (in Chinese)
2. Mengmeng, C., Cai, M., Zheng, W., Chen, X., Sun, Y., Wang, H.: Research on the application of personalized teaching mode based on big data in computer teaching. J. Phys.: Conf. Ser. **3**, 1650 (2020)
3. Yang, P.: Research on computer professional process and individualized teaching reform based on big data thinking. In: Proceedings of 2019 International Conference on Reform, Technology, Psychology in Education (ICRTPE 2019), pp. 397–400. Francis Academic Press (2019)
4. Liu, J.: Construction of teaching process evaluation system and student ability profile system based on big data. Educ. Mod. **51**, 157–160 (2019). (in Chinese)
5. Zang, R., Wang, L.: Personalized teaching model of college English based on big data. J. Phys.: Conf. Ser. **2**, 1852 (2021)
6. Ai, X., Zhang, Y.: From digital portraits to digital twins: a new approach to the construction of digital twin learners driven by the integration of digital intelligence. J. Dist. Educ. **39**, 41–50 (2021). (in Chinese)
7. Guo, N., Wei, R.: Design of a system for building multi-group user portraits based on big data. Mod. Electron. Technol. **44**, 173–177 (2021). (in Chinese)
8. Yang, X.: Research on the application of user portraits in the personalized recommendation of educational resources. Fujian Comput. **37**, 52–53 (2021). (in Chinese)
9. Huang, T., Liu, S., Li, Q.: "Internet + Education" curriculum intelligent adaptation Res. High. Eng. Educ. **01**, 181–186 (2021). (in Chinese)
10. Xu, W.: Adult education management and development innovation in the era of big data. J. Jilin Prov. Educ. Coll. **37**, 141–144 (2021). (in Chinese)

Author Index

B. J. Jansen et al. (Eds.): *International Conference on Cognitive based Information Processing and Applications (CIPA 2021)*, LNDECT 85, pp. 999–1001, 2022.
https://doi.org/10.1007/978-981-16-5854-9